Plant Breeding 2050

P. M. Priyadarshan • Rodomiro Ortiz
Editors

Plant Breeding 2050

Next-Gen Crops

Springer

Editors
P. M. Priyadarshan
Rubber Research Institute of India
(formerly)
Thiruvananthapuram, Kerala, India

Rodomiro Ortiz
Department of Plant Breeding
Swedish University of Agricultural Sciences
Alnarp, Sweden

ISBN 978-981-95-0582-1 ISBN 978-981-95-0583-8 (eBook)
https://doi.org/10.1007/978-981-95-0583-8

This Springer imprint is published by the registered company Springer Nature Singapore Pte Ltd.
The registered company address is: 152 Beach Road, #21-01/04 Gateway East, Singapore 189721, Singapore

Foreword

The goal of agricultural research has long been to develop innovations to optimize farm productivity. However, this goal was mightily disrupted starting in the 1980s by the widespread public resistance to genetically modified (GMs) crops. Now, with the growing world population and climate change exacerbating threats of food insecurity, integrating biotechnology into crop breeding has become a burgeoning reality.

A brief history of GMOs reveals that Kary Mullis and Michael Smith were awarded the Nobel Prize in Chemistry in 1993 for creating the polymerase chain reaction, which sparked numerous potential applications by geneticists worldwide. Two years later, in 1995, the U.S. Environmental Protection Agency authorized the cultivation of GM *Bacillus thuringiensis* crops – corn, cotton, potatoes, and tobacco – that were resistant to insects. Further advances were made in 1996, including the complete sequencing of the initial eukaryotic genome (*Saccharomyces cerevisiae*), and by 2000, the Arabidopsis sequencing consortium accomplished its entire genome sequencing.

Following the disruption of the normal commercialization process, intense regulatory requirements were developed to ensure that the use of GM crops was safe for both humans and the environment. That process has resulted in the current capture of significant market share by GMOs in many agricultural regions, and this trend is expected to continue. While consumer acceptance of GMOs remains low, attention is focused on the potential for GMOs to achieve multiple goals, including food and nutrition security, reduction of greenhouse gas emissions, and improved efficiency of breeding programs.

This timely book, *Plant Breeding 2050: Next-Gen Crops*, edited by P.M. Priyadarshan and Rodomiro Ortiz, covers a wide range of cutting-edge aspects of how plant breeding will contribute to global agriculture in the coming decades, including its role in addressing food insecurity under climate change. The book comprises 24 chapters and covers next-generation sequencing, pan-genomics, artificial intelligence, speed breeding, crop wild relatives, gene editing, nanobiotechnology, computational biology, systems biology, IoT, AI, robotics, remote sensing, phenomics, and phenotyping.

Key methods described in the book include CRISPR Cas-9, high-throughput phenotyping, marker-assisted selection, and genomic selection. Gene editing via CRISPR signifies a novel framework designed for creating and utilizing new genetic variations. At present, plant breeders are investigating techniques to reduce fieldwork and emphasize laboratory breeding through innovative concepts such as converting mitotic cells to undergo meiosis. This technology, combined with speed breeding, can greatly shorten generation time, enhancing crop breeding significantly. Smart farming is facilitated by precision agriculture that utilizes the internet of things (IoT), artificial intelligence (AI), robotics, and remote sensing via mobile devices.

After decades of debate, climate change is now recognized as not only inevitable but as presently impacting multiple breadbasket regions of the world. Food systems have shifted from a singular focus on maximizing agricultural production to delivering health and nutrition, decreasing contributions of greenhouse gas emissions, and resilience to increasing climate extremes. The editors and authors of the book deserve praise for compiling in-depth knowledge that will benefit both public-sector and private-sector researchers, as well as educators and students alike.

Plant Breeding 2050: Next-Gen Crops provides a clear and compelling roadmap of how plant breeding can make significant contributions to the future of global agriculture, food and nutrition security, and climate change.

NASA Goddard Institute for Space Studies,
Columbia University, World Food Prize Laureate 2022
New York, NY, USA

Cynthia Rosenzweig

Preface

In the next 25 years, world human population will increase by 25% to reach 10 billion. Traditional breeding methods have generated high-yielding, nutrient-rich crops that can feed the increasing population. Yet, the current increase in productivity for important crops like wheat, rice, and maize is insufficient to meet future needs. There is a demand to develop new cultivar with increased yield, better nutrition, resistance to pests and pathogens, and ability to withstand climate challenges. Due to farmers choosing genetically uniform, high yielding cultivars over local ones or landraces, 60% of dietary energy now comes from only three crops: rice, maize, and wheat. As a result, almost one-third of individuals experience a type of malnutrition.

Grain crops were first grown by humans 12,000 years ago. Traditional breeding is a method of selective breeding that chooses crops based on their superior performance. Pure-line selections, mass selection, backcross breeding, recurrent selection, and hybridization were among the most well-known conventional breeding techniques that are lengthy and rely heavily on the plants' phenotype having an impact of the environment. Following the discovery of DNA, plant breeding shifted toward the molecular age. Breeding based on traits less affected by the environment like genotypes, visual and genetic indicators, image analysis, and loci mapping is being done now. Most prevalent breeding methods now are genomic selection, marker-assisted breeding, high-throughput phenotyping, and CRISPR-based gene editing. Transgenic breeding introduced a new era; however, its outcomes have foreign DNA fragments inserted randomly in the genome that may lead to unintended regulatory consequences. CRISPR may work as an exception. In many countries, regulatory limitations have not been overcome by new plant breeding techniques such as cis-genesis, intra-genesis, grafting, agroinfiltration, reverse breeding, RNA-directed DNA methylation, genome elimination, oligonucleotide-directed mutagenesis (ODM), and site-directed nucleases (SDNs). Lately, metabolomics, nanogenomics, and bioinformatics have emerged as tools to assist plant breeding. Utilizing crop wild relatives and agrobiodiversity in breeding programs is of utmost importance to tap the unutilized genetic resources to create new variations.

The introduction of second- and third-generation sequencing technologies enable breeders to utilize DNA markers for selection, aiding in gene discovery, trait analysis, and predictive breeding advancements. Plant breeding usually involves only one or two generations per year, but with the implementation of "speed breeding" protocols, the generation times of spring wheat, barley, and chickpea are cut by over half. The fusion of cutting-edge technologies with rapid breeding will support endeavors to address the task of providing food for a population of 10 billion. Despite this fact, the problem of gene erosion has been worsened by plant breeding due to the loss of native landraces and wild plants.

There has been rapid advancement in plant phenomics in recent years, allowing for more precise breeding opportunities. AI capabilities such as computer vision, machine learning, and deep learning have been successfully incorporated into non-invasive imaging techniques. By utilizing machine learning for reliable image evaluation, this integration is progressively improving the effectiveness of data recording and analysis, especially in field phenotyping. This research will enhance resilience to climate change. Artificial intelligence has been employed to speed up high-throughput genomics, genomic prediction, genomic selection, and marker-assisted selection. Incorporating these technologies with traditional and speed breeding will augment the development of new cultivars.

Developing food that is nutritious and eco-friendly requires collaborating with the environment. Nature cannot be disregarded or outsmarted. Global farming exerts significant stress on global limits, creating the task of ensuring future food security without jeopardizing the resilience of the Earth's systems. Almost half of the world's food production relies on crossing planetary boundaries (biosphere integrity, land-system change, freshwater use, nitrogen flows). Areas with high levels of activity, primarily in Asia, also confront multiple breaches of local boundaries at the same time. Only 3.4 billion people could be provided with a balanced diet of 2355 kcal per capita per day if strict boundaries were followed in the current food system. Transformation toward more sustainable production and consumption patterns could support 10.2 billion people within the examined planetary limits. Essential requirements include reallocating cropland, enhancing water and nutrient management, reducing food waste, and making dietary adjustments.

The sustainability of current agriculture is being called into question by the Intergovernmental Panel on Climate Change (IPCC) and numerous experts believe unprecedented levels of uncertainty regarding future food security. Agricultural activities are responsible for approximately 23% of total greenhouse gas emissions worldwide, including CO_2, methane, and nitrous oxides. Even though this number does not take into account the CO_2 absorption from agricultural crop production, the clear message is the intricate relationship between climate, environment, and economics. Investing in innovative breeding of plants and animals could significantly cut agricultural emissions and enhance sustainability, while also boosting crop productivity and nutritional biofortification. Still, there is a significant and unchanging disparity between progress in plant science and societal approval of new genetic technologies worldwide.

This book on "Plant Breeding 2050: Next-Gen Crops" addresses all the aforesaid issues to feed the world of 10 billion. Due care is taken to eliminate errors. If "typos" are there, the same may kindly be forgiven. We are thankful to Springer Nature for agreeing to publish this book. Also, we are extremely indebted to our families for their unflinching support and warm counsel. We are indebted to Dr. Cynthia Rosenzweig, World Food Prize Laureate, 2022, for providing foreword to this book.

Thiruvananthapuram, Kerala, India P. M. Priyadarshan
Alnarp, Sweden Rodomiro Ortiz

Contents

Part V New Plant Breeding Techniques

Editors and Contributors

About the Editors

P. M. Priyadarshan is a prominent *Hevea* rubber breeder. He commenced his research career by breeding triticale and wheat. He joined the Rubber Research Institute of India (Rubber Board, Ministry of Commerce, Govt. of India) in 1990 and specialized in breeding *Hevea* rubber under suboptimal environments. In 2009, he became the Deputy Director of the institute, and managed its Central Experiment Station till 2016. As a scientist, he has been involved in breeding cereals, spices, and rubber for the past 36 years. He has published several research papers and chapters in journals and books of international repute. He was a visiting scientist at the Hardwood Tree Improvement and Regeneration Center of Purdue University, United States. He currently works as a freelance consultant in *Hevea* management and is involved in authoring and editing books on agriculture/life sciences with a global perspective.

Rodomiro Ortiz (Lima, Perú; 1958) is currently working as a Faculty Professor and Chair of Genetics and Plant Breeding at the Swedish University of Agricultural Sciences (SLU), and SLU Breeding Network Chairman. He holds a PhD in Plant Breeding and Genetics from the University of Wisconsin-Madison, and worked as a young researcher at UNALM (Perú), Centro Internacional de la Papa (CIP, Perú), and Rutgers University. He was a scientist and director of various CGIAR Centers (CIMMYT, ICRISAT, IITA), and held a Nordic professorship on plant genetic resources at the then Royal Veterinary and Agricultural University of Denmark (merged thereafter with the University of Copenhagen). He has written ca. 1000 reports, of which about 60% are journal articles or edited book chapters with an h-index of 80. He was a member of CGIAR Independent Science and Partnership Council (ISPC, 2015–2019) and of the Commission on Sustainable Agriculture Intensification (CoSAI, 2020–2021), international fellow of the Royal Swedish Academy of Agriculture and Forestry (KSLA) since 2019, member of the Royal Physiographic Society at Lund (KFS) from 2021 onwards, and fellow of the African Academy of Sciences (FAAS) since 2021.

Contributors

B. C. Ajay ICAR-Indian Institute of Groundnut Research, Sagdividi, Gujarat, India

Manickavelu Alagu Department of Genomic Science, Central University of Kerala, Periye, Kerala, India

Andrey Vega Alfaro Department of Plant and Agroecosystem Sciences, University of Wisconsin-Madison, Madison, WI, USA

U. Aswathi BioMedical Technology Wing, Sree Chitra Tirunal Institute of Medical Sciences and Technology, Thiruvananthapuram, Kerala, India

Anagha Balakrishnan Department of Bioinformatics, University of North Bengal, District-Darjeeling, West Bengal, India

K. C. Bansal Department of Biotechnology, Guru Jambheshwar University of Science and Technology, Hisar, India

Centre for Crop and Food Innovation, Murdoch University, Perth, Australia

P. S. Basavraj ICAR-National Institute of Abiotic Stress Management, Baramati, Maharashtra, India

C. Bharadwaj The UWA Institute of Agriculture, School of Natural and Agricultural Sciences, The University of Western Australia, Perth, WA, Australia

ICAR-Indian Agricultural Research Institute, New Delhi, India

Minhui Bi State Key Laboratory of Herbage Improvement and Grassland AgroEcosystems, College of Ecology, Lanzhou University, Lanzhou, Gansu, China

U. Bitla ICAR-National Institute of Abiotic Stress Management, Baramati, Maharashtra, India

Ramesh Bondada School of Biology, Indian Institute of Science Education and Research, Thiruvananthapuram (IISER TVM), Maruthamala, Vithura, Kerala, India

K. M. Boraiah ICAR-National Institute of Abiotic Stress Management, Baramati, Maharashtra, India

Creuci Maria Caetano Universidade Federal de Rondônia UNIR, Presidente Médici, Rondônia, Brazil

S. Chandramohan ICAR-Indian Institute of Groundnut Research, Sagdividi, Gujarat, India

Jianjun Chen Mid-Florida Research and Education Center, Department of Environmental Horticulture, Institute of Food and Agricultural Sciences, University of Florida, Apopka, FL, USA

Roy Chen Department of Molecular Biosciences, University of South Florida, Tampa, FL, USA

Tabsum Chhetri Department of Bioinformatics, University of North Bengal, District-Darjeeling, West Bengal, India

Carlos Iván Cardozo Conde Universidad Nacional de Colombia sede Palmira, Facultad de Ciencias Agropecuarias, Palmira, Valle del Cauca, Colombia

Grupo de Investigación en Recursos Fitogenéticos Neotropicales (GIRFIN), Universidad Nacional de Colombia sede Palmira, Facultad de Ciencias Agropecuarias, Palmira, Valle del Cauca, Colombia

Diego Felipe Conejo-Rodríguez Universidad Nacional de Colombia sede Palmira, Facultad de Ciencias Agropecuarias, Palmira, Valle del Cauca, Colombia

Grupo de estudio Fenómica, Ciencia de datos e Inteligencia Artificial, Universidad Nacional de Colombia sede Palmira, Facultad de Ciencias Agropecuarias, Palmira, Valle del Cauca, Colombia

Grupo de Investigación en Recursos Fitogenéticos Neotropicales (GIRFIN), Universidad Nacional de Colombia sede Palmira, Facultad de Ciencias Agropecuarias, Palmira, Valle del Cauca, Colombia

Genetic Resources Program, International Center for Tropical Agriculture (CIAT), Palmira, Valle del Cauca, Colombia

Andrés J. Cortés Department of Plant Breeding, Swedish University of Agricultural Sciences, Alnarp, Sweden

Facultad de Ciencias Agrarias—Departamento de Ciencias Forestales, Universidad Nacional de Colombia—Sede Medellín, Medellín, Colombia

Lumbini Devi Darapu Department of Biological Sciences, SRM University AP, Amaravati, Andhra Pradesh, India

M. Dash ICAR-National Institute of Abiotic Stress Management, Baramati, Maharashtra, India

R. Dey ICAR-National Institute of Abiotic Stress Management, Baramati, Maharashtra, India

Harsh Kumar Dikshit Division of Genetics, ICAR-Indian Agricultural Research Institute, New Delhi, India

P. Duraimurugan ICAR-Indian Institute of Oilseeds Research, Hyderabad, Telangana, India

Amanda Ortiz Escobar Universidad Nacional de Colombia sede Palmira, Facultad de Ciencias Agropecuarias, Palmira, Valle del Cauca, Colombia

Grupo de Investigación en Recursos Fitogenéticos Neotropicales (GIRFIN), Universidad Nacional de Colombia sede Palmira, Facultad de Ciencias Agropecuarias, Palmira, Valle del Cauca, Colombia

Daniel Orlando Osorio Garcia Universidad Nacional de Colombia sede Palmira, Facultad de Ciencias Agropecuarias, Palmira, Valle del Cauca, Colombia

Karla Gasparini National Institute of Science and Technology on Plant Physiology Under Stress Conditions, Departamento de Biologia Vegetal, Universidade Federal de Viçosa, Viçosa, MG, Brazil

Andrés Gatica-Arias University of Costa Rica, School of Biology, Plant Biotechnology Laboratory, San José, Costa Rica

Red Latinoamericana en Biotecnología, San José, Costa Rica

S. Geethanjali Tamil Nadu Agricultural University, Coimbatore, Tamil Nadu, India

John J. Georrge Department of Bioinformatics, University of North Bengal, District-Darjeeling, West Bengal, India

Vanesa Diaz Giraldo Grupo de estudio Fenómica, Ciencia de datos e Inteligencia Artificial, Universidad Nacional de Colombia sede Palmira, Facultad de Ciencias Agropecuarias, Palmira, Valle del Cauca, Colombia

Aiswarya Girija Institute of Biological, Environmental & Rural Sciences (IBERS), Aberystwyth University, Plas Gogerddan, Aberystwyth, Ceredigion, Wales, UK

Irwin L. Goldman Department of Plant and Agroecosystem Sciences, University of Wisconsin-Madison, Madison, WI, USA

K. Jagadis Gupta BRIC-National Institute of Plant Genome Research, New Delhi, India

Prateek Gupta Department of Biological Sciences, SRM University AP, Amaravati, Andhra Pradesh, India

Kusum Gurung Department of Bioinformatics, University of North Bengal, District-Darjeeling, West Bengal, India

Nolan T. Hartwick Plant Molecular and Cellular Biology Laboratory, The Salk Institute for Biological Studies, La Jolla, CA, USA

Alejandro Hernández-Soto Technologic Institute of Costa Rica, Biology School, Biotechnology Research Center, Cartago, Costa Rica

Saeideh Javid Botany and Mycology Unit, Finnish Museum of Natural History, University of Helsinki, Helsinki, Finland

P. Kadirvel ICAR-Indian Institute of Oilseeds Research, Hyderabad, Telangana, India

Divya Kattupalli ICAR-Indian Institute of Rice Research, Rajendra Nagar, Hyderabad, Telangana, India

Smita Kaul International Center for Agricultural Research in the Dry Areas (ICARDA), New Delhi, India

Justine K. Kitony Plant Molecular and Cellular Biology Laboratory, The Salk Institute for Biological Studies, La Jolla, CA, USA

Narendra Kumar ICAR-Indian Institute of Groundnut Research, Sagdividi, Gujarat, India

Neeraj Kumar ICAR-Indian Agricultural Research Institute, New Delhi, India

Shiv Kumar International Center for Agricultural Research in the Dry Areas (ICARDA), New Delhi, India

Sudhir Kumar ICAR-Indian Agricultural Research Institute, New Delhi, India

Maryke T. Labuschagne Department of Plant Sciences (Plant Breeding), University of the Free State, Bloemfontein, South Africa

Ana María Serna Landaeta Universidad Nacional de Colombia sede Palmira, Facultad de Ciencias Agropecuarias, Palmira, Valle del Cauca, Colombia

Felipe López-Hernández Corporación Colombiana de Investigación Agropecuaria—AGROSAVIA, CI La Selva, Rionegro, Colombia

Carlos Federico Marfil Consejo Nacional de Investigaciones Científicas y Técnicas-CONICET. Estación Experimental Agropecuaria—Mendoza—Instituto Nacional de Tecnología Agropecuaria-INTA, Mendoza, Argentina

Héctor Estrada Marin Universidad Nacional de Colombia sede Palmira, Facultad de Ciencias Agropecuarias, Palmira, Valle del Cauca, Colombia

Grupo de estudio Fenómica, Ciencia de datos e Inteligencia Artificial, Universidad Nacional de Colombia sede Palmira, Facultad de Ciencias Agropecuarias, Palmira, Valle del Cauca, Colombia

Ravi Maruthachalam School of Biology, Indian Institute of Science Education and Research, Thiruvananthapuram (IISER TVM), Maruthamala, Vithura, Kerala, India

R. K. Mathur ICAR-Indian Institute of Oilseeds Research, Hyderabad, Telangana, India

Todd P. Michael Plant Molecular and Cellular Biology Laboratory, The Salk Institute for Biological Studies, La Jolla, CA, USA

University of California San Diego, La Jolla, CA, USA

San Diego Botanical Garden, Encinitas, CA, USA

G. P. Mishra ICAR-Indian Agricultural Research Institute, New Delhi, Delhi, India

Saurav Kumar Mishra Department of Bioinformatics, University of North Bengal, District-Darjeeling, West Bengal, India

Pragati Shukla nee Misra Centre for Tissue Culture Technology, Jacob Institute of Biotechnology and Bioengineering, Sam Higginbottom University of Agriculture, Technology and Sciences, Prayagraj, Uttar Pradesh, India

Kutubuddin A. Molla ICAR-Central Rice Research Institute, Cuttack, Odisha, India

Riniya Najeeb School of Biology, Indian Institute of Science Education and Research, Thiruvananthapuram (IISER TVM), Maruthamala, Vithura, Kerala, India

J. Nikitha Department of Genomic Science, Central University of Kerala, Periye, Kerala, India

Rodomiro Ortiz Department of Plant Breeding, Swedish University of Agricultural Sciences, Alnarp, Sweden

Lillian K. Padgitt-Cobb Plant Molecular and Cellular Biology Laboratory, The Salk Institute for Biological Studies, La Jolla, CA, USA

K. K. Pal ICAR-National Institute of Abiotic Stress Management, Baramati, Maharashtra, India

Peter Poczai Botany and Mycology Unit, Finnish Museum of Natural History, University of Helsinki, Helsinki, Finland

P. M. Priyadarshan Rubber Research Institute of India (formerly), Thiruvananthapuram, Kerala, India

Revanth Kishore Pulamolu University of Illinois, Urbana-Champaign, Urbana, IL, USA

T. Radhakrishnan ICAR-Indian Institute of Groundnut Research, Sagdividi, Gujarat, India

Amara Rafi Department of Plant Protection, Faulty of Agriculture, Universiti Putra Malaysia, Serdang, Selangor, Malaysia

Kirti Rani ICAR-National Bureau of Plant Genetic Resources, New Delhi, Delhi, India

Reena Rani Agricultural Biotechnology Division, National Institute for Biotechnology and Genetic Engineering (NIBGE), Constituent College Pakistan Institute of Engineering and Applied Sciences (PIEAS), Faisalabad, Pakistan

A. L. Rathnakumar ICAR-Indian Institute of Oilseeds Research, Hyderabad, Telangana, India

K. S. Reddy ICAR-National Institute of Abiotic Stress Management, Baramati, Maharashtra, India

Sneha Priya Pappula Reddy The UWA Institute of Agriculture, School of Natural and Agricultural Sciences, The University of Western Australia, Perth, WA, Australia

School of Biological Sciences, The University of Western Australia, Perth, WA, Australia

Muhammad Rizwan Plant Breeding and Genetics Division, Nuclear Institute of Agriculture (NIA), Tandojam, Pakistan

Department of Plant Sciences, University of Saskatchewan, Saskatoon, Canada

Federico Roda Max Planck Tandem Group, Facultad de Ciencias, Universidad Nacional de Colombia, Bogotá, Colombia

Lucía Rodríguez Gargantini Consejo Nacional de Investigaciones Científicas y Técnicas-CONICET. Estación Experimental Agropecuaria—Mendoza—Instituto Nacional de Tecnología Agropecuaria-INTA, Mendoza, Argentina

Sneha Roy Department of Bioinformatics, University of North Bengal, District-Darjeeling, West Bengal, India

Mohammad Sabzehzari Division of Biotechnology, Department of Agronomy and Plant Breeding, College of Agriculture and Natural Resources, University of Tehran, Tehran, Iran

K. Sakthivel ICAR-Indian Institute of Oilseeds Research, Hyderabad, Telangana, India

Liam Dixon Scully Department of Plant and Agroecosystem Sciences, University of Wisconsin-Madison, Madison, WI, USA

Kwame W. Shamuyarira Department of Plant Sciences (Plant Breeding), University of the Free State, Bloemfontein, South Africa

Hussein Shimelis School of Agricultural, Earth and Environmental Sciences, African Centre for Crop Improvement, University of KwaZulu-Natal, Pietermaritzburg, South Africa

Pradeep Kumar Shukla Department of Biological Sciences, Sam Higginbottom University of Agriculture, Technology and Sciences, Prayagraj, Uttar Pradesh, India

Akanksha Singh International Center for Agricultural Research in the Dry Areas (ICARDA), New Delhi, India

Atul Kumar Singh Department of Plant Physiology, KL University, Guntur, Andhra Pradesh, India

Jagannath Swain BRIC-National Institute of Plant Genome Research, New Delhi, India

Shivani Tiwari Department of Biological Sciences, SRM University AP, Amaravati, Andhra Pradesh, India

Abhinav Uniyal School of Biology, Indian Institute of Science Education and Research, Thiruvananthapuram (IISER TVM), Maruthamala, Vithura, Kerala, India

Akshaya Unnikrishnan Plant Biotechnology Laboratory, Department of Biotechnology, Cochin University of Science and Technology, Kochi, Kerala, India

Milan Oldřich Urban International Center for Biosaline Agriculture (ICBA), Near Zayed University, Dubai, United Arab Emirates

Sneha Yogindran Plant Biotechnology Laboratory, Department of Biotechnology, Cochin University of Science and Technology, Kochi, Kerala, India

Masoumeh Zeinali Division of Biotechnology, Department of Agronomy and Plant Breeding, Faculty of Agricultural, University of Mohaghegh Ardabili, Ardabil, Iran

Agustin Zsögön National Institute of Science and Technology on Plant Physiology Under Stress Conditions, Departamento de Biologia Vegetal, Universidade Federal de Viçosa, Viçosa, MG, Brazil

Part I

Generalia

Voyage of Plant Breeding to 2050

1

P. M. Priyadarshan and Rodomiro Ortiz

Abstract

The world population is expected to surpass nine billion by 2050, necessitating a 70% rise in food production to satisfy the anticipated population growth. The Earth's average temperature has risen by 1.2 ± 0.1 °C compared to preindustrial times, attributed to increased CO_2 emissions into the atmosphere. Climate change is leading to various noticeable impacts on agriculture and food production. Occurrences of plant diseases present major risks to worldwide food security and ecological sustainability. We have reached Breeding 4.0, where the cost of genome resequencing might be lower than that of a replicated yield trial, and genome editing is anticipated to allow multiple accurate modifications at different sites. High-throughput phenotyping enables the evaluation of many characteristics with excellent spatiotemporal accuracy, while machine learning techniques permit the analysis and understanding of agricultural data beyond human capacity.

The emergence of transgenic plants that are now accessible worldwide occurred afterward. Recently, CRISPR-based editing of plant genomes has been utilized to identify precise breeding goals. In the last 25 years, progress in sequencing technology has initiated a new era in plant breeding. Since the sequencing of the first plant genome (*Arabidopsis thaliana*) 24 years ago, numerous crop genomes of greater complexity and size have been decoded. Research on pangenomes, concentrating on species diversity via extensive genome sequencing, has uncovered genetic differences in essential traits because of dele-

P. M. Priyadarshan (✉)
Rubber Research Institute of India (formerly), Thiruvananthapuram, Kerala, India
e-mail: rriipriya@gmail.com

R. Ortiz
Department of Plant Breeding, Swedish University of Agricultural Sciences, Alnarp, Sweden
e-mail: rodomiro.ortiz@slu.se

P. M. Priyadarshan, R. Ortiz (eds.), *Plant Breeding 2050*,
https://doi.org/10.1007/978-981-95-0583-8_1

tions, insertions, and different types of chromosomal changes. Understanding of species and alleles is advancing alongside progress in pangenome research. Plant breeding is yet to adopt new "omics" techniques fully for assessing genetic worth, mainly due to their significant expense. Genetic diversity is essential for future progress in genetics, and new technologies like CRISPR provide potential in creating useful genetic diversity that consumers willingly embrace for future genetic improvements.

Progress in next-generation sequencing (NGS) has opened doors to a new era of varied omics, including genomics, transcriptomics, and proteomics. However, crop science has extensively documented metabolomics, ionomics, and phenomics as well. High-throughput methods that combine various omics techniques have been crucial for understanding development, aging, productivity, and reactions to various stresses in multiple crops. Genetic diversity is crucial for cultivating plants. To improve diversity, our knowledge of the molecular genetic processes involved in crop domestication has to significantly progress with the emergence of next-generation sequencing. Speed breeding technology has emerged as a technique that modify the environmental conditions for crop plants to accelerate their breeding processes by enhancing flowering rates and seed yield with the aid of artificial intelligence (AI). The commencement of meiosis in mitotic cells provides a key benefit by greatly reducing the duration of the breeding cycle. Breeders must create thorough data-driven precision breeding programs that encompass genotype–phenotype–environment multiple factors. This approach will facilitate gene editing, synthetic biology, and numerous technological advancements, enabling precise and effective enhancements in developing new varieties while fostering accuracy, intelligence, and an industrial-focused seed transformation. In Plant Breeding 5, each phase from breeding design to phenotype prediction is executed automatically through Deep Learning (DL). Extra stages could be added if new technologies emerge in the future. Nonetheless, breeders encounter constraints regarding their comprehension of the process of plant development. Enabling crops to flourish naturally in their environment is a crucial element of research. Breeders have defined duties in the breeding process, yet they also require support from appropriate institutions and regulations.

Keywords

Plant breeding 5.0 · Genomics · Phenomics proteomics NGS Speed breeding · Phenotyping · Artificial intelligence

1.1 Intro duction

Water, energy, and food availability have always influenced sustainable development, and in the near future, these aspects probably will become much more crucial (Zhou et al. 2022a, b). The availability of these resources should also be examined globally because social systems and institutions are becoming more globalized. It should be highlighted that there is a strong correlation between resource availability and globalization (Abdi et al. 2020). In the context of food demand, a variety of linkages are seen between the procurement, manufacturing, and distribution of food products and raw materials (Yadav et al. 2021).

One of the biggest challenges in the future will be the global food markets' capability to satisfy the needs of a fast-expanding population. It is projected that the world's population will surpass 9 billion by 2050 (Adam 2021) and reach a peak of 9.73 billion in 2064 (Vollset et al. 2020; Bahar et al. 2020). As per the projections of United Nations, the world's population is to reach 9.4–10.1 billion in 2050 and is expected to rise to 9.4–12.7 billion by 2100 (United Nations 2019) (Fig. 1.1). Several factors, including population fertility rates, policies affecting fertility, development level, cultural aspects, and mortality rates, will influence the accuracy of these estimates (Vollset et al. 2020). China is a prime illustration of the aforementioned. Although China is highly populated, its fertility rate dropped to 1.3 in 2020, ranking below many other countries globally (Daszkiewicz 2022). As per the United Nations (UN 2019) report, by 2050, only nine nations will account for over half of the world's population growth (Democratic Republic of the Congo, Egypt, Ethiopia, India, Indonesia, Nigeria, Pakistan, United Republic of Tanzania, United States) (Daszkiewicz 2022). It is projected that the population of Sub-Saharan Africa will triple by the end of the twenty-first century, increasing from 1.03 billion in 2017 to 3.07 billion in 2100. According to Vollset et al. (2020), the population in several Asian and Central-Eastern European nations is expected to decrease.

Global food production needs to rise by 70% to keep up with the expected population growth by 2050 (Van Dijk et al. 2021; Giller et al. 2021; Gao 2021).

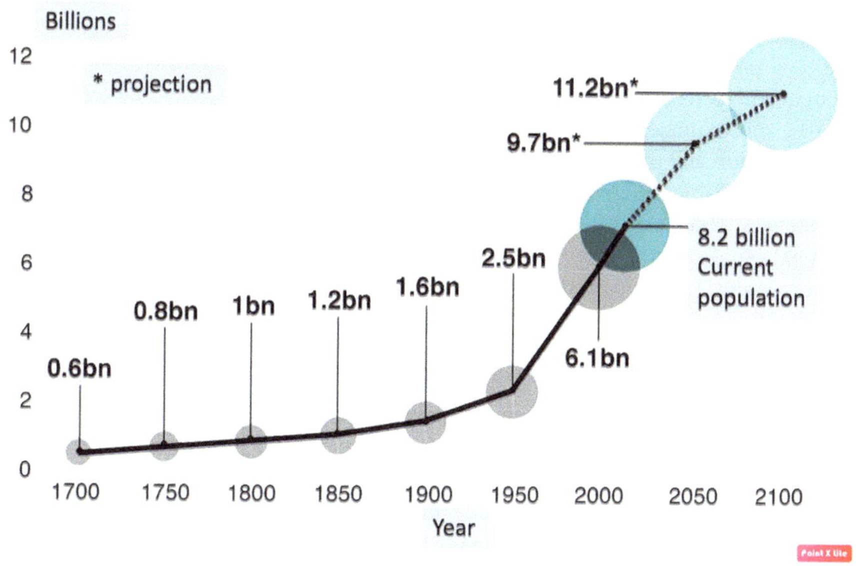

Fig. 1.1 Meeting the food demand within the next 30 years necessitates a 70% increase in production. *(Source: University College, London)*

Nevertheless, the rise in global population is not the sole reason for the escalating demand for food. Increased wealth in developing nations will also increase the need for food (Andreoli et al. 2021; Al Hasan et al. 2022). In most of these nations, the main source of protein is from plants;, however, this is expected to shift toward animal-based protein, mirroring the patterns seen in more advanced countries (Henchion et al. 2017). It is estimated that by 2030, the average person worldwide will consume fresh dairy products at a rate that increases by 1.2% annually. As per OECD/FAO, global meat consumption, however, is projected to rise by 14%, with Africa potentially seeing a 30% increase, the Asia-Pacific region at 18%, Latin America at 12%, North America at 9%, and Europe at 0.4% (Font-I-Furnols 2023).

Under all aforementioned circumstances, plant breeding has to come a long way to derive improved varieties. In this chapter, our attention is on the role of new and innovative breeding techniques in advancing plant breeding toward 2050.

1.2 Climate Change Factor

According to Song et al. (2022), the Earth's average temperature has risen by 1.2 ± 0.1 °C compared to preindustrial times, attributed to the increased release of CO_2 into the atmosphere. Between 2081 and 2100, it is projected that the average temperature will rise by 2.6–4.8 °C compared to the levels seen between 1986 and 2005 in high-emission scenarios. Surface temperatures are expected to stay consistently high for many centuries even after all human-induced CO_2 emissions stop. A significant portion of human-caused climate change caused by CO_2 emissions is permanent over hundreds to thousands of years, unless there is substantial removal of CO_2 from the air for a long time (IPCC 2014) (Fig. 1.2).

Climate change is having various and currently noticeable impacts on agriculture and food production. It results in reduced water resource availability, causing land drought and desertification in extensive regions. According to Sterk and Stoorvogel (2020), approximately 12 million hectares of land are lost annually as a result of desertification and drought, leading to a loss of 20 million tons of grain. This issue is most noticeable in Africa, Asia, and Latin America, but it is also becoming more prevalent in other areas such as Eastern Europe and Mediterranean countries like Portugal, Spain, Italy, Greece, Albania, Bosnia and Herzegovina, Croatia, Cyprus, France, Malta, Slovenia and Turkey, where there is a greater danger of soil degradation (Salvia et al. 2019).

Climate change raises the likelihood of severe weather events and atmospheric anomalies that result in natural disasters such as droughts, floods, hurricanes, typhoons, cyclones, tornadoes, and windstorms (Zandalinas et al. 2022). These events harm crops and lower yields, while also causing long-term damage to soil and reducing land fertility in areas that experience persistent extreme weather conditions. These deviations pose a risk to the security of the food supply. An example is the extended dry spell in 2020 causing a decline in crop yield potential, resulting in a 6.0% decrease in maize yields, a 3.0% decrease in winter wheat yields, a 5.4% decrease in soybean yields, and a 1.8% decrease in rice yields compared to the 1981–2010 average (Daszkiewicz 2022).

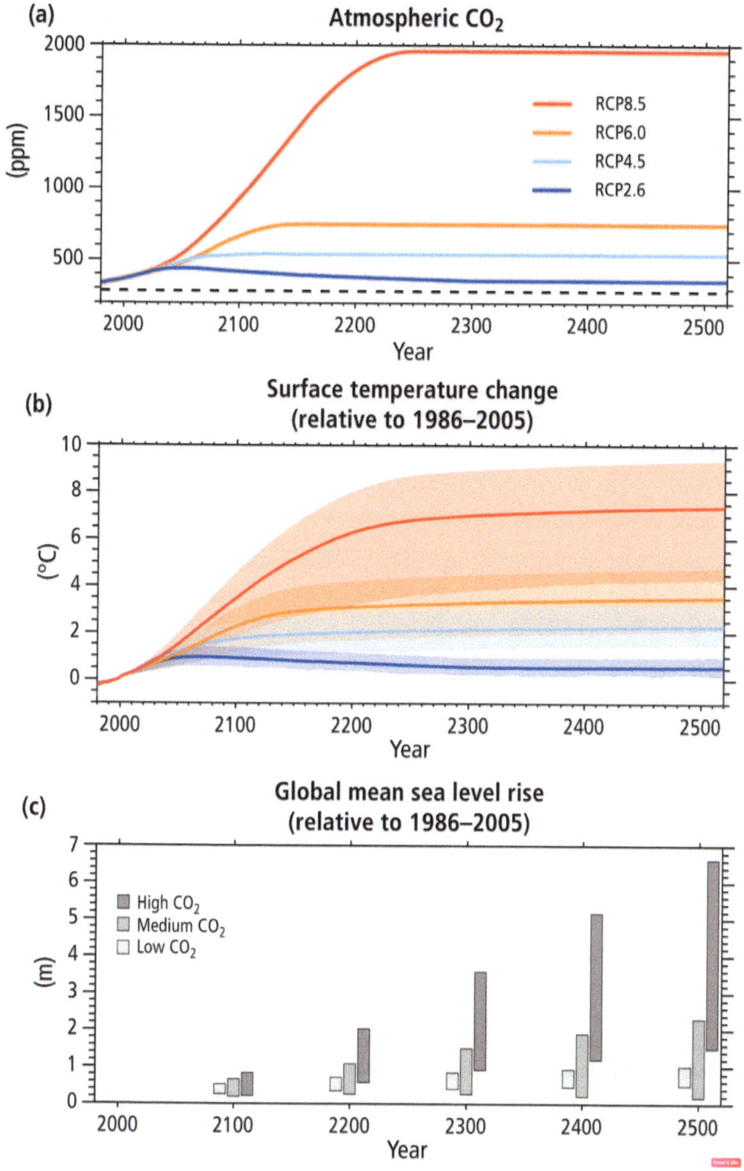

Fig. 1.2 (**a**) Atmospheric carbon dioxide (CO_2) and (**b**) anticipated changes in global mean surface temperature as modeled by Earth System Models of Intermediate Complexity (EMICs) for the four Representative Concentration Pathways (RCPs) through 2300 (compared to 1986–2005), followed by a consistent radiative forcing at the year 2300 level. A smoothing of 10 years was implemented. The dashed line in (a) represents the CO_2 concentration before industrialization. (**c**) Projections of sea level rise are classified into three categories based on greenhouse gas concentration (in CO_2-eq) for the year 2100: low: concentrations that peak and then decrease, remaining under 500 ppm, as seen in the scenario RCP2.6; medium: concentrations between 500 and 700 ppm, including RCP4.5; high: concentrations greater than 700 ppm but less than 1500 ppm, as reflected in scenarios RCP6.0 and RCP8.5. The bars in (c) represent the largest potential spread achievable with the limited available model results (and these should not be viewed as uncertainty ranges). These models *probably* undervalue the Antarctic ice sheet's contribution, leading to an underestimated forecast of sea level rise after 2100. (*Source: IPCC* 2014)

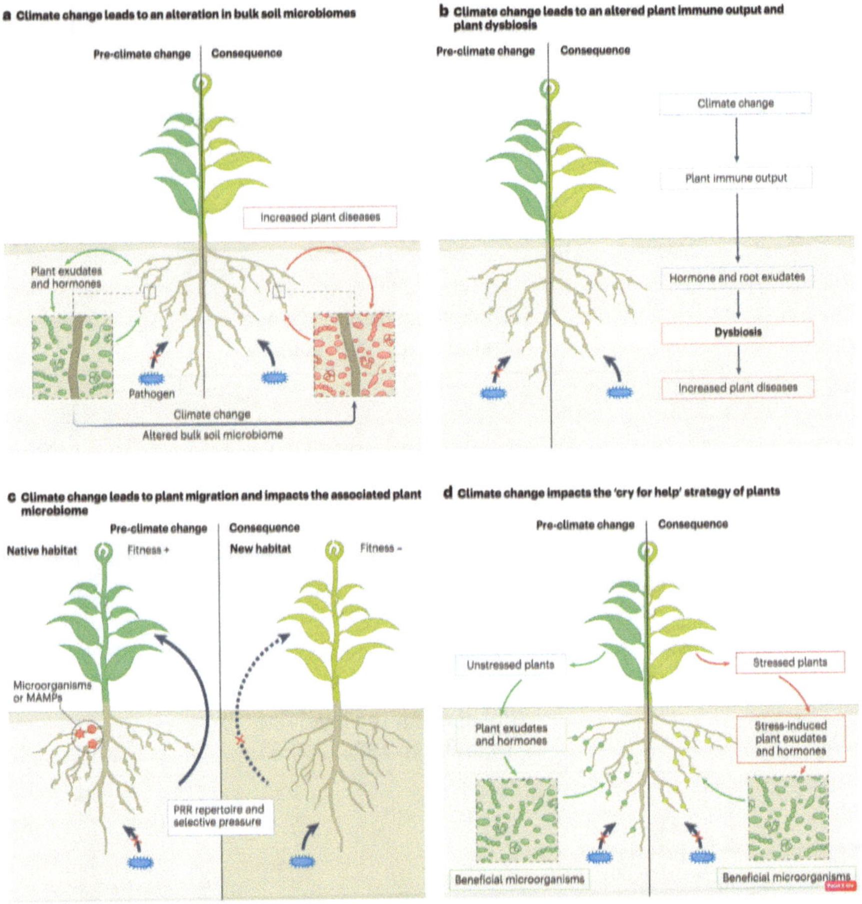

Fig. 1.3 Four potential scenarios are proposed regarding how microbiomes might respond to climate change. Green plants symbolize the strong health of vegetation prior to climate change, whereas yellow plants indicate the effects of climate change and disease intrusion on plant life. (**a**) Plants employ different strategies that depend on efficient immune responses, root exudates, and hormonal balance to nurture intricate microbiomes. Microbiomes linked to plants in the soil, particularly in the rhizosphere, act as the first line of defence against pathogens. Climate change is anticipated to alter the makeup of the microbial community in the bulk soil. These elements, together with changes in plant immunity, may influence the composition of the rhizosphere microbiome and impact initial defences, potentially allowing the pathogen to invade. (**b**) Plants maintain the equilibrium of related (leaf, root, stem, endophyte) microbiomes through rigorous and complex regulation of immune responses. Climate change may affect plant physiology and immune reactions, such as the production of root exudates, volatile organic compounds, and phytohormones, potentially hindering plants' ability to draw in and foster beneficial microbiomes and promoting dysbiosis that causes diseases. Furthermore, plant migration to novel habitats (niche range shift) could interfere with the plant immune system and its mutualistic coevolution with native soil microbiomes that support robust growth and disease resistance. As a result, the migration of plants could expose them to local pathogens to which they are susceptible, whereas in other cases, migrating plants might escape local pathogens. Similarly, a change in the niche range of pathogens (along with evolutionary adaptations) may increase the pathogen's ability to transmit its virulence.

Outbreaks of plant diseases greatly endanger food security and environmental sustainability globally (Chaloner et al. 2021). Climate change raises the chances of outbreaks by changing how pathogens evolve and interact with their hosts, as well as making it easier for new harmful strains to appear (Singh et al. 2023a, b) (Fig. 1.3). The range of pathogens may change, leading to the expansion of plant diseases in unfamiliar territories. The impact of plant pathogens can be assessed through direct reductions in crop productivity or through the associated societal, environmental, and economic expenses of controlling them (Fones et al. 2020). As with all organisms, crop pests and pathogens have specific tolerances or needs for certain environmental conditions (Chaloner et al. 2020). These tolerances establish the ecological niche, dictating where and when pests and pathogens can thrive and harm crops (Chaloner et al. 2020). Due to climate change, the ideal conditions for pest outbreaks are changing in both timing and location, leading to different challenges for farmers and adjustments needed in their management strategies (Bebber 2015). Predicting the future trends in pest and pathogen burdens is essential for ensuring future food security (Bebber et al. 2019).

1.3 On the Way to Plant Breeding 5.0

Only 1–3% of individuals in modern industrialized societies grow their own food, a substantial decrease from rates in previous cultures (Wallace et al. 2018). Plows, seeders, and fertilizers have been key management innovations that have contributed greatly to this transformation, but developments in crop genetics through breeding have also played a vital part (Nerkar et al. 2022; Potts et al. 2023). Most important agricultural traits are largely measurable, such as plant structure, development rate, nutrient content, and production, while others, such as Mendel's famous pea characteristics or important disease resistance genes, have specific genetic characteristics. Although it has been around for roughly one hundred years, the principles of quantitative genetics have been utilized in plant breeding for an extended period of time (Bernardo 2020a, b). Wallace and colleagues (2018) categorized plant breeding into four primary phases according to the techniques employed. Three have been achieved already, and one is expected to happen soon. Through a combination of traditional techniques and modern advancements, each stage enhances the one before it and improves breeding efficiency:

Fig. 1.3 (continued) Various regions where local plants that lack effective immune defences make local microflora provide little resistance (**c**). Additionally, the plant's "cry for help" strategy, which involves recruiting beneficial microorganisms during pathogen attacks, is anticipated to be affected by climate change. Climate change may restrict plants' ability to produce signaling molecules (like root exudates and volatiles) that attract beneficial microorganisms and/or change microbial diversity and traits, as well as their capacity to respond to these signals. Climate change may also reduce the effects of pathogen attacks since a shift in microbiomes has either boosted helpful microorganisms or triggered the plant's immune system (**d**). MAMP = molecular motif associated with microbes, PRR is receptor for pattern recognition. (Courtesy: Nature Reviews Microbiology; https://doi.org/10.1038/s41579-023-00900-7)

- The early phases of breeding, referred to as Breeding 1.0, began approximately 10–12 thousand years ago when various groups globally identified and cultivated around 7000 different species of edible plants (Khoshbakht and Hammer 2008). Because of the scarcity of skilled breeders, local, independent farmers depended on selecting plants according to their physical traits (Meyer et al. 2012).
- Breeding 2.0 began in the late nineteenth and early twentieth centuries with the recognition of inbreeding depression (Byers and Waller 1999), the revival of Mendelian genetics, and the development of quantitative genetics theory. Advancements were achieved via replicated field trials, controlled crosses, statistical assessments, structured experimental designs, hybrid breeding techniques, and pedigree-derived estimations of breeding values.
- About 30 years ago, Breeding 3.0 emerged as molecular markers and genomic information began to improve phenotypic data. The process started with marker-assisted backcrossing and verifying pedigrees, then advanced to the analysis of complex traits through linkage mapping. High-throughput genotyping has enabled novel approaches in quantitative genetics to examine variation in natural populations (genome-wide association) and to select based on genomic-estimated breeding values (genomic selection) (Meuwissen et al. 2001) (refer to Uffelmann et al. 2021; Lu et al. 2020; Cortes et al. 2021 for recent developments).
- We have arrived at Breeding 4.0, a phase in breeding where genome resequencing may cost less than a replicated yield trial, and genome editing is expected to enable several precise modifications at various locations (Zhang et al. 2023). High-throughput phenotyping facilitates the assessment of numerous traits with outstanding spatiotemporal precision, whereas machine learning approaches permit the analysis and interpretation of agronomic data beyond human proficiency (Xu et al. 2022; Wang et al. 2023a, b; Araújo et al. 2023).

One can imagine a higher level (Breeding 5.0) that includes the creation of genes, pathways, and traits from scratch (Lorenzo et al. 2022). It is expected to happen soon (Fang 2024).

1.4 Plant Breeding Advances and Society

When people were still reliant on hunting and gathering, 11,700 BC marked the beginning of modern human civilization (Carey 2023). After 2300 years, primitive people began domesticating sheep and other animals. Potatoes, maize, and wheat were progressively planted in distinct plots. The introduction of the plough and irrigation system occurred around 2900 BC. Following that, the domestication of other crops accelerated (Bellwood 2005). According to Zohary et al. (2021), the earliest domesticated crops were Einkorn and Emmer wheat, which were discovered in South-West Asia during the Neolithic era. All of the basic grains and crops that humans eat today were domesticated by Neolithic man before 20 or 30 centuries ago (Borlaug 1983). Currently, 2500 plant species have undergone domestication (Dirzo and Raven 2003; Pratap et al. 2023), with 250 species having reached the pinnacle of domestication (Sulaiman et al. 2023; see Purugganan 2019 for details).

Humanity has been refining plant-based crops for thousands of years in order to produce valuable commodities like phytomedicines and food. For instance, the most common methods employed during the "pre-green" revolution era were phenotype selection, incorporating preference, experience, and germplasm collections (Doebley et al. 2006). Then, the "green revolution" of the 1940s to 1960s significantly increased our capacity to manufacture goods and meals derived from plants. With the introduction of novel kinds and hybrids, technology played a crucial role in this production step-change (Khush 2001). These novel cultivars were frequently the result of physical and chemical mutagenesis experiments. Fertilizers and pesticide formulations were among the innovative chemical inputs used with the new germplasm.

1.4.1 Domestication to Biotechnology, Molecular Breeding, and Genomics

By definition, the goal of plant breeding is to produce agricultural products that will meet human needs for fuel, food, and fiber. According to Hebrew University of Jerusalem's Arie Altman (https://doi.org/10.1016/j.tplants.2021.03.013), three main pillars are the basis for modern plant breeding:

- The selection of exceptional plant phenotypes at an early stage in the domestication process, approximately 12,000 years ago, resulted from unintentional early hybrids.
- From Mendel's discovery of the genetic "rules" of intentional plant breeding in the early nineteenth century to the discovery of DNA structure in the 1950s, these developments made it possible to assign a genotype to phenotypes, formulate the "central dogma," engage in mutation breeding, and produce hybrid varieties.
- The extraordinary application of novel molecular technologies gave rise to the modern field of "molecular plant breeding." This includes the successful integration of marker-assisted selection of domesticated plants and their wild relatives into conventional plant breeding.

The creation of transgenic plants (Herrera-Estrella et al. 1983) that are presently sold all over the world came next. More recently, plant genome editing has been used to identify particular breeding targets (Doudna and Charpentier 2014), a technique that earned them the 2020 Nobel Prize in Chemistry. The onset of plant biotechnology, in the early twentieth century, is associated with the advancement of cultivating plant cells and tissues in a lab setting, enabling regeneration and cloning of new plants. These abilities became necessary for the production of transgenic and genome-edited plants, in addition to basic molecular technologies (such as cloning, sequencing, and recombinant DNA). In the near future, it is quite likely that two research approaches will coexist in parallel: the transgenic approach and the non-transgenic approach, which involve better traditional breeding.

Although agricultural biotechnology and plant breeding have demonstrated potential, Altman (https://doi.org/10.1016/j.tplants.2021.03.013) proposes that future plant breeding endeavors should concentrate on addressing the primary challenges that remain to be addressed. Among them are:

- Improving quantitative traits to bridge the genotype–phenotype gap; and automated methods for selection and screening focused on whole-plant physiology and quality characteristics.
- To bridge the gap between genetics and environment: A greater embrace of translational and interactome research across all R&D stages should ideally enhance breeding objectives, as numerous sought-after plant traits rely on the interplay of multiple genes and metabolic pathways with environmental factors, including the plant microbiome.
- Molecular epigenetic mechanisms crucial for plant adaptation to fluctuating environments deserve greater focus in the context of evolution.
- Promoting transparent dialogue and cooperation among molecular biologists, plant physiologists, farmers, breeding companies, and the public to tackle ethical, legal, sociocultural, and economic issues.

1.4.2 Decoding Genomes

In the last 25 years, significant progress in sequencing technology has led to a new era in plant breeding (Henry 2022). Since the first plant genome (*Arabidopsis thaliana*) was released 24 years ago (*Arabidopsis* genome initiative 2000), numerous crop genomes with growing complexity and size have been sequenced (Song et al. 2023). These combinations of plants led to an increase in research focused on genomes, often leading to extensive resequencing and marker identification. This resulted in a wide range of genes and variations linked to significant agricultural traits such as flowering time, disease resistance, plant structure, and environmental stressors (Amiteye 2021; Lübberstedt and Suza 2023). Therefore, the process of replicating their basic genetic makeup has become quicker and less laborious, leading to a significant improvement in our capacity to tame a wild species or enhance a crop to meet societal demands (von Wettberg et al. 2020; Gasparini et al. 2021; Lei et al. 2024).

Utilizing genomic data to its fullest extent is crucial for enhancing crop production (Bassi et al. 2024). Over the next 25 years, one of the most thrilling tasks in plant breeding will involve creating synthetic breeds through genome design informed by our knowledge of genome development and the relationships between genes and traits (Markel and Shih 2021). Yu and Li (2022) predict that the de novo breeding method will completely revolutionize the field of plant breeding.

1.4.3 Plant Breeding Under Sustainable Ecosystems

Over the last 25 years, there has been a fundamental change in plant science technologies and research philosophy, primarily influenced by the Sustainable Development Goals set by the United Nations (Khalid 2023). Public acceptance has had a significant impact on the progress of utilizing technological developments to enhance plant breeding outcomes for a more sustainable agriculture in order to meet the needs of a growing global population. The international plant science community was filled with a wave of hopeful anticipation when transgenic crops were first introduced in 1994 with Calgene's delayed-ripening tomato (Kramer and Redenbaugh 1994). The expected public backlash toward genetically modified (GM) crops after their release has hindered the worldwide acceptance of GM crops. Gradual, yet consistent expansion in biotech crops led to the development of GM crops worth $17.2 billion in 2017, according to the International Service for the Acquisition of Agri-biotech Applications (ISAAA) 2019. In 23 years, the cultivation of GM crops has covered a total area of 2.53 billion hectares. According to the latest data from 2019, 29 countries cultivated 190.4 million hectares with GMOs, with the Americas having the largest cultivated area globally (Rozas et al. 2022). Nevertheless, the extent devoted to biotech crops has leveled off in recent years. The main advantages of biotech crops include enhancing profits per hectare, as well as promoting better health and social benefits by improving the environment, ultimately resulting in more sustainable agricultural practices.

Research on pangenomes, which involves examining species diversity using complete genome sequencing, has shown genetic variations in important traits like deletions, insertions, and other types of chromosome modifications (Shi et al. 2023; Hu et al. 2024). The understanding of species and alleles is evolving with advancements in pangenome research. Plant breeding has not yet adopted new "omics" techniques for assessing genetic worth, mainly due to their expensive nature. Genetic diversity is essential for future genetic advancements, and new technologies like CRISPR provide hope for creating valuable genetic diversity that is widely accepted by consumers for future genetic improvement (Tang et al. 2023). Accelerating plant breeding will involve using rapid single seed descent to create inbred lines more quickly (Rutkoski et al. 2022; Kigoni et al. 2023) and applying genomic selection on noninbred lines to speed up selection cycles (Alemu et al. 2024). It is unavoidable that technologies like CRISPR will be superseded by more efficient technologies in the future, which will make multiple nucleotide changes simultaneously. The STU-SR-SSA (single transcript unit-surrogate reporter-single-strand annealing system) is an example (Tang et al. 2024). STU-SR-base editor systems use base editing to restore the start codon by increasing C-to-T and A-to-G base editing occurrences.

In the past 25 years, we have discovered that new plant breeding technologies cannot reach their full potential without the approval of the public. Regrettably, there is still widespread public resistance to GM crops because of deep-seated biases and incorrect beliefs about potential dangers (Ichim 2020). The opposition to GM crops is especially strong in Europe, and it has now spread to other parts of the world from Europe. New gene-editing tools have been created, promising to enhance the accuracy and speed of developing innovative crop traits (Bhuyan et al. 2023). Using CRISPR and other gene-editing tools, it is possible to create numerous new characteristics in crops without the need for introducing foreign DNA. This could help address many of the public's worries surrounding genetically modified crops (Ahmad 2023). Currently in Europe, gene-edited crops are viewed and controlled in a similar manner to transgenic GM crops. However, as more knowledge is gained about the benefits and safety of these crops, along with better science communication, public concerns may lessen in the future.

Genomic data from crops and their wild counterparts provided valuable understanding into the domestication process (Krug et al. 2023). Developing new crop varieties through de novo domestication of wild plants is a novel approach in crop breeding to address upcoming food demands (Yu and Li 2022). Pan-genomic information and the evolution of gene repositories into bio-digital assets enable the mapping of genetic variation throughout complete groups. Extensive genotyping combined with efficient and detailed phenotyping, supported by FAIR large-scale "omics" datasets, serves as the foundation for using machine learning to forecast phenotypes based on genotypes (Liao et al. 2023). These, in addition to multiscale computational models connecting genes and genetic variation to phenotypes will ultimately enable trait design from scratch (refer to Wang et al. 2023a, b). Genome editing allows for the transformation of information into practical use, transitioning from decoding and comprehending genetic sequences to altering them, by merging different genetic variations in unique ways, free from the limitations of sexual reproduction and genetic linkage.

In 25 years from now, metabolic pathways created by humans will have surpassed the inefficiencies of naturally occurring pathways, like the oxygenation reaction of Rubisco (Zhao et al. 2024). Items produced from new, specially bred plants will have taken the place of animal-based products like meat and dairy, maintaining the same taste, texture, and price. Taking livestock out of the human food system will greatly decrease the area of arable land allocated for food production, therefore creating more land available for supporting biodiversity (refer to Wang 2022). Crops with high energy content and improved nutrient and water efficiency will enable the use of bioenergy with carbon capture and storage to effectively combat human-caused climate change, without impacting land use or biodiversity (Hoover et al. 2023; Feigin et al. 2023). Therefore, plant science and plant breeding have the potential to ensure the future of upcoming generations (Bassi et al. 2024; Ortiz 2018).

Over the past 25 years, one of the biggest successes in breeding has been ensuring the protection of crops against pests and diseases, which becomes more challenging with increasing agricultural intensity due to demographic and economic changes (Kalogiannidis et al. 2022). Within the same time period, molecular markers have been effectively utilized in breeding for pest and disease resistance, which is essential for food security in resource-limited countries where crops often develop without the use of fungicides (Luo et al. 2023). In addition to being good for the environment, genetic resistance to biotic stresses also protects people from harmful agrochemicals. While transgenic crops are still debated, everyone agrees on the positive impact of the *Bt* gene in cotton. Rural communities used to face harmful insecticides that led to birth defects before the introduction of the *Bt* gene from *Bacillus thuringiensis* by Monsanto (carrying Cry 1Ac gene) (see Abbas 2018).

Yet, these technologies have failed to meet expectations in enhancing complex genetic traits. A change in mindset will be needed for this. Inheritance of resistance to biotic stresses is more closely aligned with Mendelian patterns compared to the more intricate abiotic stress factors like heat and drought (refer to Nawaz et al. 2023). The likelihood of cloning genes responsible for traits or finding molecular markers associated with them decreases with the growing number of genes in effect. This can be seen in the necessity for plants to be flexible in adapting to various environments across varying periods of time (Schneider 2022). Moreover, the advancement of genomic selection for yield over the past 25 years has highlighted the significance of both genetic background effects and environmental factors in determining which alleles influence crop performance (Alemu et al. 2024). This is the reason why efforts to genetically modify complicated plant characteristics have not succeeded beyond tightly controlled settings and are unlikely to offer solutions, especially for climate adaptation.

In the future, it may be more practical to enhance the chances of acquiring advantageous alleles by crossbreeding different sources of heritable, beneficial complex traits, and utilizing phenomic and genomic selection methods to pinpoint the top offspring (Sukumaran et al. 2022; Zhu et al. 2022). Information gained from yield trials at multiple locations can be leveraged in reverse genetics to determine causality, thanks to another major accomplishment of the past 25 years: the sequencing of crop genomes. By modeling potential impacts and relationships between traits and genes, more targeted screening of crop genetic resources can be achieved, tapping into a vast yet mostly unexplored opportunity in the next 25 years. This will enhance the development of superior germplasm through hybridization. In this situation, well-known field-trial networks like the International Wheat Improvement Network, together with advanced phenomics methods (such as android, drone, and satellite sensors), can assist in confirming effective breeding goals.

1.4.4 Current Yield and Future Demands

Meeting the needs of a rapidly expanding population through global food markets will present a significant challenge in the future. The United Nations predicts that the world's population will grow to 9.4–10.1 billion in 2050, and it is expected to rise to 9.4–12.7 billion by 2100 (United Nations 2019). Multiple factors, including population fertility rate, policies affecting fertility rates, development level, cultural influences, and mortality rates, will play a role in determining the precision of these estimates (Vollset et al. 2020). China serves as a positive illustration of the aforementioned point. Even though China is highly populated, its fertility rate dropped to 1.3 in 2020, which is lower than the global average of 2.3, causing it to fall behind other countries in this aspect (Daszkiewicz 2022). Van Dijk et al. (2021) suggest that a 70% increase in global food production is necessary to match the anticipated population growth by 2050. Nevertheless, the rise in global population is not the sole factor fueling the demand for food. The increase in wealth in developing nations will also boost the need for food (Al Hasan et al. 2022).

Recent research suggests that crop yields may not be continuing to rise in various parts of the world (Ray et al. 2012; Hickey et al. 2019). Yields are not seeing any further increases in 24–39% of our key agricultural land. Several of these regions exist in leading agricultural countries, experiencing growing populations, rising affluence, or a combination of both factors (Tilman et al. 2011). This could make it harder to achieve future crop production targets, but there are still important uncertainties when it comes to creating and implementing strategies. These uncertainties include how crop yields are evolving globally, whether productivity gains can keep up with increasing demands, and where crop yields are not keeping pace. Several research studies indicate that in order to feed a larger and wealthier global population, agricultural production needs to double by 2050 (Godfray et al. 2010), with an estimated annual crop production growth rate of ~2.4%. The average yield improvements for the top four global crops—maize, rice, wheat, and soybean—are currently only 0.9–1.6% per year, much slower than what is needed to double their production by 2050 through yield increases alone. This is due to the fact that there is a yearly yield increase of less than ~2.4% in numerous regions of our key agricultural lands. By 2050, global production of maize, rice, wheat, and soybean could rise by 67%, ~42%, ~38%, and ~ 55%, respectively, at these levels of growth. It is highly likely that the overall worldwide production growth due to yields alone would range from 34% to 101% for maize, 21% to 59% for rice, 4% to 76% for wheat, and 13% to 84% for soybean by 2050. If these growth rates do not rise, land clearing may be necessary to improve or maintain global food security. (Tilman et al. 2011) (see Fig. 1.4). Yields for maize, rice, wheat, and soybean are rising by 1.6%, 1.0%, 0.9%, and 1.3% annually, respectively. However, these rates are lower than the 2.4% needed to double global production by 2050 (Ray et al. 2013).

1.4.5 Obstacles to New Technology Adoption

Upcoming breeding technologies are closely examined by the public and policymakers due to advancements in crop breeding technology. See Table 1.1 for more information. This is especially accurate for gene editing using CRISPR (clustered regularly interspaced short palindromic repeats). While the EU is dedicated to protecting advancements in plant technologies (Anders et al. 2021), current EU regulations lumping CRISPR-produced crops with other forms of genetic engineering pose a significant obstacle to reaping the full advantages of gene editing in crop breeding (Herman et al. 2019). The strict policies in the EU against genetic engineering in crops have had significant repercussions globally, just as the EU's control over gene editing will also have far-reaching effects beyond its own borders (Callaway 2018). China and India, which are two of the most heavily populated countries in the world, have a history of neglecting the needs and desires of their agricultural communities when making policy decisions (Anders et al. 2021).

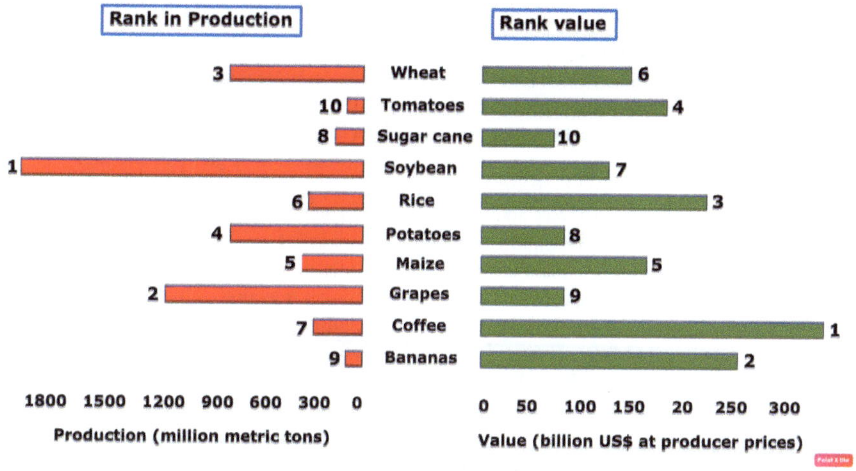

Fig. 1.4 Global Top Ten Crops, by Production Volume and Value. Food and Agriculture Organization of the United Nations (FAOSTAT), Statistical Database, published online. http://www.fao.org/faostat/en/#data. (Courtesy: *Trends in Plant Science;* https://doi.org/10.1016/j.tplants.2021.03.004)

Therefore, solutions in crop-breeding for current food security issues must take into consideration both the scientifically proven risks and benefits perceived by consumers as well as their impact on the agronomic and cultural practices of farmers. For instance, the approval for the consumption and processing of high beta-carotene (golden) rice in the Philippines in January 2020 made it the initial genetically modified crop to address a public health problem (vitamin A deficiency) in a lower-income nation. After two decades, golden rice was finally approved thanks to dedicated researchers collaborating with partners to allow public research institutions in developing countries to transfer high-beta-carotene genes into the most suitable local varieties (Potrykus 2001). Several issues were identified as the cause of the lack of acceptance among farmers in the Philippines (and other Asian countries):

- A deficiency in trust toward institutional figures in society.
- Rice farmers were more focused on agronomic factors, production, yield stability, and cost rather than the nutritional benefits that consumers can enjoy from golden rice, which they are willing to pay for.
- Farmers wanting to cultivate golden rice may feel they lack the necessary marketing knowledge and support (Chong 2003).

Certainly, CRISPR gene editing has the potential to greatly address various global challenges, such as abiotic stress and enhancing micronutrient levels in food (Shaheen et al. 2023). Addressing any of these obstacles necessitates the wholehearted collaboration of millions of farmers and consumers from varying agronomic and cultural backgrounds. Since the introduction of genetically modified (GM) plant production, many research studies have highlighted that these technologies do not present any higher risks to society compared to traditional agriculture (Zaidi

Table 1.1 Description of some of the new plant breeding techniques

Technique	Summary
Accelerated plant breeding (speed breeding)	Induction of early-flowering to accelerate crossbreeding. Also, implement in vitro nurseries, which could substantially shorten generation time through rapid cycles of meiosis and mitosis (where new genotypes are formed by in vitro production of gametes and their subsequent fusion; generation time can be reduced by quickly forming new gametes and fusing these gametes quickly)
Agroinfiltration	Use of recombinant *Agrobacterium* to achieve transient expression of genes in plant tissues. Here, a suspension of *Agrobacterium tumefaciens* is introduced into a plant leaf by direct injection or by vacuum infiltration, or brought into association with plant cells immobilized on a porous support (plant cell packs), whereafter the bacteria transfer the desired gene into the plant cells via transfer of T-DNA
Centromere-mediated genome elimination	Centromeres are points where spindle fibers are attached. Centromeres depend on an epigenetic signal, that is, a persistent DNA modification that does not depend on sequence. This largely mysterious epigenetic signal requires a variant histone H3, called CENH3. The experimental alteration of CENH3, by swapping its amino-terminal region and fusing it to green fluorescent protein (GFP) to produce "Tailswap CENH3," can lead to genome elimination. Genome elimination only occurred when a plant strain with the altered CENH3, referred to as the "Tailswap" haploid inducer, was crossed to a wild-type plant, leading to the elimination of all the Tailswap chromosomes. To date, this event has only been reported in *Arabidopsis*, but given the conserved nature of the perturbed mechanism, it is likely to also apply to crop plants
Cisgenesis	Cisgenesis is the production of genetically modified crops/plants using donor DNA fragment from the species itself or from a cross compatible species. The newly introduced gene is unchanged and includes its own introns and regulatory sequences and is free of vector DNA, except T-DNA border sequences that flank the cisgene. Such a phenotype can be achieved through conventional breeding also, but it takes a longer time. The advantage of cisgenesis is that it introduces only the desired gene, thus avoiding linkage drag (reduction in fitness in a cultivar due to deleterious genes introduced along with the beneficial gene during backcrossing). It also eliminates the hectic and time-consuming backcrossing to recover the recurrent parent genotype
Intragenesis	Intragenesis is very much similar to cisgenesis, but intragenesis allows the creation of novel combinations of DNA fragments. New genes can be created in vitro by combining functional genetic elements like promoters, coding region and terminator sequences, and this new chimeric gene can be inserted into existing varieties. Intragenesis also allows the use of antisense or RNA interference (RNAi) with the aim of silencing the gene(s). Such phenotypes cannot always be achieved through conventional breeding. While cisgenesis involves genetic modification using a complete copy of natural genes with their regulatory elements that belong exclusively to sexually compatible plants, intragenesis refers to the transference of new combinations of genes and regulatory sequences belonging to that particular species

(continued)

Table 1.1 (continued)

Technique	Summary
Grafting on GM rootstocks	Production of chimeras from GM rootstocks and non-GM scions. Here, only rootstocks are genetically modified. Use of *short interfering RNA* (siRNA) is another application which are made in the genetically modified rootstock. They are transported to the graft (scion), where they cause the desired effect. Using this technique, protein production, for example, can be regulated in the upper stem and fruit without the upper stem requiring genetic modification
Induced hypomethylation	Silencing of genes encoding DNA methylation regulators to induce epiallelic variation (variants among alleles that have undergone changes due to transcriptional regulation resulting from epigenetic differences are commonly referred to as epialleles)
Meganuclease technique	Meganuclease provides an efficient way to generate DNA double-strand breaks at specific loci of interest. They are used to knock out targeted genes; to correct targeted genes, or insert new genes at a predetermined site in the genome
Methyltransferase technique	Synthetic methyl transferases are used for targeted methylation of genomic sequences. This will further alter the protein structure and function
Oligonucleotide-directed mutagenesis (ODM)	ODM is a tool for targeted mutagenesis, employing a specific oligonucleotide, typically 20–100 bp in length, to produce a single DNA base change in the plant genome. This oligonucleotide is of a single base-pair change. In cultured plant cells, they bind to the corresponding homologous plant DNA sequence. Then, the cell's natural repair machinery recognizes this single base mismatch and undertakes required repair. Plants carrying the specific mutation are subsequently regenerated by tissue culture techniques and can be used for breeding the desirable trait into elite plant varieties
Reverse breeding	In order to assess hybrid performance, hybrid crop types are typically created by choosing and crossing parental lines. Selecting uncharacterized heterozygotes and creating parental lines from them is made possible through reverse breeding. By using these, the number of hybrids that may be screened in breeding programs is significantly increased as the chosen heterozygotes can be reproduced as F_1 hybrids. In order to guarantee that nonrecombinant chromosomes are transferred to haploid gametes, reverse breeding relies on suppressing meiotic crossovers in hybrid plants. After that, these gametes regenerate into double-haploid (DH) progeny. To recreate the original hybrid, complementing pairs can be crossed between each DH, which carries combinations of its parents' chromosomes. Chromosome number variation caused by achiasmatic meiosis and haploid production results in unique characteristics in the progeny

(continued)

Table 1.1 (continued)

Technique	Summary
RNA-directed DNA-methylation (RdDM)	RNA-directed DNA methylation (RdDM) is a biological process in which noncoding RNA molecules direct the addition of DNA methylation to specific DNA sequences. The RdDM pathway is unique to plants. To date, the RdDM pathway is best characterized within angiosperms (flowering plants), and particularly within the model plant *Arabidopsis thaliana*. The RdDM pathway closely resembles other sRNA pathways, particularly the highly conserved RNAi pathway found in fungi, plants, and animals. Both the RdDM and RNAi pathways produce sRNAs and involve conserved Argonaute, Dicer, and RNA-dependent RNA polymerase proteins. The DNA methylation added by RdDM is generally associated with transcriptional repression of the genetic sequences targeted by the pathway. Since DNA methylation patterns in plants are heritable, these changes can often be stably transmitted to progeny. Transposable element (TE) activity is thus steadily and transgenerationally suppressed, which is one of RdDM's main functions. Abiotic stress responses, pathogen defense, and the control of multiple crucial developmental transitions have all been connected to RdDM. In general, RdDM is a significant system in plants that controls several functions through the establishment and maintenance of particular DNA methylation patterns. These patterns can have transgenerational epigenetic effects on phenotypic and gene expression
Seed Production Technology	Use of transgenic maintainer lines to propagate male sterile female parental lines used in producing hybrid seeds. Hybrid seed production uses cytoplasmic male sterile lines or photoperiod/thermosensitive genic male sterile lines (PTGMS) as female parents. Cytoplasmic male sterile lines are propagated via cross-pollination by corresponding maintainer lines, whereas PTGMS lines are propagated via self-pollination under environmental conditions restoring male fertility. Alternatively, construction of the male sterility system using a nuclear gene that encodes a putative glucose–methanol–choline oxidoreductase regulating tapetum degeneration and pollen exine formation. Cross-pollination of the fertile transgenic plants to the nontransgenic male sterile plants produces male sterile seeds of high purity
TALEN technique	Transcription activator-like effector nucleases (TALEN) are restriction enzymes that can be engineered to cut specific sequences of DNA. They are made by fusing a TAL effector DNA-binding domain to a DNA cleavage domain (a nuclease that cuts DNA strands). Transcription activator-like effectors (TALEs) can be engineered to bind to practically any desired DNA sequence, so when combined with a nuclease, DNA can be cut at specific locations. TALEN is a tool in genome editing
Targeted chemical mutagenesis	Use of oligonucleotides coupled to chemical mutagens to trigger mutations at a predetermined site of the genome
Target mutagenesis with T-DNA	Use of T-DNA to replace an endogenous target gene with a homologous gene with altered DNA sequence
Transformation with wild-type *Agrobacterium*	Use of wild-type *Agrobacterium rhizogenes* for producing transformed plants

(continued)

Table 1.1 (continued)

Technique	Summary
Virus-induced Gene Silencing (VIGS)	This is a technique for using recombinant viruses to achieve transient gene silencing in plants. VIGS is a technology that exploits an RNA-mediated antiviral defence mechanism. It is one of the reverse genetics tools for the analysis of gene function that uses viral vectors carrying a target gene fragment to produce dsRNA, which trigger RNA-mediated gene silencing. Virus-derived inoculations are performed on host plants using different methods such as agro-infiltration and in vitro transcriptions. In plants infected with unmodified viruses, the mechanism is specifically targeted against the viral genome
Zinc finger nuclease technique	Zinc finger nucleases (ZFNs) are a class of engineered DNA-binding proteins that facilitate targeted editing of the genome by creating double-strand breaks in DNA at user-specified locations. Each zinc finger nuclease (ZFN) consists of two functional domains: (a) A DNA-binding domain comprised of a chain of two-finger modules, each recognizing a unique hexamer (6 bp) sequence of DNA. Two-finger modules are stitched together to form a zinc finger protein, each with specificity of ≥24 bp. (b) A DNA-cleaving domain comprised of the nuclease domain of Fok I. When the DNA-binding and DNA-cleaving domains are fused together, a highly-specific pair of "genomic scissors" is created (see chapter on Genetic Engineering)
CRISPR-cas9	CRISPR-Cas9 (Clustered Regularly Interspaced Short Palindromic Repeats) was adapted from a naturally occurring genome editing system in bacteria. The bacteria capture snippets of DNA from invading viruses and use them to create DNA segments known as CRISPR arrays. If the viruses attack again, the bacteria produce RNA segments from the CRISPR arrays to target the viruses' DNA. The bacteria then use Cas9 or a similar enzyme to cut the DNA apart, which disables the virus. A small piece of RNA with a short "guide" sequence that attaches to a specific target sequence of DNA in a genome with the Cas9 enzyme is made. The Cas9 enzyme cuts the DNA at the targeted location. Once the DNA is cut, the cell's own DNA repair machinery is used to add or delete pieces of genetic material, or to make changes to the DNA by replacing an existing segment with a customized DNA sequence
Single base editors	Scientists have developed a single-base editing system (base editor) by combining the CRISPR/Cas9 system with cytosine deaminase. Compared with the Cas9 system, this base editor can convert cytosine to thymine (C > T) at a specific site more efficiently without inducing double-strand breaks to avoid the generation of indels (insertion or deletion of bases). However, the base editor can only generate the transition of pyrimidine but cannot modify purines. Recently, a novel base editing system to convert adenine to guanine (ABEs, adenine base editors) through the fusion of Cas9 nickase to a modified deaminase has been evolved through screening of a random library based on tRNA adenine deaminase from *E. coli*

(continued)

Table 1.1 (continued)

Technique	Summary
Crop breeding, Chips, and genotyping platforms	Up to now, over 50 SNP arrays and 15 various genotyping-by-sequencing (GBS) platforms have been created for more than 25 crop species and perennial trees, a genetic screening method used to find new SNPs. It is anticipated that in the near future, major crops will have genotyping platforms due to advancements in gene discovery, understanding of quantitative traits, integration of omics data, and cost-effectiveness in genotyping. Crop breeding microchips and genotyping platforms offer unique chances to speed up the creation of plant varieties with desired yield, quality, and improved resilience to combat climate change impacts
Breeding for C4 photosynthesis	C4 plants are more effective than C3 plants at using resources and at photosynthesis, especially in hot areas where productivity potential is high. The introduction of C4 photosynthesis into C3 crops is widely acknowledged as a crucial approach to boost crop yield. In C3 crops, lowering photorespiration could considerably boost rates of photosynthesis and yield. There is evidence that various lineages of C4 photosynthesis diverged from C3 photosynthesis. However, C3-to-C4 engineering is a challenging endeavor since C3 plants must be given access to a number of traits necessary for C4 photosynthesis. The geographical division of the two components of photosynthesis—CO_2 fixation and carbohydrate synthesis—into the mesophyll and bundle sheath cells, respectively, is one of these characteristics. The Kranz anatomy, which is distinguished by a strong relationship between the mesophyll and bundle sheath (BS) cells (1:1 ratio), is another trait. These anatomical characteristics create a CO2 concentration mechanism that ensures a high photosynthetic efficiency together with a C4-specific carbon fixation enzyme phosphoenolpyruvate carboxylase (PEPC).
RNA-seq (whole transcriptome shotgun *sequencing*—WTSS)	Shotgun transcriptome sequencing is used to characterize and annotate the functional properties of genes that have been discovered through DNA sequencing as well as to find and quantify coding and noncoding RNAs. Gene-to-gene interaction networks can be created using whole transcriptome analysis by shotgun sequencing, commonly referred to as RNA sequencing (RNA seq), in order to understand the operation of many biological systems. To put it simply, whole transcriptome shotgun sequencing contributes to creating a transcriptome blueprint that contains the whole population of cellular RNAs (mRNA, tRNA, and rRNA). By doing so, it is able to ascertain the degree of gene expression as well as the pattern, status, and time of gene activation. RNA editing, alternative splicing, transcriptional networks, variable gene expression, and post-transcriptional alterations (polyadenylation and 5′ capping) are among the specific uses for these sequencing approaches

(continued)

Table 1.1 (continued)

Technique	Summary
Metabolomic profiling	Metabolomics is a useful technique for bettering our understanding of the physiology and biochemistry of organisms. This useful method is being used to study numerous aspects of plant biology in plants, such as growth and development, reactions to pressures from the environment, genetics, and nutrient requirements. As a result, it assists other omics technologies used to clarify how cells, tissues, and plant systems function. Understanding how plants produce a variety of metabolites in response to environmental challenges is crucial to understanding how they adapt, acclimatize, and defend themselves. Metabolomics also has the potential to be employed in genome editing programs to create improved next-generation crops because it is simple to apply to the phenotyping of plants
Phenomics	The systematic investigation of phenotypic traits is known as phenomics. Plant phenomics is a high-throughput field that is breaking new ground and satisfies all criteria for accurate, speedy, and comprehensive phenotypic data collection. Research in this area crosses a number of discipline boundaries, including biology, data sciences, engineering, and others. A phenome is a set of qualities (physical and biochemical attributes) that can be created by a specific organism over the course of development and in response to genetic mutation and environmental effects. Phenomics is concerned with measuring the phenotype
Genomic prediction	The goal of genomic selection is to forecast an individual's future genetic worth or phenotypic feature. The genomic selection-based technique in plant breeding that incorporates the genomic prediction (GP) model uses a set of markers distributed across the entire genome. A GP model is built on a reference population whose genotypes and accompanying phenotypes are known before being applied to related populations whose genotypes are only known. For the purpose of choosing the best parents for the following generation(s), the total genomic value, calculated from the GP model, is employed as a pseudo-phenotype. GP is difficult for complex traits since they are typically driven by a combination of additive and nonadditive (such as dominant/recessive and epistatic) allele effects
Climate-resilient breeding with artificial intelligence	AI is one tool that can help us understand what the models are saying to us and where uncertainties in climate change are coming from. This understanding may then be used to improve observation programs, enhance models, or even incorporate AI into model systems. This field of study is to create technological systems that are able to solve problems and carry out tasks or duties that would typically be carried out by the human mind. It can be characterized as a set of studies and techniques dealing with the computer science and mathematical aspects of statistical modeling

(continued)

Table 1.1 (continued)

Technique	Summary
Controlled recombination	In plant species, crossovers (COs), which promote genetic exchange across homologous chromosomes, are skewed greatly in favor of sub-telomeric areas. A long-standing objective is to manipulate the pace and locations of COs to broaden the genetic diversity available to breeders. In order to accomplish this, it is becoming more and more practical to use genome editing agents that cause double-strand breaks (DSBs) or alter the epigenome at desired recombination locations, as well as CO factor modification. These "controlled recombination" techniques offer the ability to shorten the traditional breeding process's time and cost, disclose previously hidden genetic diversity, and provide breeders more control over the inheritance of desired haplotypes. The quantity of accessible genetic variation can be increased through carefully managed recombination. Multiple crop species have shown that silencing anti-CO genes can up to threefold the rate of recombination

et al. 2019). However, the overall view of GM and other revolutionary technologies continues to be unfavorable. The rise of genome-editing technologies has sparked great anticipation among plant researchers, breeding firms, and the food sector, with the common belief being that scientific advances can alter how consumers view things. Despite CRISPR/Cas technologies being more accepted compared to GM plant production, global consumer acceptance is still affected by food technology neophobia (Farid et al. 2020). This viewpoint is influenced partly by the law that considers genome-edited crops as equivalent to early-generation GM plants. Some countries permit genome-edited crops that do not contain foreign DNA without any limitations as they are seen as traditional plants (Voigt 2023) (see Fig. 1.5). Research on adoption preferences and barriers for CRISPR/Cas technologies among farmers, particularly in developing countries, is limited, making it unclear how they perceive these technologies as unique adopters (Pixley et al. 2022).

1.4.6 Smart Breeding

In favorable environments, plants put most of their energy from photosynthesis into growth; however, in harsh conditions, they shift their focus to stress responses in order to survive. Nevertheless, under extended periods of stress, plants may need to alter their reproductive growth to ensure the survival of future offspring needs to be considered (Verslues et al. 2023). Trade-offs happen when a certain characteristic gives a benefit in one situation but comes with a drawback in the same or a different situation (Wang et al. 2024). An example that is commonly used to demonstrate trade-offs is the balance between carbon absorption and water evaporation in the process of photosynthesis. As stomata open to take in carbon dioxide, they also inevitably release water. The intelligent plant has the ability to choose between a "water-spender" approach that focuses on obtaining a lot of carbon, even if it

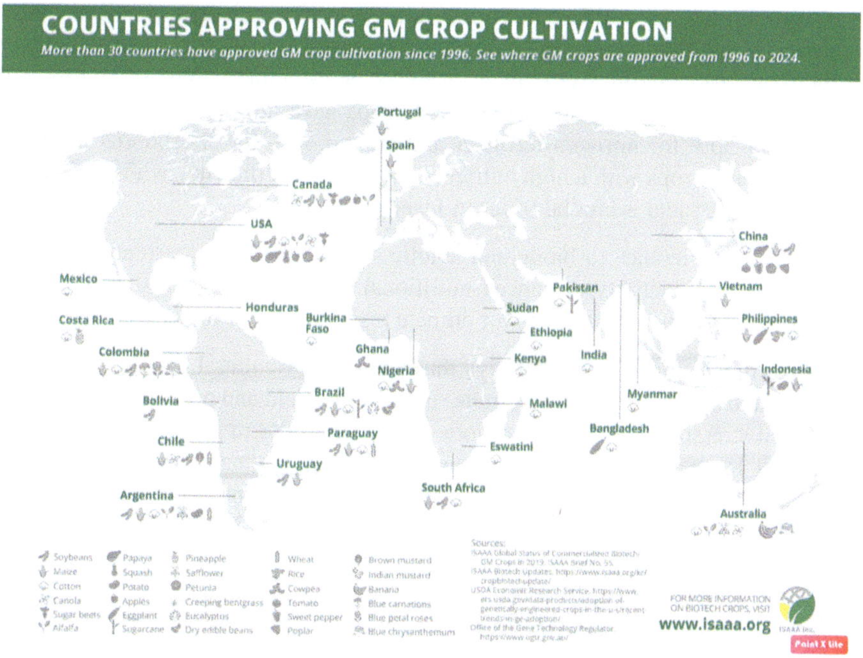

Fig. 1.5 An up-to-date global status of the regulation of GM plants (with permission of ISAAA)

requires a lot of water, and a "water-saver" approach that limits water usage to acquire less carbon, while optimizing water use efficiency (WUE). Wang et al. (2024) utilized the idea of trade-offs to explain the concept of "SMART CROP," which refers to crops capable of effectively managing the trade-off between growth/survival and stress response/yield. Intelligent crops yield higher food quantities in a faster timeframe, decreasing the reliance on chemicals such as pesticides and fungicides while contributing to eco-friendly, sustainable farming practices (Richard et al. 2021). The challenging objectives for intelligent crops go beyond just improving productivity and tolerance to stress. They also involve maintaining or enhancing nutritional content, maximizing the use of fertilizers and water, and exploring new methods of utilizing plant–microbiome interactions (Yang et al. 2022). Tian et al. (2021) proposed five factors for developing future smart crops:

- Advancing productivity: The primary goal for upcoming intelligent crop design initiatives is to create new high-yield plant types in order to provide food for ten billion individuals by 2050 (refer to Wang et al. 2024).
- Enhancing nutritional value: Around 828 million people are still undernourished and 1 billion individuals suffer from malnutrition worldwide (von Braun 2010; also cited in Webb 2021). Create plants with improved nutritional content or that can produce specific secondary compounds using techniques such as metabolic engineering and synthetic biology (refer to Mipeshwaree Devi et al. 2023).

- Improving agricultural resource utilization: The various plant mechanisms linked to increased drought resistance, regulation of leaf stomatal closure, and stimulation of root system development are crucial for minimizing water loss and enhancing WUE. Furthermore, modern crops have nutrient-use efficiencies of only 30–50% for nitrogen fertilizer and around 45% for phosphorus fertilizer. Developing crops with a high nutrient and water-use efficiency while avoiding a reduction in yield is crucial (refer to Javed et al. 2022).
- Enhancing tolerance to biotic and abiotic stress: Enhancing current crops with increased productivity, enhanced nutritional characteristics, resistance to pests and diseases, and adjustments for climate resilience (refer to Mukherjee 2022).
- Exploring the molecular pathways that regulate agricultural characteristics is a difficult task due to conflicting traits such as crop yield and resistance to environmental stressors like abiotic or biotic factors. In order to improve crop production, it is necessary to design future crops with optimal structure, enhanced disease and pest resistance, improved water and nutrient utilization efficiency, and better metabolic quality (refer to Lastochkina et al. 2022).

1.4.7 Omics in Crop Breeding

Several "omics" methodologies have become effective tools for studying plant systems in recent past.. Progress in next-generation sequencing (NGS) has opened up opportunities for a new era of various omics, including genomics, transcriptomics, and proteomics. Nevertheless, crop science has also extensively documented metabolomics, ionomics, and phenomics. High-throughput techniques combining multiple omics methods have been crucial in understanding the development, aging, productivity, and reactions to different stresses in various crops. The combination of functional genomics with other omics reveals the connections between crop genomes and phenotypes in distinct physiological and environmental scenarios (Yang et al. 2021). Among these categories, genomics can be divided into structural genomics, functional genomics, epigenomics, and pan genomics, each highlighting a distinct aspect of genomics (Yang et al. 2021). These multi-omics datasets, coupled with bioinformatics workflows, play crucial roles in elucidating the genetic origins and molecular underpinnings of intricate crop characteristics. For example, omics allows for quick identification, isolation, and functional characterization of quantitative trait loci (QTL) through cloning and high-resolution association analysis between markers and traits (Zhang et al. 2019). Analyzing transcripts in different environments helps to discover genes or networks crucial for responding to stress or adaptation (Groen et al. 2020). Furthermore, utilizing omics technology at the single-cell level allows for the discovery of unknown cell types and states, leading to a more comprehensive understanding of plant development (Sun et al. 2022).

Incorporating various omics data and extracting valuable knowledge from the combined information requires significant adjustments to the analytical approach, which involves traditional model-based statistical analysis (Zhang et al. 2022). Yan and Wang (2023) suggest that machine learning and deep learning (DL) techniques can efficiently manage the integration of multi-omics data and convert the acquired information and biological data into targeted breeding in plants. Additionally, there have been tools and platforms created for the purpose of combining multi-omics data. PAINTOMICS, as an example, is a tool available on the web that displays combined multi-omics data and their pathway annotations (Liu et al. 2022a, b). Panomics is a newly described concept that offers a framework for combining intricate omics information, such as genomics, epigenomics, transcriptomics, proteomics, PTM proteomics, metabolomics, and phenomics (Weckwerth et al. 2020). The utilization of multiple omics approaches has resulted in the discovery of commonly regulated pathways and genes that respond to drought in various crops (Subramanian et al. 2020). Because of the quick progress in high-throughput technologies and computational tools, combining multi-omics data with other biological methods will speed up the identification of functional genes and regulatory networks, leading to crop improvement suitable for changing environments (Fig. 1.6).

1.4.8 High-Throughput Phenomics Techniques

With the rapid progress of high-throughput phenotyping technologies, a new era named "phenomics" has arisen in this field (Jiang and Zhu 2024). In recent years, significant strides have been taken to create high-throughput phenotyping methods for various objectives like cells, seeds, shoots, leaves, roots, individual plants, and canopy (Yang et al. 2020a, b). High-quality imaging methods like micro-CT imaging and microscopic imaging can be employed to analyze cell count, alterations in cell structure, cell growth rate, and tissue morphology. In terms of seed phenotyping, visible light imaging has been extensively used to analyze morphological characteristics like color, coleoptile length, and germination rate. Ma et al. (2016) created a portable device using smartphones for evaluating seed morphological parameters. Additionally, X-ray imaging has been utilized for assessing seed morphometric characteristics and tissue health (De Medeiros et al. 2020). Near-infrared spectroscopy and time-domain pulsed nuclear magnetic resonance (NMR) showed benefits in evaluating protein, oil content, and fatty acids in seeds (Melchinger et al. 2018). Regarding organs, individual plants, and the canopy, the features and capabilities of different high-throughput phenotyping methods for gathering phenotypes at these levels have been recorded. Yang et al. (2013) provided an overview of high-throughput phenotyping techniques for plants, focusing on morphological, physiological, and pathological traits, as well as their applications in assessing abiotic stress, pest stress, and yield quality in rice and other crops, as summarized by Zhang and Zhang (2018). Shakoor et al. (2017) conducted a study on high-throughput phenotyping platforms and sensors. The numerous potential uses of high-throughput phenotyping methods in evaluating diseases were also explained. Furthermore, Liu et al. (2020) provided an in-depth examination of hyperspectral imaging and 3D sensing in plant phenotyping. Jang et al. (2020) and Yang et al. (2017) concentrated on examining the uses of drones in field crop phenotyping. They provided a detailed

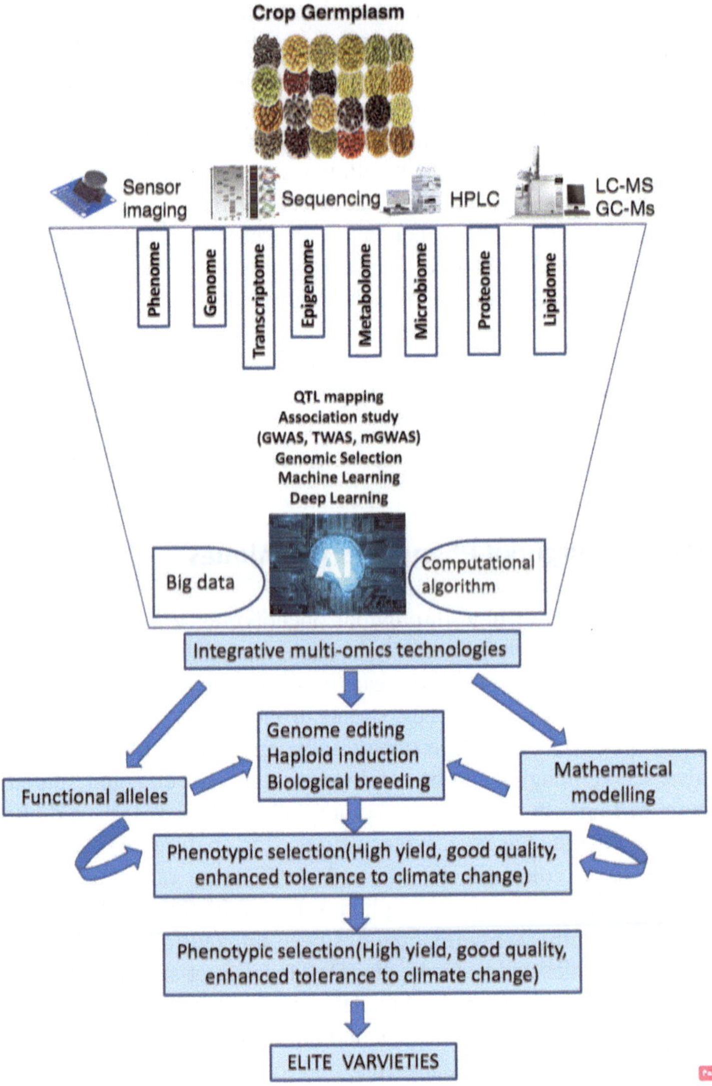

Fig. 1.6 Integration of multi-omics technologies accelerates crop improvement. MS mass spectrometry, NMR nuclear magnetic resonance, HPLC high-performance liquid chromatography, GC gas chromatography, AI artificial intelligence. (Modified from: Frontiers in Bioinformatics https://doi.org/10.3389/fbinf.2022.1027457)

summary of the sensors that can be installed on unmanned aerial vehicles. These techniques are summarized in Fig. 1.7.

Apart from the high-throughput phenotyping methods mentioned earlier, a few additional techniques have been utilized for plant phenotyping, with a few instances being documented. High-throughput phenotyping techniques can be combined to

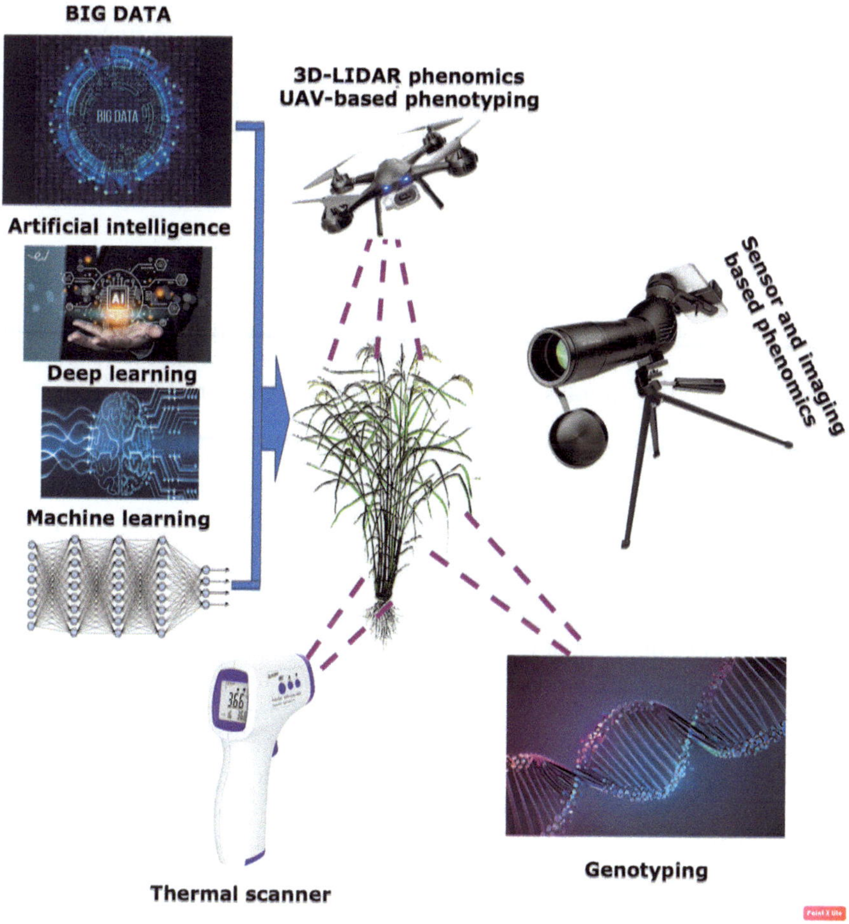

Fig. 1.7 Incorporating advanced technologies into phenomics for sustainable farming

achieve more thorough and precise outcomes. RGB (red, green, blue) cameras and multispectral sensors were combined to estimate yield, as stated by Herrero-Huerta et al. in 2020. The integration of thermal sensors, RGB cameras, and multi-spectral sensors has been utilized for yield assessment (Maimaitijiang et al. 2020), determination of biochemical parameters, and quantification of biophysical parameters (Maimaitijiang et al. 2017). In addition, ultrasonic sensors for measuring distance, along with spectral, thermal, and other sensors, have been used in studies on crop canopy analysis by Bai et al. (2016) and yield estimation by Rischbeck et al. (2016). High-throughput phenotyping is a reliable basis for discovering the genetic traits that cause plant phenotypes (Yang et al. 2020a, b). Thanks to the advancements in computer technology, deep learning proves to be highly beneficial in processing large amounts of data, according to Zhou et al. in 2019. Several studies investigating deep learning algorithms have started to appear in the field of plant phenotyping research (Mostafa et al. 2023). Moreover, deep learning has excelled in various plant phenotyping tasks such as forecasting relative moisture content (Rehman et al. 2020), identifying flowering time and rate (Wang et al. 2019a, b, c), estimating and predicting yield (Moghimi et al. 2020), and detecting and quantifying diseases (Stewart et al. 2019). Using deep learning in plant phenotyping is a predictable trend, with an increase in research expected in the future (Xiao et al. 2022).

1.4.9 Genome-Wide Association Studies

GWAS employ statistical techniques to investigate links between genetic variations in the genome, such as single nucleotide polymorphisms (SNPs) and structural variations, and variations in traits. GWAS greatly aids in examining the genetic structures linked to complex traits (Rebolledo et al. 2016) and explores the genetic foundation for the diversity of plant phenotypes (Xiao et al. 2017). The benefits of GWAS, which have been emphasized, surpass the efforts needed for constructing mapping groups in traditional genetic mapping methods. Initially, GWAS employed natural populations for research, leading to a significant decrease in research duration. Furthermore, GWAS is characterized by its high efficiency and high resolution. Thirdly, GWAS is carried out on a genome-wide level and enables the simultaneous identification of various alleles at the same locus. Natural populations exhibit extensive genetic variation; genetic information and germplasm from a single natural population can be applied to analyze numerous traits, eliminating the necessity for establishing populations repeatedly and significantly cutting down on gene sequencing expenses. Significant agricultural characteristics like productivity, crop quality, and resistance to diseases are usually determined by several genes and can be affected by constantly changing environmental factors. The genetic makeup of complex traits is more complex than traits influenced by a single gene. Hence, the use of GWAS in uncovering the genetic foundation of plant characteristics has garnered significant interest. Figure 1.6 demonstrates the detailed analytical method for high-throughput phenotyping along with GWAS.

Speed breeding combines various techniques to speed up the flowering and seed production of crop genotypes by manipulating environmental factors, and can be combined with the aforementioned technologies. Speed breeding can produce about three to nine generations annually (Ghosh et al. (2018). Speed breeding methods work efficiently alongside marker-assisted selection (MAS) and high-throughput phenotyping techniques to choose multiple desired traits at the same time. Controlled growth chambers have a significant influence on plant growth processes by providing adjustable factors such as light intensity, temperature, photoperiod, soil moisture, carbon dioxide level, and nutrient supply (Abdul Aziz and Masmoudi 2024). Speed breeding can be clubbed with omics and AI to further augment and innovate the production of new varieties (see Ćeran et al. 2024). Yet another upcoming area of research is meiosis-like induction in cells (in vitro nurseries) (see Box 1.1).

Box 1.1 Meiosis-Like Induction in Cells (In Vitro Nurseries)

The speed at which generations (vegetative and reproductive) alternate and the time it takes for plants to flower restrict the genetic advances achieved by plant breeders. Various attempts have been made to shorten the transition time from vegetative to reproductive stages in plants, but these methods are typically tailored to specific species and involve inducing flowering. In vitro nurseries involve inducing somatic plant cells to develop recombinant haploid cells, allowing for cell fusion in a Petri dish to facilitate breeding (Cook et al. 2023).

It can take months for flowering and gamete formation to occur, like in maize, or even decades in certain woody species. Speed breeding, early flowering induction through phytohormones, and dwarf plant breeding are all techniques aimed at shortening the seed-to-seed timeline. Speed breeding has likely had the biggest effect on shortening the breeding cycle by significantly reducing flowering time through the manipulation of growth conditions. These advancements in breeding still depend on fully grown plants with flower parts for the formation of gametes. On the other hand, in vitro breeding programs have been proposed as a method to enable meiosis to take place separately from gamete formation during flowering. In a mammal system, in vitro meiosis has been successfully achieved with rates of approximately 1%. Up to now, the only concepts are in vitro nurseries (IVN) or cycling of gametes in vitro (COGIV). Getting meiosis induction to occur in a laboratory setting brings us one step closer to bringing these concepts to life. In order to assist with this task, Cook et al. (2023) outlined a dual-fluorescent, single-cell system in *Arabidopsis thaliana* that enables testing of candidate factors for their abilities to induce meiosis-like reactions. They additionally confirmed the system's accuracy by utilizing large simulated meiosis induction test datasets, with over four million datapoints. Furthermore, Cook et al. (2024) also established the rates of false detection for the fluorescent cells employed in this setup and identified the most suitable tissue origin for conducting factor tests. If accomplished as a routine production system, this can revolutionize plant breeding in the years to come through augmenting speed breeding (Fig. 1.8).

Much of the existing knowledge regarding MMC specification is derived from the model organism *Arabidopsis*, where the essential role of the nuclear-localized protein SPOROCYTELESS/NOZZLE (SPL/NZZ) in forming the female germline has been identified (Mendes et al. 2020). Without SPL/NZZ, no megaspore mother cells (MMCs) are created. SPL/NZZ gathers TOPLESS/TOPLESS-RELATED (TPL/TPR) corepressors to suppress the expression of CINCINNATA (CIN)-like TEOSINTE BRANCHED 1/CYCLOIDEA/PCF (TCP) transcription factors and create conditions for MMC formation (Wei et al. 2015). In contrast, the absence of TPL1 function alongside the overexpression of TCP factors leads to plants that frequently do not form an MMC

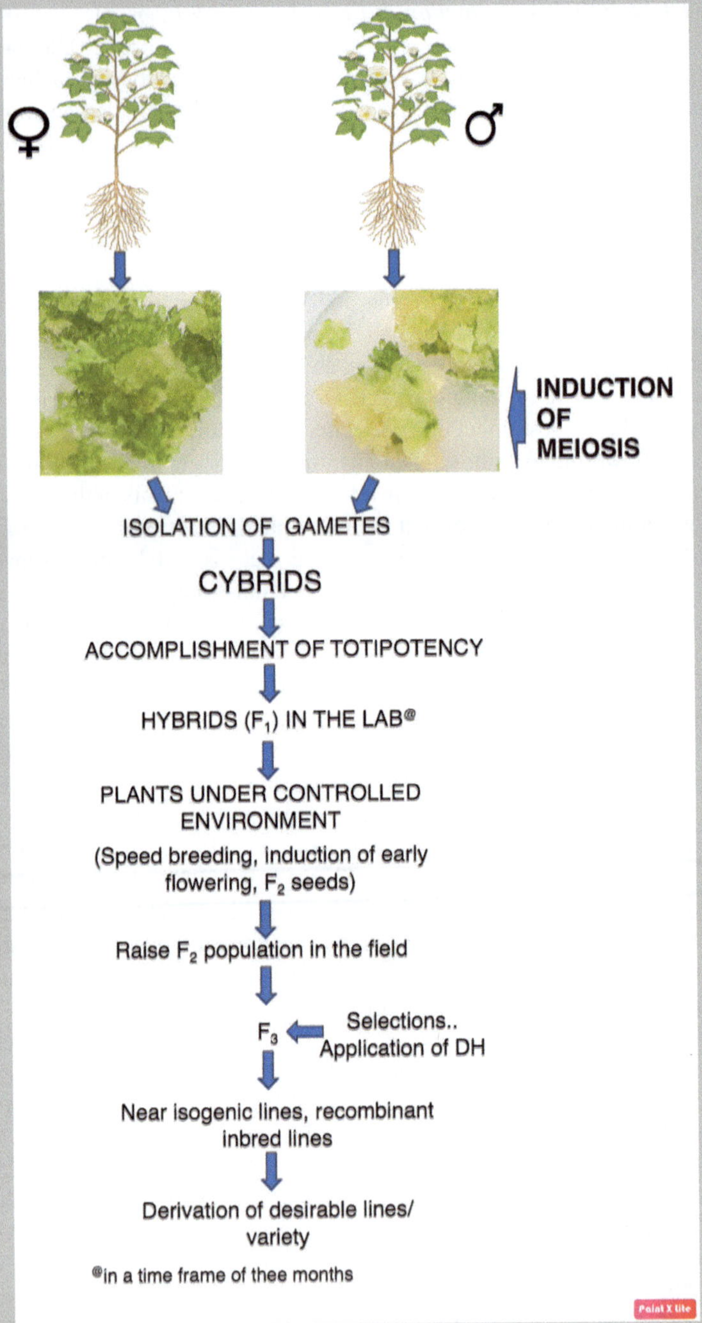

Fig. 1.8 Mitosis to meiosis phase change and its advantages to augment speed breeding. The concept of in vitro nursery

in ovule primordia and bear some similarities to the *spl/nzz* mutant phenotype (Böwer and Schnittger 2021).

Similar to the specification of the female germline, SPL/NZZ seems to play a key role in the regulation of pollen mother cell (PMC) formation. In *spl/nzz* mutants, ARs cease to differentiate and display a vacuolated appearance (Li et al. 2019a, b). The goals of SPL/NZZ related to the PMC specification are ambiguous, and WUS, regulated by SPL/NZZ throughout female germline development, is absent in anthers (Deyhle et al. 2007). WUS is a crucial transcription factor primarily recognized for its role in preserving stem cell identity in the shoot apical meristem. However, issues like molecular differences in female versus male germline development and how germline development is conserved need to be addressed.

Uniparental Genome Elimination Uniparental genome elimination (UGE) happens when the chromosome set from one parent is eliminated during embryonic development post-fertilization, resulting in the creation of haploid offspring from that single parent. This man-made occurrence was observed following the cultivation of plants from distant origins. The elimination of paternal chromosomes takes place selectively during the initial stages of development in hybrid embryos of *Hordeum vulgare x H. bulbosum* (Gernand et al. 2006). After fertilization, the CENH3 genes from both the mother and father are actively expressed. CENH3 is incorporated into the centromeres of *H. vulgare*, while this does not occur in *H. bulbosum*, likely due to the lack of synchronization of the two parental genomes during the mitotic G2 phase. This causes chromosome lagging in *H. bulbosum* because of inactive centromeres in anaphase, resulting in micronuclei formation. To summarize, the chromatin in micronucleated *H. bulbosum* will disintegrate, resulting in the formation of a haploid *H. vulgare* embryo (Sanei et al. 2011; Liu et al. 2014) (see Fig. 1.9). Typically, most of the *H. bulbosum* chromatin is swiftly removed in almost all embryos within nine days after pollination. Differences are present in the mitotic processes of the original chromosomes, since *H. bulbosum* chromatids split unevenly during anaphase.

Due to lack of genetic tools and genomic resources in species with UGE, it has been challenging to determine the molecular mechanism behind this process. Ravi and Chan demonstrated in 2010 that *Arabidopsis thaliana* is commonly used for researching the molecular mechanisms of UGE. Changing centromeres by manipulating the CENH3 gene can result in mimicking genetic results of distant hybrids within *A. thaliana* intraspecific cross (Thondehaalmath et al. 2021). The advantages of UGE in crop improvement have been harnessed through the experimental benefits of *A. thaliana* (refer to Marimuthu et al. (2021) and Zhou et al. (2022a, b) for more details at the molecular level). Both meiosis-like induction in cells and uniparental genome elimination can be brought together to significantly curtail the time taken to produce a variety (Fig. 1.10). In this way, the time taken to produce a variety can be confined to 2 years.

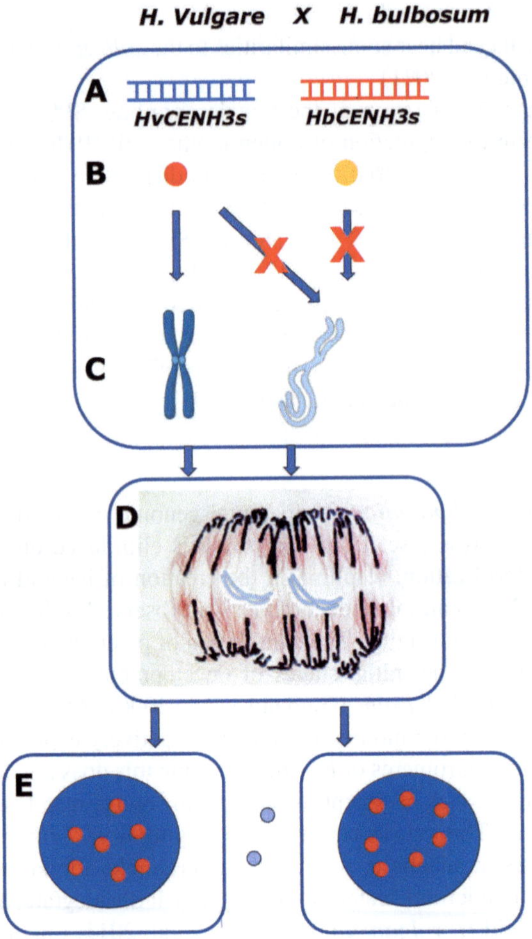

Fig. 1.9 Proposed framework for understanding the mitosis-driven mechanism of uniparental chromosome elimination in *H. vulgare* × *H. bulbosum* hybrid embryos. (**a**) Following the fertilization of the *H. vulgare* egg by the *H. bulbosum* sperm, all parental CENH3s exhibit transcriptional activity. (**b**) HvCENH3s undergo translation; it is unknown if HbCENH3s undergo translation. (**c**) In *H. vulgare*, CENH3 (red) is loaded into the centromeres, whereas in *H. bulbosum*, it is not. Due to the cell cycle asynchrony between the two parental genomes, the incorporation of CENH3 likely occurs solely in the centromeres of *H. vulgare* during the G2 phase. A lower temperature during early embryonic development encourages standard centromere function of both parental genomes. (**d**) Due to the inactivity of the centromere in unstable hybrids, anaphase chromosomes of *H. bulbosum* lag behind and later develop into micronuclei. Micronucleated *H. bulbosum* chromatin eventually breaks down leading to the development of a haploid *H. vulgare* embryo. (Modified from Sanei et al. 2011). (Courtesy: Proc Natl Acad Sci U S A. https://doi.org/10.1073/pnas.1103190108)

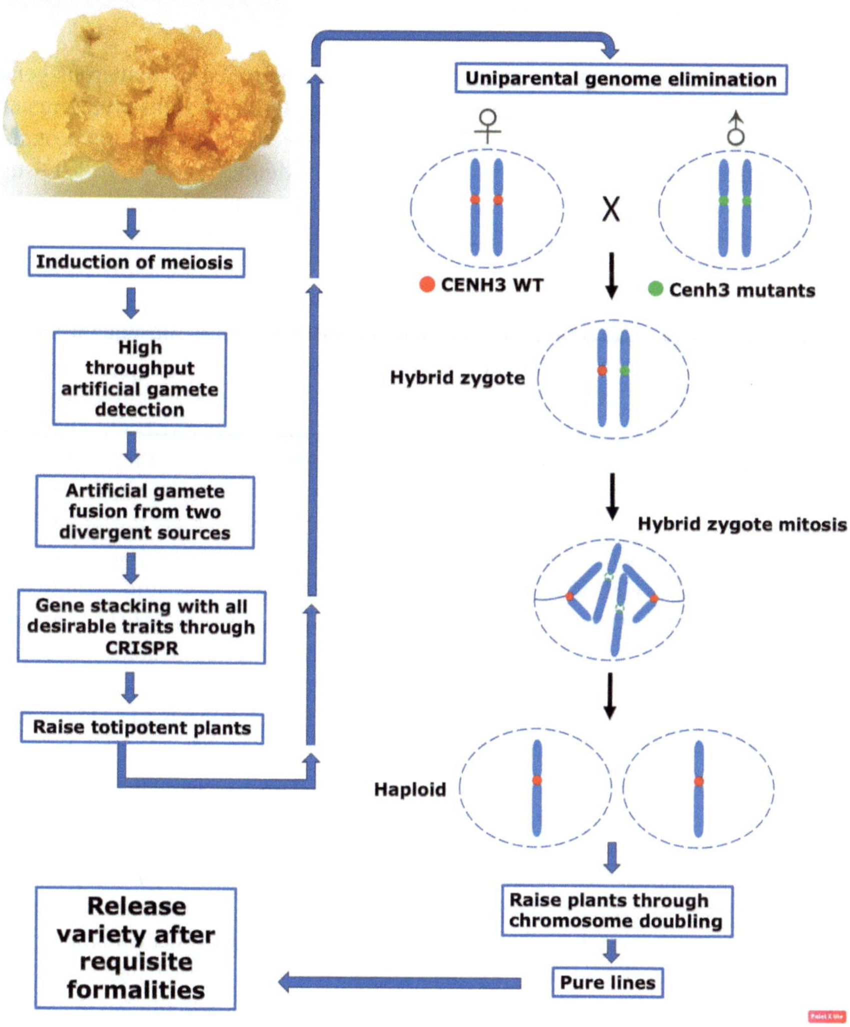

Fig. 1.10 Centromere manipulations induce ploidy changes in plants. Centromere-mediated haploid induction. CENH3 mutants can induce haploids when crossing with CENH3 wild-type line. During mitosis of the hybrid zygote, the CENH3 nucleosome can incorporate in the chromosome of WT normally, while CENH3 nucleosomes cannot load on the chromosome from the cenh3 mutant lines, whose centromeres lose the ability to bind with microtubules. Following the anaphase of hybrid zygotes, the laggard chromosomes form micronuclei and are torn into fragments, eventually leading to genome elimination and haploids (after Zhou et al. 2022a, b)

1.4.10 Genomic Selection (GS)

Using markers is challenging for quantitative traits controlled by numerous genes with small effects. Earlier strategies for handling quantitative traits were created, including a multi-marker MAS system, but they proved challenging to recognize and consider all allele effects (Bernardo 2020a, b). Meuwissen et al. (2001) suggested the concept of estimating all markers at the same time, instead of focusing only on those that are considered "significant," in order to capture all of their effects. Meuwissen et al. (2001) named this approach "Genomic Selection" (GS), which has also been known as genome-wide selection or genomic prediction (GP). The initial GS research in plant breeding was carried out in maize with subsequent studies conducted in wheat, barley, and oat (Merrick et al. 2022).

GS utilizes every available marker across the entire genome for calculating breeding values, known as genomic estimated breeding values (GEBV),[1] instead of assessing their significance. In contrast, traditional MAS only considers a predetermined subset of significant markers, disregarding the rest as having no impact (Nayaka and Isobe 2012). By utilizing highly concentrated markers through whole-genome sequencing and/or genotyping, GS can effectively capture both major and minor QTL effects (Heffner et al. 2009). In addition, there is no requirement to look for significant QTL-marker connections for every trait as opposed to QTL mapping. While QTL studies remain important for comprehending the genetic makeup of quantitative traits, breeders may find that GS can streamline the process of managing individual biparental populations and reassessing the impacts of QTLs within the breeding population. Hence, incorporating GS into plant breeding can lead to a more efficient and expedited selection of top genotypes compared to traditional phenotypic selection or MAS methods, ultimately reducing the breeding cycle of various crop species (Kim et al. 2020).

In order to conduct GS, two groups of populations are needed: a training population (TRN) and a testing population (TST) (also known as breeding population; BP). GS constructs a predictive model using the genetic and physical characteristics of individuals in the TRN (Fig. 1.11). The educated models are then utilized to forecast the breeding values of the genotyped individuals in TST without having knowledge of their phenotypes. Breeders can then choose the individuals with better predicted phenotypes using the GEBVs. In this selection process, phenotype evaluations for desired traits are unnecessary; only a genomic profile of individuals is needed, which can be assessed affordably with different genotyping platforms. In this way, GS allows for the selection of superior individuals without needing physical traits, offering breeders significant advantages. Breeding accelerates with GS due to shorter generation intervals between selection cycles. Selecting elite individuals can occur without the need for seed planting, leading to a significant decrease in both the time and cost involved in the selection process. Breeding intensifies with the ability to assess more candidates in each selection cycle (Cobb et al.

[1] Genetic effects spread across the genome are captured through genome-wide marker data to determine an individual's breeding value.

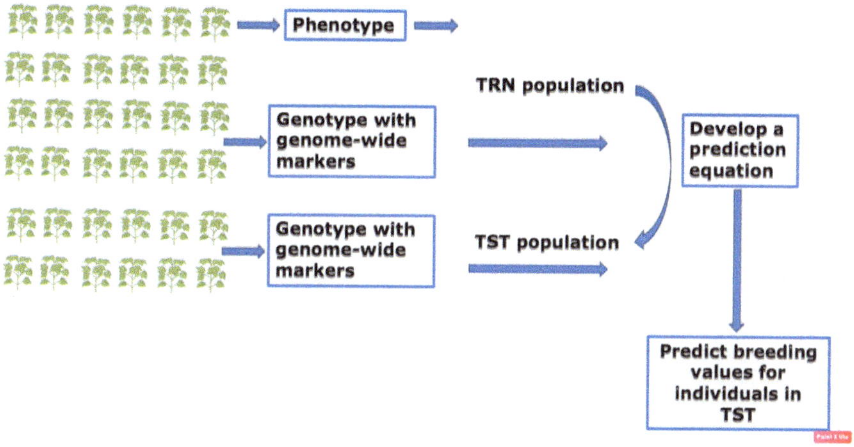

Fig. 1.11 Genomic selection (GS) necessitates a training population (TRN) that has undergone genotyping and phenotyping, as well as a testing population (TST) that has only been genotyped and not phenotyped. (Modified from Crossa et al. 2017)

2019). Before introducing exotic germplasm and wild relatives to the breeding population, it is important to test them in order to assess their potential breeding values. A forecasting model developed under varied growing conditions might identify the individuals or genomic profiles best suited for the shifting climate (Zhang et al. 2015). Plant breeders have various alternatives besides GS, like double-haploid induction and speed breeding, to shorten generation times and replicate clones and inbreds through self or vegetative propagation, limiting the use of GS in plants compared to animal breeders (Hickey et al. 2017). The phenotype of plants is greatly influenced by the interaction between genotype and environment, which is crucial for enhancing the predictive accuracy of genomic selection in plants (Vos-Fels et al. 2019a, b).

Both phenotypic selection (PS) and genomic selection (GS) can be merged to gain benefits (Crossa et al. 2017). In maize breeding, half of the available lines can be testcrossed and evaluated in initial multi-locational trials. The phenotypic data can then be used to predict the other half using genomic selection (GS). Figure 1.12 illustrates that GS is superior to PS in: (i) cutting costs by as much as 50% and (ii) saving time by directly picking lines for stage II instead of going through stage I (as done in PS). This greatly decreases the expenses of creating and assessing testcrosses in every phase of multi-location assessments. The improved time efficiency in the process could be due to the second selection cycle, where the TRN from the prior cycle is utilized to forecast the new DH lines, eliminating testcross formation and initial multi-location evaluations. According to GS, top-performing lines may skip the first stage and move directly to the second round of evaluations in multiple locations (Crossa et al. 2017).

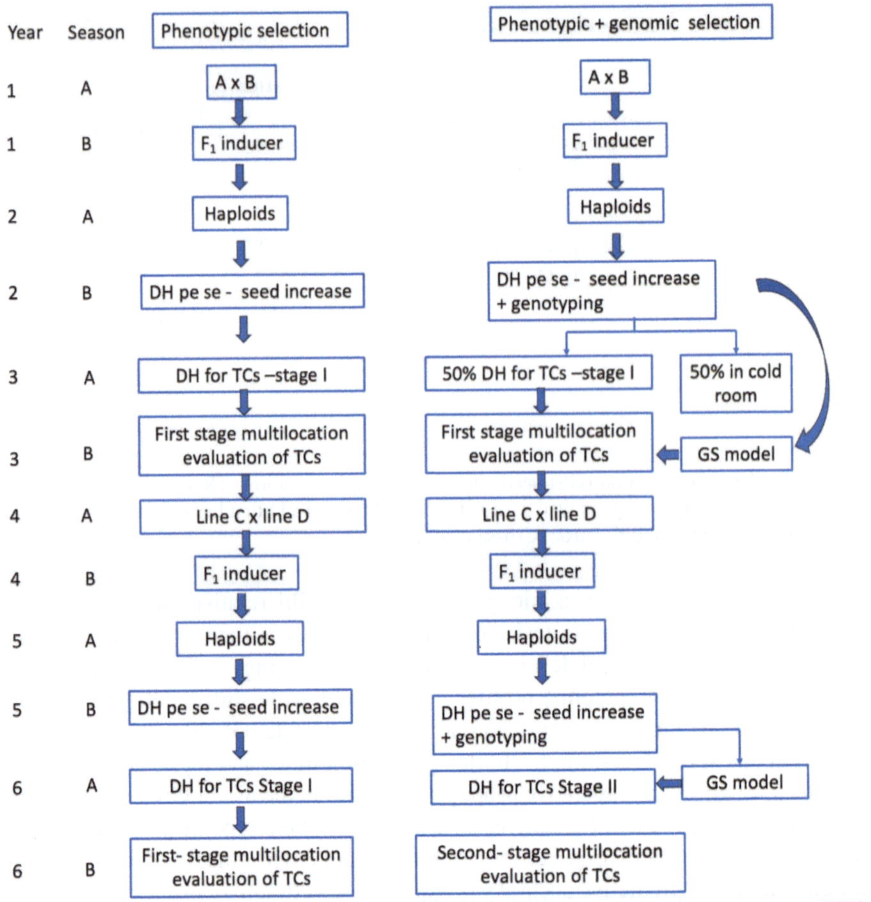

Fig. 1.12 A Scheme of Phenotypic and Genomic Selection in Maize Breeding. Reduction of cycle length through GS in maize using doubled haploids (DH) crossed with a tester (TC). (Modified from Crossa et al. 2017)

1.4.11 CRISPR Breakthrough

In the 70 years since the DNA double helix was first discovered, technologies for studying, controlling, and introducing foreign DNA have transformed the field of biological research. However, making changes to genomes specific to a certain site continued to be difficult. Zinc-finger nucleases (ZFN) and transcription-activator-like effector nucleases (TALEN) demonstrated the promise of precise gene editing, but their effectiveness was hindered by constraints in targeting capabilities and the complex design process (Westermann et al. 2021). CRISPR-Cas9 is seen as a significant breakthrough because of its ease and effectiveness (Doudna and Charpentier 2014).

CRISPR (Clustered Regularly Interspaced Short Palindromic Repeats) is utilized by bacteria in combating phage infections. Essentially, the CRISPR system allows bacteria to identify genetic sequences of intruders and use specific enzymes to destroy these sequences. The DNA endonuclease Cas9 is among the CRISPR-associated proteins known as Cas enzymes. In 2011, Charpentier's team discovered an important part of the CRISPR system, specifically an RNA molecule known as trans-activating crRNA (tracrRNA), which is responsible for identifying foreign phage sequences in bacteria. A later partnership between the Charpentier and Doudna labs led to the ground-breaking 2012 Science publication, marking the start of CRISPR-Cas9 genome editing (Jinek et al. 2012). The article demonstrated that a bacterial CRISPR-Cas9 enzyme is directed by a pair of RNA molecules as the tracrRNA:crRNA duplex, in order to target precise DNA cutting. It was demonstrated that the dual tracrRNA:crRNA can be modified into a single guide RNA (sgRNA) with key elements: a 20 nucleotide sequence for targeting and a double-stranded structure for binding to Cas9. This led to the development of a basic system with two components, allowing modifications to the 20 nucleotide guide sequence to direct CRISPR-Cas9 toward any desired DNA sequence (Doudna and Charpentier 2014).

CRISPR-Cas9 can disturb gene function by creating double-strand breaks, which are then repaired using nonhomologous end joining or gene replacement through homology-directed repair. This means that any gene can be removed in many different types of cells and organisms. Nonetheless, CRISPR technology's uses go beyond just gene disruption, as it has the capability to make precise alterations to individual bases (Komor et al. 2016). CRISPR-Cas9 has sparked numerous uses in fundamental science, agriculture, and medicine. Although it has great potential for practical use, it is important to remember that this powerful technology originated from basic research driven by curiosity. Please see Liu et al. (2022a, b) and Pixley et al. (2022) for reviews on the applications of CRISPR in crop improvement. Over the last decade, transgenic lines have been derived in a number of crops utilizing CRISPR (Son and Park 2022; Cardi et al. 2023). See Ahmad et al. (2023) for a comprehensive review on crops modified through CRISPR technology.

Gene editing techniques have transformed the area of biotechnology. CRISPR-Cas techniques employing RNA-guided enzymes are the leading gene editing tools and have generated gene-edited crops such as rice, wheat, and corn. Nonetheless, these systems exhibit certain constraints, such as the necessity for a protospacer adjacent motif sequence, the creation of unwanted double-strand breaks (DSBs), and the limitation in editing extensive genomic regions. Certain limitations are partly tackled through the advancement of second-generation editors, such as base editors (BEs) and prime editors (PEs). Gene editing technologies of the third generation, like seekRNA and bridgeRNA, can address many of these challenges (Gupta and Kumar 2025). PASTE (Prime Editing-assisted Site-Specific Targeting) and PASSIGE (Prime Editing-assisted Site-Specific Integrase Gene Editing) are cutting-edge gene editing techniques that merge the accuracy of prime editing with site-specific integrases and targeting strategies to insert large DNA sequences into specific genomic sites. PASSIGE specifically utilizes site-targeted large serine recombinases (LSRs) in conjunction with prime editing to insert extensive, multi-kilobase DNA sequences into designated regions of the mammalian genome, achieving significant efficiency following a single editing instance (Li et al. 2025). This field of knowledge shifts between reading and writing and keeping oneself updated with the advancements is a challenge.

Because of the numerous practical uses in food production, genome editing is expected to be employed in addressing agricultural issues that impact food security, as well as the significant yield losses in C_3 plants like rice and barley, caused by ineffective photorespiration. Enhancing photorespiration efficiency in crops like maize, rice, and wheat can greatly boost food production for the increasing global population without expanding cultivation areas or harming forest areas. Numerous countries have specific rules governing the cultivation of GMOs. In the years to come, a consortium of nations can collaborate to address regulatory challenges related to CRISPR-based transgenic crops like that in the field of biodiversity conservation.

1.4.12 Next-Generation Sequencing and Omics

More complex genomes demanded greater effort for sequencing. Plant sequencing projects face significant challenges due to the high levels of repetitive elements and ploidy in complex genomes. The presence of a reference genome allows for the identification of many genes related to both biotic and abiotic stresses, as well as molecular markers. Re-sequencing projects are better suited for pre-breeding efforts in order to detect genomic variations and collect data on beneficial polymorphisms. As of now, there are 30,530 eukaryotic genomes listed in the NCBI database (https://www.ncbi.nlm.nih.gov/), which includes 5119 that are complete or at a chromosome level, representing 12,205 species (Bernal-Gallardo and de Folter 2024). Sequencing efforts spanning over 20 years have increased the quantity of sequenced plant genomes that are available. The initial *Arabidopsis* plant genome was released in 2000, and to date, 4604 genomes from 1482 plant species have been published. NGS-based methods offer the chance to improve the resolution of QTLs and identify genetic variations. Various types of high-throughput Next Generation Sequencing (NGS) have been employed for mapping crop populations. Table 1.2 features a few of these studies (Naqvi et al. 2022). Due to the introduction of modern NGS technologies, the following methods are effectively employed for discovering genes and QTLs in crop plants:

- GWAS facilitate more precise mapping in a larger sample by detecting significant associations between phenotype and genotype through LD (Linkage Disequilibrium).
- RADSeq (restriction site-associated DNA sequencing) employs next-generation sequencing to examine various genetic markers in individuals within a population. A more sophisticated and economical approach is genotyping-by-sequencing (GBS).

Table 1.2 Tools and platforms sued in NGS analysis

Tool/Platform	Language	Traits
autoSNP	Perl	SNP identification
Blast2Go	Linus/windows	Genome/transcriptome annotation
CNVKit	Phyton/Linux	Variant discovery in NGS data
DAVID	Set of annotation tools	Data annotation
Galaxy	Cloud platform	NGS data analysis
GATK	Linux	Variant Discovery in NGS data
InterProScan	Online/Linux	Domain and motif analysis
KEGG	Web interface/Linux	Pathway analysis
MapMaker	Windows	QTL analysis
MapQTL	Windows	QTL mapping
MISA	Perl	SSRV detection
QTLcartographer	Windows	Composite interval QTL mapping
Qu-gene	Windows	Simulation for quantitative genetics
SnpEff	Java/Linux	SNP calling
SNPpipeline	C++, Perl, and Phyton	SNP detection
Snpsift	Java/Linux	SNP filtering
Tassel	Windows/Linux	Association mapping
TROLL	C++	SS occurrence locator
Pangenomics		
PanGP	Windows/Linux	Profiling analysis for the development of the core genome
RPAN	Linux	Rice pan-genome analysis
Splitmem	Linux	de Brujin graph-based visualization algorithm
PanViz	Linux	Pan genome visualization
PGAPPGAP-X	Linux	Pan genome cluster and evolution analysis
Pantools	Linux /Windows	Genome mapping

- GBS is an efficient and dependable approach for performing GWAS in crops. It also allows breeders to explore genetic lineage, pinpoint markers, evaluate genomic diversity, and apply genomic selection in different crop breeding programs.
- Bulk-segregant analysis sequencing (BSA-Seq) merges Bulk-segregant analysis and NGS to effectively identify markers associated with a particular trait in a breeding population.
- TILLING by Sequencing integrates Tilling and Eco-Tilling approaches, employing Next Generation Sequencing to rapidly detect induced or natural genetic changes. This approach helps in identifying rare and novel mutations.

1.4.13 Epigenomics/Epigenetics

Epigenomics is defined as the total alterations in gene expression and cellular function caused by DNA and histone modifications, as well as the production of noncoding RNA, all while maintaining the original DNA sequence (Kumar 2020). DNA methylation, histone acetylation, ubiquitination, and phosphorylation are typical instances of these alterations. One of the most researched epigenetic markers in plants is DNA methylation (Holliday 2006). Using high-throughput next-generation sequencing technologies can track and map various modifications, including this one, to gain a better understanding of how histone and DNA modifications regulate plant growth, development, and responses to stresses (Kujur et al. 2013). Changes in DNA methylation patterns can be caused by environmental factors like abiotic and biotic stressors or by genetic factors (Kawakatsu and Ecker 2019). Analyzing DNA methylation patterns and genetic information stored in epigenetic markers can help us understand how epialleles regulate gene expression and transcript levels during a plant's growth phases and its reaction to stress in various crop varieties (Kujur et al. 2013). The CG pattern, also called gene body methylation (gbM), is methylation occurring in the transcriptional regions and associated with low expression levels. The CG transposable element (TE)-like methylation pattern suppresses TE activities. The roles of TE activities in increasing genetic diversity are important for breeding purposes (Nadeem et al. 2014).

RNA interference is a recognized technique of epigenetic manipulation. RNAi is a process where gene expression is silenced after transcription by using double-stranded RNA (dsRNA) molecules (Das and Sherif 2020). Small interfering RNA molecules are employed in this technique. A particular gene shows increased activity in its genetic makeup, enabling a specific characteristic (Younis et al. 2014). Cutting double-stranded RNA (such as micro RNA and small interfering RNAs) into functional small non-coding RNAs is a prerequisite. The process is made easier by the Dicer (or similar to Dicer) multifunctional endoribonuclease (Rajput et al. 2021). These minuscule noncoding RNAs play a role in gene regulation by being integrated. Another ribonuclease system, comprised of RISC, argonaute, and other effector proteins, triggers gene silencing by degrading targeted messenger RNAs. RNAi systems can be transiently delivered through traditional transformation methods such as particle bombardment, viral delivery, and *Agrobacterium*-mediated delivery (Chaudhary et al. 2024). It is well known that epigenetic changes impact crucial characteristics in crop plants. DNA/chromatin-modifying enzymes that are directed toward a specific target can be employed to epigenetically alter a specific epigenetic marker that impacts gene activity by altering the expression of genes, both increasing and decreasing their transcription levels (Kumar 2020). RNAi, a technology widely utilized to combat various biotic and abiotic stress factors, has proven effective in addressing different levels and types of stressors (Chaudhary et al. 2024). RNAi has enabled the acquisition of characteristics by triggering atypical activation of defensive genes in response to plant pathogens and other intruding factors (Williams et al. 2024).

In order to withstand environmental changes, a stable inheritance of the adaptive epiallele is necessary. When the DNA methylation inheritance becomes stable, the epiallele will also stay stable, consequently decreasing the creation of additional epialleles. For instance, Ketumile et al. (2022) discovered that changes in gene expression in sorghum can improve crop resilience, specifically by studying the Mutator S HOMOLOG 1 (MSH1) gene. In this research, silencing of the MSH1 gene in sorghum using RNAi caused methylation repatterning, leading to intergenerational epigenomic changes (Chaudhary et al. 2024). Utilizing epigenetic tools in speed breeding can reduce generation time and create stress-tolerant crops.

1.5 Will Yield Increases Be Enough?

The narrowing of yield gaps will also be influenced by climate change, as it is predicted to reduce crop productivity in significant global regions where bread is produced (Vogel et al. 2019). Climate change will impact crop production by affecting temperature, rainfall, pests, and CO_2 levels in the atmosphere (Ostberg et al. 2018). In low-input agricultural systems with large yield gaps, it is important to note that yields will likely decrease (Rosenzweig et al. 2014). Rain-fed croplands will face increased vulnerability to unpredictable changes in rainfall and precipitation patterns (Rojas et al. 2019). Many research papers about global food security only look ahead until 2050, such as studies by Godfray et al. (2010) and Fader et al. (2013), while predictions show that climate change and growing food demand will worsen. Additionally, past research has analyzed key factors influencing worldwide food production and demand separately, overlooking the comprehensive range of elements that will impact future food security: population growth, dietary shifts, climate change, and the potential for reducing crop yield gaps sustainably using the Earth's limited freshwater resources (Beltran-Peña et al. 2020). The upsurge of new pests and diseases along with the effects exerted by climate change are to be addressed efficiently (Lahlali et al. 2024; Skendžić et al. 2021).

Global wheat yields rose by an average of 1% annually in the twentieth century, while maize yields in the United States saw a growth of 200 kg/ha/yr or 1.8% per year until a drought hit the crop in 2012 (Campbell et al. 2016). Nonetheless, there is a need for a higher rate of annual yield improvement in key food crops worldwide in order to satisfy the projected global demand (Fischer et al. 2014). Yield improvement and the adoption of new varieties are minimal in many parts of Africa and South Asia, as stated by Santantonio et al. (2020), while wheat yields in developed countries like Australia have dropped due to climate change in the twenty-first century, according to Hochman et al. (2017). The IPCC advises (FAO 2017, 2018) to avoid possible climate disaster in the worldwide food system by advocating for (i) a prompt and substantial change in eating habits moving from meat to sustainable plant-based diets, and (ii) a prompt and substantial increase in both public and private funding in agricultural research and supportive policy structures that can turn scientific progress into practical, resilient farming methods and systems. In order for these interventions to achieve success, citizens, consumers, and stakeholders in global agrifood systems and supply chains must be prepared to make significant disruptions and sacrifices. Yet, the extent of these effects, where they are happening, and possible ways to reduce them are mostly not understood.

A few major crops like wheat, maize, rice, potato, and soybean make up a significant portion of global caloric consumption and agricultural earnings (Anders et al. 2021). The lack of price responsiveness in food and land markets, growing uncertainty in yield, shortages in food supply, and misguided policy reactions to advancements in plant technologies could lead to the IPCC's forecast of a 7.6% median rise in cereal prices, potentially reaching 23% by 2050. This forecast underscores the serious socioeconomic impact of depending on a limited number of key crop species for human nutrition during a time of uncertainty caused by climate change for both consumers and farmers (FAO 2017, 2018). Even though it puts financial strain on low-income families, increasing food costs are frequently linked to advantages for farmers and a decrease in rural poverty. It is unclear if these advantages outweigh the negative economic impacts of pricier food. It is crucial to have research evidence that combines scientific advances, economic outcomes, and policy strategies for farmers and consumers. Presently, it is important to conduct a more extensive evaluation of how climate change impacts on agriculture affect economies both nationally and internationally (Fróna et al. 2021). Research in this area should address important issues regarding the implementation of new technology on farms and the development of policies that support its spread among a large number of impoverished farmers, each with their own unique needs and obstacles (Lan et al. 2018). Yet, regulatory environments, consumer demands, and farmer acceptance are limiting access to genome-based technologies. The potential benefits of genetic advances are frequently overlooked due to the ongoing "technology controversy," which has hindered the acceptance of genetic modification and gene-editing techniques worldwide (Kleter et al. 2019).

1.5.1 Crop Wild Relatives

Wild relatives serve as both the foundational material for breeding and an important source of diversity. This variation can be utilized to introduce new alleles for disease resistance and other quality characteristics into enhanced germplasm (Hufford et al. 2019; Zhang et al. 2017). Thus, these wild populations play an essential role in comprehending how the selection of traits significant to agriculture has influenced the genome of cultivated species throughout domestication and enhancement (Schreiber et al. 2018). Major obstacles in researching wild relatives encompass the scarce availability of germplasm and genetic information in numerous crop species (Zhang et al. 2017) and the chance that current wild populations may not be genuinely ancestral, but rather feral or mixed with cultivated germplasm (Wang et al. 2017). This association may grow even more intricate because of the potential for gene exchange between cultivated plants and wild, feral, or hybrid varieties over different time spans, as recorded in many crop types. Recent studies have demonstrated that cultivated crops can become wild through different mechanisms, such as adaptive introgression from their wild relatives and selection on certain genomic areas during domestication, as observed in cases like eggplant (Page et al. 2019) and weedy rice (Li et al. 2017). In the future, understanding the variety of crop wild relatives and the extent of genetic blending with cultivated plants will be crucial for identifying genomic regions that can improve crops and prevent the rise and spread of undesirable weed species. Enhancing the representation of wild relatives in crop germplasm collections and developing methods to assess hybridization between wild and cultivated populations will be essential.

1.6 Systems Biology-Driven Breeding

Advancements in technology have enabled the exploration of cellular response at all organizational levels by generating large amounts of data from genome, transcriptome, proteome, metabolome, and high-throughput phenome, a process that is now standard. Nevertheless, these data have typically been analyzed separately up until now. Combining transcriptomics, proteomics, and metabolomics would significantly help in uncovering and understanding intricate plant regulatory networks (Pazhamala et al. 2021). Systems biology has surfaced as a hopeful multidisciplinary field that combines extensive omics datasets with carefully crafted mathematical models to validate hypotheses and forecast biological systems (Fig. 1.13). This allows for a comprehensive grasp of the overall system response to growth, development, and stress adaptation, essential for advancing breeding of climate-resilient crops. Yet, the initial stage in adopting a systems biology approach is to formulate hypotheses rooted in existing knowledge. This forms the foundation for planning a systems biology experiment and is the most crucial stage (Pinu et al. 2019).

1.6.1 Integration of Data

Datasets collected from various sources are typically analyzed using multivariate techniques to identify strong correlations among biological components. PANOMICS platform offers the integration of intricate omics datasets from various fields such as genomics, epigenomics, transcriptomics, proteomics, post-translational modifications proteomics, metabolomics, and phenomics (Weckwerth

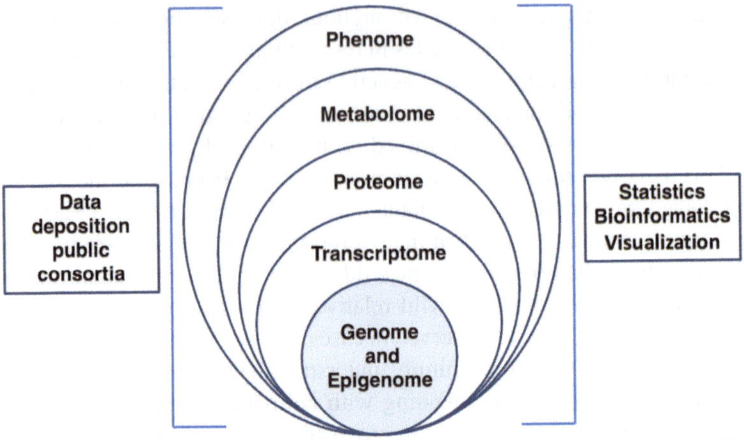

Fig. 1.13 Integration of heterogeneous multiple "omics" data. (Courtesy: Springer Nature https://doi.org/10.1007/s00425-014-2059-5)

et al. 2020). In crop plants, these methods allow for the examination of plant metabolism and the comprehension of the molecular processes that influence plant characteristics with agricultural significance. In rice, unique metabolic and regulatory characteristics related to light were discovered through transcriptomics, metabolomics, and genome-scale in silico modeling (Lakshmanan et al. 2015). Not long ago, researchers examined transcriptomics, proteomics, and metabolomics data to add to existing information that helped understand the fertility transition mechanisms in a pigeon pea thermosensitive male sterile line for possible application in two-line hybrid breeding (Pazhamala et al. 2020). In addition, multiomics data connect phenotype with genome variations to provide additional data layers for genomics-based predictions (Azodi et al. 2020). Multiple omics data were combined in prediction models to enhance accuracy in predicting outcomes in crops like maize (Azodi et al. 2020). The use of multiomics data is growing for predicting phenotypes because it goes beyond just the genome and reflects how the organism's biology is influenced by the environment (Li et al. 2019a, b).

1.6.2 Biology of Networks

Regulatory networks play a key role in controlling plant response and adaptation to stress, as noted by Gehan et al. in 2015. Analysis of transcriptome, proteome, and metabolome data can aid in the interpretation of regulatory networks and their functional relationships within biological systems, leading to the reconstruction of pathways and networks (Moreno-Risueno et al. 2010). Biological interactions for cellular functions can be illustrated as networks and graphs, showing the connections among genes, proteins, and metabolites (Weckwerth et al. 2004). In short, omics data are normalized for a similarity matrix, which is then used to create an adjacency matrix that can be converted into an undirected graph or network (Langfelder et al. 2013). Multiple platforms exist for the integration of multidimensional omics datasets.

1.6.3 Modeling Systems

The upcoming focus in network biology involves dynamic modeling to provide a holistic understanding of how gene expression influences protein actions to produce metabolites in plants in response to external signals (McCormack et al. 2016). Systems modeling is mostly used to evaluate biological hypotheses and can also be applied to forecast the overall response of a system. Computational modeling, such as dynamic simulation models and machine learning approaches, is employed to interpret the data biologically. Put simply, the biologist's hypothesis is used to simulate the biological system and is then tested over a period of time. Next, the simulated outcomes are analyzed against the actual data to assess the hypothesis and adjusted as needed to align with the experimental results. Having a precise understanding of the system's behavior under dynamic conditions and time is crucial for achieving accurate predictions (Muthuramalingam et al. 2019). For cellular modeling, four main strategies have been proposed:

- Constraint-based modeling
- Metabolic modeling
- Analysis of metabolic control
- Cybernetic modeling

Constraint-based modeling necessitates data on metabolic reactions and flux capacities and is able to function effectively with minimal experimental data. This approach outperforms all other methods and has been applied to build metabolic networks at the entire cell or genome level, a growing use alongside omics data for improved predictability. Muthuramalingam et al. (2019) discovered that this method offers valuable understanding of how cellular metabolism in rice responds to abiotic stress. On the other hand, metabolic modeling focuses on a particular aspect of metabolism by using equations to predict the fluctuating levels of metabolite concentrations. Metabolic modeling in plants is still restricted due to the lack of kinetic variables needed to model large-scale networks for most biochemical reactions (Muthuramalingam et al. 2019). Examining metabolic regulation is a common method for investigating control mechanisms in plants, assessing how system variables react to slight alterations in system parameters such as enzyme levels or activity. The cybernetic modeling framework was first employed in macroscopic input–output models that characterized substrate uptake, growth, and product formation. Afterwards, this method has been expanded to be used for more intricate, internal cellular processes. Cell Illustrator (Nagasaki et al. 2010), COBRAToolbox (Schellenberger et al. 2011), Acorn (Sroka et al. 2011), and CellNetanalyzer (Klamt et al. 2007) are among the available software applications and markup languages for systems modeling.

In plant research, the main objective still revolves around predicting phenotypes and comprehending the connection between genotype and phenotype. Anticipating plant characteristics like stress reactions and quality features, using multiomics data is crucial for enhancing crops. Regarding this matter, a research study on corn categorized DNA sequence regions into active versus inactive pseudogenes, based on characteristics like DNA methylation (Sartor et al. 2019). Another research found important genes, proteins, and metabolites that can predict the quality of potato tubers using data from transcriptomics, proteomics, and metabolomics. This research employed random forest regression to combine omics data sources (Acharjee et al. 2016). The effectiveness of predicting phenotype relies on the use of predictive modeling and analytics, a complex topic that was recently examined by Tong et al. in 2020.

Systems biology has the potential to revolutionize sustainable agriculture by uncovering the intricate relationship between different traits that connect the genotype to the phenotype. It can be utilized for simulating and examining multi-gene characteristics associated with agricultural output like plant structure, nitrogen utilization efficiency, water utilization efficiency, and tolerance to abiotic and biotic stress. Due to the increase in recent high-throughput experimental analyses and computational power, there is an opportunity to combine various fields to elucidate complex traits. Mathematical models developed from time series data help pinpoint important genes for breeding programs and aid in creating a systems biology approach to breeding (Lavarenne et al. 2018). They concluded that the main practical goal of systems biology is to understand the intricate interactions that control the multigenic trait for enhancing crops. Incorporating multiple types of omics data with detailed phenotyping will lead to the creation of advanced models that will enhance breeding programs for improving important agricultural traits in the future (Gehan et al. 2015).

1.7 Metabolomics

Since the metabolome is almost at the very end of the expression of the organism's genetically encoded design, it is thought to have a substantial number of gene–gene and gene–environment interactions. The metabolome contains "condensed information" about how well the genotype performs under specific growth conditions (Katam et al. 2022). In reality, plant metabolites can offer valuable information for agronomically important feature prediction. Metabolic indicators are quantitative traits that are represented by the intensities of MS or NMR signals, which show their concentrations (Shen et al. 2023). Usually, they are expressed in relative units. They can be associated with quantitative phenotypic features.

1.7.1 Metabolomics of Plant Stress Response

Abiotic and biotic stresses lead to alterations in the plant metabolome. The stress response is likely the domain where metabolomics has concentrated the greatest focus. Plants react to abiotic stress via phytohormone- and $Ca2^{+}$-dependent signaling, which activates transcription factors and triggers the expression of genes responsive to stress. The stress also triggers the production of reactive oxygen species (ROS) (Raza et al. 2020). The overall pattern of reaction to pathogens is alike. The reception of biotic stress signals triggers signaling pathways and hormone regulation primarily through salicylic acid (mainly in reaction to biotrophic pathogens) and jasmonate together with ethylene (primarily in response to necrotrophic pathogens), although other hormones (gibberellins, auxin, brassinolide, cytokinins, and abscisic acid) also play a role in the response (Van Esse et al. 2020). Every one of these reactions leads to ROS production and a metabolic reaction.

1.7.1.1 Metabolomics for Abiotic Stress Tolerance

A metabolic reaction to drought stress was documented long before the word "metabolomics" was introduced. In 1954, the buildup of the amino acid proline was noted in wilted rye. During the 1970–1990s, the catalog of metabolites found in plants exposed to osmotic stresses (such as drought and soil salinity) grew (Litvinov et al. 2021). The primary function of high levels of these metabolites during osmotic stress is to preserve turgor, and these metabolites are referred to as "compatible solutes" (which do not disrupt cell metabolism) and "osmolytes." Research has shown that osmolytes, besides osmotic pressures, also build up in response to various other nonbiological stresses. So far, many osmolytes have been identified. Generally, they can be categorized into three groups: soluble carbohydrates (sugars and oligosaccharides from the raffinose family, and sugar alcohols or polyols), amino acids (which include nonproteinogenic amino acids like 4-aminobutyrate), and polyamines along with betaine and its derivatives, such as glycine betaine (Litvinov et al. 2021). Transgenic sugar cane varieties NXI-1 T, NXI-4 T, and NXI-6 T exhibit heightened tolerance to water stress due to increased glycine betaine synthesis. It is believed that in stressful conditions, osmolytes help stabilize lipid membrane structures, ensure correct protein folding, and safeguard macromolecules and membranes against aggregation (Ianutsevich et al. 2023).

1.7.1.2 Metabolomics for Biotic Stress Tolerance

Plants are perpetually under threat from viruses, bacteria, fungi, nematodes, insects, and animals. To ensure their safety, plants utilize metabolites that are always present, known as phytoanticipins, which include saponins. When infected, plants trigger oxidative stress and produce antimicrobial secondary metabolites from different classes known as phytoalexins. Phytoalexins include isoprenoids (terpenoids), flavonoids, and alkaloids (Yeshi et al. 2022). Pathogens cause alterations in the levels of various classes of metabolites: amino acids, organic acids, fatty acids, phytohormones, and polyamines (Serag et al. 2023). To determine genes that are implicated in responding to a specific stress, a concurrent analysis of the transcriptome is performed. It enables the identification of correlations between the expression of specific genes and the emergence/accumulation of the targeted metabolite. The involvement of related genes in the production of metabolites is presumed based on the knowledge of biosynthetic pathways. Consequently, a recent metabolomic analysis uncovered several alterations in the metabolite concentrations when exposed to pathogens in tomato (Singh et al. 2023a, b). The metabolomic reaction varied based on the type of pathogen and elicitor. Among various alterations to the biotic stress caused by the fungus *Cladosporium fulvum*, the production of falcarindiol was elevated. Falcarindiol is an acetylenic fatty acid and an oxylipin. According to the overall knowledge of falcarindiol biosynthesis, it has been proposed that the essential enzymes needed for producing these acetylenic fatty acids are desaturases. To uncover the potential biosynthetic pathway that gets activated and leads to the production of falcarindiol, gene expression was analyzed through RNA-seq alongside a metabolic investigation. Various types of pathogens/elicitors and multiple time points were subjected to simultaneous metabolome and transcriptome analysis. Out of 40 potential desaturases identified in the tomato transcriptome, the mRNA expression of three desaturases showed the strongest correlation with falcarindiol accumulation under stress conditions.

Additionally, a suspected decarbonylase was found within the same 20 kb region, and its expression showed a positive correlation with the level of falcarindiol. The creation of tomato lines with a knockout of these genes through CRISPR/Cas9 inhibited the production of falcarindiol, and genetic complementation assays employing *Agrobacterium*-mediated transient overexpression of cDNA for these genes reinstated falcarindiol production. Unexpectedly, knockout tomato strains that could not produce falcarindiol showed greater resistance to pathogens than the original falcarindiol-producing tomato line. Nonetheless, the integration of metabolomic analysis with gene expression successfully revealed genes associated with the biosynthesis of particular metabolites. Plant metabolomics will surely play a significant role in supporting upcoming breeding programs.

1.8 Digital Devices in Breeding Plants

Digital breeding involves automating breeding design, experimental plot layout, growth stages, and selection, all while taking into account and predicting different environmental factors. Digital breeding intends to streamline the plant breeding process by reducing the need for human involvement. Advanced techniques such as next-generation sequencing, machine/deep learning, speed breeding, and advanced statistical models are incorporated in breeding technologies, facilitating predictive breeding (Jeon et al. 2023). According to Farooq et al. (2024), incorporating machine learning (ML) algorithms in prebreeding techniques, local selection, and marker-assisted selection can enhance genetic variability and accelerate the breeding of climate-resilient cultivars. These algorithms are able to both preserve and recover the dynamic processes leading to genetic diversity, which is essential for plants to adapt to shifting climates (Armstrong 2023). Genetic improvement results from the extensive genetic diversity found in crops. Over 1750 gene banks around the globe store more than seven million germplasm accessions, such as cultivars, landraces, and wild relatives, with vast untapped potential (McCouch et al. 2020). Utilizing genome-wide genotyping of germplasm resources stored in gene banks presents a hopeful opportunity to enhance our understanding and utilization of these valuable resources. Significantly, extensive genome-wide genotyping data have been produced for a large number of wheat, maize, and barley samples, amounting to more than 80,000, 4000, and 20,000, respectively (Farooq et al. 2024). These datasets can be used as a base for selectively choosing and testing accessions in certain environments with the help of AI-driven predictive genomics.

1.8.1 Artificial Intelligence

Research in OMICS has caused the surplus generation of intricate datasets. Understanding the precise significance of these intricate datasets is crucial for describing crop plants for a particular characteristic (Harfouche et al. 2009) At the same time, NGS technology has greatly hastened the access to the entire genome sequence of desired plant species, resulting in the creation of extensive datasets (Godwin et al. 2019). Moreover, in combination with transcriptomics and proteomics analysis, MAS and GWAS methods also thoroughly explore plant genotypes and phenotypes (Yang et al. 2021).

Utilizing appropriate models and definitive algorithms is imperative in efficiently extracting valuable information from breeding datasets for the implementation of Next-Gen AI in plant breeding (Najafabadi et al. 2023). AI, like neural networks (NN) and deep learning (DL), are currently being utilized to enhance the effectiveness and precision of multi-omics data (Wang et al. 2023a, b). The way in which these two AI functions are frequently hidden involves several complex hierarchical techniques to create nodes for simple categorization of datasets resembling neurons in the brain (Niazian and Niedbała 2020). On the other hand, plant breeders are envisioning a new AI system that will evaluate breeding values and offer a detailed examination of intricate traits in varying environmental conditions (Xu et al. 2022). Moreover, AI will be continuously trained and enhanced to increase the accuracy and efficiency of data mining in order to better predict factors related to disease resistance and agronomic traits, thus speeding up breeding programs. Extensive breeding and strict selection criteria have greatly changed the ability of crop plants to adapt to different environments (Abdul Aziz and Masmoudi 2024). Current breeding programs are focused on enhancing crop plant tolerance to abiotic stress by addressing the genotype–phenotype gap caused by changes in phenotype plasticity (Brooker et al. 2022).

1.8.2 Connecting DNA to Traits Through Artificial Intelligence

It is crucial to accurately and precisely connect the genotype with the crop phenotype (Khan et al. 2022). Connecting all genetic information across the entire genome

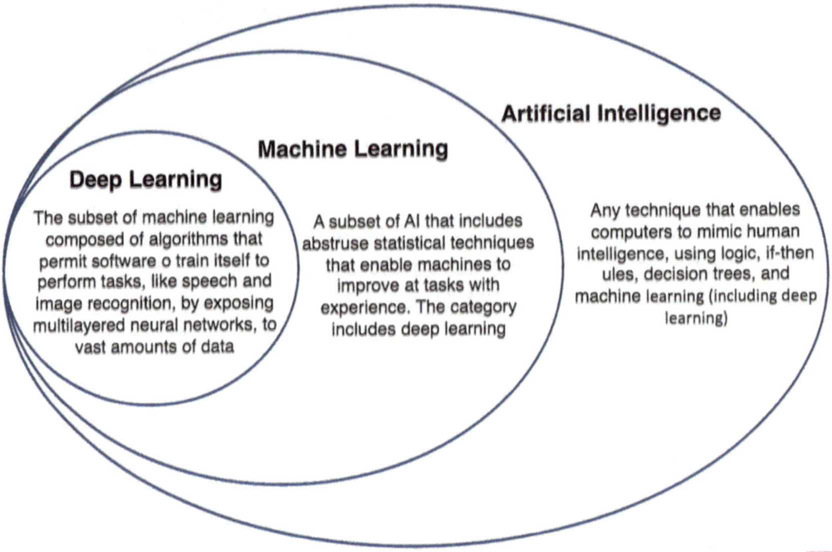

Fig. 1.14 Interrelationships of AI, ML, and DL

with high-throughput traits in advanced breeding is a significant obstacle, hindering the effective use of field phenotyping and omics techniques (Harfouche et al. 2009). Germplasm collection and mapping populations are able to effectively distinguish phenomics and genomics data using AI. Crop variety, identification of single nucleotide polymorphisms (SNPs), examination of quantitative trait loci (QTL), study of genome-wide association, and utilization of genomic selection and sequencing all produce vast amounts of data; artificial intelligence (AI) can analyze and connect phenomics and genomics data from this large dataset to enhance breeding strategies. AI involved in a computation and training framework is capable of forecasting gene functional analysis and high-throughput crop phenotyping, as well as predicting the crop's yield and traits performance (Wang et al. 2020). Hence, combining AI with phenomics and genomics tools can help quickly detect genes linked to crop traits, speeding up crop enhancement initiatives. Figure 1.7 outlines the process of integrating AI technology to connect high-throughput genomics with phenomics.

1.8.3 AI Technologies and Phenomics Challenges

There has been rapid progress in plant phenomics in recent years, allowing for more precise breeding opportunities. Recently, there has been a surge in the application of AI across various scientific disciplines. AI capabilities such as computer vision, machine learning, and deep learning have been successfully incorporated into non-invasive imaging techniques. AI, Machine Learning, and Deep Learning are all interconnected (Fig. 1.14). By employing ML for strong image analysis, it can consistently improve the effectiveness of the collection and analysis of data. Moreover, AI has assisted in the advancement of software and tools for collecting and managing data in field phenotyping (Najafabadi et al. 2023). This entails open-source devices and platforms that enable community-driven research and data sharing, supplying the vast amounts of data necessary for dependable phenotyping studies (Nabwire et al. 2021). AI plays a crucial role in managing three key aspects of phenomic data:

Algorithms and software for translating sensory data into phenotypic data
Development of models to comprehend GxPxE interactions
Database management enables the sharing of information and resources

Recent research has demonstrated that using AI for crop phenotyping has led to advancements in crop phenotyping and predictions (Jung et al. 2021). Selvaraj et al. (2020) found that ML algorithms such as k-Nearest Neighbours (kNN), RF, and SVM showed superior results in predicting root yield in cassava (*Manihot esculenta* Crantz), with R^2 values of 0.67, 0.66, and 0.64, respectively. Furthermore, AI-supported high-throughput phenotyping systems have shown successful utilization in wheat and maize for plant growth stage identification and plant image segmentation; oilseed crops for crops and weeds semantic segmentation; disease resistance phenotyping in crops; and enhancing plant productivity (Khan et al. 2022).

AI will support the algorithms and enable high-throughput analysis of crop characteristics for quantitative traits (Razzaq et al. 2021). Furthermore, AI can be seamlessly integrated with bioinformatics and genome sequencing analysis to analyze different molecular features like transcription factor binding sites (Long et al. 2020), long noncoding RNAs (lncRNAs) (Sun et al. 2015), microRNA (miRNAs), epistatic modifications, coding genes, targeted polyadenylation sites (Gao et al. 2018), and cis-regulatory elements (CREs) (Mahood et al. 2020). Additionally, utilizing AI to determine the connections between candidate genes and CREs is an innovative method for classifying and discovering new genes to enhance crop productivity. Additionally, AI approaches have a greater capacity in analyzing crop yield, assessing climate variability, conducting high-throughput crop stress phenotyping, as well as temperature, UV radiation, wind, and hail in the environment (Tong and Nikoloski 2021). AI's importance is growing in acquiring, analyzing, integrating, and overseeing genomic and phenomic data to enhance agricultural climate resilience, according to Shen et al. (2022).

NGS-based genotyping methods have enhanced gene mapping resolution and identification, and have been utilized in crop improvement through GWAS analysis (Razzaq et al. 2021). Bulk segregant analysis, along with its modified methods, is currently being applied in numerous crops, as noted by Majeed et al. in 2022. The NGS-based BSA method is gaining popularity in pinpointing potential genes linked to traits, including soybean mosaic virus, flowering time, phytophthora resistance, and powdery-mildew resistance. A new deep-learning algorithm called DeepBSA was created for QTL mapping and functional gene cloning in maize (Li et al. 2022).

1.8.4 AI and Envirotyping

Envirotyping is suggested as a third technology, along with phenotyping and genotyping, to understand how the environment affects crop breeding (Xu 2016). Envirotyping is important in crop modeling and predicting phenotypes by incorporating various components like genotype–environment interaction, environmental cues, genes, and stresses, along with integrative phenotyping. Cortes et al. (2022) examine the current status of the developing area of research called "genome–environment associations," which merges ecological climatic information with evolutionary genomics (GEAs) (Cortes et al. 2022). The writers support the idea of communities starting to gather genomic estimated adaptive values (GEAVs) for genomic prediction (GP) and multi-dimensional ML models to incorporate polygenic evolutionary adaptation. Xu et al. (2022) have recently suggested a breeding scheme that combines genomic-enviromics prediction with multiomics data, big data technology, and artificial intelligence.

Concerns about food security worldwide have been heightened by climate change, increasing global population, and pathogen pressure. In the future, it is essential to create plans to optimize the use of scarce resources for crop breeding and land management (Resende et al. 2021). Climate-resilient crops and climate-resilient soils have been modified to fit the environment, enhancing breeding programs, which could be used by breeders to develop new resilient crops for the changing climate. Genomic technologies, combined with high-throughput phenotyping, are giving researchers and farmers the necessary information to direct and inform breeding strategies for climate-smart breeding (Marsh et al. 2021). AI is essential in managing and combining the rapidly accumulating data by performing association studies to discover genomic targets linked to adaptive climate-resilient traits (Shafiekhani et al. 2017). Cortes et al. (2022) proposed a plan to employ ML, genomic prediction, and multi-trait gene editing methods to identify new abiotic stress resilience traits from wild crop relatives for developing drought-resistant crops (Cortés and López-Hernández 2021).

1.9 Contributions of Nanoscience

The CRISPR/Cas system has been used for improving major crops, including rice, sorghum, tobacco, wheat, maize, barley, cotton, tomatoes, soybeans, and rapeseed (Ahmar et al. 2021). Yet, the successful, effective, and accurate delivery of CRISPR components while saving time is still a difficult task (Rui et al. 2020). Traditional methods of delivering biomolecules in plants suffer from significant limitations, including poor gene transmission efficiency, restricted species applicability, limited cargo variety, and tissue damage. Nanotechnology methods have been suggested as cost-effective, simple, and durable approaches for delivering genes or molecules into plants efficiently and safely (Chandrasekaran et al. 2020; Demirer et al. 2021). Plant biomolecule delivery via nanoparticles is more difficult than mammalian cell delivery due to the natural barrier of the cell wall, as stated by Mao et al. in 2019. Some research indicates that NPs can make plant transformation more effective by

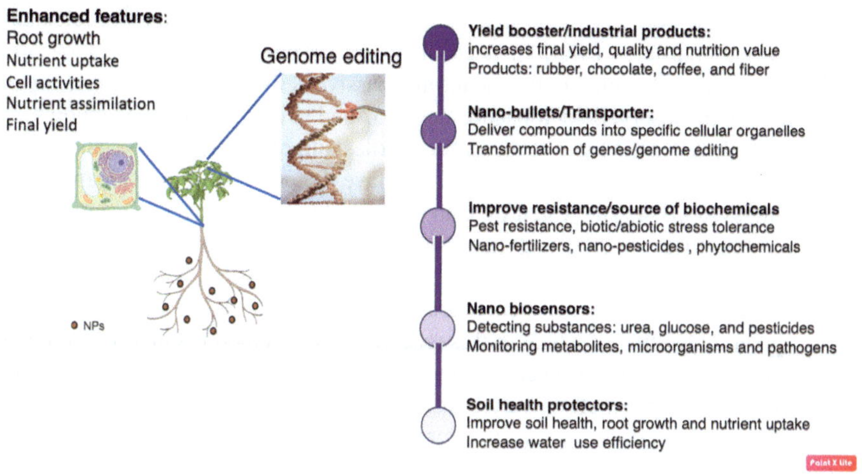

Fig. 1.15 Application of nanotechnology in plant breeding. (Courtesy: Frontiers in Plant Science; https://doi.org/10.3389/fpls.2021.663849)

shielding the genetic material from enzymes that break it down within the cell (e.g., nucleases; Finiuk et al. 2017; Joldersma and Liu 2018). Different types of NPs used for gene delivery are categorized based on the type of material they are made from, which include carbon-based NPs, silicon-based NPs, metallic NPs, and polymer-based NPs. Different genetic cargos are delivered by each type of NP. For instance, while carbon nanotubes (CNTs) have the ability to transport RNA and DNA (Karimi et al. 2015), metallic nanoparticles can exclusively distribute DNA as genetic material (Zhao et al. 2017). Moreover, silicon nanoparticles have the ability to transport DNA and proteins, while polymeric nanoparticles like PEG and polyethyleneimine, can deliver encapsulated RNA, DNA, and proteins into cells (Zhou et al. 2018).

In general, NPs need to have the ability to penetrate the cell membrane and target specific organelles for localization. Cationic nanoparticles are the top choice for delivering genes to plants since they can attach to the negatively charged cell wall and transfer genes, while carbon nanotube nanoparticles have been utilized to transport DNA into different crops (Fig. 1.15). Nevertheless, gene delivery into plant cells typically necessitates supplementary physical techniques such as magnetoinfection and electroporation when utilizing NPs. In contrast, NPs like silicon carbide whiskers (SCW) and mesoporous silica NPs (MSN) have been successfully utilized for gene transfer in plants without the need for other physical techniques (Chang et al. 2013). In this study, SCW-mediated transformation was effectively employed for the transformation of tobacco plants (Golestanipour et al. 2018). However, the SCW method generally has a downside when compared to other NP-mediated plant transformation methods: it necessitates a well-defined protocol for plant regeneration from cell cultures. Polymer nanoparticles are also capable of transporting nucleic acids into plant cells. Silva et al. (2010) utilized polymer NPs to deliver siRNA into tobacco protoplasts, offering a different method for gene knockout in plant cells. Some nanoparticles, like carbon nanotubes and mesoporous silica, are able to enter the cell wall naturally, while others, such as gold nanoparticles and magnetic nanoparticles (MNPs), need chemical or physical treatments to carry genetic materials into cells. In the meantime, passive delivery mediated by nanoparticles has been observed in tobacco, cowpea, and arugula crops per studies by Burlaka et al. (2015), Mitter et al. (2017), Golestanipour et al. (2018), and Kwak et al. (2019). New materials, including NPs, have the potential to be valuable tools for editing systems, as suggested by Gao (2021). Hamada et al. (2017) suggested a plant bombardment method targeting the shoot apical meristems of wheat within this same framework (Imai et al. 2020). Gold particles covered with the gene construct for green fluorescent protein were transported into the L2 cell layer of the shoot apical meristems of wheat in this research. This method resulted in a reliable change in wheat without using embryogenic callus culture and can be used for other crops that have not been effectively transformed using traditional techniques.

Unlike traditional methods of plant genetic engineering, which rely on *Agrobacterium*, viruses, or external biolistic/chemical assistance, a nanomaterial-based delivery system can deliver various biomolecule cargos by passively crossing intact cell walls (Fig. 1.16). This delivery system offers numerous advantages for plant genetic modification. Initially, it surpasses restrictions related to the range of hosts and thus has the potential to be used in a variety of plant species and various tissues, even those that are typically challenging to regenerate through tissue culture (a necessity for traditional transgenic techniques). Furthermore, it is simple to customize by functionalizing nanomaterials to allow for controlled or target-specific release of the cargo. Achieving this is possible through either an external gate-opening trigger or a gatekeeper design that responds to conditions. These benefits can enhance the effectiveness of this method for precise and controlled delivery of cargo to a specific destination once certain conditions are satisfied, such as pH-dependent release (Kwak et al. 2019). It also enables the introduction of external functional genes or siRNA/dsRNA to produce genetically improved plants without integrating transgenes into the genome, known as "engineered nontransgenic plants." In this scenario, the altered characteristics achieved through temporary gene activation or deactivation cannot be passed on to future generations. Ultimately, the efficiency of delivery has been increased within the system. It has the ability to carry a considerable number of biomolecules and enables the simultaneous delivery of various biomolecules to the desired locations (Wang et al. 2019a, b, c).

Fig. 1.16 (continued) exogenous plasmid DNA (pDNA) and siRNA into intact plant cells. Techniques employing nanomaterials can overcome host-range limitations and create genetically modified plants free of transgenes. The nanocarrier can be adjusted to transport cargo in a controlled, targeted, and species-independent manner. The nanomaterial-based delivery system can also be employed to carry a Cas9/single guide (sg)RNA plasmid into pollen or callus tissues for the generation of transgenic plants, especially for species that are difficult to genetically engineer. Nanomaterials are shown without scale representation. Abbreviations: CNTs carbon nanotubes, MSN mesoporous silica nanoparticles, MNP magnetic nanoparticles. (Courtesy: Trends in Plant Science; https://doi.org/10.1016/j.tplants.2019.05.004)

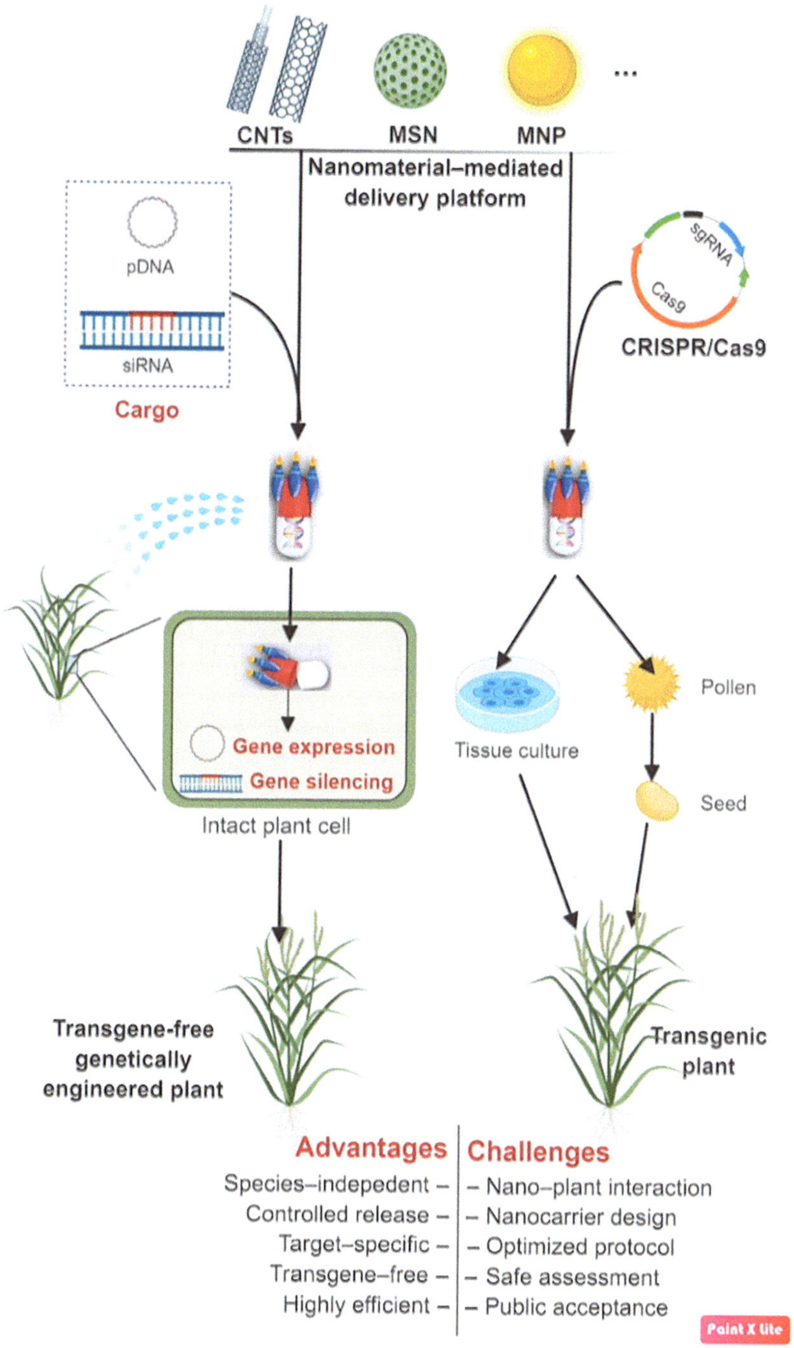

Fig. 1.16 Nanomaterial-enhanced genetic modification of plants. Functionalized nanomaterials can act as a delivery system capable of traversing barriers (e.g., multilayered cell walls) to deliver

There is a strong desire to develop genetically enhanced plants with regulated gene expression, without integrating exogenous DNA into their genome. Therefore, this method could avoid the stringent rules that control the use of genetically modified organisms. Furthermore, advancements in nanomaterial-based plant genetic engineering can aid in plant functional genetic research by quickly creating RNAi interference or overexpression lines to confirm gene functions (Ijaz et al. 2023). This will be particularly beneficial for inducing temporary changes in gene expression, either by increasing or decreasing it, within a specific timeframe. In the same way, the method could also be expanded to deliver a genome-editing tool like CRISPR/Cas9 to create genetically modified plants. It is certain that future research will necessitate cooperation between scientists working with nanomaterials and those in the field of plant biology in order to create specifically tailored nanomaterials and improve the methods of their use. Moreover, addressing the risk of using nanomaterials in agricultural systems is important to ensure safe use and public acceptance, especially for plant species that are challenging to genetically transform through traditional methods. Hence, the aim is to create an innovative category of nanomaterials that are physiologically based and more effective. The nanomaterial-based approach, which is both efficient and can be used across different species, may be further advanced as a tool for high-throughput screening to pinpoint genotypes that lead to desired traits. Nanoscience is a rapidly expanding field and one must be fully alert to keep abreast of the latest advancements.

1.10 Prospective

The world is reaching a turning point, wherein the population and food production are not in tandem. To augment food production, plant breeding needs to have a multifaceted approach. Integration of appropriate technologies at the right time is vital. Several upcoming technologies are getting prominence. Salse et al. (2024) enumerate the following for future plant breeding: (i) climate change presents a major obstacle to agricultural output and food safety; (ii) future ecosystems necessitate new strains for sustainable food cultivation;(iii) developing new varieties should take into account multiple stresses at the plant, plot, and landscape scales to preserve ecosystem services; (iv) concentrating solely on plants is inadequate; it is crucial to consider the holobiont and incorporate microorganisms linked to plants in breeding approaches for upcoming ecosystems; and (v) integrated biological systems must be used to tackle agricultural adaptation more effectively.

Recent research emphasizes the effectiveness of genome sequencing and gene editing technologies (Meemken and Bellemare 2020). Quick and affordable sequencing technologies have transformed the process of genome sequencing. Nevertheless, a small number of technologies tested in the laboratory have been proven in the field in a manner that would be valuable to the farmer (Napier et al. 2019). Findings from experiments conducted in controlled environments frequently contrast with those seen in real-world situations. The idea of domesticating wild plants from scratch has been suggested as a way to achieve food security and sustainable agriculture with minimal inputs (Fernie and Yan 2019). The new domestication of wild plants (de novo domestication) into novel crops is commendable. For instance, through the genome editing of essential genes related to domestication and agronomic characteristics, wild allotetraploid rice can be swiftly transformed into novel rice cultivars exhibiting excellent environmental adaptability and high grain production (Xie and Liu 2021). Likewise, rapid-cycle breeding programs that create enhanced varieties can also contribute to climate change adaptation. This approach involves quickly rotating different plant varieties by actively promoting new ones and removing outdated ones. Along with impacting genetic controls related to plant responses to climate change stresses, environmental triggers also highly regulate the epigenome (Iwasaki and Paszkowski 2014).

Most plant genomes are primarily made up of retrotransposons, which are a very dynamic aspect of the genome. Additionally, retrotransposons are essential for genome evolution and mutations, and they also influence plant responses to environmental stressors such as heat and drought associated with climate change, in conjunction with factors like UV radiation (Anders et al. 2021). The function of retrotransposon promoters significantly affects the expression of adjacent genes. Moreover, retrotransposon activity triggers epigenetic regulation via the RNAi pathway, resulting in post-transcriptional repression and CHG DNA methylation, which then follows the siRNA pathway to label DNA for silencing. The formation of heterochromatin may arise from the insertion of neighboring retrotransposons (Yang et al. 2023). Methylation of retrotransposons may influence nearby genes, resulting in their concurrent silencing. It is important that many crucial cereal genes are positioned close to retrotransposons. The activation of retrotransposons may play an essential role in modulating the epigenetic state in response to climate change-related stresses (Papolu et al. 2022). A more thorough understanding of how climate change-related environmental stresses affect retrotransposon replication and the resulting increase in genome size is essential for their incorporation into current plant breeding methods (Fig. 1.17). Nonetheless, it is expected that the use of retrotransposon technologies in crop breeding will help breeders achieve higher and sustainable productivity rates.

The method by which tools of New Plant Breeding Technologies (NPBTs) such as CRISPR-Cas are introduced into the plant cell might include a step forward in engineering plants. Recent advancements are aimed at utilizing DNA-free systems to provide the elements for genome editing. DNA-free systems face two significant issues, the extent of which may vary according to the plant species: (i) transport through the plant cell wall and (ii) regeneration of plants from protoplasts or callus cells (Metje-Sprink et al. 2019). Consequently, upcoming studies should also focus on creating dependable DNA-free genome editing methods for various crop types. Additionally, the integration of genome editing techniques with double haploid (DH) technology is highly significant. Homozygous pure DH lines could facilitate the desired trait enhancement within two generations, avoiding the prolonged process of multiple crossings and backcrossings adopted in traditional breeding to incorporate a preferred trait into top commercial lines (Wang et al. 2019a, b, c).

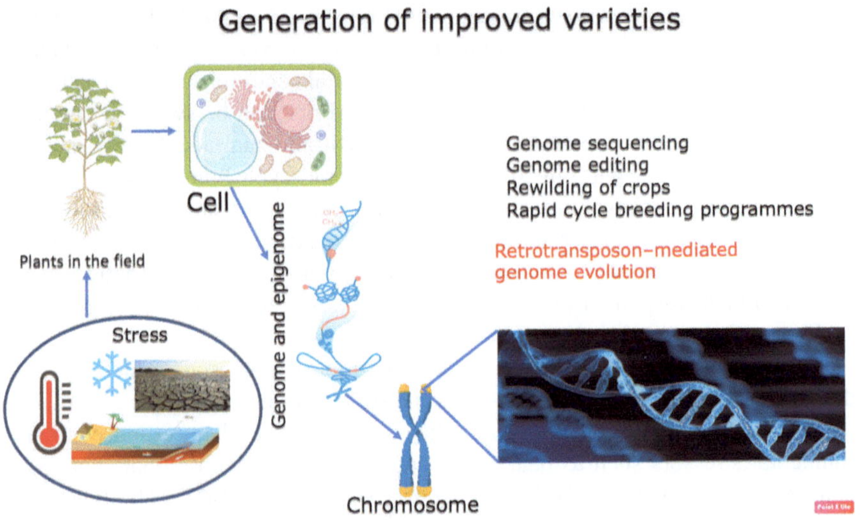

Fig. 1.17 Generation of improved varieties. (Courtesy: Trends in Plant Science; https://doi.org/10.1016/j.tplants.2021.03.004)

Genetic diversity is key to plant breeding. To increase diversity, our comprehension of the molecular genetic processes involved in crop domestication has significantly improved with the introduction of next-generation sequencing. Several studies have shown that leveraging our understanding of the domestication of genes can enable the development of multi-gene strategies aimed at improving the agronomic potential of underutilized species through gene editing or introgression methods. Recently, the potential of altering gene regulator sequences to change the expression of groups of genes simultaneously has been suggested, and initial proof-of-concept studies in this field have been conducted (Curtin et al. 2022). Similar exercises must be augmented to inculcate new diversity into breeding programs.

Speed breeding technology has emerged as a versatile collection of techniques that modify the growing conditions of crop plants to accelerate their breeding cycles by enhancing the rate of flowering and seed yield with the aid of AI (Razzaq et al. 2021; Rai 2022). Induction of meiosis in mitotic cells comes as a boon to drastically reduce the time of the breeding cycle (Cook et al. 2023). However, this approach needs refinement to achieve meiotic cells on a "turn-key" basis. AI technology has been utilized to speed up the development of new plant varieties, including high-throughput genomics and phenomics for advanced breeding (Esposito et al. 2020; Khan et al. 2022). More frequently, ML techniques have been applied in genomic prediction, genomic selection, and marker-assisted selection (Reinoso-Peláez et al. 2022). Numerous agricultural firms like Monsanto and John Deere have invested hundreds of millions in developing technologies that leverage vast data on soil types, seed varieties, and weather to aid farmers in minimizing costs and improving yields (Khan et al. 2022). Plant breeders are creating systems to enhance the comprehension of plant behavior in diverse climatic conditions (Sun et al. 2024). A new model of digital breeding that is rapidly developing is encouraging.

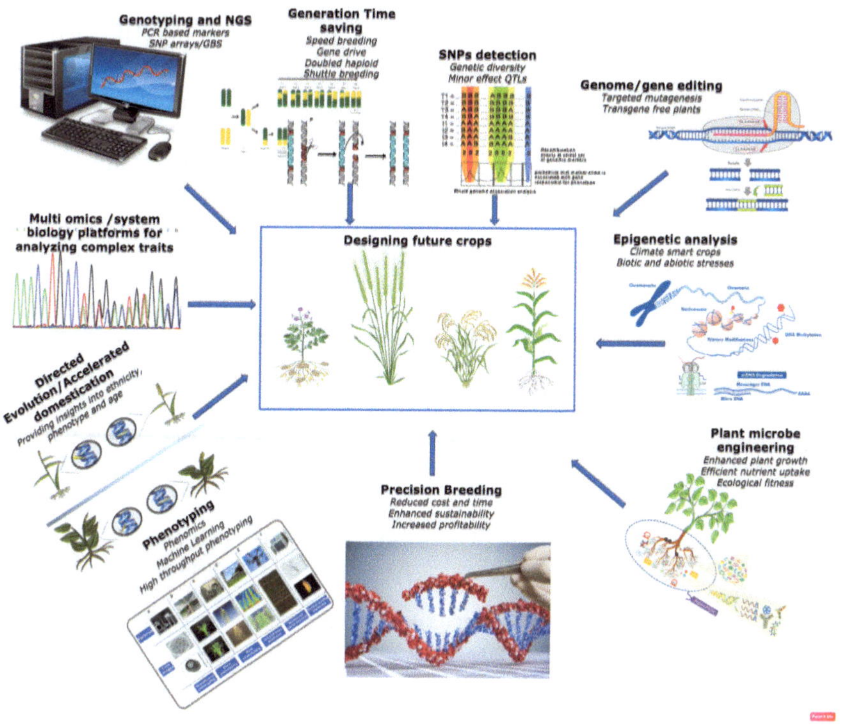

Fig. 1.18 Designing future crops (used Biorender)

The rapid progress in cutting-edge gene sequencing and phenotype collection technologies is resulting in a growing amount of genetic, phenotypic, and environmental information in crop breeding studies. The thorough integration of artificial intelligence, information technology, and conventional breeding, together with biotechnology, is shifting breeding from traditional hybrid techniques and field assessments to a data-informed intelligent breeding technology framework. A comprehensive strategy employing methods like gene drive, rapid breeding, shuttle breeding, doubled haploids, SNP identification, genome/gene modification, epigenetic study, plant-microbe engineering, precision breeding, phenotyping, directed evolution/accelerated domestication, and multi-omics/systems biology frameworks for intricate traits, along with genotyping and NGS, should be adopted in plant breeding initiatives (Fig. 1.18). In the future breeding system, technology and modern tools will be employed in all areas of breeding such as data gathering, analysis, storage, management, and utilization, resulting in a groundbreaking new breeding model where computers support or replace human cognitive tasks and intelligent devices replace manual labor. Breeders need to create comprehensive, data-oriented precision breeding programs that include genotype–phenotype–environment multi-dimensional factors. This method will aid gene editing, synthetic biology, and other technological progress, allowing for focused and efficient improvements in developing new varieties while promoting precision, intelligence, and a factory-driven seed transformation. In Plant breeding 5, every step from breeding design to phenotype prediction is conducted automatically through Deep Learning (DL) (Jeon et al. 2023). Additional phases may be included if new technologies appear in the future. However, breeders face limitations in their understanding of the progression of plant growth. Allowing crops to grow naturally in their environment is also an essential aspect of research. Breeders have specific responsibilities in the breeding process, but also need assistance from relevant institutions and policies.

References

Abbas MS (2018) Genetically engineered (modified) crops (Bacillus thuringiensis crops) and the world controversy on their safety. Egypt J Biol Pest Control 28(1):1–12. https://doi.org/10.1186/s41938-018-0051-2

Abdi H, Shahbazitabar M, Mohammadi-Ivatloo B (2020) Food, energy and water nexus: a brief review of definitions, research, and challenges. Inventions 5:56

Abdul Aziz M, Masmoudi K (2024) Molecular breakthroughs in modern plant breeding techniques. Hortic Plant J. https://doi.org/10.1016/j.hpj.2024.01.004

Acharjee A, Kloosterman B, Visser RG, Maliepaard C (2016) Integration of multi-omics data for prediction of phenotypic traits using random forest. BMC Bioinformatics 17:363–373. https://doi.org/10.1186/s12859-016-1043-4

Adam D (2021) How far will global population rise? Researchers can't agree. Nature 597:462–465

Ahmad M (2023) Plant breeding advancements with "CRISPR-Cas" genome editing technologies will assist future food security. Front Plant Sci 14:1133036. https://doi.org/10.3389/fpls.2023.1133036

Ahmar S, Mahmood T, Fiaz S, Shafique MS, Chattha MS, Jung K (2021) Advantage of nanotechnology-based genome editing system and its application in crop improvement. Front Plant Sci 12:663849. https://doi.org/10.3389/fpls.2021.663849

Al Hasan SM, Saulam J, Mikami F, Kanda K, Ngatu NR, Yokoi H, Hirao T (2022) Trends in per capita food and protein availability at the National Level of the Southeast Asian Countries: an analysis of the FAO'S food balance sheet data from 1961 to 2018. Nutrients 14:603

Alemu A, Åstrand J, Montesinos-López OA, Isidro y Sánchez, J., Fernández-Gónzalez J, Tadesse W, Vetukuri RR, Carlsson AS, Ceplitis A, Crossa J, Ortiz R, Chawade A (2024) Genomic selection in plant breeding: key factors shaping two decades of progress. Mol Plant 17(4):552–578. https://doi.org/10.1016/j.molp.2024.03.007

Amiteye S (2021) Basic concepts and methodologies of DNA marker systems in plant molecular breeding. Heliyon 7(10):e08093. https://doi.org/10.1016/j.heliyon.2021.e08093. PMID: 34765757; PMCID: PMC8569399

Anders S, Cowling W, Pareek A, Gupta KJ, Singla-Pareek SL, Foyer CH (2021) Gaining acceptance of novel plant breeding technologies. Trends Plant Sci 26(6):575–587. https://doi.org/10.1016/j.tplants.2021.03.004. Epub 2021 Apr 20. PMID: 33893048

Andreoli V, Bagliani M, Corsi A, Frontuto V (2021) Drivers of protein consumption: a cross-country analysis. Sustainability 13:7399

Araújo SO, Peres RS, Ramalho JC, Lidon F, Barata J (2023) Machine learning applications in agriculture: current trends, challenges, and future perspectives. Agronomy 13(12):2976. https://doi.org/10.3390/agronomy13122976

Armstrong EM et al (2023) One hundred important questions facing plant science: an international perspective. New Phytol 238(2):470–481. https://doi.org/10.1111/nph.18771

Azodi CB, Pardo J, VanBuren R, de Los Campos G, Shiu SH (2020) Transcriptome-based prediction of complex traits in maize. Plant Cell 32:139–151. https://doi.org/10.1105/tpc.19.00332

Bahar NH, Lo M, Sanjaya M, Van Vianen J, Alexander P, Ickowitz A, Sunderland T (2020) Meeting the food security challenge for nine billion people in 2050: what impact on forests? Glob Environ Chang 62:102056. https://doi.org/10.1016/j.gloenvcha.2020.102056

Bai G, Ge YF, Hussain W, Baenziger PS, Graef G (2016) A multi-sensor system for high throughput field phenotyping in soybean and wheat breeding. Comput Electron Agric 128:181–192. https://doi.org/10.1016/j.compag.2016.08.021

Bassi FM, Sanchez-Garcia M, Ortiz R (2024) What plant breeding may (and may not) look like in 2050? Plant Genome 17(1):e20368. https://doi.org/10.1002/tpg2.20368

Bebber DP (2015) Range-expanding pests and pathogens in a warming world. Annu Rev Phytopathol 53:335–356

Bebber DP et al (2019) Many unreported crop pests and pathogens are probably already present. Glob Change Biol 25:2703–2713

Bellwood P (2005) First farmers: the origins of agricultural societies. Blackwel Publishing, Oxford

Beltran-Peña A, Rosa L, D'Odorico P (2020) Global food self-sufficiency in the 21st century under sustainable intensification of agriculture. Environ Res Lett 15(9):95004

Bernal-Gallardo JJ, de Folter S (2024) Plant genome information facilitates plant functional genomics. Planta 259:117. https://doi.org/10.1007/s00425-024-04397-z

Bernardo R (2020a) Reinventing quantitative genetics for plant breeding: something old, something new, something borrowed, something BLUE. Heredity 125(6):375–385. https://doi.org/10.1038/s41437-020-0312-1

Bernardo R (2020b) Breeding for quantitative traits in plants, 3rd edn. Stemma Press, Woodbury. ISBN 978-0-9720724-3-4

Bhuyan SJ, Kumar M, Devde PR, Rai AC, Mishra AK, Singh PK, Siddique KH (2023) Progress in gene editing tools, implications and success in plants: a review. Front Genome Ed 5. https://doi.org/10.3389/fgeed.2023.1272678

Borlaug NE (1983) Contributions of conventional plant breeding to food production. Science 219(4585):689–693

Böwer F, Schnittger A (2021) How to switch from mitosis to meiosis: regulation of germline entry in plants. Annu Rev Genet 55:427–452. https://doi.org/10.1146/annurev-genet-112618-043553

Brooker R, Brown LK, George TS, Pakeman RJ, Palmer S, Ramsay L, Schöb C, Schurch N, Wilkinson MJ (2022) Active and adaptive plasticity in a changing climate. Trends Plant Sci 27(7):717–728. https://doi.org/10.1016/j.tplants.2022.02.004

Burlaka OM, Pirko YV, Yemets AI, Blume YB (2015) Plant genetic transformation using carbon nanotubes for DNA delivery. Cytol Genet 49:3–12. https://doi.org/10.3103/S009545271506002X

Byers D, Waller DM (1999) Do plant populations purge their genetic load? Effects of population size and mating history on inbreeding depression. Annu Rev Ecol Syst 30:479–513. https://doi.org/10.1146/annurev.ecolsys.30.1.479

Callaway E (2018) CRISPR plants now subject to tough GM laws in European Union. Nature 560(7716):16. https://doi.org/10.1038/d41586-018-05814-6. PMID: 30065322

Campbell BM et al (2016) Reducing risks to food security from climate change. Glob Food Sec 11:34–43

Cardi T, Murovec J, Bakhsh A, Boniecka J, Bruegmann T, Bull SE, Eeckhaut T, Fladung M, Galovic V, Linkiewicz A, Lukan T, Mafra I, Michalski K, Kavas M, Nicolia A, Nowakowska J, Sági L, Sarmiento C, Yıldırım K, Zlatković M, Hensel G, Van Laere K (2023) CRISPR/Cas-mediated plant genome editing: outstanding challenges a decade after implementation. Trends Plant Sci 28(10):1144–1165. https://doi.org/10.1016/j.tplants.2023.05.012

Carey J (2023) Unearthing the origins of agriculture. Proc Natl Acad Sci 120(15):e2304407120. https://doi.org/10.1073/pnas.2304407120

Ćeran M, Miladinović D, Đorđević V, Trkulja D, Radanović A, Glogovac S (2024) Genomics-assisted speed breeding for crop improvement: present and future. Front Sustain Food Syst 8:1383302. https://doi.org/10.3389/fsufs.2024.1383302

Chaloner TM, Gurr SJ, Bebber DP (2020) Geometry and evolution of the ecological niche in plant-associated microbes. Nat Commun 11:2955

Chaloner TM, Gurr SJ, Bebber DP (2021) Plant pathogen infection risk tracks global crop yields under climate change. Nat Clim Change 11(8):710–715. https://doi.org/10.1038/s41558-021-01104-8

Chandrasekaran R, Rajiv P, Abd-Elsalam KA (2020) Carbon nanotubes: plant gene delivery and genome editing. Elsevier Inc, Amsterdam

Chang FP, Kuang LY, Huang CA, Jane WN, Hung Y, Hsing YIC et al (2013) A simple plant gene delivery system using mesoporous silica nanoparticles as carriers. J Mater Chem B 1:5279–5287. https://doi.org/10.1039/c3tb20529k

Chaudhary D, Jeena AS, Rohit GS, Raj R, Mishra S, Kajal GOP, Meena MR (2024) Advances in RNA interference for plant functional genomics: unveiling traits, mechanisms, and future directions. Appl Biochem Biotechnol. https://doi.org/10.1007/s12010-023-04850-x

Chong M (2003) Acceptance of golden rice in the Philippine 'rice bowl'. Nat Biotechnol 21(9):971–972. https://doi.org/10.1038/nbt0903-971. PMID: 12949549

Cobb JN et al (2019) Enhancing the rate of genetic gain in public-sector plant breeding programs: lessons from the breeder's equation. Theor Appl Genet 132:627–645

Cook TM, Isenegger D, Dutta S, Sahab S, Kay P, Aboobucker SI, Biswas E, Heerschap S, Nikolau BJ, Dong L, Lübberstedt T (2023) Overcoming roadblocks for in vitro nurseries in plants: induction of meiosis. Front Plant Sci 14:1204813. https://doi.org/10.3389/fpls.2023.1204813

Cook TM, Biswas E, Aboobucker SI, Dutta S, Lübberstedt T (2024) A cell-based fluorescent system and statistical framework to detect meiosis-like induction in plants. Front Plant Sci 15:1386274. https://doi.org/10.3389/fpls.2024.1386274

Cortés AJ, López-Hernández F (2021) Harnessing crop wild diversity for climate change adaptation. Genes 12:783. https://doi.org/10.3390/genes12050783

Cortes LT, Zhang Z, Yu J (2021) Status and prospects of genome-wide association studies in plants. Plant Genome 14(1):e20077. https://doi.org/10.1002/tpg2.20077

Cortés AJ, López-Hernández F, Blair MW (2022) Genome–environment associations, an innovative tool for studying heritable evolutionary adaptation in orphan crops and wild relatives. Front Genet 13:910386

Crossa J et al (2017) Genomic selection in plant breeding: methods, models, and perspectives. Trends Plant Sci 22(11):961–975. https://doi.org/10.1016/j.tplants.2017.08.011

Curtin S, Qi Y, Peres LEP, Fernie AR, Zsögön A (2022) Pathways to de novo domestication of crop wild relatives. Plant Physiol 188(4):1746–1756. https://doi.org/10.1093/plphys/kiab554. PMID: 34850221; PMCID: PMC8968405

Das PR, Sherif SM (2020) Application of exogenous dsRNAs-induced RNAi in agriculture: challenges and triumphs. Front Plant Sci 11:946

Daszkiewicz T (2022) Food production in the context of global developmental challenges. Agriculture 12(6):832. https://doi.org/10.3390/agriculture12060832

De Medeiros AD, Da Silva LJ, Pereira MD, Oliveira AMS, Dias DCFS (2020) High- throughput phenotyping of brachiaria grass seeds using free access tool for analyzing x-ray images. An Acad Bras Cienc 92. https://doi.org/10.1590/0001-3765202020190209e20190209

Demirer GS, Silva TN, Jackson CT, Thomas JB, Rhee SY, Mortimer JC, Landry MP (2021) Nanotechnology to advance CRISPR–Cas genetic engineering of plants. Nat Nanotechnol 16(3):243–250. https://doi.org/10.1038/s41565-021-00854-y

Deyhle F, Sarkar AK, Tucker EJ, Laux T (2007) WUSCHEL regulates cell differentiation during anther development. Dev Biol 302:154–159

Dirzo R, Raven PH (2003) Global state of biodiversity and loss. Annu Rev Environ Resour 28:137–167

Doebley JF, Gaut BS, Smith BD (2006) The molecular genetics of crop domestication. Cell 127(7):1309–1321. https://doi.org/10.1016/j.cell.2006.12.006. PMID: 17190597

Doudna JA, Charpentier E (2014) Genome editing. The new frontier of genome engineering with CRISPR-Cas9. Science 346:1258096. https://doi.org/10.1126/science.1258096

Esposito S, Carputo D, Cardi T, Tripodi P (2020) Applications and trends of machine learning in genomics and phenomics for next-generation breeding. Plants 9(1):34. https://doi.org/10.3390/plants9010034

Fader M, Gerten D, Krause M, Lucht W, Cramer W (2013) Spatial decoupling of agricultural production and consumption: quantifying dependences of countries on food imports due to domestic land and water constraints. Environ Res Lett 8:014046

Fang J (2024) Breeding 5.0: AI-driven revolution in designed plant breeding. Mol Plant Breed 15:27–33. https://doi.org/10.5376/mpb.2024.15.0004

Farid M, Cao J, Lim Y, Arato T, Kodama K (2020) Exploring factors affecting the acceptance of genetically edited food among youth in Japan. Int J Environ Res Public Health 17:2935. https://doi.org/10.3390/ijerph17082935

Farooq MA et al (2024) Artificial intelligence in plant breeding. Trends Genet 40(10):891–908. https://doi.org/10.1016/j.tig.2024.07.001

Feigin SV, Wiebers DO, Lueddeke G, Morand S, Lee K, Knight A, Brainin M, Feigin VL, Whitfort A, Marcum J, Shackelford TK, Skerratt LF, Winkler AS (2023) Proposed solutions to anthropogenic climate change: a systematic literature review and a new way forward. Heliyon 9(10):e20544. https://doi.org/10.1016/j.heliyon.2023.e20544

Fernie AR, Yan J (2019) De novo domestication: an alternative route toward new crops for the future. Mol Plant 12:615–631

Finiuk N, Buziashvili A, Burlaka O, Zaichenko A, Mitina N, Miagkota O et al (2017) Investigation of novel oligoelectrolyte polymer carriers for their capacity of DNA delivery into plant cells. Plant Cell Tissue Organ Cult 131:27–39. https://doi.org/10.1007/s11240-017-1259-7

Fischer RA et al (2014) Crop yields and global food security: will yield increase continue to feed the world? ACIAR Monograph No. 158 Australian Centre for International Agricultural Research

Flowers JM, Hazzouri KM, Gros-Balthazard M, Mo Z, Koutroumpa K, Perrakis A, Ferrand S, Khierallah HSM, Fuller DQ, Aberlenc F et al (2019) Cross-species hybridization and the origin of North African date palms. Proc Natl Acad Sci U S A 116:1651–1658

Fones HN et al (2020) Threats to global food security from emerging fungal and oomycete crop pathogens. Nat Food 1:332–342

Font-I-Furnols M (2023) Meat consumption, sustainability and alternatives: an overview of motives and barriers. Foods 12(11):2144. https://doi.org/10.3390/foods12112144. PMID: 37297389; PMCID: PMC10252260

Food and Agriculture Organization of the United Nations (2017) The future of food and agriculture: trends and challenges. Published online. http://www.fao.org/3/a-i6583e.pdf

Food and Agriculture Organization of the United Nations (2018) The future of food and agriculture: alternative pathways to 2050. Published online. http://www.fao.org/3/I8429EN/i8429en.pdf

Fróna D, Szenderák J, Harangi-Rákos M (2021) Economic effects of climate change on global agricultural production. Nat Conserv 44:117–139. https://doi.org/10.3897/natureconservation.44.64296

Gao X, Zhang J, Wei Z, Hakonarson H (2018) DeepPolyA: a convolutional neural network approach for polyadenylation site prediction. IEEE Access 2018(6):24340–24349

Gao C (2021) Genome engineering for crop improvement and future agriculture. Cell 184:1621–1635

Gasparini K, Moreira JDR, Peres LEP, Zsögön A (2021) De novo domestication of wild species to create crops with increased resilience and nutritional value. Curr Opin Plant Biol 60:102006. https://doi.org/10.1016/j.pbi.2021.102006

Gehan MA, Greenham K, Mockler TC, McClung CR (2015) Transcriptional networks—crops, clocks, and abiotic stress. Curr Opin Plant Biol 24:39–46. https://doi.org/10.1016/j.pbi.2015.01.004

Gernand D, Rutten T, Pickering R, Houben A (2006) Elimination of chromosomes in Hordeum vulgare x H. bulbosum crosses at mitosis and interphase involves micronucleus formation and progressive heterochromatinization. Cytogenet Genome Res 114(2):169–174. https://doi.org/10.1159/000093334

Ghosh S, Watson A, Gonzalez-Navarro OE, Ramirez-Gonzalez RH, Yanes L, Simmonds J, Wells R, Rayner T, Green P, Hafeez A, Hayta S, Melton RE, Steed A, Sarkar A, Carter J, Perkins L, Lord J, Tester M, Osbourn A et al (2018) Speed breeding in growth chambers and glass-

houses for crop breeding and model plant research. Nat Protoc 13(12):2944–2963. https://doi.org/10.1038/s41596-018-0072-z
Giller KE, Delaune T, Silva JV et al (2021) The future of farming: who will produce our food? Food Sec 13:1073–1099. https://doi.org/10.1007/s12571-021-01184-6
Godfray HCJ, Beddington JR, Crute IR, Haddad L, Lawrence D, Muir JF, Pretty J, Robinson S, Thomas SM, Toulmin C (2010) Food security: the challenge of feeding 9 billion people. Science 327:812–818
Godwin ID, Rutkoski J, Varshney RK, Hickey LT (2019) Technological perspectives for plant breeding. Theor Appl Genet 132:555–557. https://doi.org/10.1007/s00122-019-03321-4
Golestanipour A, Nikkhah M, Aalami A, Hosseinkhani S (2018) Gene delivery to tobacco root cells with single-walled carbon nanotubes and cell- penetrating fusogenic peptides. Mol Biotechnol 60:863–878. https://doi.org/10.1007/s12033-018-0120-5
Groen SC, Ćalić I, Joly-Lopez Z, Platts AE, Choi JY, Natividad M, Dorph K, Mauck WM 3rd, Bracken B, Cabral CLU, Kumar A, Torres RO, Satija R, Vergara G, Henry A, Franks SJ, Purugganan MD (2020) The strength and pattern of natural selection on gene expression in rice. Nature 578(7796):572–576
Gupta PK, Kumar S (2025) Third-generation novel technologies for gene editing. Trends Biotechnol. Jul 29: S0167-7799(25)00273-2. https://doi.org/10.1016/j.tibtech.2025.07.012. Epub ahead of print. PMID: 40738762
Hamada H, Linghu Q, Nagira Y, Miki R, Taoka N, Imai R (2017) An in planta biolistic method for stable wheat transformation. Sci Rep 7:11443. https://doi.org/10.1038/s41598-017-11936-0
Harfouche AL, Jacobson DA, Kainer D, Romero JC, Harfouche AH, Mugnozza GS, Moshelion M, Tuskan GA, Keurentjes JJ, Heffner EL, Sorrells ME, Jannink J-L (2009) Genomic selection for crop improvement. CropSci 49:1–12
Heffner EL, Sorrells ME, Jannink JL (2009) Genomic Selection for Crop Improvement. Crop Sci. 2009, 49:1–12
Henchion M, Hayes M, Mullen AM, Fenelon M, Tiwari B (2017) Future protein supply and demand: strategies and factors influencing a sustainable equilibrium. Foods 6:53
Henry RJ (2022) Progress in Plant Genome Sequencing. Appl Biosci 1(2):113–128. https://doi.org/10.3390/applbiosci1020008
Herman RA, Fedorova M, Storer NP (2019) Will following the regulatory script for GMOs promote public acceptance of gene-edited crops? Trends Biotechnol 37(12):1272–1273. https://doi.org/10.1016/j.tibtech.2019.06.007. Epub 2019 Jul 13. PMID: 31307666
Herrera-Estrella L, Depicker A, Van Montagu M et al (1983) Expression of chimaeric genes transferred into plant cells using a Ti-plasmid-derived vector. Nature 303:209–213. https://doi.org/10.1038/303209a0
Herrero-Huerta M, Rodriguez-Gonzalvez P, Rainey KM (2020) Yield prediction by machine learning from UAS-based mulit-sensor data fusion in soybean. Plant Methods 16:78. https://doi.org/10.1186/s13007-020-00620-6
Hickey JM, Chiurugwi T, Mackay I, Powell W, Hickey JM, Chiurugwi T, Mackay I, Powell W, Eggen A, Kilian A et al (2017) Genomic prediction unifies animal and plant breeding programs to form platforms for biological discovery. Nat Genet 49:1297–1303
Hickey LT, Robinson H, Jackson SA, Leal-Bertioli SC, Tester M, Gao C, Godwin ID, Hayes BJ, Wulff BB (2019) Breeding crops to feed 10 billion. Nat Biotechnol 37(7):744–754. https://doi.org/10.1038/s41587-019-0152-9
Hochman Z et al (2017) Climate trends account for stalled wheat yields in Australia since 1990. Glob Chang Biol 23:2071–2081
Holliday R (2006) Epigenetics: a historical overview. Epigenetics 1:76–80
Hoover DL, Abendroth LJ, Browning DM, Saha A, Snyder K, Wagle P, Witthaus L, Baffaut C, Biederman JA, Bosch DD, Bracho R, Busch D, Clark P, Ellsworth P, Fay PA, Flerchinger G, Kearney S, Levers L, Saliendra N et al (2023) Indicators of water use efficiency across diverse agroecosystems and spatiotemporal scales. Sci Total Environ 864:160992. https://doi.org/10.1016/j.scitotenv.2022.160992

Hu H, Wang J, Nie S, Zhao J, Batley J, Edwards D (2024) Plant pangenomics, current practice and future direction. Agric Commun 2(2):100039. https://doi.org/10.1016/j.agrcom.2024.100039

Hufford MB, Berny Mier Y, Teran JC, Gepts P (2019) Crop biodiversity: an unfinished magnum opus of nature. Annu Rev Plant Biol 70:727–751

Ianutsevich EA, Danilova OA, Grum-Grzhimaylo OA, Tereshina VM (2023) The role of Osmolytes and membrane lipids in the adaptation of acidophilic fungi. Microorganisms 11(7):1733. https://doi.org/10.3390/microorganisms11071733

Ichim MC (2020) The more favourable attitude of the citizens toward GMOs supports a new regulatory framework in the European Union. GM Crops & Food 12:18–24. https://doi.org/10.1080/21645698.2020.1795525

Ijaz M, Khan F, Ahmed T, Noman M, Zulfiqar F, Rizwan M, Chen J, Siddique KH, Li B (2023) Nanobiotechnology to advance stress resilience in plants: current opportunities and challenges. Mater Today Bio 22:100759. https://doi.org/10.1016/j.mtbio.2023.100759

Imai R, Hamada H, Liu Y, Linghu Q, Kumagai Y, Nagira Y et al (2020) In planta particle bombardment (IPB): a new method for plant transformation and genome editing. Plant Biotechnol 37:171–176. https://doi.org/10.5511/PLANTBIOTECHNOLOGY.20.0206A

IPCC (2014) Climate Change 2014: Synthesis Report. Contribution of Working Groups I, II and III to the Fifth Assessment Report of the Intergovernmental Panel on Climate Change [Core Writing Team, R.K. Pachauri and L.A. Meyer (eds.)]. IPCC, Geneva, 151 pp

ISAAA (2019) Global Status of Commercialized Biotech/GM Crops in 2019: biotech crops drive socio economic development and sustainable environment in the new frontier. ISAAA, Ithaca, pp 3–13

Iwasaki M, Paszkowski J (2014) Identification of genes preventing transgenerational transmission of stress induced epigenetic states. Proc Natl Acad Sci USA 111:8547–8552

Jang G, Kim J, Yu J-K, Kim H-J, Kim Y, Kim D-W et al (2020) Review: cost-effective unmanned aerial vehicle (UAV) platform for field plant breeding application. Remote Sens 12:998. https://doi.org/10.3390/rs12060998

Javed T, Indu I, Singhal RK, Shabbir R, Shah AN, Kumar P, Jinger D, Dharmappa PM, Shad MA, Saha D, Anuragi H, Adamski R, Siuta D (2022) Recent advances in agronomic and physio-molecular approaches for improving nitrogen use efficiency in crop plants. Front Plant Sci 13:877544. https://doi.org/10.3389/fpls.2022.877544

Jeon D, Kang Y, Lee S, Choi S, Sung Y, Lee T, Kim C (2023) Digitalizing breeding in plants: a new trend of next-generation breeding based on genomic prediction. Front Plant Sci 14:1092584. https://doi.org/10.3389/fpls.2023.1092584

Jiang N, Zhu X (2024) Modern phenomics to empower holistic crop science, agronomy, and breeding research. J Genet Genom. https://doi.org/10.1016/j.jgg.2024.04.016

Jinek M, Chylinski K, Fonfara I, Hauer M, Doudna JA, Charpentier E (2012) A programmable dual-RNA-guided DNA endonuclease in adaptive bacterial immunity. Science 337:816–821. https://doi.org/10.1126/science.1225829

Joldersma D, Liu Z (2018) Plant genetics enters the nano age? J Integr Plant Biol 60:446–447. https://doi.org/10.1111/jipb.12646

Jung J, Maeda M, Chang A, Bhandari M, Ashapure A, Landivar-Bowles J (2021) The potential of remote sensing and artificial intelligence as tools to improve the resilience of agriculture production systems. Curr Opin Biotechnol 70:15–22

Kalogiannidis S, Kalfas D, Chatzitheodoridis F, Papaevangelou O (2022) Role of crop-protection technologies in sustainable agricultural productivity and management. Land 11:1680. https://doi.org/10.3390/land11101680

Karimi M, Solati N, Ghasemi A, Estiar MA, Hashemkhani M, Kiani P et al (2015) Carbon nanotubes part II: a remarkable carrier for drug and gene delivery. Expert Opin Drug Deliv 12:1089–1105. https://doi.org/10.1517/17425247.2015.1004309

Katam R, Lin C, Grant K, Katam CS, Chen S (2022) Advances in plant metabolomics and its applications in stress and single-cell biology. Int J Mol Sci 23(13):6985. https://doi.org/10.3390/ijms23136985

Kawakatsu T, Ecker JR (2019) Diversity and dynamics of DNA methylation: epigenomic resources and tools for crop breeding. Breed Sci 69:191–204

Ketumile D, Yang X, Sanchez R, Kundariya H, Rajewski J, Dweikat IM, Mackenzie SA (2022) Implementation of epigenetic variation in sorghum selection and implications for crop resilience breeding. Front Plant Sci 12:798243

Khalid AM (2023) Creating synergies among the sustainable development goals and climate action: insights from a developing economy. Sustainability 15(17):13137. https://doi.org/10.3390/su151713137

Khan MH, Wang S, Wang J, Ahmar S, Saeed S, Khan SU, Xu X, Chen H, Bhat JA, Feng X (2022) Applications of artificial intelligence in climate-resilient smart-crop breeding. Int J Mol Sci 23(19):11156. https://doi.org/10.3390/ijms231911156

Khoshbakht K, Hammer K (2008) How many plant species are cultivated? Genet Resour Crop Evol 55(7):925–928

Khush GS (2001) Green revolution: the way forward. Nat Rev Genet 2(10):815–822. https://doi.org/10.1038/35093585. PMID: 11584298

Kigoni M, Choi M, Arbelaez JD (2023) 'Single-Seed-SpeedBulks:' a protocol that combines 'speed breeding' with a cost-efficient modified single-seed descent method for rapid-generation-advancement in oat (Avena sativa L.). Plant Methods 19:92. https://doi.org/10.1186/s13007-023-01067-1

Kim KD, Kang Y, Kim C (2020) Application of genomic big data in plant breeding: past, present, and future. Plants 9(11):1454. https://doi.org/10.3390/plants9111454

Klamt S, Saez-Rodriguez J, Gilles ED (2007) Structural and functional analysis of cellular networks with CellNetAnalyzer. BMC Syst Biol 1:2. https://doi.org/10.1186/1752-0509-1-2

Kleter A et al (2019) Gene-edited crops: towards a harmonized safety assessment. Trends Biotechnol 37:443–447

Komor AC, Kim YB, Packer MS, Zuris JA, Liu DR (2016) Programmable editing of a target base in genomic DNA without double-stranded DNA cleavage. Nature 533:420–424. https://doi.org/10.1038/nature17946

Kramer MG, Redenbaugh K (1994) Commercialization of a tomato with an antisense polygalacturonase gene: the FLAVR SAVR(tm) tomato story. Euphytica 79:293–297. https://doi.org/10.1007/BF00022530

Krug AS, Van Tassel DL, Warschefsky EJ (2023) The next era of crop domestication starts now. Proc Natl Acad Sci 120(14):e2205769120. https://doi.org/10.1073/pnas.2205769120

Kujur A, Saxena MS, Bajaj D, Laxmi; Parida, S.K. (2013) Integrated genomics and molecular breeding approaches for dissecting the complex quantitative traits in crop plants. J Biosci 38:971–987

Kumar S (2020) Epigenomics for crop improvement: current status and future perspectives. J Genet Cell Biol 3:128–134

Kwak SY, Lew TTS, Sweeney CJ, Koman VB, Wong MH, Bohmert- Tatarev K et al (2019) Chloroplast-selective gene delivery and expression in planta using chitosan-complexed single-walled carbon nanotube carriers. Nat Nanotechnol 14:447–455. https://doi.org/10.1038/s41565-019-0375-4

Lahlali R, Taoussi M, Laasli S, Gachara G, Ezzouggari R, Belabess Z, Aberkani K, Assouguem A, Meddich A, El Jarroudi M, Barka EA (2024) Effects of climate change on plant pathogens and host-pathogen interactions. Crop Environ 3(3):159–170. https://doi.org/10.1016/j.crope.2024.05.003

Lakshmanan M, Lim SH, Mohanty B, Kim JK, Ha SH, Lee DY (2015) Unraveling the light-specific metabolic and regulatory signatures of rice through combined in silico modeling and multiomics analysis. Plant Physiol 169:3002–3020

Lan L et al (2018) Farm-level and community aggregate economic impacts of adopting climate smart agricultural practices in three mega environments. PLoS One 13:e0207700

Langfelder P, Mischel PS, Horvath S (2013) When is hub gene selection better than standard meta-analysis? PLoS One 8:e61505. https://doi.org/10.1371/journal.pone.0061505

Lastochkina O, Aliniaeifard S, SeifiKalhor M, Bosacchi M, Maslennikova D, Lubyanova A (2022) Novel approaches for sustainable horticultural crop production: advances and prospects. Horticulturae 8(10):910. https://doi.org/10.3390/horticulturae8100910

Lavarenne J, Guyomarc'h S, Sallaud C, Gantet P, Lucas M (2018) The spring of systems biology-driven breeding. Trends Plant Sci 23:706–720. https://doi.org/10.1016/j.tplants.2018.04.005

Lei Z, Li Z, Zhang W, He D, Zhang Y (2024) From wild to cultivated crops: general shift in morphological and physiological traits for yield enhancement following domestication. Crop Environ. https://doi.org/10.1016/j.crope.2024.03.001

Li L-F, Li Y-L, Jia Y, Caicedo AL, Olsen KM (2017) Signatures of adaptation in the weedy rice genome. Nat Genet 49:811–814

Li X, Lian H, Zhao Q, He Y (2019a) MicroRNA166 monitors SPOROCYTELESS/NOZZLE for building of the anther internal boundary. Plant Physiol 181:208–220

Li Z, Gao N, Martini JW, Simianer H (2019b) Integrating gene expression data into genomic prediction. Front Genet 10:126. https://doi.org/10.3389/fgene.2019.00126

Li Z, Chen X, Shi S, Zhang H, Wang X, Chen H, Li W, Li L (2022) DeepBSA: a deep-learning algorithm improves bulked segregant analysis for dissecting complex traits. Mol Plant 15:1418–1427

Li M, Lin Y, Cheng Q, Wei T (2025) Prime Editing: A Revolutionary Technology for Precise Treatment of Genetic Disorders. Cell Prolif. 2025 Apr;58(4): e13808. https://doi.org/10.1111/cpr.13808

Liao X, Niehues A, de Visser C, et al (2023) FAIR Data Cube, a FAIR data infrastructure for integrated multi-omics data analysis. medRxiv; 2023. https://doi.org/10.1101/2023.04.23.23289000

Litvinov DY, Karlov GI, Divashuk MG (2021) Metabolomics for crop breeding: general considerations. Genes 12(10):1602. https://doi.org/10.3390/genes12101602

Liu D, Zhang H, Zhang L, Yuan Z, Hao M, Zheng Y (2014) Distant hybridization: a tool for interspecific manipulation of chromosomes. In: Pratap A, Kumar J (eds) Alien gene transfer in crop plants, Volume 1. Springer, New York. https://doi.org/10.1007/978-1-4614-8585-8_2

Liu H, Bruning B, Garnett T, Berger B (2020) Hyperspectral imaging and 3D technologies for plant phenotyping: from satellite to close-range sensing. Comput Electron Agric 175:105621. https://doi.org/10.1016/j.compag.2020.105621

Liu H, Chen W, Li Y, Sun L, Chai Y, Chen H, Nie H, Huang C (2022a) CRISPR/Cas9 technology and its utility for crop improvement. Int J Mol Sci 23(18):10442. https://doi.org/10.3390/ijms231810442

Liu T, Salguero P, Petek M, Martinez-Mira C, Balzano-Nogueira L, Ramšak Ž, McIntyre L, Gruden K, Tarazona S, Conesa A (2022b) PaintOmics 4: new tools for the integrative analysis of multi-omics datasets supported by multiple pathway databases. Nucleic Acids Res 50(W1):W551–W559

Long P, Zhang L, Huang B, Chen Q, Liu H (2020) Integrating genome sequence and structural data for statistical learning to predict transcription factor binding sites. Nucleic Acids Res 48:12604–12617

Lorenzo CD, Debray K, Herwegh D, Develtere W, Impens L, Schaumont D, Vandeputte W, Aesaert S, Coussens G, Boe YD, Demuynck K, Hautegem TV, Pauwels L, Jacobs TB, Ruttink T, Nelissen H, Inzé D (2022) BREEDIT: a multiplex genome editing strategy to improve complex quantitative traits in maize. Plant Cell 35(1):218–238. https://doi.org/10.1093/plcell/koac243

Lu S, Liu Y, Yu X, Li Y, Yang Y, Wei M, Zhou Q, Wang J, Zhang Y, Zheng W, Chen S (2020) Prediction of genomic breeding values based on pre-selected SNPs using ssGBLUP, WssGBLUP and BayesB for Edwardsiellosis resistance in Japanese flounder. Genet Sel Evol 52:49. https://doi.org/10.1186/s12711-020-00566-2

Lübberstedt T, Suza W (2023) Molecular plant breeding concepts. In: Suza WP, Lamkey KR (eds) Molecular plant breeding. Iowa State University Digital Press

Luo K, He D, Guo J, Li G, Li B, Chen X (2023) Molecular advances in breeding for durable resistance against pests and diseases in wheat: opportunities and challenges. Agronomy 13(3):628. https://doi.org/10.3390/agronomy13030628

Ma ZH, Mao YH, Gong L, Liu CL (2016) Smartphone-based visual measurement and portable instrumentation for crop seed phenotyping. Ifac Papersonline 49:259–264. https://doi.org/10.1016/j.ifacol.2016.10.048

Mahood EH, Kruse LH, Moghe GD (2020) Machine learning: a powerful tool for gene function prediction in plants. Appl Plant Sci 8:e11376

Maimaitijiang M, Ghulam A, Sidike P, Hartling S, Maimaitiyiming M, Peterson K et al (2017) Unmanned aerial system (UAS)-based phenotyping of soybean using multi-sensor data fusion and extreme learning machine. ISPRS J Photogramm Remote Sens 134:43–58. https://doi.org/10.1016/j.isprsjprs.2017.10.011

Maimaitijiang M, Sagan V, Sidike P, Hartling S, Esposito F, Fritschi FB (2020) Soybean yield prediction from UAV using multimodal data fusion and deep learning. Remote Sens Environ. 237:111599. https://doi.org/10.1016/j.rse.2019.111599

Majeed A, Johar P, Raina A, Salgotra RK, Feng X, Bhat JA (2022) Harnessing the potential of bulk segregant analysis sequencing and its related approaches in crop breeding. Front Genet 13:944501

Mao Y, Botella JR, Liu Y, Zhu JK (2019) Gene editing in plants: progress and challenges. Natl Sci Rev 6:421–437. https://doi.org/10.1093/nsr/nwz005

Marimuthu MPA, Maruthachalam R, Bondada R, Kuppu S, Tan EH, Britt A, Chan SWL, Comai L (2021) Epigenetically mismatched parental centromeres trigger genome elimination in hybrids. Sci Adv 7(47):eabk1151. https://doi.org/10.1126/sciadv.abk1151

Markel K, Shih PM (2021) From breeding to genome design: a genomic makeover for potatoes. Cell 184(15):3843–3845. https://doi.org/10.1016/j.cell.2021.06.027

Marsh JI, Hu H, Gill M, Batley J, Edwards D (2021) Crop breeding for a changing climate: integrating phenomics and genomics with bioinformatics. Theor Appl Genet 134:1677–1690

McCormack ME, Lopez JA, Crocker TH, Mukhtar MS (2016) Making the right connections: network biology and plant immune system dynamics. Curr Plant Biol 5:2–12. https://doi.org/10.1016/j.cpb.2015.10.002

McCouch S et al (2020) Mobilizing crop biodiversity. Mol Plant 13(10):1341–1344. https://doi.org/10.1016/j.molp.2020.08.011

Meemken E-M, Bellemare MF (2020) Smallholder farmers and contract farming in developing countries. Proc Natl Acad Sci USA 117:259–264

Melchinger AE, Boehm J, Utz HF, Mueller J, Munder S, Mauch FJ (2018) High- throughput precision phenotyping of the oil content of single seeds of various oilseed crops. Crop Sci 58:670–678. https://doi.org/10.2135/cropsci2017.07.0429

Mendes MA, Petrella R, Cucinotta M, Vignati E, Gatti S et al (2020) The RNA-dependent DNA methylation pathway is required to restrict SPOROCYTELESS/NOZZLE expression to specify a single female germ cell precursor in Arabidopsis. Development 147(23):dev194274

Merrick LF, Herr AW, Sandhu KS, Lozada DN, Carter AH (2022) Optimizing Plant Breeding Programs for Genomic Selection. Agronomy 12:714. https://doi.org/10.3390/agronomy12030714

Metje-Sprink J, Menz J, Modrzejewski D, Sprink T (2019) DNA-free genome editing: past, present and future. Front Plant Sci 9:1957. https://doi.org/10.3389/fpls.2018.01957

Meuwissen TH, Hayes BJ, Goddard ME (2001) Prediction of total genetic value using genome-wide dense marker maps. Genetics 157(4):1819–1829

Meyer RS, DuVal AE, Jensen HR (2012) Patterns and processes in crop domestication: an historical review and quantitative analysis of 203 global food crops. New Phytol 196(1):29–48

Mipeshwaree Devi A, Khedashwori Devi K, Premi Devi P, Lakshmipriyari Devi M, Das S (2023) Metabolic engineering of plant secondary metabolites: prospects and its technological challenges. Front Plant Sci 14:1171154. https://doi.org/10.3389/fpls.2023.1171154

Mitter N, Worrall EA, Robinson KE, Li P, Jain RG, Taochy C et al (2017) Clay nanosheets for topical delivery of RNAi for sustained protection against plant viruses. Nat Plants 3:1–10

Moghimi A, Yang C, Anderson JA (2020) Aerial hyperspectral imagery and deep neural networks for high-throughput yield phenotyping in wheat. Comput Electron Agric 172:105299. https://doi.org/10.1016/j.compag.2020.105299

Moreno-Risueno MA, Busch W, Benfey PN (2010) Omics meet networks—using systems approaches to infer regulatory networks in plants. Curr Opin Plant Biol 13:126–131. https://doi.org/10.1016/j.pbi.2009.11.005

Mostafa S, Mondal D, Panjvani K, Kochian L, Stavness I (2023) Explainable deep learning in plant phenotyping. Front Artif Intell 6:1203546. https://doi.org/10.3389/frai.2023.1203546

Mukherjee D (2022) New paradigm for higher crop productivity through climate smart strategies. In: Mahdi SS, Singh R (eds) Innovative approaches for sustainable development. Springer, Cham. https://doi.org/10.1007/978-3-030-90549-1_4

Muthuramalingam P, Jeyasri R, Krishnan SR, Pandian STK, Sathishkumar R, Ramesh M (2019) Integrating the bioinformatics and omics tools for systems analysis of abiotic stress tolerance in Oryza sativa (L.). In: Sathishkumar R, Kumar SR, Hema J, Baskar V (eds) Advances in plant transgenics: methods and applications. Springer, pp 59–77

Nabwire S, Suh HK, Kim MS, Baek I, Cho BK (2021) Review: application of artificial intelligence in phenomics. Sensors 21:4363

Nadeem SM, Ahmad M, Zahir ZA, Javaid A, Ashraf M (2014) The role of mycorrhizae and plant growth promoting rhizobacteria (PGPR) in improving crop productivity under stressful environments. Biotechnol Adv 32:429–448

Nagasaki M, Saito A, Jeong E, Li C, Kojima K, Ikeda E, Miyano S (2010) Cell illustrator 4.0: a computational platform for systems biology. In Silico Biol 10:5–26. https://doi.org/10.3233/ISB-2010-0415

Najafabadi MY, Hesami M, Eskandari M (2023) Machine learning-assisted approaches in modernized plant breeding programs. Genes 14(4). https://doi.org/10.3390/genes14040777

Nakaya A, Isobe SN (2012) Will genomic selection be a practical method for plant breeding? Ann Bot 110:1303–1316

Napier JA et al (2019) The challenges of delivering genetically modified crops with nutritional enhancement traits. Nat Plants 5:563–567

Naqvi RZ, Siddiqui HA, Mahmood MA, Najeebullah S, Ehsan A, Azhar M, Farooq M, Amin I, Asad S, Mukhtar Z, Mansoor S, Asif M (2022) Smart breeding approaches in post-genomics era for developing climate-resilient food crops. Front Plant Sci 13:972164. https://doi.org/10.3389/fpls.2022.972164

Nawaz M, Sun J, Shabbir S, Khattak WA, Ren G, Nie X, Bo Y, Javed Q, Du D, Sonne C (2023) A review of plants strategies to resist biotic and abiotic environmental stressors. Sci Total Environ 900:165832. https://doi.org/10.1016/j.scitotenv.2023.165832

Nerkar G, Devarumath S, Purankar M, Kumar A, Valarmathi R, Devarumath R, Appunu C (2022) Advances in crop breeding through precision genome editing. Front Genet 13. https://doi.org/10.3389/fgene.2022.880195

Niazian M, Niedbała G (2020) Machine learning for plant breeding and biotechnology. Agriculture 10(10):436. https://doi.org/10.3390/agriculture10100436

Ortiz R (2018) Role of plant breeding to sustain food security under climate change. pp 145–158. https://doi.org/10.1002/9781119180661.ch8

Ostberg S, Schewe J, Childers K, Frieler K (2018) Changes in crop yields and their variability at different levels of global warming. Earth Syst Dyn 9:479–496

Page A, Gibson J, Meyer RS, Chapman MA (2019) Eggplant domestication: pervasive gene flow, feralization, and transcriptomic divergence. Mol Biol Evol 36:1359–1372

Papolu PK, Ramakrishnan M, Mullasseri S, Kalendar R, Wei Q, Zou H, Ahmad Z, Vinod KK, Yang P, Zhou M (2022) Retrotransposons: how the continuous evolutionary front shapes plant genomes for response to heat stress. Front Plant Sci 13:1064847. https://doi.org/10.3389/fpls.2022.1064847

Pazhamala LT, Chaturvedi P, Bajaj P, Srikanth S, Ghatak A, Chitikineni A, Bellaire A, Hingane A, Sameer Kumar CV, Saxena KB, Weckwerth W, Saxena RK, Varshney RK (2020) Multiomics approach unravels fertility transition in a pigeon- pea line for a two-line hybrid system. Plant Genome 13:e20028. https://doi.org/10.1002/tpg2.20028

Pazhamala LT, Kudapa H, Weckwerth W, Millar AH, Varshney RK (2021) Systems biology for crop improvement. Plant Genome 14(2):e20098. https://doi.org/10.1002/tpg2.20098

Pinu FR, Beale DJ, Paten AM, Kouremenos K, Swarup S, Schirra HJ, Wishart D (2019) Systems biology and multi-omics integration: viewpoints from the metabolomics research community. Metabolites 9:76. https://doi.org/10.3390/metabo9040076

Pixley KV, Falck-Zepeda JB, Paarlberg RL, Phillips PW, Slamet-Loedin IH, Dhugga KS, Campos H, Gutterson N (2022) Genome-edited crops for improved food security of smallholder farmers. Nat Genet 54(4):364–367. https://doi.org/10.1038/s41588-022-01046-7

Potrykus I (2001) The 'Golden rice' tale. In Vitro Cell Dev Biol-Plant 37:93–100. https://doi.org/10.1007/s11627-001-0019-9

Potts J, Jangra S, Michael VN, Wu X (2023) Speed breeding for crop improvement and food security. Crops 3(4):276–291. https://doi.org/10.3390/crops3040025

Pratap A, Somta P, Smýkal P, Gupta S (2023) Editorial: genetics of domestication and diversification towards evolution of crop plants. Front Genet 14:1175931. https://doi.org/10.3389/fgene.2023.1175931

Purugganan MD (2019) Evolutionary insights into the nature of plant domestication. Curr Biol 29(14):R705–R714. https://doi.org/10.1016/j.cub.2019.05.053

Rai KK (2022) Integrating speed breeding with artificial intelligence for developing climate-smart crops. Mol Biol Rep 49:11385–11402. https://doi.org/10.1007/s11033-022-07769-4

Rajput M, Choudhary K, Kumar M, Vivekanand V, Chawade A, Ortiz R, Pareek N (2021) RNA interference and CRISPR/Cas gene editing for crop improvement: paradigm shift towards sustainable agriculture. Plants 10:1914

Ravi M, Chan SW (2010) Haploid plants produced by centromere-mediated genome elimination. Nature 464(7288):615–618. https://doi.org/10.1038/nature08842

Ray DK, Ramankutty N, Mueller ND, West PC, Foley JA (2012) Recent patterns of crop yield growth and stagnation. Nat Commun 3:1293. https://doi.org/10.1038/ncomms2296

Ray DK, Mueller ND, West PC, Foley JA (2013) Yield trends are insufficient to double global crop production by 2050. PLoS One 8(6):e66428. https://doi.org/10.1371/journal.pone.0066428

Raza A, Ashraf F, Zou X, Zhang X, Tosif H (2020) Plant adaptation and tolerance to environmental stresses: mechanisms and perspectives. In: Hasanuzzaman M (ed) Plant ecophysiology and adaptation under climate change: mechanisms and perspectives I: general consequences and plant responses. Springer, Singapore, pp 114–117

Razzaq A, Kaur P, Akhter N, Wani SH, Saleem F (2021) Next-generation breeding strategies for climate-ready crops. Front Plant Sci 12:620420. https://doi.org/10.3389/fpls.2021.620420

Rebolledo MC, Pena AL, Duitama J, Cruz DF, Dingkuhn M, Grenier C et al (2016) Combining image analysis, genome wide association studies and different field trials to reveal stable genetic regions related to panicle architecture and the number of spikelets per panicle in rice. Front Plant Sci 7:1384. https://doi.org/10.3389/fpls.2016.01384

Rehman TU, Ma D, Wang L, Zhang L, Jin J (2020) Predictive spectral analysis using an end-to-end deep model from hyperspectral images for high-throughput plant phenotyping. Comput Electron Agric 177:105713. https://doi.org/10.1016/j.compag.2020.105713

Reinoso-Peláez EL, Gianola D, González-Recio O (2022) Genome-enabled prediction methods based on machine learning. In: Ahmadi N, Bartholomé J (eds) Genomic prediction of complex traits. Methods in molecular biology, vol 2467. Humana, New York. https://doi.org/10.1007/978-1-0716-2205-6_7

Resende RT, Piepho HP, Rosa GJM, Silva-Junior OB, ESilva FF, de Resende MDV, Grattapaglia D (2021) Enviromicsin breeding: applications and perspectives on envirotypic-assisted selection. Theor Appl Genet 134:95–112

Richard B, Qi A, Fitt L (2021) Control of crop diseases through integrated crop management to deliver climate-smart farming systems for low- and high-input crop production. Plant Pathol 71(1):187–206. https://doi.org/10.1111/ppa.13493

Rischbeck P, Elsayed S, Mistele B, Barmeier G, Heil K, Schmidhalter U (2016) Data fusion of spectral, thermal and canopy height parameters for improved yield prediction of drought stressed spring barley. Eur J Agron 78:44–59. https://doi.org/10.1016/j.eja.2016.04.013

Rojas M, Lambert F, Ramirez-Villegas J, Challinor AJ (2019) Emergence of robust precipitation changes across crop production areas in the 21st century. Proc Natl Acad Sci USA 116:6673–6678

Rosenzweig C et al (2014) Assessing agricultural risks of climate change in the 21st century in a global gridded crop model intercomparison. Proc Natl Acad Sci USA 111:3268–3273

Rozas P, Kessi-Pérez EI, Martínez C (2022) Genetically modified organisms: adapting regulatory frameworks for evolving genome editing technologies. Biol Res 55. https://doi.org/10.1186/s40659-022-00399-x

Rui Y et al (2020) Poly(Beta-Amino Ester) nanoparticles enable nonviral delivery of CRISPR-Cas9 plasmids for gene knockout and gene deletion. Mol Ther Nucleic Acids 20:661–672. https://doi.org/10.1016/j.omtn.2020.04.005

Rutkoski JE, Krause MR, Sorrells ME (2022) Breeding methods: line development. In: Reynolds MP, Braun HJ (eds) Wheat improvement. Springer, Cham. https://doi.org/10.1007/978-3-030-90673-3_5

Salse J, Barnard RL, Veneault-Fourrey C, Rouached H (2024) Strategies for breeding crops for future environments. Trends Plant Sci 29(3):303–318. https://doi.org/10.1016/j.tplants.2023.08.007

Salvia R, Egidi G, Vinci S, Salvati L (2019) Desertification risk and rural development in southern Europe: permanent assessment and implications for sustainable land management and mitigation policies. Land 8:191

Sanei M, Pickering R, Kumke K, Nasuda S, Houben A (2011) Loss of centromeric histone H3 (CENH3) from centromeres precedes uniparental chromosome elimination in interspecific barley hybrids. Proc Natl Acad Sci USA 108(33):E498–E505. https://doi.org/10.1073/pnas.1103190108

Santantonio N et al (2020) Strategies for effective use of genomic information in crop breeding programs serving Africa and South Asia. Front Plant Sci 11:353

Sartor RC, Noshay J, Springer NM, Briggs SP (2019) Identification of the expressome by machine learning on omics data. Proc Natl Acad Sci 116:18119–18125. https://doi.org/10.1073/pnas.1813645116

Schellenberger J, Que R, Fleming RM, Thiele I, Orth JD, Feist AM, Zielinski DC, Bordbar A, Lewis NE, Rahmanian S, Kang J, Hyduke DR, Palsson BØ (2011) Quantitative prediction of cellular metabolism with constraint-based models: the COBRA toolbox v2. 0. Nat Protoc 6:1290–1307. https://doi.org/10.1038/nprot.2011.308

Schneider HM (2022) Characterization, costs, cues and future perspectives of phenotypic plasticity. Ann Bot 130(2):131–148. https://doi.org/10.1093/aob/mcac087

Schreiber M, Stein N, Mascher M (2018) Genomic approaches for studying crop evolution. Genome Biol 19:140

Selvaraj MG, Valderrama M, Guzman D, Valencia M, Ruiz H, Acharjee A (2020) Machine learning for high-throughput field phenotyping and image processing provides insight into the association of above and below-ground traits in cassava (Manihot esculenta Crantz). Plant Methods 16:87

Serag A, Salem MA, Gong S, Wu L, Farag MA (2023) Decoding metabolic reprogramming in plants under pathogen attacks, a comprehensive review of emerging metabolomics technologies to maximize their applications. Metabolites 13(3):424. https://doi.org/10.3390/metabo13030424

Shafiekhani A, Kadam S, Fritschi FB, De Souza GN (2017) Vinobot and vinoculer: two robotic platforms for high-throughput field phenotyping. Sensors 17:214

Shaheen N, Ahmad S, Alghamdi SS, Rehman HM, Javed MA, Tabassum J, Shao G (2023) CRISPR-Cas system, a possible "savior" of rice threatened by climate change: an updated review. Rice (N Y) 16(1):39. https://doi.org/10.1186/s12284-023-00652-1

Shakoor N, Lee S, Mockler TC (2017) High throughput phenotyping to accelerate crop breeding and monitoring of diseases in the field. Curr Opin Plant Biol 38:184–192. https://doi.org/10.1016/j.pbi.2017.05.006

Shen Y, Zhou G, Liang C, Tian Z (2022) Omics-based interdisciplinarity is accelerating plant breeding. Curr Opin Plant Biol 66:102167

Shen S, Zhan C, Yang C, Fernie AR, Luo J (2023) Metabolomics-centered mining of plant metabolic diversity and function: past decade and future perspectives. Mol Plant 16(1):43–63. https://doi.org/10.1016/j.molp.2022.09.007

Shi J, Tian Z, Lai J, Huang X (2023) Plant pan-genomics and its applications. Mol Plant 16(1):168–186. https://doi.org/10.1016/j.molp.2022.12.009

Silva AT, Nguyen A, Ye C, Verchot J, Moon JH (2010) Conjugated polymer nanoparticles for effective siRNA delivery to tobacco BY-2 protoplasts. BMC Plant Biol 10:291. https://doi.org/10.1186/1471-2229-10-291

Singh BK, Egidi E, Guirado E, Leach JE, Liu H, Trivedi P (2023a) Climate change impacts on plant pathogens, food security and paths forward. Nat Rev Microbiol 21(10):640–656. https://doi.org/10.1038/s41579-023-00900-7

Singh DP et al (2023b) Untargeted metabolomics of Alternaria solani-challenged wild tomato species Solanum cheesmaniae revealed key metabolite biomarkers and insight into altered metabolic pathways. Metabolites 13(5):585. https://doi.org/10.3390/metabo13050585

Skendžić S, Zovko M, Živković IP, Lešić V, Lemić D (2021) The impact of climate change on agricultural insect pests. Insects 12(5). https://doi.org/10.3390/insects12050440

Son S, Park SR (2022) Challenges facing CRISPR/Cas9-based genome editing in plants. Front Plant Sci 13:902413. https://doi.org/10.3389/fpls.2022.902413

Song F, Zhang GJ, Ramanathan V, Leung LR (2022) Trends in surface equivalent potential temperature: a more comprehensive metric for global warming and weather extremes. Proc Natl Acad Sci USA 119:e2117832119

Song B, Ning W, Wei D, Jiang M, Zhu K, Wang X, Edwards D, Odeny DA, Cheng S (2023) Plant genome resequencing and population genomics: current status and future prospects. Mol Plant 16(8):1252–1268. https://doi.org/10.1016/j.molp.2023.07.009

Sroka J, Bieniasz-Krzywiec Ł, Gwóźdź S, Leniowski D, Łącki J, Markowski M, Avignone-Rossa C, Bushell ME, McFadden J, Kierzek AM (2011) Acorn: a grid computing system for constraint based modeling and visualization of the genome scale metabolic reaction networks via a web interface. BMC Bioinformatics 12:196. https://doi.org/10.1186/1471-2105-12-196

Sterk G, Stoorvogel JJ (2020) Desertification-scientific versus political realities. Land 9:156

Stewart EL, Wiesner-Hanks T, Kaczmar N, Dechant C, Wu H, Lipson H et al (2019) Quantitative phenotyping of northern leaf blight in UAV images using deep learning. Remote Sens 11:2209. https://doi.org/10.3390/rs11192209

Subramanian I, Verma S, Kumar S, Jere A, Anamika K (2020) Multi-omics data integration, interpretation, and its application. Bioinform Biol Insights 14. https://doi.org/10.1177/1177932219899051

Sukumaran S, Rebetzke G, Mackay I, Bentley AR, Reynolds MP (2022) Pre-breeding strategies. In: Reynolds MP, Braun HJ (eds) Wheat improvement. Springer, Cham. https://doi.org/10.1007/978-3-030-90673-3_25

Sulaiman N, Aziz MA, Stryamets N, Mattalia G, Zocchi DM, Ahmed HM, Manduzai AK, Shah AA, Faiz A, Sõukand R et al (2023) The importance of becoming tamed: wild food plants as possible novel crops in selected food-insecure regions. Horticulturae 9:171

Sun L, Liu H, Zhang L, Meng J (2015) lncRScan-SVM: a tool for predicting long non-coding RNAs using support vector machine. PLoS One 10:e0139654

Sun G, Xia M, Li J, Ma W, Li Q, Xie J, Bai S, Fang S, Sun T, Feng X, Guo G, Niu Y, Hou J, Ye W, Ma J, Guo S, Wang H, Long Y, Zhang X, Zhang J, Zhou H, Li B, Liu J, Zou C, Wang H, Huang J, Galbraith DW, Song CP (2022) The maize single-nucleus transcriptome comprehensively describes signaling networks governing movement and development of grass stomata. Plant Cell 34:1890–1911

Sun L, Lai M, Ghouri F, Nawaz MA, Ali F, Baloch FS, Nadeem MA, Aasim M, Shahid MQ (2024) Modern plant breeding techniques in crop improvement and genetic diversity: from molecular markers and gene editing to artificial intelligence—a critical review. Plants 13(19):2676. https://doi.org/10.3390/plants13192676

Tang Q, Wang X, Jin X, Peng J, Zhang H, Wang Y (2023) CRISPR/Cas technology revolutionizes crop breeding. Plants 12(17):3119. https://doi.org/10.3390/plants12173119

Tang X, Ren Q, Yan X, Zhang R, Liu L, Han Q, Zheng X, Qi Y, Song H, Zhang Y (2024) Boosting genome editing in plants with single transcript unit surrogate reporter systems. Plant Commun 5(6):100921. https://doi.org/10.1016/j.xplc.2024.100921. Epub 2024 Apr 15. PMID: 38616491; PMCID: PMC11211634

The Arabidopsis Genome Initiative (2000) Analysis of the genome sequence of the flowering plant Arabidopsis thaliana. Nature 408:796–815. https://doi.org/10.1038/35048692

Thondehaalmath T, Kulaar DS, Bondada R, Maruthachalam R (2021) 2021. Understanding and exploiting uniparental genome elimination in plants: insights from Arabidopsis thaliana. J Exp Bot 72(13):4646–4662. https://doi.org/10.1093/jxb/erab161

Tian Z, Wang JW, Li J, Han B (2021) Designing future crops: challenges and strategies for sustainable agriculture. Plant J 105:1165–1178

Tilman D, Balzer C, Hill J, Befort BL (2011) Global food demand and the sustainable intensification of agriculture. Proc Natl Acad Sci USA 108:20260–20264

Tong H, Nikoloski Z (2021) Machine learning approaches for crop improvement: leveraging phenotypic and genotypic big data. J Plant Physiol 257:153354

Tong H, Madison I, Long TA, Williams CM (2020) Computational solutions for modeling and controlling plant response to abiotic stresses: a review with focus on iron deficiency. Curr Opin Plant Biol 57:8–15. https://doi.org/10.1016/j.pbi.2020.05.006

Uffelmann E, Huang QQ, Munung NS, De Vries J, Okada Y, Martin AR, Martin HC, Lappalainen T, Posthuma D (2021) Genome-wide association studies. Nat Rev Methods Primers 1(1):1–21. https://doi.org/10.1038/s43586-021-00056-9

United Nations (2019) Department of Economic and Social Affairs, population division. In: World population prospects 2019: highlights (ST/ESA/SER.A/423). UN, New York, pp 5–12

van Dijk M, Morley T, Rau ML, Saghai Y (2021) A meta-analysis of projected global food demand and population at risk of hunger for the period 2010–2050. Nat Food 2:494–501

Van Esse HP, Reuber TL, van der Does D (2020) Genetic modification to improve disease resistance in crops. New Phytol 225:70–86

Verslues PE et al (2023) Burning questions for a warming and changing world: 15 unknowns in plant abiotic stress. Plant Cell 35:67–108

Vogel E, Donat MG, Alexander LV, Meinshausen M, Ray DK, Karoly D, Meinshausen N, Frieler K (2019) The effects of climate extremes on global agricultural yields. Environ Res Lett 14:054010

Voigt B (2023) EU regulation of gene-edited plants—A reform proposal. Front Genome Ed 5:1119442. https://doi.org/10.3389/fgeed.2023.1119442

Vollset SE, Goren E, Yuan C-W, Cao J, Smith AE, Hsiao T, Bisignano C, Azhar GS, Castro E, Chalek J et al (2020) Fertility, mortality, migration, and population scenarios for 195 countries and territories from 2017 to 2100: a forecasting analysis for the Global Burden of Disease Study. Lancet 396:1285–1306

von Braun J (2010) Food insecurity, hunger and malnutrition: necessary policy and technology changes. New Biotechnol 27:449–452

von Wettberg E, Davis TM, Smýkal P (2020) Editorial: wild plants as source of new crops. Front Plant Sci 11:591554. https://doi.org/10.3389/fpls.2020.591554

Voss-Fels KP, Stahl A, Hickey LT (2019a) Q&A: modern crop breeding for future food security. BMC Biol 17:18. https://doi.org/10.1186/s12915-019-0638-4

Voss-Fels KP, Cooper M, Hayes BJ (2019b) Accelerating crop genetic gains with genomic selection. Theor Appl Genet 132:669–686

Wallace JG, Rodgers-Melnick E, Buckler ES (2018) On the road to breeding 4.0: Unraveling the good, the bad, and the boring of crop quantitative genomics. Annu Rev Genet 52:421–444. https://doi.org/10.1146/annurev-genet-120116-024846

Wang X (2022) Managing land carrying capacity: key to achieving sustainable production systems for food security. Land 11(4):484. https://doi.org/10.3390/land11040484

Wang H, Vieira FG, Crawford JE, Chu C, Nielsen R (2017) Asian wild rice is a hybrid swarm with extensive gene flow and feralization from domesticated rice. Genome Res 27:1029–1038

Wang X, Xuan H, Evers B, Shrestha S, Pless R, Poland J (2019a) High-throughput phenotyping with deep learning gives insight into the genetic architecture of flowering time in wheat. GigaScience 8:giz120. https://doi.org/10.1093/gigascience/giz120

Wang B, Zhu L, Zhao B, Zhao Y, Xie Y, Zheng Z et al (2019b) Development of a haploid-inducer mediated genome editing system for accelerating maize breeding. Mol Plant 12:597–602. https://doi.org/10.1016/j.molp.2019.03.006

Wang P, Zhao F, Kopittke PM (2019c) Engineering crops without genome integration using nanotechnology. Trends Plant Sci 24(7):574–577. https://doi.org/10.1016/j.tplants.2019.05.004

Wang H, Cimen E, Singh N, Buckler E (2020) Deep learning for plant genomics and crop improvement. Curr Opin Plant Biol 54:34–41

Wang X, Zeng H, Lin L, Huang Y, Lin H, Que Y (2023a) Deep learning-empowered crop breeding: intelligent, efficient and promising. Front Plant Sci 14:1260089. https://doi.org/10.3389/fpls.2023.1260089

Wang X, Han L, Li J, Shang X, Liu Q, Li L, Zhang H (2023b) Next-generation bulked segregant analysis for Breeding 4.0. Cell Rep 42(9):113039. https://doi.org/10.1016/j.celrep.2023.113039. Epub 2023 Aug 30. PMID: 37651230

Wang P, Li Z, Li H, Zhang D, Wang W, Xu X, Xie Q, Duan Z, Xia X, Guo G, Shaheen A, Zhou Y, Wang D, Guo S, Hu Z, Galbraith DW, Song C (2024) Smart crops. New Crops 1:100007. https://doi.org/10.1016/j.ncrops.2023.12.002

Webb P (2021) The 2020 Nobel peace prize rewards the persistent vision of a world without hunger, famine, or malnutrition. Am J Clin Nutr 113(2):290–293. https://doi.org/10.1093/ajcn/nqaa355

Weckwerth W, Loureiro ME, Wenzel K, Fiehn O (2004) Differential metabolic networks unravel the effects of silent plant phenotypes. Proc Natl Acad Sci 101:7809–7814. https://doi.org/10.1073/pnas.0303415101

Weckwerth W, Ghatak A, Bellaire A, Chaturvedi P, Varshney RK (2020) PANOMICS meets germplasm. Plant Biotechnol J 18:1507–1525. https://doi.org/10.1111/pbi.13372

Wei B, Zhang J, Pang C, Yu H, Guo D et al (2015) The molecular mechanism of sporocyteless/nozzle in controlling Arabidopsis ovule development. Cell Res 25:121–134

Westermann L, Neubauer B, Köttgen M (2021) Nobel prize 2020 in chemistry honors CRISPR: a tool for rewriting the code of life. Pflugers Arch 473(1):1–2. https://doi.org/10.1007/s00424-020-02497-9

Williams K, Subramani M, Lofton LW, Penney M, Todd A, Ozbay G (2024) Tools and techniques to accelerate crop breeding. Plants 13:1520. https://doi.org/10.3390/plants13111520

Xiao Y, Liu H, Wu L, Warburton M, Yan J (2017) Genome-wide association studies in maize: praise and stargaze. Mol Plant 10:359–374. https://doi.org/10.1016/j.molp.2016.12.008

Xiao Q, Bai X, Zhang C, He Y (2022) Advanced high-throughput plant phenotyping techniques for genome-wide association studies: a review. J Adv Res 35:215–230. https://doi.org/10.1016/j.jare.2021.05.002. PMID: 35003802; PMCID: PMC8721248

Xie X, Liu YG (2021) De novo domestication towards new crops. Natl Sci Rev 8:nwab033. https://doi.org/10.1093/nsr/nwab033. PMID: 34691626; PMCID: PMC8288345

Xu Y (2016) Envirotyping for deciphering environmental impacts on crop plants. TAG Theor Appl Genet 129:653–673

Xu Y, Zhang X, Li H, Zheng H, Zhang J, Olsen MS, Varshney RK, Prasanna BM, Qian Q (2022) Smart breeding driven by big data, artificial intelligence, and integrated genomic-enviromic prediction. Mol Plant 15(11):1664–1695. https://doi.org/10.1016/j.molp.2022.09.001

Yadav K, Geli HME, Cibils AF, Haye M, Fernald A, Peach J, Sawalhah MN, Tidwell VC, Johnson LE, Zaied AJ et al (2021) An integrated food, energy, and water nexus, human well-being, and resilience (FEW-WISE) framework: New Mexico. Front Environ Sci 9:667018

Yan J, Wang X (2023) Machine learning bridges omics sciences and plant breeding. Trends Plant Sci 28:199–210

Yang W, Duan L, Chen G, Xiong L, Liu Q (2013) Plant phenomics and high- throughput phenotyping: accelerating rice functional genomics using multidisciplinary technologies. Curr Opin Plant Biol 16:180–187. https://doi.org/10.1016/j.pbi.2013.03.005

Yang G, Liu J, Zhao C, Li Z, Huang Y, Yu H et al (2017) Unmanned aerial vehicle remote sensing for field-based crop phenotyping: current status and perspectives. Front Plant Sci 8:1111. https://doi.org/10.3389/fpls.2017.01111

Yang W, Feng H, Zhang X, Zhang J, Doonan JH, Batchelor WD et al (2020a) Crop phenomics and high-throughput phenotyping: past decades, current challenges, and future perspectives. Mol Plant 13:187–214. https://doi.org/10.1016/j.molp.2020.01.008

Yang Q, Jin H, Yu X, Fu X, Zhi H, Yuan F (2020b) Rapid identification of soybean resistance genes to soybean mosaic virus by SLAF-seq bulked segregant analysis. Plant Mol Biol Report 38:666–675

Yang Y, Saand MA, Huang L, Abdelaal WB, Zhang J, Wu Y, Li J, Sirohi MH, Wang F (2021) Wang Applications of multi-omics technologies for crop improvement. Front Plant Sci 12:563953. https://doi.org/10.3389/fpls.2021.563953

Yang L, Zhang P, Wang Y, Hu G, Guo W, Gu X, Pu L (2022) Plant synthetic epigenomic engineering for crop improvement. Sci China Life Sci 65:2191–2204

Yang H, Shi X, Chen C, Hou J, Ji T, Cheng J, Birchler JA (2023) Genomic imbalance modulates transposable element expression in maize. Plant Commun 4(2):100467. https://doi.org/10.1016/j.xplc.2022.100467

Yeshi K, Crayn D, Ritmejerytė E, Wangchuk P (2022) Plant secondary metabolites produced in response to abiotic stresses has potential application in pharmaceutical product development. Molecules 27(1):313. https://doi.org/10.3390/molecules27010313

Younis A, Siddique MI, Kim C-K, Lim K-B (2014) RNA interference (RNAi) induced gene silencing: a promising approach of hi-tech plant breeding. Int J Biol Sci 10:1150–1158

Yu H, Li J (2022) Breeding future crops to feed the world through de novo domestication. Nat Commun 13(1):1–4. https://doi.org/10.1038/s41467-022-28732-8

Zaidi et al (2019) New plant breeding technologies for food security. Science 363,1390–1391(2019).https://doi.org/10.1126/science.aav6316

Zandalinas SI, Balfagón D, Mittler R (2022) Plant responses to climate change: metabolic changes under combined abiotic stresses. J Exp Bot 73(11):3339–3354. https://doi.org/10.1093/jxb/erac073

Zhang Y, Zhang N (2018) Imaging technologies for plant high-throughput phenotyping: a review. Front Agric Sci Eng 5:406–419. https://doi.org/10.15302/j-fase-2018242

Zhang X et al (2015) Genomic prediction in biparental tropical maize populations in water-stressed and well-watered environments using low-density and GBS SNPs. Heredity 114:291–299

Zhang H, Mittal N, Leamy LJ, Barazani O, Song B-H (2017) Back into the wild-apply untapped genetic diversity of wild relatives for crop improvement. Evol Appl 10:5–24

Zhang CY, Si YS, Lamkey J, Boydston RA, Garland-Campbell KA, Sankaran S (2018) High-throughput phenotyping of seed/seedling evaluation using digital image analysis. Agronomy-Basel 8:63. https://doi.org/10.3390/agronomy8050063

Zhang H, Wang X, Pan Q, Li P, Liu Y, Lu X, Zhong W, Li M, Han L, Li J, Wang P, Li D, Liu Y, Li Q, Yang F, Zhang YM, Wang G, Li L (2019) QTG-seq accelerates QTL fine mapping through QTL partitioning and whole-genome sequencing of bulked segregant samples. Mol Plant 12:426–437

Zhang R, Zhang C, Yu C, Dong J, Hu J (2022) Integration of multi-omics technologies for crop improvement: status and prospects. Frontiers. Bioinformatics 2:1027457. https://doi.org/10.3389/fbinf.2022.1027457

Zhang C, Jiang S, Tian Y, Dong X, Xiao J, Lu Y, Liang T, Zhou H, Xu D, Zhang H, Luo M, Xia Z (2023) Smart breeding driven by advances in sequencing technology. Mod Agric 1(1):43–56. https://doi.org/10.1002/moda.8

Zhao X, Meng Z, Wang Y, Chen W, Sun C, Cui B et al (2017) Pollen magnetofection for genetic modification with magnetic nanoparticles as gene carriers. Nat Plants 3:956–964. https://doi.org/10.1038/s41477-017-0063-z

Zhao L, Cai Z, Li Y, Zhang Y (2024) Engineering rubisco to enhance CO2 utilization. Synth Syst Biotechnol 9(1):55–68. https://doi.org/10.1016/j.synbio.2023.12.006

Zhou Y, Quan G, Wu Q, Zhang X, Niu B, Wu B et al (2018) Mesoporous silica nanoparticles for drug and gene delivery. Acta Pharm Sin B 8:165–177. https://doi.org/10.1016/j.apsb.2018.01.007

Zhou L, Zhang C, Liu F, Qiu ZJ, He Y (2019) Application of deep learning in food: a review. Compr Rev Food Sci Food Saf 18:1793–1811. https://doi.org/10.1111/1541-4337.12492

Zhou J, Liu Y, Guo X, Birchler JA, Han F, Su H (2022a) Centromeres: from chromosome biology to biotechnology applications and synthetic genomes in plants. Plant Biotechnol J 20(11):2051–2063. https://doi.org/10.1111/pbi.13875

Zhou Y, Wei B, Zhang R, Li H (2022b) Evolution of water-energy-food-climate study: current status and future prospects. J Water Clim Chang 13:463

Zhu X, Maurer HP, Jenz M, Hahn V, Ruckelshausen A, Leiser WL, Würschum T (2022) The performance of phenomic selection depends on the genetic architecture of the target trait. Theor Appl Genet 135:653–665. https://doi.org/10.1007/s00122-021-03997-7

Zohary D, Hopf M, Weiss E (2021) Domestication of plants in the old world: the origin and spread of domesticated plants in Southwest Asia, Europe, and the Mediterranean Basin. Oxford University Press

From Deficiency to Sufficiency: Tackling Hidden Hunger with Biofortified Staples

2

Akanksha Singh, Smita Kaul, Harsh Kumar Dikshit, and Shiv Kumar

Abstract

Biofortification has become a key approach in combating micronutrient deficiencies, which disproportionately affect populations in regions facing malnutrition. It involves increasing the key essential nutrients such as iron (Fe), zinc (Zn), and vitamin A in staple crops such as rice, wheat, maize, sweet potatoes, cassava, and legumes. Biofortification provides a long-term solution to improve nutritional status without requiring substantial dietary changes. This approach can be achieved through conventional plant breeding, agronomic practices, and modern breeding tools. The HarvestPlus initiative has been instrumental in promoting biofortified crops, having released more than 443 biofortified varieties across 41 countries. These efforts have had a substantial impact on addressing "hidden hunger" by reducing deficiencies in critical micronutrients. For the future, biofortification strategies need to be incorporated into mainstream breeding programs to ensure that nutrient-rich varieties become the norm. Advanced techniques such as gene stacking (combining multiple nutrient traits in one crop), introgression breeding (incorporating beneficial traits from other varieties), and integrating conventional breeding with metabolic engineering are expected to enhance the nutritional content of crops. Efforts must also focus on improving the bioavailability of these nutrients, ensuring that they can be efficiently absorbed by the human body, along with ensuring widespread access to these crops. Research advancements, combined with supportive policy interventions, will be essential to overcoming barriers and achieving nutritional security glob-

A. Singh · S. Kaul · S. Kumar (✉)
International Center for Agricultural Research in the Dry Areas (ICARDA), New Delhi, India
e-mail: SK.Agrawal@cgiar.org

H. K. Dikshit
Division of Genetics, ICAR-Indian Agricultural Research Institute, New Delhi, India

P. M. Priyadarshan, R. Ortiz (eds.), *Plant Breeding 2050*,
https://doi.org/10.1007/978-981-95-0583-8_2

ally. This approach represents a cost-effective way to address malnutrition, particularly in low- and middle-income countries where access to diverse and nutrient-rich diets may be limited.

Keywords

Biofortification · Genetic · Agronomic · Transgenic · Genome editing

2.1 Introduction

The Food and Agriculture Organization (FAO) was established in 1945 through the collaborative efforts of representatives from 44 nations. The main aim was to improve food production and distribution of agricultural produce along with improved nutrition. Initially, the focus was on enhancing human nutrition, with agricultural supply being a secondary concern. However, in the early 1960s, the growing global population and the threat of famine necessitated a shift in priorities from improving nutrition to increasing food production. In recent decades, global efforts to enhance agricultural productivity per unit area have significantly increased food production, contributing to a notable reduction in global hunger. This emphasis on maximizing the food output has played a crucial role in feeding a growing population and stabilizing food supply chains worldwide. However, this progress has come with an unintended consequence: the prioritization of food quantity over nutritional quality. While the world has succeeded in producing enough food, there is a decline in the density of essential minerals and vitamins over time as yield increases (Davis et al. 2004; Fan et al. 2008). The focus on yield has overlooked the hidden costs of nutrient deficiencies in regions that predominantly rely on staple crops (Stein and Qaim 2007; Ambikapathi et al. 2022). These staple foods provide up to 70% of the calories consumed in some countries, especially by individuals with low incomes. Currently, micronutrient deficiencies impact around one-third of the world's population, primarily in Asia and Africa (Kumar et al. 2022). Hidden hunger, or micronutrient deficiency (MND), disproportionately impacts pregnant women and young children under 5 years of age. Poor maternal nutrition during pregnancy and breastfeeding can lead to impaired physical growth and cognitive development (Marshall et al. 2022). Certainly, the current estimate of over two billion people suffering from "hidden hunger" is an understatement, as Fe deficiency alone is expected to affect more than 50% of the world population (Qaim et al. 2007; Petry et al. 2016). The HarvestPlus initiative of CGIAR has been implemented to enhance the micronutrient content of staple food crops through a process called "biofortification." By 2030, HarvestPlus aims to feed one billion people and ensure that each meal nourishes, empowers, and changes people's lives (Bouis and Saltzman 2017). Biofortification efforts have focused primarily on addressing vitamin A, Fe, and Zn deficiencies, which collectively account for the greatest unaddressed burden of diseases associated with "hidden hunger" in low- and middle-income countries (LMICs). Improving human health by delivering a range of essential nutrients through improved agricultural practices is recognized as a significant challenge. Biofortification involves three key approaches: agronomic biofortification (where micronutrients are added to crops during their growth through soil or foliar applications), conventional biofortification (identifying genetic variations for desired traits within existing germplasm and then incorporating into breeding programs to develop biofortified crop varieties without compromising yield), and the third strategy is biotechnological interventions such as transgenics and gene editing to target traits that are difficult or impossible to find in natural germplasm, but it often involves regulatory hurdles and public acceptance challenges. In this chapter, we focus on biofortification through conventional plant breeding, agronomic, and biotechnological approaches for improving vitamin A, Fe, and Zn in staple crops.

2.2 Role of Fe, Zn, and Vitamin A in Human Health

Micronutrient deficiencies represent one of the most pervasive public health challenges worldwide. The global impact of these deficiencies continues to be profound, affecting approximately two billion people globally (IFPRI 2016). Among these, Fe and Zn deficiencies are the most widespread (WHO 2017). Fe deficiency, for instance, leads to anemia, impacting about 42% of children aged 6 months to 5 years and one-third of women of reproductive age (WHO 2017). Iron (Fe) plays a vital role in key metabolic functions, including electron transport, DNA replication, and the formation of oxygen-carrying proteins such as hemoglobin and myoglobin (Abbaspour et al. 2014). Dietary deficiency of Fe can cause anemia, fatigue, cognitive decline, and significantly increase the risk of maternal mortality during childbirth (Winkler et al. 2011).

Zn is essential to support immune function, to protect against viral infections, including the coronavirus (Skalny et al. 2020). It is also involved in enzymatic activity, tissue growth, taste acuity, bone mineralization, and fetal development (Gammoh and Rink 2017). Vitamin A, important for vision, immune function, and cell division, also has cancer-fighting properties by protecting cells from oxidative damage. Deficiency in vitamin A can lead to impaired growth, blindness, and higher susceptibility to infections like diarrhea (Hodge and Taylor 2023). These micronutrients—Fe, Zn, and vitamin A—are essential for human health, and deficiencies in any of these can have severe public health consequences, especially in vulnerable populations like children and pregnant women. Addressing these deficiencies remains a critical goal in global health initiatives.

2.3 Breeding Target for Biofortification

Targeted breeding (TB) for micronutrient content, initiated by HarvestPlus, prioritizes the traits of staple crops in specific countries. TB focuses on consistently enhancing the germplasm and early-stage breeding products to improve nutrient content in crop varieties. This involves collaboration with both public and private sectors to build breeding capacity. Breeding objectives are determined by the dietary intake of staple crops within particular agro-ecological regions. These goals also consider nutrient degradation during storage and processing, along with the bioavailability of nutrients present in the crops. Breeding targets for primary staple crops were initially proposed by Hotz and McClafferty (2007) and subsequently revised with the availability of more data (Bouis and Saltzman 2017; Andersson et al. 2017). The baseline levels of key micronutrients and vitamins were measured for various crops (Govindaraj et al. 2019; Beebe 2020). Targeted breeding measures improve these crop-specific baselines. The target increase for Zn was set at +12 ppm for wheat, rice, and maize. However, for Fe the target is set to +44 ppm for beans and +30 ppm for pearl millet. For provitamin A, the improvement benchmarks were 15 ppm in both cassava and maize, and 70 ppm in sweet potato (Bouis and Saltzman 2017). More than 40 million people in developing nations with poor incomes who cannot afford a varied diet today consume biofortified crops that are grown in more than 41 countries. To date, many biofortified varieties have been released and adopted globally, and these varieties have exhibited a quantifiable effect enhancing the status of micronutrient status in the resource-poor population in developing countries.

2.4 Strategies for Biofortification

Strategies for biofortification include conventional plant breeding, transgenic technologies, and agronomic practices. In recent years, genome-editing technology has also been employed to enhance the nutritional quality of food crops (Nagamine and Ezura 2022). A global program (HarvestPlus) involving CG centers and national programs is concentrating on Fe, Zn, and vitamin A, and majorly focusing on Asia and Africa (Bouis and Welch 2010). A list of biofortified varieties released globally is given (Kumar et al. 2024).

2.4.1 Biofortification Through Conventional Breeding

Breeding programs for biofortification using conventional plant breeding methods focus on crop productivity, micronutrient enrichment, stability of micronutrient enrichment trait across the environments, nutritional efficacy, and consumers' acceptance. Biofortified seeds with increased micronutrients improve the viability and overall performance of seedlings and are likely to enhance crop yields and nutritional quality. Conventional breeding schemes like transfer of any other economic traits involve four steps, namely, the selection of parents, crossing parents with desired traits, the selection of recombinants combining traits of interest in an appropriate agronomic background, and large-scale evaluation of improved breeding lines in the target population of environments. This process is further strengthened with the generation of knowledge on the number of genes/QLTs involved in controlling traits, their mode of inheritance, heritability of the traits, their genetic correlations with other desirable traits, etc. The process is very complex, time consuming, and becomes cumbersome with an increase in the number of genes responsible for micronutrient uptake and tissue transport ability (Velu et al. 2014; Kumar et al. 2022). Biofortified varieties developed via traditional breeding approach do not require any regulatory barriers and are widely accepted by farmers and consumers, and is a sustainable way to feed nutritious food to the targeted populations (Winkler 2011). The cultivated species and crop wild relatives hold genetic variation for micronutrient content, which is utilized to breed micronutrient-dense crops.

Rice: Biofortified rice varieties rich in Zn fulfill 40% of the daily Zn requirement of the population dependent on rice as a staple food. Zulfiqar et al. (2023) reported that among 14 genotypes, Kissan Basmati exhibited high Fe content at 44.65 ppm, while Super Basmati 2019 had the highest Zn content at 34.51 ppm. Similarly, Bollinedi et al. (2020) found that the Fe content varied between 6.2 and 23.1 ppm in brown rice, and 0.8 to 12.3 ppm in milled rice, whereas the Zn content was recorded from 11.0 to 47.0 ppm in brown rice, and 8.2 to 40.8 ppm in milled rice. Significant variation has also been recorded by other researchers in rice (Banerjee et al. 2010; Jahan et al. 2013).

Wheat: In developing countries where micronutrient deficiencies are widespread and wheat consumption is high, this major staple food crop is an ideal candidate for biofortification. Conventional breeding approaches have been extensively utilized under various globally funded programs to biofortify wheat with Zn and Fe. High Fe (up to 88 mg/kg) and Zn contents (14 to 190 mg/kg) have been found among wild wheat, especially wild emmer wheat (Cakmak et al. 2010). Further, many reports suggest substantial variability for micronutrient content (Singh et al. 2017; Khokhar et al. 2018; Goel et al. 2018). Substantial genetic variation for Fe and Zn was reported in a panel of 245 bread wheat lines derived from a cross between landraces from Watkin collections and Paragon (a wheat cultivar from the United Kingdom) (Khokhar et al. 2020).

Maize: Menkir (2008) reported a range of 0.7 to 4.7 μg/g for kernel β-carotene in maize. In Chinese maize inbreds, it varied from 0.01 to 1.72 μg/g (Chander et al. 2008). Vignesh et al. (2012) characterized 105 maize inbreds of diverse pedigrees from India and CIMMYT and reported a range of 0.02 to 16.50 μg/g for kernel β-carotene content. Significant variation has also been recorded by various researchers (Prasanna et al. 2011; Queiroz et al. 2011).

Sweet potato: Orange-fleshed sweet potato (OFSP) is a rich source of provitamin A, particularly β-carotene, with concentrations reaching up to 276.98 μg per gram (Tumuhimbise et al. 2009). A 125 g serving of boiled OFSP can provide enough vitamin A to fulfill the daily requirement for preschool children, thereby reducing the risk of night blindness (Mitra et al. 2012). Alam et al. (2016) analyzed the total carotenoid concentration of nine varieties of orange-fleshed sweet potato grown in Bangladesh. Total carotenoids varied from 0.38 to 7.24 mg/100 g fresh weight of sweet potato. The result exhibited that Bangladesh Agricultural Research Institute (BARI) Sweet Potato 8 (SP8) contained the highest total carotenoid content, whereas BARI SP6 contained the lowest amount of carotenoid content. Islam et al.

(2016) observed that among seven sweet potato varieties, Kamalasundari (BARI SP-2) exhibited the highest carotenoid content in both raw and boiled forms (61.94 ± 2.37 μg/g vs 58.32 ± 0.47 μg/g), followed by BARI SP-4 (36.23 ± 1.67 μg/g vs 32.29 ± 0.93 μg/g) and BARI SP-5 (19.31 ± 0.78 μg/g vs 18.24 ± 0.32 μg/g). The other varieties were found to have relatively low amounts of carotenoids. Other researchers, including Bhuiyan et al. (2008) and Tumwegamire et al. (2014), have also reported the high carotenoid contents in the Kamalasundari variety. Similar advancements in breeding for enhanced β-carotene levels have been made in other regions as well. For instance, the Indian variety BHU Sona has been developed with a high β-carotene content of 14.0 mg/100 g, significantly surpassing the 2.0–3.0 mg/100 g found in other popular Indian varieties (Yadava et al. 2017).

Legumes: Legumes, particularly lentil, mung bean, cowpea, and common bean, have shown significant variation in Fe and Zn contents, indicating strong potential for their enhancement. Several studies on lentils have reported a wide variability for grain Fe and Zn contents (Rasheed et al. 2020; Podder et al. 2021).In lentil, Singh et al. (2017) reported Fe from 34.4 to 19.5 mg/kg and Zn from 12.3 to 78.75 mg/kg in seed. Kumar et al. (2019b) reported that grain Fe content varied from 42.8 to 110.63 mg/kg and Zn by 38.18–81.68 mg/kg. In mung bean, the Fe content varied from 35 to 87 mg/kg, while the Zn content ranged from 21 to 62 mg/kg across different genotypes (Nair et al. 2015). Large-scale studies on common beans at CIAT found Fe levels ranging from 34 to 92 mg/kg in cultivated beans, with some wild beans showing up to 96 mg/kg (Islam et al. 2002). Twelve large-scale studies conducted across ten countries involving 50 to 1512 genotypes, consistently identified bean varieties with high Fe content, with some exceeding 150 mg/kg (Delfini et al. 2020; Araújo et al. 2022). Araújo et al. (2022) reported mean Fe and Zn levels of 51.1 mg/kg and 46.8 mg/kg, respectively. Another study explored Fe and Zn diversities in cowpea, showcasing broad ranges: the Fe content spanned 26.48 mg/kg to 140.89 mg/kg, while the Zn content ranged from 36.83 mg/kg to 203.42 mg/kg (Ringo et al. 2022). These findings underscore the potential to enhance nutrient content in legumes, contributing to nutritional security.

2.4.1.1 Variation of Fe and Zn in Crop Wild Relatives

Crop wild relatives, rich in genetic diversity, hold valuable genes/QTLs that can significantly enhance the nutritional profile of cultivated varieties. By incorporating these genes/QTLs into modern varieties, breeders have made important strides in biofortification, particularly in increasing the levels of essential micronutrients like Fe and Zn in staple crops. However, there is a limited genetic variability in modern wheat varieties, whereas wheat's wild relatives and its progenitors, such as einkorn wheat, wild emmer wheat, and landraces, exhibit higher variation for grain Fe and Zn contents (Gupta et al. 2021). Through cytogenetic techniques, the addition of chromosomes from the wild *Aegilops* has significantly raised the Fe content of wheat grains, virtually doubling the original values (Tiwari et al. 2010; Neelam et al. 2011). Synthetic wheats have shown an increase in the Zn content in both Zn-deficient and sufficient soil (Cakmak et al. 1997; Kumar et al. 2019a). In a related study, synthetic hexaploid wheat demonstrated increased Fe and Zn contents with conventional wheat varieties, with nearly equivalent yield to traditional varieties (Calderini and Ortiz-Monasterio 2003). Screening of 22 wild emmer wheat accessions identified lines with high grain Fe (85 mg/kg) and Zn (125 mg/kg) contents in Zn-deficient soils (Peleg et al. 2008). In another study involving 825 wild emmer accessions under field conditions, the Fe content in wheat kernels ranged from 15 to 109 mg/kg, with an average of 46 mg/kg, while the Zn content ranged from 30 to 118 mg/kg, with an average of 61 mg/kg (Cakmak et al. 2004). Wide variations in Fe and Zn contents have also been reported by other researchers. For instance, Oury et al. (2006) found that the Fe content ranged from 20 to 88 ppm, and Zn content ranged from 15 to 43 ppm. Similarly, Peleg et al. (2008) reported Fe content ranging from 20 to 88 ppm and Zn content ranging from 15 to 43 ppm. Successful examples of harnessing wheat wild relatives for developing high Zn varieties in India include "Zinc Shakti" (Chitra), "WB 02" and "HPBW 01" (HarvestPlus 2017), and the Fe-rich variety, "HD 3171" (Gupta et al. 2018).

Wild rice species, such as *Oryza nivara*, *O. longistaminata*, *O. barthii*, *O. rufipogon*, and African cultivated rice *O. glaberrima*, have been reported to possess Zn concentrations 2–3 times higher than cultivated varieties, with the unpolished grain Zn content ranging between 37 and 55 mg/kg(Cheng et al. 2005; Banerjee et al. 2010). This makes them promising genetic resources for improving Zn content in rice.

In lentils, mineral content was analyzed in both cultivated and wild species. The variability in cultivated lentils for Fe, Zn, Ca, Cu, and Mg ranges from 26 to 92, 17 to 51, 97 to 536, 3 to 12, and 272 to 892 mg/kg, respectively. However, mineral contents in wild species were recorded as follows: *L. lamottei* (Fe: 64–80, Zn: 26–40, Ca: 311–434, Cu: 2–6, Mg: 754–839 mg/kg), *L. nigricans* (Fe: 60–70, Zn: 33–39, Ca: 508–590, Cu: 3–4, Mg: 445–738 mg/kg), and *L. ervoides* (Fe: 65, Zn: 37, Ca: 339, Cu: 6, Mg: 638 mg/kg) (Gupta et al. 2016).

In chickpea, wild species exhibited substantial variation for seed protein and mineral content. Among these, *Cicer chorassanicum* was found to have a higher level of seed calcium and protein. *C. judaicum* exhibits high Fe, Cu, and Mg contents, *C. yamashiatae* shows high Zn and Fe, and *C. pinnatifidum* is rich in Mn (Sharma et al. 2021).

2.4.2 Agronomic Biofortification of Food Crops

Agronomic biofortification is a rapid and complementary approach for developing biofortified food crops with increased nutrient content in a crop season. The strategy is based on the application of nutrients through fertilizer formulations. The importance of agronomic biofortification has grown with the development of innovative techniques like nutripriming, foliar spraying, soilless cultivation, and mechanized delivery methods (Bhardwaj et al. 2022). Cereal crops possess very limited amounts of Fe and Zn, and growing these crops in Fe- and Zn-deficient soil reduces their content in grain. Zn content in cereals has been improved by agronomic biofortification (Cakmak and Kutman 2018). In comparison to Zn, the application of iron fertilizers in improving the content of Fe in cereal crops has been less investigated. Due to the potentially limited effectiveness of foliar Fe fertilizers in boosting grain Fe content, multiple studies have suggested that a comprehensive approach involving integrated agricultural strategies, such as plant breeding, genetic modification, and agronomic practices, is essential for enhancing grain Fe content in cereals (Cakmak and Kutman 2018).

2.4.2.1 Agronomic Biofortification in Cereals

Extensive research has been conducted on agronomic biofortification for enhancing Fe and Zn contents in major cereals. Low levels of these micronutrients in grains like maize and rice are often attributed to their limited availability in the soil. Consequently, various fertilizer application methods, including soil, seed, and foliar treatments, have been explored to increase the micronutrient content in these cereal grains. Zn and other nutrients work together to impact the accumulation of Zn in rice grains. Grain Zn content was found to increase with the appropriate application of sulfur (S) to the soil or with the use of nitrogen (N) fertilizers (Khampuang et al. 2023). Saha et al. (2017) reported improved Zn content, improved yield, and greater bioavailability in cooked grains when using soil-applied Zn along with two foliar applications. Integrating Zn and Fe through soil application has shown notable advantages in rice cultivation. For instance, Rao et al. (2020) observed a benefit-cost ratio of 2.13 when applying 25 kg/ha of $ZnSO_4$ and 30 kg/ha of $FeSO_4$. Integrating zinc and iron through soil application has shown notable advantages in rice cultivation, for instance, Rao et al. (2020), observed a benefit-cost ratio of 2.13 when applying 25 kg/ha of $ZnSO_4$ and 30 kg/ha of $FeSO_4$. In rice, Zulfiqar et al. (2021) reported that the most effective way for enhancing grain Fe accumulation is foliar spray. Additional studies revealed that foliar fortification is an effective strategy for increasing Fe and Zn in rice (Mabesa et al. 2013; Yuan et al. 2013; Boonchuay et al. 2013; Zulfiqar et al. 2021). According to Guo et al. (2016), the Zn content in rice grown in Zn-deficient soil can be effectively increased by combining foliar application with soil application of Zn fertilizer.

Similar research trends have been observed in maize. Varying the application of NPK along with Zn and Fe led to a significant increase in Fe (23–107%) and Zn contents (28–122%) in maize grains. Higher content was recorded with foliar application compared to soil application (Salakinkop and Kumar 2018). Saleem et al. (2016) reported significant Zn and Fe fortification levels by foliar application. Similar results were reported by Stewart et al. (2021) and Abdelgalil et al. (2022). Integrated nutrient management, comprising 50% NPK, 50% farmyard manure, and foliar application of Fe and Zn (0.5%) during the active vegetative and grain filling stages, improved vegetative growth, yield components, and grain quality in maize varieties (Augustine and Imayavaramban 2022).

Foliar application of Fe and Zn has also been reported to enhance micronutrient content in wheat (Ram et al. 2016; Ciccolini et al. 2017; Sattar et al. 2022). Fe biofortification of wheat grains has been achieved using both chemical and organic fertilizers (Ramesh et al. 2014; Ramzani et al. 2016). In their study, Yadav et al. (2023) conducted a review on the impact of fertilizer management on the agronomic biofortification of wheat with Zn to boost the bioavailability of Zn in the grain. The findings of the study indicated that the administration of Zn salts through foliar or combined soil and foliar methods can lead to an increase in Zn content of up to 60 mg/kg in wheat grains. Nano biofortification has also gained popularity as a successful alternative method to improve nutrition in wheat (Khan et al. 2021).

2.4.2.2 Agronomic Biofortification in Legumes

Agronomic biofortification has been utilized to increase nutrient content in cowpea, chickpea, common bean, lentil, and mung bean (Fassinou Hotegni et al. 2024). Fertilizer application through foliar spray has enhanced Zn content in legumes. For instance, Ibrahim and Ramadan (2015) demonstrated that the application of foliar Zn fertilizer significantly improved the content of Zn in beans. Study by Ram et al. (2016) in beans showed that the uptake of copper (Cu), manganese (Mn), and Zn was improved by applying organic and synthetic fertilizers, demonstrating the effect of combined inputs in enhancing micronutrient content.

Foliar application of zinc sulfate heptahydrate (ZnSHH), zinc–ethylene diamine tetraacetic acid (Zn-EDTA), sodium selenate, and sodium selenite significantly increased Zn and Se contents in chickpea grains as demonstrated by Shivay et al. (2015). Furthermore, (Jahan and Tar'an 2023) application of 10 kg/ha, and 30 kg/ha of Fe-EDDHA results in an increase in Fe content in chickpea cultivars X05TH20-2 and CDC Frontier. Similarly, Poblaciones and Rengel (2016) reported that the foliar application of Zn alone or in a combination with soil application enhanced Zn content in field peas.

In cowpea, Dhaliwal et al. (2022) observed significant improvements in yield, length of root, nodules per plant, root length, and nutrient uptake after the application of molybdenum (Mo) to the soil along with foliar spray of 0.5% of $FeSO_4 \cdot 7H_2O$ and $ZnSO_4 \cdot 7H_2O$. This integrated approach to biofortification demonstrated the value of combining soil and foliar treatments for enhancing the nutritional value and productivity of cowpea. Further supporting the efficacy of foliar application, López-Morales et al. (2020) demonstrated that in cowpea, a combined treatment of 0.250 g edaphic and 12.5 mM foliar doses of Zn sulfate significantly increased size and weight of grain, Zn content, total phenolic content, and antioxidant capacity.

In addition, selenium (Se) biofortification has also been explored in legume crops. Yang et al. (2003) reported that Se-enriched soybean cultivation was facilitated using Se complex fertilizers applied via foliar sprays. This approach resulted in an increase in the Se content in soybean seeds, contributing to improved nutritional profiles and addressing Se deficiencies in Se-deficient regions. Foliar application of iron nanoparticles (FeNPs), chelated Fe, and Fe sulfate fertilizers significantly improved yield and related growth parameters such as number of branches, plant height, number of pods, leaf area, and seed weight in broad beans (Mahmoud et al. 2022). Additionally, FeNPs boosted the levels of crude protein, amino acids, carbohydrates, minerals such as Fe, Cu, Zn, and Mn in broad bean seeds, highlighting the effectiveness of FeNPs in improving both growth and nutritional quality. These studies collectively highlighted the potential of foliar biofortification strategies, particularly when combined with soil application to enhance the micronutrient content in legume crops, offering a sustainable solution to combat nutrient deficiencies and improve food security.

2.4.3 Genetic Engineering for Enhancement of Fe, Zn, and Vitamin A Content

Genetic engineering has been deployed to increase Fe, Zn, and Vit A in staple crops. This approach is a viable option, especially when crop genetic resources exhibit limited or no genetic variation for micronutrients (Kumar et al. 2019b). The technology enables the incorporation of genes that enhance micronutrient content, improve bioavailability, and reduce antinutritional factors in staple crops. By increasing efficiency or reconstructing biochemical pathways, genetic modifications can increase the level of micronutrients in the edible parts of crops. Key targets for biofortification include genes involved in mineral metabolism (e.g., ferritin and nicotianamine synthase), essential amino acids (e.g., PSY and albumin), vitamins (e.g., carotene desaturase), and essential fatty acids (e.g., Δ6 desaturase) (Garg et al. 2018).

Rice is a primary target crop for biofortification to address Fe deficiency-related anemia. Transgenic rice was developed by Wairich et al. (2022) co-expressing nicotianamine synthase (NAS) and other genes, whereas previous studies reported enhanced Fe content by expressing NAS1 and NAS2 genes in rice grains (Trijatmiko et al. 2016). Kawakami et al. (2022) explored the expression of ZmYS1 and OsTOM1 in biofortified rice, with ferritin from common bean and soybean, as well as Fe transporters like OsIrT1, contributing to enhanced Fe uptake (Lucca et al. 2001; Vasconcelos et al. 2003). Zn content in genetically modified rice has been increased through the overexpression of OsIRT1 and genes involved in mugineic acid synthesis (Masuda et al. 2009). In wheat, TaVIT2 (Connorton et al. 2017) and NAS2 (Beasley et al. 2019) have been targeted to increase Fe content.

For vitamin A biofortification, rice lacks pro-vitamin A, but humans can convert dietary precursors into vitamin A. Golden Rice was a pioneering project that introduced genes such as phytoene synthase (PSY) and carotene desaturase to enhance the accumulation of β-carotene levels (Beyer et al. 2002; Paine et al. 2005). In wheat, maize and sorghum pro-vitamin A content has been enhanced through bacterial carotenoid biosynthesis genes (Wang et al. 2014) and (Aluru et al. 2008). Multivitamin corn has also been developed by engineering multiple metabolic pathways (Naqvi et al. 2009). Table 2.1 summarizes key genes and transgenic strategies employed to biofortify major cereals with enhanced levels of Fe, Zn, and vitamin A.

2.5 CRISPR-Cas Genome Editing Approach

The CRISPR-based genome-editing technique is widely recognized as one of the most efficient and preferred methods in genetic engineering, enabling rapid, precise, and multiplexed genetic modifications (Zheng et al. 2021). Many crops edited

Table 2.1 List of transgenic biofortified crops and their targeted genes for enhanced Fe, Zn, and vitamin A contents

Nutrient	Crop	Gene/protein	References
Fe	Rice	*OsYSL2*	Ishimaru et al. (2010)
		OsIDEF1	Kobayashi et al. (2009)
		Lactoferrin	Suzuki et al. (2001)
		Metallothionein-like protein	(Lucca et al. 2001)
		OsNAC5	Ricachenevsky et al. (2013)
		SoyferH2, *HvNAS1*, and *OsYSL2*	Masuda et al. (2012)
		HvNAS1, *HvNAAT-A*, *HvNAAT-B*	Masuda et al. (2013)
		Pvferritin, *AtNAS1*, and *Afphytase*	Wirth et al. (2009)
	Wheat	TaVIT2	Connorton et al. (2017)
		GmFerritin	Borg et al. (2012)
	Barley	*AtZIP1*	Ramesh et al. (2004)
	Maize	Ferritin and lactoferrin	Drakakaki et al. (2005)
	Pea	*brz* and *dgl*	Rogers and Guerinot (2002)
Zn	Rice	*OsHMA1*	Lee et al. (2007)
		OsNAS1, OsNAS2, and *OsNAS3*	Johnson et al. (2011)
	Wheat	*NAM-B1*	Connorton et al. (2017)
Fe and Zn	Rice	*HvNAS1*	Masuda et al. (2009)
		fumigates phytase	Lucca et al. (2001)
		AtIRT1, *Pvferritin*, and *AtNAS1*	Boonyaves et al. (2017)
	Wheat	*phy* A	Abid et al. (2017)
		phy A	Brinch-Pedersen et al. (2000)
	Soybean	Phytase	Gao et al. (2007)
Vit A	Wheat	*CrtB* and *CrtI*	Wang et al. (2014)
	Rice	*PSY* and carotene desaturase	Beyer et al. (2002), Paine et al. (2005)
	Maize	Maize *PSY* and *CrtI*, *Zmpsy*, *PacrtI*, *Gllycb*, *Glbch*, *ParacrtW*	Aluru et al. (2008), Zhu et al. (2008)
		crtB	Manjeru et al. (2019)

with CRISPR-Cas have been biofortified with carotenoids, such as golden-flesh melons (Tzuri et al. 2015), golden bananas (Kaur et al. 2020), and golden rice (Dong et al. 2020). A notable example is the development of Golden Rice, where a 5.2-kb carotenogenesis cassette containing the *CrtI* and maize *PSY* genes was introduced into the Kitaake variety. The variety has a β-carotene content of 7.9 μg/g dry weight (DW) in its endosperm (Dong et al. 2020). The CRISPR-Cas9 technology has been utilized to enhance the level of micronutrients like Fe and Zn in crops. In rice, the *OsVIT2* gene was edited to enhance Fe availability (Ludwig and Slamet-Loedin 2019). In wheat, Ibrahim et al. (2022) reported an increase in Fe and Zn contents while decreasing phytic acid and demonstrated that CRISPR-Cas9 disrupted the *inositol pentakisphosphate 2-kinase 1* (*TaIPK1*) genes, which reduced phytic acid content and increased Fe and Zn levels in the grains.

2.6 Status of Micronutrient Bioavailability in Food Crops

The first step in making food crops more nutritious for humans is to make these crops rich in micronutrient content. However, as not all nutrients in food are absorbed, the absolute or total concentration of a particular nutrient in a food is not a reliable indicator of its nutritional quality. Therefore, the concept of bioavailability forms a conceptual framework for studying nutritional studies. The amount of a nutrient in a person's diet that is efficiently absorbed and used by the body for different metabolic processes is known as bioavailability.

The amount of bioavailable nutrients can be measured in human beings and in animals through blood (Human), storage tissue (animal), and excretion in isotope studies. Bioaccessibility is the release of an amount of nutrients which is available for absorption (Van Lieshout et al. 2001; Tanumihardjo et al. 2010). Bioefficiency and bioconversion are the terms applied mostly to provitamin A carotenoids. Bioefficiency refers to the absorption and conversion of retinol and is evaluated by recording the vitamin A stores in the body. Bioconversion is that part of absorbed provitamin A, which is converted into retinol.

There is a micronutrient deficiency in the vegetarian population due to the limited bioavailability of micronutrients in the edible portion of a plant. Therefore, one of the most important steps in any biofortification program is to assess the amount of bioavailable grain nutrients in the newly developed varieties. According to Bouis and Welch (2010), bioavailability of micronutrients is low in staple crops (5% for Fe and 25% for Zn). Many factors, such as the host that intakes the micronutrient, the digestive environment of the host, and the presence of inhibitors and promoters (Table 2.2) for the absorption of minerals in food affect the bioavailability of micronutrients. For instance, promoters related to the absorption of minerals, like ascorbic acid, amino acids (sulfur-containing), fibers, prebiotics, carbohydrates, and carotenoids, increase the absorption of these micronutrients in the digestive system. However, polyphenols and phytic acid are the most prominent inhibitors of micronutrients such as Fe and Zn in a plant-based diet. The bioavailability of micronutrients can be enhanced by increasing the promoters or decreasing the inhibitors like phytic acid and polyphenols. Some of the dietary components, like organic acids,

Table 2.2 Bioavailability of the food matrix (Welch 2002)

Nutrient	Food matrix component
	Promoters
Fe, Zn	Provitamin A carotenoids-beta carotene
Ca, Mg, Se	Prebiotics (fructans)
Fe, Zn	Ascorbic acid
Fe, Zn, Ca, Mg	Sulfur containing amino acids (cysteine)
Fe, Zn, Ca, Mg	Nondigestible carbohydrates (inulin)
Fe	Phytoferritin
	Inhibitors
Fe, Zn, Ca, Mg	Phytic acid
Fe, Zn, Mg	Phenolic compounds

act as enhancers, while some other components, like tannins, calcium, phytic acid, and dietary fiber, act as inhibitors, which affect the bioavailability of Fe and Zn contents and influence the absorption of these nutrients.

2.7 Socioeconomic Constraints of Biofortification

Biofortification, while promising as a solution to micronutrient deficiencies, faces several socioeconomic challenges that limit its widespread adoption and impact. These constraints are deeply rooted in various economic, cultural, and social factors.

- Lack of awareness and consumer acceptance: The primary challenge is that farmers, consumers, and policymakers do not fully comprehend biofortified crops. In contrast to that, farmers and consumers are unaware of the possible health benefits of biofortified cultivars. Due to the lack of knowledge, there may not be as much demand for biofortified seeds, which would hinder their adoption and assimilation into regional food systems. Furthermore, cultural dietary habits sometimes influence consumer preferences, and biofortified crops might not immediately fit into conventional diets, which would limit their acceptability (Bouis and Welch 2010).
- Cost and market accessibility: Another significant barrier is the cost of biofortified seeds. The initial cost of biofortified seeds may be more for smallholder farmers than for conventional varieties. This problem is further exacerbated by resource-poor farmers, making it difficult for them to invest in biofortified crops. Furthermore, biofortified crops often face challenges in gaining market access and competing with well-established varieties. These market entry barriers exist both domestically and internationally, making it difficult for biofortified crops to gain a place in the competitive agricultural market.
- Inadequate seed system and infrastructure support: In many underdeveloped nations, inadequate seed systems impede the spread and availability of biofortified crops. The limited infrastructure for seed production, delivery, and marketing makes it difficult to reach isolated farming areas where biofortified crops could have the most impact (Muthayya et al. 2013). Furthermore, farmers may have limited or no access to extension services that educate them on the benefits and production practices of biofortified cultivars.
- Regulatory and institutional framework: Biofortification programs rely on strong regulatory and institutional frameworks that encourage the production, distribution, and use of biofortified crops. However, biofortification programs frequently receive insufficient support from government institutions or have policies in place to encourage biofortified cultivars (Bouis and Welch 2010). Furthermore, incorporating biofortified crops into national agricultural policies and programs necessitates collaboration across various sectors, including agriculture, health, and nutrition, which can be challenging to accomplish.
- Economic viability for farmers: Farmers' decisions to grow biofortified crops are often influenced by economic incentives. For biofortified crops to be viable, they

must offer competitive yields, marketability, and prices compared to nonbiofortified varieties. If biofortified crops are perceived to offer lower yields or market value, farmers may be reluctant to adopt them, regardless of their nutritional benefits. Economic viability is crucial for scaling biofortification in areas where farming is the primary source of livelihood.

- Gender inequity: Gender dynamics can also pose socioeconomic challenges for biofortification. Women contribute significantly to food production and family nutrition, yet they often lack access to the same resources, such as land, credit, and agricultural inputs, as men. Ensuring that biofortification programs are inclusive and address gender disparities is critical for their success, as empowering women can lead to improved nutritional outcomes for entire households.

2.8 Prospects

A viable and efficient way to address micronutrient deficiencies worldwide, especially in areas with poor dietary diversity, is through the biofortification of staple crops. Without necessitating major adjustments to dietary habits, biofortification can benefit public health by increasing the nutritional value of staple crops. Advances in breeding techniques, genetic engineering, and gene-editing technologies have accelerated the development of crops enriched with essential micronutrients like iron, zinc, and vitamin A. The genetic diversity present in wild relatives and landraces further enhances the potential for nutritional improvements. While biofortification holds promise as a complementary strategy to improve nutritional quality in staple crops, it cannot be viewed as a standalone solution. Issues such as limited crop reach, nutrient bioavailability, farmer and consumer acceptance, and uneven nutritional benefits across populations must be tackled to unlock the true potential of biofortification. It must be integrated into a comprehensive strategy that includes dietary diversification, education, and other agricultural and health initiatives to be fully effective. Moving forward, the vision for biofortification must include collaborative research to enhance nutrient uptake and bioavailability, efforts to increase crop diversity, and initiatives to educate and engage both farmers and consumers. Addressing these challenges will be crucial for the global success of biofortification in reducing malnutrition and improving food security.

References

Abbaspour N, Hurrell R, Kelishadi R (2014) Review on iron and its importance for human health. J Res Med Sci 19(2):164

Abdelgalil A, Mustafa AA, Ali SAM, Yassin OM (2022) Effect of different water deficit and foliar spray of zinc and silicon treatments of chemical composition of maize. Curr Chem Lett 11:191–198. https://doi.org/10.5267/j.ccl.2022.2.001

Abid N, Khatoon A, Maqbool A, Irfan M, Bashir A, Asif I, Shahid M, Saeed A, Brinch-Pedersen H, Malik KA (2017) Transgenic expression of phytase in wheat endosperm increases bioavailability of iron and zinc in grains. Transgenic Res 26(1):109–122. https://doi.org/10.1007/s11248-016-9983-z

Alam MK, Rana ZH, Islam SN (2016) Comparison of the proximate composition, total carotenoids and total polyphenol content of nine orange-fleshed sweet potato varieties grown in Bangladesh. Foods 5(3):64. https://doi.org/10.3390/FOODS5030064

Aluru M, Xu Y, Guo R, Wang Z, Li S, White W, Wang K, Rodermel S (2008) Generation of transgenic maize with enhanced provitamin A content. J Exp Bot 59(13):3551–3562

Ambikapathi R, Schneider KR, Davis B, Herrero M, Winters P, Fanzo JC (2022) Global food systems transitions have enabled affordable diets but had less favourable outcomes for nutrition, environmental health, inclusion and equity. Nat Food 3(9):764–779. https://doi.org/10.1038/s43016-022-00588-7

Andersson MS, Saltzman A, Virk PS, Pfeiffer WH (2017) Progress update: crop development of biofortified staple food crops under harvestplus. Afr J Food Agric Nutr Dev 17(2):11905–11935. https://doi.org/10.18697/ajfand.78.HarvestPlus05

Araújo MDS, de Aragão WFL, dos Santos SP, Freitas TKT, da Costa Saraiva V, Damasceno-Silva KJ, dos Santos Dias LA, de Moura Rocha M (2022) Evaluation of adaptability and stability for iron, zinc and protein content in cowpea genotypes using GGE biplot approach. Heliyon 8(12):e11832. https://doi.org/10.1016/J.HELIYON.2022.E11832

Augustine R, Imayavaramban V (2022) Agronomic biofortification through integrated nutrient management on maize (*Zea mays* L.) hybrids. Agric Sci Dig 42(2):187–191. https://doi.org/10.18805/AG.D-5457

Banerjee S, Sharma DJ, Verulkar SB, Chandel G, Gandhi I, Vishwavidyalaya K (2010) Use of in silico and semiquantitative RT-PCR approaches to develop nutrient rich rice (*Oryza sativa* L.). Indian J Biotechnol 9:203–212

Beasley JT, Bonneau JP, Sánchez-Palacios JT, Moreno-Moyano LT, Callahan DL, Tako E, Glahn RP, Lombi E, Johnson AAT (2019) Metabolic engineering of bread wheat improves grain iron concentration and bioavailability. Plant Biotechnol J 17(8):1514–1526. https://doi.org/10.1111/PBI.13074

Beebe S (2020) Biofortification of common bean for higher iron concentration. Front Sustain Food Syst 4. https://doi.org/10.3389/fsufs.2020.573449

Beyer P, Al-Babili S, Ye X, Lucca P, Schaub P, Welsch R, Potrykus I (2002) Golden rice: introducing the β-carotene biosynthesis pathway into rice endosperm by genetic engineering to defeat vitamin A deficiency. J Nutr 132(3):506S–510S. https://doi.org/10.1093/jn/132.3.506S

Bhardwaj AK, Chejara S, Malik K, Kumar R, Kumar A, Yadav RK (2022) Agronomic biofortification of food crops: an emerging opportunity for global food and nutritional security. Front Plant Sci 13:1055278. https://doi.org/10.3389/FPLS.2022.1055278/BIBTEX

Bhuiyan, MKR, Hossain, MM, Alam, MS, Islam ATMT, Hossain MI, Hossain M (2008) Characteristics of some high yielding sweet potato and aroid varieties (in Bangal). Gazipur-1701: Tuber Crops Research Centre (TCRC) of Bangladesh Agricultural Research Institute (BARI), Joydebpur, Bangladesh.

Bollinedi H, Yadav AK, Vinod KK, Gopala Krishnan S, Bhowmick PK, Nagarajan M, Neeraja CN, Ellur RK, Singh AK (2020) Genome-wide association study reveals novel marker-trait associations (MTAs) governing the localization of Fe and Zn in the rice grain. Front Genet 11:497543. https://doi.org/10.3389/FGENE.2020.00213/BIBTEX

Boonchuay P, Cakmak I, Rerkasem B, Prom-U-Thai C (2013) Effect of different foliar zinc application at different growth stages on seed zinc concentration and its impact on seedling vigor in rice. Soil Sci Plant Nutr 59(2):180–188. https://doi.org/10.1080/00380768.2013.763382

Boonyaves K, Wu TY, Gruissem W, Bhullar NK (2017) Enhanced grain iron levels in iron-regulated metal transporter, nicotianamine synthase, and ferritin gene cassette. Front Plant Sci 8(FEBRUARY):130. https://doi.org/10.3389/fpls.2017.00130

Borg S, Brinch-Pedersen H, Tauris B, Madsen LH, Darbani B, Noeparvar S, Holm PB (2012) Wheat ferritins: improving the iron content of the wheat grain. J Cereal Sci 56(2):204–213. https://doi.org/10.1016/j.jcs.2012.03.005

Bouis HE, Saltzman A (2017) Improving nutrition through biofortification: a review of evidence from HarvestPlus, 2003 through 2016. Glob Food Sec 12:49–58

Bouis HE, Welch RM (2010) Biofortification—a sustainable agricultural strategy for reducing micronutrient malnutrition in the global south. Crop Sci 50:S-20. https://doi.org/10.2135/CROPSCI2009.09.0531

Brinch-Pedersen H, Olesen A, Rasmussen SK, Holm PB (2000) Generation of transgenic wheat (*Triticum aestivum* L.) for constitutive accumulation of an *Aspergillus phytase*. Mol Breed 6(2):195–206. https://doi.org/10.1023/A:1009690730620

Cakmak I, Kutman UB (2018) Agronomic biofortification of cereals with zinc: a review. EuJSS 69(1):172–180. https://doi.org/10.1111/EJSS.12437

Cakmak I, Derici R, Torun B, Tolay I, Braun HJ, Schlegel R (1997) Role of rye chromosomes in improvement of zinc efficiency in wheat and triticale. In: Plant nutrition for sustainable food production and environment, pp 237–241. https://doi.org/10.1007/978-94-009-0047-9_65

Cakmak I, Torun A, Özkan H, Millet E, Feldman M, Fahima T, Korol A, Nevo E, Braun HJ (2004) Triticum dicoccoides: an important genetic resource for increasing zinc and iron concentration in modern cultivated wheat. Soil Sci Plant Nutr 50(7):1047–1054. https://doi.org/10.1080/00380768.2004.10408573

Cakmak I, Kalayci M, Kaya Y, Torun AA, Aydin N, Wang Y, Arisoy Z, Erdem H, Yazici A, Gokmen O, Ozturk L, Horst WJ (2010) Biofortification and localization of zinc in wheat grain. J Agric Food Chem 58(16):9092–9102. https://doi.org/10.1021/JF101197H

Calderini DF, Ortiz-Monasterio I (2003) Are synthetic hexaploids a means of increasing grain element concentrations in wheat? Euphytica 134(2):169–178. https://doi.org/10.1023/B:EUPH.0000003849.10595.AC/METRICS

Chander S, Meng Y, Zhang Y, Yan J, Li J (2008) Comparison of nutritional traits variability in selected eighty-seven inbreds from Chinese maize (*Zea mays* L.) germplasm. J Agric Food Chem 56(15):6506–6511. https://doi.org/10.1021/JF7037967/SUPPL_FILE/JF7037967-FILE002.PDF

Cheng CH, Chung MC, Liu SM, Chen SK, Kao FY, Lin SJ, Hsiao SH, Tseng IC, Hsing YIC, Wu HP, Chen CS, Shaw JF, Wu J, Matsumoto T, Sasaki T, Chen HH, Chow TY (2005) A fine physical map of the rice chromosome 5. Mol Gen Genomics 274(4):337–345. https://doi.org/10.1007/S00438-005-0039-Y/METRICS

Ciccolini V, Pellegrino E, Coccina A, Fiaschi AI, Cerretani D, Sgherri C, Quartacci MF, Ercoli L (2017) Biofortification with iron and zinc improves nutritional and nutraceutical properties of common wheat flour and bread. J Agric Food Chem 65(27):5443–5452. https://doi.org/10.1021/ACS.JAFC.7B01176

Connorton JM, Jones ER, Rodríguez-Ramiro I, Fairweather-Tait S, Uauy C, Balk J (2017) Wheat vacuolar iron transporter *TaVIT2* transports Fe and Mn and is effective for biofortification. Plant Physiol 174(4):2434–2444. https://doi.org/10.1104/pp.17.00672

Davis DR, Epp MD, Riordan HD, Davis DR (2004) Changes in USDA food composition data for 43 garden crops, 1950 to 1999. J Am Coll Nutr 23(6):669–682. https://doi.org/10.1080/07315724.2004.10719409

Delfini J, Moda-Cirino V, dos Santos NJ, Buratto JS, Ruas PM, Gonçalves LSA (2020) Diversity of nutritional content in seeds of Brazilian common bean germplasm. PLoS One 15(9):e0239263. https://doi.org/10.1371/JOURNAL.PONE.0239263

Dhaliwal SS, Sharma V, Shukla AK, Kaur J, Verma V, Kaur M, Singh P, Brestic M, Gaber A, Hossain A (2022) Interactive effects of molybdenum, zinc and iron on the grain yield, quality, and nodulation of cowpea (*Vigna unguiculata* (L.) Walp.) in North-Western India. Molecules 27(11):3622. https://doi.org/10.3390/MOLECULES27113622

Dong OX, Yu S, Jain R, Zhang N, Duong PQ, Butler C, Li Y, Lipzen A, Martin JA, Barry KW, Schmutz J, Tian L, Ronald PC (2020) Marker-free carotenoid-enriched rice generated through targeted gene insertion using CRISPR-Cas9. Nat Commun 11(1). https://doi.org/10.1038/S41467-020-14981-Y

Drakakaki G, Marcel S, Glahn RP, Lund EK, Pariagh S, Fischer R, Christou P, Stoger E (2005) Endosperm-specific co-expression of recombinant soybean ferritin and Aspergillus phytase

in maize results in significant increases in the levels of bioavailable iron. Plant Mol Biol 59(6):869–880. https://doi.org/10.1007/s11103-005-1537-3
Fan MS, Zhao FJ, Fairweather-Tait SJ, Poulton PR, Dunham SJ, McGrath SP (2008) Evidence of decreasing mineral density in wheat grain over the last 160 years. J Trace Elem Med Biol 22(4):315–324. https://doi.org/10.1016/j.jtemb.2008.07.002
Fassinou Hotegni NV, Sohindji FS, Salaou MAB, Agbandou PC, Azonhoumon LWS, Tchokponhoué D, Houdegbe C, Adjé CAO, Achigan-Dako EG (2024) Agronomic biofortification of cereals and legumes with iron, zinc, calcium and magnesium for food and nutrition security: available options for farmers in Sub-Saharan Africa. J Agric Food Res 18:101391
Gammoh N, Rink L (2017) Zinc in infection and inflammation. Nutrients 9(6):624. https://doi.org/10.3390/nu9060624
Gao XR, Wang GK, Su Q, Wang Y, An LJ (2007) Phytase expression in transgenic soybeans: stable transformation with a vector-less construct. Biotechnol Lett 29(11):1781–1787. https://doi.org/10.1007/s10529-007-9439-x
Garg M, Sharma N, Sharma S, Kapoor P, Kumar A, Chunduri V, Arora P (2018) Biofortified crops generated by breeding, agronomy, and transgenic approaches are improving lives of millions of people around the world. Front Nutr 5:12. https://doi.org/10.3389/fnut.2018.00012
Goel S, Singh B, Grewal S, Jaat RS, Singh NK (2018) Variability in Fe and Zn content among indian wheat landraces for improved nutritional quality. Indian J Genet Plant Breed 78(4):426–432. https://doi.org/10.31742/IJGPB.78.4.4
Govindaraj M, Rai KN, Cherian B, Pfeiffer WH, Kanatti A, Shivade H (2019) Breeding biofortified pearl millet varieties and hybrids to enhance millet markets for human nutrition. Agriculture (Switzerland) 9(5):106. https://doi.org/10.3390/agriculture9050106
Guo JX, Feng XM, Hu XY, Tian GL, Ling N, Wang JH, Shen QR, Guo SW (2016) Effects of soil zinc availability, nitrogen fertilizer rate and zinc fertilizer application method on zinc biofortification of rice. J Agric Sci 154(4):584–597. https://doi.org/10.1017/S0021859615000441
Gupta A, Singh C, Kumar V, Tyagi BS, Tiwari V, Chatrath, R, Singh, GP (2018) Wheat varieties notified in India since 1965. Karnal: ICAR-Indian Institute of Wheat and Barley Research pp 101
Gupta PK, Balyan HS, Sharma S, Kumar R (2021) Biofortification and bioavailability of Zn, Fe and Se in wheat: present status and future prospects. Theor Appl Genet 134(1):1–35. https://doi.org/10.1007/S00122-020-03709-7/METRICS
HarvestPlus (2017) Biofortification addresses the serious public health problem of mineral and vitamin deficiencies
Hodge C, Taylor C (2023) Vitamin A deficiency. Clin Exp Optom 77(2):76–77. https://doi.org/10.1111/j.1444-0938.1994.tb02378.x
Hotz C, McClafferty B (2007) From harvest to health: challenges for developing biofortified staple foods and determining their impact on micronutrient status. Food Nutr Bull 28:S271–S279
Ibrahim EA, Ramadan WA (2015) Effect of zinc foliar spray alone and combined with humic acid or/and chitosan on growth, nutrient elements content and yield of dry bean (*Phaseolus vulgaris* L.) plants sown at different dates. Sci Hortic 184:101–105. https://doi.org/10.1016/J.SCIENTA.2014.11.010
Ibrahim S, Saleem B, Rehman N, Zafar SA, Naeem MK, Khan MR (2022) CRISPR/Cas9 mediated disruption of Inositol Pentakisphosphate 2-Kinase 1 (*TaIPK1*) reduces phytic acid and improves iron and zinc accumulation in wheat grains. J Adv Res 37:33–41. https://doi.org/10.1016/J.JARE.2021.07.006
Ishimaru Y, Masuda H, Bashir K, Inoue H, Tsukamoto T, Takahashi M, Nakanishi H, Aoki N, Hirose T, Ohsugi R, Nishizawa NK (2010) Rice metal-nicotianamine transporter, *OsYSL2,* is required for the long-distance transport of iron and manganese. Plant J 62(3):379–390
International Food Policy Research Institute (2016) Global Food Policy Report. International Food Policy Research Institute, Washington, DC
Islam FMA, Basford KE, Jara C, Redden RJ, Beebe S (2002) Seed compositional and disease resistance differences among gene pools in cultivated common bean. Genet Resour Crop Evol 49(3):285–293. https://doi.org/10.1023/A:1015510428026/METRICS

Islam SN, Nusrat T, Begum P, Ahsan M (2016) Carotenoids and β-carotene in orange fleshed sweet potato: a possible solution to Vitamin A deficiency. Food Chem 199:628–631. https://doi.org/10.1016/j.foodchem.2015.12.057

Jahan TA, Tar'an B (2023) Agronomic approach to iron biofortification in chickpea. Agronomy 13(12):2894

Jahan GS, Hassan L, Begum SN, Islam SN (2013) Identification of iron rich rice genotypes in Bangladesh using chemical analysis. J Bangladesh Agric Univ 11(1):73–78. https://doi.org/10.22004/AG.ECON.209753

Johnson AAT, Kyriacou B, Callahan DL, Carruthers L, Stangoulis J, Lombi E, Tester M (2011) Constitutive overexpression of the *OsNAS* gene family reveals single-gene strategies for effective iron- and zinc-biofortification of rice endosperm. PLoS One 6(9):e24476. https://doi.org/10.1371/journal.pone.0024476

Kaur N, Alok A, Shivani KP, Kaur N, Awasthi P, Chaturvedi S, Pandey P, Pandey A, Pandey AK, Tiwari S (2020) CRISPR/Cas9 directed editing of lycopene epsilon-cyclase modulates metabolic flux for β-carotene biosynthesis in banana fruit. Metab Eng 59:76–86. https://doi.org/10.1016/J.YMBEN.2020.01.008

Kawakami Y, Gruissem W, Bhullar NK (2022) Novel rice iron biofortification approaches using expression of *ZmYS1* and *OsTOM1* controlled by tissue-specific promoters. J Exp Bot 73(16):5440–5459. https://doi.org/10.1093/JXB/ERAC214

Khampuang K, Chaiwong N, Yazici A, Demirer B, Cakmak I, Prom-U-Thai C (2023) Effect of sulfur fertilization on productivity and grain zinc yield of rice grown under low and adequate soil zinc applications. Rice Sci 30(6):632–640. https://doi.org/10.1016/J.RSCI.2023.07.003

Khan M, Pandey A, Hamurcu M, Gezgin S, Athar T, Rajput V, Gupta O, Minkina T (2021) Insight into the prospects for nanotechnology in wheat biofortification. Biology (Basel) 10(11):1123. https://doi.org/10.3390/biology10111123

Khokhar JS, Sareen S, Tyagi BS, Singh G, Wilson L, King IP, Young SD, Broadley MR (2018) Variation in grain Zn concentration, and the grain ionome, in field-grown Indian wheat. PLoS One 13(1):e0192026. https://doi.org/10.1371/JOURNAL.PONE.0192026

Khokhar JS, King J, King IP, Young SD, Foulkes MJ, De Silva J, Weerasinghe M, Mossa A, Griffiths S, Riche AB, Hawkesford M, Shewry P, Broadley MR (2020) Novel sources of variation in grain Zinc (Zn) concentration in bread wheat germplasm derived from Watkins landraces. PLoS One 15(2):e0229107. https://doi.org/10.1371/JOURNAL.PONE.0229107

Kobayashi T, Itai RN, Ogo Y, Kakei Y, Nakanishi H, Takahashi M, Nishizawa NK (2009) The rice transcription factor IDEF1 is essential for the early response to iron deficiency, and induces vegetative expression of late embryogenesis abundant genes. Plant J 60(6):948–961. https://doi.org/10.1111/j.1365-313X.2009.04015.x

Kumar A, Kapoor P, Chunduri V, Sharma S, Garg M (2019a) Potential of *Aegilops* sp. for improvement of grain processing and nutritional quality in wheat (*Triticum aestivum*). Front Plant Sci 10:422039. https://doi.org/10.3389/FPLS.2019.00308/BIBTEX

Kumar S, Palve A, Joshi C, Srivastava RK, Rukhsar (2019b) Crop biofortification for iron (Fe), zinc (Zn) and vitamin a with transgenic approaches. Heliyon 5:e01914

Kumar S, Dikshit HK, Mishra GP, Singh A (2022) Biofortification of staple crops. Springer Nature

Kumar S, Singh A, Dikshit HK, Mishra GP, Aski M, Bansal R, Gupta P (2024) Global scenario of biofortification. In: Biofortification for nutrient-rich crops. pp 26–42. https://doi.org/10.1201/9781032690636-3/GLOBAL-SCENARIO-BIOFORTIFICATION-SHIV-KUMARA-AKANKSHA-SINGH-DIKSHIT-GYAN-PRAKASH-MISHRA-ASKI-RUCHI-BANSAL-PRIYANKA-GUPTA

Lee S, Kim YY, Lee Y, An G (2007) Rice P1B-type heavy-metal ATPase, OsHMA9, is a metal efflux protein. Plant Physiol 145(3):831–842. https://doi.org/10.1104/pp.107.102236

López-Morales D, de la Cruz-Lázaro E, Sánchez-Chávez E, Preciado-Rangel P, Márquez-Quiroz C, Osorio-Osorio R (2020) Impact of agronomic biofortification with zinc on the nutrient content, bioactive compounds, and antioxidant capacity of cowpea bean (*Vigna unguiculata* L. Walpers). Agronomy 10:1460. https://doi.org/10.3390/AGRONOMY10101460

Lucca P, Hurrell R, Potrykus I (2001) Approaches to improving the bioavailability and level of iron in rice seeds. J Sci Food Agric 81(9):828–834. https://doi.org/10.1002/jsfa.886

Ludwig Y, Slamet-Loedin IH (2019) Genetic biofortification to enrich rice and wheat grain iron: from genes to product. Front Plant Sci 10:450726. https://doi.org/10.3389/FPLS.2019.00833/BIBTEX

Mabesa RL, Impa SM, Grewal D, Johnson-Beebout SE (2013) Contrasting grain-Zn response of biofortification rice (Oryza sativa L.) breeding lines to foliar Zn application. Field Crops Res Complete 149:223–233. https://doi.org/10.1016/J.FCR.2013.05.012

Mahmoud AWM, Ayad AA, Abdel-Aziz HSM, Williams LL, El-Shazoly RM, Abdel-Wahab A, Abdeldaym EA (2022) Foliar application of different iron sources improves morpho-physiological traits and nutritional quality of broad bean grown in sandy soil. Plants 11(19):2599. https://doi.org/10.3390/PLANTS11192599/S1

Manjeru P, Van Biljon A, Labuschagne M (2019) The development and release of maize fortified with provitamin A carotenoids in developing countries. Crit Rev Food Sci Nutr 59(8):1284–1293. https://doi.org/10.1080/10408398.2017.1402751

Marshall NE, Abrams B, Barbour LA, Catalano P, Christian P, Friedman JE, Hay WW, Hernandez TL, Krebs NF, Oken E, Purnell JQ, Roberts JM, Soltani H, Wallace J, Thornburg KL (2022) The importance of nutrition in pregnancy and lactation: lifelong consequences. Am J Obstet Gynecol 226(5):607–632. https://doi.org/10.1016/j.ajog.2021.12.035

Masuda H, Usuda K, Kobayashi T, Ishimaru Y, Kakei Y, Takahashi M, Higuchi K, Nakanishi H, Mori S, Nishizawa NK (2009) Overexpression of the barley nicotianamine synthase gene *HvNAS1* increases iron and zinc concentrations in rice grains. Rice 2(4):155–166. https://doi.org/10.1007/s12284-009-9031-1

Masuda H, Ishimaru Y, Aung MS, Kobayashi T, Kakei Y, Takahashi M, Higuchi K, Nakanishi H, Nishizawa NK (2012) Iron biofortification in rice by the introduction of multiple genes involved in iron nutrition. Sci Rep 2. https://doi.org/10.1038/srep00543

Masuda H, Kobayashi T, Ishimaru Y, Takahashi M, Aung MS, Nakanishi H, Mori S, Nishizawa NK (2013) Iron-biofortification in rice by the introduction of three barley genes participated in mugineic acid biosynthesis with soybean ferritin gene. Front Plant Sci 4:132. https://doi.org/10.3389/fpls.2013.00132

Menkir A (2008) Genetic variation for grain mineral content in tropical-adapted maize inbred lines. Food Chem 110(2):454–464. https://doi.org/10.1016/j.foodchem.2008.02.025

Mitra S, Chandra B, Viswavidyalaya K (2012) Citation: Mitra S (2012) Nutritional status of orange-fleshed sweet potatoes in alleviating vitamin A malnutrition through a food-based approach. J Nutr Food Sci 2(8):160. https://doi.org/10.4172/2155-9600.1000160

Muthayya S, Rah JH, Sugimoto JD, Roos FF, Kraemer K, Black RE (2013) The global hidden hunger indices and maps: an advocacy tool for action. PLoS One 8(6):e67860. https://doi.org/10.1371/journal.pone.0067860

Nagamine A, Ezura H (2022) Genome editing for improving crop nutrition. Front Genome Ed 4:850104. https://doi.org/10.3389/fgeed.2022.850104

Nair RM, Thavarajah D, Thavarajah P, Giri RR, Ledesma D, Yang RY, Hanson P, Easdown W, Hughes J, Keatinge JDHD (2015) Mineral and phenolic concentrations of mungbean [*Vigna radiata* (L.) R. Wilczek var. *radiata*] grown in semi-arid tropical India. J Food Compos Anal 39:23–32. https://doi.org/10.1016/J.JFCA.2014.10.009

Naqvi S, Zhu C, Farre G, Ramessar K, Bassie L, Breitenbach J, Conesa DP, Ros G, Sandmann G, Capell T, Christou P (2009) Transgenic multivitamin corn through biofortification of endosperm with three vitamins representing three distinct metabolic pathways. Proc Natl Acad Sci USA 106(19):7762–7767. https://doi.org/10.1073/PNAS.0901412106

Neelam K, Rawat N, Tiwari VK, Kumar S, Chhuneja P, Singh K, Randhawa GS, Dhaliwal HS (2011) Introgression of group 4 and 7 chromosomes of *Ae. peregrina* in wheat enhances grain iron and zinc density. Mol Breed 28(4):623–634. https://doi.org/10.1007/S11032-010-9514-1/METRICS

Oury FX, Leenhardt F, Rémésy C, Chanliaud E, Duperrier B, Balfourier F, Charmet G (2006) Genetic variability and stability of grain magnesium, zinc and iron concentrations in bread wheat. Eur J Agron 25(2):177–185. https://doi.org/10.1016/J.EJA.2006.04.011

Paine JA, Shipton CA, Chaggar S, Howells RM, Kennedy MJ, Vernon G, Wright SY, Hinchliffe E, Adams JL, Silverstone AL, Drake R (2005) Improving the nutritional value of Golden Rice through increased pro-vitamin A content. Nat Biotechnol 23(4):482–487

Peleg Z, Saranga Y, Yazici A, Fahima T, Ozturk L, Cakmak I (2008) Grain zinc, iron and protein concentrations and zinc-efficiency in wild emmer wheat under contrasting irrigation regimes. Plant Soil 306(1–2):57–67. https://doi.org/10.1007/S11104-007-9417-Z/METRICS

Petry N, Olofin I, Hurrell RF, Boy E, Wirth JP, Moursi M, Angel MD, Rohner F (2016) The proportion of anemia associated with iron deficiency in low, medium, and high human development index countries: a systematic analysis of national surveys. Nutrients 8:693

Poblaciones MJ, Rengel Z (2016) Soil and foliar zinc biofortification in field pea (*Pisum sativum* L.): grain accumulation and bioavailability in raw and cooked grains. Food Chem 212:427–433. https://doi.org/10.1016/J.FOODCHEM.2016.05.189

Podder R, Glahn RP, Vandenberg A (2021) Iron- and zinc-fortified lentil (*Lens culinaris* Medik.) demonstrate enhanced and stable iron bioavailability after storage. Front Nutr 7:614812. https://doi.org/10.3389/FNUT.2020.614812/BIBTEX

Prasanna BM, Mazumdar S, Chakraborti M, Hossain F, Manjaiah KM, Agrawal PK, Guleria SK, Gupta HS (2011) Genetic variability and genotype × environment interactions for kernel iron and zinc concentrations in maize (Zea mays) genotypes. Ind J Agric Sc 81 (8)

Qaim M, Stein AJ, Meenakshi JV (2007) Economics of biofortification. In: Agricultural economics, pp 119–133

Queiroz VAV, de Oliveira Guimarães PE, Queiroz LR, Guedes EO, Vasconcelos VDB, Guimarães LJ, de Aquino Ribeiro PE, Schaffert RE (2011) Iron and zinc availability in maize lines. Food Sci Technol 31(3):577–583. https://doi.org/10.1590/S0101-20612011000300005

Ram H, Rashid A, Zhang W, Duarte AP, Phattarakul N, Simunji S, Kalayci M, Freitas R, Rerkasem B, Bal RS, Mahmood K, Savasli E, Lungu O, Wang ZH, de Barros VLNP, Malik SS, Arisoy RZ, Guo JX, Sohu VS, Zou CQ, Cakmak I (2016) Biofortification of wheat, rice and common bean by applying foliar zinc fertilizer along with pesticides in seven countries. Plant Soil 403(1–2):389–401. https://doi.org/10.1007/S11104-016-2815-3/METRICS

Ramesh SA, Choimes S, Schachtman DP (2004) Over-expression of an Arabidopsis zinc transporter in Hordeum vulgare increases short-term zinc uptake after zinc deprivation and seed zinc content. Plant Mol Biol 54(3):373–385. https://doi.org/10.1023/B:PLAN.0000036370.70912.34

Ramesh A, Sharma SK, Sharma MP, Yadav N, Joshi OP (2014) Inoculation of zinc solubilizing *Bacillus aryabhattai* strains for improved growth, mobilization and biofortification of zinc in soybean and wheat cultivated in Vertisols of Central India. Appl Soil Ecol 73:87–96. https://doi.org/10.1016/J.APSOIL.2013.08.009

Ramzani PMA, Khalid M, Naveed M, Ahmad R, Shahid M (2016) Iron biofortification of wheat grains through integrated use of organic and chemical fertilizers in pH affected calcareous soil. Plant Physiol Biochem 104:284–293. https://doi.org/10.1016/J.PLAPHY.2016.04.053

Rao SG, Balachandrakaumar V, Immanuel RR, Nambi J, Raj ST (2020) Influence of zinc and iron fortified micronutrients on the growth, yield and economics of rice (*Oryza sativa* L.). Crop Res 55(5 & 6). https://doi.org/10.31830/2454-1761.2020.029

Rasheed N, Maqsood MA, Aziz T, Jabbar A (2020) Characterizing lentil germplasm for zinc biofortification and high grain output. J Soil Sci Plant Nutr 20(3):1336–1349. https://doi.org/10.1007/S42729-020-00216-Y/METRICS

Ricachenevsky FK, Menguer PK, Sperotto RA, Williams LE, Fett JP (2013) Roles of plant metal tolerance proteins (MTP) in metal storage and potential use in biofortification strategies. Front Plant Sci 4:144

Ringo GD, Sibiya J, Odindo A, Kimaro D (2022) Genetic diversity in selected cowpea accessions using grain protein, iron and zinc contents. SSRN Electr J. https://doi.org/10.2139/SSRN.4180082

Rogers EE, Guerinot ML (2002) FRD3, a member of the multidrug and toxin efflux family, controls iron deficiency responses in Arabidopsis. Plant Cell 14(8):1787–1799. https://doi.org/10.1105/tpc.001495

Saha S, Chakraborty M, Padhan D, Saha B, Murmu S, Batabyal K, Seth A, Hazra GC, Mandal B, Bell RW (2017) Agronomic biofortification of zinc in rice: influence of cultivars and zinc application methods on grain yield and zinc bioavailability. Field Crop Res 210:52–60. https://doi.org/10.1016/J.FCR.2017.05.023

Salakinkop SR, Kumar N (2018) Agronomic biofortification of maize with zinc and iron micronutrients. 1(5). https://doi.org/10.31031/MCDA.2018.01.000522

Saleem I, Javid S, Bibi F, Ehsan S, Niaz A, Ashfaq Ahmad Z, Hasanuzzaman M (2016) Biofortification of maize grain with zinc and iron by using fertilizing approach. J Agric Ecol Res Int 7(4):1–6. https://doi.org/10.9734/JAERI/2016/24532

Sattar A, Wang X, Ul-Allah S, Sher A, Ijaz M, Irfan M, Abbas T, Hussain S, Nawaz F, Al-Hashimi A, Al Munqedhi BM, Skalicky M (2022) Foliar application of zinc improves morpho-physiological and antioxidant defense mechanisms, and agronomic grain biofortification of wheat (*Triticum aestivum* L.) under water stress. Saudi J Biol Sci 29(3):1699–1706. https://doi.org/10.1016/J.SJBS.2021.10.061

Sen Gupta D, Thavarajah D, McGee RJ, Coyne CJ, Kumar S, Thavarajah P (2016) Genetic diversity among cultivated and wild lentils for iron, zinc, copper, calcium and magnesium concentrations. Aust J Crop Sci 10(10):1381–1387. https://doi.org/10.21475/ajcs.2016.10.10.pne6

Sharma S, Lavale SA, Nimje C, Singh S (2021) Characterization and identification of annual wild *Cicer* species for seed protein and mineral concentrations for chickpea improvement. Crop Sci 61(1):305–319. https://doi.org/10.1002/CSC2.20413

Shivay YS, Prasad R, Pal M, Shivay YS, Prasad R, Pal M (2015) Effects of source and method of zinc application on yield, zinc biofortification of grain, and Zn uptake and use efficiency in chickpea (*Cicer arietinum* L.). CSSPA 46(17):2191–2200. https://doi.org/10.1080/00103624.2015.1069320

Singh A, Sharma V, Dikshit HK, Aski M, Kumar H, Thirunavukkarasu N, Patil BS, Kumar S, Sarker A (2017) Association mapping unveils favorable alleles for grain iron and zinc concentrations in lentil (*Lens culinaris* subsp. *culinaris*). PLoS One 12(11):e0188296. https://doi.org/10.1371/JOURNAL.PONE.0188296

Stein AJ, Qaim M (2007) The human and economic cost of hidden hunger. Food Nutr Bull 28(2):125–134. https://doi.org/10.1177/156482650702800201

Skalny AV, Rink L, Ajsuvakova, OP, Aschner M, Gritsenko VA, Alekseenko SI, Svistunov AA, Petrakis D, Spandidos DA, Aaseth J and Tsatsakis A (2020) Zinc and respiratory tract infections: Perspectives for COVID-19. International journal of molecular medicine, 46(1), pp.17-26

Stewart ZP, Paparozzi ET, Wortmann CS, Jha PK, Shapiro CA (2021) Effect of foliar micronutrients (B, Mn, Fe, Zn) on maize grain yield, micronutrient recovery, uptake, and partitioning. Plants 10(3):1–25. https://doi.org/10.3390/PLANTS10030528

Suzuki YA, Shin K, Lönnerdal B (2001) Molecular cloning and functional expression of a human intestinal lactoferrin receptor. Biochemistry 40(51):15771–15779. https://doi.org/10.1021/bi0155899

Tanumihardjo SA, Palacios N, Pixley KV (2010) Provitamin A carotenoid bioavailability: what really matters? Int J Vitam Nutr Res 80(45):336–350. https://doi.org/10.1024/0300-9831/a000042

Tiwari VK, Rawat N, Neelam K, Kumar S, Randhawa GS, Dhaliwal HS (2010) Substitutions of 2S and 7U chromosomes of Aegilops kotschyi in wheat enhance grain iron and zinc concentration. Theor Appl Genet 121(2):259–269. https://doi.org/10.1007/S00122-010-1307-8/METRICS

Trijatmiko KR, Duenãs C, Tsakirpaloglou N, Torrizo L, Arines FM, Adeva C, Balindong J, Oliva N, Sapasap MV, Borrero J, Rey J, Francisco P, Nelson A, Nakanishi H, Lombi E, Tako E, Glahn RP, Stangoulis J, Chadha-Mohanty P, Johnson AAT, Tohme J, Barry G, Slamet-Loedin IH (2016) Biofortified indica rice attains iron and zinc nutrition dietary targets in the field. Sci Rep 6:19792. https://doi.org/10.1038/SREP19792

Tumuhimbise GA, Namutebi A, Muyonga JH (2009) Microstructure and in vitro beta carotene bioaccessibility of heat processed orange fleshed sweet potato. Plant Foods Hum Nutr 64(4):312–318. https://doi.org/10.1007/s11130-009-0142-z

Tumwegamire S, Mwanga ROM, Andrade M, Low J, Ssemakula GN, Laurie S, Chipungu FP, Nidirigue J, Agili S, Karanja L, Chiona M, Njoku JC, Mtunda N, Ricardo J, Adofo K, Carey T, Gruneberg W (2014) Orange-fleshed Sweetpotato for Africa: catalogue 2014. International Potato Center (CIP)

Tzuri G, Zhou X, Chayut N, Yuan H, Portnoy V, Meir A, Sa'Ar U, Baumkoler F, Mazourek M, Lewinsohn E, Fei Z, Schaffer AA, Li L, Burger J, Katzir N, Tadmor Y (2015) A "golden" SNP in CmOr governs the fruit flesh color of melon (*Cucumis melo*). Plant J 82(2):267–279. https://doi.org/10.1111/TPJ.12814

Van Lieshout M, West CE, Muhilal PD, Wang Y, Xu X, Van Breemen RB, Creemers AFL, Verhoeven MA, Lugtenburg J (2001) Bioefficacy of β-carotene dissolved in oil studied in children in Indonesia. Am J Clin Nutr 73(5):949–958. https://doi.org/10.1093/AJCN/73.5.949

Vasconcelos M, Datta K, Oliva N, Khalekuzzaman M, Torrizo L, Krishnan S, Oliveira M, Goto F, Datta SK (2003) Enhanced iron and zinc accumulation in transgenic rice with the ferritin gene. Plant Sci 164(3):371–378. https://doi.org/10.1016/S0168-9452(02)00421-1

Velu G, Ortiz-Monasterio I, Cakmak I, Hao Y, Singh RP (2014) Biofortification strategies to increase grain zinc and iron concentrations in wheat. J Cereal Sci 59:365–372

Vignesh M, Hossain F, Nepolean T, Saha S, Agrawal PK, Guleria SK, Prasanna BM, Gupta HS (2012) Genetic variability for kernel β-carotene and utilization of crtRB1 3'TE gene for biofortification in maize (Zea mays L.). Indian J Genet Plant Breed 72(2):189–194

Wairich A, Ricachenevsky FK, Lee S (2022) A tale of two metals: biofortification of rice grains with iron and zinc. Front Plant Sci 13:944624. https://doi.org/10.3389/FPLS.2022.944624/BIBTEX

Wang C, Zeng J, Li Y, Hu W, Chen L, Miao Y, Deng P, Yuan C, Ma C, Chen X, Zang M, Wang Q, Li K, Chang J, Wang Y, Yang G, He G (2014) Enrichment of provitamin A content in wheat (*Triticum aestivum* L.) by introduction of the bacterial carotenoid biosynthetic genes CrtB and CrtI. J Exp Bot 65(9):2545–2556. https://doi.org/10.1093/jxb/eru138

Welch RM (2002) Breeding strategies for biofortified staple plant foods to reduce micronutrient malnutrition globally. J Nutr 132(3):495S–499S. https://doi.org/10.1093/JN/132.3.495S

Wirth J, Poletti S, Aeschlimann B, Yakandawala N, Drosse B, Osorio S, Tohge T, Fernie AR, Günther D, Gruissem W, Sautter C (2009) Rice endosperm iron biofortification by targeted and synergistic action of nicotianamine synthase and ferritin. Plant Biotechnol J 7(7):631–644. https://doi.org/10.1111/j.1467-7652.2009.00430.x

World Health Organization. Anaemia in Women and Children: WHO Global Anaemia Estimates, (2021) Edition World Health Organization Geneva, Switzerland, Available online: https://www.who.int/data/gho/data/themes/topics/anaemia_in_women_and_children (accessed on 5 August 2025)

Winkler JT, (2011) Biofortification: improving the nutritional quality of staple crops. Access not excess, pp.100-112

Yadav AK, Seth A, Kumar V, Datta A (2023) Agronomic biofortification of wheat through proper fertilizer management to alleviate zinc malnutrition: a review. Commun Soil Sci Plant Anal 54(2):154–177. https://doi.org/10.1080/00103624.2022.2110892

Yadava D, Choudhury P, Hossain F, Kumar D (2017) Biofortified varieties: sustainable way to alleviate malnutrition. Indian Council of Agricultural Research, New Delhi

Yang F, Chen L, Hu Q, Pan G (2003) Effect of the application of selenium on selenium content of soybean and its products. Biol Trace Elem Res 93(1–3):249–256. https://doi.org/10.1385/BTER:93:1-3:249

Yuan L, Wu L, Yang C, Lv Q (2013) Effects of iron and zinc foliar applications on rice plants and their grain accumulation and grain nutritional quality. J Sci Food Agric 93(2):254–261. https://doi.org/10.1002/JSFA.5749

Zheng X, Kuijer HNJ, Al-Babili S (2021) Carotenoid biofortification of crops in the CRISPR era. Trends Biotechnol 39(9):857–860. https://doi.org/10.1016/J.TIBTECH.2020.12.003

Zhu C, Naqvi S, Breitenbach J, Sandmann G, Christou P, Capell T (2008) Combinatorial genetic transformation generates a library of metabolic phenotypes for the carotenoid pathway in maize. Proc Natl Acad Sci 105(47):18232–18237. https://doi.org/10.1073/pnas.0809737105

Zulfiqar U, Hussain S, Maqsood M, Zamir SI, Ishfaq M, Ali N, Ahmad M, Maqsood MF (2021) Enhancing the accumulation and bioavailability of iron in rice grains via agronomic interventions. Crop Pasture Sci 73(2):32–43. https://doi.org/10.1071/CP21140

Zulfiqar A, Naseer S, Saleem A, Sabar M, Ahmed S, Sardar R, Shahzadi F, Raza Q (2023) Genetic diversity studies for grain iron and zinc content analysis for Elite rice (*Oryza sativa* L.) genotype by using SSR markers. Journal of Food Composition and Analysis 115:104816. https://doi.org/10.1016/J.JFCA.2022.104816

Part II

Genetic Resources and Breeding for Climate Change

3 Breeding Climate-Resilient Varieties: Focus on Cereals and Pulses for Future Food Security

Aiswarya Girija, U. Aswathi, and Divya Kattupalli

Abstract

Climate change poses a significant threat to global food security, particularly for staple crops like cereals (wheat, rice, maize) and pulses (chickpeas, lentils, beans), which are crucial sources of nutrition for billions of people. Breeding climate-resilient varieties has become a critical focus of agricultural research to address challenges such as drought, heat, salinity, and emerging pest and disease pressures. This chapter of the book provides an overview of various aspects of breeding climate-resilient cereals and pulses with the integration of traditional breeding methods and modern biotechnological approaches, including marker-assisted selection (MAS), genomic selection (GS), gene editing (CRISPR-Cas9), soil microbiome, and speed breeding. Additionally, the topics also cover the role of emerging technologies such as the application of artificial intelligence (AI) in precision breeding, emphasizing their contribution to identifying key genetic traits and predicting crop performance under future climate scenarios. This integrated roadmap aims to enhance food security by 2050 and beyond, with a focus on sustainability and resilience in global agriculture.

A. Girija (✉)
Institute of Biological, Environmental & Rural Sciences (IBERS), Aberystwyth University, Plas Gogerddan, Aberystwyth, Ceredigion, Wales, UK
e-mail: aig15@aber.ac.uk

U. Aswathi
BioMedical Technology Wing, Sree Chitra Tirunal Institute of Medical Sciences and Technology , Thiruvananthapuram, Kerala, India

D. Kattupalli
ICAR-Indian Institute of Rice Research, Rajendra Nagar, Hyderabad, Telangana, India

P. M. Priyadarshan, R. Ortiz (eds.), *Plant Breeding 2050*,
https://doi.org/10.1007/978-981-95-0583-8_3

Keywords

Artificial intelligence · Cereals · Climate smart · Diversity · Gene editing · Plant breeding · Pea · Pulses · Rice · Speed breeding · Wheat

3.1 Introduction

Climate resilience in cereals and pulses is vital for ensuring global food security in the face of increasingly unpredictable weather patterns and environmental stressors (Considine et al. 2017). These climatic shifts disrupt traditional growing seasons, alter crop yields, and exacerbate vulnerabilities in already marginalized regions (Fatima et al. 2020). Consequently, maintaining and enhancing the resilience of cereals and pulses to climate change emerges as a critical imperative for ensuring future food security. Cereals such as wheat, rice, maize, and pulses such as lentils, chickpeas, and beans form the backbone of diets worldwide, particularly in developing regions where they are primary sources of protein and carbohydrates. However, these crops are vulnerable to climate-related challenges such as drought, extreme temperatures, flooding, and pest outbreaks (Brás et al. 2021).

Enhancing climate resilience means breeding and cultivating varieties that can withstand these stresses, ensuring stable yields despite environmental uncertainties. This is particularly critical as global warming intensifies, causing more frequent and severe weather events. Climate-resilient crops help farmers avoid devastating losses, reduce the risk of food shortages, and stabilize agricultural economies (Benitez-Alfonso et al. 2023). Developing and promoting resilient varieties will be crucial in adapting agricultural systems to a rapidly changing climate, helping to feed a growing population while preserving ecosystems. By prioritizing climate resilience in cereals and pulses, we can build more robust food systems capable of withstanding future environmental challenges, ensuring that nutritious and affordable food remains accessible for all. This chapter explores the current challenges posed by climate change, the adaptive strategies being employed, and the innovative technologies and policies needed to safeguard these essential crops.

3.1.1 Impact of Climate Change on Cereal and Pulse Production

Climate change has profound implications for agricultural productivity, particularly impacting cereals (such as wheat, rice, maize, and barley) and pulses (including beans, lentils, and chickpeas). Key climate-related challenges include temperature extremes, drought, flooding, and salinity, each exerting distinct pressures on crop growth, yield, and quality.

Temperature Extremes Temperature extremes, including heatwaves and cold spells, disrupt physiological processes critical for crop development. Higher temperatures accelerate crop maturation, potentially reducing yields by shortening the growing period. For instance, heat stress in wheat can decrease grain size and quality, impacting its market value and nutritional content (Farhad et al. 2023). Conversely, cold spells during sensitive growth stages can cause frost damage, compromising yield and quality. Additionally, cold spells can stunt pulse crops, affecting their ability to flower and set pods (Staniak et al. 2021).

Drought Drought, exacerbated by climate change, significantly impacts cereals and pulses by reducing soil moisture essential for plant growth. Prolonged drought conditions impair seed germination, root development, and nutrient uptake, leading to reduced yields. Drought stress in rice, for example, lowers water availability for proper grain filling, resulting in reduced yields (Yang et al. 2022). Pulses, known for their ability to fix nitrogen from the atmosphere, also suffer as drought limits their ability to establish symbiotic relationships with soil microbes, crucial for nitrogen fixation (Díaz-Pereira et al. 2019).

Flooding Conversely, increased frequency and intensity of rainfall events can lead to flooding, particularly in low-lying areas. Excessive waterlogged conditions impede root respiration and nutrient uptake, promoting diseases like root rot and reducing yields. Floods can also wash away topsoil, crucial for nutrient retention and beneficial microbial activity. In rice cultivation, prolonged submergence due to flooding can lead to complete crop loss if not managed promptly (Michael et al. 2023). Pulses, which prefer well-drained soils, are similarly vulnerable to waterlogging, affecting their ability to fix nitrogen and causing rotting of seeds (Zaman et al. 2019).

Salinity Rising sea levels and poor irrigation practices contribute to soil salinization, a critical issue for both cereals and pulses. High soil salinity disrupts water uptake and nutrient absorption, limiting growth and reducing yields. Salinity stress in wheat and barley inhibits germination and restricts root development, resulting in stunted growth and lower grain production (Alhudhaibi et al. 2024; Mwando et al. 2020). Pulses, sensitive to saline conditions, exhibit reduced nodulation and nitrogen fixation, compromising their overall productivity (Bouzroud et al. 2023). In conclusion, climate change such as temperature extremes, drought, flooding, and salinity poses significant challenges to global food security, particularly for cereals and pulses. These crops, which are vital for nutrition and livelihoods, require adaptive strategies and resilient varieties to mitigate the adverse impacts of changing climatic conditions.

3.2 Genetic Diversity in Cereals and Pulses for Climate Resilience

3.2.1 The Role of Genetic Diversity in Enhancing Resilience

Genetic diversity plays a critical role in enhancing resilience, offering a buffer against environmental stresses and contributing to sustainable agricultural practices. Genetic diversity refers to the range of genetic material present within a species or population, including variations in genes that influence traits such as drought tolerance, disease resistance, and yield stability (Swarup et al. 2021). In cereals and pulses, genetic diversity is crucial for maintaining productivity under variable and changing environmental conditions. It provides breeders with the raw materials needed to develop new crop varieties that can thrive in diverse climates and cope with pests, diseases, and other stressors (Badaeva et al. 2022). Without a robust pool of genetic diversity, crops become more susceptible to climate-induced vulnerabilities, threatening food security. In response to these challenges, there has been growing interest in harnessing the genetic diversity of cereals like wheat, rice, maize, and barley, along with pulses such as lentils, chickpeas, and peas. By exploring the genetic potential within these crops, breeders can develop varieties that are more resilient to climate extremes, thus enhancing agricultural sustainability and food security in the face of climate change.

3.2.2 Sources of Genetic Diversity: Wild Relatives, Landraces, and Modern Cultivars

The genetic diversity in cereals and pulses can be traced back to three primary sources: wild relatives, landraces, and modern cultivars. Each of these sources has unique attributes that contribute to climate resilience.

Wild Relatives Wild relatives of domesticated crops represent a valuable reservoir of untapped genetic variation. These species, which have evolved in diverse and often harsh environments, possess traits that can enhance climate resilience in cultivated varieties. For example, wild relatives of cereals and pulses may carry genes for tolerance to drought, salinity, or heat, as well as resistance to pests and diseases. By introgressing genes from wild relatives into domesticated crops, breeders can introduce new traits that improve resilience to climate-related stressors. One notable example is the use of wild barley (*Hordeum spontaneum*) in breeding programs for cultivated barley (*Hordeum vulgare*), which has led to the development of varieties with improved drought tolerance (Lakew et al. 2011). Similarly, wild relatives of wheat, such as *Aegilops* species, have been used to transfer genes for disease resistance and abiotic stress tolerance into modern wheat varieties (Yu et al. 2022).

Landraces Landraces are traditional, locally adapted crop varieties that have been cultivated over generations by farmers in specific regions. These varieties have evolved under natural and human selection pressures, resulting in a wide range of adaptations to local environmental conditions. Landraces often exhibit high levels of genetic diversity and possess traits that enhance resilience to climate variability, including tolerance to drought, heat, and poor soil conditions. For instance, landraces of maize grown in arid regions of Mexico have shown remarkable tolerance to water stress (Barbosa et al. 2021), while landraces of chickpeas cultivated in the Mediterranean basin are better adapted to drought conditions than modern varieties

(Rocchetti et al. 2020). Incorporating the genetic diversity of landraces into breeding programs can provide a rich source of traits for improving the climate resilience of modern cereals and pulses.

Modern Cultivars Modern cultivars, developed through scientific breeding programs, have focused primarily on increasing yields, uniformity, and disease resistance. However, this has often come at the cost of reducing genetic diversity (Sinha et al. 2023). The narrow genetic base of many modern cereal and pulse varieties can make them vulnerable to climate-related stressors, as they may lack the diversity needed to adapt to changing environmental conditions. Nevertheless, modern cultivars can still be a valuable source of genetic diversity, particularly when combined with traits from wild relatives and landraces. Advances in molecular breeding and genomic technologies have made it possible to identify and incorporate beneficial genes from diverse sources into modern cultivars, allowing breeders to develop varieties that combine high yields with improved resilience to climate change (Abdul Aziz and Masmoudi 2024).

3.2.3 Strategies for Conserving and Utilizing Genetic Resources in Cereals and Pulses

Ensuring the availability and sustainable use of genetic diversity in cereals and pulses requires coordinated efforts to conserve genetic resources and integrate them into breeding programs. Several strategies are essential for achieving this goal:

In Situ and Ex Situ Conservation In situ conservation involves the preservation of genetic diversity within natural ecosystems and traditional farming systems. By maintaining populations of wild relatives and landraces in their native habitats, in situ conservation allows these crops to continue evolving in response to changing environmental conditions (Salgotra and Chauhan 2023). This dynamic form of conservation ensures that genetic diversity remains available for future breeding efforts. Protected areas, such as national parks and reserves, play a critical role in in situ conservation, particularly for wild relatives of cereals and pulses. On-farm conservation, where farmers continue to grow traditional landraces, is another key strategy. In many regions, farmers serve as custodians of genetic diversity, selecting and maintaining locally adapted varieties that are resilient to climate fluctuations. Ex situ conservation involves the storage and preservation of genetic material outside its natural environment, typically in gene banks. Gene banks house collections of seeds, plant tissues, and other genetic resources that can be used in breeding programs (Salgotra and Chauhan 2023). Ex situ conservation is critical for safeguarding genetic diversity and ensuring its availability for future use. International organizations, such as the Consultative Group on International Agricultural Research (CGIAR), maintain large gene banks that store genetic material from cereals and pulses around the world (Turner-Hissong et al. 2020). These collections are invaluable for breeding programs seeking to improve climate resilience, as they provide access to a vast array of genetic resources that can be used to develop new crop varieties. In situ and ex situ conservation efforts, coupled with participatory breeding and advanced genomic approaches, offer pathways to harnessing genetic diversity for the benefit of future generations. As climate change continues to impact agriculture, maintaining and utilizing genetic diversity will be essential for building resilient and sustainable food systems.

3.3 Omics Technologies in Breeding Climate-Resilient Cereals and Pulses

The advancement in high-throughput technologies has demystified the application of omics technologies such as phenomics, genomics, metabolomics, transcriptomics, and proteomics in crop improvement and also enables the identification of key genetic, molecular, and metabolic regulators associated with stress tolerance (Anusha and Amir 2024).

3.3.1 Recent Phenotyping Technologies

Phenomics focuses on the measurement of plant phenotypes such as root architecture, canopy size, or photosynthetic efficiency under different environmental conditions (Ninomiya 2022). High-throughput phenotyping technologies, such as drones, imaging systems, and sensors, can monitor plant responses to stress in real-time, enabling breeders to rapidly assess the performance of different genotypes in the field (Moreira et al. 2020). The significance of phenotyping lies in its ability to capture complex interactions between a plant's genetic makeup and its environment (G x E interactions), which are fundamental for improving resilience to climate variability. In cereals like wheat, maize, and rice, traits such as deep rooting systems, enhanced stomatal regulation, and improved grain filling under abiotic stress are key indicators of resilience. Similarly, in pulses like chickpeas, lentils, and beans, traits like early flowering, nitrogen fixation efficiency, and drought tolerance are essential for ensuring yield stability in arid and semi-arid regions. Accurate phenotyping is crucial for breeding programs because it helps in validating the effectiveness of genotypic selection, especially when combined with molecular breeding techniques like marker-assisted selection (MAS) or genomic selection (GS). For example, drought-resilient varieties of rice and wheat were developed by identifying phenotypic traits that enable survival under water stress, such as increased root depth and leaf rolling to reduce transpiration (Ahmed et al. 2020; Shafi et al. 2023). However, the phenotyping process can be time-consuming and labor-intensive when done manually, thus limiting the speed of breeding cycles and the development of climate-resilient varieties. Recent advances in high-throughput phenotyping (HTP) technologies have transformed the field of plant breeding, allowing for the rapid collection of large-scale phenotypic data under field conditions. These technologies use automated systems, sensors, and imaging tools to measure multiple traits simultaneously, providing detailed insights into plant performance under various environmental stresses (Fig. 3.1).

Noninvasive Imaging Technologies Imaging techniques, such as multispectral, hyperspectral, and thermal imaging, allow for noninvasive monitoring of plant traits such as canopy temperature, chlorophyll content, and photosynthetic efficiency (Silva-Perez et al. 2018). Thermal imaging, for instance, has been widely used to assess heat tolerance in cereals like wheat by measuring leaf temperature, which correlates with a plant's ability to maintain transpiration under high temperatures (Anderegg et al. 2024). Similarly, hyperspectral imaging has been applied to pulses like chickpeas to assess drought tolerance by monitoring changes in leaf water content (Pappula-Reddy et al. 2024). Hyperspectral imaging has been instrumental in screening large populations of wheat and maize for drought tolerance by detecting changes in reflectance patterns associated with water stress (Ram et al. 2024). These imaging technologies are valuable in identifying early signs of stress that may not be visible to the naked eye. By capturing subtle changes in plant physiology, they enable breeders to make informed decisions about which varieties are more likely to thrive in stressful conditions.

Drone and Satellite-Based Phenotyping Unmanned aerial vehicles (UAVs), or drones, and satellite-based remote sensing technologies have also become integral to modern phenotyping. These platforms can capture large-scale data over vast agricultural fields, providing detailed spatial information about crop performance. Drones equipped with multispectral cameras can monitor traits like canopy cover, plant height, and biomass accumulation, which are crucial for assessing the impact of climate stressors on nutrient use, growth, and yield (Banerjee et al. 2020; Gano et al. 2024). In cereals such as maize and rice, drone-based phenotyping has proven effective in mapping field-scale variability in response to drought or flooding (Bongomin et al. 2024; Zaman-Allah et al. 2015). In rice, near-infrared (NIR) imaging is used for determining water content and photosynthetic efficiency (Kim et al. 2020). For pulses, UAVs are useful for monitoring traits like pod formation and canopy senescence under heat stress. Satellite-based remote sensing, though less granular than drone data, offers the advantage of covering large geographic areas and tracking long-term crop performance, making it an essential tool for breeding programs targeting climate resilience on a regional scale.

Robotics and Automation Robotic platforms and ground-based sensors have further enhanced the precision of phenotyping efforts. Field robots equipped with various sensors can autonomously navigate breeding trials, collecting continuous data on plant traits such as leaf area, stem diameter, and nutrient content (Atefi et al. 2021). These robots can operate under diverse environmental conditions, making them ideal for collecting data on stress responses in cereals and pulses grown in challenging climates. In wheat, rice, and barley, robotic phenotyping platforms have been used to measure plant height and tiller number under drought stress, providing breeders with valuable data on yield potential (Nagy et al. 2018). In pulse crops like

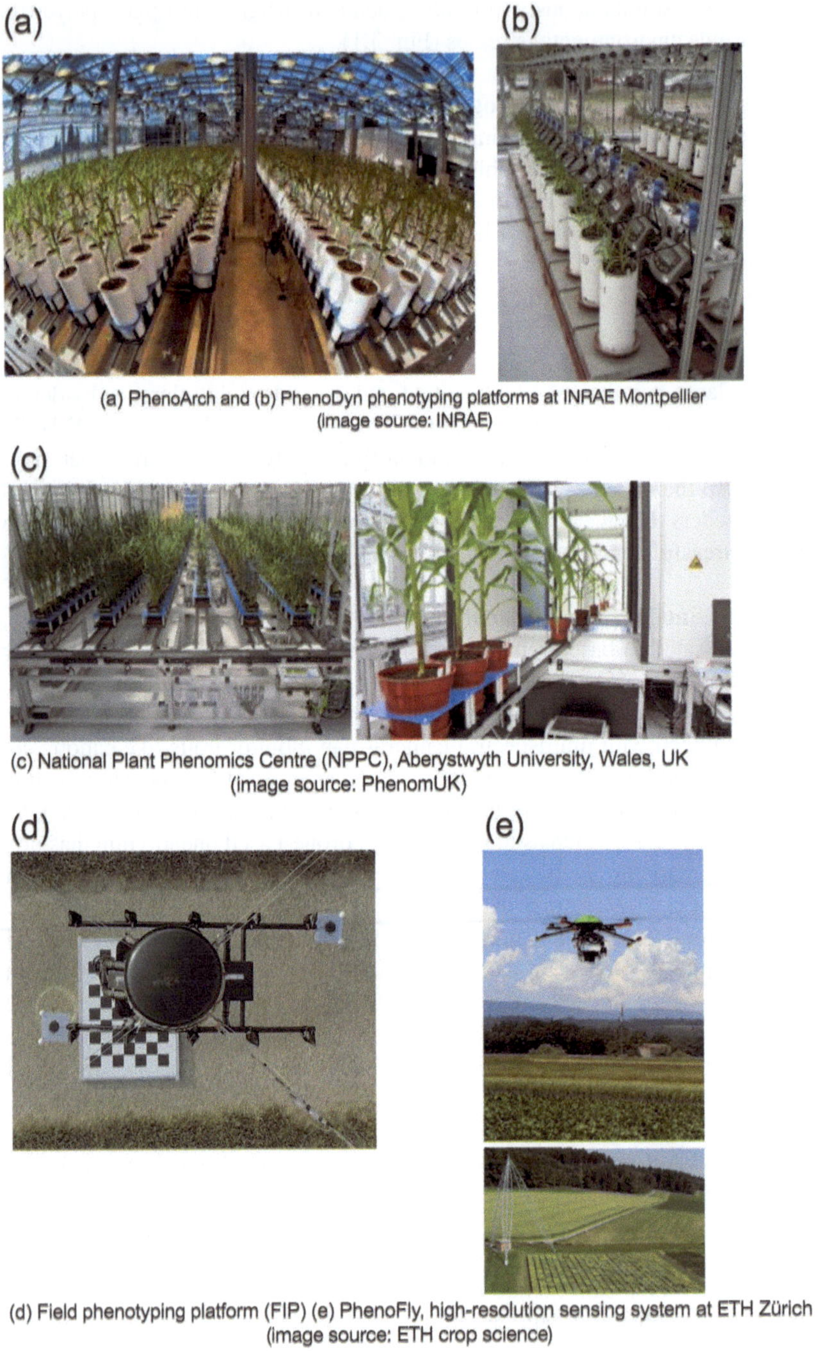

Fig. 3.1 Different high-throughput phenotyping platforms and technologies for precision phenotyping in crops

cowpea and soybean, robotic systems have helped to assess traits like pod number and root growth, which are key indicators of climate resilience in water-limited environments (Mekonnen et al. 2022).

3.3.2 Genome Data-Driven Marker-Assisted Genomic Selection

In plant breeding, genomics aids researchers identify genes and genetic markers associated with key traits, such as drought tolerance, heat resistance, or salinity tolerance. Additionally, genomics has expanded the availability of high-quality reference genomes for a wide range of model and nonmodel plant species (Li et al. 2012). This has aided quantitative trait loci (QTL) mapping, genome-wide association studies (GWAS), and gene editing in crops. Genomic data help breeders identify genetic markers linked to stress tolerance traits, enabling Marker-Assisted Selection (MAS) and Genomic Selection (GS) to predict plant performance in specific environments. This reduces the time and cost of field trials. In cereals, MAS has been widely used to incorporate genes conferring resistance to biotic and abiotic stresses, such as the *Sub1A* gene in rice, which enables submergence tolerance during floods (Fukao et al. 2011). *Sub1 g*ene was identified through genomics and incorporated into popular varieties using MAS, leading to the rapid adoption of flood-tolerant rice in Southeast Asia (Oladosu et al. 2020). In pulses, MAS has been instrumental in breeding varieties with improved drought tolerance by selecting for traits like deeper root systems or water use efficiency (Torres et al. 2010). The advantage of GS lies in its ability to increase selection accuracy and reduce breeding cycles, significantly shortening the time required to develop improved varieties (Mangal et al. 2024). In Southeast Asia, intermittent drought periods can cause significant yield losses in rice. To address this, scientists have used MAS to introduce the *Drought1 (DT1)* gene in rice to promote deeper rooting systems (Singh et al. 2024). In conjunction with MAS, CRISPR-based modifications to rice have targeted genes involved in stomatal regulation, improving the plant's water-use efficiency (Rathnasamy et al. 2023). Genomic selection has been utilized to breed wheat varieties that exhibit better tolerance to heat stress (Molero et al. 2023). By using GS models based on genomic data from heat-tolerant varieties, breeders can predict which wheat lines will perform better under high temperatures. A notable success is the breeding of wheat lines in India that can withstand heat stress during the grain-filling period. Genomic selection has reduced the conventional breeding cycle of 15–20 years for these lines (Bellundagi et al. 2022). For cereals like wheat, rice, and maize, genomic selection has shown great potential in predicting grain yield under drought conditions. In pulses, genomic selection is gaining traction, particularly for complex traits such as heat tolerance, which are difficult to improve using traditional phenotypic selection methods (Tables 3.1, 3.2, and 3.3).

Table 3.1 Examples of genes known for resistance to pathogens and insect–pests in crops

Crop	Disease	Genes	Reference
Rice	Blast (*Magnaporthe grisea*)	*bsr-d1, Pb1, Pb2, Pb3, Pib, Pid2, Pid4, Pi5, Pit, pi21, Pish/Pi35, Pi36, Pi37, Pi50, Pi54rh, Pi56, Pi56(t), Ptr, Pita2, Pi63, Pi64, Pizh, Pigm, Pi9/ Pi2=Piz5/Piz-t/ Pi50/ PigmR(e)/ Pizh, Pid3/Pi25/Pid3-I1, Pia/PiCO39, Pi54rh/ Pi54=Pikh, Pik/ Pikm/ Pik-p/ Pi1/ Pi7, Pita, Ptr = Pita2*	Liu et al. (2021), Simon et al. (2023)
	Bacterial leaf blight (*Xanthomonas oryzae* pv. *oryzae*)	*Xa3=Xa26, Xa4(b), xa5, Xa7, Xa7(a), Xa10, Xa21, Xa23, Xa27, xa41(t), Xa47(t), Xa1/ Xa2=Xa31 (t)/Xa14/ Xa45(t)/ CGS-Xo1, xa13/ OsSWEET11/ Os8N3, xa41(t)/ OsSWEET14/ Os11N3, Xa23, Xa47(t), xa25/ OsSWEET13/ OsMtN3*	Liu et al. (2021)
	Brown planthopper (*Nilaparvata lugens*)	*BPH14, BPH30, BPH17(b), BPH6, BPH29, BPH32=BPH3©, BPH1=BPH10=BPH18=BPH21/ BPH2=BPH26/ BHP7/BPH9(d)*	Simon et al. (2023)
	Virus	Rice tungro virus (tsv1), Rice Yellow Mottle Virus 1 (*RYMV1*), *RYMV2*, *RYMV3*, Rice stripe virus (*STV11*)	Simon et al. (2023)
Wheat	Rust (*Puccinia triticina, P. graminis, P. striiformis*)	*Lr1, Lr9, Lr10, Lr13/Ne2, Lr14a, Lr21, Lr22a, Lr42, Lr47, Lr58; Sr22, Sr22b, Sr26, Sr27, Sr33, Sr35, Sr43, Sr45, Sr46, Sr50, Sr60* (WKS2), *Sr61, Sr62, SrTA1662* (*Sr66*); *Yr5, Yr5a* (*Yr5*), *Yr5b* (*YrSP*), *Yr7, Yr10* (*Yr10cg*), *Yr15, Yr27, Yr28* (*YrAs2388*), *Yr36* (*WKS1*), *YrNAM, YrU1, Lr34/Yr18/Sr57, Lr67/Yr46/Sr55, Yr46/Lr67/Sr55, Yr18/Lr34/Sr57*	Li et al. (2023), Geethanjali et al. (2024), Tong et al. (2024)

(continued)

Table 3.1 (continued)

Crop	Disease	Genes	Reference
Maize	Northern corn leaf blight (*Exserohilum turcicum*)	*ZmWAK-RLK1, remorin* (*ZmREM6.3*),	Zhu et al. (2021), Chen et al. (2022)
	Southern corn leaf blight (*Bipolaris maydis*)	*ZmCCoAOMT2, rhm1, RppK*	
	Banded leaf and sheath blight (*Rhizoctonia solani*)	*ZmFBL41*	
	Gray leaf spot (*Cercospora zeae-maydis*)	*ZmCCoAOMT2*	
	Common rust (*Puccinia sorghi*)	*Rp1-D*	
	Southern rust (*Puccinia polysora*)	*ZmREM1.3*	
	Head smut (*Sphacelotheca reiliana*)	*ZmWAK*	
	Gibberella ear rot (*Gibberella zeae*)	*ZmAuxRP1*	
	Fusarium ear rot (*Fusarium verticillioides*)	*ZmLOX3, ZmLOX12*	
	Aspergilus ear rot (*Aspergillus flavus*)	*ZmLOX3*	
	Gibberella stalk rot (*Gibberella acuminate*)	*ZmCCT, ZmAuxRP1*	
	Anthracnose stalk rot (*Colletotrichum graminicola*)	*Rcg1*	
	Maize rough dwarf disease	*ZmGDIα-hel*	
	Maize dwarf mosaic disease	*ZmTrxh, ZmABP1*	

Table 3.2 Examples of heterologous expression of nonplant and plant-derived genes conferring biotic stress resistance in transgenic plants

S. No	Transgenic crop	Transgene Introduced	Source of transgene	Biotic stress resistance to	Gene transfer method	Reference
	Nonplant sources of biotic stress-resistant genes					
1	Rice	*Cry1Ac, Cry2A*	*Bacillus thuringenesis*	Yellow stem borer Striped stem borer	Particle bombardment	Gahakwa et al. (2000)
2	Maize	*vip3Aa* and *Cry1Ab*	*Bacillus thuringenesis*	Fall armyworm (*Spodoptera frugiperda*) Asian corn borer (*Ostrinia furnacalis*)	*Agrobacterium tumefaciens*	Chen et al. (2024)
3	Pigeonpea	*Cry1Ac and Cry2Aa*	*Bacillus thuringenesis*	*Helicoverpa armigera*	*Agrobacterium*-mediated	Gosh et al. (2017)
4	Cowpea	*Vip3ba*	*Bacillus thuringenesis*	*Legume pod borer (Maruca vitrata)*	*Agrobacterium*-mediated	Bett et al. (2017)
5	Chickpea	*Cry1Ac*	*Bacillus thuringenesis*	*Helicoverpa armigera*	*Agrobacterium tumefaciens*	Chakraborthy et al. (2016)
6	Groundnut	*Cry1EC*	*Bacillus thuringenesis*	*Tobacco cutworm (Spodoptera litura)*	*Agrobacterium*-mediated	Tiwari et al. (2008)
7	Cotton	*Cry1Ia12*	*Bacillus thuringenesis*	*Fall Armyworm (Spodoptera frugiperda) and Cotton Boll Weevil (Anthonomus grandis)*	*Agrobacterium tumefaciens*	de Oliveira et al. (2016)
8	Cotton	*Cry1Ia12*	*Bacillus thuringenesis*	*Pink bollworm (Pectinophora gossypiella)*	*Agrobacterium tumefaciens*	Tariq et al. (2022)

9	Cotton	*Cry3Bb1 and Cry3*	*Bacillus thuringenesis*	Pink bollworm (*Pectinophora gossypiella*), army bollworm (*Helicoverpa armigera*)	Agrobacterium tumefaciens	Zafar et al. (2022)
10	Sugarcane	*Cry1Ac*	*Bacillus thuringenesis*	*Stem borer (Diatraea saccharalis*	*Particle bombardment*	Zhou et al. (2018)
11	Potato	*Cry3A*, SN-*19*	*Bacillus thuringenesis*	Potato tuber moth, *Phthorimaea operculella Colorado potato beetle (Leptinotarsa decemlineata*	*Agrobacterium tumefaciens*	Salim et al. (2024) Mi et al. (2015)
12	Brinjal	*Cry1Aa3, Cry1Ac, Cry1Fa1*	*Bacillus thuringenesis*	*Eggplant fruit and shoot borer Leucinodes orbonalis)*	*Agrobacterium tumefaciens*	Rai et al. (2013) Shrivastava et al. (2011); Kumar et al. (1998)
13	Tobacco	*Cry1Ah*	*Bacillus thuringenesis*		*Agrobacterium tumefaciens*	Sun et al. (2012)
	Plant-derived biotic stress-resistant genes					
14	Wheat	*CERK1-V*	*Haynaldia villosa*	*Powdery mildew, Yellow rust, and Fusarium head blight pathogens*	*Agrobacterium tumefaciens*	Fan et al. (2022)
15	Rice	*ZmNBS25*	*Zea mays*	*Rice sheath blight pathogen (Rhizoctonia solani)*	*Agrobacterium tumefaciens*	Xu et al. (2018)
16	Sugarcane	*Kunitz trypsin inhibitor (SKTI) and Bowman–Birk inhibitor (SBBI)*	*Glycine max*	*Sugarcane borer (Diatraea saccharalis)*	*Particle bombardment*	Falco et al. (2003)

(continued)

Table 3.2 (continued)

S. No	Transgenic crop	Transgene Introduced	Source of transgene	Biotic stress resistance to	Gene transfer method	Reference
17	Potato	*Oryzacystatin I and II (OCI and OCII)*	*Oryza sativa*	*Colorado potato beetle (Leptinotarsa decemlineata)*	*Agrobacterium tumefaciens*	Cingel et al. (2017)
18	Tomato	*Mi-1.2* gene	*Solanum lycopersicum*	*Root-knot nematode (Meloidogyne javanica)*	*Agrobacterium tumefaciens*	Goggin et al. (2006)
19	Tomato	*Proteinase inhibitor PI-II and carboxypeptidase inhibitors (PCI)*	*Potato*	*Heliothis obsoleta Liriomyza trifolii*	*Agrobacterium tumefaciens*	Abdeen et al. (2005)
20	Tomato	*Cysteine proteinase inhibitor (CeCPI)*	*Colocasia esculenta*	*Root knot nematode (Meloidogyne incognita)*	*Agrobacterium tumefaciens*	Chan et al. (2010)
21	Tomato	*ferredoxin-I protein*	*Sweet pepper*	*Ralstonia solanacearum*	*Agrobacterium tumefaciens*	Huang et al. (2007)
22	Tobacco	*AhMPK3*	*Arachis hypogaea*	*Spodoptera litura*	*Agrobacterium tumefacien*	Kumar et al. (2009)
23	Tobacco	*Chitinase gene (SmchiC)*	*Serratia marcescens*	*Necrotrophic fungus (Botrytis cinerea) Spodoptera frugiperda*	*Agrobacterium tumefacien*	Navarro-González et al. (2019)
24	Tobacco	*Taxane biosynthesis genes—taxadiene synthase, taxane 5α-hydroxylase, taxane 13α-hydroxylase*		*Fall armyworm*	*Agrobacterium tumefacien*	Xu et al. (2024)
25	Tobacco	*Transcription binding factor AtNPR1*	*Arabidopsis thaliana*	*Spodoptera litura*	*Agrobacterium tumefaciens*	Meur et al. (2008)

26	Tobacco	Kunitz-type (PKPI—(*PKI1*, *PKI2*) and Potato type I protease inhibitor genes (*PPI3A2*, *PPI3B2*, *PPI2C4*)	*Potato*	*Botrytis cinerea* *Alternaria alternata*	*Agrobacterium tumefaciens*	Turra et al. (2020)
27	Arabidopsis	*Kunitz-type protease inhibitor GsKTI*	*Glycine soja*	*Helicoverpa armigera*	*Agrobacterium tumefaciens*	Zhang et al. (2023)
28	Arabidopsis	*PtrWRKY18 and PtrWRKY35*	*Populus trichocarpa*	*Botrytis cinerea*	*Agrobacterium tumefaciens*	Guo et al. (2020)
29	Arabidopsis	*Constitutive disease-resistance genes PtCDR2* and *PtCDR8*	*Poncirus trifoliata*	*Pseudomonas syringae*	*Agrobacterium tumefaciens*	Ying et al. (2020)
30	Arabidopsis	*Pathogenesis-related protein AsPR5*	*Allium sativum*	*Fusarium oxysporum f.* sp. *Cepae* *Botrytis cinerea*	*Agrobacterium tumefaciens*	Rout et al. (2016)

Table 3.3 RNAi-based host-induced gene silencing to combat pest and pathogen in crop plants

Crop	*Pest/pathogen*	*Targeted pest/pathogenic genes for silencing*	Effect of host-induced gene silencing on pest/pathogen	RNAi approach for HIGS	Reference
Rice	*Rice stripe virus* (RSV)	*RSV nucleocapsid protein (NCP) gene*	Tenfold increase in virus-derived siRNAs, imparting resistance to rice stripe virus	siRNA	Li et al. (2016)
Rice	Striped stem borer (*Chilo suppressalis*)	*csu-novel-miR15*	Growth inhibition and delayed pupation	dsRNA	Jiang et al. (2017)
Rice	Rice blast fungus *Magnaporthe oryzae*	*Pathogenic genes MoABC1, MoMAC1 and MoPMK1,*	Reduction in transcription levels of targeted fungal genes	siRNA	Zhu et al. (2017)
Rice	Striped stem borer (*Chilo suppressalis*)	*Csu-miR-14 / CsSpo and CsEcR (Spook and Ecdysome Receptors)*	Impairment of ecdysone biosynthesis led to high mortality and developmental defects in RSB	amiRNA	He et al. (2019)
Rice	Striped stem borer (*Chilo suppressalis*)	*Csu-novel-miR260 disembodied*	Negative regulation of ecdysteroid biosynthesis	amiRNA	Zheng et al. (2021)
Rice	Striped stem borer (SSB; *Chilo suppressalis*)	*CsFAR (Fatty acyl-CoA reductase)*	Downregulation of FAR increased mortality rate by 80% and darkening of epidermis	siRNA	Sun et al. (2022)
Rice	Striped stem borer (*Chilo suppressalis*)	*CssHsp (heat shock protein)*	Larval mortaliy (60–90%) by the time of pupation	ds RNAi	Mao et al. (2022)

Rice (Transgenic line with Cry1ca)	Striped stem borer (*Chilo suppressalis*)	*microRNA-7322-5p/p38/ Hsp19 heat shock protein 19*	Enhanced cry1ca sequestration and increased mortality of SSB	dsRNA	Wu et al. (2023)
Rice	Striped stem borer (*Chilo suppressalis*)	*CsmiR1579-CsKr-h1 (Krüppel homologue 1)*	Negative regulation of transcription factor Kr-h1 leading to a decline in larval growth and survival.	miRNA	Tang et al. (2024)
Rice	Brown plant hopper (*Nilaparvata lugens*)	*Glutathione S-transferase NlGST1-1 gene*	Retarded growth rate of nymphs and insect female fecundity	dsRNA	Yang et al. (2021)
Wheat	*Fusarium graminearum*	*Chitin synthase Chs3b*	Reduced fungal growth		Cheng et al. (2015)
Wheat	*Grain aphid (Sitobion avenae)*	*Salivary sheath protein (SHP)*	Decline in growth, reproduction and survival rates of the aphid	dsRNA	Abdellatef et al. (2015)
Wheat	*Grain aphid (Sitobion avenae)*	*Lipase maturation factor 2-like gene*	Reduced survival rate and reproductive capacity		Xu et al. (2017)
Wheat	*Grain aphid (Sitobion avanae)*	*Chitin synthase 1 (CHS1)*	Impaired insect molting leading to reduced survival rate	dsRNA	Zhao et al. (2018)
Wheat	*Sitobion avenae*	*Zinc finger protein (SaZFP)*	Reduced fecundity and survival of aphids	dsRNA	Sun et al. (2019)
Wheat	*Puccinia striiformis f.* sp. *tritici*	*Glycine-serine-rich effector (PstGSRE1)*	Reduction in fungal biomass and virulence of Pst	siRNA	Qi et al. (2019)
Wheat	*Blumeria graminis* f. sp. *tritici*	*β2-tub*, *$SvrPm3^{a1/f1}$*	Impaired fungal haustorium formation in seedlings and reduced growth in adult plants	ckRNAi	Schaefer et al. (2020)

(continued)

Table 3.3 (continued)

Crop	*Pest/pathogen*	*Targeted pest/pathogenic genes for silencing*	Effect of host-induced gene silencing on pest/pathogen	RNAi approach for HIGS	Reference
Wheat	*Sitobion miscanthi*	*putative salivary protein gene SmDSR33*	Reduced fecundity and survival	dsRNA	Zhang et al. (2023)
Maize	Western corn rootworm (*Diabrotica virgifera virgifera)*	*Vacuolar -ATPase A*	Reduced feeding	dsRNA	Baum et al. (2007)
Maize	Western corn rootworm (*Diabrotica virgifera virgifera)*	*DvSnf7*	Downregulation leads to corn rootworm mortality and reduced emergence	dsRNA	Head et al. (2017)
Maize	*Diabrotica virgifera virgifera*	*boule (bol) gene*	Reduction in egg production and hatching rate	dsRNA	Niu et al. (2017)
Maize	Maize stem borer *(Chilo partellus)*	*Chitinase CpCHI*	Phenotypic abnormalities and mortality of stemborer	dsRNA	Adeyinka et al. (2023)
Soybean	*Heterodera glycines* *Fusarium oxysporum*	*Chitin synthase (SCN-CHS)*	Downregulation leads to suppression of cyst and egg formation in nematodes Reduced mycelial growth of the pathogen	dsRNA	Kong et al. (2022)
Soybean	*Alfalfa mosaic virus, Bean pod mottle virus, and Soybean mosaic virus*	Replicase genes of *Alfalfa mosaic virus, Bean pod mottle virus, and Soybean mosaic virus* AMV, BPMV, and SMV (ABS)	Inhibition of viral replication and clearance of the virus within a short time period	dsRNA	Zhang et al. (2011)

Sugarcane	*Sphenophorus levis*	*V-ATPase subunit E*	Downregulation of the gene resulted in reduced larval growth and survival	dsRNA	Mohan et al. (2021)
Sugarcane	*Sporisorium scitamineum*	*Beta-1,6-glucanase precursor encoding gene SsGlcP*	Reduced fungal biomass of the smut pathogen	dsRNA	Wu et al. (2025)
Cotton	Cotton bollworm (*Helicoverpa armigera*)	*P450 monooxygenase*	Silencing impairs larval tolerance to Gossypol	ds-RNA	Mao et al. (2007)
Cotton	*Aphis gossypii*	*AgCYP6CY3-P1*	Delayed growth and development of the aphid	dsRNA	Zhang et al. (2023)
Mustard	Mustard aphid (*Lipaphis erysimi*)	*Sucrase 1*	Impaired parthenogenetic reproduction	dsRNA	Dhatwalia et al. (2022)
Brassica napus (rape)	Green peach aphid, (*Myzus persicae*)	*Partitioning-defective protein 6 (Par6)*	Reduced survival rate and impaired fecundity	dsRNA	Zhang et al. (2024)
Potato	*Leptinotarsa decemlineata*	*Ecdysone receptor (EcR)*	Reduction in larval weight	ds RNA	Hussain et al. (2019)
Potato	*Henosepilachna vigintioctopunctata*	*β-Actin*	Complete larval mortality	dsRNA	Xu et al. (2023)
Tomato	Western flower thrips (*Frankliniella occidentalis*)	*Coatomer protein subunit epsilon (CopE)* *Toll-like receptor 6 (TLR6)*	Increased larval mortality	dsRNA	Venkatesh et al. (2023)
Tomato	*Myzus persicae*	*Acetylcholinesterase 1 gene (Ace 1)*	Reduction in aphid colonization	amiRNA	Faisal et al. (2021)
Tomato, tobacco	*Helicoverpa armigera*	*Chitinase (HaCHI) gene*	Impaired insect molting and metamorphosis resulted in overall reduction in growth and survival	siRNA	Mamta et al. (2016)
Tomato	*Helicoverpa armigera*	*Ecdysone Receptor (HaECR)*	Reduction in growth and survival of the insect	amiRNA	Yogindran and Rajam (2021)

3.3.3 Metabolomics, Proteomics, and Transcriptomics for Understanding Stress Responses

Metabolites are direct indicators of physiological responses to environmental stress. Given the functional role of metabolites, the profiling of metabolome associated with stress tolerance will help to identify metabolite QTL (mQTLs) regulating agronomic traits. Metabolomics-based mGWAS studies will provide deep insights into metabolite variation under climatic stresses, further allowing breeders to select elite varieties (Razzaq et al. 2022). Proteomics, on the other hand, is the large-scale study of proteins, which are the primary functional molecules in cells. Since proteins mediate plant responses to stress, understanding the entire set of proteins expressed in a plant provides insight into how plants adapt to environmental changes (Kosová et al. 2018). By identifying stress-responsive proteins, breeders can pinpoint critical molecular pathways and markers involved in heat, drought, or salinity responses and select varieties that express beneficial proteins under these conditions. Similarly, the integration of transcript information through transcriptomics enables the detection of differential genes responsible for yield, stress, or disease resistance (Fernie and Schauer 2009). By combining phenotypic data with omics information, breeders can identify the most promising genotypes for specific stress conditions, improving the accuracy of selection.

3.3.4 Tailoring the Soil and Crop Microbiome to Enhance Resilience

The use of agrochemicals and synthetic fertilizers has significantly increased in recent decades, aiming to maximize agricultural production and minimize pest attack (Foong et al. 2020). The use of these agrochemicals not only pollutes the environment, but over time, it becomes harmful for humans and livestock (Fu et al. 2022). In addition to this, they kill beneficial microbes and reduce nutrient availability, which are essential elements for plant growth and productivity (Meena et al. 2020). Plant microbiome engineering is one of the emerging biotechnological approaches for crop production under extreme climatic conditions. The manipulation of the bacterial microbiome and the production of bioinoculants have enabled scientists to control and properly monitor plant health and production (Adesemoye et al. 2009; Alok et al. 2020). This includes various methods such as manipulating plant mycobiome—the enhancement and optimization of advantageous plant–fungal interactions (Pozo et al. 2021), in situ microbiome—modification of the natural microbial community in their native setting (Tosi et al. 2020) and synthetic microbiome—inoculating genetically designed microorganisms to host plant (Glick and Gamalero 2021). There are bacterial consortia for phosphate solubilization, biofilm formation, root colonization, and plant growth-promoting rhizobacteria (PGPR) providing significant biocontrol and diverse growth characteristics (Magallon-Servin et al. 2020; Syed et al. 2020). Different agricultural systems, farming practices, and soil treatments will also influence the microbial communities, which will affect soil health (Lupatini et al. 2017). For example, the application of pelletized spent mushroom substrate in combination with mineral fertilizer improved plant performance and soil health in *Hordeum vulgare* L (Obermeier et al. 2020). A study in alfalfa with the application of organic amendments effectively improved soil fertility, promoting beneficial soil microbiome and inducing plant growth, with high yields (Bonanomi et al. 2020). Substantial organic additions enhance the decomposition process, triggering symbiotic microbial colonization, and indirectly result in pathogen suppression (Bonanomi et al. 2020). Therefore, improving soil health with beneficial microbiota can complement improved growth and performance in cereals and pulses and could be a transformative technique for increasing climate resilience and performance (Cernava 2024).

3.4 Future Perspectives and Innovations in Climate-Resilient Breeding for Cereals and Pulses

3.4.1 Role of Digital Tools and Big Data in Precision Breeding for Cereals and Pulses

Breeding efforts for cereals and pulses are undergoing a transformation by integrating advanced digital tools, big data, cutting-edge technologies such as artificial intelligence (AI), machine learning (ML), and speed breeding.

Data-Driven Decision Making Big data analytics are now essential in processing the enormous datasets generated by high-throughput phenotyping and genotyping technologies. Data analytics platforms like Field Book (Rife and Poland 2014), Breeding Insight, and the Integrated Breeding Platform (IBP) allow breeders to store, manage, and analyze these data in real-time, enabling more accurate selection decisions (Sinha et al. 2023). These tools also facilitate data sharing across institutions, fostering collaboration and accelerating crop improvement efforts. For example, in maize breeding, platforms like the Genomic Open-source Breeding Informatics Initiative (GOBII) integrate data from multiple sources to facilitate large-scale genomic selection for drought tolerance (Santantonio et al. 2020).

Using AI and Machine Learning for Predictive Breeding Artificial intelligence (AI) and machine learning (ML) are increasingly used to predict complex trait outcomes and make breeding more efficient. AI algorithms can analyze large datasets of genotype–phenotype interactions and environmental factors to identify patterns and predict plant performance under different conditions (Farooq et al. 2024). AI can enhance genomic selection by processing large genomic datasets and identifying complex gene networks associated with stress tolerance. For example, AI models can analyze thousands of molecular markers and environmental variables to predict the best parental combinations for breeding programs. AI-driven genomic selection has been used in wheat to improve heat tolerance, with models predicting the best crosses based on genome-wide marker data and environmental factors

(Dreisigacker et al. 2023). Machine learning algorithms such as random forests, support vector machines, and deep learning models can predict crop performance based on environmental variables, genetic markers, and historical data. These models help breeders prioritize which genotypes are most likely to perform well under specific climate scenarios. In rice, ML models have been used to predict yield stability under varying water availability, allowing breeders to develop drought-tolerant varieties more quickly (Hashim et al. 2024). In Australia, the application of digital agriculture tools in wheat breeding has led to the rapid development of varieties that can thrive in water-limited environments. By integrating these phenotypic data with genomic selection models, Australian breeders have accelerated the development of drought-tolerant wheat varieties. The International Crops Research Institute for the Semi-Arid Tropics (ICRISAT) has been leveraging digital agriculture tools to improve chickpea breeding in India and sub-Saharan Africa. Data from high-throughput phenotyping platforms, such as ground-based sensors and drones, are integrated into a machine learning model to predict elite varieties for the future (Jangra et al. 2021).

3.4.2 Emerging Technologies

Gene Editing Gene editing technologies, particularly CRISPR-Cas9, are allowing breeders to make precise modifications to the DNA of crops, speeding up the development of varieties with enhanced stress tolerance. Clustered, regularly interspaced short palindromic repeats (CRISPR) CRISPR-associated (Cas) protein 9 is a popular tool for targeted mutagenesis and has been used to edit genes to improve desired traits in crops. In the last 5 years, CRISPR-Cas9 rice (*Oryza sativa*) mutants have been developed and these included targeting *Drought and salt tolerance (OsDST), CHLOROPHYLL A OXYGENASE 1, LAZY1, MAP kinase, Semi-Rolled Leaf1,2, Enhanced response to aba1, Hexokinase gene* which improved abiotic tolerance, disease resistance, drought and cold tolerance, and yield (Ahmar et al. 2023). In wheat, *dehydration-responsive element binding protein 2 (TaDREB2)* and *wheat ethylene responsive factor 3 (TaERF3)* increased drought tolerance (Abdallah et al. 2022); *photoperiod-1 (TaPpd-1)* increased the number of spikelets with improved grain traits (Errum et al. 2023). CRISPR is widely applied in other cereals such as maize and barley (Ahmar et al. 2023). In pulses, CRISPR is applied in alfalfa, soybeans, and chickpeas to enhance traits such as nitrogen fixation, nodule formation, and flowering time, which are critical for productivity and quality (Baloglu et al. 2022).

Speed Breeding and Accelerated Development Speed breeding, which allows for up to six crop generations per year, is a key tool for accelerating the breeding cycle (Watson et al. 2018). Future research should focus on optimizing speed breeding protocols for different cereals and pulses, ensuring that this technique can be applied across a range of crop species. Recently, a speed vernalization protocol was developed to reduce winter wheat and winter barley generation time (Cha et al.

2022). Additionally, efforts should be made to develop low-cost, energy-efficient growth chambers to make speed breeding accessible to a wider range of breeding programs globally (Cha et al. 2023). Combining speed breeding with gene editing offers a powerful approach for rapidly developing climate-resilient crops. Research should explore how to best integrate these techniques, allowing for the quick assessment of gene-edited crops across multiple generations. This will be critical for shortening the time from the discovery of key genetic traits to the commercial release of resilient crop varieties.

Integration of High-throughput Phenotyping and Machine Learning into Cereals and Pulses Breeding Integrating high-throughput phenotyping data into breeding programs has revolutionized the way of developing climate-resilient varieties. When phenotypic data on climate-resilient traits such as root architecture, transpiration efficiency, or stay-green traits are integrated with genomic selection models, breeders can better identify and select high-performing individuals before they undergo extensive field testing. For example, in wheat and maize, genomic selection models incorporating phenotypic data on drought tolerance have enabled breeders to predict yield stability under water-limited conditions with higher accuracy (Das et al. 2021; Guo et al. 2020). In pulses like chickpeas and lentils, integrating phenotypic data with heat tolerance traits such as early flowering and pod set has accelerated the development of varieties capable of withstanding high temperatures during reproductive stages (Choukri et al. 2022). Modern breeding programs are increasingly utilizing data-driven platforms that combine phenotypic, genomic, and environmental data to make more informed breeding decisions. These platforms use machine learning algorithms to analyze vast datasets and identify correlations between phenotypic traits and genetic markers associated with climate resilience (Parmley et al. 2019). This approach allows breeders to prioritize traits that are most important for adapting to specific climate conditions and regions. In cereals, these platforms have been instrumental in the development of rice varieties that can tolerate both drought and submergence, a combination of stresses that are increasingly common due to erratic rainfall patterns. For pulses, data-driven approaches have been applied to breed lentil varieties with improved water-use efficiency and heat tolerance, ensuring higher yields even in arid environments (Noor et al. 2024).

3.4.3 The Role of Interdisciplinary Research in Advancing Breeding Efforts

Climate-resilient breeding requires the collaboration of multiple scientific disciplines, including genetics, genomics, data science, agronomy, and climatology. Interdisciplinary research teams are essential for integrating advanced technologies into breeding programs and for addressing the complex interactions between plants and their environments (Beans 2020). The integration of genetics with data science has been crucial for developing predictive models that guide breeding decisions. Collaborations between geneticists, agronomists, climatologists, and computer scientists have led to the development of machine learning algorithms that can analyze vast genomic datasets and predict how plants will respond to climate variables like temperature, rainfall, and humidity.

3.5 Visions for the Future: 2050 and Beyond

By 2050, it is expected that the global population will exceed nine billion, and the demand for food will increase significantly (Tripathi et al. 2019). As we face the escalating challenges of climate change, food security, and population growth, a clear and actionable roadmap is essential to guide future research and development (R&D) in breeding climate-resilient cereals and pulses. This roadmap envisions a strategic and multidisciplinary approach to crop improvement, incorporating the latest technological innovations and focusing on sustainable practices (Fig. 3.2).

Expanding Genomic Resources and Digital Platforms for Breeding Future research must continue to expand genomic resources for under-researched cereals and pulses like cowpea, mung bean, and pigeon pea, which are crucial for food security in marginal regions. Comprehensive genome mapping of these crops will allow breeders to identify key genetic markers for traits like drought and heat tolerance, as well as pest and disease resistance. Genomic databases should be expanded and updated regularly to include new marker-trait associations that can guide breeding efforts. Gene editing technologies, such as CRISPR-Cas9, will remain a corner-

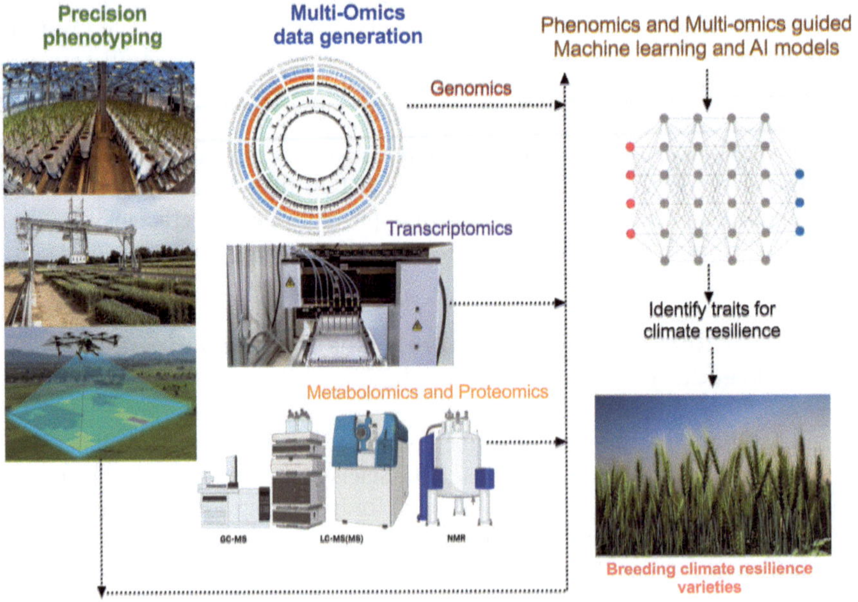

Fig. 3.2 A Road map to precision breeding for climate resilient varieties

stone of future R&D. The focus should be on enhancing the precision, efficiency, and accessibility of gene editing tools, making them more widely available to breeders in developing countries. Research should also prioritize the development of multiplex genome editing techniques, which can modify several genes simultaneously, accelerating the breeding of crops with multiple climate-resilient traits. Artificial intelligence (AI) and machine learning (ML) should be fully integrated into breeding programs to assist in predictive breeding. This will be crucial for anticipating future climate impacts and ensuring the development of resilient varieties that can thrive under changing environmental conditions.

Sustainable and Regenerative Agriculture Future breeding efforts will focus not only on resilience to abiotic and biotic stresses but also on promoting sustainability and regeneration in agricultural systems. Research should prioritize breeding for traits that enhance soil health, carbon sequestration, and nutrient use efficiency. This includes developing crops that are better adapted to no-till farming, intercropping systems, and agroforestry practices, which can improve sustainability while mitigating climate impacts. Breeding programs must also focus on developing low-input crop varieties that require minimal fertilizers, pesticides, and water. This will be especially important in regions where resources are scarce, and it will contribute to more sustainable agricultural practices. Low-input, climate-resilient varieties will help smallholder farmers in developing countries adapt to climate change while reducing their environmental footprint.

Fostering Interdisciplinary Collaboration and Capacity Building Interdisciplinary collaboration will be critical for addressing the complex challenges of breeding climate-resilient crops. Geneticists, data scientists, agronomists, climatologists, and social scientists must work together to create holistic breeding strategies. Future R&D should encourage collaborative projects that bring together expertise from diverse fields to tackle issues like climate modeling, genomic prediction, and sustainable farming systems. To ensure that all regions benefit from advancements in breeding technology, it is essential to invest in capacity building in developing countries. Training programs for breeders, agronomists, and data scientists should be expanded, with a focus on improving access to advanced tools such as genomic selection, ML, and speed breeding. In addition, partnerships between international research organizations and local agricultural institutions should be strengthened to ensure that breeding programs are tailored to the specific challenges faced by farmers in these regions.

To ensure the timely release of climate-resilient crops, it will be important to streamline regulatory approvals for new crop varieties, especially those developed using gene editing technologies like CRISPR. This should include studies on the biosafety and environmental impacts of edited crops, providing the data needed to facilitate regulatory decisions. Public acceptance of new breeding technologies is crucial for their widespread adoption, and this can be done by including outreach and engagement efforts aimed at educating farmers and consumers about the benefits of gene editing, speed breeding, and other advanced breeding technologies.

Transparent communication about the safety and advantages of these technologies will be essential for overcoming skepticism and ensuring their adoption.

3.6 Conclusion

The future of climate-resilient breeding for cereals and pulses is bright, but it requires a concerted effort from the global research community to fully realize its potential. By expanding genomic resources, integrating digital tools, leveraging AI, and promoting interdisciplinary collaboration, future breeding programs can address the complex challenges posed by climate change. With a focus on sustainability, capacity building, and public engagement, this roadmap can ensure that by 2050 and beyond, global agriculture remains resilient, productive, and capable of feeding a growing population in an increasingly volatile world.

References

Abdallah NA, Elsharawy H, Abulela HA, Thilmony R, Abdelhadi AA, Elarabi NI (2022) Multiplex CRISPR/Cas9-mediated genome editing to address drought tolerance in wheat. GM Crops Food 16:1–17. https://doi.org/10.1080/21645698.2022.2120313

Abdul Aziz M, Masmoudi K (2024) Molecular breakthroughs in modern plant breeding techniques. Hortic Plant J. https://doi.org/10.1016/j.hpj.2024.01.004

Adesemoye AO, Torbert HA, Kloepper JW (2009) Plant growth-promoting rhizobacteria allow reduced application rates of chemical fertilizers. Microb Ecol 58(4):921–929. https://doi.org/10.1007/s00248-009-9531-y

Ahmar S, Hensel G, Gruszka D (2023) CRISPR/Cas9-mediated genome editing techniques and new breeding strategies in cereals—current status, improvements, and perspectives. Biotechnol Adv 69:108248. https://doi.org/10.1016/j.biotechadv.2023.108248

Ahmed K, Shabbir G, Ahmed M, Shah KN (2020) Phenotyping for drought resistance in bread wheat using physiological and biochemical traits. Sci Total Environ 729:139082. https://doi.org/10.1016/j.scitotenv.2020.139082

Alhudhaibi AM, Ibrahim MAR, Abd-Elaziz SMS, Farag HRM, Elsayed SM, Ibrahim HA, Hossain ABMS, Alharbi BM, Haouala F, Elkelish A, Srour HAM (2024) Enhancing salt stress tolerance in wheat (Triticum aestivum) seedlings: insights from trehalose and mannitol. BMC Plant Biol 24(1):472. https://doi.org/10.1186/s12870-024-04964-2

Alok D, Annapragada H, Singh S, Murugesan S, Singh NP (2020) Symbiotic nitrogen fixation and endophytic bacterial community structure in Bt-transgenic chickpea (Cicer arietinum L). Sci Rep 10(1):5453. https://doi.org/10.1038/s41598-020-62199-1

Anderegg J, Kirchgessner N, Aasen H, Zumsteg O, Keller B, Zenkl R, Walter A, Hund A (2024) Thermal imaging can reveal variation in stay-green functionality of wheat canopies under temperate conditions [original research]. Front Plant Sci 15. https://doi.org/10.3389/fpls.2024.1335037

Atefi A, Ge Y, Pitla S, Schnable J (2021) Robotic technologies for high-throughput plant phenotyping: contemporary reviews and future perspectives. Front Plant Sci 12:611940. https://doi.org/10.3389/fpls.2021.611940

Badaeva ED, Konovalov FA, Knüpffer H, Fricano A, Ruban AS, Kehel Z, Zoshchuk SA, Surzhikov SA, Neumann K, Graner A, Hammer K, Filatenko A, Bogaard A, Jones G, Özkan H, Kilian B (2022) Genetic diversity, distribution and domestication history of the neglected

GGAtAt genepool of wheat. Theor Appl Genet 135(3):755–776. https://doi.org/10.1007/s00122-021-03912-0

Baloglu MC, Celik Altunoglu Y, Baloglu P, Yildiz AB, Türkölmez N, Özden Çiftçi Y (2022) Gene-editing technologies and applications in legumes: progress, evolution, and future prospects [review]. Front Genet 13. https://doi.org/10.3389/fgene.2022.859437

Banerjee BP, Joshi S, Thoday-Kennedy E, Pasam RK, Tibbits J, Hayden M, Spangenberg G, Kant S (2020) High-throughput phenotyping using digital and hyperspectral imaging-derived biomarkers for genotypic nitrogen response. J Exp Bot 71(15):4604–4615. https://doi.org/10.1093/jxb/eraa143

Barbosa PAM, Fritsche-Neto R, Andrade MC, Petroli CD, Burgueño J, Galli G, Willcox MC, Sonder K, Vidal-Martínez VA, Sifuentes-Ibarra E, Molnar TL (2021) Introgression of maize diversity for drought tolerance: subtropical maize landraces as source of new positive variants. Front Plant Sci 12:691211. https://doi.org/10.3389/fpls.2021.691211

Beans C (2020) Inner workings: crop researchers harness artificial intelligence to breed crops for the changing climate. Proc Natl Acad Sci USA 117(44):27066–27069. https://doi.org/10.1073/pnas.2018732117

Benitez-Alfonso Y, Soanes BK, Zimba S, Sinanaj B, German L, Sharma V, Bohra A, Kolesnikova A, Dunn JA, Martin AC, Khashi u Rahman M, Saati-Santamaría Z, García-Fraile P, Ferreira EA, Frazão LA, Cowling WA, Siddique KHM, Pandey MK, Farooq M et al (2023) Enhancing climate change resilience in agricultural crops. Curr Biol 33(23):R1246–R1261. https://doi.org/10.1016/j.cub.2023.10.028

Bonanomi G, De Filippis F, Zotti M, Idbella M, Cesarano G, Al-Rowaily S, Abd-ElGawad A (2020) Repeated applications of organic amendments promote beneficial microbiota, improve soil fertility and increase crop yield. Appl Soil Ecol 156:103714. https://doi.org/10.1016/j.apsoil.2020.103714

Bongomin O, Lamo J, Guina JM, Okello C, Ocen GG, Obura M, Alibu S, Owino CA, Akwero A, Ojok S (2024) UAV image acquisition and processing for high-throughput phenotyping in agricultural research and breeding programs. Plant Phenome J 7(1):e20096. https://doi.org/10.1002/ppj2.20096

Bouzroud S, Henkrar F, Fahr M, Smouni A (2023) Salt stress responses and alleviation strategies in legumes: a review of the current knowledge. *3*. Biotech 13(8):287. https://doi.org/10.1007/s13205-023-03643-7

Brás TA, Seixas J, Carvalhais N, Jägermeyr J (2021) Severity of drought and heatwave crop losses tripled over the last five decades in Europe. Environ Res Lett 16(6):065012. https://doi.org/10.1088/1748-9326/abf004

Cernava T (2024) Coming of age for microbiome gene breeding in plants. Nat Commun 15(1):6623. https://doi.org/10.1038/s41467-024-50700-7

Cha J-K, O'Connor K, Alahmad S, Lee J-H, Dinglasan E, Park H, Lee S-M, Hirsz D, Kwon S-W, Kwon Y, Kim K-M, Ko J-M, Hickey LT, Shin D, Dixon LE (2022) Speed vernalization to accelerate generation advance in winter cereal crops. Mol Plant 15(8):1300–1309. https://doi.org/10.1016/j.molp.2022.06.012

Cha J-K, Park H, Choi C, Kwon Y, Lee S-M, Oh K-W, Ko J-M, Kwon S-W, Lee J-H (2023) Acceleration of wheat breeding: enhancing efficiency and practical application of the speed breeding system. Plant Methods 19(1):118. https://doi.org/10.1186/s13007-023-01083-1

Choukri H, El Haddad N, Aloui K, Hejjaoui K, El-Baouchi A, Smouni A, Thavarajah D, Maalouf F, Kumar S (2022) Effect of high temperature stress during the reproductive stage on grain yield and nutritional quality of lentil (Lens culinaris Medikus). Front Nutr 9:857469. https://doi.org/10.3389/fnut.2022.857469

Considine MJ, Siddique KHM, Foyer CH (2017) Nature's pulse power: legumes, food security and climate change. J Exp Bot 68(8):1815–1818. https://doi.org/10.1093/jxb/erx099

Das RR, Vinayan MT, Seetharam K, Patel M, Phagna RK, Singh SB, Shahi JP, Sarma A, Barua NS, Babu R, Zaidi PH (2021) Genetic gains with genomic versus phenotypic selection for drought and waterlogging tolerance in tropical maize (Zea mays L.). Crop J 9(6):1438–1448. https://doi.org/10.1016/j.cj.2021.03.012

Díaz-Pereira E, Marín Sanleandro P, Asencio AD (2019) Effects of drought and water pulses on microbial functionality and the role of Cyanoprokaryota in the rhizospheres of gypsophytes. Sci Total Environ 691:919–932. https://doi.org/10.1016/j.scitotenv.2019.07.145

Dreisigacker S, Pérez-Rodríguez P, Crespo-Herrera L, Bentley AR, Crossa J (2023) Results from rapid-cycle recurrent genomic selection in spring bread wheat. G3 Genes|Genomes|Genetics 13(4). https://doi.org/10.1093/g3journal/jkad025

Emediegwu LE, Wossink A, Hall A (2022) The impacts of climate change on agriculture in sub-Saharan Africa: A spatial panel data approach. World Dev 158:105967. https://doi.org/10.1016/j.worlddev.2022.105967

Errum A, Rehman N, Uzair M, Inam S, Ali GM, Khan MR (2023) CRISPR/Cas9 editing of wheat Ppd-1 gene homoeologs alters spike architecture and grain morphometric traits. Funct Integr Genomics 23(1):66. https://doi.org/10.1007/s10142-023-00989-2

Farhad M, Kumar U, Tomar V, Bhati PK, Krishnan JN, Kishowar-E-Mustarin, Barek V, Brestic M, Hossain A (2023) Heat stress in wheat: a global challenge to feed billions in the current era of the changing climate [review]. Front Sustain Food Syst 7. https://doi.org/10.3389/fsufs.2023.1203721

Farooq MA, Gao S, Hassan MA, Huang Z, Rasheed A, Hearne S, Prasanna B, Li X, Li H (2024) Artificial intelligence in plant breeding. Trends Genet 40(10):891–908. https://doi.org/10.1016/j.tig.2024.07.001

Fatima Z, Ahmed M, Hussain M, Abbas G, Ul-Allah S, Ahmad S, Ahmed N, Ali MA, Sarwar G, Haque E, u., Iqbal, P., & Hussain, S. (2020) The fingerprints of climate warming on cereal crops phenology and adaptation options. Sci Rep 10(1):18013. https://doi.org/10.1038/s41598-020-74740-3

Fernie AR, Schauer N (2009) Metabolomics-assisted breeding: a viable option for crop improvement? Trends Genet 25(1):39–48. https://doi.org/10.1016/j.tig.2008.10.010

Foong SY, Ma NL, Lam SS, Peng W, Low F, Lee BHK, Alstrup AKO, Sonne C (2020) A recent global review of hazardous chlorpyrifos pesticide in fruit and vegetables: prevalence, remediation and actions needed. J Hazard Mater 400:123006. https://doi.org/10.1016/j.jhazmat.2020.123006

Fu H, Tan P, Wang R, Li S, Liu H, Yang Y, Wu Z (2022) Advances in organophosphorus pesticides pollution: current status and challenges in ecotoxicological, sustainable agriculture, and degradation strategies. J Hazard Mater 424(Pt B):127494. https://doi.org/10.1016/j.jhazmat.2021.127494

Gano B, Bhadra S, Vilbig JM, Ahmed N, Sagan V, Shakoor N (2024) Drone-based imaging sensors, techniques, and applications in plant phenotyping for crop breeding: a comprehensive review. Plant Phenome J 7(1):e20100. https://doi.org/10.1002/ppj2.20100

Glick BR, Gamalero E (2021) Recent developments in the study of plant microbiomes. Microorganisms 9(7). https://doi.org/10.3390/microorganisms9071533

Guo X, Svane SF, Füchtbauer WS, Andersen JR, Jensen J, Thorup-Kristensen K (2020) Genomic prediction of yield and root development in wheat under changing water availability. Plant Methods 16(1):90. https://doi.org/10.1186/s13007-020-00634-0

Hashim N, Ali MM, Mahadi MR, Abdullah AF, Wayayok A, Mohd Kassim MS, Jamaluddin A (2024) Smart farming for sustainable rice production: an insight into application, challenge, and future prospect. Rice Sci 31(1):47–61. https://doi.org/10.1016/j.rsci.2023.08.004

Jangra S, Chaudhary V, Yadav RC, Yadav NR (2021) High-throughput phenotyping: a platform to accelerate crop improvement. Phenomics 1(2):31–53. https://doi.org/10.1007/s43657-020-00007-6

Kim SL, Kim N, Lee H, Lee E, Cheon KS, Kim M, Baek J, Choi I, Ji H, Yoon IS, Jung KH, Kwon TR, Kim KH (2020) High-throughput phenotyping platform for analyzing drought tolerance in rice. Planta 252(3):38. https://doi.org/10.1007/s00425-020-03436-9

Kosová K, Vítámvás P, Urban MO, Prášil IT, Renaut J (2018) Plant abiotic stress proteomics: the major factors determining alterations in cellular proteome. Front Plant Sci 9:122. https://doi.org/10.3389/fpls.2018.00122

Lakew B, Eglinton J, Henry RJ, Baum M, Grando S, Ceccarelli S (2011) The potential contribution of wild barley (Hordeum vulgare ssp. spontaneum) germplasm to drought tolerance of cultivated barley (H. vulgare ssp. vulgare). Field Crop Res 120(1):161–168. https://doi.org/10.1016/j.fcr.2010.09.011

Lupatini M, Korthals GW, de Hollander M, Janssens TKS, Kuramae EE (2017) Soil microbiome is more heterogeneous in organic than in conventional farming system [original research]. Front Microbiol 7. https://doi.org/10.3389/fmicb.2016.02064

Magallon-Servin P, Antoun H, Taktek S, de-Bashan LE (2020) Designing a multi-species inoculant of phosphate rock-solubilizing bacteria compatible with arbuscular mycorrhizae for plant growth promotion in low-P soil amended with PR. Biol Fertil Soils 56(4):521–536. https://doi.org/10.1007/s00374-020-01452-1

Meena RS, Kumar S, Datta R, Lal R, Vijayakumar V, Brtnický M, Sharma MP, Yadav GS, Jhariya MK, Jangir CK, Pathan SI, Dokulilová T, Pecina V, Marfo TD (2020) Impact of agrochemicals on soil microbiota and management: a review. Land 9(2):34

Mekonnen TW, Gerrano AS, Mbuma NW, Labuschagne MT (2022) Breeding of vegetable cowpea for nutrition and climate resilience in Sub-Saharan Africa: progress, opportunities, and challenges. Plants (Basel) 11(12). https://doi.org/10.3390/plants11121583

Michael PS, Mwakyusa L, Sanga HG, Shitindi MJ, Kwaslema DR, Herzog M, Meliyo JL, Massawe BHJ (2023) Floods stress in lowland rice production: experiences of rice farmers in Kilombero and Lower-Rufiji floodplains, Tanzania [original research]. Front Sustain Food Syst 7. https://doi.org/10.3389/fsufs.2023.1206754

Moreira FF, Oliveira HR, Volenec JJ, Rainey KM, Brito LF (2020) Integrating high-throughput phenotyping and statistical genomic methods to genetically improve longitudinal traits in crops. Front Plant Sci 11:681. https://doi.org/10.3389/fpls.2020.00681

Mwando E, Angessa TT, Han Y, Li C (2020) Salinity tolerance in barley during germination-homologs and potential genes. J Zhejiang Univ Sci B 21(2):93–121. https://doi.org/10.1631/jzus.B1900400

Nagy É, Lehoczki-Krsjak S, Lantos C, Pauk J (2018) Phenotyping for testing drought tolerance on wheat varieties of different origins. S Afr J Bot 116:216–221. https://doi.org/10.1016/j.sajb.2018.03.009

Ninomiya S (2022) High-throughput field crop phenotyping: current status and challenges. Breed Sci 72(1):3–18. https://doi.org/10.1270/jsbbs.21069

Noor MMA, Tahjib-Ul-Arif M, Alim SMA, Islam MM, Hasan MT, Babar MA, Hossain MA, Jewel ZA, Murata Y, Mostofa MG (2024) Lentil adaptation to drought stress: response, tolerance, and breeding approaches [review]. Front Plant Sci 15. https://doi.org/10.3389/fpls.2024.1403922

Obermeier MM, Minarsch E-ML, Durai Raj AC, Rineau F, Schröder P (2020) Changes of soil-rhizosphere microbiota after organic amendment application in a Hordeum vulgare L. short-term greenhouse experiment. Plant Soil 455(1):489–506. https://doi.org/10.1007/s11104-020-04637-7

Pappula-Reddy S-P, Kumar S, Pang J, Chellapilla B, Pal M, Millar AH, Siddique KHM (2024) High-throughput phenotyping for terminal drought stress in chickpea (Cicer arietinum L.). Plant Stress 11:100386. https://doi.org/10.1016/j.stress.2024.100386

Parmley KA, Higgins RH, Ganapathysubramanian B, Sarkar S, Singh AK (2019) Machine learning approach for prescriptive plant breeding. Sci Rep 9(1):17132. https://doi.org/10.1038/s41598-019-53451-4

Pozo MJ, Zabalgogeazcoa I, Vazquez de Aldana BR, Martinez-Medina A (2021) Untapping the potential of plant mycobiomes for applications in agriculture. Curr Opin Plant Biol 60:102034. https://doi.org/10.1016/j.pbi.2021.102034

Ram BG, Oduor P, Igathinathane C, Howatt K, Sun X (2024) A systematic review of hyperspectral imaging in precision agriculture: analysis of its current state and future prospects. Comput Electron Agric 222:109037. https://doi.org/10.1016/j.compag.2024.109037

Razzaq A, Wishart DS, Wani SH, Hameed MK, Mubin M, Saleem F (2022) Advances in metabolomics-driven diagnostic breeding and crop improvement. Meta 12(6):511. https://www.mdpi.com/2218-1989/12/6/511

Rife TW, Poland JA (2014) Field book: an open-source application for field data collection on android. Crop Sci 54(4):1624–1627. https://doi.org/10.2135/cropsci2013.08.0579

Rocchetti L, Bellucci E, Cortinovis G, Di Vittori V, Lanzavecchia G, Frascarelli G, Nanni L, Del Gatto A, Pieri S, Mangoni L, Bitocchi E, Papa R (2020) The development of a European and Mediterranean Chickpea Association Panel (EMCAP). Agronomy 10(9):1417. https://www.mdpi.com/2073-4395/10/9/1417

Salgotra RK, Chauhan BS (2023) Genetic diversity, conservation, and utilization of plant genetic resources. Genes (Basel) 14(1). https://doi.org/10.3390/genes14010174

Santantonio N, Atanda SA, Beyene Y, Varshney RK, Olsen M, Jones E, Roorkiwal M, Gowda M, Bharadwaj C, Gaur PM, Zhang X, Dreher K, Ayala-Hernández C, Crossa J, Pérez-Rodríguez P, Rathore A, Gao SY, McCouch S, Robbins KR (2020) Strategies for effective use of genomic information in crop breeding programs serving Africa and South Asia. Front Plant Sci 11:353. https://doi.org/10.3389/fpls.2020.00353

Shafi S, Shafi I, Zaffar A, Zargar SM, Shikari AB, Ranjan A, Prasad PVV, Sofi PA (2023) The resilience of rice under water stress will be driven by better roots: evidence from root phenotyping, physiological, and yield experiments. Plant Stress 10:100211. https://doi.org/10.1016/j.stress.2023.100211

Silva-Perez V, Molero G, Serbin SP, Condon AG, Reynolds MP, Furbank RT, Evans JR (2018) Hyperspectral reflectance as a tool to measure biochemical and physiological traits in wheat. J Exp Bot 69(3):483–496. https://doi.org/10.1093/jxb/erx421

Sinha D, Maurya AK, Abdi G, Majeed M, Agarwal R, Mukherjee R, Ganguly S, Aziz R, Bhatia M, Majgaonkar A, Seal S, Das M, Banerjee S, Chowdhury S, Adeyemi SB, Chen JT (2023) Integrated genomic selection for accelerating breeding programs of climate-smart cereals. Genes (Basel) 14(7). https://doi.org/10.3390/genes14071484

Staniak M, Czopek K, Stępień-Warda A, Kocira A, Przybyś M (2021) Cold stress during flowering alters plant structure, yield and seed quality of different soybean genotypes. Agronomy 11(10):2059. https://www.mdpi.com/2073-4395/11/10/2059

Swarup S, Cargill EJ, Crosby K, Flagel L, Kniskern J, Glenn KC (2021) Genetic diversity is indispensable for plant breeding to improve crops. Crop Sci 61(2):839–852. https://doi.org/10.1002/csc2.20377

Syed S, Tollamadugu N, Lian B (2020) Aspergillus and Fusarium control in the early stages of Arachis hypogaea (groundnut crop) by plant growth-promoting rhizobacteria (PGPR) consortium. Microbiol Res 240:126562. https://doi.org/10.1016/j.micres.2020.126562

Tosi M, Mitter EK, Gaiero J, Dunfield K (2020) It takes three to tango: the importance of microbes, host plant, and soil management to elucidate manipulation strategies for the plant microbiome. Can J Microbiol 66(7):413–433. https://doi.org/10.1139/cjm-2020-0085

Tripathi AD, Mishra R, Maurya KK, Singh RB, Wilson DW (2019) Chapter 1 – Estimates for world population and global food availability for global health. In: Singh RB, Watson RR, Takahashi T (eds) The role of functional food security in global health. Academic Press, pp 3–24. https://doi.org/10.1016/B978-0-12-813148-0.00001-3

Turner-Hissong SD, Mabry ME, Beissinger TM, Ross-Ibarra J, Pires JC (2020) Evolutionary insights into plant breeding. Curr Opin Plant Biol 54:93–100. https://doi.org/10.1016/j.pbi.2020.03.003

Watson A, Ghosh S, Williams MJ, Cuddy WS, Simmonds J, Rey M-D, Asyraf Md Hatta M, Hinchliffe A, Steed A, Reynolds D, Adamski NM, Breakspear A, Korolev A, Rayner T, Dixon LE, Riaz A, Martin W, Ryan M, Edwards D et al (2018) Speed breeding is a powerful tool to accelerate crop research and breeding. Nat Plants 4(1):23–29. https://doi.org/10.1038/s41477-017-0083-8

Yang Y, Yu J, Qian Q, Shang L (2022) Enhancement of heat and drought stress tolerance in Rice by genetic manipulation: a systematic review. Rice 15(1):67. https://doi.org/10.1186/s12284-022-00614-z

Yu G, Matny O, Champouret N, Steuernagel B, Moscou MJ, Hernández-Pinzón I, Green P, Hayta S, Smedley M, Harwood W, Kangara N, Yue Y, Gardener C, Banfield MJ, Olivera PD, Welchin C, Simmons J, Millet E, Minz-Dub A et al (2022) Aegilops sharonensis genome-assisted iden-

tification of stem rust resistance gene Sr62. Nat Commun 13(1):1607. https://doi.org/10.1038/s41467-022-29132-8
Zaman MSU, Malik AI, Kaur P, Ribalta FM, Erskine W (2019) Waterlogging tolerance at germination in field pea: variability, genetic control, and indirect selection [original research]. Front Plant Sci 10. https://doi.org/10.3389/fpls.2019.00953
Zaman-Allah M, Vergara O, Araus JL, Tarekegne A, Magorokosho C, Zarco-Tejada PJ, Hornero A, Albà AH, Das B, Craufurd P, Olsen M, Prasanna BM, Cairns J (2015) Unmanned aerial platform-based multi-spectral imaging for field phenotyping of maize. Plant Methods 11(1):35. https://doi.org/10.1186/s13007-015-0078-2

Crop Wild Relatives for Breeding Climate-Smart and Highly Productive Cultivars

4

Masoumeh Zeinali, Mohammad Sabzehzari, Saeideh Javid, and Peter Poczai

Abstract

Crop improvement for the remainder of this century should adopt current perspectives and methods to meet changing demands, such as improving yield and quality, along with adapting to increasing environmental stresses due to climate change. Crop wild relatives (CWRs) include a variety of adaptive characteristics that are beneficial to improve stress resilience and maintain food security. Advances in high-throughput phenotyping and omics have accelerated data production and reduced the cost per experiment. This has resulted in large amounts of digital information on morphological, bio-physiological, and molecular diversities in CWRs. However, conservation and use of CWRs in breeding practices are challenging. This chapter focuses on advances in breeding approaches for the exploration and utilization of genes or traits of CWRs for the establishment of climate-smart and highly productive cultivars. Moreover, the opportunities and challenges in the conservation of CWRs, along with their access and benefit-sharing, will be discussed.

M. Zeinali
Division of Biotechnology, Department of Agronomy and Plant Breeding, Faculty of Agricultural, University of Mohaghegh Ardabili, Ardabil, Iran

M. Sabzehzari
Division of Biotechnology, Department of Agronomy and Plant Breeding, College of Agriculture and Natural Resources, University of Tehran, Tehran, Iran

S. Javid (✉) · P. Poczai
Botany and Mycology Unit, Finnish Museum of Natural History, University of Helsinki, Helsinki, Finland
e-mail: saeideh.javid@helsinki.fi

P. M. Priyadarshan, R. Ortiz (eds.), *Plant Breeding 2050*,
https://doi.org/10.1007/978-981-95-0583-8_4

Keywords

Crop wild relatives · Digital imaging · Germplasm · Machine learning

4.1 Introduction

A food gap >50% is estimated between 2005 and 2050 due to global population growth toward 10 B by the 2050s (Bassi et al. 2024). To address this challenge, the yield should be increased, or arable lands should be extended. In developing countries, 20% of productivity increment originates from expanding arable lands; thus, 80% must come from improvement in cropping intensity and yield (Bohra et al. 2022). This highlights the necessity of adopting breeding strategies that improve crop performance in the field. This task is hindered considerably by climate change-derived stresses. With every degree of temperature increase, a 6% decrease is estimated in the field yield of wheat (Asseng et al. 2015). In addition to global warming, increases in the number and severity of biotic and abiotic stresses highlight the serious consequences of climate change (Zhang and Batley 2020).

Opportunities for improvement of crop yield and development of climate-smart crops include exploring natural (epi)genetic variations from crop wild relatives (CWRs), identifying genomic regions involved in yield and tolerance to stressful conditions, alteration of endogenous gene regulatory or relevant protein function, or improvement of crop health via interactions with symbiotic fungi or bacteria (Bailey-Serres et al. 2019). To achieve these goals, readers should follow current advances in breeding approaches for the exploration and utilization of genes or traits of CWRs to establish climate-smart and highly productive cultivars.

4.2 Genetic and Epigenetic Variations

Approximately 75% of the genetic variation available in landraces has been missed in the past century (Khoury et al. 2022). Accordingly, there is a need to conserve CWRs along with exploring and using their diversity. CWRs usually retain more genetic variability than elite cultivars due to domestication bottlenecks, including the founder impact (a decrease in genetic diversity due to establishing a new colony from a small subset of a large population), fixation of domestication syndrome (changes in syndrome traits, including loss of shattering), and post-domestication (such as local preference underlying cultural selection) (Alam and Purugganan 2024; Purugganan and Fuller 2009). Such diversity in CWRs can be used to achieve some breeding goals, including resistance to environmental stresses. Achievements in the improvement of crop yield and stress tolerance by CWRs have been discussed previously (Flint-Garcia et al. 2023).

Epigenetic diversity harbors gene regulatory mechanisms that occur beyond the DNA sequence changes of an individual. From epigenetic models, DNA methylators or demethylators, histone modifiers, and chromatin remodelers interact with noncoding RNAs to regulate crop adaptation to their surroundings (Sabzehzari and Naghavi 2019a, b; Lämke and Bäurle 2017). Recently, epigenetic breeding has been suggested for introducing epigenetic variants from CWRs into crops (Gahlaut et al. 2020). However, CWR epigenetic structures remain largely unclear, which prevents broad utilization of CWR accessions in epigenetic breeding (Varotto et al. 2022). In addition, inheritance of epidiversity and its consistency across multiple generations are other potential restrictions in employing epidiversity for epibreeding and enrichment of crop gene pools by epialleles (Lieberman-Lazarovich et al. 2022).

To drive crop improvement, methylations should be sustainably inherited to subsequent generations. Some differentially methylated regions are linked with causal shifts and are not inherited, while others are related to proximal structural diversity and are faithfully inherited along with the genetic polymorphisms (Springer and Schmitz 2017). The transgenerational stability of methylation inheritance is also another key to the use of epibreeding (Lieberman-Lazarovich et al. 2022). In addition, methylations should exhibit significant natural variety within CWRs and affect agronomic characteristics. To influence crop traits, natural diversity available in DNA methylations should change the transcriptome (Springer and Schmitz 2017). Recently, experimental evidence suggests that domestication-induced DNA methylations can affect the expression level of distal or nearby genes, possibly by changing chromatin loops and accessibility (Cao et al. 2024). Advances in plant biotechnology have provided a promising opportunity to uncover the function of epigenetic diversity in the control of agronomic traits. This knowledge can be exploited to alter the intended characteristics for crop improvement (Gahlaut et al. 2020). As a milestone, the development of epi-RIL in model crops has enabled the mapping of epigenetic quantitative trait loci (epiQTL) and the precise genetic exploration of epigenetic variety (Gahlaut et al. 2020).

Epigenetic mechanisms, like chromatin remodeling, RNAi, and DNA methylation, can silence or induce (or both) transposable elements during abiotic or biotic stressful conditions (Ito et al. 2011). These transposable elements, in turn, participate in the formation of eco-geographical epigenetic and genetic variation in various wild populations (such as emmer wheat) toward local adaptation (Ito et al. 2011).

Data from epigenetic research should be analyzed and interpreted by efficient bioinformatics algorithms and tools. The large number of articles, tools, and servers can confuse newcomers in bioinformatics and bioanalysis. Fortunately, multiple benchmark studies and reviews have been published in recent years to help navigate through available algorithms and tools, such as histone modification and remodeling by ChIP-seq (Nakato and Sakata 2021), DNA methylation sequencing (Omony et al. 2020), and RNA-seq (Conesa et al. 2016).

4.3 Pre-breeding Approaches

Utilization of genetic or epigenetic diversity hidden in CWRs is discouraged in plant breeding by the following three factors: (i) linkage drag, or the transfer of unwanted characteristics physically associated with the target traits; (ii) failure in the development of fertile offspring due to crossing barriers between CWRs and relevant crops; and (iii) weak agronomic performance of CWRs (Tirnaz et al. 2022). To overcome these limitations, the pre-breeding term was suggested in which breeders try to detect and transfer the intended alleles or traits from genetic material that cannot be employed directly in crop improvement (Renzi et al. 2022).

In pre-breeding, the approaches for incorporating CWR alleles or traits into crops are grouped as "cross first" or "choose first" (Renzi et al. 2022). In the former, elite cultivars are crossed with one or more CWR accessions, and the progenies are then assessed. This method is favorable when the intended trait cannot be evaluated directly in CWR donors. In "choose first," the selected accessions are expected to harbor a characteristic of interest. Based on the phenotypic or genotypic traits, CWRs are selected from a larger number of accessions, and these accessions are then used in a crossing design (Renzi et al. 2022).

4.4 FIGS: Germplasm Screening for Traits or Alleles of Interest

CWRs contribute annually to the global economy by approximately $185 billion (Tyack et al. 2020). Landraces and their wild ancestors are a valuable reservoir of tolerance to stressful conditions, which remain vital for the adaptation of crop varieties to changing climates. However, the identification of CWRs carrying traits or alleles of interest in a large ex situ or in situ collection is akin to "searching for a needle in a haystack." To identify CWR accessions harboring the intended genes or traits, it is necessary to efficiently narrow the search space (Kashyap et al. 2022). As a solution, the focused identification of germplasm strategy (FIGS) facilitates evaluation of a large germplasm collection. FIGS is based on the hypothesis that natural selection on genetic variation leads to adaptive evolution of crop characteristics (Warschefsky et al. 2014). This eco-geographic approach utilizes environmental information related to each location to evaluate the environment–trait relationship, which provides a priori facts on "best-bet germplasm" with a high possibility of harboring the trait of interest (Stenberg and Ortiz 2021). The performance of FIGS was best demonstrated via an allele-mining experiment in which researchers tried to identify the genes resistant to powdery mildew in a collection of 1320 bread wheat landraces (Bhullar et al. 2009).

Advances in computer science, such as machine learning (ML), have taken a step forward in ML-assisted FIGS (Danilevicz et al. 2022). To date, numerous FIGS-based ML algorithms have been developed and utilized to identify germplasms of interest. An example is a survey of geographical, phenotypic, or genotypic data of wheat accessions for stem rust from 2000 locations (1988–1994) that indicated a relationship between environmental variables and the geographical distribution of resistant accessions (Bari et al. 2012).

4.5 High-Throughput Phenotyping

To screen germplasms for traits or alleles of interest, the high volume of accessions in local and global gene banks should be characterized in a rapid, precise, and economic manner. Advances in artificial intelligence, cloud services, 5G networks, automation, machine learning, sensors, and digital imaging have reshaped phenotyping in crop research (Yang et al. 2020a, b). Now, international efforts are trying to overcome the next-generation phenotyping challenges. Phenotyping projects of the European community, such as EPPN2020, have uncovered the importance of connecting omics and phenotyping researchers to improve crop breeding (Van Tassel et al. 2022). The International Crop Phenome Project should explore the hidden value in CWRs.

4.6 High-Throughput Genotyping

Genomic selection is an innovative method that uses molecular markers to develop marker-based models for genetic assessment, which improves genetic gain of complex traits per cost and unit time. In this approach, the training population is used to calculate genomic breeding values (Brozynska et al. 2016). Multiple bottlenecks should be addressed when employing genomic selection to an elite × CWR cross. One bottleneck is describing the structure of the training subset in a manner that can maximize predicting precision in CWR and elite accessions. The second bottleneck is avoiding germplasm base narrowing that occurs when accessions with the largest breeding values are picked to cross in each generation (Brozynska et al. 2016). A solution is to harness optimum contribution selection (OCS), which improves the selection of exotic material for use as a parent while maintaining a logical level of variation in a whole population (Cowling et al. 2017). In an alternative approach, origin-specific genomic selection (OSGS), the genes of interest are first partitioned based on the donating parent (CWR or elite), marker impacts are figured, and the breeding value of the candidate line is then predicted by all markers (Yang et al. 2020a, b). Unlike OCS, which is focused on selecting the best CWR accessions for crossing, OSGS directly applies genomic selection into the elite × CWR cross toward improving the "introgression capability" of the CWR parent (Yang et al. 2020a, b). Both approaches described can assist transgressive breeding, which accelerates the development of crop cultivars with a predictable collection of economically important traits (Kochetov et al. 2021). Moreover, the utility of genomic selection in the discovery of valuable traits has been documented by the large genotyping of approximately 80,000 wheat accessions, including 3903 CWRs under the "Seeds of Discovery" project (Sansaloni et al. 2020).

4.7 Introgression Populations

Upon identification of CWRs carrying traits or alleles of interest, developing populations with introgression from CWRs can bring desired genes to a breeding pool. The most popular introgression populations include Multiparent Advanced Generation Intercross (MAGIC), Nested Association Mapping (NAM), and Chromosome Segment Substitution Lines (CSSL) populations (Ortiz et al. 2024). Ideally, the entire genome of the CWR donor is segment-wise distributed in the whole population of substitution lines, meaning each line can harbor a single chromosome segment of the CWR. CSSL has been used in the pre-breeding of several crops, such as soybean, maize, barley, wheat, and rice, to identify and transfer valuable alleles linked with drought or heat tolerance (Balakrishnan et al. 2019). NAM is a set of biparental populations with a common parent. In a NAM population, lines are classed into families with as many families as donor parents; therefore, the haplotype variation is restricted by the common parent, and the genome of NAM lines is a mosaic of only two parents (Pratap et al. 2021). The MAGIC population has a more complex crossing design, usually including 4, 8, or 16 founders. The MAGIC line genomes are a mosaic of all parents, and haplotype variation increases with the number of parents, although it is limited by the number of population lines (Pratap et al. 2021).

An introgression library offers some features over other mapping populations, as its genetic architecture efficiently removes background noise of single QTLs (Renzi et al. 2022). The accessibility of these libraries makes it possible to reveal QTLs with both minor and major effects, validate QTL effects, perform fine mapping and cloning, and elucidate the genetic interaction between environment and QTLs [environment × gene effect] between the elite genetic background and a QTL (epistasis) or among multiple QTLs (Renzi et al. 2022). The resources and time needed for the development of introgression libraries are the main factors limiting the broad utilization of these libraries in plant research.

4.8 Introgression Challenges and Their Solutions

Crop breeders face several challenges when developing introgression populations or libraries and introgressing a trait of interest from CWRs into elite varieties (Flint-Garcia et al. 2023). A main challenge is location-specific diversity, which can only be seen in a certain ecological condition. In other words, beneficial alleles may not be expressed when growing CWRs outside their natural distribution range (Wang et al. 2017a, b). This discourages further utility of CWRs as a resource of genetic diversity in an improvement program. Therefore, breeders need innovative methods to efficiently discover ecological location-specific alleles (Dempewolf et al. 2017). As another bottleneck, a strong linkage between desirable and undesirable alleles (the so-called linkage drag) adversely affects introgression output, such as yield penalty related to introgression of the *Pch1* resistance gene to eyespot from *Aegilops ventricosa* Tausch in wheat (Summers and Brown 2013). In these cases, marker-assisted backcrossing is efficient in the removal of linkage between targeted and untargeted genes or alleles (Brozynska et al. 2016).

Accurate selection for target loci can help reduce linkage drag by meiotic crossover-mediated recombination (Choi et al. 2013). However, meiotic crossovers are not evenly distributed along eukaryotic chromosomes (usually two per chromosome) (Fernandes et al. 2018). In *Arabidopsis thaliana* (L.) Heynh., 75% of meiotic crossovers occur in only 25% of the genome, with recombination cold spots avoiding crossover while recombination hotspots harbor crossover clusters (Choi et al. 2013). An approach for increasing the meiotic crossovers is to bypass pathways suppressing meiotic recombination in crops. These mechanisms rely on the activity of RecQ helicase 4 (RECQ4), Fidgetin-like protein 1 (FIGL1), and FANCM helicase, which induce the creation of nonmeiotic crossovers from unrepaired double-strand breaks in the genome (Tourrette et al. 2021). Meiotic recombination may be induced by altering the above pathways through a change in ploidy level, creation of knockout mutations, or both (Bohra et al. 2022). Although the problem of linkage drag may be addressed by enhancing the genome-wide recombination rate, the ability to target recombination to an intended genomic location would be a challenging step forward (Bohra et al. 2022). To compare the utility of the above pathways to enhance meiotic crossovers, the output was achieved only in cases where the recombination rate was enhanced in cold spots containing the locus of interest. In contrast, enhancing the recombination rate in hotspots harboring target loci appeared detrimental (Tourrette et al. 2021).

4.9 AB-QTL: Concurrent Detection and Introgression of Traits or Alleles of Interest

AB-QTL allows concurrent detection and introgression of intended genes from CWRs to an elite genetic background. This approach attempts to select early against pre-domestication characteristics along with negative epistasis that occurs between elite background and donor QTL (Tanksley and Nelson 1996). AB-QTL can accelerate the recovery of useful QTL near-isogenic lines (NIL) by trying to bridge the gap between uncovering and transferring of intended QTL. NILs can then be used directly as parents in crop improvement or serve as a source for isolating the genes of interest (Ho et al. 2002). This approach has been used to exploit CWR diversity to improve agronomic traits in various crops, such as cotton (Wang et al. 2017a, b), oat (Herrmann et al. 2014), barley (Pillen et al. 2003), maize (Ho et al. 2002), wheat (Buerstmayr et al. 2011), and rice (Lu et al. 2014). The deployment of AB-QTL to crop improvement is limited by the following factors: reduced recombination rate, bias toward recurrent parent alleles, the use of exotic QTLs without sufficient validation of their effects, and the creation of wide linkage disequilibrium. These bottlenecks can be overcome by a promising design that merges the nested association mapping with backcrossing (AB-NAM), in which a common elite parent is crossed to diverse CWR accessions (Nice et al. 2016).

4.10 Pangenomics: An Avenue Toward Detection of Presence-Absence and Copy-Number Variation

Thanks to advanced sequencing technologies, reference genomes have provided an opportunity to perform genomic analyses, such as characterization and cloning of intended genes, uncovering of trait-gene associations (GWAS), and genome-wide mapping of genetic variants (Flint-Garcia et al. 2023). However, the increasing volume of high-quality data from CWR genome sequencing has shown that reference genome-mediated approaches are inadequate to reveal structural variations, such as copy-number variation and presence-absence variation that can distinguish accessions from each other (Bohra et al. 2022).

Resequencing of CWR genomes has enabled the development of pangenomes (i.e., genomic diversity of a species) in various crops (Bohra et al. 2022). Evidence from pangenome research supports a key role in structural variation in domestication and crop improvement (Varshney et al. 2019) (Table 4.1). Therefore, pangenomics represents new opportunities to reveal genetic variation that has only scarcely been brought into the gene pool, selected against, or lost during the domestication process (Schreiber et al. 2024). From pangenome studies, some dispensable genes are linked with stressful responses, making them interesting since crop breeders explore routes to improve climate resilience in novel cultivars (Zhao et al. 2018).

In addition to the structural variants, pangenomics provides an efficient means to explore diverse populations to discover single-nucleotide polymorphisms (SNPs) (Ortiz et al. 2024). To reveal the "contribution of CWR variation" by the number of SNPs per accession, researchers observed that sorghum inbred lines contain one SNP every 1543 bp, landraces one per 1282 bp, and wild species one per 763 bp

Table 4.1 Typical pangenome research in crops and their CWRs

Crops	Number of accessions	Dispensable genes	Core genes	References
Wheat	18 diverse accessions and Chinese Spring	26,711	89,795	Montenegro et al. (2017)
Rice	66 of *Oryza rufipogon* and *O. sativa*	16,208	26,372	Zhao et al. (2018)
Maize	26	70,981	32,052	Hufford et al. (2021)
Barley	20 (wild, landrace, cultivars)	14,948	25,228	Jayakodi et al. (2020)
Sorghum	354 genetically diverse	18,898	16,821	Ruperao et al. (2021)
Soybean	176	33,312	18,738	Liu et al. (2020)
Cotton	1961	25,093	65,595	Li et al. (2021)
Sunflower	287 lines, 17 landraces, 189 wild	16,525	61,205	Hübner et al. (2019)
Tomato	725	1266	3606	Gao et al. (2019)
Cabbage	9	11,484	49,895	Golicz et al. (2016)
Pepper	383	37,424	51,757	Ou et al. (2018)

(Mace et al. 2021). This is evidence that through breeding or domestication, genetic diversity has declined, emphasizing that CWRs can be used to broaden crop variation (Bailey-Serres et al. 2019). Moreover, pangenomics allows more accurate prediction of genetic regions linked with phenotypic traits, such as double-stranded break or repair, recombination, and transposable elements (Song et al. 2020).

4.11 Speed Cloning of CWR Genes

Tolerance to pests and diseases accounts for nearly 80% of features introduced from CWRs into crop cultivars (Bohra et al. 2022). Access to a variation of resistance (R) genes in CWR accessions opened novel avenues to pyramid or stack various genes in a single variety to improve the permanence of the tolerance and to induce broad-spectrum resistance to a range of pathogens (Chen et al. 2024). The most plentiful group of R genes in crops is the family NBS-LRR, which has been detected and cloned by innovative approaches in recent years (Chen et al. 2024). An important approach is resistance gene enrichment sequencing (RenSeq), which presents an economical alternative to whole-genome sequencing without positional cloning. This approach includes target gene enrichment by the bait library of NBS-LRR, followed by enriched sequences de novo assembly. The assembled sequence is then utilized as a reference to discover each variant between the resistant and the sensitive accessions (Jupe et al. 2013). Advances in long-read sequencing can overcome the limitations of RenSeq, where copy-number variation and high similarity of sequences between alleles and paralogs make it difficult to align short-read sequences precisely (Jupe et al. 2013). An important breakthrough occurred when researchers combined SMRT sequencing with RenSeq and successfully cloned the gene conferring tolerance to late blight disease from a diploid wild potato, *Solanum americanum* Mill. (Witek et al. 2016).

4.12 Reclaiming Lost Genetic Diversity: De Novo Domestication

Domestication and breeding endeavors created today's crops that have high yields in the current climate. However, some useful traits have been lost during these processes, such as those contributing to yield consistency (Gasparini et al. 2021). De novo domestication provides a way to directly engineer domestication-associated genes or traits in CWR accessions instead of transferring their alleles into elite cultivars by backcrossing (Curtin et al. 2022). The process includes two steps, namely the determination of the target trait and then the implementation of genome editing, cis-genesis, or both (Wu et al. 2021). To identify the traits for de novo domestication of wheat ancestors, researchers evaluated adaptability to European climate in six modern cultivars, 38 landraces, and 111 wild accessions from different wheat species (*Triticum aestivum L.*, *T. araraticum* Jakubz, *T. monococcum* L., *T. monococcum* subsp. *aegilopoides* (Link) Thell. (Syn.: *T. boeoticum* Boiss.), *T. turgidum* subsp. *dicoccoides* (Asch. & Graebn.) Thell. (Syn.: *T. dicoccoides* (Ssch. & Graebn.) Schweing.), *T. turgidum* subsp. *durum* (Desf.) Husn. (Syn.: *T. durum* Desf.), and *T. urartu* Thumanjan ex Ganilyan). They found that the thermal time to flowering and heading has high adaptability in wheat wild relatives. The values of thousand kernel weight, number of tillers, and grain weight per plant were lower in the CWRs than in the landraces, suggesting that these agronomic characteristics can be addressed for de novo domestication-based breeding (Zeibig et al. 2024).

The idea of de novo domestication can be implemented by different approaches, including genome editing and cis-genesis. These two approaches are of interest in the case of strong hybridization barriers in the secondary and tertiary gene pools. Cis-genesis genetically transforms a crop with genes from the CWR without transferring selectable markers or reporters from other species, thus preventing linkage drag (Jacobsen and Schouten 2007). This approach was successfully used in several crops, such as rice, potato, bread wheat, barley, and durum wheat (EFSA Panel on Genetically Modified Organisms 2022). The increasing popularity of genome editing via CRISPR/Cas9 is due to its flexibility, precision, and efficiency for the targeted introduction of a wide range of genetic and epigenetic changes (Sabzehzari et al. 2020a, b). Utility for gene editing-mediated de novo domestication was documented by creating agronomically important traits in tomato (Li et al. 2018) and rice (Yu et al. 2021) wild relatives.

4.13 Conservation of CWRs

CWR accessions can be conserved against climate change, human activities, and natural disasters by ex situ or in situ conservation. In situ conservation tries to protect CWRs in their native habitats, whereas ex situ ensures CWRs are available for education, farming, and research activities that eventually support reintroduction efforts toward preventing genetic material from becoming extinct (Engels and Thormann 2020).

The utility of in situ is that CWRs are dynamic genetically and retain the ability to evolve in response to artificial and natural selection, leading to improved acclimation to the environment in which they are grown. However, in situ collection is sensitive to some factors, such as habitat destruction (Engels and Thormann 2020).

There is increasing interest in ex situ management of CWR accessions in light of naturally occurring risks, including climate change, mining, pollution, invasive species, urbanization, competition from biofuel crops, and land-use change (Castañeda-Álvarez et al. 2016). Ex situ collections can be kept in gene banks, botanical gardens, and labs. CWR accessions account for approximately 6% of gene bank collections globally (Tyack et al. 2020). A major challenge is to preserve and propagate CWRs outside their zone of adaptation, which results in the loss of genetic diversity over time. Moreover, ex situ conservation of CWRs is often less successful and more expensive than for landraces, in part because of challenges related to viability, dormancy, seed shattering, pollination requirements, mating habits, and flowering time

(Engels and Thormann 2020). Taken together, this suggests that ex situ management procedures should be improved.

To overcome the challenges that conservation approaches are facing, genome editing offers the possibility to remove deleterious alleles that reduce fitness and introduce new alleles that confer resistance to climate change (Johnsson et al. 2019; Varshney et al. 2018). This is helpful for those species that are seriously threatened by a general lack of adaptation capacity, stressful environmental conditions, or both. Despite its utility, genome editing suffers from some risks that must be addressed carefully before using it for in situ or ex situ conservation (Wambugu and Henry 2022).

4.14 Access and Benefit-Sharing

Sustainable farming needs accessibility to CWRs and benefit-sharing derived from this availability. The principle of "accessibility and benefit-sharing" under public international law fairly distributes benefits from CWRs or landraces between the provider countries of those resources and their users (i.e., companies or universities) (Bohra et al. 2022). These principles are codified in The Nagoya Protocol on Access and Benefit-sharing 2011, The International Treaty on Plant Genetic Resources for Food and Agriculture 2001, and the Convention on Biological Diversity 1992.

4.15 Conclusion

Crop breeders may be hesitant to use CWR accessions in improvement programs due to problems including poor agronomic performance, linkage drag, and cross-ability barriers. Progress in biotechnology has provided novel opportunities for overcoming several of these bottlenecks. Innovative approaches to use and conserve CWRs include pre-breeding, epigenetic breeding, FIGS, high-throughput phenotyping and genotyping, introgression populations, AB-QTL, epi-RIL, GWAS, pangenomics, RenSeq, de novo domestication, ex situ or in situ conservation, and their accessibility and benefit sharing.

References

Alam O, Purugganan MD (2024) Domestication and the evolution of crops: variable syndromes, complex genetic architectures, and ecological entanglements. Plant Cell 36(5):1227–1241

Asseng S, Ewert F, Martre P, Rötter RP, Lobell DB, Cammarano D, Kimball BA, Ottman MJ, Wall GW, White JW, Reynolds MP (2015) Rising temperatures reduce global wheat production. Nat Clim Chang 5(2):143–147

Bailey-Serres J, Parker JE, Ainsworth EA, Oldroyd GE, Schroeder JI (2019) Genetic strategies for improving crop yields. Nature 575(7781):109–118

Balakrishnan D, Surapaneni M, Mesapogu S, Neelamraju S (2019) Development and use of chromosome segment substitution lines as a genetic resource for crop improvement. Theor Appl Genet 132:1–25

Bari A, Street K, Mackay M, Endresen DTF, De Pauw E, Amri A (2012) Focused identification of germplasm strategy (FIGS) detects wheat stem rust resistance linked to environmental variables. Genet Resour Crop Evol 59:1465–1481

Bassi FM, Sanchez-Garcia M, Ortiz R (2024) What plant breeding may (and may not) look like in 2050? Plant Genome 17(1):e20368

Bhullar NK, Street K, Mackay M, Yahiaoui N, Keller B (2009) Unlocking wheat genetic resources for the molecular identification of previously undescribed functional alleles at the Pm3 resistance locus. Proc Natl Acad Sci 106(23):9519–9524

Bohra A, Kilian B, Sivasankar S, Caccamo M, Mba C, McCouch SR, Varshney RK (2022) Reap the crop wild relatives for breeding future crops. Trends Biotechnol 40(4):412–431

Brozynska M, Furtado A, Henry RJ (2016) Genomics of crop wild relatives: expanding the gene pool for crop improvement. Plant Biotechnol J 14(4):1070–1085

Buerstmayr M, Lemmens M, Steiner B, Buerstmayr H (2011) Advanced backcross QTL mapping of resistance to Fusarium head blight and plant morphological traits in a Triticum macha× T. aestivum population. Theor Appl Genet 123:293–306

Cao S, Sawettalake N, Li P, Fan S, Shen L (2024) DNA methylation variations underlie lettuce domestication and divergence. Genome Biol 25(1):158

Castañeda-Álvarez NP, Khoury CK, Achicanoy HA, Bernau V, Dempewolf H, Eastwood RJ, Guarino L, Harker RH, Jarvis A, Maxted N, Müller JV (2016) Global conservation priorities for crop wild relatives. Nat Plants 2(4):1–6

Chen R, Gajendiran K, Wulff BB (2024) R we there yet? Advances in cloning resistance genes for engineering immunity in crop plants. Curr Opin Plant Biol 77:102489

Choi K, Zhao X, Kelly KA, Venn O, Higgins JD, Yelina NE, Hardcastle TJ, Ziolkowski PA, Copenhaver GP, Franklin FCH, McVean G (2013) Arabidopsis meiotic crossover hot spots overlap with H2A. Z nucleosomes at gene promoters. Nat Genet 45(11):1327–1336

Conesa A, Madrigal P, Tarazona S, Gomez-Cabrero D, Cervera A, McPherson A, Szcześniak MW, Gaffney DJ, Elo LL, Zhang X, Mortazavi A (2016) A survey of best practices for RNA-seq data analysis. Genome Biol 17:1–19

Cowling WA, Li L, Siddique KH, Henryon M, Berg P, Banks RG, Kinghorn BP (2017) Evolving gene banks: improving diverse populations of crop and exotic germplasm with optimal contribution selection. J Exp Bot 68(8):1927–1939

Curtin S, Qi Y, Peres LE, Fernie AR, Zsögön A (2022) Pathways to de novo domestication of crop wild relatives. Plant Physiol 188(4):1746–1756

Danilevicz MF, Gill M, Anderson R, Batley J, Bennamoun M, Bayer PE, Edwards D (2022) Plant genotype to phenotype prediction using machine learning. Front Genet 13:822173

Dempewolf H, Baute G, Anderson J, Kilian B, Smith C, Guarino L (2017) Past and future use of wild relatives in crop breeding. Crop Sci 57(3):1070–1082

EFSA Panel on Genetically Modified Organisms (GMO), Mullins E, Bresson JL, Dalmay T, Dewhurst IC, Epstein MM, Firbank LG, Guerche P, Hejatko J, Moreno FJ, Naegeli H (2022) Updated scientific opinion on plants developed through cisgenesis and intragenesis. EFSA J 20(10):e07621

Engels JM, Thormann I (2020) Main challenges and actions needed to improve conservation and sustainable use of our crop wild relatives. Plants 9(8):968

Fernandes JB, Séguéla-Arnaud M, Larchevêque C, Lloyd AH, Mercier R (2018) Unleashing meiotic crossovers in hybrid plants. Proc Natl Acad Sci 115(10):2431–2436

Flint-Garcia S, Feldmann MJ, Dempewolf H, Morrell PL, Ross-Ibarra J (2023) Diamonds in the not-so-rough: wild relative diversity hidden in crop genomes. PLoS Biol 21(7):e3002235

Gahlaut V, Zinta G, Jaiswal V, Kumar S (2020) Quantitative epigenetics: a new avenue for crop improvement. Epigenomes 4(4):25

Gao L, Gonda I, Sun H, Ma Q, Bao K, Tieman DM, Burzynski-Chang EA, Fish TL, Stromberg KA, Sacks GL, Thannhauser TW (2019) The tomato pan-genome uncovers new genes and a rare allele regulating fruit flavor. Nat Genet 51(6):1044–1051

Gasparini K, dos Reis Moreira J, Peres LEP, Zsögön A (2021) De novo domestication of wild species to create crops with increased resilience and nutritional value. Curr Opin Plant Biol 60:102006

Golicz AA, Bayer PE, Barker GC, Edger PP, Kim H, Martinez PA, Chan CKK, Severn-Ellis A, McCombie WR, Parkin IA, Paterson AH (2016) The pangenome of an agronomically important crop plant *Brassica oleracea*. Nat Commun 7(1):13390

Herrmann MH, Yu J, Beuch S, Weber WE (2014) Quantitative trait loci for quality and agronomic traits in two advanced backcross populations in oat (*Avena sativa* L.). Plant Breed 133(5):588–601

Ho J, McCouch S, Smith M (2002) Improvement of hybrid yield by advanced backcross QTL analysis in elite maize. Theor Appl Genet 105:440–448

Hübner S, Bercovich N, Todesco M, Mandel JR, Odenheimer J, Ziegler E, Lee JS, Baute GJ, Owens GL, Grassa CJ, Ebert DP (2019) Sunflower pan-genome analysis shows that hybridization altered gene content and disease resistance. Nat Plants 5(1):54–62

Hufford MB, Seetharam AS, Woodhouse MR, Chougule KM, Ou S, Liu J, Ricci WA, Guo T, Olson A, Qiu Y, Della Coletta R (2021) De novo assembly, annotation, and comparative analysis of 26 diverse maize genomes. Science 373(6555):655–662

Ito H, Gaubert H, Bucher E, Mirouze M, Vaillant I, Paszkowski J (2011) An siRNA pathway prevents transgenerational retrotransposition in plants subjected to stress. Nature 472(7341):115–119

Jacobsen E, Schouten HJ (2007) Cisgenesis strongly improves introgression breeding and induced translocation breeding of plants. Trends Biotechnol 25(5):219–223

Jayakodi M, Padmarasu S, Haberer G, Bonthala VS, Gundlach H, Monat C, Lux T, Kamal N, Lang D, Himmelbach A, Ens J (2020) The barley pan-genome reveals the hidden legacy of mutation breeding. Nature 588(7837):284–289

Johnsson M, Gaynor RC, Jenko J, Gorjanc G, De Koning DJ, Hickey JM (2019) Removal of alleles by genome editing (RAGE) against deleterious load. Genet Sel Evol 51:1–18

Jupe F, Witek K, Verweij W, Śliwka J, Pritchard L, Etherington GJ, Maclean D, Cock PJ, Leggett RM, Bryan GJ, Cardle L (2013) Resistance gene enrichment sequencing (RenSeq) enables reannotation of the NB-LRR gene family from sequenced plant genomes and rapid mapping of resistance loci in segregating populations. Plant J 76(3):530–544

Kashyap A, Garg P, Tanwar K, Sharma J, Gupta NC, Ha PTT, Bhattacharya RC, Mason AS, Rao M (2022) Strategies for utilization of crop wild relatives in plant breeding programs. Theor Appl Genet 135(12):4151–4167

Khoury CK, Brush S, Costich DE, Curry HA, De Haan S, Engels JM, Guarino L, Hoban S, Mercer KL, Miller AJ, Nabhan GP (2022) Crop genetic erosion: understanding and responding to loss of crop diversity. New Phytol 233(1):84–118

Kochetov AA, Mirskaya GV, Sinyavina NG, Egorova KV (2021) Transgressive breeding: a methodology for accelerated creation of new forms of plants with a predictable complex of economically valuable traits. Russ Agric Sci 47(Suppl 1):S40–S50

Lämke J, Bäurle I (2017) Epigenetic and chromatin-based mechanisms in environmental stress adaptation and stress memory in plants. Genome Biol 18:1–11

Li T, Yang X, Yu Y, Si X, Zhai X, Zhang H, Dong W, Gao C, Xu C (2018) Domestication of wild tomato is accelerated by genome editing. Nat Biotechnol 36(12):1160–1163

Li J, Yuan D, Wang P, Wang Q, Sun M, Liu Z, Si H, Xu Z, Ma Y, Zhang B, Pei L (2021) Cotton pan-genome retrieves the lost sequences and genes during domestication and selection. Genome Biol 22:1–26

Lieberman-Lazarovich M, Kaiserli E, Bucher E, Mladenov V (2022) Natural and induced epigenetic variation for crop improvement. Curr Opin Plant Biol 70:102297

Liu Y, Du H, Li P, Shen Y, Peng H, Liu S, Zhou GA, Zhang H, Liu Z, Shi M, Huang X, Li Y, Zhang M, Wang Z, Zhu B, Han B, Liang C, Tian Z (2020) Pan-genome of wild and cultivated soybeans. Cell 182(1):162-176.e13. https://doi.org/10.1016/j.cell.2020.05.023

Lu CHAI, Zhang J, Pan XB, Zhang F, Zheng TQ, Zhao XQ, Wang WS, Jauhar A, Xu JL, Li ZK (2014) Advanced backcross QTL analysis for the whole plant growth duration salt tolerance in rice (*Oryza sativa* L.). J Integr Agric 13(8):1609–1620

Mace ES, Cruickshank AW, Tao Y, Hunt CH, Jordan DR (2021) A global resource for exploring and exploiting genetic variation in sorghum crop wild relatives. Crop Sci 61(1):150–162

Montenegro JD, Golicz AA, Bayer PE, Hurgobin B, Lee H, Chan CKK, Visendi P, Lai K, Doležel J, Batley J, Edwards D (2017) The pangenome of hexaploid bread wheat. Plant J 90(5):1007–1013

Nakato R, Sakata T (2021) Methods for ChIP-seq analysis: a practical workflow and advanced applications. Methods 187:44–53

Nice LM, Steffenson BJ, Brown-Guedira GL, Akhunov ED, Liu C, Kono TJ, Morrell PL, Blake TK, Horsley RD, Smith KP, Muehlbauer GJ (2016) Development and genetic characterization of an advanced backcross-nested association mapping (AB-NAM) population of wild × cultivated barley. Genetics 203(3):1453–1467

Omony J, Nussbaumer T, Gutzat R (2020) DNA methylation analysis in plants: review of computational tools and future perspectives. Brief Bioinform 21(3):906–918

Ortiz R, Bassi FM, Rao M, Rubiales D (2024) Trends and perspectives for the use of crop wild relatives in crop breeding. Front Plant Sci 15:1424160

Ou L, Li D, Lv J, Chen W, Zhang Z, Li X, Yang B, Zhou S, Yang S, Li W, Gao H (2018) Pangenome of cultivated pepper (*Capsicum*) and its use in gene presence–absence variation analyses. New Phytol 220(2):360–363

Pillen K, Zacharias A, Léon J (2003) Advanced backcross QTL analysis in barley (*Hordeum vulgare* L.). Theor Appl Genet 107:340–352

Pratap A, Das A, Kumar S, Gupta S (2021) Current perspectives on introgression breeding in food legumes. Front Plant Sci 11:589189

Purugganan MD, Fuller DQ (2009) The nature of selection during plant domestication. Nature 457:843

Renzi JP, Coyne CJ, Berger J, von Wettberg E, Nelson M, Ureta S, Hernández F, Smýkal P, Brus J (2022) How could the use of crop wild relatives in breeding increase the adaptation of crops to marginal environments? Front Plant Sci 13:886162

Ruperao P, Thirunavukkarasu N, Gandham P, Selvanayagam S, Govindaraj M, Nebie B, Manyasa E, Gupta R, Das RR, Odeny DA, Gandhi H (2021) Sorghum pan-genome explores the functional utility for genomic-assisted breeding to accelerate the genetic gain. Front Plant Sci 12:666342

Sabzehzari M, Naghavi MR (2019a) Phyto-miRNAs-based regulation of metabolites biosynthesis in medicinal plants. Gene 682:13–24

Sabzehzari M, Naghavi MR (2019b) Phyto-miRNA: a molecule with beneficial abilities for plant biotechnology. Gene 683:28–34

Sabzehzari M, Zeinali M, Naghavi MR (2020a) CRISPR-based metabolic editing: next-generation metabolic engineering in plants. Gene 759:144993

Sabzehzari M, Zeinali M, Naghavi MR (2020b) Alternative sources and metabolic engineering of Taxol: advances and future perspectives. Biotechnol Adv 43:107569

Sansaloni C, Franco J, Santos B, Percival-Alwyn L, Singh S, Petroli C, Campos J, Dreher K, Payne T, Marshall D, Kilian B (2020) Diversity analysis of 80,000 wheat accessions reveals consequences and opportunities of selection footprints. Nat Commun 11(1):4572

Schreiber M, Jayakodi M, Stein N, Mascher M (2024) Plant pangenomes for crop improvement, biodiversity and evolution. Nat Rev Genet:1–15

Song JM, Guan Z, Hu J, Guo C, Yang Z, Wang S, Liu D, Wang B, Lu S, Zhou R, Xie WZ (2020) Eight high-quality genomes reveal pan-genome architecture and ecotype differentiation of Brassica napus. Nat Plants 6(1):34–45

Springer NM, Schmitz RJ (2017) Exploiting induced and natural epigenetic variation for crop improvement. Nat Rev Genet 18(9):563–575

Stenberg JA, Ortiz R (2021) Focused identification of germplasm strategy (FIGS): polishing a rough diamond. Curr Opin Insect Sci 45:1–6

Summers RW, Brown JKM (2013) Constraints on breeding for disease resistance in commercially competitive wheat cultivars. Plant Pathol 62:115–121

Tanksley SD, Nelson JC (1996) Advanced backcross QTL analysis: a method for the simultaneous discovery and transfer of valuable QTLs from unadapted germplasm into elite breeding lines. Theor Appl Genet 92:191–203

Tirnaz S, Zandberg J, Thomas WJ, Marsh J, Edwards D, Batley J (2022) Application of crop wild relatives in modern breeding: an overview of resources, experimental and computational methodologies. Front Plant Sci 13:1008904

Tourrette E, Falque M, Martin OC (2021) Enhancing backcross programs through increased recombination. Genet Sel Evol 53:1–13

Tyack N, Dempewolf H, Khoury CK (2020) The potential of payment for ecosystem services for crop wild relative conservation. Plants 9(10):1305

Van Tassel DL, DeHaan LR, Diaz-Garcia L, Hershberger J, Rubin MJ, Schlautman B, Turner K, Miller AJ (2022) Re-imagining crop domestication in the era of high throughput phenomics. Curr Opin Plant Biol 65:102150

Varotto S, Krugman T, Aiese Cigliano R, Kashkush K, Kondić-Špika A, Aravanopoulos FA, Pradillo M, Consiglio F, Aversano R, Pecinka A, Miladinović D (2022) Exploitation of epigenetic variation of crop wild relatives for crop improvement and agrobiodiversity preservation. Theor Appl Genet 135(11):3987–4003

Varshney RK, Singh VK, Kumar A, Powell W, Sorrells ME (2018) Can genomics deliver climate-change ready crops? Curr Opin Plant Biol 45:205–211

Varshney RK, Thudi M, Roorkiwal M, He W, Upadhyaya HD, Yang W, Bajaj P, Cubry P, Rathore A, Jian J, Doddamani D (2019) Resequencing of 429 chickpea accessions from 45 countries provides insights into genome diversity, domestication and agronomic traits. Nat Genet 51(5):857–864

Wambugu PW, Henry R (2022) Supporting in situ conservation of the genetic diversity of crop wild relatives using genomic technologies. Mol Ecol 31(8):2207–2222

Wang B, Draye X, Zhuang Z, Zhang Z, Liu M, Lubbers EL, Jones D, May OL, Paterson AH, Chee PW (2017a) QTL analysis of cotton fiber length in advanced backcross populations derived from a cross between Gossypium hirsutum and G. mustelinum. Theor Appl Genet 130:1297–1308

Wang C, Hu S, Gardner C, Lübberstedt T (2017b) Emerging avenues for utilization of exotic germplasm. Trends Plant Sci 22(7):624–637

Warschefsky E, Penmetsa RV, Cook DR, Von Wettberg EJ (2014) Back to the wilds: tapping evolutionary adaptations for resilient crops through systematic hybridization with crop wild relatives. Am J Bot 101(10):1791–1800

Witek K, Jupe F, Witek AI, Baker D, Clark MD, Jones JD (2016) Accelerated cloning of a potato late blight–resistance gene using RenSeq and SMRT sequencing. Nat Biotechnol 34(6):656–660

Wu D, Lao S, Fan L (2021) De-domestication: an extension of crop evolution. Trends Plant Sci 26(6):560–574

Yang CJ, Sharma R, Gorjanc G, Hearne S, Powell W, Mackay I (2020a) Origin specific genomic selection: a simple process to optimize the favorable contribution of parents to progeny. G3: Genes, Genomes, Genetics 10(7):2445–2455

Yang W, Feng H, Zhang X, Zhang J, Doonan JH, Batchelor WD, Xiong L, Yan J (2020b) Crop phenomics and high-throughput phenotyping: past decades, current challenges, and future perspectives. Mol Plant 13(2):187–214

Yu H, Lin T, Meng X, Du H, Zhang J, Liu G, Chen M, Jing Y, Kou L, Li X, Gao Q (2021) A route to de novo domestication of wild allotetraploid rice. Cell 184(5):1156–1170

Zeibig F, Kilian B, Özkan H, Pantha S, Frei M (2024) Phenotyping and identification of target traits for de novo domestication of wheat wild relatives. Food Energy Secur 13(1):e497

Zhang F, Batley J (2020) Exploring the application of wild species for crop improvement in a changing climate. Curr Opin Plant Biol 56:218–222

Zhao Q, Feng Q, Lu H, Li Y, Wang A, Tian Q, Zhan Q, Lu Y, Zhang L, Huang T, Wang Y (2018) Pan-genome analysis highlights the extent of genomic variation in cultivated and wild rice. Nat Genet 50(2):278–284

The Power of *De Novo Domestication* of Wild Solanaceae for a Changing Climate

5

Karla Gasparini, Lucía Rodríguez Gargantini,
Carlos Federico Marfil, and Agustin Zsögön

Abstract

Changing climate is putting a severe strain on crop production. New breeding methods are urgently needed to cope with abiotic stresses such as drought, high temperature, and soil salinity. Crop wild relatives (CWRs) have long been used as sources of genetic variation to improve stress resistance. However, the multigenic nature of resistance traits makes it very difficult to use conventional breeding (*i.e.*, wide crosses followed by selection). The advent of gene editing (GE) techniques now makes it possible to manipulate CWRs to improve their agronomic value, through a platform known as de novo domestication. The Solanaceae family has been intensely studied due to their profusion of valuable CWRs with cultivation potential. Here, we summarize and discuss the prospects for de novo domestication in CWRs via GE for two major solanaceous crops: tomato (*Solanum lycopersicum*) and potato (*S. tuberosum*).

Keywords

Crop breeding · de novo domestication · Gene editing · Ideotype · Solanaceae

The original version of the chapter has been revised. A correction to this chapter can be found at https://doi.org/10.1007/978-981-95-0583-8_24

K. Gasparini · A. Zsögön (✉)
National Institute of Science and Technology on Plant Physiology Under Stress Conditions, Departamento de Biologia Vegetal, Universidade Federal de Viçosa, Viçosa, MG, Brazil
e-mail: agustin.zsogon@ufv.br

L. Rodríguez Gargantini · C. F. Marfil
Consejo Nacional de Investigaciones Científicas y Técnicas-CONICET. Estación Experimental Agropecuaria—Mendoza—Instituto Nacional de Tecnología Agropecuaria-INTA, Mendoza, Argentina

P. M. Priyadarshan, R. Ortiz (eds.), *Plant Breeding 2050*,
https://doi.org/10.1007/978-981-95-0583-8_5

5.1 Introduction

Gene editing (GE) is rapidly emerging as a supporting tool to traditional breeding techniques, particularly in crops where abundant genetic and genomic information is available (Gao 2021). The key advantage of this approach is that it allows the precise molecular design of plant traits in a predictable and efficient manner. Unlike conventional breeding, which often relies on empirical methods and the chance generation of specific allelic combinations, GE requires a detailed understanding of plant physiology and molecular genetics. With this knowledge, ideal plant types or "ideotypes" can be modeled and then targeted modifications can be made to specific genes to rapidly create agronomically valuable traits (Čermák et al. 2021; Ahn et al. 2023).

The success of GE across molecular biology fields has been driven by the customizability and efficiency of Clustered Regularly Interspaced Short Palindromic Repeats (CRISPR/Cas9), which stands out for its simplicity and efficiency (Doudna and Charpentier 2014). The ability to design a guide RNA to target virtually any DNA sequence has made it a versatile tool across many organisms and cell types, enabling researchers to quickly and accurately manipulate gene sequences (Chen et al. 2019). The main use of CRISPR/Cas9 thus far has been the generation of precise double-strand breaks (DSBs) in DNA (Shen et al. 2017; Huang et al. 2022). Once DSBs are introduced, they are repaired either through nonhomologous end joining (NHEJ) or homology-directed repair (HDR), leading to gene inactivation via small insertions or deletions (InDels) or the precise introduction of desired genetic changes, respectively (Xue and Greene 2021).

The strategy of targeted deletion compares favorably to crop breeding methods such as crossing and selection, chemical or physical mutagenesis and traditional transgenesis mainly due to lower time and labor requirements (Fig. 5.1). Conventional plant breeding methods, such as controlled mating and selection, have been widely used for centuries. These strategies exploit natural genetic variation to combine desirable traits, such as yield or disease resistance, into new cultivars. While effective and low-cost, the process is time-consuming, limited by available genetic diversity, and in some cases does not achieve specific desired traits (Acquaah 2012). When the selection of natural variability reaches a limit, induced mutagenesis can be an alternative in breeding programs. Induced mutagenesis introduces genetic variation through chemical or physical agents, such as radiation or mutagenic chemicals, causing random mutations throughout the genome, some of which can result in useful traits. Although several crop cultivars have been successfully obtained using this approach (Ahloowalia et al. 2004), the stochastic nature of mutations requires extensive screening, making it labor-intensive and unpredictable.

In the late twentieth century, transgenic technology introduced new possibilities in plant breeding. The insertion of foreign genes from unrelated species directly into a plant's genome allowed the introduction of new traits like pest resistance (e.g., Bt crops) or herbicide tolerance (James 2015). This technology allows more targeted trait modifications than conventional breeding, but regulatory norms and public acceptance are still significant challenges (Kausch et al. 2019). In the early years of

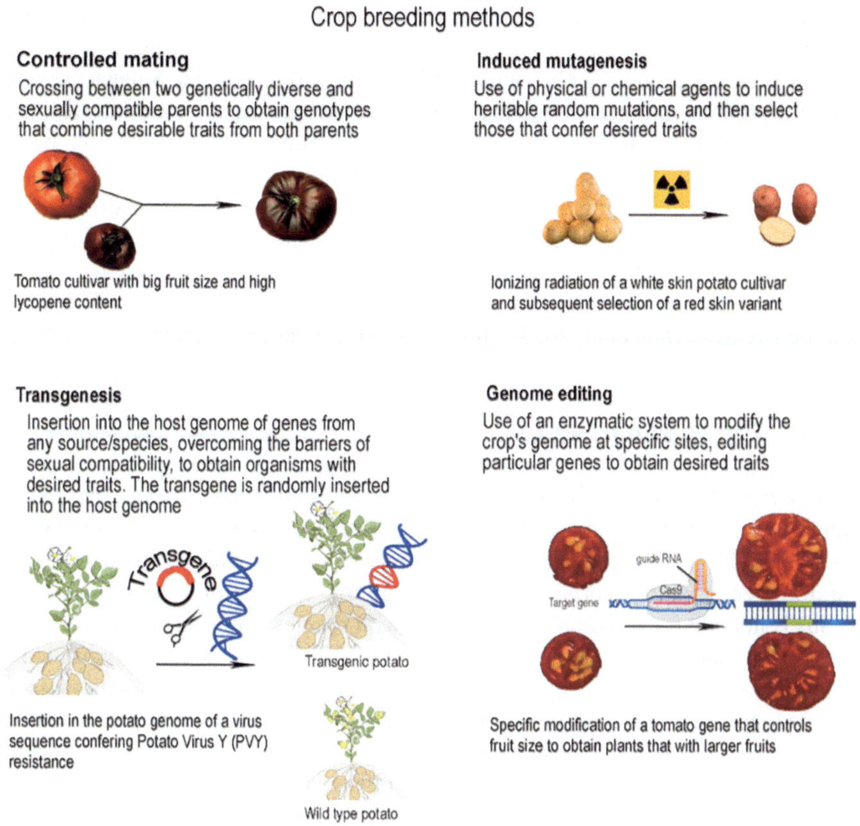

Fig. 5.1 A schematic summary of the most common plant breeding methods

this century, the development of sequence-specific nucleases opened new avenues for crop improvement. GE technologies based on Zinc Finger Nucleases (ZFN), Transcription Activation-Like Effector Nucleases (TALENs) or CRISPR/Cas9, allow precise modifications of host genome target sequences to improve specific traits (Kausch et al. 2019). Because of its versatility, low cost, and higher time-efficiency, CRISPR/Cas9 GE represents a breakthrough in biology in general, and in plant breeding in particular (Kausch et al. 2019; Gao 2021) (Fig. 5.1). One of the plant families where GE has been deployed more intensively is the Solanaceae, which we discuss next.

5.2 The Solanaceae: Biosystematic Background

The Solanaceae, or nightshade family, is a diverse angiosperm group consisting of over 2700 species, including several important crop species that have also contributed as models to classical and molecular genetics and genomics studies, such as tomato (*Solanum lycopersicum*), tobacco (*Nicotiana tabacum*), petunia (*Petunia hybrida*), potato (*S. tuberosum*), pepper (*Capsicum annuum*), and eggplant (*S. melongena*) (Olmstead et al. 2008; Gebhardt 2016). Economically, the Solanaceae rank as the third most important botanical family and include the most productive vegetable crops globally (FAOSTAT 2022). Due to its cosmopolitan distribution, members of this family demonstrate exceptional adaptability to a wide range of climates, from rainforests to arid regions, and display immense phenotypic diversity, ranging from small annual herbs to large trees (Knapp 2002). However, the domestication of crops like tomato, potato, and pepper has significantly narrowed their genetic diversity, often resulting in the loss of valuable traits that confer tolerance to biotic and abiotic stresses (Tang et al. 2022; Zhou et al. 2022; Chen et al. 2024).

5.2.1 The *Solanum* Genus

The *Solanum* genus is the largest and most diverse clade within the family Solanaceae, comprising approximately 1500 species distributed worldwide (Weese and Bohs 2007). This genus includes a variety of species with significant agricultural and economic importance, as well as a large number of wild relatives. The genus is notable for its complex taxonomy, with species exhibiting a wide range of floral, fruit, and leaf morphologies (Knapp et al. 2004). This complexity has led to extensive research on the evolutionary relationships within the genus, with molecular phylogenetics revealing insights into its classification and evolutionary history (Särkinen et al. 2013). Species within *Solanum* are also known for their various economic uses, including food, ornamental plants, and their pharmaceutical value due to the presence of bioactive secondary metabolites such as tomatine, solanine, and capsaicin in some species (Faria-Silva et al. 2022).

In addition to its economic value, the genetic diversity within the genus has made it a target for breeding programs, particularly in efforts to improve traits like disease resistance, abiotic stress tolerance, and nutritional content in crops (Jansky and Peloquin 2006; Firdaus et al. 2012; Conesa et al. 2017). Research on wild relatives of *Solanum* has been instrumental in dissecting the genetic basis of valuable traits (Alseekh et al. 2013). These traits could be introduced into cultivated species to help broaden their genetic base and address climate change challenges (Zsögön et al. 2022).

5.2.2 Tomato and Its Wild Relatives

The cultivated tomato is closely related to at least 13 wild species (some of which are depicted in Fig. 5.2), which were previously classified under a separate genus, *Lycopersicon* (Taylor 1986). However, molecular phylogeny analyses led to their reclassification within the monophyletic *Solanum* genus (Peralta and Spooner 2001). These species share common features, including herbaceous growth habit, presence of glandular trichomes, laterally dehiscent anthers and pinnate leaves. All

Fig. 5.2 Leaves and inflorescences of some representative wild relatives of the tomato. From left to right: *S. peruvianum* (LA0153), *S. neorickii* (LA0247), *S. pimpinellifolium* (LA0373), *S. esculentum* var. *cerasiforme* (LA0292), *S. chilense* (LA1930), *S. chmielewskii* (LA1028), *S. pennellii* (LA706), *S. esculentum* cv. M82 (LA3475)

species in the *Lycopersicon* section are diploids with 12 chromosomes (2n = 24) and exhibit a high degree of synteny (Zhou et al. 2022). They are distributed across a range of environments from southern Ecuador —including the Galápagos Islands— down through Peru to northern Chile (Moyle 2008). These regions include diverse habitats, such as deserts, high-altitude areas with cold nights, and coastal regions affected by salinity (Nakazato et al. 2008, 2010).

Each wild species is adapted to its particular environment, making them valuable resources for broadening the genetic diversity of cultivated tomatoes (Warnock 1991). For instance, *S. galapagense*, which grows on the Galápagos Islands, thrives in salty soils close to the shoreline and offers a vast array of traits for salt tolerance (Pailles et al. 2017; Pailles et al. 2020). *S. habrochaites* is endemic to central Peru, where it grows between 500 and 3500 m above sea level, enduring nighttime temperatures as low as 5 °C, making it a source of cold tolerance (Arms et al. 2015; Chen et al. 2015). Additionally, *S. habrochaites* is noted for its resistance to arthropods due to its high density of glandular trichomes (Gasparini et al. 2023). However, progress in incorporating these traits into cultivated tomatoes has been limited due to the multigenic nature of trichome development and the associated metabolic pathways (Therezan et al. 2021; Vendemiatti et al. 2022).

5.2.3 Potato and Its Wild Relatives

Potato plants are classified within the section *Petota* and form a diverse group that encompasses hundreds of wild and cultivated taxonomic species. These species share a close relationship with the species of the section *Etuberosum*, which do not develop tubers, and with the section *Lycopersicon*, which includes cultivated and wild tomato species. The species of the section *Petota* have the capacity to develop tubers from stolons, specialized belowground branches that undergo a process of thickening and transformation.

Wild potato species are native to the Americas and distributed along a latitudinal transect of about 10,000 km from South America in Argentina and Chile (latitude 41 °S) to Southwestern United States (latitude 38 °N) (Peralta et al. 2021). The wild potatoes comprise an euploid series from diploid ($2n = 2x = 24$) to hexaploid ($2n = 6x = 72$) species. Two diversity centers of wild potatoes have been identified, one in the Andes of Peru, Bolivia, and north-west Argentina, and the other in Mexico, regions where the greatest number of taxonomic species occur per degree of latitude (Hawkes and Hjerting 1969; Hijmans and Spooner 2001).

The species grows in a wide range of environmental conditions. This widespread geographic distribution of the wild potatoes is explained by the great ecological diversity and the high adaptive value to diverse and contrasting environmental conditions. Some species can tolerate sub-zero temperatures above 4000 m a.s.l. as *S. acaule*, some that grow in Yungas grassland or temperate forests with high humidity, like *S. vernei*, and others that are adapted to desert areas with high temperatures and low humidity as *S. kurtzianum* (Hawkes and Hjerting 1969; Kozub et al. 2019; Ibañez et al. 2021) (Fig. 5.3).

The wide ecological adaptability of wild potatoes is explained in part by the great genetic variability present in all possible hierarchical analyses: within the individual (heterozygosity), among individuals within populations, among populations, and among taxonomical species (Marfil and Masuelli 2015; Hardigan et al. 2017, Tang et al. 2022). Wild potato germplasm has been a source of genes for resistance to biotic stresses and genes related to tuber quality for more than a century (Plaisted and Hoopes 1989; Bradshaw 2009; Bethke et al. 2017). However, the potential value of many wild potato species has not yet been evaluated (Castañeda-Álvarez et al. 2015). Although the significance of wild potatoes' contribution to modern cultivars is controversial (Jansky et al. 2021), only a dozen of the more than one

Fig. 5.3 Schematic representation of the diversity of wild potato species, genus *Solanum* section *Petota*. *Solanum acaule*, a tetraploid species with rosette to semi-rosette reaching the high puna plateau and possessing frost tolerance. Herbaceous low-growing rosette plants of the diploid species *Solanum boliviense*. The ubiquitous species *Solanum brevicaule* shows three ploidy levels. *Solanum kurtzianum*, a diploid species adapted to desert and semi-desert regions. The nematode-resistant diploid species *Solanum vernei*. All of these species, except *S. boliviense*, are present in the pedigree of modern potato cultivars

undred wild taxonomic species are present in the pedigree of modern cultivars—and few or unique clones of each species have been used in breeding programs (Plaisted and Hoopes 1989; Bradshaw 2009).

Biological barriers and limited knowledge about this exotic germplasm pose significant challenges to the use of wild potato species in breeding (Bethke et al. 2017). To overcome these hurdles, multiple successive backcrosses are necessary to break linkage between desirable and undesirable wild genes (Jansky et al. 2021). GE using CRISPR/Cas9 offers a novel approach for modifying undesirable or harmful traits in wild potatoes (Zsögön et al. 2017; Fernie and Yan 2019; Gasparini et al. 2021). Additionally, the limited understanding of beneficial traits in wild potato germplasm continues to hinder the advancement of this new breeding focus. Therefore, characterizing and selecting wild genotypes for editing is a crucial step in the improvement process (Ibañez 2019).

5.3 Processes of Domestication

5.3.1 Domestication of Tomato and Potato

The domestication of the tomato was a complex evolutionary process that began in the Andean region of South America, where wild relatives of tomatoes were originally found. Early domesticated forms are thought to have been small berry-like fruits resembling modern wild tomatoes of the species *S. pimpinellifolium* (PIM) (Blanca et al. 2012). Farmers in the Americas selected for traits like larger fruit size, improved flavor, and reduced bitterness (Pereira et al. 2021). This selective breeding led to significant changes in the fruit's morphology and genetic makeup, creating larger, fleshier tomatoes better suited for human consumption (Tanksley 2004). An elegant model has suggested that tomato domestication occurred through two significant successive transitions: from wild to a semidomesticated intermediate (*S. lycopersicum* var. *cerasiforme*—CER) in South America, and from this to the fully domesticated *S. lycopersicum* var. *lycopersicum* (LYC) in Mesoamerica (Blanca et al. 2015; Gao et al. 2019). While this model continues to be generally accepted, the specific patterns are still debated. First, Razifard et al. (2020) suggested that the CER intermediate form emerged prior to domestication, and that migration of these protodomesticates northward entailed the loss of cultivated traits. A re-domestication then occurred in Mesoamerica, whereby valuable traits appeared in convergent fashion prior to the tomato's worldwide expansion during the Columbian Exchange. Blanca et al. (2022b) subsequently proposed that PIM evolved into CER during a slow migration from Peru to Mesoamerica and that Peruvian and Ecuadorian CERs are hybrid populations originating from Mesoamerican LYCs and Peruvian and Ecuadorian PIMs. Subsequently, Peruvian domesticated CERs migrated north to Mexico and then evolved into LYC (Blanca et al. 2022b). As tomatoes spread beyond their native regions, particularly after their introduction to Europe following the Columbian Exchange, further domestication and breeding took place. In Europe, especially in Mediterranean climates, additional selection for traits such as enhanced flavor, reduced acidity, and disease resistance occurred (Blanca et al. 2022a). This global spread of tomatoes and subsequent breeding resulted in a wide variety of cultivars adapted to different climates and culinary uses. Modern tomato breeding continues to focus on improving traits such as yield, shelf life, and resistance to pathogens, while also revisiting wild relatives to reintroduce beneficial traits lost during early domestication efforts.

Wild potatoes were domesticated in high Andean valleys near Lake Titicaca, around 7000–10,000 years ago. The emergence of agriculture in South America had potatoes as the main source of carbohydrates, which, along with the protein obtained from domesticated camelids (llama, *Lama glama,* and alpaca, *Vicugna pacos*), allowed the demographic expansion of the Andean peoples to reach 10–12 million inhabitants at the apogee of the Inca Empire (King 1987). Although wild potatoes were used as food by the first hunter-gatherers who populated America (Ugent et al. 1982; Louderback and Pavlik 2017), these plants generally produce tubers that are toxic for human consumption because they contain high concentrations of steroidal glycoalkaloids. The selection of those tubers with reduced content of glycoalkaloids was an essential part of crop domestication (Johns and Alonso 1990; Friedman 2006). The domesticated species *S. tuberosum* evolved through a complex process involving hybridization and polyploidization, with the diploid *S. candolleanum* considered its ancestral wild relative. This domestication also resulted in the emergence of plants with larger tubers and shorter stolons, modifications in the regulation of carbohydrate biosynthesis and transport, as well as reduced sexual fertility (Jansen et al. 2001; Bradshaw et al. 2006; Hardigan et al. 2017) (Fig. 5.4).

The results of this first domestication event led to the emergence of Andean landraces, which are distributed from Venezuela to Northwest Argentina. The Andean Landraces include diploid, triploid, tetraploid, and pentaploid species and cultivars adapted to short-day flowering and tuberization. They are taxonomically categorized into four cultivated species: *S. tuberosum* group Andigenum, *S. ajanhuiri, S. curtilobum,* and *S. juzepczukii* (Spooner et al. 2007). In the southward migration of Andean landraces, these cultivated potatoes naturally incorporated wild alleles through hybridization and introgression, enabling tuber setting in long-day conditions (Hardigan et al. 2017).

A second domestication process arose from this germplasm, giving rise to *S. tuberosum* group Chilotanum, the lowland tetraploid landraces adapted to long-day photoperiod and distributed in Southern Chile Chonos and Guaitecas Arquipelagos (Peralta et al. 2021) (Fig. 5.4). Sequence analysis of a diversity panel of wild species, landraces, and North American modern cultivars indicates that varieties derived from modern breeding programs and grown around the world are closely related to the Chilotanum group (Hardigan et al. 2017). The introgression of wild potato genes and the different selection strategies implemented in highland Andigenum and lowland Chilotanum populations (Hardigan et al. 2017) clearly demonstrate that the ancestral domestication was a combination of natural events and human-directed practices (Vavilov 1926).

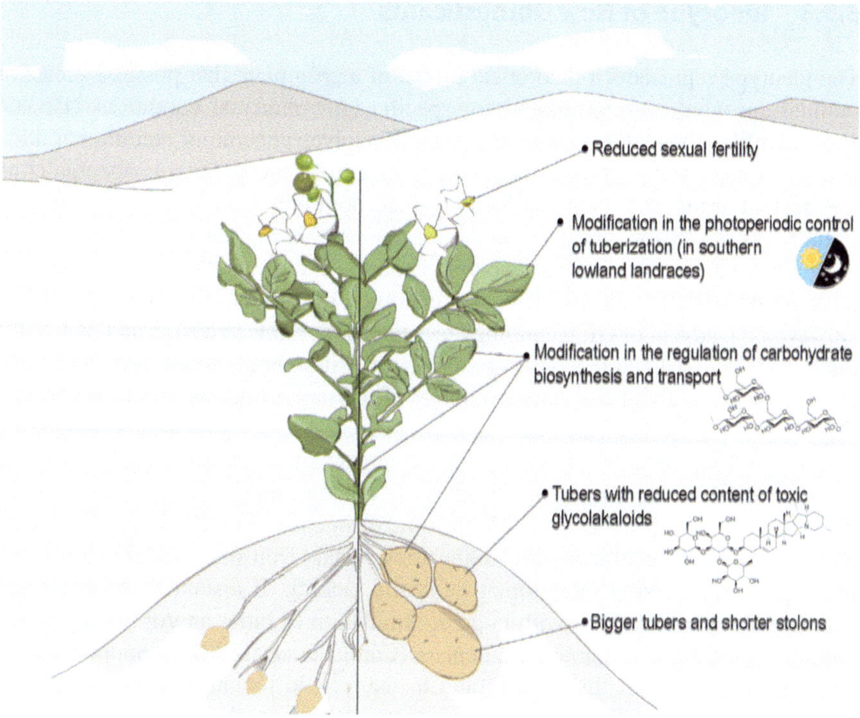

Fig. 5.4 Schematic representation of the ancestral potato domestication. Phenotypic traits associated with the ancestral potatoes landraces (right half) domesticated from wild ancestors (left half)

5.3.2 Justification for New Domestication

Crops are highly dynamic entities and their use and abandonment are determined by a balance between economic considerations, practical aspects of growth and harvest, and climate conditions. Climate change is placing a serious burden on traditional crops, which are generally not well suited to excessive temperatures or water deficit. The domestication of crop wild relatives (CWRs) could be an essential contribution to "future-proof" crops (Lehretz et al. 2020), and to extend agriculture to currently unsuitable marginal lands in developing countries (Shahid and Al-Shankiti 2013). Previous work has shown that targeting a few major genes in CWRs using CRISPR/Cas9 can rapidly and reliably bring about domestication and breeding traits in tomato (Zsögön et al. 2018) and rice (Yu et al. 2021). It should also be noted that totally new domestication is not the only available option, but accelerated improvement of semi-domesticated orphan crops (Lemmon et al. 2018) and new modes of use of existing crops are also feasible using this approach (Smartt 1990).

5.3.3 Ideotype of New Domesticants

The ideotype represents a theoretical model of a crop plant that possesses a set of optimal agronomic traits tailored for specific environmental conditions (Donald 1968). Unlike conventional breeding, which involves generating random variation and selecting advantageous traits, ideotype breeding involves first conceptualizing the desired traits and then deliberately incorporating them into crop varieties. Initially proposed for wheat (*Triticum aestivum*), this approach faced limitations due to its reliance on existing phenotypic variation and the challenges of integrating multiple beneficial traits into a single genotype or variety (Buxton et al. 1993). However, the advent of GE technology has made it possible to design and realize the rapid stacking of advantageous agronomic traits in a single plant, provided sufficient knowledge about the genetic control of the target traits is available (Zsögön et al. 2017).

Tomato domestication entailed primarily gigantism of the fruit, probably based on selection of mutations in major genes of "predomesticated" plants with edible, but small fruits (Vicente et al. 2023). In spite of intense research effort in this area, and the identification of many loci influencing tomato fruit size, a clear roadmap for predictably engineering large tomatoes is still lacking. If distant relatives are targeted, green, unedible fruit with large accumulation of bitter and/or toxic glycoalkaloids would pose important challenges (Cárdenas et al. 2019). Suppression of seed dormancy, a trait that is common to many wild tomato relatives is also an important feature of domesticated tomatoes (Wang et al. 2018). Phenological changes, including the removal of the short-day photoperiodic flowering requirement, would permit extension of cultivation to higher latitudes and can be achieved via discrete genetic changes in major loci (Song et al. 2020). Novel genes controlling flowering in response to water availability or other environmental cues could also be targeted (Robledo et al. 2020; Moreira et al. 2022). The suppression of side branching has been a long-sought goal to complete the domestication and improvement of tomato, as it demands a high amount of labor to remove sucker branches, which can divert energy away from the primary body of the plant, potentially reducing growth, health, and productivity. Although the subject of intense research effort, the genetic pathways that need to be targeted to preclude a tomato plant from branching are still not fully understood (Brophy et al. 2018; Nicolas et al. 2022). However, as a deeper understanding of the genetic basis of domestication is achieved, powerful GE tools may be used to rapidly engineer multiple traits simultaneously in wild relatives (Zsögön et al. 2018; Čermák et al. 2021).

The ideotype of the potato for the future should address the following strategic goals: (i) tolerance to biotic and abiotic stresses, (ii) enhanced tuberization, (iii) biofortification, improved nutrient composition, and industrial performance, and (iv) reproductive strategy in productive systems.

Solanum tuberosum sp. *tuberosum* is a drought-sensitive crop due to its shallow root system, which impacts tuber yield and quality. Therefore, a root architecture that promotes soil exploration, combined with reduced leaf area and a canopy structure with low leaf: stem ratio, are highly desirable traits to improve water-use efficiency (Schittenhelm et al. 2006; Hill et al. 2021). Furthermore, stomatal control that reduces water loss through transpiration, while providing continuous access to CO_2 for photosynthesis and subsequent tuber filling, along with mechanisms that protect proteins and membranes from oxidative damage, are also crucial (Hill et al. 2021).

Leaf and stem trichomes modifications could contribute toward the ideotype of pest and disease-resistant potatoes. The presence of glandular trichomes in wild potato species plays a role in aphid resistance, common vectors of potato viruses (Cho et al. 2017). Furthermore, trichomes, especially type A, are also associated with a lower incidence of potato late blight (*Phytophthora infestans*) (Lai et al. 2000).

Deciphering mechanisms and factors that regulate tuber induction is key to developing more efficient and sustainable potato productive systems. Adapting wild species that tuberize under short-day conditions to more flexible photoperiods is a key strategy in de novo domestication. Cytokinins promote potato tuberization by establishing the stolon tip as a sink (Roitsch and Ehneß 2000). In addition, specific tuberization signals are required. SP6A, an ortholog of the FLOWERING LOCUS T (FT) protein in *Arabidopsis*, has been identified as a fundamental player in tuber formation under short-day conditions (Navarro et al. 2011). On the other hand, the *SP5G* gene, which is expressed under long-day conditions, acts as a repressor of tuberization, blocking the action of *SP6A* under noninductive conditions (Kloosterman et al. 2013; Navarro et al. 2015). Therefore, inhibition of *SP5G* would enable short-day potatoes to be cultivated at higher latitudes. Another source of molecular manipulation for potato improvement is the *StCDF1* locus. The allelic composition and dosage at this locus affect tuberization, potato maturity, and yield (Kloosterman et al. 2013; Hoopes et al. 2022). Evolutionary studies of the section *Petota* have shown that the gene *StCDF1*, a positive regulator of tuberization through its interaction with the protein *CONSTANS*, contains deletions in clones that tuberize under long-day conditions. Genetic and phylogenetic analyses suggest that these deletions in *StCDF1* locus are mutations induced by transposable elements, leading to the production of truncated proteins. Moreover, clones that tuberize under long-day have been observed to contain alleles for shortened *StCDF1* proteins, which were introgressed from the wild potato *S. microdontum* (Hardigan et al. 2017). By manipulating these genes using GE, potato germplasm that tuberizes in short days could be adapted to tuberize in long days, allowing for cultivation at higher latitudes around the world.

Tuber nutritional quality is a crucial factor in feeding the world's population. Potatoes are vital food source, especially in developing countries, where micronutrient deficiency—often referred to as "hidden hunger"—presents a significant challenge (Singh et al. 2021). Biofortification is a promising strategy to mitigate these deficiencies (Burlingame et al. 2009), and GE of wild potatoes could enhance the utilization of this valuable source of genetic variability.

Another key aspect of tuber quality is their postharvest preservation. Optimal quality is achieved at harvest time. During storage, sprouting can deteriorate tubers, resulting in weight loss and increased susceptibility to physical damage, which leads to starch mobilization. In this sense, the control of tuber dormancy (suppression of tuber sprouting) is a prominent task in the commercial potato production. Tuber dormancy in potato is a physiological process regulated by a complex genetic network (Tai et al. 2020; Dogramaci et al. 2024). The polygenic nature of tuber dormancy is a major hurdle for potato breeding programs aimed at enhancing postharvest storage and extending the shelf life of tubers (Di et al. 2024). De novo domestication of wild potato species with long dormancy periods could be an alternative to obtain varieties with extended postharvest life. A wide range of dormancy periods has been described in wild potatoes (Horgan et al. 2013; Marfil et al. 2015; Janiszewska et al. 2023), but those with longer dormancies tend to originate from semi-arid regions (Dogramaci et al. 2024). In this context, species adapted to arid regions, such as *S. jamesii*, native to northern Mexico and the southwestern USA (Bamberg 2010), or *S. kurtzianum*, endemic from Argentina (Ibañez et al. 2021), are promising candidates to capitalize longer dormancy periods in new varieties developed via de novo domestication.

The lowering of reducing sugars content is another demand for tuber quality from the chip processing industry. The low postharvest temperatures activate a process known as cold-induced sweetening, in which starch is hydrolyzed into reducing sugars. This increase in the reducing sugar levels induced by low temperatures can lead to nonenzymatic browning, resulting in Maillard reaction products that generate undesirable dark colors (Alamar et al. 2017). In the species *S. berthaultii*, native to Bolivia and northern Argentina, the *SbAI* gene has been identified as inhibiting the activity of amylases that break down starch in tubers stored under cold conditions (Zhang et al. 2014). The use of this trait would reduce postharvest losses and ensure higher quality for consumers. In addition, potato CWRs could be a wide source of starches, an important component across diverse food and industrial applications (Bachmann-Pfabe and Dehmer 2020: Ahmad et al. 2024). Given the great intra- and interspecific variability in starch contents and qualities (Jansen et al. 2001), the characterization and selection of genotypes suitable for the ideotype of interest will be an interesting challenge.

In the last decades, diploid F_1 hybrid potato has emerged as a great breeding alternative to convert the third most important food crop worldwide into a true seed crop (Jansky et al. 2016). This strategy relies on the generation of a series of inbred lines through self-fertilization. However, most diploid potatoes are self-incompatible and obligate outcrossers, a genetic system controlled by the *S*-locus. This challenge has been met in two ways. On one hand, thanks to CRISPR/Cas9 technology, self-compatible diploid lines have been obtained by deactivating the *S-RNase* gene (Enciso-Rodriguez et al. 2019). On the other hand, the discovery in *S. chacoense* of the *Sli* gene, a dominant self-incompatibility inhibitor, has shown that homozygosity does not always imply low vigor, as is evident in inbred lines of *S. chacoense* (Jansky et al. 2016). These advances open the door to the domestication of new potato varieties from wild inbred lines, and allow seed propagation as an alternative to asexual propagation.

5.4 Conclusions

Plant domestication was a long and protracted process achieved by selective cross-breeding that resulted in the plethora of extant modern crops. Climate change, however, threatens the sustainability of our current agricultural systems, and radically new ideas and disruptive breeding platforms are required. De novo domestication is one such approach, based on harnessing the knowledge about the genetic basis of domestication to introduce agronomic traits in CWRs using GE techniques. The Solanaceae family contains crops that are key contributors to food security and includes a wide range of CWRs with enormous potential for domestication and breeding upon minimal intervention. In this chapter, we have summarized the prospects and potential for the development of new tomato and potato crops via de novo domestication.

Acknowledgments Support was provided by the Conselho Nacional de Desenvolvimento Científico e Tecnológico (CNPq) [Grants 406455/2022-8 and 402294/2024-6]; Fundação de Amparo à Pesquisa do Estado de Minas Gerais (FAPEMIG) [RED-00060-23; APQ-01942-22 and BPD-00823-22]; Coordenação de Aperfeiçoamento de Pessoal de Nível Superior (CAPES, Brazil); and Agencia I + D + I, Argentina (PICT-2020-03991).

References

Acquaah G (2012) Principles of plant genetics and breeding. John Wiley & Sons

Ahloowalia BS, Maluszynski M, Nichterlein K (2004) Global impact of mutation-derived varieties. Euphytica 135:187–204

Ahmad D, Ying Y, Bao J (2024) Understanding starch biosynthesis in potatoes for metabolic engineering to improve starch quality: a detailed review. Carbohydr Polym 346:122592. https://doi.org/10.1016/j.carbpol.2024.122592

Ahn E, Botkin J, Curtin SJ, Zsögön A (2023) Ideotype breeding and genome engineering for legume crop improvement. Curr Opin Biotechnol 82:102961. https://doi.org/10.1016/j.copbio.2023.102961

Alamar MC, Tosetti R, Landahl S, Bermejo A, Terry Leon A (2017) Assuring potato tuber quality during storage: a future perspective. Front Plant Sci 8. https://doi.org/10.3389/fpls.2017.02034

Alseekh S, Ofner I, Pleban T, Tripodi P, Di Dato F, Cammareri M, Mohammad A, Grandillo S, Fernie AR, Zamir D (2013) Resolution by recombination: breaking up Solanum pennellii introgressions. Trends Plant Sci 18:536–538. https://doi.org/10.1016/j.tplants.2013.08.003

Arms EM, Bloom AJ, St. Clair DA (2015) High-resolution mapping of a major effect QTL from wild tomato Solanum habrochaites that influences water relations under root chilling. Theor Appl Genet 128:1713–1724. https://doi.org/10.1007/s00122-015-2540-y

Bachmann-Pfabe S, Dehmer KJ (2020) Evaluation of wild potato germplasm for tuber starch content and nitrogen utilization efficiency. Plants 9(7):833. https://doi.org/10.3390/plants9070833

Bamberg JB (2010) Tuber dormancy lasting eight years in the wild potato Solanum jamesii. Am J Pot Res 87:226–228. https://doi.org/10.1007/s12230-009-9124-9

Baur S, Frank O, Hausladen H, Hückelhoven R, Hofmann T, Eisenreich W, Dawid C (2021) Biosynthesis of α-solanine and α-chaconine in potato leaves (Solanum tuberosum L.)—a 13CO2 study. Food Chem 365:130461. https://doi.org/10.1016/j.foodchem.2021.130461

Bethke PC, Halterman DA, Jansky S (2017) Are we getting better at using wild potato species in light of new tools? Crop Sci 57:1241–1258

Blanca J, Cañizares J, Cordero L, Pascual L, Diez MJ, Nuez F (2012) Variation revealed by SNP genotyping and morphology provides insight into the origin of the tomato. PLoS One 7:e48198. https://doi.org/10.1371/journal.pone.0048198

Blanca J, Montero-Pau J, Sauvage C, Bauchet G, Illa E, Díez MJ, Francis D, Causse M, van der Knaap E, Cañizares J (2015) Genomic variation in tomato, from wild ancestors to contemporary breeding accessions. BMC Genomics 16:257. https://doi.org/10.1186/s12864-015-1444-1

Blanca J, Pons C, Montero-Pau J, Sanchez-Matarredona D, Ziarsolo P, Fontanet L, Fisher J, Plazas M, Casals J, Rambla JL, Riccini A, Palombieri S, Ruggiero A, Sulli M, Grillo S, Kanellis A, Giuliano G, Finkers R, Cammareri M, Grandillo S, Mazzucato A, Causse M, Díez MJ, Prohens J, Zamir D, Cañizares J, Monforte AJ, Granell A (2022a) European traditional tomatoes galore: a result of farmers' selection of a few diversity-rich loci. J Exp Bot 73:3431–3445. https://doi.org/10.1093/jxb/erac072

Blanca J, Sanchez-Matarredona D, Ziarsolo P, Montero-Pau J, van der Knaap E, Díez MJ, Cañizares J (2022b) Haplotype analyses reveal novel insights into tomato history and domestication driven by long-distance migrations and latitudinal adaptations. Hortic Res 19:uhac030. https://doi.org/10.1093/hr/uhac030

Bradshaw JE (2009) Potato breeding at the Scottish plant Breeding Station and the Scottish Crop Research Institute: 1920–2008. Potato Res 52:141–172

Bradshaw JE, Bryan G, Ramsay G (2006) Genetic resources (including wild and cultivated Solanum species) and progress in their utilisation in potato breeding. Potato Res 49:49–65

Brophy JAN, LaRue T, Dinneny JR (2018) Understanding and engineering plant form. Semin Cell Dev Biol 79:68–77. https://doi.org/10.1016/j.semcdb.2017.08.051

Burlingame B, Mouillé B, Charrondière R (2009) Nutrients, bioactive non-nutrients and anti-nutrients in potatoes. J Food Compos Anal 22:494–502. https://doi.org/10.1016/j.jfca.2009.09.001

Buxton DR, Shibles R, Forsberg RA, Blad BL, Asay KH, Paulsen GM, Wilson RF, Hamblin J (1993) The Ideotype concept: useful or outdated? International Crop Science I. Crop Science Society of America, pp 589–597

Cárdenas PD, Sonawane PD, Heinig U, Jozwiak A, Panda S, Abebie B, Kazachkova Y, Pliner M, Unger T, Wolf D, Ofner I, Vilaprinyo E, Meir S, Davydov O, Gal-on A, Burdman S, Giri A, Zamir D, Scherf T, Szymanski J, Rogachev I, Aharoni A (2019) Pathways to defense metabolites and evading fruit bitterness in genus Solanum evolved through 2-oxoglutarate-dependent dioxygenases. Nat Commun 10:5169. https://doi.org/10.1038/s41467-019-13211-4

Čermák T, Gasparini K, Kevei Z, Zsögön A (2021) Genome editing to achieve the crop Ideotype in tomato. In: Tripodi P (ed) Crop breeding: genetic improvement methods. Springer, New York, pp 219–244

Castañeda-Álvarez NP, de Haan S, Juárez H, Khoury CK, Achicanoy HA, et al. (2015) Ex Situ Conservation Priorities for the Wild Relatives of Potato (Solanum L. Section Petota). PLOS ONE 10(4): e0122599. https://doi.org/10.1371/journal.pone.0122599

Chen H, Chen X, Chen D, Li J, Zhang Y, Wang A (2015) A comparison of the low temperature transcriptomes of two tomato genotypes that differ in freezing tolerance: Solanum lycopersicum and Solanum habrochaites. BMC Plant Biol 15:132. https://doi.org/10.1186/s12870-015-0521-6

Chen K, Wang Y, Zhang R, Zhang H, Gao C (2019) CRISPR/Cas genome editing and precision plant breeding in agriculture. Annu Rev Plant Biol 70:667–697. https://doi.org/10.1146/annurev-arplant-050718-100049

Chen W, Wang X, Sun J, Wang X, Zhu Z, Ayhan DH, Yi S, Yan M, Zhang L, Meng T, Mu Y, Li J, Meng D, Bian J, Wang K, Wang L, Chen S, Chen R, Jin J, Li B, Zhang X, Deng XW, He H, Guo L (2024) Two telomere-to-telomere gapless genomes reveal insights into capsicum evolution and capsaicinoid biosynthesis. Nat Commun 15:4295. https://doi.org/10.1038/s41467-024-48643-0

Cho KS, Kwon M, Cho JH, Im JS, Park YE, Hong SY, Hwang IT, Kang JH (2017) Characterization of trichome morphology and aphid resistance in cultivated and wild species of potato. Hortic Environ Biotechnol 58:450–457. https://doi.org/10.1007/s13580-017-0078-4

Conesa MÀ, Muir CD, Roldán EJ, Molins A, Perdomo JA, Galmés J (2017) Growth capacity in wild tomatoes and relatives correlates with original climate in arid and semi-arid species. Environ Exp Bot 141:181–190. https://doi.org/10.1016/j.envexpbot.2017.04.009

Di X, Wang Q, Zhang F, Feng H, Wang X, Cai C (2024) Advances in the modulation of potato tuber dormancy and sprouting. Int J Mol Sci 25(10):5078. https://doi.org/10.3390/ijms25105078

Dogramaci M, Dobry EP, Fortini EA, Sarkar D, Eshel D, Campbell MA (2024) Physiological and molecular mechanisms associated with potato tuber dormancy. J Exp Bot erae182. https://doi.org/10.1093/jxb/erae182

Donald CM (1968) The breeding of crop ideotypes. Euphytica 17:385–403. https://doi.org/10.1007/BF00056241

Doudna JA, Charpentier E (2014) The new frontier of genome engineering with CRISPR-Cas9. Science 346:1258096–1258096. https://doi.org/10.1126/science.1258096

Enciso-Rodriguez F, Manrique-Carpintero NC, Nadakuduti Satya S, Buell CR, Zarka D, Douches D (2019) Overcoming self-incompatibility in diploid potato using CRISPR-Cas9. Front Plant Sci 10. https://doi.org/10.3389/fpls.2019.00376

FAOSTAT (2022) Crops and livestock products

Faria-Silva C, de Sousa M, Carvalheiro MC, Simões P, Simões S (2022) Alpha-tomatine and the two sides of the same coin: an anti-nutritional glycoalkaloid with potential in human health. Food Chem 391:133261. https://doi.org/10.1016/j.foodchem.2022.133261

Fernie AR, Yan J (2019) De novo domestication: an alternative route toward new crops for the future. Mol Plant 12:615–631

Firdaus S, van Heusden AW, Hidayati N, Supena EDJ, Visser RGF, Vosman B (2012) Resistance to Bemisia tabaci in tomato wild relatives. Euphytica 187:31–45. https://doi.org/10.1007/s10681-012-0704-2

Friedman M (2006) Potato glycoalkaloids and metabolites: roles in the plant and in the diet. J Agric Food Chem 54:8655–8681

Gao C (2021) Genome engineering for crop improvement and future agriculture. Cell 184:1621–1635. https://doi.org/10.1016/j.cell.2021.01.005

Gao L, Gonda I, Sun H, et al (2019) The tomato pan-genome uncovers new genes and a rare allele regulating fruit flavor. Nature Genetics 51:1044–1051. https://doi.org/10.1038/s41588-019-0410-2

Gasparini K, dos Reis Moreira J, Pereira Peres LE, Zsögön A (2021) De novo domestication of wild species to create crops with increased resilience and nutritional value. Curr Opin Plant Biol 60:102006

Gasparini K, Gasparini J, Therezan R, Vicente MH, Sakamoto T, Figueira A, Zsögön A, Peres LEP (2023) Natural genetic variation in the HAIRS ABSENT (H) gene increases type-VI glandular trichomes in both wild and domesticated tomatoes. J Plant Physiol 280:153859. https://doi.org/10.1016/j.jplph.2022.153859

Gebhardt C (2016) The historical role of species from the Solanaceae plant family in genetic research. Theor Appl Genet 129:2281–2294. https://doi.org/10.1007/s00122-016-2804-1

Hardigan MA, Laimbeer FPE, Newton L, Crisovan E, Hamilton JP, Vaillancourt B, Wiegert-Rininger K, Wood JC, Douches DS, Farré EM et al (2017) Genome diversity of tuber-bearing Solanum uncovers complex evolutionary history and targets of domestication in the cultivated potato. Proc Natl Acad Sci USA 114:E9999–E10008

Hijmans RJ, Spooner DM (2001) Geographic distribution of wild potato species†. Am J Bot 88:2101–2112. https://doi.org/10.2307/3558435

Hill D, Nelson D, Hammond J, Bell L (2021) Morphophysiology of potato (Solanum tuberosum) in response to drought stress: paving the way forward. Front Plant Sci 11:597554. https://doi.org/10.3389/fpls.2020.597554

Hawkes, J. G. & Hjerting, J. P. (1969). Potatoes of Argentina, Brazil, Paraguay, and Uruguay: a biosystematic study. Annals of Botany Memoirs, No. 3. Oxford: Clarendon Press. 525 pp., with illustrations and plates. ISBN: 0198543646

Hoopes G, Meng X, Hamilton JP, Achakkagari SR, de Alves Freitas Guesdes F, Bolger ME, Coombs JJ, Esselink D, Kaiser NR, Kodde L, Kyriakidou M, Lavrijssen B, van Lieshout N,

Shereda R, Tuttle HK, Vaillancourt B, Wood JC, de Boer JM, Bornowski N, Bourke P, Douches D, van Eck HJ, Ellis D, Feldman MJ, Gardner KM, Hopman JCP, Jiang J, De Jong WS, Kuhl JC, Novy RG, Oome S, Sathuvalli V, Tan EH, Ursum RA, Vales MI, Vining K, Visser RGF, Vossen J, Yencho GC, Anglin NL, Bachem CWB, Endelman JB, Shannon LM, Strömvik MV, Tai HH, Usadel B, Buell CR, Finkers R (2022) Phased, chromosome-scale genome assemblies of tetraploid potato reveal a complex genome, transcriptome, and predicted proteome landscape underpinning genetic diversity. Mol Plant 15(3):520–536. https://doi.org/10.1016/j.molp.2022.01.003

Horgan FG, Quiring DT, Lagnaoui A, Pelletier Y (2013) Tuber production, dormancy and resistance against Phthorimaea operculella (Zeller) in wild potato species. J Appl Entomol 137:739–750

Huang J, Rowe D, Subedi P, Zhang W, Suelter T, Valent B, Cook DE (2022) CRISPR-Cas12a induced DNA double-strand breaks are repaired by multiple pathways with different mutation profiles in Magnaporthe oryzae. Nat Commun 13:7168. https://doi.org/10.1038/s41467-022-34736-1

Ibañez VN (2019) Variabilidad genética y epigenética en poblaciones naturales de la especie silvestre de papa *Solanum kurtzianum* distribuidas a lo largo de un gradiente altitudinal. PhD Thesis, UNCuyo (in Spanish)

Ibañez VN, Kozub PC, Gonzalez CV, Jerez DN, Masuelli RW, Berli FJ, Marfil CF (2021) Response to water deficit of semi-desert wild potato Solanum kurtzianum genotypes collected from different altitudes. Plant Sci 110911:110911

James C (2015) Global status of commercialized biotech/GM crops: 2015. ISAAA Brief No. 51

Janiszewska M, Smyda-Dajmund P, Sobkowiak S, Michałowska D, Śliwka J (2023) Tuber flesh colour, enzymatic discolouration, dormancy and late blight resistance of 29 tuber-bearing accessions of Solanum spp. Potato Res 66:1–21

Jansen G, Flamme W, Schüler K, Vandrey M (2001) Tuber and starch quality of wild and cultivated potato species and cultivars. Potato Res 44:137–146

Jansky SH, Peloquin SJ (2006) Advantages of wild diploid solanum species over cultivated diploid relatives in potato breeding programs. Genet Resour Crop Evol 53:669–674. https://doi.org/10.1007/s10722-004-2949-7

Jansky SH, Charkowski AO, Douches DS, Gusmini G, Richael C, Bethke PC, Spooner DM, Novy RG, De Jong H, De Jong WS, Bamberg JB, Thompson AL, Bizimungu B, Holm DG, Brown CR, Haynes KG, Sathuvalli VR, Veilleux RE, Miller JC, Bradeen JM, Jiang J (2016) Reinventing potato as a diploid inbred line–based crop. Crop Sci 56:1412–1422. https://doi.org/10.2135/cropsci2015.12.0740

Jansky, S.H., De Jong, W.S., Douches, D.S., Haynes, K.G., Holm, D.G. (2021). Cultivar Improvement with Exotic Germplasm: An Example from Potato. In: Carputo, D., Aversano, R., Ercolano, M.R. (eds) The Wild Solanums Genomes. Compendium of Plant Genomes. Springer, Cham. https://doi.org/10.1007/978-3-030-30343-3_12

Johns T, Alonso JG (1990) Glycoalkaloid change during the domestication of the potato, *Solanum* section *Petota*. Euphytica 50:203–210

Kausch AP, Nelson-Vasilchik K, Hague J, Mookkan M, Quemada H, Dellaporta S, Fragoso C, Zhang ZJ (2019) Edit at will: genotype independent plant transformation in the era of advanced genomics and genome editing. Plant Sci 281:186–205. https://doi.org/10.1016/j.plantsci.2019.01.006

King S (1987) Four endemic Andean tuber crops: promising food resources for agricultural diversification. Mt Res Dev 7:43–51

Kloosterman B, Abelenda JA, Gomez Mdel M, Oortwijn M, de Boer JM, Kowitwanich K, Horvath BM, van Eck HJ, Smaczniak C, Prat S, Visser RG, Bachem CW (2013) Naturally occurring allele diversity allows potato cultivation in northern latitudes. Nature 495(7440):246–250. https://doi.org/10.1038/nature11912

Knapp S (2002) Tobacco to tomatoes: a phylogenetic perspective on fruit diversity in the Solanaceae. J Exp Bot 53:2001–2022. https://doi.org/10.1093/jxb/erf068

Knapp S, Bohs L, Nee M, Spooner DM (2004) Solanaceae — a model for linking genomics with biodiversity. CFG 5:285–291. https://doi.org/10.1002/cfg.393

Kozub PC, Ibañez VN, Digilio A, Atencio HM, Garavano ME, Sánchez ME, Marfil CF (2019) Wild potato genetic reserves in protected areas: prospection notes from Los Cardones National Park, Salta, Argentina. Rev FCA UNCUYO 51:461–474

Lai A, Cianciolo V, Chiavarini S, Sonnino A (2000) Effects of glandular trichomes on the development of Phytophthora infestans infection in potato (S. tuberosum). Euphytica 114:165–174. https://doi.org/10.1023/A:1003924318577

Lehretz GG, Sonnewald S, Lugassi N, Granot D, Sonnewald U (2020) Future-proofing potato for drought and heat tolerance by overexpression of hexokinase and SP6A. Front Plant Sci 11:614534. https://doi.org/10.3389/fpls.2020.614534

Lemmon ZH, Reem NT, Dalrymple J, Soyk S, Swartwood KE, Rodriguez-Leal D, Van Eck J, Lippman ZB (2018) Rapid improvement of domestication traits in an orphan crop by genome editing. Nat Plants 4:766–770. https://doi.org/10.1038/s41477-018-0259-x

Louderback LA, Pavlik BM (2017) Starch granule evidence for the earliest potato use in North America. Proc Natl Acad Sci USA 114(29):7606–7610

Marfil CF, Hidalgo V, Masuelli RW (2015) In situ conservation of wild potato germplasm in Argentina: example and possibilities. Glob Ecol Conserv 3:461–476. https://doi.org/10.1016/j.gecco.2015.01.009

Moreira J dos R, Quiñones A, Lira BS, Robledo JM, Curtin SJ, Vicente MH, Ribeiro DM, Ryngajllo M, Jiménez-Gómez JM, Peres LEP, Rossi M, Zsögön A (2022) SELF PRUNING 3C is a flowering repressor that modulates seed germination, root architecture, and drought responses. J Exp Bot 73:6226–6240. https://doi.org/10.1093/jxb/erac265

Moyle LC (2008) Ecological and evolutionary genomics in the wild tomatoes (solanum sect. Lycopersicon). Evolution 62:2995–3013. https://doi.org/10.1111/j.1558-5646.2008.00487.x

Nakazato T, Bogonovich M, Moyle LC (2008) Environmental factors predict adaptive phenotypic differentiation within and between two wild Andean tomatoes. Evolution 62:774–792. https://doi.org/10.1111/j.1558-5646.2008.00332.x

Nakazato T, Warren DL, Moyle LC (2010) Ecological and geographic modes of species divergence in wild tomatoes. Am J Bot 97:680–693. https://doi.org/10.3732/ajb.0900216

Navarre DA, Pillai SS, Shakya R, Holden MJ (2011) HPLC profiling of phenolics in diverse potato genotypes. Food Chem 127:34–41. https://doi.org/10.1016/j.foodchem.2010.12.080

Navarro C, Abelenda JA, Cruz-Oró E, Cuéllar CA, Tamaki S, Silva J, Shimamoto K, Prat S (2011) Control of flowering and storage organ formation in potato by FLOWERING LOCUS T. Nature 478(7367):119–122. https://doi.org/10.1038/nature10431. PMID: 21947007

Navarro C, Cruz-Oró E, Prat S (2015) Conserved function of Flowering Locus T (FT) homologues as signals for storage organ differentiation. Curr Opin Plant Biol 23:45–53

Naves ER, Scossa F, Araújo WL, Nunes-Nesi A, Fernie AR, Zsögön A (2022) Heterosis for capsacinoids accumulation in chili pepper hybrids is dependent on parent-of-origin effect. Sci Rep 12:14450. https://doi.org/10.1038/s41598-022-18711-w

Nicolas M, Torres-Pérez R, Wahl V, Cruz-Oró E, Rodríguez-Buey ML, Zamarreño AM, Martín-Jouve B, García-Mina JM, Oliveros JC, Prat S, Cubas P (2022) Spatial control of potato tuberization by the TCP transcription factor BRANCHED1b. Nat Plants 8:281–294. https://doi.org/10.1038/s41477-022-01112-2

Olmstead RG, Bohs L, Migid HA, Santiago-Valentin E, Garcia VF, Collier SM (2008) A molecular phylogeny of the Solanaceae. Taxon 57:1159–1181. https://doi.org/10.1002/tax.574010

Pailles Y, Ho S, Pires IS, Tester M, Negrão S, Schmöckel SM (2017) Genetic diversity and population structure of two tomato species from the Galapagos Islands. Front Plant Sci 8

Pailles Y, Awlia M, Julkowska M, Passone L, Zemmouri K, Negrão S, Schmöckel SM, Tester M (2020) Diverse traits contribute to salinity tolerance of wild tomato seedlings from the Galapagos Islands. Plant Physiol 182:534–546. https://doi.org/10.1104/pp.19.00700

Peralta IE, Spooner DM (2001) Granule-bound starch synthase (GBSSI) gene phylogeny of wild tomatoes (Solanum L. section Lycopersicon [Mill.] Wettst. subsection Lycopersicon). Am J Bot 88:1888–1902

Peralta IE, Clausen AM, Zorrilla C, Ames M, Digilio A, Rodriguez F (2021) Wild and cultivated potato species diversity, taxonomy, and conservation. In: Carputo D et al (eds) The

wild Solanums genomes, compendium of plant genomes. Springer, Cham. https://doi.org/10.1007/978-3-030-30343-3_4
Pereira L, Zhang L, Sapkota M, Ramos A, Razifard H, Caicedo AL, van der Knaap E (2021) Unraveling the genetics of tomato fruit weight during crop domestication and diversification. Theor Appl Genet 134:3363–3378. https://doi.org/10.1007/s00122-021-03902-2
Plaisted RL, Hoopes RW (1989) The past record and future prospects for the use of exotic potato germplasm. Am Potato J 66:603–627
Razifard H, Ramos A, Della Valle AL, Bodary C, Goetz E, Manser EJ, Li X, Zhang L, Visa S, Tieman D, van der Knaap E, Caicedo AL (2020) Genomic evidence for complex domestication history of the cultivated tomato in Latin America. Mol Biol Evol 37:1118–1132. https://doi.org/10.1093/molbev/msz297
Robledo JM, Medeiros D, Vicente MH, Azevedo AA, Thompson AJ, Peres LEP, Ribeiro DM, Araújo WL, Zsögön A (2020) Control of water-use efficiency by florigen. Plant Cell Environ 43:76–86. https://doi.org/10.1111/pce.13664
Roitsch T, Ehneß R (2000) Regulation of source/sink relations by cytokinins. Plant Growth Regul 32:359–367. https://doi.org/10.1023/A:1010781500705
Särkinen T, Bohs L, Olmstead RG, Knapp S (2013) A phylogenetic framework for evolutionary study of the nightshades (Solanaceae): a dated 1000-tip tree. BMC Evol Biol 13:214–214. https://doi.org/10.1186/1471-2148-13-214
Schittenhelm S, Sourell H, Löpmeier FJ (2006) Drought resistance of potato cultivars with contrasting canopy architecture. Eur J Agron 24:193–202
Shahid SA, Al-Shankiti A (2013) Sustainable food production in marginal lands—case of GDLA member countries. Int Soil Water Conserv Res 1:24–38. https://doi.org/10.1016/S2095-6339(15)30047-2
Shen H, Strunks GD, Klemann BJPM, Hooykaas PJJ, de Pater S (2017) CRISPR/Cas9-induced double-strand break repair in Arabidopsis nonhomologous end-joining mutants. G3 Genes|Genomes|Genetics 7:193–202. https://doi.org/10.1534/g3.116.035204
Singh B, Goutam U, Kukreja S, Sharma J, Sood S, Bhardwaj V (2021) Potato biofortification: an effective way to fight global hidden hunger. Physiol Mol Biol Plants 10:2297–2313. https://doi.org/10.1007/s12298-021-01081-4
Smartt J (1990) 5. Vavilov's law of homologous series and de novo crop plant domestication. Biol J Linn Soc 39:27–38. https://doi.org/10.1111/j.1095-8312.1990.tb01610.x
Song J, Zhang S, Wang X, Sun S, Liu Z, Wang K, Wan H, Zhou G, Li R, Yu H, Cui X (2020) Variations in both FTL1 and SP5G, two tomato FT paralogs, control day-neutral flowering. Mol Plant 13:939–942. https://doi.org/10.1016/j.molp.2020.05.004
Spooner DM, Núñez J, Trujillo G, del Rosario Herrera M, Guzmán F, Ghislain M (2007) Extensive simple sequence repeat genotyping of potato landraces supports a major reevaluation of their gene pool structure and classification. Proc Natl Acad Sci USA 104:19398–19403
Tai HH, Lagüe M, Thomson S, Aurousseau F, Neilson J, Murphy A, Bizimungu B, Davidson C, Deveaux V, Bègue Y et al (2020) Tuber transcriptome profiling of eight potato cultivars with different cold-induced sweetening responses to cold storage. Plant Physiol Biochem 146:163–176. https://doi.org/10.1016/J.PLAPHY.2019.11.001
Tang D, Jia Y, Zhang J, Li H, Cheng L, Wang P, Bao Z, Liu Z, Feng S, Zhu X, Li D, Zhu G, Wang H, Zhou Y, Zhou Y, Bryan GJ, Buell CR, Zhang C, Huang S (2022) Genome evolution and diversity of wild and cultivated potatoes. Nature 606:535–541. https://doi.org/10.1038/s41586-022-04822-x
Tanksley SD (2004) The genetic, developmental, and molecular bases of fruit size and shape variation in tomato. Plant Cell 16:S181–S189
Taylor IB (1986) Biosystematics of the tomato. In: Atherton JG, Rudich J (eds) The tomato crop: a scientific basis for improvement. Chapman & Hall, London, pp 1–34
Therezan R, Kortbeek R, Vendemiatti E, Legarrea S, de Alencar SM, Schuurink RC, Bleeker P, Peres LEP (2021) Introgression of the sesquiterpene biosynthesis from Solanum habrochaites to cultivated tomato offers insights into trichome morphology and arthropod resistance. Planta 254:11. https://doi.org/10.1007/s00425-021-03651-y

Ugent D, Pozorski S, Pozorski T (1982) Archaeological potato tuber remains from the Casma Valley of Peru. Econ Bot 36:182–192

Vavilov NI (1926) Studies on the origin of cultivated plants. Bulletin of Applied Botany and Plant Breeding, Leningrad 16:2

Vendemiatti E, Therezan R, Vicente MH, Pinto M de S, Bergau N, Yang L, Bernardi WF, Alencar SM de, Zsögön A, Tissier A, Benedito VA, Peres LEP (2022) The genetic complexity of type-IV Trichome development reveals the steps towards an insect-resistant tomato. Plants 11:1309. https://doi.org/10.3390/plants11101309

Vicente MH, MacLeod K, Zhu F, Rafael DD, Figueira A, Fernie AR, Mohareb F, Kevei Z, Thompson AJ, Zsögön A, Peres LEP (2023) The ORGAN SIZE (ORG) locus modulates both vegetative and reproductive gigantism in domesticated tomato. Ann Botany 132:mcad150. https://doi.org/10.1093/aob/mcad150

Wang M, Li W, Fang C, Xu F, Liu Y, Wang Z, Yang R, Zhang M, Liu S, Lu S, Lin T, Tang J, Wang Y, Wang H, Lin H, Zhu B, Chen M, Kong F, Liu B, Zeng D, Jackson SA, Chu C, Tian Z (2018) Parallel selection on a dormancy gene during domestication of crops from multiple families. Nat Genet 50:1435–1441. https://doi.org/10.1038/s41588-018-0229-2

Warnock SJ (1991) Natural habitats of Lycopersicon species. HortScience 26:466–471

Weese TL, Bohs L (2007) A three-gene phylogeny of the genus Solanum (Solanaceae). Syst Bot 32:445–463. https://doi.org/10.1600/036364407781179671

Xue C, Greene EC (2021) DNA repair pathway choices in CRISPR-Cas9-mediated genome editing. Trends Genet 37:639–656. https://doi.org/10.1016/j.tig.2021.02.008

Yu H, Lin T, Meng X, Du H, Zhang J, Liu G, Chen M, Jing Y, Kou L, Li X, Gao Q, Liang Y, Liu X, Fan Z, Liang Y, Cheng Z, Chen M, Tian Z, Wang Y, Chu C, Zuo J, Wan J, Qian Q, Han B, Zuccolo A, Wing RA, Gao C, Liang C, Li J (2021) A route to de novo domestication of wild allotetraploid rice. Cell 184:1156–1170.e14. https://doi.org/10.1016/j.cell.2021.01.013

Zhang H, Liu J, Hou J, Yao Y, Lin Y, Ou Y, Song B, Xie C (2014) The potato amylase inhibitor gene *SbAI* regulates cold-induced sweetening in potato tubers by modulating amylase activity. Plant Biotechnol J 12:984–993

Zhou Y, Zhang Z, Bao Z, Li H, Lyu Y, Zan Y, Wu Y, Cheng L, Fang Y, Wu K, Zhang J, Lyu H, Lin T, Gao Q, Saha S, Mueller L, Fei Z, Städler T, Xu S, Zhang Z, Speed D, Huang S (2022) Graph pangenome captures missing heritability and empowers tomato breeding. Nature 606:527–534. https://doi.org/10.1038/s41586-022-04808-9

Zsögön A, Cermak T, Voytas D, Peres L (2017) Genome editing as a tool to achieve the crop ideotype and de novo domestication of wild relatives: case study in tomato. Plant Sci 256:120–130

Zsögön A, Čermák T, Naves ER, Notini MM, Edel KH, Weinl S, Freschi L, Voytas DF, Kudla J, Peres LEP (2018) De novo domestication of wild tomato using genome editing. Nat Biotechnol 36:1211–1216. https://doi.org/10.1038/nbt.4272

Zsögön A, Peres LEP, Xiao Y, Yan J, Fernie AR (2022) Enhancing crop diversity for food security in the face of climate uncertainty. Plant J 109:402–414. https://doi.org/10.1111/tpj.15626

Biotic Stress Buildup Under Climate Change and Breeding Innovations

6

A. L. Rathnakumar, S. Geethanjali, P. Kadirvel, K. Sakthivel, P. Duraimurugan, and R. K. Mathur

Abstract

The global population is rising rapidly and is projected to hit 10 billion by 2050, requiring a 70% increase in food production compared to current levels. Human activities, primarily through the emission of greenhouse gases, have clearly led to global warming, with temperatures rising by 1.1°C above pre-industrial levels. Human induced increases in CO_2 levels in the atmosphere occurred just over the past 30 years. Alterations in global climate are reflected in various ways like more frequent and extreme heatwaves, drought, very high precipitation, water-logging, and associated changes in soil moisture impacting crop yields and food security globally. These severe weather conditions could influence the behavior of plant pathogens or insect pests or cause buildup of specific pathogen(s) or an insect pest leading to potential risks and varying degrees of crop yield losses. Extreme weather events can impact plant pathogens by expanding their geographic range, enhancing their overwintering survival, and increasing the number of generations they produce in case of insect pests, greater risk of invasion by aggressive crop pests and diseases, decline in pollinator populations and biocontrol agents. To tackle these challenges, it is essential to investigate the complex interactions between climate factors and the emergence, severity, and spatial distribution of plant diseases and insect pests in relation to shifts in the global temperature, CO_2 levels, precipitation, and humidity exacerbate virulence of pathogen and alter disease dynamics. The present chapter reviews the subject briefly and additionally discusses the breeding innovations and biotechnological and molecular approaches currently available and utilized in various crop species to mitigate the climate change challenges.

A. L. Rathnakumar (✉) · P. Kadirvel · K. Sakthivel · P. Duraimurugan · R. K. Mathur
ICAR-Indian Institute of Oilseeds Research, Hyderabad, Telangana, India

S. Geethanjali
Tamil Nadu Agricultural University, Coimbatore, Tamil Nadu, India

P. M. Priyadarshan, R. Ortiz (eds.), *Plant Breeding 2050*,
https://doi.org/10.1007/978-981-95-0583-8_6

Keywords

Climate change · Global warming · Food security · Biotic stresses · Plant pathogens · Insect pests

6.1 Introduction

The global population is rapidly increasing and is projected to grow by approximately 25–30% over the next 30 years, reaching around 10 billion and would require food production to increase by 70% from the present level (Ray et al. 2012; Hickey et al. 2019; Lam 2025). With the paradigm shift in dietary preferences due to higher per capita income levels, especially towards meat and high calorie food and its products, the production of which also require more feed by 100–110% increase by 2050 (Palmgren and Shabala 2024). Per capita agricultural land area is also shrinking every year across countries (Ritchie and Roser 2023). Confounding to these, climate change is further worsening the situation, with global temperatures projected to increase by 2°C by 2050 (Zaidi et al. 2020; IPCC 2023) and between 2.6°C and 4.8°C by 2100 if no mitigation measures are implemented (IPCC 2014).

Over the past century, the earth's climate has experienced a temperature increase of 0.3–0.6 °C (IPCC 1996). The pace of global warming has differed significantly across regions, with the most rapid increases occurring between 1925 and 1944 and 1978 and 1997 (Jones et al. 1999). More specifically, the global surface temperature was 1.09 °C higher in the most recent decade (2011–2020) than over 50 years starting from 1850 to 1906. The increase in temperature was larger (1.59°C) in the land than over the ocean (0.88 °C) (IPCC 2021). The global surface temperature between 2001 and 2020 was 0.99 °C higher compared to the 1850–1900 period, and the rate of warming since 1970 has been faster than during any other 50-year span in at least the past 2000 years. The estimated range of global surface temperature rise solely due to human activities between 1850 and 1900 and 2010 and 2019 is between 0.8°C and 1.3°C. During the same timeframe, well-mixed greenhouse gases (GHGs) likely contributed to a warming of between 1.0°C and 2.0°C (IPCC 2023). It has been forecasted that a global average increase in temperature would be 2–4 °C (IPCC 2021) toward the close of the century.

Likewise, the total historical net CO_2 emissions from 1850 to 2019 amounted to 2400 ± 240 $GtCO_2$, with over half (58%) released between 1850 and 1989, and approximately 42% emitted within the three decades from 1990 to 2019. As of 2019, atmospheric CO_2 levels (410 ppm) were the highest recorded in at least two million years, while methane (1866 ppb) and nitrous oxide (332 ppb) concentrations exceeded levels observed in at least the past 800,000 years (IPCC 2023) indicating the vulnerability of the all living systems to the overall changes in the global climate. Global climate change is expected to cause an increase in the frequency and intensity of heatwaves, droughts, waterlogging, and related shifts in soil moisture (IPCC 2021). As a result, by 2030, an estimated 5 billion people may reside in water-scarce areas where drought could jeopardize agricultural production (UN-DESA 2011; Palmgren and Shabala 2024).

In addition, fluctuations in climate have affected crop yields and food security by altering rainfall patterns and temperatures, further intensified by potential increases or shifts in the spread of diseases across various regions (Zhao et al. 2017a, b; Li et al. 2019; IPCC 2019; Jägermeyr et al. 2021; Zhang et al. 2022c; Eckardt et al. 2023). Projections indicate that by 2040, average global temperatures may rise by 1.5–2.0°C, significantly threatening the stability of the global food system and food security, with potential reductions in global food production ranging from 30% to 40% (Ray et al. 2015; IPCC 2018).

Moreover, climate change is most evident through rising CO_2 levels, increasing temperatures, intensified rainfall, and a higher likelihood of extreme weather events (Wheeler and von Braun 2013; Seneviratne et al. 2021). By 2040, the chance of experiencing what was once a 1-in-100-year extreme weather event is projected to rise to 1 in 30 (Bailey and van de Pol 2015). Specifically, occurrences of extreme heat, severe drought, and heavy rainfall are expected to become more common and widespread. In Asian countries, where multiple cropping is practiced in a year during different seasons, due to the drastic change in climate, the *kharif* crops may face delayed onset and early withdrawal of monsoon, intermittent or prolonged drought spells, leading to low productivity. About 40% of global land area covering over 70 countries is regularly hit by drought costing farming over US$ 80 billion per annum (UNCCD 2022b). Drought impacts countries across all inhabited continents, regardless of their development status. Between 2007 and 2017, it affected at least 1.5 billion people and resulted in global economic losses of around US$125 billion (UNCCD 2022a). Projections suggest that by 2050, over 75% of the global population could be exposed to drought conditions. Other predictions indicate over 50% yield loss at least in major crops by 2050 (Razzaq et al. 2021). The rise in temperature by 1°C during nights for *rabi* crops would cost 3–6% drop in crop yields (Zhao et al. 2017a).

During *kharif* season, heavy rainfall events leading to flash flooding, river surges, rise in ground water table and water logging may lead to crop lodging and consequent yield losses. Besides, other factors like reduction in number of rainy days, high amount of precipitation on a single rainy day (cloud bursting), short or prolonged monsoon or early retrieval of monsoon, unprecedented rainfall at harvest, etc., confound the problems. The crop losses due to water logging are estimated around US$ 74 billion annually (Li et al. 2019c; Kaur et al. 2020; Liu et al. 2020). Similarly, *rabi* crops may face problems like delayed sowing due to delay in harvest of preceding *kharif* crop in Asian countries thus exposing the seeds and seedlings of *rabi* crops to a low to very low temperatures at germination, emergence, and seedling establishment; high temperature at flowering, or at important crop growth stages and harvest, interspersed with low diurnal variation/mean average temperatures, shortened winter, and thus, affecting the productivity of different crops.

This apart, enhanced CO_2 levels in the atmosphere, ozone holes and intense infrared radiations, salinity ingress in coastal areas, reducing water tables in arid and semi-arid regions further aggravate the situation. Published research results indicate that increased CO_2 levels can offer several benefits for C_3 crops, such as enhanced yields, greater efficiency in resource use, improved competitiveness against C_4 weeds, decreased sensitivity to ozone damage, and in some instances, increased resistance to pests and diseases (Wang et al. 2020). However, surface ozone (O_3) pollution represents a major threat to global crop production and food security. Worldwide estimates of yield losses attributed to surface ozone exposure are approximately 3.6 ± 1.1% for maize, 2.6 ± 0.8% for rice, 6.7 ± 4.1% for soybean, and 7.2 ± 7.3% for wheat (Tai et al. 2021). The anticipated adverse effects of global warming on agricultural productivity and the resulting socio-economic consequences for food systems have been thoroughly examined in existing literature (e.g., Yuan et al. 2024; Cushman et al. 2022; Slattery and Ort 2019; Vermeulen et al. 2012; Hatfield et al. 2011; Fischer et al. 2005), However, there is limited understanding of the important interactions and trade-offs among various climate change-related biotic stresses affecting plant productivity (Rivero et al. 2022; Leisner et al. 2023), despite the fact that the influence of climate on plant disease development has been recognized for more than 2000 years (Luck et al. 2011; Chakraborty 2013).

6.2 Biotic Stress Buildup and Crop Losses Associated with Climate Change

A key concern regarding potential climate change is the rising frequency of extreme weather events, as highlighted in other sections of this chapter. These extreme weather events can also impact plant pathogens or insect pests, potentially leading to the proliferation of certain pathogens or pests and resulting in increased risks and yield losses from the associated diseases or infestations (Rosenzweig et al. 2001). Nevertheless, this topic is rarely explored in detail and new knowledge on the subject has started accumulating progressively (Seidel 2017). For instance, the effects of moisture deficit stress on crop diseases have received relatively limited research attention (Wakelin et al. 2018) compared to diseases of trees (Desprez-Loustau et al. 2006). Drought conditions can affect the stomatal openings in plant leaves (Mittler 2006), which in turn may alter the infection success of foliar pathogens (Bostock et al. 2014), as stomata serve as primary entry points for many pathogen species (Qi et al. 2018). Additionally, plant stress can influence the virulence or aggressiveness of pathogens (Bidzinski et al. 2016). As a result, climate change research must address the intricate interactions between abiotic and biotic stress factors (Ramegowda and Kumar 2015), more so under field conditions (Aung et al. 2018). Therefore, adopting an integrated approach that combines multiple, complementary research methods may offer the most effective strategy to address this complex challenge.

Assessing and predicting the future impacts of climate change on plant pathogen epidemics is highly complex due to the presence of numerous nonlinear interactions among the host, pathogen, vector, and environment. These dynamics are influenced by factors such as the specific pathosystem, severity of plant damage, type of cultivation system, and the broader socioeconomic context (Savary et al. 2019; Savary et al. 2006). Many reviews (more than 100 have been written on life cycles, inoculum levels, interactions with hosts, expression of host resistances, and epidemiologies (see Jones and Barbetti 2012; Fisher et al. 2012; Juroszek et al. 2020 for further details on the published reviews), information on how climate change parameters affect diseases caused by fungal, viral, or bacterial plant pathogens remains limited and insufficiently documented in the existing literature (Coakley et al. 1999; Chakraborty 2013).

Approximately 40% of global crop yields are lost each year due to pests and diseases. Annually, these losses result in a global economic cost exceeding US$220 billion, with invasive insects alone accounting for up to US$70 billion (Brajesh et al. 2023). Across various crops, the estimated global potential losses caused by insect pests range from around 50% in wheat to over 80% in cotton production. The estimated crop losses caused by pest attacks for other major crops are approximately 26–29% for soybean, wheat, and cotton; and 31, 37, and 40% for maize, rice, and potatoes, respectively (Oerke 2006). Obtaining standardized and quantitative data on crop losses at the global level is cumbersome and compare across crops, agroecosystems, and region-wise especially in connection with climate change. Crop insect pests and diseases cause significant global crop losses, ranging from moderate to high. For example, wheat, rice, maize, potato, and soybean experience yield reductions of 21.5%, 30.0%, 22.5%, 17.2%, and 21.4%, respectively, due to 137 identified pathogens and pests (Savary et al. 2019).

Smallholder farmers in some regions face severe pre-harvest losses. Survey indicated that nearly 50% of Central American bean and maize farmers and about 50% of potato farmers of South America experienced huge losses (Delgado et al. 2017). Substantial post-harvest losses estimated to the extent of 6–12% were also reported from developing tropical countries lacking infrastructure (Paterson and Lima 2010). Historically, plant diseases have even led to major societal and political upheavals, as evidenced in the Irish potato famine (Fry 2008) and the Bengal famine (Padmanabhan 1973; Chakraborty and Newton 2011).

Besides yield losses in food crop species, plant diseases can decrease species diversity, enhanced cost for control measures, and adverse impacts on human health (Ristainoa et al. 2021). These problems are particularly acute in food-insecure regions with rapidly growing populations, where new or re-emerging pests and diseases are common. Climate change is expected to worsen this situation, highlighting the urgent need for strategic crop health management to ensure sustainable agricultural practices and societal well-being (Fanzo et al. 2024).

Under global climate change, insect pests and pathogens also evolve or migrate for their survival. Changes in the climate can accelerate their evolution leading to development of new variants more suited to the altered climate conditions and changes in geographic distributions resulting in enhanced species diversity. It has been observed that during the twentieth century, distribution of crop pests and pathogens were mainly through growing agricultural production and trade, global warming has been reported to be a major factor for their poleward movement (Chakraborty 2013). Since 1960s, over 600 crop pests and pathogens have shifted poleward at an average rate of 2.7 km/year (Daniel et al. 2013; Barford 2013). Different groups and species have shown varied rate of shift with fungi, beetles, true bugs, mites, butterflies, and moths exhibiting movements to higher latitudes, whereas viruses and nematodes worms settling at lower latitudes. Notably, fungi and oomycetes demonstrate rapid movement (at 7 and 6 km/year, respectively) (Barford 2013).

Climate change over the years has drastically impacted mainly the driving forces of disease in plants, causing devastating effects on the health of crop, food security, and global agriculture. Climate is the key determinant of crop disease dynamics, influencing the occurrence, spread, and severity of plant pathogens (Mirzabaev et al. 2023). Factors like temperature, humidity, rainfall, wind, and extreme weather events can intensify or suppress disease outbreaks. Elevated temperature, high relative humidity, altered precipitation patterns, rising CO_2 levels, and other propounded occurrence of extreme weather events may provide conducive conditions for several plant pathogens (Lahlali et al. 2024; Hossain et al. 2024). Warm and humid environments, for example, often promote the growth of fungal and bacterial pathogens, while extended droughts may stress plants, causing them to be more susceptible to plant pathogens (Chakraborty and Newton 2011).

In addition, disruption in typical weather patterns due to climate change may result in unforeseen modifications in disease epidemiology, like spread of diseases to different areas or the emergence of more aggressive strains. These changes impose considerable problems to global food security, as disease-related crop losses are expected to rise (Velásquez et al. 2018) as discussed elsewhere in this chapter. For example, *Stemphylium* blight, which has been reported from Bangladesh on Chickpea as early as 1987, spread to Nepal and was found in India infecting both lentil and chickpea mainly in the North-Eastern Plain zone, Terai region and Pusa (Ghosh et al. 2012). Conversely, *Ascochyta* blight of chickpea, a major problem in North-Eastern India has by and large disappeared and moved to southern India (Chattopadhyay et al. 2019).

Plant viruses are responsible for nearly 50% of new plant diseases globally (Jones and Naidu 2019). Pandemics and epidemics caused by viruses affect staple food crops of world threatening global food security predominantly in developing tropical or subtropical countries. The threat is intensifying due to challenges in managing virus diseases under climate change conditions. Losses due to virus diseases in different crops vary widely worldwide ranging from complete crop failure to minor damage, leading to an estimated annual global economic loss to the tune of over US$30 billion (Thresh 1980; Jones and Naidu 2019).

Climate change greatly impacts agricultural production and insect pests. The models predicting crop yield losses due to insect pests' infestation, largely consider direct crop losses incurred due to the pest infestations and pesticide use but ignore the climate change impacts (Oerke 2006; Gregorc et al. 2012). In future, the change in climate may cause insect pest species to differ in their responses like a minor insect pest may become the serious pest geographically and cause relative impacts among crops (Zvereva and Kozlov 2006; Deutsch et al. 2008). Climate change has both direct and indirect effects both on crops and their corresponding insect pests. Direct impacts include effect on pests' reproduction, development, survival, and movement, whereas indirect effects mainly affect the interaction between pests, their environment and natural enemies, competitors, vectors, and mutualists

(Prakash et al. 2014). Climate induced changes can expand the geographic distribution of an insect pest, enhance its overwintering survival, increase number of cycles, disrupt plants and pest synchrony, modify interspecific interactions, raise the risk of migratory pests, invasion, increase insect-transmitted plant diseases, and reduce effectiveness of biological control and natural enemies (Subedi et al. 2023).

It has also been shown that agriculture, climate change, and their interactions result in significant decline in insect pollinators. Studies suggest that the tropical regions will face the highest risk to crop production due to pollinator losses. The threat is expected to be localized and predicted to be the highest in sub-Saharan Africa, northern South America, and Southeast Asia. Loss of pollinator can modify agricultural land use pattern and may pose a risk to human well-being (Devkota et al. 2024; Millard et al. 2023; Potts et al. 2010) and food security.

Hence, understanding the intricate interplay of climatic conditions and diseases and insect pests' incidence is critical for establishing effective mitigation methods, increasing crop resilience, and sustainable agricultural practices under changing climate. In this context, it would be appropriate to anticipate more climate-resilient strains of existing pathogens and insect pests. The effect on different pathogen groups will be based on their biology, host-specialization, survival, and dispersal mechanisms besides their ability to adapt to change in climate. A brief review of important and specific cases on how climate change impacts plant diseases and insect pests is provided below.

6.3 Climate Change Impact on Diseases and Insect Pests of Crops

6.3.1 Elevated Temperature

Temperature is a crucial climatic variable in plant disease incidence, and its alteration due to climate change has significant implications for agricultural ecosystems (Raza and Bebber 2022; Brajesh et al. 2023; Waheed et al. 2023). Between 2014 and 2019, disease outbreaks to the extent of 17.7% were linked to high temperatures, while 9.9% were due to low temperatures. The impact of higher temperatures on various plant pathogens and diseases vary widely. High temperatures (daytime at 35 °C and night time at 27 °C) significantly reduced the resistance of rice IRBB near-isogenic lines (NILs) carrying the Xa4 gene against bacterial leaf blight caused by *Xanthomonas oryzae* pv. *oryzae* (Webb et al. 2010). This reduction in plant resistance at higher temperatures is often associated with enhanced RNA-silencing-mediated resistance and reduced effector-triggered immunity (Liu et al. 2015). Similarly, the effect of high temperatures on rice blast caused by *Magnaporthe oryzae* varies widely by growing region. Since this pathogen cannot form blast lesions on rice below 14 °C or above 33 °C (Qiu et al. 2022), temperatures beyond this optimal range reduce the risk in the endemic areas (Wilson and Talbot 2009). However, warmer climates could lead to more severe outbreaks under favorable conditions due to earlier infection by the pathogen and colonization.

Across the globe, Wheat growers loose US$4.3–5.0 billion annually due to the incidence of stripe rust (*Puccinia striiformis* f. sp. *tritici or Pst*). The disease can occur at any growth stage right from one-leaf stage and can cause total loss of resistance in some varieties. Out of many environmental conditions affecting pathogen behavior and spread, temperature is the major factor associated with stripe rust outbreaks. Although low temperatures and high humidity favor the disease, the rust fungus exhibited adaptation to elevated temperature regimes also (Hovmøller et al. 2016; de Vallavieille-Pope et al. 2018). Recently, virulent and high temperature tolerant strains viz., *Pst Pst1*, *Pst2*, and *Warrier* have emerged in Europe, northern Africa, eastern Africa, and central Asia (Bhattacharya 2017; Wamalwa et al. 2022) and in southern central United States (Milus et al. 2009) and Australia (Loladze et al. 2014) causing severe disease outbreaks. Microsatellite genotyping and virulence phenotyping have detected greater degree of variability within *Pst1* and *Pst2* lineages, pointing out to the ongoing evolutionary potential of this pathogen across the globe (Nnadi and Carter 2021).

New dangerous races of wheat stem rust fungus caused by *Puccinia graminis* f.sp. *tritici* (*Pgt*) such as *Ug99* from Uganda (sub-Saharan Africa) as well as strains from Germany and Italy have evolved due to recent changes in climate and destroyed tens of thousands of hectares of wheat crop thus raising concerns among wheat producers globally (Singh et al. 2015) although several resistant varieties are available against this disease (Figueroa et al. 2016). In many countries. It has been reported that warmer climates will be advantageous to all other forms of rust which may lead to earlier infections, increased spread, and heightened disease severity due to evolution and better adaptation of the rust fungi (Caubel et al. 2017). Conversely, some diseases like potato late blight (*Phytophthora infestans*) and wheat powdery mildew (*Blumeria graminis f.* sp. *tritici*) usually infect the crop relatively at lower temperatures (Matić et al. 2018; Sparks et al. 2014), thus suggesting that these two diseases may be negatively influenced due to rise in temperature.

Incidence of Head blight (FHB) disease of wheat and other cereal crops caused by *Fusarium graminearum* species complex (FGSC) (Sarver et al. 2011) reduces yield and quality including mycotoxin contamination in the seeds consequently affecting food security (Dweba et al. 2017; Alisaac and Mahlein 2023). Warm and humid weather conditions favor outbreak of this disease leading to yield losses over 75% (Leplat et al. 2013).

Soil-borne RNA viruses like Wheat Yellow Mosaic Virus (WYMV) and Chinese Wheat Mosaic Virus (CWMV) can significantly decrease wheat yield, particularly during winter in China, often co-infecting and causing severe symptoms (Yang et al. 2022). WYMV typically causes 10–30% yield loss, but can reach even up to 70% during epidemics (Sun et al. 2013; Chen et al. 2014; Lei et al. 2023). CWMV infection is highly temperature-dependent and may be worsened further by climate change (Yang et al. 2022; Lei et al. 2023).

Temperature controls crucial aspects of the pathogens like, rate and duration of infection process, latency period, sporulation, dispersal of inoculum, and host resistance dynamics (Kweku et al. 2019). For instance, temperatures above 20 °C can deactivate temperature-sensitive stem rust resistance genes (*Pg*3 and *Pg*4) in oats.

High temperature boosts the multiplication and transmission of pathogens (especially fungi and bacteria), while simultaneously weakening the host defense mechanisms (Devi et al. 2022). Rising global temperatures can directly affect pathogens life cycles, host plant physiology, and their interactions with vectors (Velásquez et al. 2018; Legg et al. 2015). However, optimal temperature ranges for pathogen-host systems vary, with some pathogens being susceptible to temperature extremes (Porras et al. 2023). Warmer conditions can also shorten pathogen life cycle, allowing for multiple disease cycles within a single crop season, and expand the geographic range of diseases (Kumar and Mukhopadhyay 2025) allowing pathogens to establish in previously unsuitable regions. Conversely, incidence of plant diseases may also decline in areas where temperatures exceed their optimal range (Caminade et al. 2019; El-Sayed and Kamel 2020). Besides, elevated temperatures can stress plants, weakening their natural defenses, and making them more prone to susceptibility to infections. As climate change continues to determine temperature fluctuations, it is crucial to understand these impacts for developing adaptive strategies.

Climate change has positive influence on over-wintering pathogen resisting stage. For instance, a change in temperature from −6 °C to −10 °C has increased the survival of rust pathogen, *Puccinia graminis* infecting perennial ryegrass and tall fescue (Pfender and Vollmer 1999; Laine 2023). Mild winter conditions have been shown to increase the success of pathogen overwintering and subsequent disease prevalence, as observed with the powdery mildew fungus (*Podoshaera plantaginis*) infecting *Plantago lanceolata* (Penczykowski et al. 2015). Contrasting response to climate change-induced shifts in both tropical and temperate bacterial pathogens' geographical distribution and host range have been recorded. Higher temperature increased the geographical distribution of wilt-causing bacteria (*Ralstonia solanacearum, A. avenai* sub sp. *avenae, X. campestris* pv. *Vesicatoria, B. glumae*); and field distribution of soft rot-causing bacteria such as *E. chrysanthemi* and *Bacillus* sp. and in storage conditions. Conversely, the distribution of rot-causing bacteria, *E. carotovora* subsp. *Atroseptica,* and *Clostridium* in the temperate climate have decreased (Jones and Barbetti 2012) due to elevated temperature conditions.

Under future climatic conditions, the general trend in case of insect pests is their abundance and distribution. Each additional degree of temperature increase could lead to a further 10–25% yield reduction due to pest attacks (Shrestha et al. 2019). The pest survival and dispersal are also influenced by the change in climate. For instance, rising temperature accelerate the attainment of minimum flight temperatures faster in insects such as aphids and moths thereby aid in increased dispersal capability. Locust swarms alter their migratory paths due to climate induced vegetative changes (Cease 2024). Similarly, with every rise in a degree of temperature, the life table of insects will diminish. While metabolic rate of insects may tend to roughly double for every increase of 10 °C (Dukes et al. 2009) leading to increased herbivory and consequent altered insect pests' population dynamics. For instance, a rise in ambient temperature due to climate change could result in 2–4 additional generations of lepidopteran pests like *Helicoverpa armigera*, with their life cycle shortened by 5–6 days across locations, leading to increased crop losses (Rao et al. 2023).

The effects of increased temperatures are more pronounced in case of above ground insects than the subterranean insects, as soil acts as thermal insulation and minimize their impact (Bale et al. 2002). Tropical insect populations are predicted to experience a decrease in growth rate due to global warming because current temperatures are already near optimal for pest development. Conversely, insect pests in temperate zones are expected to see an increase in growth rate. A study examining pest population changes in wheat, rice, and maize under various climate change scenarios also supported this: warming will accelerate pest population growth in temperate wheat, while tropical rice pests may see a decrease in growth. Maize, grown in both regions, shows mixed responses, including range expansion and rapid reproduction of insect vectors, and reduced effectiveness of bio-control agents and predators (Deutsch et al. 2018). However, due to narrow ecological niche requirements, physiological tolerances, and varied temperature effects on insect phenology and life table, warming may not uniformly increase pests' proliferation and economic crop losses globally (Lehmann et al. 2020). The same authors, in their analysis, presented mixed responses among different insect species to warming. Analyses showed about 60% of all species exhibited reduced harmful impact and the rest with increased pest severity. In earlier studies which involved 1100 insect species, it has been predicted that by 2050, global warming would cause extinction of 15–37% of these species (Thomas et al. 2004; Hance et al. 2007).

Recent studies suggest that resistant crop varieties maintain their efficacy for extended periods but eventually become susceptible under changing climatic conditions. For instance, elevated temperatures have been found to reduce the resistance of the rice variety IR 62 against the brown planthopper, *Nilaparvata lugens* (Horgan et al. 2021). Studies on warmer temperature and rainfall on yellow stem borer (YSB) *Scirpophaga incertulas* indicated reduced risk of YSB. However, separate effect of warmer temperature and rainfall resulted in enhanced growth rate of YSB. Risk of YSB was high during abundant rainfall (Ali et al. 2020).

6.3.2 Rainfall Pattern

Climate change alters rainfall patterns which significantly impact incidence of crop disease, influence both the occurrence and severity of plant pathogens. Altered rainfall regimes, including increased intensity, irregular distribution, or prolonged droughts, can create conditions that either favor or hinder the development of diseases (Parker and Gilbert 2004; Luck et al. 2011; Brajesh et al. 2023). Excessive rainfall coupled with high humidity often support the growth and spread of fungal and bacterial pathogens, as moisture facilitates spore germination, dispersal, and infection. High moisture favors infection and virulence of many soil borne fungi and oomycete pathogens like root rot (*Pythium* and *Phytophthora*) (Delgado et al. 2020); *Myrtaceae* rust caused by *Puccinia psidii* (Meiklejohn et al. 2022); *Septoria* leaf blotch (*Zymoseptoria tritici*) (Boixel et al. 2022), and bacterial leaf blight disease caused by *Xanthomonas oryzae* pv. *oryzae* (Ahmad et al. 2023). Diseases like wheat rust, rice blast, and late blight of potatoes thrive better under wet conditions

(Hossain et al. 2024; Zhou et al. 2024). Recent research finding based on Bayesian model established that disease incidence was positively influenced by temperature, only when the relative humidity is high (Chiu et al. 2022). High rainfall or extended wet conditions create favorable environments for the proliferation of soil borne plant pathogens like *Phytophthora* and *Fusarium*, which can thrive well in moist conditions (Delai et al. 2024). For extensive sporulation in case of *Cercospora zeae-maydis* (Paul and Munkvold 2005) and the emergence of highly virulent *Phytophthora* strains (Granke, and Hausbeck 2020), high humidity is crucial. Gray mold disease of castor, reported as early as 1918 in the USA, limits global castor production (Domingo and Crooks 1945). Sparks Soares (2012) reviewed the disease and reported that incidence of gray mold is observed when temperature was in the range of 14.6–28 °C (optimal temperature is 25–28 °C). Though several factors contribute to gray mold incidence, two factors, namely, number of rainy days and temperature play a significant role. In India, castor is an important crop and the recent rise in rainfall, temperature coupled with high humidity during cultivation, the crop is prone to severe incidence of the disease causing huge economic loss to the resource poor farmers of SAT regions of the country (Manjula et al. 2024).

On the other hand, drought conditions, which have become recurrent due to climate change, create moisture deficit stress in plants, weaken their immune systems, and making them more vulnerable for infection (Seleiman et al. 2021). Additionally, erratic rainfall patterns can disrupt traditional farming practices, such as planting and harvesting schedules as discusses earlier, further exacerbating the risk of disease outbreaks. As climate change continues to alter global precipitation patterns, understanding its effects on crop disease dynamics is essential for developing adaptive management strategies, such as improving irrigation systems, selecting drought-resistant crop varieties, and implementing integrated disease management practices to mitigate the impact on agricultural productivity as well as food security.

Heavy storms and flooding significantly increase the incidence of bacterial leaf blight in rice (Amin et al. 2022) and make potatoes more susceptible to pathogens like *Pectobacterium* spp. and *Ralstonia solanacearum* (Haverkort and Verhagen 2008). Flooding reduces plant fitness and alters plant-associated microbiota (Francioli et al. 2022) leading to long-term negative effects on soil health and crop productivity even after recession of floods. Notably, the same authors reported that flooded conditions suppress mutualistic fungi, but promote proliferation of pathogenic fungi like *Gibberella intricans*, *Mycosphaerella graminicola*, *Typhula incarnata*, and *Olpidium brassicae*. This necessitates a clear-cut knowledge on the possible effects of climate change-induced abiotic stresses on various biotic stresses, which can pose a great threat on sustainable agriculture (Hossain et al. 2024).

Insect pests are also greatly influenced both positively and negatively by rainfall, affecting their populations as well as behavior, with some species thriving very well in wet conditions while others are negatively impacted. For example, the leaf miner (GLM), *Aproaerema modicella* is one of the key pests of groundnut in many parts of India and in few other countries of Asia. Hot dry conditions during post-rainy season favor this pest. Heavy and persistent rains, high relative humidity (RH) coupled with low temperatures cause decline in pest numbers, whereas dry weather, bright sunshine, and occasional rains promote its buildup (Rao et al. 2010). Heavy downpour, wash away small pests such as aphids, jassids, whiteflies, mites, etc. (Pathak et al. 2012; Skendžić et al. 2021). Incidence of *Helicoverpa armigera* is reported to be high during November (which is the deviation of normal precipitation pattern) under high rainfall conditions. But some species are unaffected by alterations in rainfall pattern. For example, *Coccoidea lecanopsis formicarum* remained unaffected by altered rainfall pattern. Whereas some other species like *A. lineatus* are positively impacted by increased rainfall. Heavy incidence of *Spodoptera litura* has been reported in groundnut, cotton, chillies, and coriander under high rainfall conditions (Shrestha 2019). Subterranean insects such as wireworms (*Agriotes lineatus* L.), the pest of potato, sugar beet, and corn become serious because of high rainfall during summer and under drought in grasslands (Staley et al. 2007).

6.3.3 Relative Humidity

Relative humidity assumes a crucial role in the growth and distribution of crop diseases similar to temperature and rainfall, and its alteration due to climate change has significant implications for agricultural systems (Velásquez et al. 2018; Lahlali et al. 2024). High humidity creates a favorable environment for many plant pathogens, particularly fungi and bacteria, as moisture is essential for spore germination, infection, and disease progression. Diseases like powdery mildew, downy mildew, and bacterial blight thrive in humid conditions, where water droplets on plant surfaces facilitate pathogen growth and transmission (Strange and Scott 2005; IPCC 2021). Diseases that infect the floral parts of plants, particularly during the post-anthesis stage as in case of wheat blast (Hossain 2022) and *Fusarium* head blight (Cowger et al. 2009), rainfall is particularly critical. Elevated humidity during grain maturation stage enhances mycotoxin contamination. For example, high moisture conditions enhance the risk of mycotoxin (deoxynivalenol) contamination in wheat grains infected by *F. graminearum* (Hossain 2022; Pande et al. 2025; Ferreira et al. 2025). Fluctuating humidity with periods of high humidity followed by dry spells, can worsen the spread of diseases like anthracnose in beans (Poloni and Schirawski 2014). The combination of intermittent rains and resultant high humidity variations result in double danger to our food supply. Not only does it become harder to grow enough food (food security), but the food that is produced might also become unsafe to eat due to the mycotoxin contaminants (food safety) under climate change situations.

Climate change is expected to increase relative humidity in some regions due to rising temperatures and altered precipitation patterns, potentially leading to escalation in the frequency and severity of disease epidemics. Conversely, in areas where low humidity prevails diseases such as rice blast (*M. oryzae*), *Fusarium* wilt in tomatoes and bananas, and potato scab (*Streptomyces turgidiscabies and S. europaeiscabiei*) may decline (Sing et al. 2023; Hossain et al. 2024). But plants could experience increased stress, making them increasingly vulnerable to other pathogens or abiotic factors (Singh et al. 2016; Bebber et al. 2014). Moreover, fluctuations in humidity levels can also disrupt the balance between host plants and pathogens, leading to unpredictable disease dynamics. As climate change continues to alter humidity patterns, understanding the impact of such changes is essential for developing mitigating strategies.

As far as insect pests are concerned, high temperatures along with high humidity correlates with whitefly outbreak in cotton (Pathania et al. 2020). *B. tabaci* was initially regarded as a minor pest in cotton cultivation. However, in 2015, it emerged as a significant threat, becoming the dominant pest affecting cotton in north-western India (Dhillon and Sidhu 2016). In Punjab, the pest inflicted severe damage on approximately 60% of the cotton crop that year, resulting in a substantial decline in lint yield (Dhillon and Sidhu 2016). The study revealed that *B. tabaci* is a serious pest of cotton, with its population growth and dynamics being significantly affected by abiotic factors like elevated temperatures, relative humidity, solar radiation, and wind speed.

6.3.4 Elevated Carbon Dioxide (CO_2)

Rising atmospheric carbon dioxide (CO_2) levels pose complex and multifaceted effects on crop diseases by influencing both plant physiology and pathogen behavior. Increased CO_2 concentrations can enhance plant growth and photosynthesis, a phenomenon known as the "CO_2 fertilization effect," which may improve plant vigor and, in some cases, increase resistance to certain diseases (Taub 2010). However, this potential benefit is often offset by indirect effects that can increase plant vulnerability. For instance, elevated CO_2 can alter plant tissue composition, increasing the carbon-to-nitrogen ratio. While this shift may reduce nutrient availability for pathogens, it can also weaken plant defenses, making crops more vulnerable to certain pests and diseases (Navarro et al. 2020). Additionally, CO_2-driven changes in plant defense mechanisms may impair the ability of plants to keep away infections. From a pathogen perspective, higher CO_2 levels can create favorable microclimatic conditions that promote the growth and reproduction of certain fungi, bacteria, and viruses, particularly in humid environments. For example, fungal pathogens such as rusts and powdery mildews may thrive better due to denser plant canopies and increased humidity associated with CO_2-enriched atmospheres. Susceptibility of soybean to downy mildew disease caused by *Peronospora manshurica* got declined due to elevated CO_2. This calls for a thorough understanding of host-pathogen interaction to draw valid conclusions and draw suitable strategies. Majority of the studies were stimulated under chambers and in reality, the interplay of climate factors may be complex and may not be synonymous with findings of simulated conditions.

Moreover, climate change-induced shifts in temperature and precipitation patterns, coupled with rising atmospheric CO_2 can influence the spatial distribution and severity of crop diseases, potentially introducing pathogens to newer regions (Velásquez et al. 2018). Understanding the interplay between rising CO_2 levels and crop disease dynamics is crucial for developing adaptive strategies. Breeding disease-resistant crop varieties, optimizing crop management practices, and implementing integrated pest management systems will be essential to mitigate as well as for sustaining global food security due to climate change.

Some pests like general feeders, switch to more nutritious host thereby affecting the crop-pest dynamics. Elevated CO_2 levels affect the plant pest interactions, it may increase the nutritional quality of plants for some pests while for other pests it becomes tough and weakens the defense mechanisms. The C_3 plants (wheat, rice, cotton, etc.) require high CO_2 and thus positively gets affected by elevated CO_2 levels (increased photosynthetic activity) but negatively affected by insect response (less palatability) and vice-versa for C_4 plants (sorghum, corn, etc.) (Lincoln et al. 1984). Therefore, CO_2 indirectly affect the insects in quantity and quality of food consumption. Abundant accumulation of sugars and starches cause less palatability to insect pests because of changes in C:N ratio (Cotrufo et al. 1998; Shrestha 2019). In some cases, CO_2 concentration is directly proportional to the plant damage. Since nitrogen is essential for their growth and development, insects may consume more plant material to meet their nutritional needs under elevated CO_2 conditions. As higher CO_2 levels often reduce plant nitrogen content, certain pest species are prompted to feed more as observed in case of foliage feeders (Bezemer et al. 1998).

The cotton bollworm (*Helicoverpa armigera* Hübner) is recognized as a major global pest among crop insects due to its substantial economic impact and rapid spread into new areas. Moreover, *H. armigera,* which is considered to be the major global pest of diverse agricultural and horticultural crops, is reported to cause crop losses to the tune of US$ 7 billion annually (Jones et al. 2019). There are ten studies published in literature on the effects of elevated CO_2 on *H. armigera* over four plant families, viz., *Fabaceae*, *Malvaceae*, *Poaceae* and *Solanaceae*. Eight studies indicated that elevated CO_2 impaired many performance parameters of the pest mainly attributed to lower plant nitrogen content of the hosts under elevated CO_2 (Johnson et al. 2020) conditions. Such a reduction in nutritional quality may also indicate decreased concentrations of plant defense compounds like flavonoids, which play a crucial role in plant defense mechanisms. It has also been reported that elevated CO_2 reduced the Jasmonic acid (JA) levels and thereby resistance against *H. armigera* in tomato (Guo et al. 2012). Enhanced levels of sugars like glucose, sucrose, and fructose in leaves of soybean plants grown under elevated CO_2 is considered to be a preferential factor for Japanese beetle, *P. japonica* (Jason et al. 2005). Depending on the type of feeding also, response to elevated CO_2 vary in insect pests. Thrips (whole-cell feeders) exhibited an increase in population size (Bezemer et al. 1998), while aphids and whiteflies (phloem-feeding insect pests) have decreased population density and increased population growth rates under elevated CO_2 (Sutherst et al. 2011).

6.4 Impact of Climate Change on the Emergence of Few New and Serious Insect Pests

6.4.1 Fall Armyworm, *Spodoptera frugiperda* (J.E. Smith) in Maize

The fall armyworm (FAW) is native to the Americas (Sparks 1979; Mitchell et al. 1991), and considered to be a serious and destructive pest of tropical–subtropical region globally (Yan et al. 2022). The FAW has been first reported from Georgia as early as 1797 (Smith and Abbott 1987), which later spread to Latin America (Overton et al. 2021) where it remained as a major crop pest for many decades. But the pest was reported from Africa in 2016 and proliferated to sub-Saharan Africa (Goergen et al. 2016); Middle East (EFSA); Asian countries such as India (Ganiger et al. 2018; Sharanabasappa et al. 2018), Sri Lanka, Thailand, Vietnam, Bangladesh and Myanmar (Cock et al. 2017; EFSA 2023), and China (Guo et al. 2018; Jiang et al. 2019) during 2016–2018. Of late, incidence of FAW was also reported from Australia, Canary Islands, and New Caledonia (EFSA 2020; Kenis et al. 2022).

FAW is a polyphagous species, feeding on over 353 host plants (of 76 botanical families), which include a wide range of cultivated and wild plants Montezano et al. (2018). Reported occurrence on crop species most importantly belonging to the families of *Poaceae*, *Asteraceae,* and *Fabaceae*. However, it is considered as a serious pest of maize and sorghum mainly (Overton et al. 2021; Kenis et al. 2023).

Significant economic losses have been reported by several countries by reducing crop yields and increasing costs on pest control as well as management. Mitchell (1979) reported annual value of yield loss due to FAW incidence as around US$300 million in the USA, which rises to over US$500 million during major outbreak years (Overton et al. 2021). In Africa, losses owing to FAW damage varied widely between countries. Ghana and Zambia reported maize yield losses estimated to be US$284 and US$198 million, respectively. Estimated losses across 12 African countries ranged between US$2.5 and 6.3 billion in 2017 (Day et al. 2017). Across sub-Saharan Africa, it has been estimated that losses due to FAW is to the tune of US$13 billion per annum among the major crops like maize, rice, sorghum, and sugarcane (Abrahams et al. 2018).

Change in climate significantly influences prevalence and impact of the FAW on maize crops. Higher temperatures accelerate the pest's rate of growth and expand its geographical distribution, leading to increased infestations in previously unaffected regions (Yan et al. 2022). As a result, FAW infestations have increased in intensity and frequency, posing a major threat to maize production worldwide. For instance, in Southern Africa and Mozambique, due to increase in air temperature, FAW has caused significant damage to maize, a staple crop for many smallholder farmers (Sumila et al. 2024). Unlike other insect pests, FAW lacks a diapause phase, making it highly sensitive to minimum temperatures. In China, its survival and migration are greatly influenced by seasonal variations in temperature. During winter, low temperatures in southern regions restrict its survival and breeding, but as temperature rises in summer, the pest migrates northward, with peak infestations occurring in July (Urhausen et al. 2024). Further, climate change worsens the pesticide resistance in fall army worm populations. Warmer temperatures may expand the pest's range, leading to more frequent and widespread pesticide applications. This increased exposure can accelerate the development of resistance, making chemical control methods less effective over time. Similar trends have also been observed in South America, where prolonged warmer seasons have led to increased pesticide application rates and accelerating resistance evolution (Ma et al. 2021).

Additionally, climate change-induced shifts in rainfall patterns can also affect fall armyworm populations. In regions experiencing prolonged droughts, moisture stressed maize crop becomes more vulnerable to infestations as plant defenses weaken. Conversely, excessive rainfall and high humidity can create favorable conditions for fungal and bacterial infections that may indirectly impact fall armyworm populations, either reducing their numbers or influencing their feeding behavior (Sokame et al. 2025). Studies conducted in Africa and China have underscored that both low and high rainfall extremes contribute significantly to the variability in infestation levels, highlighting the need to incorporate precipitation patterns into predictive models and integrated pest management strategies (Day et al. 2017; Sokame et al. 2025) in future.

FAW thrive in warm temperatures, with the optimal range for their development being between 26 °C and 30 °C; while they can survive in a broader range, very cold temperatures significantly hinder their growth and survival, meaning cooler conditions can limit their populations and slow their lifecycle (Savadatti et al. 2023). However, most of the publications have focused on revealing the impact of FAW at selected specific temperature regimes, though the impact of various temperature regimes has not been explored in detail (Kumara et al. 2022).

FAW can adjust its gene expression to adapt to the changing environment, allowing it to survive in a wider range of conditions and potentially increasing its ability to invade new areas, posing a greater threat to agriculture (Yan et al. 2022). Multi-omics analysis revealed that FAW possesses high temperature tolerance by modulation of chitin-related genes (Yan et al. 2024). Researches on the effect of elevated temperature on FAW paid attention mainly on the physiological characteristics (Kebede 2018; Lu et al. 2019; Plessis et al. 2020; Ashok et al. 2021), while few other reports focused on molecular mechanisms. But it has been reported that only a few genes (heat shock protein genes and TPS genes) regulate thermal tolerance of FAW (Yao et al. 2020; Zhou et al. 2020; Liang et al. 2021; Samanta et al. 2021; Shang et al. 2021), bringing out the dire need of understanding the physiology as well as mechanisms of thermal tolerance to control and manage FAW.

6.4.2 Climate Change and Locust *Schistocerca gregaria* (Forsskål) Outbreak

Desert locusts (*Schistocerca gregaria*) is one of the most destructive migratory pests globally threatening food security in vulnerable regions of Africa, the Middle East, and South Asia (Liu et al. 2024). A swarm of just 1 km^2 contains 800 lakh locusts which can devour food crops sufficient to feed about 35,000 people in a day (Roussi 2020; FAO 2020). The recent (2019–2020) locust crisis was the most severe even in the past 25 years in East African countries which further migrated and invaded countries like Yemen, Somalia, Ethiopia, and India simultaneously (Salih et al. 2020; Roussi 2020). According to a report of Food and Agriculture Organization (FAO 2020), the locust swarm invaded 13 other countries prior to arrival in India and damaged the crops grown in about 0.28 million hectares.

Hydro-climatic factors significantly influence the biological activities and behavior patterns of desert locusts, and rainfall plays the most predominant role in the locust life cycle. It enhances soil moisture and supports complete development of locusts' eggs. Rainfall also promotes vegetation growth to feed hoppers. This is clearly evidenced by the consecutive cyclones bringing above-normal rainfalls (Meynard et al. 2020) near the Horn of Africa, facilitating locust outbreaks. Locust abundance is also linked to long-term dry and wet conditions, as well as short-term fluctuations in precipitation, air temperature, wind speed, soil moisture, and inundation of flood waters. These factors impact breeding, incubation, development, swarming, and migration (Liu et al. 2024; Muhammad et al. 2025). Ten countries have also been identified as hotspots for locusts which are: Northwest Africa covering Morocco and Algeria; Sudan and Kenya of East Africa; Mauritania and Niger of the African Transition Zone; Saudi Arabia and Yemen located in the Middle East; and India and Pakistan (Liu et al. 2024).

The same authors based on their robust analyses of data reported that the transition between El Niño and La Niña phases of the El Niño-Southern Oscillation (ENSO) significantly affects locust abundance and distribution. El Niño years, with above-average rainfall in certain regions, lead to 65% larger active locust areas compared to La Niña years. Specifically, In East Africa and the Middle East, locust dynamics was more active during El Niño years compared to La Niña years. This is mainly contributed by the above-average rainfall in the southern Red Sea, the Horn of Africa, and northern Somalia owing to El Niño effects, which creates favorable conditions for locust breeding (WMO 2016).

The overall locust habitat has expanded since 1985 and may continue to expand at least by 5% by the end of twenty-first century, even under mitigated carbon emission situations (Guan et al. 2021). Under climate change scenario, the historical hotspots continue to be at high risk and the emerging risk is projected to shift northeastward, affecting western India and west-central Asia during the latter half of the twenty-first century (Liu et al. 2024).

Recent locust outbreaks highlight that the impacts of climate change which extend beyond shifts in average temperature, encompassing a rise in extreme and unprecedented events. These extremes are further intensified through a "butterfly effect" triggered by non-climatic factors such as political instability, armed conflicts, inadequate governance, fragile early warning systems, and limited financial capacity—disproportionately affecting the most vulnerable communities, particularly across the African continent (Salih et al. 2020).

6.5 Climate Change Effects on Pollinators and Predators

Climate change is significantly impacting beneficial insects, including pollinators and natural enemies of pests, which are crucial for agricultural productivity and ecosystem balance.

Pollinators like bees (60%), flies (19%), bats (7%), wasps (5%), beetles (5%), birds (4%), butterflies, and moths (4%) aid essentially in reproduction of about 60% of cultivated crops globally, contributing to nearly 35% of global food production. Pollination through insects alone contributes nearly over one third of global food production (FAO 2008). Changes in climate has weakened considerably the proliferation, geographic distribution, and pollination activities of important pollinator species such as bees, moths, and butterflies.

Agricultural productivity in many countries depends on pollinators and other beneficial arthropods. Naturally occurring and managed insects provide pollination services for horticultural fruit and vegetable crops, and for seed production. Within gene banks, insect pollinators are often the key to successfully being able to produce ample quality seed during regeneration activities. In addition, predatory arthropods are used to manage plant pests through integrated pest management. Critical plant-pollinator relationships, as well as the balance between plants, pests, and beneficial arthropods can easily become disrupted by altered climatic conditions (Volk et al. 2023).

In India, Basavarajappa (2010) observed that high temperatures at Mysore resulted in 2% decline in rock bee, *Apis dorsata* colonies every year. In cotton crop, elevated CO_2 concentrations altered the preference of lady beetle to aphid prey and enhanced their efficiency of biological control of aphids by lady beetle (Chen et al. 2005). This study was the first evidence to indicate that varied CO_2 levels alter the feeding preferences of the predator.

Pollinators as well as beneficial arthropods are likely to experience direct and indirect climate change effects. Pollination and biological control approaches using arthropods might also suffer from increased use of agrochemicals used to mitigate the rise in damaging insect pests and diseases. It is important to point out that not all climate change effects are a cause for alarm, as some research shows that many insect pollinators and biological control insects have the capacities for plasticity and resilience (Sloggett 2021).

6.6 Breeding Innovations

Changing climate affects global food security by decreasing productivity of agricultural crops across the globe. Food security is a great concern especially for all policymakers in different countries as the global population is likely to surpass 10 billion in the near future. Although substantial yield enhancement in many crops could be achieved globally through conventional breeding approaches but the yields of crops in many countries remained stagnant (Gerber et al. 2024). However, modern breeding techniques or breeding innovations coupled with efficient improved agronomic practices help for fulfilling future food demand. A brief review of genomics assisted breeding and genetic engineering tools that would help to mitigate ill effects of climate change are presented.

6.6.1 Genomics-Assisted Breeding for Biotic Stress Resistance/Tolerance Traits

6.6.1.1 Genome Mapping of Biotic Stress Resistance/Tolerance Traits

DNA markers refer to the variations in DNA sequence among the individuals of a species, which arise due to mutations such as insertions, deletions, and substitution or errors in replication of DNA. Several DNA marker techniques have emerged over the last 50 years (Kadirvel et al. 2015); however, following the introduction of next generation sequencing technologies, it is now relatively easy to develop high throughput DNA markers especially simple sequence repeats (SSRs) (Geethanjali et al. 2024) and single nucleotide polymorphism (SNP) (Senthilvel et al. 2019) instantly and rapidly with less cost and time. These high throughput DNA markers are used in gene mapping and marker-assisted selection (MAS) of important traits in plant breeding programs.

Genome mapping mainly involves two strategies: (i) linkage-based QTL mapping and (ii) linkage Disequilibrium (LD)-based association mapping or genome-wide association study (GWAS). The QTL mapping uses specialized mapping populations, such as recombinant inbred lines (RILs), doubled haploids lines (DHLs), backcross inbred lines (BILs), $F_{2:3}$ lines, etc., derived from two phenotypically contrasting parents while the GWAS uses the germplasm accessions to determine marker-trait associations (MTAs). These strategies have been powerful to discover genes/QTLs associated with traits important for crop productivity and climate resilience.

Genome mapping coupled with positional cloning strategies have helped to identify several resistance (R) genes, which are used in molecular breeding to develop disease or insect resistant crop cultivars. A few examples of cloned genes or mapped genes, which are implicated in biotic stresses resistance in crops are listed in Table 6.1.

6.6.1.2 Marker-Assisted Breeding (MAB) Methods

Conventionally, plant breeders select desirable plants based on phenotype, which is often slow, tedious, time-consuming, and misleading. The MAS helps to overcome these issues when gene-specific or tightly linked DNA markers are available for predicting traits. The DNA markers serve as proxy to phenotypic selection, shorten breeding cycle, enable pyramiding of multiple genes and provide rapid, cost-effective selection options. Over the past 50 years, significant advancements have been achieved in mapping genomes and identifying molecular markers to select genes or QTLs associated with biotic stress resistant or tolerant traits in crops. Gene/QTL specific DNA markers are integrated with the standard breeding methods to enhance the effectiveness and genetic progress of breeding programs (Collard and Mackill 2008).

Table 6.1 Examples of genes known for resistance to pathogens and insect-pests in crops

Crop	Disease	Genes	Reference
Rice	Blast (*Magnaporthe grisea*)	*bsr-d1*, *Pb1*, *Pb2*, *Pb3*, *Pib*, *Pid2*, *Pid4*, *Pi5*, *Pit*, *pi21*, *Pish*/*Pi35*, *Pi36*, *Pi37*, *Pi50*, *Pi54rh*, *Pi56*, *Pi56(t)*, *Ptr*, *Pita2*, *Pi63*, *Pi64*, *Pizh*, *Pigm*, *Pi9*/ *Pi2=Piz5*/*Piz-t*/ *Pi50*/ *PigmR(e)*/ *Pizh*, *Pid3*/*Pi25*/*Pid3-I1*, *Pia*/*PiCO39*, *Pi54rh*/ *Pi54=Pikh*, *Pik*/ *Pikm*/ *Pik-p*/ *Pi1*/ *Pi7*, *Pita*, *Ptr* = *Pita2*	Liu et al. (2021) and Simon et al. (2023)
	Bacterial leaf blight (*Xanthomonas oryzae* pv. *oryzae*)	*Xa3* = *Xa26*, *Xa4(b)*, *xa5*, *Xa7*, *Xa7(a)*, *Xa10*, *Xa21*, *Xa23*, *Xa27*, *xa41(t)*, *Xa47(t)*, *Xa1*/ *Xa2* = *Xa31 (t)*/*Xa14*/ *Xa45(t)*/ *CGS-Xo1*, *xa13*/ *OsSWEET11*/ *Os8N3*, *xa41(t)*/ *OsSWEET14*/ *Os11N3*, *Xa23*, *Xa47(t)*, *xa25*/ *OsSWEET13*/ *OsMtN3*	Liu et al. (2021)
	Brown planthopper (*Nilaparvata lugens*)	*BPH14*, *BPH30*, *BPH17(b)*, *BPH6*, *BPH29*, *BPH32* = *BPH3*©, *BPH1* = *BPH10* = *BPH18* = *BPH21*/ *BPH2* = *BPH26*/*BHP7*/*BPH9(d)*	Simon et al. (2023)
	Virus	Rice tungro virus (tsv1), Rice Yellow Mottle Virus 1 (*RYMV1*), *RYMV2*, *RYMV3*, Rice stripe virus (*STV11*)	Simon et al. (2023)
Wheat	Rust (*Puccinia triticina*, *P. graminis*, *P. striiformis*)	*Lr1*, *Lr9*, *Lr10*, *Lr13*/*Ne2*, *Lr14a*, *Lr21*, *Lr22a*, *Lr42*, *Lr47*, *Lr58*; *Sr22*, *Sr22b*, *Sr26*, *Sr27*, *Sr33*, *Sr35*, *Sr43*, *Sr45*, *Sr46*, *Sr50*, *Sr60* (WKS2), *Sr61*, *Sr62*, *SrTA1662* (*Sr66*); *Yr5*, *Yr5a* (*Yr5*), *Yr5b* (*YrSP*), *Yr7*, *Yr10* (*Yr10cg*), *Yr15*, *Yr27*, *Yr28* (*YrAs2388*), *Yr36* (*WKS1*), *YrNAM*, *YrU1*, *Lr34*/*Yr18*/*Sr57*, *Lr67*/*Yr46*/*Sr55*, *Yr46*/*Lr67*/*Sr55*, *Yr18*/*Lr34*/*Sr57*	Li et al. (2023), Geethanjali et al. (2024), and Tong et al. (2024)

Maize	Northern corn leaf blight (*Exserohilum turcicum*)	*ZmWAK-RLK1, remorin* (*ZmREM6.3*),	Zhu et al. (2021) and Chen et al. (2022)
	Southern corn leaf blight (*Bipolaris maydis*)	*ZmCCoAOMT2, rhm1, RppK*	
	Banded leaf and sheath blight (*Rhizoctonia solani*)	*ZmFBL41*	
	Gray leaf spot (*Cercospora zeae-maydis*)	*ZmCCoAOMT2*	
	Common rust (*Puccinia sorghi*)	*Rp1-D*	
	Southern rust (*Puccinia polysora*)	*ZmREM1.3*	
	Head smut (*Sphacelotheca reiliana*)	*ZmWAK*	
	Gibberella ear rot (*Gibberella zeae*)	*ZmAuxRP1*	
	Fusarium ear rot (*Fusarium verticillioides*)	*ZmLOX3, ZmLOX12*	
	Aspergillus ear rot (*Aspergillus flavus*)	*ZmLOX3*	
	Gibberella stalk rot (*Gibberella acuminate*)	*ZmCCT, ZmAuxRP1*	
	Anthracnose stalk rot (*Colletotrichum graminicola*)	*Rcg1*	
	Maize rough dwarf disease	*ZmGDIα-hel*	
	Maize dwarf mosaic disease	*ZmTrxh, ZmABP1*	

6.6.1.2.1 Marker-Assisted Backcrossing (MABC)

Traditionally, backcross method is used to incorporate one or a few genes into an elite variety for improvement. To restore the elite genetic back ground with the desirable trait, at least three backcrosses are necessary. The backcross breeding method can be efficiently managed with the use of gene or QTL specific DNA markers (Collard and Mackill 2008). The MABC involves foreground selection (selection of progenies carrying the target gene/QTL by genotyping of backcross population using gene/QTL specific DNA markers) followed by background selection (using genome-wide markers to assess the proportion of recurrent parent genome recovery) to identify the backcross plants that carry the target gene/QTL with maximum proportion of recurrent parent genome. The MABC has been extensively used for improving disease or insect resistance traits in various crops.

MABC has been extensively used to develop gene pyramids involving R genes to achieve broad spectrum resistance in rice for resistance against blast disease (*Pi37* + *Pid3*, *Pi5* + *Pi54*, *Pi54* + *Pid3*, *Pigm* + *Pi37*, *Pi9* + *Pi54*, *Pigm* + *Pi1*, *Pigm* + *Pi33*, *Pigm* + *Pi54*, *Pi2* + *Pi46* + *Pita*, *Pi2* + *Pi46* + *Pigm*, and *Pib* + *Pi25* + *Pi54*), bacterial blight disease (*Xa4* + *xa5* + *Xa21*, *xa5* + *xa13* + *Xa21*, *xa5* + *Xa21*, *Xa21* + *Xa33*, and *Xa23* with other genes, as well as *Xa4* + *xa5* + *Xa7* + *xa13* + *Xa21*) and resistance to both blast and bacterial blight (*Pi2* + *Xa7* and *xa5* + *xa13* + *Pi54* + *qSBR7–1* + *qSBR11–1* + *qSBR11–2*) (Liu et al. 2021).

6.6.1.2.2 Marker-Assisted Forward Breeding (MAFB)

Pyramiding of genes through MABC is a long and tedious process when the goal of the breeding programs is to improve varieties for multiple traits. Moreover, MABC is mostly used to improve the existing varieties, which are released under the category of essentially derived varieties. Alternatively, the MAFB strategy is followed to pyramid multiple resistance/tolerance genes/QTLs to different biotic stresses and come up with new varieties. The strategy involves development of multi-parent based inter-cross populations and exercising MAS for various traits at different filial generations so that breeding lines with combination of genes/QTLs are selected and stabilized. The MAFB strategy has been extensively used to pyramid multiple biotic and abiotic stress resistance/tolerance traits in rice (Dixit et al. 2020).

6.6.1.2.3 Marker-Assisted Recurrent Selection (MARS)

The MABC and MAFB strategies require prior information on the markers linked to gene/QTL, which are discovered through gene/QTL mapping. These strategies have been highly successful for relatively large effect genes/QTLs while ineffective for minor genes. Therefore, MARS is considered a potential strategy to pyramid minor genes in the breeding population specially to improve quantitative traits. Traditionally, recurrent selection (RS) is used for developing improved breeding lines through a population approach involving several cycles of inter-mating of progenies followed by selection. The MARS is an upgraded version of the traditional RS, wherein QTL detection and MAS are used in the breeding scheme. The procedure involves (i) selection of parents (two lines carrying complementary traits of interest), (ii) developing segregating populations (F_3), QTL detection, and (iii) selection of the best progenies carrying maximum number of desirable QTLs by MAS and (iv) inter-mating of the selected progenies and MAS. Maximum genetic gain in the population would require three recombination cycles. The MARS is effective for improving quantitative traits; however, it has been used only to a limited extent, especially in private sector breeding programs in the crops like maize, sunflower, and soybean (Nayak et al. 2017).

6.6.1.3 Genomic Selection

Genomic selection (GS) denotes identifying elite lines based on genomic-estimated breeding values (GEBVs). The GEBVs are derived utilizing genotypic (high density) and phenotypic data from "training population." The training population is subjected to high throughput genotyping (either array-based or GBS) and phenotyped for important traits. Subsequently, prediction models such as random regression best linear unbiased predictor (RR-BLUP), least absolute shrinkage and selection operator (LASSO), Bayesian ridge regression (RR), Bayesian LASSO, BayesA, BayesB, BayesCp, and random forest regression (RFR) are created using genotypic and phenotypic data of the training population. This model is then applied to a validation set (which is only genotyped) to determine the best fit. The most accurate model can then be used to calculate the GEBVs of each plant in the segregating populations (based on genotypic data only) and superior plants with high GEBVs can be selected. The GS allows combining the desirable QTLs including minor and epistatic QTLs in a single genotype, which has been a limitation to date; therefore, GS is considered to be the most effective strategy for improving quantitative traits. Merrick et al. (2021) demonstrated that GS can accurately predict quantitative resistance to stripe rust disease in wheat.

6.6.2 Genetic Engineering

6.6.2.1 Transgenic Technology

While conventional breeding methods were the key to evolve high-yielding, stress-tolerant, and climate-resilient cultivars in cultivated crop species, several factors—such as limited genetic variability, the presence of favorable alleles in the wild gene pool, hybridization barriers that make the introgression of donor segments difficult, and linkage drag—slow down the progress of crop improvement. Transgenic technology is a straightforward approach that helps overcome these barriers, making it possible to introduce novel sources of resistance genes from both related and foreign organisms into desirable genotypes. This approach involves cloning of desirable genes, construction of expression vectors, genetically transforming the recipient genotypes through direct or indirect gene transfer methods, selecting positive transformants and evaluating them for the target trait expression (Anjanappa and Gruissem 2021; Yu et al. 2022). Among direct and indirect gene transfer methods, biolistic, and *Agrobacterium* mediated transformation, respectively, have been widely used for producing transgenic crops.

Table 6.2 Examples of heterologous expression of non-plant and plant derived genes conferring biotic stress resistance in transgenic plants

S. No	Transgenic crop	Transgene introduced	Source of transgene	Biotic stress resistance to	Gene transfer method	Reference
	Non plant sources of biotic stress resistant genes					
1	Rice	*Cry1Ac, Cry2A*	*Bacillus thuringiensis*	Yellow stem borer Striped stem borer	Particle bombardment	Gahakwa et al. (2000)
2	Maize	*vip3Aa* and *Cry1Ab*	*Bacillus thuringiensis*	Fall armyworm (*Spodoptera frugiperda*) Asian corn borer (*Ostrinia furnacalis*)	*Agrobacterium tumefaciens*	Chen et al. (2024)
3	Pigeonpea	*Cry1Ac and Cry2Aa*	*Bacillus thuringiensis*	*Helicoverpa armigera*	*Agrobacterium*-mediated	Ghosh et al. (2017)
4	Cowpea	*Vip3ba*	*Bacillus thuringiensis*	*Legume pod borer (Maruca vitrata)*	*Agrobacterium*-mediated	Bett et al. 2017
5	Chickpea	*Cry1Ac*	*Bacillus thuringiensis*	*Helicoverpa armigera*	*Agrobacterium tumefaciens*	Chakraborty et al. (2016)
6	Groundnut	*Cry1EC*	*Bacillus thuringiensis*	*Tobacco cutworm (Spodoptera litura)*	*Agrobacterium*-mediated	Tiwari et al. (2008)
7	Cotton	*Cry1Ia12*	*Bacillus thuringiensis*	*Fall Armyworm (Spodoptera frugiperda) and Cotton Boll Weevil (Anthonomus grandis)*	*Agrobacterium tumefaciens*	de Oliveira et al. (2016)
8	Cotton	*Cry1Ia12*	*Bacillus thuringiensis*	*Pink bollworm (Pectinophora gossypiella)*	*Agrobacterium tumefaciens*	Tariq et al. (2022)

9	Cotton	*Cry3Bb1 and Cry3*	*Bacillus thuringiensis*	Pink bollworm (*Pectinophora gossypiella*), army bollworm (*Helicoverpa armigera*)	Agrobacterium tumefaciens	Zafar et al. (2022)
10	Sugarcane	*Cry1Ac*	*Bacillus thuringiensis*	*Stem borer (Diatraea saccharalis*	*Particle bombardment*	Zhou et al. (2018)
11	Potato	*Cry3A*, SN-*19*	*Bacillus thuringiensis*	Potato tuber moth, *Phthorimaea operculella Colorado potato beetle (Leptinotarsa decemlineata*	*Agrobacterium tumefaciens*	Salim et al. (2024) and Mi et al. (2015)
12	Brinjal	*Cry1Aa3, Cry1Ac, Cry1Fa1*	*Bacillus thuringiensis*	*Egg plant fruit and shoot borer Leucinodes orbonalis)*	*Agrobacterium tumefaciens*	Rai et al. (2013), Shrivastava et al. (2011), and Kumar et al. (1998)
13	Tobacco	*Cry1Ah*	*Bacillus thuringiensis*		*Agrobacterium tumefaciens*	Sun et al. (2012)
	Plant derived biotic stress resistant genes					
14	Wheat	*CERK1-V*	*Haynaldia villosa*	*Powdery mildew, Yellow rust, and Fusarium head blight pathogens*	*Agrobacterium tumefaciens*	Fan et al. (2022)
15	Rice	*ZmNBS25*	*Zea mays*	*Rice sheath blight pathogen (Rhizoctonia solani)*	*Agrobacterium tumefaciens*	Xu et al. (2018)

(continued)

Table 6.2 (continued)

S. No	Transgenic crop	Transgene introduced	Source of transgene	Biotic stress resistance to	Gene transfer method	Reference
16	Sugarcane	*Kunitz trypsin inhibitor (SKTI) and Bowman–Birk inhibitor (SBBI)*	*Glycine max*	*Sugarcane borer (Diatraea saccharalis)*	*Particle bombardment*	Falco and Silva-Filho (2003)
17	Potato	*Oryzacystatin I and II (OCI and OCII)*	*Oryza sativa*	*Colorado potato beetle (Leptinotarsa decemlineata)*	*Agrobacterium tumefaciens*	Cingel et al. (2017)
18	Tomato	*Mi-1.2* gene	*Solanum lycopersicum*	*Root-knot nematode (Meloidogyne javanica)*	*Agrobacterium tumefaciens*	Goggin et al. (2006)
19	Tomato	*Proteinase inhibitor PI-II and carboxypeptidase inhibitors (PCI)*	*Potato*	*Heliothis obsoleta Liriomyza trifolii*	*Agrobacterium tumefaciens*	Abdeen et al. (2005)
20	Tomato	*Cysteine proteinase inhibitor (CeCPI)*	*Colocasia esculenta*	*Root knot nematode (Meloidogyne incognita)*	*Agrobacterium tumefaciens*	Chan et al. (2010)
21	Tomato	*Ferredoxin-I protein*	*Sweet pepper*	*Ralstonia solanacearum*	*Agrobacterium tumefaciens*	Huang et al. (2007)
22	Tobacco	*AhMPK3*	*Arachis hypogaea*	*Spodoptera litura*	*Agrobacterium tumefacien*	Kumar et al. (2009)
23	Tobacco	*Chitinase gene (SmchiC)*	*Serratia marcescens*	*Necrotrophic fungus (Botrytis cinerea) Spodoptera frugiperda*	*Agrobacterium tumefacien*	Navarro-González et al. (2019)
24	Tobacco	*Taxane biosynthesis genes - taxadiene synthase, taxane 5α-hydroxylase, taxane 13α-hydroxylase*		*Fall armyworm*	*Agrobacterium tumefacien*	Xu et al. (2024)

25	Tobacco	*Transcription binding factor AtNPR1*	*Arabidopsis thaliana*	*Spodoptera litura*	*Agrobacterium tumefaciens*	Meur et al. (2008)
26	Tobacco	Kunitz-type (PKPI- (*PKI1*, *PKI2*) and potato type I protease inhibitor genes (*PPI3A2*, *PPI3B2*, *PPI2C4*)	*Potato*	*Botrytis cinerea* *Alternaria alternata*	*Agrobacterium tumefaciens*	Turrà et al. (2020)
27	Arabidopsis	*Kunitz-type protease inhibitor GsKTI*	*Glycine soja*	*Helicoverpa armigera*	*Agrobacterium tumefaciens*	Zhang et al. (2023)
28	Arabidopsis	*PtrWRKY18 and PtrWRKY35*	*Populus trichocarpa*	*Botrytis cinerea*	*Agrobacterium tumefaciens*	Guo et al. (2020)
29	Arabidopsis	*Constitutive disease resistance genes PtCDR2* and *PtCDR8*	*Poncirus trifoliata*	*Pseudomonas syringae*	*Agrobacterium tumefaciens*	Ying et al. (2020)
30	Arabidopsis	*Pathogenesis related protein AsPR5*	*Allium sativum*	*Fusarium oxysporum f.* sp. *Cepae* *Botrytis cinerea*	*Agrobacterium tumefaciens*	Rout et al. (2016)

6.6.2.1.1 Transgenic Approaches for Biotic Stress Resistance

Development of transgenic crops for resistance to biotic stresses involves three different approaches, namely, (1) heterologous expression of genes obtained from a genetically distant species, (2) overexpression of endogenous genes obtained from the same or sexually compatible species, and (3) RNAi may be used to switch off activity of harmful genes found naturally in an organism or those belonging to disease causing agents or pest or through genome editing technologies (Yu et al. 2022).

Heterologous Gene Expression

Heterologous gene expression involves transferring a desirable protein-coding gene from a donor species into a recipient species, enabling the host cells to produce the foreign protein (Watts et al. 2021). Appropriate selection of suitable expression systems and several other factors, including codon usage, GC content, Kozal and Shine-Dalgarno sequences, gene splicing, and mRNA stability highly influence the levels of protein expression in host cells of recipient species (Kaur et al. 2017; Watts et al. 2021). To optimize and ensure stable protein expression, various tools such as Synthetic gene designer (Wu et al. 2006), Optimizer (Puigbo et al. 2007), Gene Designer (Villalobos et al. 2006), Visual gene developer (Jung and Karen McDonald 2011) are available to modify or synthesize native genes, accounting for these factors.

Using this approach, genes from microorganisms and non-plant sources have been successfully expressed in transgenic plants to enhance their tolerance to biotic stresses, particularly insect resistance. An integrated analysis of multiple studies by Klumper and Qaim (2014) revealed that the cultivation of insecticidal transgenic crops reduced pesticide usage by 41.67%, lowered the cultivation cost incurred due to pesticide by 43.43%, increased crop yield by 24.85%, and enhanced farmer profitability by 68.78%. A vast majority of these transgenic events involves heterologous expression of insecticidal *Cry* genes (coding for crystal proteins that are δ–endotoxins) from various strains of *Bacillus thuringiensis* (Bt) to impart resistance to chewing insect pests from lepidopteran, coleopteran, and hemipteran orders in a number of crop species involving cotton, cowpea, eggplant, maize, potato, rice, soybean, sugarcane, and tomato (Table 6.1). Several *Cry* genes and their variants, and their combinations with vegetative insecticidal protein (vip3) gene from *Bacillus thuringiensis* have been deployed in gene stacking to enhance protection against these insect pests (Chen et al. 2017; Yu et al. 2022).

Particularly, in maize, private companies like Monsanto and Dow AgroSciences have pioneered the development of US approved Genuity® SmartStax corn, which incorporates multiple resistance genes (*cp4, cry1Fa2, cry2Ab2, cry34Ab1, cry35Ab1, cry3Bb1, cry1A.105*) to protect against major lepidopteran pests including corn earworm (*Helicoverpa zea*) and coleopteran pests, European corn borer (*Ostrinia nubilalis*), Southwestern corn borer (*Diatraea grandiosella*), Northern corn rootworm (*Diabrotica virgifera zeae*), Western corn rootworm (*Diabrotica virgifera virgifera*), and Mexican corn rootworm (*Diabrotica barberi*) (Sanchis 2011; Sheoran et al. 2023). The Trecepta™ technology from Bayer integrates three insecticidal genes (*Cry1A.105, Cry2Ab2,* and *Vip3Aa20*), each with distinct modes of action, to enhance resistance against maize borers, corn earworm, black cutworm (*Agrotis ipsilon*), fall armyworm, and Western bean cutworm (*Striacosta albicosta*) (Buntin 2024). Besides insectidal proteins from *Bacillus thuringiensis* and other non-plant sources, heterologous expression of plant-derived genes from species such as *Glycine max*, *Glycine soja*, *Populus trichocarpa*, *Poncirus trifoliata*, and *Allium sativum* in *Arabidopsis thaliana* has proven to be effective against herbivorous insects, fungal pathogens, and bacterial infections (Zhang et al. 2023; Guo et al. 2020; Ying et al. 2020; Rout et al. 2016). These findings underscore the potential of harnessing plant-derived resistance genes to develop transgenic crops with improved resilience against biotic stresses. A few examples of heterologous expression of non-plant and plant derived genes imparting resistance to biotic stresses in transgenic plants as reported in literature are provided in Table 6.2.

Overexpression of Endogenous Genes

The second strategy for enhancing biotic stress resistance involves the overexpression of endogenous genes through genetic engineering. This is accomplished by driving the target genes with strong or constitutive promoters, leading to elevated protein production beyond natural levels. Overexpression of protease inhibitors and genes involved in secondary metabolite biosynthesis (such as phenols and terpenoids) has been shown to strengthen plant defense against insect herbivory (Clemente et al. 2019; Singh et al. 2020; Al-Khayri et al. 2023). Similarly, resistance genes, including nucleotide-binding leucine-rich repeat (NBS-LRR) proteins and pathogenesis-related (PR) genes, have been upregulated to induce systemic acquired resistance (SAR) against fungal and bacterial pathogens (Jain and Kumar 2015; Zhang et al. 2022a, b, c). Additionally, enhanced expression of key genes in the RNA silencing pathway, such as those encoding Dicer and Argonaute proteins, has been explored to confer viral resistance by targeting and degrading viral RNA, thereby preventing its replication. For example, overexpression of rice *OsLCB2a1* gene, encoding a subunit of serine palmitoyltransferase (SPT) in the sphingolipid biosynthesis pathway, provided improved defense against brown plant hopper *Nilaparvata lugens* (Begum et al. 2016). Overexpression of *WRKY30* in rice increased resistance to *Rhizoctonia solani* and *Magnaporthe grisea* by activating pathogenesis-related (*PR3*, *PR10*) and jasmonic acid (JA) biosynthesis-related genes (*LOX*, *AOS2*) leading to increased endogenous JA accumulation during fungal infection. Similarly, endogenous *ZmDEF1* gene, encoding the defensin 1 protein overexpressed in maize cultivars LC1 and LVN99 enhanced resistance against weevils by inhibiting insect gut α-amylase (Vi et al. 2019). Overexpression of *PtaMYB186* resulted in a fuzzy phenotype with increased trichome density that enhanced the resistance to white-marked tussock moth larvae in Populus (Plett et al. 2010). Synergistic expression of endogenous stress related genes along with *cry1Ac* gene resulted in a dramatic improvement of stemborer resistance in sugarcane (Zhou et al. 2018). Overexpression of endogenous genes ensures better genetic compatibility, regulatory acceptance, and minimal unintended effects compared to traditional transgenic approaches involving foreign genes. It enhances stress resistance while preserving the plant's natural physiology and agronomic traits. However,

Table 6.3 RNAi-based host induced gene silencing to combat pest and pathogen in crop plants

Crop	Pest/Pathogen	Targeted pest/pathogenic genes for silencing	Effect of host induced gene silencing on pest/pathogen	RNAi approach for HIGS	Reference
Rice	*Rice stripe virus* (RSV)	*RSV nucleocapsid protein (NCP) gene*	Tenfold increase in virus derived siRNAs, imparting resistance to rice stripe virus	siRNA	Li et al. (2016)
Rice	Striped stem borer (*Chilo suppressalis*)	*csu-novel-miR15*	Growth inhibition and delayed pupation	dsRNA	Jiang et al. (2017)
Rice	Rice Blast Fungus *Magnaporthe oryzae*	*Pathogenic genes MoABC1, MoMAC1 and MoPMK1,*	Reduction in transcription levels of targeted fungal genes	siRNA	Zhu et al. (2017)
Rice	Striped stem borer (*Chilo suppressalis*)	*Csu-miR-14 / CsSpo and CsEcR (Spook and Ecdysome Receptors)*	Impairment of ecdysone biosynthesis led to high mortality and developmental defects in RSB	amiRNA	He et al. (2019)
Rice	Striped stem borer (*Chilo suppressalis*)	*Csu-novel-miR260 disembodied*	Negative regulation of ecdysteroid biosynthesis	amiRNA	Zheng et al. (2021)
Rice	Striped stem borer (SSB; *Chilo suppressalis*)	*CsFAR (fatty acyl-CoA reductase)*	Downregulation of FAR increased mortality rate by 80% and darkening of epidermis	siRNA	Sun et al. (2022)
Rice	Striped stem borer (*Chilo suppressalis*)	*CssHsp (heat shock protein)*	Larval mortality (60–90%) by the time of pupation	ds RNAi	Mao et al. (2022)
Rice (transgenic line with Cry1ca)	Striped stem borer (*Chilo suppressalis*)	*microRNA-7322-5p/p38/ Hsp19 heat shock protein 19*	Enhanced cry1ca sequestration and increased mortality of SSB	dsRNA	Wu et al. (2023)

Rice	Striped stem borer (*Chilo suppressalis*)	*CsmiR1579-CsKr-h1 (Krüppel homologue 1)*	Negative regulation of transcription factor Kr-h1 leading to decline in larval growth and survival.	miRNA	Tang et al. (2024)
Rice	Brown plant hopper (*Nilaparvata lugens*)	*Glutathione S-transferase NlGST1-1 gene*	Retarded growth rate of nymphs and insect female fecundity	dsRNA	Yang et al. (2021)
Wheat	*Fusarium graminearum*	*Chitin synthase Chs3b*	Reduced fungal growth		Cheng et al. (2015)
Wheat	*Grain aphid (Sitobion avenae)*	*Salivary sheath protein (SHP)*	Decline in growth, reproduction, and survival rates of the aphid	dsRNA	Abdellatef et al. (2015)
Wheat	*Grain aphid (Sitobion avenae)*	*Lipase maturation factor 2-like gene*	Reduced survival rate and reproductive capacity		Xu et al. (2017)
Wheat	*Grain aphid (Sitobion avanae)*	*Chitin synthase 1 (CHS1)*	Impaired insect molting leading to reduced survival rate	dsRNA	Zhao et al. (2018)
Wheat	*Sitobion avenae*	*Zinc finger protein (SaZFP)*	Reduced fecundity and survival of aphids	dsRNA	Sun et al. (2019)
Wheat	*Puccinia striiformis f.* sp. *tritici*	*Glycine-serine-rich effector (PstGSRE1)*	Reduction in fungal biomass and virulence of Pst	siRNA	Qi et al. (2019)
Wheat	*Blumeria graminis* f. sp. *tritici*	*β2-tub, SvrPm3$^{a1/f1}$*	Impaired fungal haustorium formation in seedlings and reduced growth in adult plants	ckRNAi	Schaefer et al. (2020)
Wheat	*Sitobion miscanthi*	*Putative salivary protein gene SmDSR33*	Reduced fecundity and survival	dsRNA	Zhang et al. (2023)

(continued)

Table 6.3 (continued)

Crop	Pest/Pathogen	Targeted pest/pathogenic genes for silencing	Effect of host induced gene silencing on pest/pathogen	RNAi approach for HIGS	Reference
Maize	Western corn rootworm (*Diabrotica virgifera virgifera*	*Vacuolar -ATPase A*	Reduced feeding	dsRNA	Baum et al. (2007)
Maize	Western corn rootworm (*Diabrotica virgifera virgifera*	*DvSnf7*	Downregulation leads to corn root worm mortality and reduced emergence	dsRNA	Head et al. (2017)
Maize	*Diabrotica virgifera virgifera*	*boule (bol) gene*	Reduction in egg production and hatching rate	dsRNA	Niu et al. (2017)
Maize	Maize stem borer *(Chilo partellus)*	*Chitinase CpCHI*	Phenotypic abnormalities and mortality of stemborer	dsRNA	Adeyinka et al. (2023)
Soybean	*Heterodera glycines* *Fusarium oxysporum*	*Chitin synthase (SCN-CHS)*	Down regulation leads to suppression of cyst and egg formation in nematode Reduced mycelial growth of pathogen	dsRNA	Kong et al. (2022)
Soybean	*Alfalfa mosaic virus, Bean pod mottle virus, and Soybean mosaic virus*	Replicase genes of *Alfalfa mosaic virus, Bean pod mottle virus, and Soybean mosaic virus* AMV, BPMV, and SMV (ABS)	Inhibition of viral replication and clearance of virus within a short time period	dsRNA	Zhang et al. (2011)
Sugarcane	*Sphenophorus levis*	*V-ATPase subunit E*	Down regulation of the gene resulted in reduced larval growth and survival	dsRNA	Mohan et al. (2021)

Sugarcane	*Sporisorium scitamineum*	*Beta-1,6-glucanase precursor encoding gene SsGlcP*	Reduced fungal biomass of the smut pathogen	dsRNA	Wu et al. (2025)
Cotton	Cotton bollworm (*Helicoverpa armigera*)	*P450 monooxygenase*	Silencing impairs larval tolerance to Gossypol	ds-RNA	Mao et al. (2007)
Cotton	*Aphis gossypii*	*AgCYP6CY3-P1*	Delayed growth and development of the aphid	dsRNA	Zhang et al. (2023)
Mustard	Mustard aphid (*Lipaphis erysimi*)	*Sucrase 1*	Impaired parthenogenetic reproduction	dsRNA	Dhatwalia et al. (2022)
Brassica napus (rape)	Green peach aphid, (*Myzus persicae*)	*Partitioning-defective protein 6 (Par6)*	Reduced survival rate and impaired fecundity	dsRNA	Zhang et al. (2024)
Potato	*Leptinotarsa decemlineata*	*Ecdysone receptor (EcR)*	Reduction in larval weight	ds RNA	Hussain et al. (2019)
Potato	*Henosepilachna vigintioctopunctata*	*β-Actin*	Complete larval mortality	dsRNA	Xu et al. (2023)
Tomato	Western Flower Thrips (*Frankliniella occidentalis*)	*Coatomer protein subunit epsilon (CopE)* *Toll-like receptor 6 (TLR6)*	Increased larval mortality	dsRNA	Venkatesh et al. (2023)
Tomato	*Myzus persicae*	*Acetylcholinesterase 1 gene (Ace 1)*	Reduction in aphid colonization	amiRNA	Faisal et al. (2021)
Tomato, tobacco	*Helicoverpa armigera*	*Chitinase (HaCHI) gene*	Impaired insect molting and metamorphosis resulted in overall reduction in growth and survival;	siRNA	Mamta et al. (2016)
Tomato	*Helicoverpa armigera*	*Ecdysone Receptor (HaECR)*	Reduction in growth and survival of the insect	amiRNA	Yogindran and Rajam (2021)

under selection pressure, insects co-evolve with plants and adapt by aggressive feeding on transgenic plants more than non-transgenic ones, discouraging the use of overexpressing endogenous genes in host plants (Cloutier et al. 2000; Winterer and Bergelson 2001; Abdeen et al. 2005: Singh et al. 2020). Further, while these approaches have demonstrated significant potential in controlled studies, their translation into commercially viable crop varieties remains a challenge, requiring further validation and field-level applications (Stuiver and Custers 2001; Yu et al. 2022).

RNA Interference Technology

RNA interference (RNAi) is a natural conserved defense mechanism found in all eukaryotic organisms regulating genes involved in plant growth and response to stress. Since its discovery by Fire and Mello in *Caenorhabditis elegans* in 1998, RNA interference (RNAi) is widely recognized as an effective approach for gene silencing (Fire et al. 1998). Introducing double-stranded RNA (dsRNA) into a cell, activates the RNAi pathway, resulting in the silencing or suppression of specific target genes (Mamta and Rajam 2018). For stable gene silencing, RNA interference (RNAi) constructs are designed to express dsRNA in transgenic plants. These constructs contain partial sequences of the target gene in sense and antisense orientations, separated by a small intron as a spacer, and are driven by a suitable promoter. The dsRNA once expressed is cleaved by the RNAse III enzyme Dicer to produce small interfering RNAs (siRNAs) and microRNAs (miRNAs). These small RNAs associate with Argonaute (AGO) proteins and accessory factors to form the RNA-Induced Silencing Complex (RISC). Following degradation of sense strand, the activated RISC, guided by the antisense strand, binds to the complementary target mRNA. Depending on sequence complementarity, RISC either induces mRNA cleavage (perfect match) or translation repression (partial match), leading to effective gene silencing (Rajam 2020; Kaur et al. 2021).

RNA interference (RNAi) has become a potential strategy for developing crops resistant to both pests and pathogens (Kanakala and Ghanim 2016; Kaur et al. 2021; Bharathi et al. 2023). Using host plant-mediated RNAi, plants are designed to express dsRNA that silence key genes in insects and pathogens (Table 6.3). When ingested or absorbed, the dsRNA triggers gene silencing, disrupting vital processes required for survival, growth, or infection (Gordon and Waterhouse 2007; Kunte et al. 2020). By blocking these critical genes crucial for the pest's development or the pathogen's virulence, RNAi effectively protects crops from biotic stresses, enhancing their resistance to pests and diseases.

Modified pre-miRNAs (Bally et al. 2020) and artificial miRNAs (amiRNAs) (Guo et al. 2018) are also currently being used to replace native miRNAs in order to achieve efficient transient transgene silencing against sucking and chewing pests (Yogindran and Rajam 2016; Yogindran and Rajam 2021). Artificial miRNAs (amiRNAs) have been shown to be more effective than native dsRNA-based RNAi in overcoming resistance mechanisms in certain insect pests. For example, *Chilo suppressalis*, the rice striped stem borer, showed resistance to RNAi triggered by dsRNA but responded more effectively to amiRNA-induced silencing (Zheng et al.

2020). In addition to RNAi, several other RNA silencing mechanisms like rasiRNA (repeat-associated small interfering RNAs), tasiRNA (transacting siRNA), and virus and transgene exosiRNA (exogenic siRNA) pathways are being explored for crop improvement (Guo et al. 2014).

Unlike the traditional transgenic approach, RNAi provides targeted pest and pathogen resistance without producing transgenic proteins. Additionally, since RNAi results in gene knockdown rather than complete knockout, it offers a more controlled and reversible approach compared to the recent genome editing techniques. These advantages make RNAi a valuable tool for sustainable agriculture, contributing to improved crop resilience and reduced reliance on chemical pesticides.

Despite the advantages of RNA interference (RNAi) in pest management, several factors influence its efficacy in insects. Key challenges include dsRNA stability, cellular uptake efficiency, the integrity of the core RNAi machinery, systemic spread of RNAi signals, and the susceptibility of target genes to silencing (Cooper et al. 2019). Additionally, applying RNAi-mediated host-induced gene silencing (HIGS) for developing biotic stress-resistant crops faces significant obstacles. These include high costs associated with dsRNA design and synthesis, potential off-target effects due to non-specific binding within target and non-target genomes, and the risk of RNAi resistance evolving in insect pests and pathogens (Yan et al. 2020).

The case of RNAi-based control of Western corn rootworm (WCR) exemplifies these challenges. MON87411, a key component of Bayer's SmartStax PRO technology, is the first genetically modified (GM) maize developed using RNAi and is authorized for commercial cultivation in over 15 countries. As a third-generation transgenic maize, it provides targeted protection against WCR (*Diabrotica virgifera virgifera*) (Head et al. 2017). One of the critical targets for RNAi-based pest control in WCR is the vacuolar ATPase (vATPase) subunit genes. The dsRNAs designed to silence *vATPaseA* and *vATPaseE* in WCR have also been shown to reduce the fitness of *Leptinotarsa decemlineata* (Colorado potato beetle) due to sequence similarity between the target genes in both species. Interestingly, dsRNAs derived from *L. decemlineata* were more effective in suppressing gene expression than those designed specifically for WCR. Further, WCR has also evolved resistance to *DvSnf7* dsRNA in transgenic maize, primarily due to impaired dsRNA uptake in the gut lumen. This resistance is non-specific, leading to cross-resistance against multiple dsRNAs (Khajuria et al. 2018). Another major hurdle is that, while coleopterans are generally susceptible to RNAi, insects of other orders—such as Lepidoptera, Diptera, Hymenoptera, and Hemiptera—exhibit variable and often unpredictable responses to dsRNA-induced gene silencing. This variability poses a significant challenge to the widespread adoption of RNAi-based pest control strategies (Cooper et al. 2019). To overcome these challenges, non-transgenic RNAi approaches that involve delivery of dsRNA through foliar spray, irrigation water, injections, nanoparticles, and oral feeds are being experimented, although their effects are expected to be short lived (Yan et al. 2020).

Breeding varieties resistant to biotic stresses and resilient to climate change through these transgenic approaches are subjected to stringent regulatory policies that require extensive biosafety assessments, lengthy approval processes, and public scrutiny. These regulations increase the cost and time needed for commercialization, making transgenic crops less attractive compared to newer, genome-editing techniques that face fewer regulatory hurdles across the globe. Examples of heterologous expression of non-plant and plant derived genes conferring biotic stress resistance as reported in case of transgenic plants in the recent literature are provided in Table 6.3.

6.6.2.2 Genome Editing

One of the key hurdles in breeding resistant varieties is the limited availability of suitable donor sources in the gene pool of an organism. Screening large collections of germplasm accessions revealed that biotic stress response has been skewed towards susceptibility, particularly for pests and pathogens with a wide host range. For example, identification of high degree of resistance to diseases like gray mold caused by in Castor (Senthilvel et al. 2024), dry root rot caused by *Macrophomina* in pulses (Sadhana et al. 2024) has been a daunting task. While mutation breeding helps to create new variations, the physical and chemical mutagens cause random mutations, with most mutations being lethal than beneficial. Moreover, identifying desirable macromutations is like searching a needle in a haystack and involves screening large mutant populations under controlled and field conditions.

Unlike random mutagenesis, site directed mutagenesis aims to create mutations within a target region in the genome. Performing such targeted mutational edits in the genome book of an organism is known as genome editing or gene editing. The methodology involves introducing double strands breaks in the target region of a genome and repairing the broken ends through error prone non homologous end joining or by high-fidelity homologous recombination mechanism. The genome editing approach uses sequence specific nucleases like ZFN (Zinc-finger nucleases), TALEN (Transcription activator like effector nucleases), CRISPR/Cas9 (Clustered regularly interspaced short palindromic repeats/associated nuclease Cas9) to induce site directed mutations for precise genetic manipulation, without the need for introducing foreign genes as done in conventional genetic modification methods.

The gene editing technique is a boon to correct or alter a defect causing gene without interfering with the rest of the genome. It helps in tinkering the genome by allowing deletions, insertions, and substitutions at nucleotide, gene, and chromosomal level with high fidelity. Creation of such novel, precise, and heritable functional variations are useful in broadening the gene pool in plants. Further, gene edited crops with modified endogenous genes do not suffer strict regulatory issues like genetically modified transgenic crops, and can be readily released for commercial cultivation. These advantages have expedited the use of gene editing techniques in crop improvement. Specifically, the CRISPR /Cas9 system originally identified as a natural bacterial defense mechanism against virus infections, has been skyrocketing in developing gene edited plants, precisely manipulated for plant architecture, flowering time, male sterility, nutrient use efficiency, and resistance to biotic and abiotic stresses (Kumar et al. 2019; Awan et al. 2022).

6.6.2.2.1 Approaches to Develop Biotic Stress Tolerant Plants through CRISPR Technologies

Conventional CRISPR/Cas9 system comprises two essential components: A guide RNA (gRNA), which leads the Cas9 nuclease to a specific DNA sequence, and the Cas9 enzyme, which induces a double-strand break (DSB) at target site. This process requires the presence of a Protospacer Adjacent Motif (PAM), a short DNA sequence adjacent to the target site, which is essential for Cas9 binding and cleavage. The cell subsequently repairs the double-strand break by its inherent repair pathway, either through non-homologous end joining (NHEJ), which can bring in small insertions or deletions, or homology-directed repair (HDR), which enables precise genetic modifications (Sander and Joung 2014; Gallagher and Haber 2018).

While NHEJ is efficient, it introduces random insertions or deletions effects, leading to gene disruptions, whereas HDR enables precise DNA modifications but is less efficient and limited in non-dividing cells. Additionally, Cas9-induced DSBs can cause unintended mutations, reducing editing accuracy. To address these challenges, alternative nucleases such as fCas9 (fidelity-enhanced Cas9), Cas12, Cas13a, Cas13b (Bhatia and Yadav 2023; Liu et al. 2022; Kondrateva et al. 2021), deactivated Cas9 (dCas9) for gene regulation, RNA-processing Cas9 (RCas9) for RNA-targeted editing (Bravo-Vázquez et al. 2024) are being explored for enhancing precision and versatility.

Further advancements like base editing and prime editing offer accurate, targeted genetic modifications without generating DSBs, minimizing off-target effects and enhancing genome-editing efficiency. DNA base editing process utilizes a Cas enzyme for accurate binding to the target DNA sequence along with a single-stranded DNA-modifying enzyme to edit specific sequence. Two types of base editors are currently available—cytosine base editors (CBEs) and adenine base editors (ABEs)—which are capable of inducing all possible transition mutations (Kantor et al. 2020). In CBE, a deamination reaction converts cytosine to uracil, which is recognized as thymine by the DNA replication machinery ultimately leading to C-G to T-A transition (Komor et al. 2016). In ABE, through a similar deamination reaction, adenine (A) is converted to inosine (I), which is interpreted as guanine (G) by the DNA replication machinery, thereby facilitating accurate A-T to G-C conversions (Gaudelli et al. 2017). To extend beyond the transition mutations, a highly programmable system called prime editing has been developed, enabling precise genome modification by locating target sites and directly replacing them with the desired sequence (Anzalone et al. 2019; Anzalone et al. 2022).

6.6.2.2.2 Development of Biotic Stress Tolerant Crops through CRISPR Technology

CRISPR-mediated approaches have been employed to develop climate-smart, pest- and disease-resistant crops by (i) deleting susceptible genes, (ii) disrupting genes through insertions/deletions (INDELs) in coding or promoter regions, and (iii) replacing faulty genes with functionally resistant candidates (also referred as biomimicking) (Zaidi et al. 2020). Several plant susceptibility (S) genes have been knocked out to enhance resistance against several fungal and oomycete pathogens causing numerous diseases including rust, mildews, and root rot (reviewed by Zaidi et al. 2020; Rato et al. 2021). In case of wheat, TaCIPK14 [Calcineurin B-like protein (CBL)-interacting protein kinases (CIPKs)] gene governs susceptibility reaction to stripe rust (*Puccinia striformis*) infection. Broad spectrum and non-race specific/partial/horizontal resistance to stripe rust was achieved by simultaneously knocking down three homologues of TaCIPK14 gene using CRISPR/Cas9 technology. The knock down phenotypes carried frame shift mutations of C terminal region of predicted amino acid sequences that impair with function of the TaCIPK14 gene (He et al. 2023). Mutating mildew resistance locus (MLO) and enhanced disease resistance locus (EDR) through CRISPR gene editing mechanism has proven to trigger an immune defense against powdery mildew in wheat by hampering the successful host pathogen interactions (Wang et al. 2014; Zhang et al. 2017). In case of rice, disruption of an ethylene responsive transcription factor gene (OsERF922) decreased the blast lesions caused by *Magnaporthe oryzae* (Wang et al. 2016). In a similar way, CRISPR-based genome editing has been effectively utilized to improve resistance to bacterial diseases in crops like rice, tomato, and citrus. In rice, targeted editing of promoter regions of *OsSWEET11, OsSWEET13,* and *OsSWEET14*—members of SWEET transporter gene family—resulted in resistance to bacterial blight caused by different strains of *Xanthomonas oryzae* pv. *oryzae* (Kim et al. 2019; Oliva et al. 2019). Editing of *SlJAZ2,* coronatine co-receptor enhanced resistance to bacterial speck disease (caused by *Pseudomonas syringae* pv. *tomato*) by suppressing stomatal opening and subsequent bacterial entry and colonization (Ortigosa et al. 2019). Targeted editing of transcription factors like *CsWRKY22* and *CsLOB1* increased resistance to canker disease in Citrus species caused by *Xanthomonas citri* (Wang et al. 2019; Peng et al. 2017).

Unlike fungal and bacterial diseases, where resistance is often achieved by modifying plant defense genes, developing virus-resistant plants through a CRISPR-mediated approach primarily involves directly targeting viral genomes within infected plants. Particularly, CRISPR has been successfully applied to confer resistance against an important class of plant infecting viruses called potyviruses. Potyviruses rely on the host plant's translation machinery to replicate, and one key component they use is the eukaryotic initiation factor 4 (eIF4) complex, particularly eIF4E and eIF4G, that encode for cap binding proteins. These proteins help the viral genome to bind to host ribosomes for protein synthesis. By targeting the *eIF4E* gene using CRISPR, resistance to potyviruses affecting a number of host species including cassava and cucumber has been achieved (Gomez et al. 2019; Chandrasekaran et al. 2016). Biomimicking of the *eIF4G* allele in the susceptible rice cultivar IR64 has resulted in widening the gene pool through generation of novel allelic variants, carrying in-frame mutations that impart resistance to rice tungro spherical virus (Macovei et al. 2018). Viral genome targeting approaches conferred resistance to black streaked dwarf virus in rice (Zhang et al. 2019; Wang et al. 2021) and yellow leaf curl virus in tomato (Tashkandi et al. 2018).

Evolving insect resistant plants through gene editing also involves two approaches. The first approach involves targeting endogenous plant genes to impart resistance to the pest. Examples of this approach, includes targeted mutagenesis of calcium dependent protein kinase GmCDPK38 (Li et al. 2022) for cut worm resistance and GmUGT gene for resistance to *Helicoverpa armigera* and *Spodoptera litura*, in soybean (Zhang et al. 2022a, b, c). Similarly, deletion of CYP71A1 gene in rice resulted in tryptamine conversion to serotonin that induced resistance to *Nilaparvata lugens* (Brown plant hopper) and *Chilo suppressalis* (striped stemborer), by reducing the growth and development of these pests on the gene edited host plants (Lu et al. 2018). Gene editing to impart leaf color changes (green to red leaf) by altering anthocyanin pathway in crops like tobacco opens up the possibility of exploiting host non preference by insects such as *Helicoverpa armigera* and *Spodoptera litura* (Moon et al. 2022). Besides editing one or few genes, a large mutant library targeting 502 endogenous cotton genes involved in host-insect interaction has been recently created through CRISPR/Cas9 technology. Screening this mutant library, based on insect bioassays of gene edited plants, identified a defense and stress responsive gene *GhMLP423* that conferred broad spectrum insect resistance against both sucking and chewing pests by activating salicylic acid pathway and eliciting wound signaling cascades (Sun et al. 2024).

The second approach involves editing target genes in insect pests that interrupt the herbivory mechanism by impairing host recognition, mating pair recognition, insect development, fecundity, and fertility. Knocking down odor receptor genes in locusts and *Spodoptera litura*, affected the pheromone reception, choice of mating partners and host plants (Li et al. 2016; Koutroumpa et al. 2016). Knocking out vital genes essential for growth and development has led to severe malformed phenotypes affecting pest survival, such as the *White* and *paired* genes in *Ceratitis capitata*, causing embryo deformities (Meccariello et al. 2017), and the *abd-A* (*Abdominal A*) gene in *Plutella xylostella, Spodoptera litura, Spodoptera frugiperda, and Ostrinia furnacalis* resulting in abnormal embryonic development, embryonic lethality, anomalous body segmentation, and defective prolegs (Huang et al. 2016; Bi et al. 2016; Wu et al. 2018; Bi et al. 2022). CRISPR/Cas technology has become an effective tool for explaining the functional roles of gene mutations associated with insect resistance to insecticides and insecticidal proteins expressed in transgenic crops (reviewed by Moon et al. 2022). Understanding these resistance mechanisms can facilitate the futuristic development of more effective and sustainable pest management strategies.

6.7 Future Directions

(i) Climate resilience in crops should focus broadening the genetic base through introgression of novel traits for biotic and abiotic stresses tolerance from wild relatives in to a well-adapted genetic background. This highlights the need for identification of specific germplasm possessing trait of interest and their utilization as a donor for resistance/tolerance.

(ii) Identification of major QTLs associated with major biotic and abiotic stresses tolerance for even minor QTLs associated with traits associated with tolerance of these stresses and pyramiding them in to a high yielding background may be a viable alternative.

(iii) Studies on Latest Smart Agricultural Technologies (SAT) including remote sensing, thermal sensing, satellite imaging to identify intensity of heat stresses, airborne infrared/visible imaging spectrometer to detect heat stresses, are coming up though not wide-spread, owing to their high-cost involvement. Cost effectiveness of these technologies is in the offing wherein these technologies will soon become a routine practice in climate resilience in many agricultural crops.

(iv) In the crop protection front, major threat due to climate change is that once been a minor disease or insect pest of a crop are emerging as major ones. To cite few examples are *Botrytis* blight, also known as gray mold in castor, *Macrophomina* root rot or dry root rot in safflower and sesame; *Sclerotium rolfsii* causing stem rot in groundnut and the sucking pests like thrips, and leaf hoppers in many other crops. Soil application of efficient strains of bio-control agents, integrated pest management modules and spray application of cost-effective new molecules should be deployed.

(v) Understanding the host-parasite relationship and inheritance of resistance, and adopting appropriate phenotyping procedures to identify resistant donors and differentials are the keys in managing the emerging new pests and diseases complex. QTLs related with resistance to few key insect pests and diseases need to be targeted and transferred and if need be, genome editing techniques need to be adopted for better combating the menace due to climate changes. "Omics" approaches (transcriptome and proteomics) to decode gene function especially under stress conditions imposed by climate changes may also throw light on understanding the stress principles per se so that these can be effectively tackled either through recombination or molecular breeding.

(vi) Although there are many national and regional plant protection agencies that undertake surveillance of key pests and diseases, a harmonized global fore-warning and reporting mechanisms are still missing.

(vii) Monitoring and surveillance for emerging plant diseases and insect pests face multiple challenges. These include underreporting of incident plant diseases and insect pests which is mainly influenced by socio-political and economic factors both at national and international levels. Political concerns on sovereignty and trade often discourages countries to report any epidemic of diseases and insect pests due to fear of repercussions like trade restrictions or ban on export-import of produce. As food production and security is linked with human and animal well being, there is a need to promote data sharing and enable real-time monitoring cutting across countries for taking effective policy frame work for managing the disease/insect pest out breaks.

References

Abdeen A, Virgos A, Olivella E, Villanueva J, Aviles X, Gabarra R, Prat S (2005) Multiple insect resistance in transgenic tomato plants over-expressing two families of plant proteinase inhibitors. Plant Mol Biol 57:189–202

Abdellatef E, Will T, Koch A, Imani J, Vilcinskas A, Kogel KH (2015) Silencing the expression of the salivary sheath protein causes transgenerational feeding suppression in the aphid *Sitobion avenae*. Plant Biotechnol J 13(6):849–857

Abrahams P, Bateman M, Beale T, Clottey V, Cock M, Colemenarez Y, Corniani N, Day R, Early R, Godwin J, Gomez J, Moreno PG, Murphy ST, Oppong-Mensah B, Phiri N, Pratt C, Richards G, Silvestri S, Witt A (2018) Fall armyworm: impacts and implications for Africa. Evidence note (2). Outlooks Pest Manag 28:196–201

Adeyinka OS, Nasir IA, Tabassum B (2023) Host-induced silencing of the CpCHI gene resulted in developmental abnormalities and mortality in maize stem borer (Chilo partellus). PLoS One 18(2):e0280963

Ahmad TM, Haider MS, Randhir TO, Randhir R, Ahmad SR (2023) Spatial analysis of factors influencing bacterial leaf blight in rice production. Brazilian J Biol 83. https://doi.org/10.1590/1519-6984.264249

Ali MP, Bari MN, Haque SS, Kabir MMM, Nowrin F, Choudhury TR, Mankin RW, Ahmed N (2020) Response of a rice insect pest, *Scirpophaga incertulas* (*Lepidoptera*: *Pyralidae*) in warmer world. BMC Zool 5:1–8

Alisaac E, Mahlein AK (2023) *Fusarium* head blight on wheat: biology, modern detection and diagnosis and integrated disease management. Toxins 2023(15):192. https://doi.org/10.3390/toxins15030192

Al-Khayri JM, Rashmi R, Toppo V, Chole PB, Banadka A, Sudheer WN, Nagella P, Shehata WF, Al-Mssallem MQ, Alessa FM, Almaghasla MI (2023) Plant secondary metabolites: the weapons for biotic stress management. Meta 13(6):716

Amin T, Gupta V, Singh A, Khusboo SS (2022) Effect of weather parameters on the severity of bacterial leaf blight of rice (cv. Basmati-370). Biol Forum 14(4):123–133

Anjanappa RB, Gruissem W (2021) Current progress and challenges in crop genetic transformation. J Plant Physiol 261:153411

Anzalone AV, Randolph PB, Davis JR, Sousa AA, Koblan LW, Levy JM, Chen PJ, Wilson C, Newby GA, Raguram A, Liu DR (2019) Search-and-replace genome editing without double-strand breaks or donor DNA. Nature 576:149–157

Anzalone AV, Gao XD, Podracky CJ, Nelson AT, Koblan LW, Raguram A, Levy JM, Mercer JA, Liu DR (2022) Programmable deletion, replacement, integration and inversion of large DNA sequences with twin prime editing. Nat Biotechnol 40(5):731–740

Arianne J., Cease (2024) How Nutrients Mediate the Impacts of Global Change on Locust Outbreaks Annual Review of Entomology 69(1):527–550. https://doi.org/10.1146/ento.2024.69.issue-1 https://doi.org/10.1146/annurev-ento-120220-110415

Ashok KV, Balasubramani JS, Kennedy V, Geethalakshmi P, Jeyakumar N, Sathiah (2021) Effect of elevated temperature on the population dynamics of fall armyworm. J Environ Biol 42:1098–1105

Aung K, Jiang Y, He SY (2018) The role of water in plant–microbe interactions. Plant J 93:771–780

Awan MJA, Pervaiz K, Rasheed A, Amin I, Saeed NA, Dhugga KS, Mansoor S (2022) Genome edited wheat-current advances for the second green revolution. Biotechnol Adv 60:108006

Bailey LD, van de Pol M (2015) Tackling extremes: challenges for ecological and evolutionary research on extreme climatic events. J Animal Ecol 85:85–96

Bale JS, Masters GJ, Hodkinson ID, Awmack C, Bezemer TM, Brown VK, Butterfield J, Buse A, Coulson JC, Farrar J et al (2002) Herbivory in global climate change research: direct effects of rising temperature on insect herbivores. Glob Chang Biol 8:1–16

Bally J, Fishilevich E, Doran RL, Lee K, de Campos SB, German MA, Narva KE, Waterhouse PM (2020) Plin-amiR, a pre-microRNA-based technology for controlling herbivorous insect pests. Plant Biotechnol J 18:1925–1932

Barford E (2013) Crop pests advancing with global warming. Nature. https://doi.org/10.1038/nature.2013.13644

Basavarajappa S (2010) Studies on the impact of anthropogenic interference on wild honeybees in Mysore district, Karnataka, India. Afr J Agri Res 5(4):298–305

Baum JA, Bogaert T, Clinton W, Heck GR, Feldmann P, Ilagan O, Johnson S, Plaetinck G, Munyikwa T, Pleau M, Vaughn T (2007) Control of coleopteran insect pests through RNA interference. Nat Biotechnol 25(11):1322–1326

Bebber DP, Ramotowski MAT, Gurr SJ (2014) Crop pests and pathogens move pole- wards in a warming world. Nat Clim Chang. https://doi.org/10.1038/nclimate1990

Begum MA, Shi XX, Tan Y, Zhou WW, Hannun Y, Obeid L, Mao C, Zhu ZR (2016) Molecular characterization of rice OsLCB2a1 gene and functional analysis of its role in insect resistance. Front Plant Sci 7:1789

Bezemer TM, Jones TH, Knight KJ (1998) Long-term effects of elevated CO_2 and temperature on populations of the peach potato aphid Myzus persicae and its parasitoid. Aphidius Matricariae Oecologia 116:128–135

Bharathi JK, Anandan R, Benjamin LK, Muneer S, Prakash MAS (2023) Recent trends and advances of RNA interference (RNAi) to improve agricultural crops and enhance their resilience to biotic and abiotic stresses. Plant Physiol Biochem 194:600–618

Bhatia S, Yadav SK (2023) CRISPR-Cas for genome editing: classification, mechanism, designing and applications. Int J Biol Macromol 238:124054

Bhattacharya S (2017) Deadly new wheat disease threatens Europe's crops. Nature 542:145–146. https://doi.org/10.1038/nature.2017.21424

Bi HL, Xu J, Tan AJ, Huang YP (2016) CRISPR/Cas9-mediated targeted gene mutagenesis in *Spodoptera litura*. Insect Sci 23:469–477

Bi H, Merchant A, Gu J, Li X, Zhou X, Zhang Q (2022) CRISPR/Cas9-mediated mutagenesis of abdominal-A and ultrabithorax in the Asian corn borer. Ostrinia furnacalis Insects 13:384

Bidzinski P, Ballini E, Ducasse A, Michel C, Zuluaga P, Genga A, Chiozzotto R, Morel JB (2016) Transcriptional basis of drought-induced susceptibility to the Rice blast fungus *Magnaporthe oryzae*. Front Plant Sci 7:1558. https://doi.org/10.3389/fpls.2016.01558

Boixel AL, Gelisse S, Marcel TC, Suffert F (2022) Differential tolerance of *Zymoseptoria tritici* to altered optimal moisture conditions during the early stages of wheat infection. J Plant Pathol 104:495–507. https://doi.org/10.1007/s42161-021-01025-7

Bostock RM, Pye MF, Roubtsova TV (2014) Predisposition in plant disease: exploiting the nexus in abiotic and biotic stress perception and response. Annu Rev Phytopathol 52:517–549

Brajesh K, Singh Manuel, Delgado-Baquerizo Eleonora, Egidi Emilio, Guirado Jan E, Leach Hongwei, Liu Pankaj, Trivedi (2023) Climate change impacts on plant pathogens food security and paths forward Nature Reviews Microbiology 21(10):640–656. https://doi.org/10.1038/s41579-023-00900-7

Bravo-Vázquez LA, Méndez-García A, Chamu-García V, Rodríguez AL, Bandyopadhyay A, Paul S (2024) The applications of CRISPR/Cas-mediated microRNA and lncRNA editing in plant biology: shaping the future of plant non-coding RNA research. Planta 259:32

Bosibori, Bett Stephanie, Gollasch Andy, Moore William, James Joel, Armstrong Tom, Walsh Robert, Harding Thomas JV, Higgins (2017) Transgenic cowpeas (Vigna unguiculata L. Walp) expressing Bacillus thuringiensis Vip3Ba protein are protected against the Maruca pod borer (Maruca vitrata) Plant Cell Tissue and Organ Culture (PCTOC) 131(2):335–345. https://doi.org/10.1007/s11240-017-1287-3

Bosibori, Bett Stephanie, Gollasch Andy, Moore William, James Joel, Armstrong Tom, Walsh Robert, Harding Thomas JV, Higgins (2017a) Transgenic cowpeas (Vigna unguiculata L. Walp) expressing Bacillus thuringiensis Vip3Ba protein are protected against the Maruca pod borer (Maruca vitrata) Plant Cell Tissue and Organ Culture (PCTOC) 131(2):335–345. https://doi.org/10.1007/s11240-017-1287-3

Buntin GD (2024) Evaluation of Bt corn products for corn earworm control. Arthropod Management Tests 49. p.tsae096

Caminade C, Mc Intyre KM, Jones AE (2019) Impact of recent and future climate change on vector-borne diseases. Ann N Y Acad Sci 1436(1):157–173

Caubel J, Launay M, Ripoche D, Gouache D, Buis S, Huard F, Huber L, Brun F, Bancal MO (2017) Climate change effects on leaf rust of wheat: implementing a coupled crop-disease model in a French regional application. Eur J Agron 90:53–66. https://doi.org/10.1016/j.eja.2017.07.004

Chakraborty S (2013) Migrate or evolve: options for plant pathogens under climate change. Glob Chang Biol 19(7):1985–2000. https://doi.org/10.1111/gcb.12205

Chakraborty S, Newton AC (2011) Climate change, plant diseases and food security: an overview. Plant Pathol 60:2–14

Chakraborty J, Sen S, Ghosh P, Sengupta A, Basu D, Das S (2016) Homologous promoter derived constitutive and chloroplast targeted expression of synthetic cry1Ac in transgenic chickpea confers resistance against *Helicoverpa armigera*. Plant Cell Tissue Organ Cult 125:521–535

Chan YL, Yang AH, Chen JT, Yeh KW, Chan MT (2010) Heterologous expression of taro cystatin protects transgenic tomato against Meloidogyne incognita infection by means of interfering sex determination and suppressing gall formation. Plant Cell Rep 29:231–238

Chandrasekaran J, Brumin M, Wolf D, Leibman D, Klap C, Pearlsman M, Sherman A, Arazi T, Gal-On A (2016) Development of broad virus resistance in non-transgenic cucumber using CRISPR/Cas9 technology. Mol Plant Pathol 17:1140–1153

Chattopadhyay C, Birah A, Bushan LJ (2019) Climate change: impact on biotic stresses afflicting crop plants. In: Peshin R, Dhawan AK (eds) Natural resource management: ecological perspectives, sustainability in plant and crop protection. Springer Nature Switzerland AG, pp 133–146. https://doi.org/10.1007/978-3-319-99768-1_8

Chen F, Ge F, Parajulee MN (2005) Impact of elevated CO_2 on tri-trophic interaction of Gossypium hirsutum, *Aphis gossypii*, and *Leis axyridis*. Environ Entomol 34(1):37–46

Chen M, Sun L, Wu H, Chen J, Ma Y, Zhang X et al (2014) Durable field resistance to wheat yellow mosaic virus in transgenic wheat containing the antisense virus polymerase gene. Plant Biotechnol J 12:447–456. https://doi.org/10.1111/pbi.12151

Chen WB, Lu GQ, Cheng HM, Liu CX, Xiao YT, Xu C, Shen ZC, Wu KM (2017) Transgenic cotton co-expressing Vip3A and Cry1Ac has a broad insecticidal spectrum against lepidopteran pests. J Invertebr Pathol 149:59–65

Chen G, Zhang B, Ding J, Wang H, Deng C, Wang J, Yang Q, Pi Q, Zhang R, Zhai H, Dong J, Huang J, Hou J, Wu J, Que J, Zhang F, Li W, Min H, Tabor G, Li B, Liu X, Zhao J, Yan J, Lai Z (2022) Cloning southern corn rust resistant gene RppK and its cognate gene AvrRppK from *Puccinia polysora*. Nat Commun 13:4392. https://doi.org/10.1038/s41467-022-32026-4

Chen Y, Lv W, Yue Q, Wen N, Wang Y, Lang Z, Xu W, Li S (2024) Utilizing the fungal Bicistronic system for multi-gene expression to generate insect-resistant and herbicide-tolerant maize. Int J Mol Sci 25(24):13408

Cheng W, Song XS, Li HP, Cao LH, Sun K, Qiu XL, Xu YB, Yang P, Huang T, Zhang JB, Qu B (2015) Host-induced gene silencing of an essential chitin synthase gene confers durable resistance to *Fusarium* head blight and seedling blight in wheat. Plant Biotechnol J 13(9):1335–1345

Chiu M, Chen C, Chen C, Lin H (2022) Weather fluctuation can override the effects of integrated nutrient management on fungal disease incidence in the rice fields in Taiwan. Sci Rep 12. https://doi.org/10.1038/s41598-022-08139-7

Cingel A, Savić J, Lazarević J, Ćosić T, Raspor M, Smigocki A, Ninković S (2017) Co-expression of the proteinase inhibitors oryzacystatin I and oryzacystatin II in transgenic potato alters Colorado potato beetle larval development. Insect Sci 24(5):768–780

Clemente M, Corigliano MG, Pariani SA, Sánchez-López EF, Sander VA, Ramos-Duarte VA (2019) Plant serine protease inhibitors: biotechnology application in agriculture and molecular farming. Int J Mol Sci 20:1345

Cloutier C, Jean C, Fournier M, Yelle S, Michaud D (2000) Adult Colorado potato beetles, *Leptinotarsa decemlineata* compensate for nutritional stress on oryzacystatin I-transgenic potato plants by hypertrophic behavior and over-production of insensitive proteases. Arch Insect Biochem Physiol 44:69–81

Coakley SM, Scherm H, Chakraborty S (1999) Climate change and plant disease management. Annu Rev Phytopathol 37:399–426. https://doi.org/10.1146/annurev.phyto.37.1.399

Cock MJW, Beseh PK, Buddie AG et al (2017) Molecular methods to detect *Spodoptera frugiperda* in Ghana, and implications for monitoring the spread of invasive species in developing countries. Sci Rep 7:4103

Collard BCY, Mackill DJ (2008) Marker-assisted selection: an approach for precision plant breeding in the twenty-first century. Philos Trans R Soc B 363:557–572

Cooper AM, Silver K, Zhang J, Park Y, Zhu KY (2019) Molecular mechanisms influencing efficiency of RNA interference in insects. Pest Manag Sci 75:18–28

Cotrufo MF, Ineson P, Scott A (1998) Elevated CO_2 reduces the nitrogen concentration of plant tissues. Glob Chang Biol 1998(4):43–54

Cowger C, Patton-Ozkurt J, Brown-Guedira G, Perugini L (2009) Post-anthesis moisture increased Fusarium head blight and deoxynivalenol levels in North Carolina winter wheat. Phytopathology 99:320–327. https://doi.org/10.1094/PHYTO-99-4-0320

Cushman JC, Denby K, Mittler R (2022) Plant responses and adaptations to a changing climate. Plant J 109:319–322

Daniel P, Bebber Mark AT, Ramotowski Sarah J, Gurr (2013) Crop pests and pathogens move polewards in a warming world Nature Climate Change 3(11):985–988. https://doi.org/10.1038/nclimate1990

Day R, Abrahams P, Bateman M, Beale T, Clottey V, Cock M, Colmenarez Y et al (2017) Fall armyworm: impacts and implications for Africa. Outlooks Pest Manag 28(5):196–201. https://doi.org/10.1564/v28_oct_02

Devkota K, Ferreira AB, Timberlake TP, & dos Santos CF (2024) The impact of pollinator decline on global protein production: Implications for livestock and plant-based products. Global Ecology and Conservation, 50:e02815

de Oliveira RS, Oliveira-Neto OB, Moura HF, de Macedo LL, Arraes FB, Lucena WA, Lourenço-Tessutti IT, de Deus Barbosa AA, da Silva MC, Grossi-de-Sa MF (2016) Transgenic cotton plants expressing Cry1Ia12 toxin confer resistance to fall armyworm (*Spodoptera frugiperda*) and cotton boll weevil (*Anthonomus grandis*). Front Plant Sci 7:165

de Vallavieille-Pope C, Bahri B, Leconte M, Zurfluh O, Belaid Y, Maghrebi E et al (2018) Thermal generalist behaviour of invasive *Puccinia striiformis f. sp. tritici* strains under current and future climate conditions. Plant Pathol 67(6):1307–1320

Delai C, Abid R, Tian T, Liu R, Xiong Y, Ma S, Ghorbani A (2024) A comprehensive review of integrated management strategies for damping-off disease in chili. Front Microbiol 15:1479957

Delgado L, Schuster M, Torero M (2017) The reality of food losses: a new measurement methodology, IFPRI Discussion Paper 1686. International Food Policy Research Institute, Washington, DC. http://ebrary.ifpri.org/cdm/ref/collection/p15738coll2/id/131530

Delgado BM, Guerra CA, Cano-Díaz C et al (2020) The proportion of soil-borne pathogens increases with warming at the global scale. Nat Clim Chang 10:550–554. https://doi.org/10.1038/s41558-020-0759-3

Desprez-Loustau M, Marcois B, Nageleisen LM, Piou D, Vannini A (2006) Interactive effects of drought and pathogens in forest trees. Ann For Sci 63:597–612

Deutsch CA, Tewksbury JJ, Huey RB, Sheldon, KS, Ghalambor CK, Haak D C, & Martin PR (2008) Impacts of climate warming on terrestrial ectotherms across latitude. Proceedings of the National Academy of Sciences of the United States of America, 105(18), 6668–6672

Deutsch CA, Tewksbury JJ, Tigchelaar M, Battisti DS, Merrill SC, Huey RB, Naylor RL (2018) Increase in crop losses to insect pests in a warming climate. Science 361:916–919

Devi R, Kaur T, Kour D, Yadav A, Yadav AN, Suman A, Ahluwalia AS, Saxena AK (2022) Minerals solubilizing and mobilizing microbiomes: a sustainable approach for managing minerals' deficiency in agricultural soil. J Appl Microbiol 133:1245–1272. https://doi.org/10.1111/JAM.15627

Dhatwalia D, Aminedi R, Kalia V, Pande V, Bhattacharya R (2022) Host-mediated attenuation of gut sucrase in mustard aphid Lipaphis erysimi impaired its parthenogenetic reproduction on Indian mustard Brassica juncea. Pest Manag Sci 78(2):803–811

Dhillon BS, Sidhu RS (2016) Successful management of whitefly incidence in cotton through joint efforts of farm experts and farmers. Progress Farm 52:5–7

Dixit S, Singh UM, Singh AK, Alam S, Venkateshwarlu C, Nachimuthu VV, Yadav S, Abbai R, Selvaraj R, Devi N, Ramayya PJ, Badri J, Ram T, Lakshmi J, Lakshmidevi G, Vidhya J, Padmakumari AP, Laha GS, Prasad MS, Seetalam M, Singh VK, Kumar A (2020) Marker assisted forward breeding to combine multiple biotic-abiotic stress resistance/tolerance in rice. Rice (N.Y) 13:29. https://doi.org/10.1186/s12284-020-00391-7

Domingo WE, Crooks DM (1945) Investigations with the castor bean plant: I. Adaptation and variety tests1. J Am Soc Agro 37(9):750

Dukes JSDS, Pontius J, Orwig D, Garnas JRGR, Rodgers VL, Brazee N, Cooke B, Theoharides KATA, Stange EESE, Harrington R et al (2009) Responses of insect pests, pathogens, and invasive plant species to climate change in the forests of northeastern North America: what can we predict? This article is one of a selection of papers from NE forests 2100: a synthesis of climate change impacts on forests of the northeastern US and Eastern Canada. Can J For Res 39:231–248

Dweba C, Figlan S, Shimelis H, Motaung T, Sydenham S, Mwadzingeni L et al (2017) Fusarium head blight of wheat: pathogenesis and control strategies. Crop Prot 91:114–122

Eckardt NA, Ainsworth EA, Bahuguna RN, Broadley MR, Busch W, Carpita NC, Castrillo G, Chory J, DeHaan LR, Duarte CM et al (2023) Climate change challenges, plant science solutions. Plant Cell 35:24–66

EFSA (European Food Safety Authority), Kinkar M, Delbianco A, Graziosi I, Mattion G, Vos S, 2023. Pest survey card on Spodoptera frugiperda. EFSA supporting publication 2021:EN-6911. https://efsa.europa.eu/plants/planthealth/monitoring/surveillance/spodoptera-frugiperda

El-Sayed A, Kamel M (2020) Climatic changes and their role in emergence and re-emergence of diseases. Environ Sci Pollut Res 27:22336–22352

Faisal M, Abdel-Salam EM, Alatar AA (2021) Artificial microRNA-based RNA interference and specific gene silencing for developing insect resistance in Solanum lycopersicum. Agronomy 11(1):136

Falco MC, Silva-Filho MC (2003) Expression of soybean proteinase inhibitors in transgenic sugarcane plants: effects on natural defense against *Diatraea saccharalis*. Plant Physiol Biochem 41(8):761–766

Fan A, Wei L, Zhang X, Liu J, Sun L, Xiao J, Wang Y, Wang H, Hua J, Singh RP, Wang Z (2022) Heterologous expression of the Haynaldia villosa pattern-recognition receptor CERK1-V in wheat increases resistance to three fungal diseases. Crop J 10(6):1733–1745

Fanzo, J., Schneider, K.R., and Wood, S. 2024. Food insecurity and malnutrition', in Barry S. Levy, and Jonathan A. Patz (eds), Climate change and public health, 2 (New York., 2024; online edn, Oxford Academic, 22 Feb. 2024). https://doi.org/10.1093/oso/9780197683293003.0008, accessed 28 Mar. 2025

FAO (2008) Rapid Assessment of Pollinators Status. Viale delle Terme di Caracalla 00153, Rome, Italy. p64

FAO (2020). https://www.fao.org/locusts/en/

FAO. Locust Watch; www.fao.org/ag/locusts/en/info/info/index.html

Ferreira SS, Pandey S, Hemminger J, Bozdag S, Antunes MS (2025) Early changes in microRNA expression in Arabidopsis plants infected with the fungal pathogen *Fusarium graminearum*. PLoS One 20(2):e0318532. https://doi.org/10.1371/journal.pone.0318532

Figueroa M, Upadhyaya NM, Sperschneider J, Park RF, Szabo LJ, Steffenson B, Ellis JG, Dodds PN (2016) Changing the game: using integrative genomics to probe virulence mechanisms of the stem rust pathogen *Puccinia graminis* f. sp. *Tritici*. Front Plant Sci 7:205

Fire A, Xu S, Montgomery MK, Kostas SA, Driver SE, Mello CC (1998) Potent and specific genetic interference by double-stranded RNA in *Caenorhabditis elegans*. Nature 391:806–811

Fischer G, Shah M, Tubiello FN, van Velhuizen H (2005) Socio-economic and climate change impacts on agriculture: an integrated assessment, 1990–2080. Philos Trans R Soc B 360:2067–2083. https://doi.org/10.1098/rstb.2005.1744

Fisher MC, Henk DA, Briggs CJ, Brownstein JS, Madoff LC, Mccraw SL, Gurr SJ (2012) Emerging fungal threats to animal, plant and ecosystem health. Nature 484:186–194

Francioli D, Cid G, Hajirezaei MR, Kolb S (2022) Response of the wheat mycobiota to flooding revealed substantial shifts towards plant pathogens. Front Plant Sci 13:1028153. https://doi.org/10.3389/fpls.2022.1028153

Fry W (2008) Phytophthora infestans: the plant (and R gene) destroyer. Mol Plant Pathol 9:385–402

Gahakwa D, Maqbool SB, Fu X, Sudhakar D, Christou P, Kohli A (2000) Transgenic rice as a system to study the stability of transgene expression: multiple heterologous transgenes show similar behaviour in diverse genetic backgrounds. Theor Appl Genet 10:388–399

Gallagher DN, Haber JE (2018) Repair of a site-specific DNA cleavage: old-school lessons for Cas9-mediated gene editing. ACS Chem Biol 13:397–405

Ganiger PC, Yeshwanth HM, Muralimohan K, Vinay N, Kumar VAR, Chandrashekara K (2018) Occurrence of the new invasive pest, fall armyworm, Spodoptera frugiperda (JE Smith) (*Lepidoptera: Noctuidae*), in the maize fields of Karnataka, India. Curr Sci 115:621–623

Gaudelli NM, Komor AC, Rees HA, Packer MS, Badran AH, Bryson DI, Liu DR (2017) Programmable base editing of A•T to G•C in genomic DNA without DNA cleavage. Nature 551:464–471

Geethanjali S, Kadirvel P, Periyannan S (2024) Wheat improvement through advances in single nucleotide polymorphism (SNP) detection and genotyping with a special emphasis on rust resistance. Theor Appl Genet 137:224. https://doi.org/10.1007/s00122-024-04730-w

Gerber JS, Ray DK, David Makowski D, Butler EE, Mueller ND, West PC, Johnson JA, Polasky S, Samberg LH, Siebert S, Sloat L (2024) Global spatially explicit yield gap time trends reveal regions at risk of future crop yield stagnation. Nat Food 5:125–135

Ghosh A, Das A, Wamalwa BK, Kumar V, Kumar A, Meena PD, Chattopadhyay C (2012) Impact of climate change on pests and diseases–the scenario unfolding in Indian agriculture. SATSA Mukhopatra Annual Technical Issue 16:15–29

Ghosh G, Ganguly S, Purohit A, Chaudhuri RK, Das S, Chakraborti DB (2017) Transgenic pigeon pea events expressing Cry1Ac and Cry2Aa exhibit resistance to *Helicoverpa armigera*. Plant Cell Rep 36:1037–1051

Goergen G, Kumar P, Sankung S, Togola A, Tamo M (2016) First report of outbreaks of the fall armyworm *Spodoptera frugiperda* (J E Smith) (*Lepidoptera, Noctuidae*), a new alien invasive pest in West and Central Africa. PLoS One 11. https://doi.org/10.1371/journal.pone.0165632

Goggin FL, Jia L, Shah G, Hebert S, Williamson VM, Ullman DE (2006) Heterologous expression of the Mi-1.2 gene from tomato confers resistance against nematodes but not aphids in eggplant. Mol Plant-Microbe Interact 19(4):383–388

Gomez MA, Lin ZD, Moll T, Chauhan RD, Hayden L, Renninger K, Beyene G, Taylor NJ, Carrington JC, Staskawicz BJ, Bart RS (2019) Simultaneous CRISPR/Cas9-mediated editing of cassava eIF 4E isoforms nCBP-1 and nCBP-2 reduces cassava brown streak disease symptom severity and incidence. Plant Biotechnol J 17:421–434

Gordon KH, Waterhouse PM (2007) RNAi for insect-proof plants. Nat Biotechnol 25:1231–1232

Granke LL, Hausbeck MK (2020) Effects of temperature, humidity, and wounding on development of Phytophthora rot of cucumber fruit. Plant Dis 94:1417–1424. https://doi.org/10.1094/PDIS-04-10-0258

Gregorc A, Evans JD, Scharf M, Ellis JD (2012) Gene expression in honey bee (*Apis mellifera*) larvae exposed to pesticides and Varroa mites (Varroa destructor). J Insect Physiol 58(8):1042–1049. https://doi.org/10.1016/j.jinsphys.2012.03.015

Guan J, Li M, Ju X, Lin J, Wu J, Zheng J (2021) The potential habitat of desert locusts is contracting: predictions under climate change scenarios. Peer J 9:e12311. https://doi.org/10.7717/peerj.12311

Guo H, Sun Y, Ren Q, Zhu-Salzman K, Kang L et al (2012) Elevated CO_2 reduces the resistance and tolerance of tomato plants to *Helicoverpa armigera* by suppressing the JA signaling pathway. PLoS One 7(7):e41426. https://doi.org/10.1371/journal.pone.0041426

Guo H, Song X, Wang G, Yang K, Wang Y, Niu L, Chen X, Fang RX (2014) Plant-generated artificial small RNAs mediated aphid resistance. PLoS One 9(5):e97410. https://doi.org/10.1371/journal.pone.0097410

Guo JF, Zhao JZ, He KL, Zhang F, Wang ZY (2018) Potential invasion of the crop-devastating insect pest fall armyworm *Spodoptera frugiperda* to China. Plant Prot 44:1–10

Guo L, Li C, Jiang Y, Luo K, Xu C (2020) Heterologous expression of poplar WRKY18/35 paralogs in *Arabidopsis* reveals their antagonistic regulation on pathogen resistance and abiotic stress tolerance via variable hormonal pathways. Int J Mol Sci 21:5440

Hance T, Van Baaren J, Vernon P, Boivin G (2007) Impact of extreme temperatures on parasitoids in a climate change perspective. Annu Rev Entomol 52:107–126

Hatfield JL, Boote KJ, Kimball BA, Ziska LH, Izaurralde RC, Ort DR, Thomson AM, Wolfe D (2011) Climate impacts on agriculture: implications for crop production. Publications from USDA-ARS/UNL Faculty. 1350. https://digitalcommons.unl.edu/usdaarsfacpub/1350

Haverkort AJ, Verhagen A (2008) Climate change and its repercussions for the potato supply chain. Potato Res 51:223–237. https://doi.org/10.1007/s11540-008-9107-0

He K, Xiao H, Sun Y, Ding S, Situ G, Li F (2019) Transgenic micro RNA-14 rice shows high resistance to rice stem borer. Plant Biotechnol J 17(2):461–471

He F, Wang C, Sun H, Tian S, Zhao G, Liu C, Wan C, Guo J, Huang X, Zhan G, Yu X (2023) Simultaneous editing of three homoeologues of TaCIPK14 confers broad-spectrum resistance to stripe rust in wheat. Plant Biotechnol J 21:354–368

Head GP, Carroll MW, Sean PE, Rule DM, Willse, Alan RW et al (2017) Evaluation of SmartStax and SmartStax PRO Maize against Western Corn Rootworm and Northern Corn Rootworm: efficacy and resistance management. Faculty Publications: Department of Entomology. 682. https://www.digitalcommons.unl.edu/entomologyfacpub/682

Hickey LT, Hafeez NA, Robinson H, Jackson SA, Leal-Bertioli SCM, Tester M, Gao C, Godwin ID, Hayes BJ, Wulff BBH (2019) Breeding crops to feed 10 billion. Nat Biotechnol 37(7):744–754. https://doi.org/10.1038/s41587-019-0152-9. Epub 2019 Jun 17

Horgan FG, Arida A, Ardestani G, Almazan MLP (2021) Elevated temperatures diminish the effects of a highly resistant rice variety on the brown plant hopper. Sci Rep 11(1):262

Hossain MM (2022) Wheat blast: a review from a genetic and genomic perspective. Front Microbiol 13:983243. https://doi.org/10.3389/fmicb.2022.983243

Hossain MM, Sultana F, Mostafa M et al (2024) Plant disease dynamics in a changing climate: impacts, molecular mechanisms, and climate-informed strategies for sustainable management. Discov Agric 2:132. https://doi.org/10.1007/s44279-024-00144-w

Hovmøller MS, Walter S, Bayles RA (2016) Replacement of the European wheat yellow rust population by new races from the Centre of diversity in the near-Himalayan region. Plant Pathol 65(3):402–411

Huang HE, Liu CA, Lee MJ, Kuo CG, Chen HM, Ger MJ, Tsai YC, Chen YR, Lin MK, Feng TY (2007) Resistance enhancement of transgenic tomato to bacterial pathogens by the heterologous expression of sweet pepper ferredoxin-I protein. Phytopathology 97(8):900–906

Huang Y, Chen Y, Zeng B, Wang Y, James AA, Gurr GM, Yang G, Lin X, Huang Y, You M (2016) CRISPR/Cas9 mediated knockout of the abdominal-A homeotic gene in the global pest, diamondback moth (*Plutella xylostella*). Insect Biochem Mol Biol 75:98–106

Hussain T, Aksoy E, Çalışkan ME, Bakhsh A (2019) Transgenic potato lines expressing hairpin RNAi construct of molting-associated EcR gene exhibit enhanced resistance against Colorado potato beetle (Leptinotarsa decemlineata, Say). Transgenic Res 28:151–164

IPCC (1996) Climate change 1995: the science of climate change. Contribution of working group I to the second assessment report of the IPCC. Cambridge University Press, New York

IPCC (2014) Core writing team. In: Pachauri RK, Meyer LA (eds) Climate change 2014: synthesis report. Contribution of working groups I, II and III to the fifth assessment report of the intergovernmental panel on climate change. Geneva, Switzerland

IPCC (2018) Global Warming of 1.5°C. An IPCC Special Report on the impacts of global warming of 1.5°C above pre-industrial levels and related global greenhouse gas emission pathways, in the context of strengthening the global response to the threat of climate change, sustainable development, and efforts to eradicate poverty [Masson-Delmotte, V., P. Zhai, H.-O. Pörtner, D. Roberts, J. Skea, P.R. Shukla, A. Pirani, W. Moufouma-Okia, C. Péan, R. Pidcock, S. Connors, J.B.R. Matthews, Y. Chen, X. Zhou, M.I. Gomis, E. Lonnoy, T. Maycock, M. Tignor, and T. Waterfield (eds.)]. In Press

IPCC (2019) Climate Change and Land: An IPCC Special Report on Climate Change, Desertification, Land Degradation, Sustainable Land Management, Food Security, and Greenhouse Gas Fluxes in Terrestrial Ecosystems. [P.R. Shukla, J. Skea, E. Calvo Buendia, V. Masson-Delmotte, H.-O. Pörtner, D. C. Roberts, P. Zhai, R. Slade, Connors, S. R., van Diemen, M., Ferrat, E. Haughey, S., Luz, S., Neogi, M., Pathak, J., Petzold, J., Portugal Pereira, P., Vyas, E. Huntley, K. Kissick, M. Belkacemi, J. Malley, (eds.)]. In press

IPCC (2021) "Summary for Policymakers" (PDF). The Physical Science Basis. Contribution of Working Group I to the Sixth Assessment Report of the Intergovernmental Panel on Climate Change. ISBN 978-92-9169-158-6

IPCC (2023) Summary for policymakers. In: climate change 2023: synthesis report. Contribution of working groups I, II and III to the sixth assessment report of the intergovernmental panel on climate change [Core writing team, H. Lee and J. Romero (eds.)]. IPCC, Geneva, Switzerland, pp 1–34. https://doi.org/10.59327/IPCC/AR6-9789291691647.001

Potts SG, Biesmeijer JC, Kremen C, Neumann P, Schweiger O, & Kunin WE (2010) Global pollinator declines: Trends, impacts and drivers. Trends in Ecology & Evolution, 25(6):345–353. https://doi.org/10.1016/j.tree.2010.01.007

Jägermeyr J, Müller C, Ruane AC, Elliott J, Balkovic J, Castillo O, Faye B, Foster I, Folberth C, Franke JA et al (2021) Climate impacts on global agriculture emerge earlier in new generation of climate and crop models. Nat Food 2:873–885

Jain S, Kumar A (2015) The pathogenesis related class 10 proteins in plant defense against biotic and abiotic stresses. Adv Plants Agric Res 3:00077

Jason G, Hamilton JG, Dermody O, Aldea M, Zangerl AR, Rogers A, Berenbaum MR, Delucia EH (2005) Anthropogenic changes in tropospheric composition increase susceptibility of soybean to insect herbivory. Environ Entomol 34(2):479–485. https://doi.org/10.1603/0046-225X-34.2.479

Jiang S, Wu H, Liu H, Zheng J, Lin Y, Chen H (2017) The overexpression of insect endogenous small RNAs in transgenic rice inhibits growth and delays pupation of striped stem borer (*Chilo suppressalis*). Pest Manag Sci 73(7):1453–1461

Jiang YY, Liu J, Xie MC, Li YH, Yang JJ, Zhang ML, Qiu K (2019) Observation on law of diffusion damage of Spodoptera frugiperda in China in 2019. Plant Prot 2019(45):10–19

Johnson SN, Waterman JM, Hall CR (2020) Increased insect herbivore performance under elevated CO2 is associated with lower plant defence signalling and minimal declines in nutritional quality. Sci Rep 10:14553. https://doi.org/10.1038/s41598-020-70823-3

Jones RAC, Barbetti MJ (2012) Influence of climate change on plant disease infections and epidemics caused by viruses and bacteria. CAB Rev 7:022

Jones RAC, Naidu RA (2019) Global dimensions of plant virus diseases: current status and future perspectives. Annu Rev Virol 6:387–409

Jones PD, New M, Parker DE, Martin S, Rigor IG (1999) Surface air temperature and its changes over the past 150 years. Rev Geophys 37:173–199

Jones CM, Parry H, Tay WT, Reynolds DR, Chapman JW (2019) Movement ecology of pest *Helicoverpa*: implications for ongoing spread. Annu Rev Entomol 64:277–295

Jung SK, Karen McDonald K (2011) Visual gene developer: a fully programmable bioinformatics software for synthetic gene optimization. BMC Bioinfor 12:340. https://doi.org/10.1186/1471-2105-12-340

Juroszek P, Racca P, Link S, Farhumand J, Kleinhenz B (2020) Overview on the review articles published during the past 30 years relating to the potential climate change effects on plant pathogens and crop disease risks. Plant Pathol 69:179–193. https://doi.org/10.1111/ppa.13119

Kadirvel P, Senthilvel S, Geethanjali S, Sujatha M, Varaprasad KS (2015) Genetic markers, trait mapping and marker-assisted selection in plant breeding. In: Bahadur B et al (eds) Plant biology and biotechnology: volume II: plant genomics and biotechnology. Springer. https://doi.org/10.1007/978-81-322-2283-5_4

Kanakala S, Ghanim M (2016) RNA interference in insect vectors for plant viruses. Viruses 8:329

Kantor A, McClements ME, Mac Laren RE (2020) CRISPR-Cas9 DNA base-editing and prime-editing. Int J Mol Sci 21:6240

Kaur J, Kumar A, Kaur J (2017) Strategies for optimization of heterologous protein expression in E. Coli: roadblocks and reinforcements. Int J Biol Macromol 106:803–822. https://doi.org/10.1016/j.ijbiomac.2017.08.080

Kaur G, Nelson KA, Motavalli PP, Singh G (2020) Adaptation to early-season soil waterlogging using different nitrogen fertilizer practices and corn hybrids. Agronomy 10:378. https://doi.org/10.3390/agronomy10030378

Kaur R, Choudhury A, Chauhan S, Ghosh A, Tiwari R, Rajam MV (2021) RNA interference and crop protection against biotic stresses. Physiol Mol Biol Plants 27:2357–2377. https://doi.org/10.1007/s12298-021-01064-5

Kebede M (2018) Outbreak, distribution and management of fall armyworm, J.E. Smith in Africa: the status and prospects. Acad Agric J 3:551–568

Kenis M, Benelli G, Biondi A, Calatayud PA, Day R, Desneux N, Harrison RD et al (2022) Invasiveness, biology, ecology, and management of the fall armyworm. Spodoptera frugiperda Entomologia Generalis 102198. https://doi.org/10.1127/entomologia/2022/1659

Khajuria C, Ivashuta S, Wiggins E, Flagel L, Moar W, Pleau M, Miller K, Zhang Y, Ramaseshadri P, Jiang C, Hodge T (2018) Development and characterization of the first dsRNA-resistant insect population from western corn rootworm, *Diabrotica virgifera* virgifera LeConte. PloS One 13:e0197059

Kim YA, Moon H, Park CJ (2019) CRISPR/Cas9-targeted mutagenesis of Os8N3 in rice to confer resistance to *Xanthomonas oryzae* pv. oryzae. Rice 12:1–13

Klumper W, Qaim M (2014) A meta-analysis of the impacts of genetically modified crops. PLoS One 9:e111629. https://doi.org/10.1371/journal.pone.0111629

Komor AC, Kim YB, Packer MS, Zuris JA, Liu DR (2016) Programmable editing of a target base in genomic DNA without double-stranded DNA cleavage. Nature 533:420–424

Kondrateva E, Demchenko A, Lavrov A, Smirnikhina S (2021) An overview of currently available molecular Cas-tools for precise genome modification. Gene 769:145225

Kong L, Shi X, Chen D, Yang N, Yin C, Yang J, Wang G, Huang W, Peng H, Peng D, Liu S (2022) Host-induced silencing of a nematode chitin synthase gene enhances resistance of soybeans to both pathogenic *Heterodera glycines* and *Fusarium oxysporum*. Plant Biotechnol J 20(5):809

Koutroumpa FA, Monsempes C, François MC, de Cian A, Royer C, Concordet JP, Jacquin-Joly E (2016) Heritable genome editing with CRISPR/Cas9 induces anosmia in a crop pest moth. Sci Rep 6:29620

Kumar D, Mukhopadhyay R (2025) Climate change and plant pathogens: understanding dynamics, risks and mitigation strategies. Plant Pathol 74(1):59–68

Kumar PA, Mandaokar A, Sreenivasu K, Chakrabarti SK, Bisaria S, Sharma SR, Kaur S, Sharma RP (1998) Insect-resistant transgenic brinjal plants. Mol Breed 4:33–37

Kumar KRR, Srinivasan T, Kirti PB (2009) A mitogen-activated protein kinase gene, AhMPK3 of peanut: molecular cloning, genomic organization, and heterologous expression conferring resistance against Spodoptera litura in tobacco. Mol Gen Genomics 282:65–81

Kumar R, Kaur A, Pandey A, Mamrutha HM, Singh GP (2019) CRISPR-based genome editing in wheat: a comprehensive review and future prospects. Mol Biol Rep 46:3557–3569

Kumara NT, Muhandiram AMKG, Ranaweera GKMMK, Buddhika A, Sarathchandra SR, Mubarak ANM (2022) Association between temperature and life table development of fall armyworm Spodoptera frugiperda under control condition. Research Square. https://doi.org/10.21203/rs.3.rs-1295288/v1

Kunte N, McGraw E, Bell S, Held D, Avila LA (2020) Prospects, challenges and current status of RNAi through insect feeding. Pest Manag Sci 76:26–41

Kweku D, Bismark O, Maxwell A, Desmond K, Danso K, Oti-Mensah E, Quachie A, Adormaa B (2019) Greenhouse effect: greenhouse gases and their impact on global warming. J Sci Res Rep 17:1–9

Lahlali R, Taoussi M, Laasli SE, Gachara G, Ezzouggari R, Belabess Z, Aberkani K, Assouguem A, Meddich A, El Jarroudi M, Barka EA (2024) Effects of climate change on plant pathogens and host-pathogen interactions. Crop Environ 1:3(3):159–170

Laine AL (2023) Plant disease risk is modified by multiple global change drivers. Curr Biol 33(11):R574–R583

Lam D (2025) The next 2 billion: can the world support 10 billion people? Popul Dev Rev 51. https://doi.org/10.1111/padr.12685/

Legg JP, Kumar PL, Makeshkumar T, Tripathi L, Ferguson M, Kanju E, Cuellar W (2015) Cassava virus diseases: biology, epidemiology, and management. Adv Virus Res 91:85–142

Lehmann P, Ammunét T, Barton M, Battisti A, Eigenbrode SD, Jepsen JU, Kalinkat G, Neuvonen S, Niemelä P, Terblanche JS et al (2020) Complex responses of global insect pests to climate warming. Front Ecol Environ 18:141–150

Lei J, Liu S, Ye Z et al (2023) Pathogenic function of the natural variation of CP in WYMV and CWMV. Phytopathol Res 5:13. https://doi.org/10.1186/s42483-023-00170-4

Leisner CP, Potnis N, Sanz-Saez A (2023) Crosstalk and trade-offs: plant responses to climate change-associated abiotic and biotic stresses. Plant Cell Environ 46:2946–2963. https://doi.org/10.1111/pce.14532

Leplat J, Friberg H, Abid M, Steinberg C (2013) Survival of *Fusarium graminearum*, the causal agent of *Fusarium* head blight. A review. Agron Sustain Dev 33(1):97–111

Li Y, Zhang J, Chen D, Yang P, Jiang F, Wang X, Kang L (2016) CRISPR/Cas9 in locusts: successful establishment of an olfactory deficiency line by targeting the mutagenesis of an odorant receptor co-receptor (Orco). Insect Biochem Mol Biol 79:27–35

Li K, Guan GD, Schnitkey E, DeLucia BP (2019) Excessive rainfall leads to maize yield loss of a comparable magnitude to extreme drought in the United States. Glob Chang Biol 25:2325–2337

Li X, Hu D, Cai L, Wang H, Liu X, Du H, Yang Z, Zhang H, Hu Z, Huang F, Kan G (2022) Calcium-dependent protein kinase38 regulates flowering time and common cutworm resistance in soybean. Plant Physiol 190:480–499

Li H, Hua L, Zhao S, Hao M, Song R, Pang S, Liu Y, Chen H, Zhang W, Shen T, Gou JY, Mao H, Wang G, Hao X, Li J, Song B, Lan C, Li Z, Deng XW, Dubcovsky J, Wang X, Chen S (2023) Cloning of the wheat leaf rust resistance gene Lr47 introgressed from *Aegilops speltoides*. Nat Comm 14:6072. https://doi.org/10.1038/s41467-023-41833-2

Liang YJ, Zhang T, Li C, Zhi JR (2021) Molecular cloning and temporal-spatial expression profiling of trehalose-6-phosphate synthase gene in *Spodoptera frugiperda* (*Lepidoptera*: *Noctuidae*) and its response to temperature stress. Acta Entomol Sini 64:1417–1426

Lincoln DE, Sionit N, Strain BR (1984) Growth and feeding response of *Pseudoplusia includens* (*Lepidoptera*: *Noctuidae*) to host plants grown in controlled carbon dioxide atmospheres. Environ Entomol 13:1527–1530

Liu J, Feng L, Li J, He Z (2015) Genetic and epigenetic control of plant heat responses. Front Plant Sci. https://doi.org/10.3389/fpls.2015.00267

Liu K, Harrison MT, Ibrahim A et al (2020) Genetic factors increasing barley grain yields under-soil waterlogging. Food Energy Secur 9:e238. https://doi.org/10.1002/fes3.238

Liu Z, Zhu Y, Shi H, Qiu J, Ding X, Kou Y (2021) Recent progress in rice broad-spectrum disease resistance. Int J Mol Sci 22(21):11658

Liu G, Lin Q, Jin S, Gao C (2022) The CRISPR-Cas toolbox and gene editing technologies. Mol Cell 82:333–347

Liu X, Dongxiao Zhang D, He X (2024) Unveiling the role of climate in spatially synchronized locust outbreak risks. Sci Adv 10:eadj1164

Loladze A, Druml T, Wellings C (2014) Temperature adaptation in Australasian populations of *Puccinia striiformis* f. sp. *tritici*. Plant Pathol 63(3):572–580

Lu HP, Luo T, Fu HW, Wang L, Tan YY, Huang JZ, Wang Q, Ye GY, Gatehouse AM, Lou YG, Shu QY (2018) Resistance of rice to insect pests mediated by suppression of serotonin biosynthesis. Nat Plants 4:338–344

Lu ZH, He SQ, Yan NS, Zhao WJ, Yao WF, Chen YP, Yang T, Jiang YY, Gui FR (2019) Effects of temperatures on the development and reproduction of fall armyworm (*Spodoptera frugiperda* Smith). Plant Prot 45:27–31+53

Luck GW, Davidson P, Boxall D, Smallbone L (2011) Relations between urban bird and plant communities and human well-being and connection to nature. Conserv Biol 25:816–826. https://doi.org/10.1111/j.1523-1739.2011.01685.x

Ma CS, Zhang W, Peng Y et al (2021) Climate warming promotes pesticide resistance through expanding overwintering range of a global pest. Nat Commun 12:5351. https://doi.org/10.1038/s41467-021-25505-7

Macovei A, Sevilla NR, Cantos C, Jonson GB, Slamet-Loedin I, Čermák T, Voytas DF, Choi IR, Chadha-Mohanty P (2018) Novel alleles of rice eIF4G generated by CRISPR/Cas9-targeted mutagenesis confer resistance to Rice tungro spherical virus. Plant Biotechnol J 16:1918–1927

Mamta B, Rajam MV (2018) RNA interference: a promising approach for crop improvement. In: Gosal S, Wani S (eds) Biotechnologies of crop improvement, volume 2: transgenic approaches. Springer, Cham, pp 41–65. https://doi.org/10.1007/978-3-319-90650-8_3

Mamta, Reddy KR, Rajam MV (2016) Targeting chitinase gene of Helicoverpa armigera by host-induced RNA interference confers insect resistance in tobacco and tomato. Plant Mol Biol 90:281–292

Manjula CP, Prathibha B, Yamanura D, Palanna KB, Punith G (2024) Studies on the characterization, physiology and transmission of Botryotinia ricini (Godfrey) whetzel causes grey mold of castor. Int J Adv Biochem Res SP-8(7):27–37

Mao YB, Cai WJ, Wang JW, Hong GJ, Tao XY, Wang LJ, Huang YP, Chen XY (2007) Silencing a cotton bollworm P450 monooxygenase gene by plant-mediated RNAi impairs larval tolerance of gossypol. Nat Biotechnol 25(11):1307–1313

Mao C, Zhu X, Wang P, Sun Y, Huang R, Zhao M, Hull JJ, Lin Y, Zhou F, Chen H, Ma W (2022) Transgenic double-stranded RNA rice, a potential strategy for controlling striped stem borer (*Chilo suppressalis*). Pest Manag Sci 78(2):785–792

Matić S, Cucu MA, Garibaldi A, Gullino ML (2018) Combined effect of CO_2 and temperature on wheat powdery mildew development. Plant Pathol J 34(4):316–326. https://doi.org/10.5423/ppj.oa.11.2017.0226

Meccariello A, Monti SM, Romanelli A, Colonna R, Primo P, Inghilterra MG, Del Corsano G, Ramaglia A, Iazzetti G, Chiarore A, Patti F (2017) Highly efficient DNA-free gene disruption in the agricultural pest Ceratitis capitata by CRISPR-Cas9 ribonucleoprotein complexes. Sci Rep 7:10061

Meiklejohn NA, Staples TL, Fensham RJ (2022) Modelling climatic suitability for myrtle rust with a widespread host species. Biol Invasions 24:831–844

Merrick LF, Burke AB, Chen X, Carter AH (2021) Breeding with major and minor genes: genomic selection for quantitative disease resistance. Front Plant Sci 12:713667. https://doi.org/10.3389/fpls.2021.713667

Meur G, Budatha M, Srinivasan T, Rajesh Kumar KR, Dutta Gupta A, Kirti PB (2008) Constitutive expression of Arabidopsis NPR1 confers enhanced resistance to the early instars of *Spodoptera litura* in transgenic tobacco. Physiol Plant 133(4):765–775

Meynard CN, Lecoq M, Chapuis P, Piou C (2020) On the relative role of climate change and management in the current desert locust outbreak in East Africa. Glob Chang Biol 26:3753–3755

Mi X, Ji X, Yang J, Liang L, Si H, Wu J, Zhang N, Wang D (2015) Transgenic potato plants expressing cry3A gene confer resistance to Colorado potato beetle. C R Biol 338(7):443–450

Millard J, Outhwaite CL, Ceaușu S, Carvalheiro LG, da Silva e Silva FD, Dicks LV, Ollerton J, & Newbold T (2023) Key tropical crops at risk from pollinator loss due to climate change and land use. Science Advances, 9(41)

Milus EA, Kristensen K, Hovmøller MS (2009) Evidence for increased aggressiveness in a recent widespread strain of Puccinia striiformis f. sp. tritici causing stripe rust of wheat. Phytopathology 99:89–94

Mirzabaev A, Kerr RB, Hasegawa T, Pradhan P, Wreford A, Von Der Pahlen MCT, Gurney H (2023) Severe climate change risks to food security and nutrition. Clim Risk Manag 39:100473. https://doi.org/10.1016/j.crm.2022.100473

Mitchell ER (1979) Fall armyworm symposium: preface. Fla Entomol 62:81

Mitchell ER, McNeil JN, Westbrook JK, Silvain JF, Lalanne-Cassou B, Chalfant RB, Pair SD, Waddill VH, Sotomayor-Rios A, Proshold FI (1991) Seasonal periodicity of fall armyworm, (*Lepidoptera: Noctuidae*) in the Caribbean basin and northward to Canada. J Entomol Sci 26(1):39–50

Mittler R (2006) Abiotic stress, the field environment and stress combination. Trends Plant Sci 11:15–19

Mohan C, Shibao PYT, de Paula FFP, Toyama D, Vieira MAS, Figueira A, Scotton D, Soares-Costa A, Henrique-Silva F (2021) hRNAi-mediated knock-down of *Sphenophorus levis* V-ATPase E in transgenic sugarcane (*Saccharum* spp interspecific hybrid) affects the insect growth and survival. Plant Cell Rep 40:507–516

Montezano DG, Sosa-Gómez DR, Specht A, Roque-Specht VF, Sousa-Silva JC, Paula-Moraes SV, Peterson JA, Hunt TE (2018) Host plants of *Spodoptera frugiperda* (*Lepidoptera: Noctuidae*) in the Americas. Afr Entomol 26(2):286–300

Moon TT, Maliha IJ, Khan AAM, Chakraborty M, Uddin MS, Amin MR, Islam T (2022) CRISPR-Cas genome editing for insect pest stress management in crop plants. Stress 2(4):493–514

Muhammad J, Zeinab FZ, Saad Moussa S (2025) Biological control of *Schistocerca gregaria* and *Locusta migratoria migratorioides* using Entomopathogenic bacteria. Sci Rep 15:4601. https://doi.org/10.1038/s41598-025-87513-7

Navarro CE, Lam SK, Trębicki P (2020) Elevated carbon dioxide and nitrogen impact wheat and its aphid Pest. Front Plant Sci 11:605337. https://doi.org/10.3389/fpls.2020.605337

Navarro-González SS, Ramírez-Trujillo JA, Peña-Chora G, Gaytán P, Roldán-Salgado A, Corzo G, Lina-García LP, Hernández-Velázquez VM, Suárez-Rodríguez R (2019) Enhanced tolerance against a fungal pathogen and insect resistance in transgenic tobacco plants overexpressing an endochitinase gene from Serratia marcescens. Int J Mol Sci 20(14):3482

Nayak SN, Singh VK, Varshney RK (2017) Marker-assisted selection. In: Thomas B, Murray BG, Murphy DJ (eds) Encyclopedia of applied plant sciences, vol 2. Academic, Waltham, pp 183–197

Niu X, Kassa A, Hu X, Robeson J, McMahon M, Richtman NM, Steimel JP, Kernodle BM, Crane VC, Sandahl G, Ritland JL (2017) Control of western corn rootworm (*Diabrotica virgifera virgifera*) reproduction through plant-mediated RNA interference. Sci Rep 7(1):12591

Nnadi NE, Carter DA (2021) Climate change and the emergence of fungal pathogens. PLoS Pathog 17(4):e1009503. https://doi.org/10.1371/journal.ppat.1009503

Oerke EC (2006) Crop losses to pests. J Agric Sci 144:31–43

Oliva R, Ji C, Atienza-Grande G, Huguet-Tapia JC, Perez-Quintero A, Li T, Eom JS, Li C, Nguyen H, Liu B, Auguy F (2019) Broad-spectrum resistance to bacterial blight in rice using genome editing. Nat Biotechnol 37:1344–1350

Ortigosa A, Gimenez-Ibanez S, Leonhardt N, Solano R (2019) Design of a bacterial speck resistant tomato by CRISPR/Cas9-mediated editing of Sl JAZ 2. Plant Biotechnol J 17:665–673

Overton K, Maino JL, Day R, Umina PA, Bett B, Carnovale D, Ekesi S, Meagher R, Olivia LR (2021) Global crop impacts, yield losses and action thresholds for fall armyworm (*Spodoptera frugiperda*): a review. Crop Prot 145. https://doi.org/10.1016/j.cropro.2021.105641

Padmanabhan SY (1973) The great Bengal famine. Annu Rev Phytopathol 11:11–26

Palmgren M, Shabala S (2024) Adapting crops for climate change: regaining lost abiotic stress tolerance in crops. Front Sci 2:1416023. https://doi.org/10.3389/fsci.2024.1416023

Pande A, Paliwal J, Fuji Jian F, Matthew G, Bakker MG (2025) Mechanisms and application of mycotoxin decontamination techniques in stored grains. J Stored Prod Res 111:102486

Parker IM, Gilbert GS (2004) The evolutionary ecology of novel plant-pathogen interactions. Annu Rev Ecol Evol Syst 35(1):675–700

Paterson RRM, Lima N (2010) How will climate change affect mycotoxins in food? Food Res Int 43:1902–1914

Pathak H, Aggarwal PK, Singh SD (2012) Climate change impact, adaptation and mitigation in agriculture: methodology for assessment and applications. Indian Agricultural Research Institute, New Delhi, p 302

Pathania M, Verma A, Singh M, Arora PK, Kaur N (2020) Influence of abiotic factors on the infestation dynamics of whitefly, *Bemisia tabaci* (Gennadius 1889) in cotton and its management strategies in North-Western India. Int J Trop Insect Sci 2020(40):v969–v981

Paul PA, Munkvold GP (2005) Influence of temperature and relative humidity on sporulation of *Cercospora zeae-maydis* and expansion of gray leaf spot lesions on maize leaves. Plant Dis 89:624–630. https://doi.org/10.1094/PD-89-0624

Pest survey card on Spodoptera frugiperda. https://efsa.europa.eu/plants/planthealth/monitoring/surveillance/spodoptera-frugiperda

Penczykowski RM, Walker E, Soubeyrand S, Laine AL (2015) Linking winter conditions to regional disease dynamics in a wild plant-pathogen metapopulation. New Phytol 205(3):1142–1152

Peng A, Chen S, Lei T, Xu L, He Y, Wu L, Yao L, Zou X (2017) Engineering canker-resistant plants through CRISPR/Cas9-targeted editing of the susceptibility gene Cs LOB 1 promoter in citrus. Plant Biotechnol J 15:1509–1519

Pfender WF, Vollmer SS (1999) Freezing temperature effect on survival of *Puccinia graminis* subsp. *graminicola* in *Festuca arundinacea* and *Lolium perenne*. Plant Dis 83:1058–1062

Plessis DH, Schlemmer ML, Van den Berg J (2020) The effect of temperature on the development of *Spodoptera frugiperda* (*Lepidoptera: Noctuidae*). Insects 11(4):228. https://doi.org/10.3390/insects11040228. PMID: 32272548; PMCID: PMC7240686

Plett JM, Wilkins O, Campbell MM, Ralph SG, Regan S (2010) Endogenous overexpression of *Populus* MYB186 increases trichome density, improves insect pest resistance, and impacts plant growth. Plant J 64:419–432

Poloni A, Schirawski J (2014) Addressing the problem of soil-borne diseases in legumes. Front Plant Sci 5:76. https://doi.org/10.3389/fpls.2014.00076

Porras MF, Navas CA, Agudelo-Cantero GA, Santiago-Martínez MG, Loeschcke V, Sørensen JG, Crandall SG, Biddinger D, Rajotte EG (2023) Extreme heat alters the performance of hosts and pathogen front. Ecol Evol 11:Article: 1186452

Prakash A, Rao J, Mukherjee AK, Berliner J, Pokhare SS, Adak T, Munda S, Shashank PR (2014) Climate change: impact on crop pests; applied zoologists research association (AZRA). Central Rice Research Institute, Odisha. ISBN 81-900947-2-7

Puigbo P, Guzmen E, Romeu A, Garcia-Vallve S (2007) OPTIMIZER: a web server for optimizing the codon usage of DNA sequences. Nucleic Acids Res 35:W126–W131. https://doi.org/10.1093/nar/gkm219

Qi J, Song CP, Wang B, Zhou J, Kangasjärvi J, Zhu JK, Gong Z (2018) Reactive oxygen species signaling and stomatal movement in plant responses to drought stress and pathogen attack. J Integr Plant Biol 60:805–826

Qi T, Guo J, Liu P, He F, Wan C, Islam MA, Tyler BM, Kang Z, Guo J (2019) Stripe rust effector PstGSRE1 disrupts nuclear localization of ROS-promoting transcription factor TaLOL2 to defeat ROS-induced defense in wheat. Mol Plant 12(12):1624–1638

Qiu J, Xie J, Chen Y, Shen Z, Shi H, Naqvi NI, Qian Q, Liang Y, Kou Y (2022) Warm temperature compromises JA-regulated basal resistance to enhance Magnaporthe oryzae infection in rice. Mol Plant 15(4):723–739. https://doi.org/10.1016/j.molp.2022.02.014

Rai NP, Rai GK, Kumar S, Kumari N, Singh M (2013) Shoot and fruit borer resistant transgenic eggplant (*Solanum melongena* L.) expressing cry1Aa3 gene: development and bioassay. Crop Prot 53:37–45

Rajam MV (2020) RNA silencing technology: a boon for crop improvement. J Biosci 45:118

Ramegowda V, Kumar SM (2015) The interactive effects of simultaneous biotic and abiotic stresses on plants: mechanistic understanding from drought and pathogen combination. J Plant Physiol 176:47–54

Rao MS, Rao GVR, Venkateswarlu B (2010) Impact of climate change on insect pests. In: Lead papers, national symposium on climate change and rainfed agriculture, February, 18–20, 2010, CRIDA, Hyderabad, pp 43–54

Rao SM, Rao RCA, Raju BMK, Rao SAVM, Gayatri DLA, Islam A, Prasad TV, Navya M, Srinivas K, Pratibha G, Srinivas I, Prabhakar M, Yadav SK, Bhaskar S, Singh VK, Chaudhari SK (2023) Pest scenario of Helicoverpa armigera (Hub.) on pigeon pea during future climate change periods under RCP based projections in India. Sci Rep 13(1):6788. https://doi.org/10.1038/s41598-023-32188-1. PMID: 37100788; PMCID: PMC10133267

Rato C, Carvalho MF, Azevedo C, Oblessuc PR (2021) Genome editing for resistance against plant pests and pathogens. Transgenic Res 30:427–459

Ray D, Ramankutty N, Mueller N et al (2012) Recent patterns of crop yield growth and stagnation. Nat Commun 3:1293. https://doi.org/10.1038/ncomms2296

Ray DK, Gerber JS, MacDonald GK, West PC (2015) Climate variation explains a third of global crop yield variability. Nat Commun 6:5989

Raza MM, Bebber DP (2022) Climate change and plant pathogens. Curr Opin Microbiol 1(70):102233

Razzaq A, Wani S, Saleem F, Yu M, Zhou M, Shabala S (2021) Rewilding crops for climate resilience: economic analysis and de novo domestication strategies. J Exp Bot 72. https://doi.org/10.1093/jxb/erab276

Ristainoa JB, Pamela KA, Daniel PB, Qingshan W et al (2021) The persistent threat of emerging plant disease pandemics to global food security. PNAS 118(23):1–9. e2022239118

Ritchie H, Roser M (2023) Half of the world's habitable land is used for agriculture. Published online at OurWorldinData.org. Retrieved from: https://ourworldindata.org/global-land-for-agriculture

Rivero RM, Mittler R, Blumwald E, Zandalinas SI (2022) Developing climate-resilient crops: improving plant tolerance to stress combination. Plant J 109:373–389

Rosenzweig C, Iglesias A, Yang XB, Epstein PR, Chivian E (2001) Climate change and extreme weather events. Implications for food production, plant disease, and pests. Glob Chang Hum Health 2:90–104

Roussi A (2020) Why gigantic locust swarms are challenging governments and researchers. Nature 579:330

Rout E, Nanda S, Joshi RK (2016) Molecular characterization and heterologous expression of a pathogen induced PR5 gene from garlic (*Allium sativum* L.) conferring enhanced resistance to necrotrophic fungi. Eur J Plant Pathol 144:345–360

Sadhana NH, Geethanjali S, Mirchandani R, Natesan S, Senthil Kumar M (2024) Navigating towards dry root rot resistance in mungbean: impacts, mechanisms, and management strategies. Plant Physiol Rep 29:439–460

Salih AA, Baraibar M, Mwangi KK, Artan G (2020) Climate change and locust outbreak in East Africa. Nat Clim Chang 10:584–585

Salim M, Bakhsh A, Naqqash MN, Gökçe A (2024) Heterologous expression of distinct insecticidal genes in potato cultivars encodes resistance against potato tuber moth, *Phthorimaea operculella* (Zeller)(*Lepidoptera: Gelechiidae*). J Plant Dis Prot 131(5):1493–1505

Samanta S, Barman M, Chakraborty S, Banerjee A, Tarafdar J (2021) Involvement of small heat shock proteins (sHsps) in developmental stages of fall armyworm, and its expression pattern under abiotic stress condition. Heliyon 7:e06906

Sanchis V (2011) From microbial sprays to insect-resistant transgenic plants: history of the biospesticide *Bacillus thuringiensis*. A review. Agron Sustain Dev 31:217–231

Sander JD, Joung JK (2014) CRISPR-Cas systems for editing, regulating and targeting genomes. Nat Biotechnol 32:347–355

Sarver BA, Ward TJ, Gale LR, Broz K, Kistler HC, Aoki T et al (2011) Novel *Fusarium* head blight pathogens from Nepal and Louisiana revealed by multilocus genealogical concordance. Fung Genet Biol 48(12):1096–1107. https://doi.org/10.1016/j.fgb.2011.09.002

Savadatti E, Sreenivas AG, Lakshmikanth Mariyanna L, Hosamani A, Rao B, Desai KR, Subbanna AD, Ashoka Jalamangala A (2023) Temperature effects on the development of life stages of fall armyworm, *Spodoptera frugiperda* (J.E. Smith) (*Lepidoptera: Noctuidae*) on Maize. Int J Environ Climate Change 13(9):1884–1892. https://doi.org/10.9734/ijecc/2023/v13i92419

Savary S, Paul T, Willocquet L, Forrest N Jr (2006) Quantification and modeling of crop losses: a review of purposes. Annu Rev Phytopathol 44:89–112. https://doi.org/10.1146/annurev.phyto.44.070505.143342

Savary S, Willocquet L, Pethybridge SJ et al (2019) The global burden of pathogens and pests on major food crops. Nat Ecol Evol 3:430–439. https://doi.org/10.1038/s41559-018-0793-y

Schaefer LK, Parlange F, Buchmann G, Jung E, Wehrli A, Herren G, Müller MC, Stehlin J, Schmid R, Wicker T, Keller B (2020) Cross-kingdom RNAi of pathogen effectors leads to quantitative adult plant resistance in wheat. Front Plant Sci 11:253

Seidel P (2017) Extreme weather events and their effects on plant pests infecting potato, sugar beet, rape and grassland. J Cultiv Plants 69:125–136

Seleiman MF, Al-Suhaibani N, Ali N, Akmal M, Alotaibi M, Refay Y, Dindaroglu T, Abdul-Wajid HH, Battaglia ML (2021) Drought stress impacts on plants and different approaches to alleviate its adverse effects. Plants 10(2):259

Seneviratne SI, Zhang X, Adnan M, Badi W, Dereczynski C, Di Luca A, Ghosh S, Iskandar I et al (2021) Weather and climate extreme events in a changing climate. In: Climate change 2021: the physical science basis. Contribution of working group I to the sixth assessment report of the intergovernmental panel on climate change. Cambridge University Press, Cambridge/New York, pp 1513–1766. https://doi.org/10.1017/9781009157896.013

Senthilvel S, Ghosh A, Shaik M, Shaw RK, Bagali PG (2019) Development and validation of an SNP genotyping array and construction of a high-density linkage map in castor. Sci Rep 9:3003. https://doi.org/10.1038/s41598-019-39967-9

Senthilvel S, Prasad RD, Sathishkumar R, Shaik M, Mohanrao MD (2024) Genomic region linked to gray mold resistance in castor (*Ricinus communis* L.). Ind Crop Prod 222:120080

Shang D, Yang MF, Yu XF, Chen YC, Tian TA (2021) Spatiotemporal expression of CYP4G15 and CYP4L4 and their response to high and low temperature stress in Spodoptera frugiperda. Chin J App Entomol 58:664–671

Sharanabasappa SD, Kalleshwaraswamy CM, Maruthi MS, Pavithra HB (2018) Biology of invasive fall army worm *Spodoptera frugiperda* (J.E. Smith) (*Lepidoptera*: *Noctuidae*) on maize. Indian J Entomol 80(3):540–543

Sheoran S, Saini M, Ramtekey V, Gupta M, Kyum M, Kumar P (2023) Genetic engineering to improve biotic and abiotic stress tolerance in maize (*Zea mays* L.). In: Wani SH, Dar ZA, Singh GP (eds) Maize improvement. Springer, Cham. https://doi.org/10.1007/978-3-031-21640-4_10

Shrestha S (2019) Effects of climate change in agricultural insect pest. Acta Sci Agric 3:74–80

Shrivastava D, Dalal M, Nain V, Sharma PC, Kumar PA (2011) Targeted integration of *Bacillus thuringiensis*-endotoxin cry1Fa1 in Brinjal (*Solanum melongena* L.). Curr Trends Biotechnol Pharm 5(2):1149–1156

Simon EV, Hechanova SL, Hernandez JE, Li CP, Tülek A, Ahn EK, Jairin J, Choi IR, Sundaram RM, Jena KK, Kim SR (2023) Available cloned genes and markers for genetic improvement of biotic stress resistance in rice. Front Plant Sci 14:1247014. https://doi.org/10.3389/fpls.2023.1247014

Singh RP, Hodson DP, Jin Y, Lagudah ES, Ayliffe MA, Bhavani S, Rouse MN, Pretorius ZA, Szabo LJ, Huerta-Espino J et al (2016) Emergence and spread of new races of wheat stem rust fungus: continued threat to food security and prospects of genetic control. Phytopathology 105:872–884

Singh S, Singh A, Kumar S, Mittal P, Singh IK (2020) Protease inhibitors: recent advancement in its usage as a potential biocontrol agent for insect pest management. Insect Sci 27:186–201

Singh BK, Delgado-Baquerizo M, Egidi E et al (2023) Climate change impacts on plant pathogens, food security and paths forward. Nat Rev Microbiol 21:640–656. https://doi.org/10.1038/s41579-023-00900-7

Skendžić S, Zovko M, Živković IP, Lešić V, Lemić D (2021) The impact of climate change on agricultural insect pests. Insects 12:440. https://doi.org/10.3390/insects12050440
Slattery RA, Ort DR (2019) Carbon assimilation in crops at high temperatures. Plant Cell Environ 42:2750–2758
Sloggett JJ (2021) *Aphidophagous* ladybirds (*Coleoptera: Coccinellidae*) and climate change: a review. Insect Conserv Diver 14:709–722. https://doi.org/10.1111/icad.12527
Smith JE, Abbott J (1987) The Natural History of the Rarer Lepidopterous Insects of Georgia Including Their Systematic Characters, the Particulars of Their Several Metamorphoses, and the Plants on Which They Feed; Collected from the Observation of Mr. John Abbot, Many years Resident in That Country; The Lepidopterists' News: London, UK, 1987
Sokame BM, Kipkorir B, Agboka KM, Niassy S, Belayneh Y, Elkahky M, Tonnang HEZ (2025) A system dynamics model for predicting African armyworm occurrence and population dynamics. Agric Ecosyst Environ 380:109378. ISSN 0167-8809. https://doi.org/10.1016/j.agee.2024.109378
Sparks AN (1979) A review of the biology of the fall armyworm. Florida Entomol 62(2):82–87
Sparks Soares DJ (2012) The gray mold of castorbean: a review. In: Cumagun CJR (ed) Plant pathology. InTech Publisher, Rijeka, pp 219–240
Sparks AH, Forbes GA, Hijmans RJ, Garrett KA (2014) Climate change may have limited effect on global risk of potato late blight. Glob Chang Biol 20:3621–3631. https://doi.org/10.1111/gcb.12587
Staley JT, Hodgson CJ, Mortimer SR, Morecroft MD, Masters GJ, Brown VK, Taylor ME (2007) Effects of summer rainfall manipulations on the abundance and vertical distribution of herbivorous soil macro-invertebrates. Eur J Soil Biol 43:189–198. https://doi.org/10.1016/j.ejsobi.2007.02.010
Strange RN, Scott PR (2005) Plant disease: a threat to global food security. Annu Rev Phytopathol 43:3.1–3.34. https://doi.org/10.1146/annurev.phyto.43.113004.133839
Stuiver MH, Custers JH (2001) Engineering disease resistance in plants. Nature 411:865–868
Subedi B, Poudel A, Aryal S (2023) The impact of climate change on insect pest biology and ecology: implications for pest management strategies, crop production, and food security. J Agric Food Res 14. https://doi.org/10.1016/j.jafr.2023.100733
Sumila TCA, Ferraz SET, Durigon A (2024) Climate change impact on *Spodoptera frugiperda* (*Lepidoptera*: *Noctuidae*) life cycle in Mozambique. PLOS Clim 3(1):e0000325. https://doi.org/10.1371/journal.pclm.0000325
Sun H, Lang Z, Zhu L, Huang D (2012) Acquiring transgenic tobacco plants with insect resistance and glyphosate tolerance by fusion gene transformation. Plant Cell Rep 31:1877–1887
Sun BJ, Sun LY, Tugume AK, Adams MJ, Yang J, Xie LH et al (2013) Selection pressure and founder effects constrain genetic variation in differentiated populations of soil borne bymovirus wheat yellow mosaic virus (Potyviridae) in China. Phytopathology 103:949–959
Sun Y, Sparks C, Jones H, Riley M, Francis F, Du W, Xia L (2019) Silencing an essential gene involved in infestation and digestion in grain aphid through plant-mediated RNA interference generates aphid-resistant wheat plants. Plant Biotechnol J 17(5):852
Sun Y, Gong Y, He Q, Kuang S, Gao Q, Ding W, He H, Xue J, Li Y, Qiu L (2022) FAR knockout significantly inhibits Chilo suppressalis survival and transgene expression of double-stranded FAR in rice exhibits strong pest resistance. Plant Biotechnol J 20(12):2272–2283
Sun L, Alariqi M, Wang Y, Wang Q, Xu Z, Zafar MN, Yang G, Jia R, Hussain A, Chen Y, Ding X (2024) Construction of host plant insect-resistance mutant library by high-throughput CRISPR/Cas9 system and identification of a broad-spectrum insect resistance gene. Adv Sci 11:2306157
Sutherst RW, Constable F, Finlay KJ, Harrington R, Luck J, Zalucki MP (2011) Adapting to crop pest and pathogen risks under a changing climate. Wiley Interdiscip Rev Clim Chang 2:220–237
Tai APK, Mehliyar S, Jacky YSP, Yung DHY, Feng Z (2021) Impacts of surface ozone pollution on global crop yields: comparing different ozone exposure metrics and incorporating co-effects of CO_2. Front Sustain Food Syst 5:534616. https://doi.org/10.3389/fsufs.2021.534616

Tang Y, Wu S, He H, Gao Q, Ding W, Xue J, Qiu L, Li Y (2024) The CsmiR1579-CsKr-h1 module mediates rice stem borer development and reproduction: an effective target for transgenic insect-resistant rice. Int J Biol Macromol 254:127752

Tariq M, Tabassum B, Bakhsh A, Farooq AM, Qamar Z, Akram F, Naz F, Rao AQ, Malik K, Nasir IA (2022) Heterologous expression of cry1Ia12 insecticidal gene in cotton encodes resistance against pink bollworm, *Pectinophora gossypiella* (*Lepidoptera: Gelechiidae*); an alternate insecticidal gene for insect pest management. Mol Biol Rep 49(11):10557–10564

Tashkandi M, Ali Z, Aljedaani F, Shami A, Mahfouz MM (2018) Engineering resistance against Tomato yellow leaf curl virus via the CRISPR/Cas9 system in tomato. Plant Signal Behav 13:e1525996

Taub D (2010) Effects of rising atmospheric concentrations of carbon dioxide on plants. Nat Educ Knowl 1(8)

Thomas CD, Cameron A, Green RE, Bakkenes M, Beaumont LJ, Collingham YC, Erasmus BFN, De Siqueira MF, Grainger A, Hannah L et al (2004) Extinction risk from climate change. Nature 427:145–148

Thresh JM (1980) The origins and epidemiology of some important plant virus diseases. Appl Biol 5:1–65

Tiwari S, Mishra DK, Singh A, Singh PK, Tuli R (2008) Expression of a synthetic cry1EC gene for resistance against Spodoptera litura in transgenic peanut (*Arachis hypogaea* L.). Plant Cell Rep 27:1017–1025

Tong J, Zhao C, Liu D, Jambuthenne DT, Sun M, Dinglasan E, Periyannan SK, Hickey LT, Hayes BJ (2024) Genome-wide atlas of rust resistance loci in wheat. Theor Appl Genet 137(8):179. https://doi.org/10.1007/s00122-024-04689-8

Turrà D, Vitale S, Marra R, Woo SL, Lorito M (2020) Heterologous expression of PKPI and Pin1 proteinase inhibitors enhances plant fitness and broad-Spectrum resistance to biotic threats. Front Plant Sci 11:461. https://doi.org/10.3389/fpls.2020.00461

UNCCD (2022a). https://www.unccd.int/land-and-life/drought/overview#

UNCCD (2022b) Intergovernmental Working Group (IWG) on Drought Task Group #2: Vulnerability Assessment. https://www.unccd.int/sites/default/files/2022-02/TASK%20GROUP%202%20REPORT%20%20VULNERABILITY%20ASSESSMENT.pdf

UN-DESA (2011) United Nations, Department of Economic and Social Affairs, Population Division, "World population prospects: the 2010 revision, volume I: comprehensive tables" (ST/ESA/SER.A/313), United Nations

Urhausen S, Bradshaw CD, Davie J et al (2024) Climate-related risk to maize crops in China from Fall Armyworm, Spodoptera frugiperda. J Pest Sci 98:959–972. https://doi.org/10.1007/s10340-024-01817-7

Velásquez AC, Castroverde CDM, He SY (2018) Plant-pathogen warfare under changing climate conditions. Curr Biol 28:R619–R634. https://doi.org/10.1016/j.cub.2018.03.054

Venkatesh J, Kim SJ, Siddique MI, Kim JH, Lee SH, Byoung-Cheorl KANG (2023) CopE and TLR6 RNAi-mediated tomato resistance to western flower thrips. J Integr Agric 22(2):471–480

Vermeulen SJ, Campbell BM, Ingram JSI (2012) Climate change and food systems. Annu Rev Environ Resour 37:195–222

Vi TXT, Nguyen TNL, Pham TTN, Nguyen HQ, Nguyen THY, Tu QT, Le VS, Chu HM (2019) Overexpression of the ZmDEF1 gene increases the resistance to weevil larvae in transgenic maize seeds. Mol Biol Rep 46:2177–2185

Villalobos A, Ness JE, Gustafsson C, Minshull J, Govindarajan S (2006) Gene designer: a synthetic biology tool for constructing artificial DNA segments. BMC Bioinfor 7:285. https://doi.org/10.1186/1471-2105-7-285

Volk GM, Carver D, Irish BM, Marek L, Frances A, Greene S, Khoury CK, Bamberg J, del Rio A, Warburton ML, Bretting PK (2023) Safeguarding plant genetic resources in the United States during global climate change. Crop Sci:1–23. https://doi.org/10.1002/csc2.21003

Waheed A, Haxim Y, Islam W, Ahmad M, Muhammad M, Alqahtani FM, Hashem M, Salih H, Zhang D (2023) Climate change reshaping plant-fungal interaction. Environ Res 30:117282

Wakelin SA, Gomez-Gallego M, Jones E, Smaill S, Lear G, Lambie S (2018) Climate change induced drought impacts on plant diseases in New Zealand. Australas Plant Pathol 47:101–114

Wamalwa MN, Wanyera R, Algaba JR, Boyd LA, Owuoche J, Ogendo J, Bhavani S, Uauy C, Justesen AF, Hovmøller M (2022) Distribution of Puccinia striiformis f. sp. tritici races and virulence in wheat growing regions of Kenya from 1970 to 2014. Plant Dis 106(2):701–710

Wang Y, Cheng X, Shan Q, Zhang Y, Liu J, Gao C, Qiu JL (2014) Simultaneous editing of three homoeoalleles in hexaploid bread wheat confers heritable resistance to powdery mildew. Nat Biotechnol 32:947–951

Wang F, Wang C, Liu P, Lei C, Hao W, Gao Y, Liu YG, Zhao K (2016) Enhanced rice blast resistance by CRISPR/Cas9-targeted mutagenesis of the ERF transcription factor gene OsERF922. PLoS One 11:e0154027

Wang L, Chen S, Peng A, Xie Z, He Y, Zou X (2019) CRISPR/Cas9-mediated editing of CsWRKY22 reduces susceptibility to Xanthomonas citri subsp. citri in Wanjincheng orange (*Citrus sinensis* (L.) Osbeck). Plant Biotechnol Rep 13:501–510

Wang P, Ma J, Wang X, Tan Q (2020) Rising atmospheric CO_2 levels result in an earlier cyanobacterial bloom-maintenance phase with higher algal biomass. Water Res 185:116267. ISSN 0043-1354. https://doi.org/10.1016/j.watres.2020.116267

Wang W, Ma S, Hu P, Ji Y, Sun F (2021) Genome editing of rice eIF4G loci confers partial resistance to rice black-streaked dwarf virus. Viruses 13:2100

Watts A, Sankaranarayanan S, Watts A, Raipuria RK (2021) Optimizing protein expression in heterologous system: strategies and tools. Meta Gene 29:100899

Webb KM, Ona I, Bai J, Garrett KA, Mew T, Vera Cruz CM, Leach JE (2010) A benefit of high temperature: increased effectiveness of a rice bacterial blight disease resistance gene. New Phytol 185(2):568–576. https://doi.org/10.1111/j.1469-8137.2009.03076.x

Wheeler T, von Braun J (2013) Climate change impacts on global food security. Science 341(6145):508–513. https://doi.org/10.1126/science.1239402

Wilson RA, Talbot NJ (2009) Under pressure: investigating the biology of plant infection by *Magnaporthe oryzae*. Nat Rev Microbiol 7:185–195. https://doi.org/10.1038/nrmicro2032

Winterer J, Bergelson J (2001) Diamondback moth compensatory consumption of protease inhibitor-transformed plants. Mol Ecol 10:1069–1074

World Meteorological Organization and Food and Agriculture Organization of the United Nations (2016) Weather and Desert Locusts. WMO-No. 1175, Geneva 2, Switzerland, pp 47

Wu G, Bashir-Bello N, Freeland SJ (2006) The synthetic gene designer: a flexible web platform to explore sequence manipulation for heterologous expression. Protein Expr Purif 47:441–445

Wu K, Shirk PD, Taylor CE, Furlong RB, Shirk BD, Pinheiro DH, Siegfried BD (2018) CRISPR/Cas9 mediated knockout of the abdominal-A homeotic gene in fall armyworm moth (*Spodoptera frugiperda*). PLoS One 13:e0208647

Wu Y, Weng Z, Yan H, Yao Z, Li Z, Sun Y, Ma K, Hull JJ, Zhang D, Ma W, Hua H (2023) The microRNA-7322-5p/p38/Hsp19 axis modulates Chilo suppressalis cell-defences against Cry1Ca: an effective target for a stacked transgenic rice approach. Plant Biotechnol J 21(9):1827–1838

Wu H, Qiu J, Zhang P, Lu S, Meng J, Huang X, Li R, Chen B (2025) Host-induced gene silencing of *Sporisorium scitamineum* enhances resistance to smut in sugarcane. Plant Biotechnol J 23:1067–1069. https://doi.org/10.1111/pbi.14562

Xu L, Hou Q, Zhao Y, Lu L, Li B, Ni Z, Liang R (2017) Silencing of a lipase maturation factor 2-like gene by wheat-mediated RNAi reduces the survivability and reproductive capacity of the grain aphid, Sitobion avenae. Arch Insect Biochem Physiol 95(3):e21392

Xu Y, Liu F, Zhu S, Li X (2018) The maize NBS-LRR gene ZmNBS25 enhances disease resistance in Rice and *Arabidopsis*. Front Plant Sci 9:1033. https://doi.org/10.3389/fpls.2018.01033

Xu W, Zhang M, Li Y, He W, Li S, Zhang J (2023) Complete protection from Henosepilachna vigintioctopunctata by expressing long double-stranded RNAs in potato plastids. J Integr Plant Biol 65(4):1003–1011

Xu Z, Zhai Y, Chang H, Yan D, Ge P, Ren G, Zhang L, Yuan Y, Wang R, Li W, Li F (2024) Heterologous expression of taxane genes confers resistance to fall armyworm in *Nicotiana benthamiana*. Plant Cell Rep 43(4):94

Yan S, Ren B, Zeng B, Shen J (2020) Improving RNAi efficiency for pest control in crop species. Bio Techniques 68:283–290

Yan XR, Wang ZY, Feng SQ, Zhao ZH, Li ZH (2022) Impact of temperature change on the Fall Armyworm, *Spodoptera frugiperda* under global climate change. Insects 13:981. https://doi.org/10.3390/insects13110981

Yan X, Zhao Z, Feng S, Zhang Y, Wang Z, Li Z (2024) Multi-omics analysis reveal the fall armyworm *Spodoptera frugiperda* tolerate high temperature by mediating chitin-related genes. Insect Biochem Mol Biol 174. https://doi.org/10.1016/j.ibmb.2024.104192

Yang J, Sun XQ, Zhu-Salzman K, Qin QM, Feng HQ, Kong XD, Zhou XG, Cai QN (2021) Host-induced gene silencing of brown planthopper glutathione S-transferase gene enhances rice resistance to sap-sucking insect pests. J Pest Sci 94:769–781

Yang J, Liu P, Zhong K, Ge T, Chen L, Hu H, Zhang T, Zhang H, Guo J, Sun B, Chen J (2022) Advances in understanding the soil-borne viruses of wheat: from the laboratory bench to strategies for disease control in the field. Phytopathol Res 4:27

Yao MS, Meng JY, Yang CL, Zhang CY (2020) cDNA cloning and expression profiling of the heat shock protein Hsc70 gene and its response to different environmental stresses in fall armyworm *Spodoptera frugiperda*. J Plant Prot 47:797–806

Ying X, Redfern B, Gmitter FG Jr, Deng Z (2020) Heterologous expression of the constitutive disease resistance 2 and 8 genes from *Poncirus trifoliata* restored the hypersensitive response and resistance of Arabidopsis cdr1 mutant to bacterial pathogen Pseudomonas syringae. Plants 9:821

Yogindran S, Rajam MV (2016) Artificial miRNA-mediated silencing of ecdysone receptor (EcR) affects larval development and oogenesis in *Helicoverpa armigera*. Insect Biochem Mol Biol 77:21–30

Yogindran S, Rajam MV (2021) Host-derived artificial miRNA-mediated silencing of ecdysone receptor gene provides enhanced resistance to *Helicoverpa armigera* in tomato. Genomics 113:736–747

Yu H, Wang Y, Fu F, Li W (2022) Transgenic improvement for biotic resistance of crops. Int J Mol Sci 23:14370

Yuan X, Li S, Chen J, Yu H, Yang T, Wang C, Huang S, Chen H, Ao X (2024) Impacts of global climate change on agricultural production: a comprehensive review. Agronomy 14(7):1360. https://doi.org/10.3390/agronomy14071360

Zafar MM, Mustafa G, Shoukat F, Idrees A, Ali A, Sharif F, Shakeel A, Mo H, Youlu Y, Ali Q, Razzaq A (2022) Heterologous expression of cry3Bb1 and cry3 genes for enhanced resistance against insect pests in cotton. Sci Rep 12(1):10878

Zaidi SSEA, Mahas A, Vanderschuren H, Mahfouz MM (2020) Engineering crops of the future: CRISPR approaches to develop climate-resilient and disease-resistant plants. Genome Biol 21:289

Zhang X, Sato S, Ye X, Dorrance AE, Morris TJ, Clemente TE, Qu F (2011) Robust RNAi-based resistance to mixed infection of three viruses in soybean plants expressing separate short hairpins from a single transgene. Phytopathology 101(11):1264–1269

Zhang Y, Bai Y, Wu G, Zou S, Chen Y, Gao C, Tang D (2017) Simultaneous modification of three homoeologs of Ta EDR 1 by genome editing enhances powdery mildew resistance in wheat. Plant J 91:714–724

Zhang H, Tan X, He Y, Xie K, Li L, Wang R, Hong G, Li J, Li J, Taliansky M, MacFarlane S, Yan F, Chen J, Sun Z (2019) Rice black-streaked dwarf virus P10 acts as either a synergistic or antagonistic determinant during superinfection with related or unrelated virus. Mol Plant Pathol 20(5):641–655

Zhang H, Ye Z, Liu Z, Sun Y, Li X, Wu J, Zhou G, Wan Y (2022a) The cassava NBS-LRR genes confer resistance to cassava bacterial blight. Front Plant Sci 13:790140

Zhang Y, Guo W, Chen L, Shen X, Yang H, Fang Y, Ouyang W, Mai S, Chen H, Chen S, Hao Q (2022b) CRISPR/Cas9-mediated targeted mutagenesis of GmUGT enhanced soybean resistance against leaf-chewing insects through flavonoids biosynthesis. Front Plant Sci 13:802716

Zhang T, He Y, DePauw R, Jin Z, Garvin D, Yue X, Anderson W, Li T, Dong X, Zhang T, Yang X (2022c) Climate change may outpace current wheat breeding yield improvements in North America. Nat Commun 13:5591

Zhang Y, Guo W, Cao D, Chen L, Yang H, Chen H, Chen S, Hao Q, Qiu D, Shan Z, Yang Z (2023) Heterologous expression of the Glycine soja Kunitz-type protease inhibitor GsKTI improves resistance to drought stress and *Helicoverpa armigera* in transgenic Arabidopsis lines. Plant Physiol Biochem 202:107915

Zhang Q, Zhan W, Li C, Chang L, Dong Y, Zhang J (2024) Host-induced silencing of MpPar6 confers Myzus persicae resistance in transgenic rape plants. J Integr Agric 23(1):187–194

Zhao C, Liu B, Piao S, Wang X, Lobell DB, Huang Y, Huang M, Yao Y, Bassu S, Ciais P et al (2017a) Temperature increase reduces global yields of major crops in four independent estimates. Proc Natl Acad Sci USA 114(35):9326–9331. https://doi.org/10.1073/pnas.1701762114

Zhao Z, Zhang G, Zhou S, Ren Y, Wang W (2017b) The improvement of salt tolerance in transgenic tobacco by overexpression of wheat F-box gene TaFBA1. Plant Sci 259:71–85

Zhao Y, Sui X, Xu L, Liu G, Lu L, You M, Xie C, Li B, Ni Z, Liang R (2018) Plant-mediated RNAi of grain aphid CHS1 gene confers common wheat resistance against aphids. Pest Manag Sci 74(12):2754–2760

Zheng X, Weng Z, Li H, Kong Z, Zhou Z, Li F, Ma W, Lin Y, Chen H (2021) Transgenic rice overexpressing insect endogenous microRNA csu-novel-260 is resistant to striped stem borer under field conditions. Plant Biotechnol J 19:421

Zhou D, Liu X, Gao S, Guo J, Su Y, Ling H, Wang C, Li Z, Xu L, Que Y (2018) Foreign cry1Ac gene integration and endogenous borer stress-related genes synergistically improve insect resistance in sugarcane. BMC Plant Biol 18:1–14

Zhou L, Meng JY, Yang CL, Li J, Hu CX, Zhang CY (2020) Cloning of heat shock protein gene SfHsp90 and its expression under high and low temperature and UV-A stresses in Spodoptera frugiperda (Lepidoptera: Noctuidae). Acta Entomol Sini 63:533–544

Zhou J, Schaefer X, Qu Z, Zhang C, Wang F, Gao T, Yao Y, Liang J (2024) Progress in research on prevention and control of crop fungal diseases in the context of climate change. Agriculture 14(7):1108. https://doi.org/10.3390/agriculture14071108

Zhu L, Zhu J, Liu Z, Wang Z, Zhou C, Wang H (2017) Host-induced gene silencing of rice blast fungus *Magnaporthe oryzae* pathogenicity genes mediated by the brome mosaic virus. Genes 8(10):241

Zhu M, Tong L, Xu M, Zhong T (2021) Genetic dissection of maize disease resistance and its applications in molecular breeding. Mol Breed 41:32. https://doi.org/10.1007/s11032-021-01219-y

Zvereva EL, and Kozlov MV (2006) Consequences of simultaneous elevation of carbon dioxide and temperature for plant–herbivore interactions: a meta-analysis. Global Change Biology, 12:(1):27–41

Potential Breeding Strategies for Abiotic Stress Tolerance in Crops

7

T. Radhakrishnan, G. P. Mishra, Kirti Rani, B. C. Ajay, K. K. Pal, R. Dey, S. Chandramohan, M. Dash, K. M. Boraiah, P. S. Basavraj, U. Bitla, K. S. Reddy, and Narendra Kumar

Abstract

The adverse effects of changing climate are making abiotic stresses in crop plants more frequent and severe. Major stresses due to water deficit, waterlogging, changes in light quality, nutritional imbalances, salinity, acidity, and accumulation of heavy metals reduce crop productivity and quality. Information on how the neglected and underutilised species respond to abiotic stress is important for crop breeding and diversification. Breeding efforts should focus on the use of advanced biotechnological tools to combine different drought tolerance mechanisms into single crops. Both rising and freezing temperatures adversely affect the overall structure and function of growth and biomass of crop plants, necessitating the assessment of various physiological and biochemical traits in heat-tolerant genotypes. A better understanding of the genetics of heat tolerance can make targeted approaches to genetic improvement. In this context, utilisation of the molecular markers of heat tolerance in different crops is vital. The quantity and quality of light, which varies daily are important in the development of plants. One solution to food supply challenges is developing crop varieties suitable for vertical farming, indoor growing, and urban agriculture. Significant progress has been made in identifying genes linked to shade tolerance and avoidance, benefiting other organisms as well. Growing shade-tolerant plants can be

T. Radhakrishnan (✉) · B. C. Ajay · S. Chandramohan · N. Kumar
ICAR-Indian Institute of Groundnut Research, Sagdividi, Gujarat, India

G. P. Mishra
ICAR-Indian Agricultural Research Institute, New Delhi, Delhi, India

K. Rani
ICAR-National Bureau of Plant Genetic Resources, New Delhi, Delhi, India

K. K. Pal · R. Dey · M. Dash · K. M. Boraiah · P. S. Basavraj · U. Bitla · K. S. Reddy
ICAR-National Institute of Abiotic Stress Management, Baramati, Maharashtra, India

P. M. Priyadarshan, R. Ortiz (eds.), *Plant Breeding 2050*,
https://doi.org/10.1007/978-981-95-0583-8_7

difficult, but it has advantages in low-light conditions. Salt stress negatively affects seed germination, root and shoot growth, and fruit development. Many QTLs linked to salinity tolerance have been found in various crops and can be used in marker-assisted back-crossing to improve high-yield but vulnerable genotypes. As soil pH drops, toxic metals increase, leading to growth issues; aluminium toxicity is especially harmful to crops. Genes related to aluminium and iron toxicity have been identified, which may aid in developing tolerant genotypes. While heavy metals can be useful in small amounts, excess levels harm plant growth. Research indicates that plant-specific peroxidases help plants cope with heavy metal stress. Deployment of nutrient-use-efficient genotypes and improving nutrient efficiency are crucial for higher productivity, especially in nutrient-deficient soils. Advancements in genomics and proteomics have led to identifying key genes, which may improve nitrogen efficiency and crop yields. Breeding for improved biological nitrogen fixation in leguminous crops may also boost yields. Additionally, identifying phosphorus-efficient genotypes tolerant of low phosphorus is essential, with several QTLs already discovered. Gene banks play a vital role in crop enhancement initiatives, holding many beneficial genetic resources. The Indian Gene Centre is notable for its agricultural biodiversity, featuring a wide variety of cultivated plants and wild crop relatives. The complex nature of abiotic stress tolerance makes breeding for resilience challenging and calls for a clear understanding of molecular mechanisms of plant responses to stress. While controlled environment studies provide insights, open-field studies are limited. Improved phenotyping approaches can help close this gap and have selections of tolerant/resistant genotypes. Future research should focus on identifying genetic pathways for stress tolerance and utilizing next-generation sequencing data to link candidate genes. Incorporating stable quantitative trait loci offers a promising strategy for improving stress. Developing transgenic varieties may help create heat stress-resistant pulse genotypes. Many questions remain about target genes and their stress response impacts.

Keywords

Abiotic stress · QTLs · Water stress · Nutritional deficit stress · Salinity and soil stresses · Breeding approaches · Phenotyping · Genomics and proteomics · Candidate genes

7.1 Introduction

To achieve global food and nutritional security for a population projected to surpass 9.5 billion by the year 2050, it is crucial to optimise agricultural productivity on limited land resources while simultaneously tackling the challenges presented by abiotic stresses. These stresses are expected to become more frequent and severe due to the impacts of climate change. The adverse effects of abiotic stresses on agriculture and related sectors—including crops, horticulture, livestock, poultry, and fisheries—are significantly undermining farmers' incomes and jeopardising their livelihoods. The growth and development of plants are closely linked to the surrounding temperatures, and climate change is likely to exacerbate conditions such as drought and extreme temperature events, leading to water shortages and heat stress. Factors such as waterlogging, variations in light quality and intensity, nutrient imbalances, salinity, acidity, and the presence of heavy metals are critical issues that contribute to reduced productivity and quality in agricultural outputs. Salinity, in particular, poses a significant global challenge to crop productivity, resulting in water stress, cytotoxic effects, and nutritional imbalances due to excessive ion absorption (Bhoite et al. 2024). Studies reveal that nearly half of the yield reductions in crops can be attributed to abiotic stresses, with high temperatures contributing to 40%, salinity to 20%, drought to 17%, and low temperatures to 15%, while other factors account for the remaining losses. It is noteworthy that only about 9% of the earth's land is deemed suitable for crop cultivation, with the rest subjected to various stresses. To address the anticipated global food requirements, there is a pressing necessity to improve annual genetic gains by harnessing greater genetic diversity in breeding programs aimed at enhancing crop resilience to the challenging conditions posed by climate change. This chapter provides a concise overview of the advancements and future directions in the genetic enhancement of crop plants for improved tolerance to diverse stress factors.

7.2 Water Stress

7.2.1 Drought (Water Deficit Stress)

The efficiency of agricultural production and the guarantee of food security are greatly undermined during drought conditions, particularly in arid and semi-arid regions, leading to a notable decline in agricultural yields. Different crop species exhibit varying degrees of response to this environmental stress. Drought resistance in plants refers to their ability to sense signals of water deficiency and activate response mechanisms that facilitate their survival and reproductive success. There are three primary strategies employed by crop plants to achieve drought resistance: escape, avoidance, and tolerance.

Drought escape strategies assist plants in coping with limited water availability by allowing them to complete their life cycle before the onset of severe water shortages, thereby enhancing water absorption, conservation, and improving osmotic adjustment along with cell wall flexibility.

Drought avoidance focuses on sustaining cell turgor and volume through proactive water uptake or minimising water loss. While this may restrict above-ground growth, it can promote root development, thereby enhancing water absorption. These strategies enable plants to retain adequate water to function effectively in conditions of low water availability. Root characteristics such as depth, distribution, and density play a vital role in drought resistance. A greater root length density allows plants to access additional water in the soil, which is essential for maintaining yield across varying drought scenarios.

Drought tolerance is achieved through plant adaptations to stress, which involve reducing metabolic activities and undergoing osmotic adjustments, along with the accumulation of specific metabolites and proteins. These adaptations are essential for ensuring plant survival and productivity in stressful conditions.

7.2.1.1 Breeding for Drought Tolerance

To develop drought-tolerant crops, we can use natural genes from crop wild relatives and landraces. This can be done through traditional breeding methods as well as molecular techniques. Another approach is to study underutilised plant species that are known to be more resilient to drought than traditional crops. These species may also have other beneficial traits. Introducing them into existing cropping systems can help make the overall system more adaptable to water shortages. In the future, breeding efforts should combine various disciplines such as plant breeding, genomics, statistics, experimental design, and genetic diversity management. This integrated approach can help identify specific genetic regions associated with drought tolerance, use marker-assisted selection (MAS), and potentially employ genetic engineering (Khan et al. 2016a; Sallam et al. 2019).

Conventional Breeding

Several strategies are being used in the traditional approach to increase drought tolerance:

1. Develop short-duration genotypes that incorporate mechanisms for drought escape, avoidance, and tolerance tailored to specific environmental conditions.
2. Develop genotypes capable of sustaining high yields with minimal yield loss during drought periods (Sallam et al. 2019; Hurd 1971; Mitra 2001). Nevertheless, this selection strategy faces challenges due to low heritability, genetic interactions, genotype-environment interactions, and polygenic influences, necessitating comprehensive phenotypic evaluations and leading to gradual advancements (Mohammadi 2018; Roy and Murty 1970).
3. Identify genotypes that achieve high yields in optimal, non-water-stressed environments. These genotypes should also exhibit resilience under drought conditions (Johnson and Frey 1967). However, the notion of realising maximum genetic potential in ideal conditions is contentious (Blum 1973), as genotype-environment interactions may limit the performance of high-yielding genotypes in drought scenarios.

The effectiveness of any breeding program is significantly influenced by the availability of screening methods, particularly for drought resistance. The challenges faced in developing drought-tolerant varieties are exacerbated by the unpredictable nature of drought conditions during field trials and the diverse tolerance mechanisms present in various crop species and genotypes. The selection of a drought-resistant genotype in a given year and environment frequently does not fulfill the broader criteria required for a cultivar to be versatile across different years and environments. This discrepancy arises because the testing environment is a random sample from a heterogeneous array of target environments, and the conditions of the test year are also uncertain in the future. An alternative strategy could involve enhancing drought resistance in high-yielding genotypes by integrating morphological and physiological drought resistance mechanisms (Guadarrama-Escobar et al. 2024; Blum 2000). Nevertheless, many physiological parameters pose challenges, as they are either destructive during sampling or not conducive to breeding due to their high costs, lengthy procedures, and complex analytical requirements. Consequently, the feasibility of incorporating such traits into breeding programs that require screening of large populations each generation remains questionable. In this context, high throughput screening methods gain relevance. The response of plant growth and development to water scarcity can alter a plant species' tolerance to drought (Cuellar-Ortiz et al. 2008). Imaging techniques can be used to measure plant responses to water stress, root properties, and other desirable plant traits (Li et al. 2014; McAusland et al. 2013; Takahashi et al. 2015; Kim et al. 2015; Parent et al. 2015; Cai et al. 2015; Kim et al. 2021). RGB imaging is one of the most widely used imaging technologies (Humplík et al. 2015), and the traits detected through imaging can be correlated with plant yield. Non-destructive 3D reconstruction systems are now available to better evaluate root systems (Liang et al. 2024; Kumar et al. 2015). High-throughput phenotyping (HTP) platforms, utilising multi-spectral and hyperspectral imaging, chlorophyll fluorescence, RGB, thermal infrared, and other sensor equipment, enable the collection of trait data from a large number of plant samples under various conditions (Banan et al. 2018; Junker et al. 2014). In addition to ground-based imaging, crop responses to drought can also be monitored using aerial imagery from unmanned aerial vehicles (UAVs) (Kim et al. 2021; Bhandari et al. 2021; Francesconi et al. 2021).

Molecular-Breeding

Conventional breeding methods are often labour-intensive, lengthy, and frequently do not yield the intended results, particularly with abiotic stress tolerance. Drought tolerance is a complex quantitative trait controlled by multiple genes, pleiotropic effects, and unpredictable gene-environment interactions (Varshney et al. 2021a). Because of this complexity, it is difficult to rely on a single genetic intervention strategy to develop new cultivars with drought tolerance (Khan et al. 2019; Yang et al. 2010a). The limited number of studies identifying genomic regions linked to specific drought-response traits has led to an emphasis on genomic analysis for drought tolerance, focusing on traits that affect yield under drought conditions (Gupta et al. 2017). In crops such as wheat, the quantitative trait loci (QTLs) linked to drought tolerance are important; however, their utilisation in breeding programs remains constrained (Wang et al. 2024a, b; Ballesta et al. 2019; Sallam et al. 2019).

Molecular-breeding technologies such as marker-assisted selection, genome selection, gene transformation, and genome editing are currently employed to develop drought-resistant genotypes in a variety of crops. Leveraging "omics" data (genomics, transcriptomics, proteomics, metabolomics) is pivotal but requires bioinformatics to analyse the large amounts of genotypic and phenotypic information for QTL mapping, SNP discovery, and genomic selection (Upadhyay et al. 2017; Sharma and Sarkar 2013). The growing availability of bioinformatics tools, both open-source and commercial, facilitates the interpretation of this complex data at multiple levels.

Marker-Assisted Breeding

Drought stress is controlled by many minor genes (polygenes) that have additive effects in their expression (Thi Lang and Chi Buu 2008), and the location of such genes is referred to as quantitative trait loci (QTL). The natural genetic variation of a crop can be utilised by directly selecting under stressful conditions, whether simulated or natural or by mapping QTL (polygenes) and then using marker-assisted selection (Ashraf et al. 2008). QTL mapping allows for the assessment of the locations, numbers, magnitude of phenotypic effects, and pattern of gene action (Wang et al. 2024a, b; Vinh and Paterson 2005; Asíns 2002). (*the compilation of quantitative trait loci (QTLs) linked to drought resistance in crop plants is quite extensive; therefore, it is not included here to maintain brevity*). The role of polygenes in controlling a trait has been widely assessed by traditional means, but the use of DNA markers and QTL mapping has made it convenient to dissect complex traits (Humphreys and Humphreys 2005). The common strategy of marker-assisted backcrossing (MABC) is to introgress major QTLs for drought-resistant donors into high-yielding but drought-sensitive recipient parents. Since several minor and epigenetic QTLs are important in governing drought tolerance, MABC has not been very successful in improving drought resistance. To increase the frequency of favourable marker alleles in a population before inbred line extract, the marker-assisted recurrent selection (MARS) approach has been applied, involving selecting individuals with the most significant effects in each selection cycle (Gokidi et al. 2016; Bernardo 2008). MARS was successfully used to develop superior varieties of sweet corn, sunflower, soybean (Eathington et al. 2007), wheat (Sunilkumar et al. 2023), and sorghum (Mwamahonje et al. 2021). Genome-wide selection or genomic selection is another way to develop superior lines with overall excellent performance and drought resistance, but so far, this approach has not generated any drought-resistant crop products. MARS and genome-wide selection provide more powerful tools for pyramiding multiple QTLs or integrating multiple traits for drought resistance (Hu and Xiong 2014).

Recent advancements in genomics, especially genomic selection (GS) combined with high-throughput phenotyping, are being leveraged to achieve various important objective traits. GS is acknowledged as a promising method for selecting quantitative traits that are influenced by many genes with small effects. GS models are developed using genome-wide markers and phenotypic data from a training population, and these optimised models are then used to predict the genome-estimated breeding values (GEBVs) of individuals in a test population (Yeon et al. 2024; Selga et al. 2022; Xu et al. 2021b), aiding in the identification of superior lines (Varshney et al. 2021a, b, c). Unlike traditional marker-assisted selection (MAS), which relies on previously identified markers or genes that significantly influence the genetic control of target traits, GS represents a modern advancement of MAS, showcasing enhanced capabilities in discovering valuable genetic resources (Mohammadi 2018). In recent years, next-generation sequencing-based genotyping methods, such as genotyping by sequencing (GBS), complexity reduction of polymorphic sequences (CRoPS), RAD-seq, and diversity arrays technology (DArT), have emerged as leading technologies for identifying molecular markers within a partial representation of the genome through enzyme restriction (Gurgul et al. 2019; Ray and Satya 2014).

Genetic Engineering

In response to drought conditions, plants have evolved a complex network of signalling pathways that modulate various genes associated with drought response, resulting in the synthesis of diverse proteins, including transcription factors, enzymes, molecular chaperones, and other functional proteins. Understanding the genetics of drought tolerance (Daszkowska-Golec et al. 2018) has led to the discovery of numerous genes associated with this trait (Guadarrama-Escobar et al. 2024; Gupta et al. 2017). These genes have been utilised in a range of methodologies, like gene-editing technologies (Shan et al. 2013), gene silencing techniques (Yin et al. 2014), and overexpression methods (Saad et al. 2013), to create materials that may demonstrate drought resistance (Nuccio et al. 2015). Genetic engineering has enabled the alteration of plant genomes, facilitating the investigation of gene structure and function (Mohammadi 2018). However, this strategy has yet to produce crop varieties that can withstand drought conditions effectively (Sallam et al. 2019; Blum 2014).

The experimental findings have indicated that genetic modification can significantly improve drought resistance in crops. The overexpression of the maize transcription factor ZmNF-YBs has led to enhancements in several stress-related metrics, including chlorophyll levels, stomatal conductance, leaf temperature, and the preservation of photosynthetic activity, which ultimately resulted in increased grain yields in conditions of water scarcity (Wang et al. 2018). The overexpression of phytochrome-interacting factor 3 (PIF3) has demonstrated an ability to bolster tolerance against dehydration and salinity stress, with transgenic ZmPIFs plants showing enhanced relative water and chlorophyll content, chlorophyll fluorescence, and stability of cell membranes under stress conditions (Wu et al. 2019; Gao et al. 2015). The transgenic peanuts overexpressing mtlD have also shown improvements in drought-related traits (Patel et al. 2017). Genetically modified maize lines that overexpress the trehalose-6-phosphate phosphatase (TPP) gene have achieved yield levels comparable to their non-modified counterparts, even in the face of mild to severe drought conditions (Nuccio et al. 2015). Furthermore, the farnesyltransferase ERA1, which is involved in ABA signalling, has been downregulated using an antisense construct driven by a drought-inducible promoter, resulting in enhanced drought tolerance in transgenic plants (Nuccio et al. 2015). Another approach targets the transcriptional regulation of stress-responsive genes by controlling the functions of ABA-dependent transcription factors like MYB2, MYC2, CBF, and AREB, as well as ABA-independent factors such as DREB 1, DREB2, DREB3, and ZHDH (Bhalani et al. 2019; Khan et al. 2016b; Joshi et al. 2016; Zhang et al. 2016; Sarkar et al. 2014, 2016, 2019a).

Future investigations should continue to delve into the exploration, characterisation, identification, and application of favourable traits while also aiding in the development of crops and cropping systems that demonstrate resilience to drought conditions. A comprehensive understanding of the morpho-anatomical, physiological, and genetic factors that influence drought stress responses in crop wild relatives, as well as in neglected and underutilised species, is essential for recognising their potential contributions to crop breeding and diversification. Breeding initiatives aimed at enhancing drought resistance should aim to incorporate multiple mechanisms of drought tolerance within a single crop, utilising cutting-edge biotechnological tools.

7.2.2 Waterlogging (Submergence)

The increasing frequency of waterlogging events linked to climate change presents a considerable risk to global food security (Yamauchi et al. 2018; Singh 2015). Major crops affected by waterlogging include cereals such as rice, wheat, sorghum, maize, and barley, which are predominantly cultivated in southern and south-eastern Asia (Zhang et al. 2023; Pais et al. 2022; De Castro et al. 2022; Panda and Barik 2021; Tian et al. 2019; Samad et al. 2001). Waterlogging tolerance is defined as the ability of plants to maintain high growth rates and productivity under waterlogged conditions (Setter and Waters 2003). The effects of waterlogging are multifaceted, leading to various stresses, including limited gas exchange, mineral nutrient deficiencies, and toxicities from microelements (Setter and Waters 2003; Setter et al. 2009). Plants subjected to waterlogged environments encounter oxygen deficits in their rhizospheres, which disrupts the diffusion of cellular O2 and gas exchange, ultimately impeding growth and causing metabolic disturbances (Xu et al. 2022; Guo et al. 2020; Yan et al. 2018; Voesenek and Bailey-Serres 2013, 2015; Liu et al. 2012; Perata and Alpi 1993; Vartapetian and Jackson 1997). In response to waterlogging, plants have developed a range of adaptive strategies, including morphological modifications, changes in respiratory pathways, enhanced antioxidant defences, and the regulation of endogenous hormones (Habibi et al. 2023; Pucciariello and Perata 2017; Voesenek and Bailey-Serres 2015; Bailey-Serres and Colmer 2014).

Breeding for Water Logging Tolerance

The genetic variability in waterlogging tolerance among different plant species or genotypes within a single species provides valuable genetic resources for enhancing waterlogging resistance in crops, as highlighted by various studies (Xu et al. 2022; Ma et al. 2021; Borrego-Benjumea et al. 2020; Luan et al. 2018a, b). Variability is observed in the timing, duration, and intensity of waterlogging stress, as noted by Setter and Waters (2003), along with the physical and chemical alterations that occur in waterlogged soils, as discussed by Ponnamperuma (1972), Kozlowski (1984), Marschner (1984), and Kirk et al. (2003). Additionally, there is a range of diversity concerning the abiotic factors linked to environments susceptible to waterlogging, as highlighted by Laanbroek (1990) and Drew and Lynch (1980).

In the case of rice, certain landraces and traditional varieties have demonstrated the ability to endure flooding, a trait first identified in the 1950s and further explored in the 1970s. Notable Indian accessions such as FR13A and FR43B, along with Sri Lankan varieties, such as Kurkaruppan, Goda Heenati, and Thavalu, are recognised for their resilience to complete submergence, with FR13A exhibiting full tolerance for up to 7 days under complete submersion of 10-day-old seedlings (Keimeso 2022). Through traditional breeding methods, a submergence-tolerant rice line, IR49830-7-1-2-2, was developed from the FR13A lineage, incorporating the submergence-tolerant gene into high-yielding, short-to-intermediate stature varieties (Mishra et al. 1996). Subsequently, the high-yielding line IR 49830-7-2-2, known for its pest and disease resistance, has served as a donor parent, leading to the development of the variety "Sudhir" in India (crosses FR13A x "Biraj") (Dana et al. 2013). Later on, the line IR49830-7-1-2-1-3 and the landrace Rashpanjor (IC 575321) from India have been identified as an exceptional donor for submergence tolerance while also maintaining high yield potential.

In wheat, one of the major food crops, the Ducula sister lines, specifically Ducula-1 to Ducula-4, exhibited markedly enhanced tolerance to prolonged waterlogging conditions exceeding 3 months, as documented by Sayre et al. (1994) and van Ginkel et al. (1992). Conversely, the Chinese varieties Mikn Yang #11 and Zhen 7853 displayed comparatively low tolerance to waterlogging over a 42-day duration, which extended from 10 days after emergence to the mid-boot stage; however, they did demonstrate some degree of tolerance during the anthesis to grain filling phase (Sayre et al. 1994). Other wheat varieties recognised for their tolerance include Ning 8675 from China, Nonglin 46 from Japan, Yang 85–85 from China, Pato from Argentina, and Triticum macha from the Soviet Union (Cao and Cai 1991), Zhemani No. 2 and Zhengzhou 761 from China, along with Nonglin No. 46 from Japan (Lin et al. 1994) as tolerant.

In maize, the lines HZ32 (noted for high waterlogging tolerance) and K12 (recognised as highly sensitive to waterlogging) were utilised in traditional breeding and marker discovery efforts (Tang et al. 2005; Zhang et al. 2003).

The wild barley genotype Tam407227 and its dihaploid line TF58 were identified as waterlogging tolerant (Li et al. 2008). The cultivars Deder 2, Hayakiso 2, and Akashinriki have also been reported to be tolerant to waterlogging (Abiko et al. 2023).

Tolerance to waterlogging in plants is a quantitative trait influenced by multiple genes that exhibit additive effects (Zhou et al. 2007). However, some studies have indicated that it is governed by a single dominant gene in common wheat (Cao et al. 1995), Makha wheat (Triticum aestivum macha) (Fang et al. 1997), and maize (Sachs 1993). Numerous QTLs related to waterlogging tolerance have been documented across various plant species, including rice (Oryza sativa), wheat (Triticum aestivum), barley (Hordeum vulgare), and maize (Zea mays) (Table 7.1). However, the majority of these QTLs are confined to specific environmental conditions and populations (Tong et al. 2021), which limit their wider applicability in modern

Table 7.1 Some traits and QTLs associated with tolerance to water logging in major cereals

Trait	QTL	References
Rice		
Coleoptile length	*LOC_Os01g04430, LOC_Os01g04530, LOC_Os03g22720*	Kong et al. (2022)
Elongation	*qEL-NS-1.1*	Chattopadhyay et al. (2021)
Grain weight	*qGW-NS-10.1*	Chattopadhyay et al. (2021)
Plant height	*qPH-NS.3.1*	Chattopadhyay et al. (2021)
Panicle number	*qPN-NS-4.1*	Chattopadhyay et al. (2021)
Susceptibility index	*qSSI-PH-6.1, qSSI-GW-1.1, qSTI-EL-11.1, qSTI-GW-10.1*	Chattopadhyay et al. (2021)
Anaerobic germination	*qAG1-2, qAG6-2, qAG7-4, qAG10-1*	Ghosal et al. (2020)
Survival rate	*qSUB1.1, qSUB2.1, qSUB9.1 qSUB12.1*	Septiningsih et al. (2012)
Anaerobic germination	*qAG-1-2), qAG-3-1, qAG-7-2, qAG-9-1, qAG-9-2*	Angaji et al. (2010)
Internode elongation	*qTIL12C9285, qNEI12C9285, qLEI12C9285*	Hattori et al. (2007, 2008)
Green leaf recovery	*Sub1*	Xu and Mackill (1996)
Wheat		
Thousand-kernel weight	*QWTC.caas-4AL, QWTC.caas-5AS*, and *QWTC.caas-7DL*	Ding et al. (2023)
Root survival	*IWA2092, IWB11193, IWA3325, IWB22137, IWB10994*	Xu et al. (2022)
Formation aerenchyma	*IWB11078, IWA8342, IWB41008, IWB7934, IWB33938, TWB16121*	Xu et al. (2022)
Chlorophyll content	*QSpad3.ua-1D.5*	Ballesteros et al. (2015)
Germination index	*Xcdo1283, Xfbb283, Xfbb264, Xfba337, Xrz2*	Yu et al. (2014)
Survival rate	*Xbarc169, Xcdo312, Xbarc147, Xbcd1431, Xbcd508, Xgwm626, Xbcd1338*	Yu et al. (2014)
Chlorophyll content index	*Xpsr11, Xgwm71, Xwg583, XksuG12*	Yu et al. (2014)
Root length index	*Xbarc206, Xgwm16, Xgwm333, Xfba301, Xbcd1872 2D*	Yu et al. (2014)
Plant height index	*Xbarc228, Xgwm547, Xbcd310, Xbcd1338, Xabc455 2B*	Yu et al. (2014)
Root dry weight index	*Xbcd1119, Xfba67 1A*	Yu et al. (2014)
Shoot dry weight index	*Xgwm357, Xgwm135, Xfba67*	Yu et al. (2014)
Total dry weight index	*Xfba67*	Yu et al. (2014)

(continued)

Table 7.1 (continued)

Trait	QTL	References
Maize		
Adventitious root and aerenchyma formation		Manik et al. (2022)
Leaf chlorosis and senescence	*Subtol6*	Campbell et al. (2015)
Grain yield		Zaidi et al. (2015)
Leaf chlorosis		Mano and Omori (2013)
Root/shoot dry weight	*sdw4-3, sdw9-4, tdw9-2, tdw9-3*	Qiu et al. (2007)
Aerenchyma formation	QTLs, *Qaer1.02-3*, *Qaer1.07*, *Qaer5.09* and *Qaer8.06-7*	Mano et al. (2007)
Barley		
Biomass	QWSL.1H, QWSL.1H-(1-5), QBIO.2H, QSP.2H, QPH2.H-1, QPH.2H-2, QWLS.2H, QKWP.2H, QKWP.4H, QWLS.4H-1, QWLS.H2, QSP.5H-1, QSP.5H-2, QSP.5H-3, QWLS.6H, QPH.7H	Borrego-Benjumea et al. (2021)
Superoxide anion radicals and hydrogen peroxide content	*QSO.TxNn.2H, QHP.TxNn.2H*	Gill et al. (2019)
Root porosity	Q_{WL}*.YeFr.4H*	Broughton et al. (2015)
Leaf chlorosis	*tfy1.1-1, tfy1.1-2, tfy1.1-3, tfy1.2-1, tfy2.1-1, tfy2.1-2*	
Waterlogging tolerance	Q_{WL}*.YeFr.4H*, Q_{WL}*.YeFr.2H.2* Q_{WL}*.YeFr.2H.1*, Q_{WL}*.YeFr.3H*	Zhou et al. (2012)
Plant biomass reduction	*tfmas, yfmas,*	Li et al. (2008)
Plant survival	*tfsur-1, tfsur-2, yfsur-1, yfsur-2*	Li et al. (2008)
Leaf yellowing	*yfy1.1-1, yfy1.1-2, yfy2.1-1 yfy2.1-2, yfy2.1-3, yfy2.2-1 yfy2.2-2, yfy2.2-*	Li et al. (2008)

breeding programs. Therefore, it is essential to develop accessible molecular markers for beneficial QTLs to facilitate their incorporation into breeding strategies. The use of molecular markers enhances the precision of selection in contemporary breeding practices and has proven successful in pinpointing genetic loci. In rice, a significant locus for submergence tolerance, known as Sub1, has been identified (Xu and Mackill 1996) and is currently utilised in marker-assisted selection (MAS) breeding. A notable instance of the application of marker-assisted backcrossing (MABC) in creating stress-tolerant varieties is the development of a submergence-tolerant rice variety through the transfer of the major QTL, Sub1, which was later integrated into other cultivars (Singh et al. 2013).

Recently, notable progress has been made in the identification of genes associated with waterlogging tolerance, with many of these genes undergoing validation

(Xu et al. 2024). These validated genes can be effectively employed in focused breeding programs aimed at improving waterlogging tolerance. A high-throughput platform combining state-of-the-art imaging technologies, accurate sensor systems, fully automated robotic support, and comprehensive data processing and analytical software can enable the non-invasive collection and analysis of phenotypic data from plants from large and diverse populations (Yang et al. 2021.

7.3 High- and Low-Temperature Stress

High temperatures and heat stress are becoming major abiotic stresses due to climate change. Sustained high soil and air temperatures cause irreversible damage to plant development and physiology. This heat stress significantly affects pollen viability, stigma receptivity, and pollen tube growth, leading to ovule abortion, irregular pod filling, shrivelled seeds, and severe yield losses (Dhanasekar and Souframanien 2023). It also impacts chloroplast thylakoid membranes, electron transport activity, respiration, photosystems I and II, nitrogen fixation, microspore formation, embryogenesis, seed set, and seed development (Punniyamoorthy and Jegadeesan 2023). Thus, rising temperatures affect the overall morphology, anatomy, and physiology of plants. Researchers use various physiological and biochemical traits, such as photosynthetic activity, membrane stability, pollen viability, and phenolic compounds, to identify heat-tolerant genotypes (Sita et al. 2017a). Studying the association of crop phenology with temperature changes is essential for understanding plant adaptation strategies (Chaudhary et al. 2022). Knowledge of the genetics underlying key heat tolerance traits helps breeders make precise genetic improvements. Recently, molecular markers have been used to study the genetics of heat tolerance in several crops. Some of the high temperature tolerant genotypes of major cereals are listed in Table 7.2.

7.3.1 Heat Stress in Major Cereals

Wheat

Wheat is sensitive to high temperatures (Mason et al. 2010), and a temperature rise of 1–2 °C can shorten the grain-filling period, reducing the productive tillers by approximately 15.38%, leading to 53.57% grain yield reduction (Nahar et al. 2010). As heat stress intensifies, the chlorophyll content index (CCI) decreases, with healthy plants typically showing a CCI between 40 and 60 (Mamrutha et al. 2017). Prolonged maintenance of canopy greenness, also known as the stay-green trait, is a physiological adaptation to heat stress. This stay-green trait has been associated with sustained yield levels under heat stress, and specific QTL regions that regulate this trait have been identified (Vijayalakshmi et al. 2010).

Early-maturing genotypes complete growth phases before late heat stress, reducing its negative effects. Research shows that these genotypes have a higher

Table 7.2 Important heat-tolerant genotypes of major cereals

Genotype	Screening/other	References
Wheat		
CWI # 59788, 60155, 60391	Leaf chlorophyll content (LCC) and canopy temperature depression (CTD)	Hede et al. (1999)
Raj 4014 × WH730 (HT) RIL population (113)	1000-grain weight (TGW)	Pandey et al. (2014)
Aegilops tauschii Coss.	Cell membrane stability and TTC-based cell viability	Gupta et al. (2010)
A. speltoides Tausch; A. geniculate Roth	Spikelet fertility	Pradhan et al. (2012a)
ALTAR 84/AO'S′; ALTAR 84/A. tauschii	Leaf chlorophyll, grain weight, and grain yield	Pradhan et al. (2012b)
Moomal-2000, Mehran-89	Germination-related traits	Buriro et al. (2011)
Jimai-22	Photosynthesis, PS II, carboxylation, and grain yield	Feng et al. (2014)
CB # 367, 333, 335	Grain development and survival	Rehman et al. (2009)
WH # 1021, 730	Grain yield	Dhanda and Munjal (2012)
SYN # 11, 36, 44	1000-grain weight	Sharma et al. (2015)
Rice		
Dular, Todorokiwase, Milyang23, IR2006-P12-12-2-2, Giza178	Spikelet fertility and seed set	Tenorio et al. (2013)
N22, Bala, Co 39; CG14 (Oryza glaberrima)	Spikelet fertility and seed set	Jagadish et al. (2008)
N22, NH219	Spikelet fertility and pollen viability	Poli et al. (2013)
Bala (HT) × Azucena RIL population	Spikelet fertility	Jagadish et al. (2010); Yan et al. (2003)
Oryza meridionalis	Growth rate and photosynthesis	Scafaro et al. (2010)
N-22	Spikelet fertility	Prasad et al. (2006)
Nipponbare, Akitakomachi	Spikelet fertility	Matsui and Omasa (2002)
Maize		
ZPBL 1304 (HT); (ZPBL 1304 × ZPL 389) F2 population (160)	Heat shock protein (HSP)	Ristic et al. (1998)
B76, Tx205, C273A, BR1, B105C, C32B, S1W, C2A554-4	Leaf firing and tassel blast	Chen et al. (2012)

(continued)

Table 7.2 (continued)

Genotype	Screening/other	References
Hybrids: YH-1898, KJ. Surabhi, FH-793, ND-6339, NK-64017	Grain filling duration, kernel dry weight, starch, protein, and oil contents yield	Wilhelm et al. (1999); Rahman et al. (2013)
Pearl millet		
9444, Nandi 32, ICMB 05666, ICMB 92777; ICMB 02333	Seed set	Gupta et al. (2015)
F1's: H77/29-2 × CVJ-2-5-3-1-3; H77/833-2 × 96 AC-93	Seedling thermotolerance index (STI), seed to seedling thermotolerance index (SSTI), and membrane thermostability (MTS)	Yadav et al. (2014)
CVJ-2-5-3-1-3; 77/371 × BSECT CP 1	STI and SSTI	Yadav et al. (2011)

grain-filling rate, improving heat stress tolerance in durum wheat (Al-Karaki 2012). Early maturation also helps cultivars adapt to higher temperatures by extending the post-heading phase and grain-filling duration. Given the importance of photosynthesis for yield under heat stress, developing phenotyping methods to identify genotypes with better photosynthetic efficiency is essential (Yadava et al. 2022).

Temperatures exceeding 34 °C in the northern plains of India result in significant yield losses (Lobell et al. 2012). In South Asia, elevated night temperatures during February and March have negatively affected wheat production, posing a threat to food security (de Lima et al. 2021). Although frost injury is generally less severe than heat stress for wheat, mountainous and sub-mountainous regions are increasingly experiencing cold or frost damage due to sudden temperature drops (Yadava et al. 2022).

Rice (Oryza sativa L.)

Heat and cold stress have severe effects on rice growth, development, and yield (Yadava et al. 2022; Cao et al. 2008). High temperature stress can lead to poor panicle exertion and spikelet sterility, causing significant yield losses, low grain quality, increased chalkiness, shrivelled grains, and decreased storage protein levels.

There is an urgent need to develop heat-tolerant rice cultivars. Under heat stress, rice plants exhibit various morphological and physiological symptoms, making precise phenotyping important for selecting elite germplasm and identifying thermotolerance genes. In response to heat stress, rice triggers a cascade of events and activates complex transcriptional regulatory networks. Maintaining protein homeostasis under heat stress is vital for rice thermotolerance, involving protein quality control, the effective removal of toxic proteins, and translational regulation. Although some agronomic and genetic strategies have been implemented to improve heat tolerance in rice, the molecular mechanisms underlying rice's response to heat stress remain elusive, and progress in breeding for thermotolerance has been limited (Xu et al. 2021a). Though rice is generally adapted to tropical climates, it is sensitive to low temperatures. Low temperature tolerance is a critical trait in hilly regions, where sudden temperature drops can lead to significant productivity losses (Yadava et al. 2022).

Maize (Zea mays L.)

Maize is particularly vulnerable to heat stress during the reproductive and early grain-filling stages. Heat stress tolerance is considered a complex trait, and leveraging the genetic diversity found in maize landraces and wild accessions is a promising strategy for identifying novel donors, traits, quantitative trait loci (QTLs), and genes that can be introgressed into elite cultivars (Djalovic et al. 2024). Several QTLs associated with heat stress tolerance in maize have been mapped using genomic approaches (Longmei et al. 2021). Frey et al. (2016) identified two QTL hotspots on chromosomes 2 and 3 that are linked to heat tolerance, explaining 7–13% of the phenotypic variance (PVE). In these QTL hotspot regions, three heat tolerance genes (GRMZM2G115658 and GRMZM2G537291 on chromosome 2 and GRMZM2G324886 on chromosome 3) were identified. The gene GRMZM2G324886 encodes a calcyclin-binding protein involved in calcium signalling in response to stress conditions. Additionally, QTLs for the heat susceptibility index related to leaf scorching were found on chromosome 9, while QTLs associated with the heat susceptibility index for grain yield were located on chromosomes 2 and 3, indicating that different genomic regions regulate the genetic mechanisms for leaf scorching and grain yield under heat stress (Frey et al. 2016).

Inghelandt et al. (2019) identified six QTLs related to the heat susceptibility index for five traits at the seedling stage, which explained 7–9% of the phenotypic variance (PVE). Qin et al. (2007) reported that the increased expression of the ZmDREB2A gene in maize could induce heat-responsive genes, enhancing the plant's adaptability to high-temperature stress. Further, QTLs controlling leaf temperature, anthesis-silking interval (ASI), and grain yield have also been mapped on chromosome 7 in maize (Sanguineti et al. 1999). A transcriptome study in maize responding to heat stress identified 1029 upregulated and 828 downregulated differentially expressed genes (DEGs), highlighting the critical role of protein processing in the endoplasmic reticulum under heat stress conditions (Qian et al. 2019). Heat stress also induces endoplasmic reticulum stress, leading to the upregulation of ZmbZIP60 (Li et al. 2018a). Genome-wide association studies (GWAS) are a key approach for identifying potential QTLs and dominant genes for crop improvement. Recent advancements in molecular techniques, including clustered, regularly interspaced short palindromic repeats (CRISPR), may accelerate the development of thermotolerant maize genotypes (Djalovic et al. 2024).

Maize is highly sensitive to cold stress during seed germination and the seedling stage, which can result in reduced plant vigour and lower grain production. Cold-tolerant varieties generally have a greater ability to maintain traits related to seed germination, root development, and shoot photosynthesis under cold conditions. These morphological and physiological traits are commonly used to assess the cold tolerance of maize varieties in genetic studies. Significant progress has been made in recent years in understanding the mechanisms of cold tolerance in maize. Numerous QTL, GWAS, and transcriptomic analyses have been conducted on various maize genotypes and populations with substantial variation in cold tolerance, leading to the identification of hundreds of candidate genes involved in cold regulation. However, only a few of these candidate genes have been functionally characterised so far (Zhou et al. 2022).

Pearl Millet (Pennisetum glaucum)

Pearl millet serves as an important cereal crop in the arid and semiarid regions of Sub-Saharan Africa and South Asia, where it encounters challenges such as elevated temperatures and erratic rainfall patterns. Its inherent capacity to withstand high temperatures, drought, salinity, and suboptimal soil conditions enables pearl millet to flourish in adverse environments. The ideal temperature for its normal growth is 33–34 °C, and high temperatures are harmful to seedlings and reproductive stages. Hence, heat tolerance at the seedling stage becomes a priority. Genetic variations in seedling survival under high soil surface temperatures have been observed (Peacock et al. 1993), and selecting for better seedling emergence under artificial screening techniques (Soman and Peacock 1985) has been effective (Lynch 1994). Extensive field screenings over several years and in multiple locations during the summer season have revealed that the stigma is more sensitive to heat than pollen. Pearl millet is an open-pollinated crop with a 75–80% outcrossing rate. Significant genetic variation was found among breeding lines and within open-pollinating populations. It was observed that the boot-leaf stage is more heat-sensitive than the panicle-emergence stage, and heat tolerance acts as a dominant trait. These screenings assisted in identifying heat-tolerant breeding materials, which were then utilised to further enhance heat tolerance. This process also facilitated combining heat tolerance in high-yielding hybrids (Gupta et al. 2016, 2019a, b).

Conventional and molecular breeding techniques can be used for genetic improvement. Recurrent selection and pedigree selection are the two common methods for genetic improvement. Pearl millet's natural out-crossing and protogynous flowering behaviour enables random mating, which is essential for recurrent selection. Various forms of recurrent selection are suitable for enhancing the genetics of pearl millet (Sanjana 2015). According to Yadav et al. (2011), breeding for heterosis can lead to the development of heat-tolerant pearl millet varieties. Wahid et al. (2007) also noted that while some plants have shown improved heat tolerance through traditional breeding methods, success in genetic transformation approaches has been limited. The International Crops Research Institute for the Semi-Arid Tropics in Hyderabad identified 424,085 genomic structural variations in the pearl millet genome from eight major geographical regions (Nature 2023). Genes near these variations are more likely to respond to heat stress, making them potential targets for breeding stress-tolerant pearl millet. These genetic variations impact the expression of specific genes that have helped the plant adapt to high temperatures during domestication. However, the molecular mechanisms governing heat stress responses (HSRs) in pearl millet have not been extensively studied compared to other major crops (Huang et al. 2021; Sun et al. 2020) and thus remain poorly understood. Yan et al. (2023) proposed a potential mechanism involving the endoplasmic reticulum (ER) and identified an RWP-RK transcription factor as a key regulator of heat tolerance, along with the ER pathway, through integrated multi-omics analyses. These findings advance our understanding of heat tolerance in pearl millet and could aid in genomics-assisted breeding for heat tolerance. Currently, molecular breeding, particularly the application of QTLs and other markers specific to high-temperature tolerance, is limited. However, exploiting common QTLs and candidate genes relevant for both drought and high temperature could be highly beneficial.

7.3.2 Heat Stress in Pulses

Chickpea (Cicer arietinum)

Chickpea is a cool-season legume known for its nutritional value, but it is sensitive to high temperatures (>32/20 °C max/min) during its reproductive and seed-filling stages. Heat stress significantly reduces leaf sucrose production, which negatively impacts the accumulation of various seed constituents, leading to poor nutritional quality (Devi et al. 2022). Interactions between temperature, day length, and genotype are crucial in determining the flowering time of chickpea. The reproductive structures of chickpea are particularly sensitive and vulnerable to heat stress (Jeffrey et al. 2021). Heat stress causes structural abnormalities in anthers, resulting in small, shrunken, and empty pollen grains and ovaries, leading to ovule abnormalities and decreased stigma receptivity (Kaushal et al. 2013). It is widely believed that day/night temperatures of 35/20 °C are critical for reducing pollen viability and pod set in chickpea. High temperature stress at anthesis is thought to primarily cause sterility due to reduced carbohydrate accumulation and altered sink-source dynamics in pollen grains and stigmatic tissue (Devasirvatham et al. 2010). Stomatal conductance plays a vital role during heat stress (Kaushal et al. 2013). Chickpeas exposed to 35/20 °C temperatures can show a 50% reduction in pod set compared to the optimum temperature of 28/16 °C (Devasirvatham et al. 2010). Twenty-two heat shock factors (HSFs) were identified in both desi and kabuli genomes during podding and seedling growth (Chidambaranathan et al. 2018), 482 heat-responsive proteins were correlated with heat response (Parankusam et al. 2017), and five putative genes encoding heat shock proteins (HSPs) or heat shock transcription factors were identified as contributing to thermotolerance (Agarwal et al. 2016).

Pigeonpea (Cajanus cajan)

Pigeonpea is a climate-resistant grain legume, and its optimal growing temperature is 25–35 °C. However, its wild relatives can thrive in temperatures ranging from 18 to 45 °C (Tiwari et al. 2022). Heat stress (HS) reduces plant biomass and results in lower yields, especially in tropical and subtropical regions (Ramakrishna et al. 2021). Previous analyses indicate that a 1 °C increase in maximum temperature during the growing season could lead to a 20.8% decrease in pigeonpea yield. HS causes the dissociation of crucial protein complexes and the production of reactive oxygen species (ROS) (Becklin et al. 2016). In response to HS, plants upregulate genes encoding molecular chaperones and signalling molecules, triggering a series of events that mediate HS responses (Mishra et al. 2017).

Pea (Pisum sativum)

Pea crop has relatively low heat tolerance, and yield declines when the maximum day-time air temperature during flowering exceeds 25 °C (Sadras et al. 2012). Several studies have addressed the impact of high temperature on pea (Mohapatra et al. 2020). Lamichaney et al. (2021) observed a reduction in the seed yield (24–60%), seed germination (4–8%), seed set (7–14%), and 100-seed weight (6–16%) under elevated ambient temperature during vegetative and reproductive stages. The ambient temperature above 25 °C negatively affects plant growth and crop duration (Bueckert et al. 2015). The photosynthetic activity was detained at 40 °C, and at 45 °C, net photosynthesis reduction reached up to 80% (Haldimann and Feller 2005). The reduction in the different yield attributes like seed germination, seed set, seed weight, and overall grain yield has been noticed under heat stress (Lamichaney et al. 2021). The exposure of field pea to elevated temperature during the flowering stage has a higher negative impact on grain yield than the heat stress at the grain filling period, which occurs due to the reduced pods/ plant and seeds per pod. The medium-tall genotypes had longer grain-filling duration, higher pod-bearing nodes and yield, and lower yield losses as compared to semi-dwarf and dwarf genotypes under heat stress, suggesting higher heat tolerance capacity of tall-type plants (Parihar et al. 2023). The high-temperature exposure also reduced maturity duration prominently in long-duration genotypes (Parihar et al. 2022).

Black Gram (Vigna mungo)

Black gram thrives well under 25–35 °C temperature, and a drastic reduction in yield under high temperatures (>35 °C) has been reported (Sen Gupta et al. 2021). A method to assess the thermotolerance in black gram was developed (Partheeban et al. 2017), wherein the temperature induction response technique (TIR) (induction at 36–46 °C for 3 h and lethal temperature at 52 °C for 3 h) was used to study the cellular responses. Similarly, Pavithra et al. (2023) carried out screening of 100 genotypes for TIR (induction temperature: 36–46 °C for 3 h; lethal temperature: 54 °C for 3 h). Sen Gupta et al. (2021) evaluated a panel under field and laboratory for heat tolerance and >40 °C during flowering resulted in deformation/abortion of flower parts leading to poor yield (Sen Gupta et al. 2021). Sen Gupta et al. (2021) characterised 97 urdbean genotypes and identified 8 heat tolerant and 35 heat sensitive genotypes. Higher proline, SOD and POX activity, and enzymatic activities in heat tolerant genotypes and more MDA accumulation in the susceptible genotypes have been reported (Pavithra et al. 2023). Sen Gupta et al. (2021) used 21 genetic markers and observed that the heat tolerant genotype (UPU 85–86) got distinctly clustered from the heat sensitive genotype (PKGU-1).

Mungbean (Vigna radiata)

The threshold temperatures for the mungbean plant and its responses are inconsistent and not yet clearly identified (Aski et al. 2023; Aski et al. 2021). Traits such as upright growing habit, lower canopy temperature and small seeded nature have shown their utility for better adaptation under heat stress (Devi et al. 2022). Identifying variations in mungbean root traits will help breeding lines with enhanced root characteristics, improving water uptake and helping to alleviate heat and drought stresses (Mishra et al. 2009). Screening of the AVRDC mungbean mini core collection (296 genotypes) for 14 root-related traits displayed wide variations, and genotypes with favourable root traits were identified as donors for breeding cultivars with enhanced adaptation to water-deficit stress (Aski et al. 2021).

Lentil (Lens culinaris)

Lentil is a sensitive crop to the rising temperature, and it requires 18–30 °C for its growth (Choudhury et al. 2012). The maximum temperature of 32/20 °C (max/min) at the flowering and pod filling stage in lentil can sharply deteriorate the quality and yield (Delahunty et al. 2015). The consecutive heat waves (35 °C) for 6 days resulted in a 70% reduction in yield across southeastern Australia (Delahunty et al. 2015). The heat stress also results in reduced germination (%), abnormal seedling growth, degenerated nodules, early flowering, increased lipid peroxidation, etc. (Sehgal et al. 2017). The genotypes tolerant to heat stress exhibit superiority in functionality of pollen and enhanced level of antioxidants (Sita et al. 2017b), reduced reproductive period, and forced maturity (Bhandari et al. 2016). Screening for heat stress is done by growing them in late sown conditions so that the temperature at the reproductive stage is above 35 °C (Sita et al. 2017b). However, precise phenotyping can be done in laboratory and hydroponics conditions for traits like rate of photosynthesis, pollen viability and membrane stability (Kumar et al. 2018). The best way to identify the heat tolerant genotypes is by combined screening under field and controlled conditions (Chen et al. 2019).

Cowpea (Vigna unguiculata L.)

In cowpea, with every 1 °C increment in night temperatures beyond 16 °C, pod filling and seed yield were found to be severely impacted to the extent of 4–14% (Hall 2004). Elevated temperatures with flowering and pod filling stages acutely affect its yield (Ntare 1991). At temperatures beyond 40 °C, cowpeas exhibit poor pod setting, diminished harvest index, increased floral drop, male sterility, poor pollen shedding, tanned testa and decreased biological nitrogen fixation (Jha et al. 2017). Five QTLs (Cht-1, -2, -3, -4, -5; PVE: 11.5–18.1%) were tagged with 48 SNPs, 4 of which were contributed by CB27, while 1 was from IT82E-18 (Lucas et al. 2013). In addition, three QTLs (Hbs-1, -2, -3) regulating heat-induced tanning of cowpea testa, which explained 77.3%, 12.3%, and 6.8% PVE, respectively (Pottorff et al. 2014). The transcription factors NAC1/NAC2 were found to impart heat tolerance in transgenic cowpea. The inheritance of heat stress tolerance indicated single dominant, single recessive, or double dominant genes and was later confirmed as QTLs (Lucas et al. 2013). Selections of heat tolerant genotypes based on the abundance of blooming and pod set under high nocturnal temperatures and long-day growing conditions (Marfo and Hall 1992) or based on depressed leaf-electrolyte-leakage (Hall 2004) were found promising. Heat stress tolerant genotypes can also be identified based on stomatal function, photosynthetic rate, pollen viability, relative water content, etc. (Kumar et al. 2020).

Grass pea (Lathyrus sativus L.)
Grass pea has comparatively better tolerance to drought, flooding, salinity, temperature extremities (high and low) and problematic soils than other pulse crops (Dixit et al. 2016; Sarkar et al. 2019b). Breeding efforts in grasspea focus on developing varieties resilient to heat stress (Gonçalves et al. 2022). Traits associated with heat stress tolerance include deep root systems and efficient water and nutrient uptake, which are targeted for genetic improvement. Molecular techniques aid in identifying heat stress-resilience genes and markers, enabling precise breeding strategies. Ultimately, the goal is to enhance the adaptability of *Lathyrus* cultivars to adverse environmental conditions, ensuring sustainable production in challenging environments. A cytogenetic study investigated cytomictic variations induced by abiotic stresses (temperature, heavy metal Pb, water stress) on meiotic cells of *Lathyrus sativus* L. (Kumar and Tripathi 2008).

7.3.2.1 Cold Tolerance

Chickpea (Cicer arietinum)
Most of the chickpea cultivars are susceptible to chilling temperatures at flowering (Croser et al. 2003). Freezing temperature reduces growth vigour and vegetative biomass while chilling temperature at flowering results in flower and pod abortion. International Center for Agricultural Research in the Dry Areas (ICARDA) has identified several cold tolerant genotypes from the cultivated and wild gene pools (Singh et al. 1995; Robertson et al. 1995). When mean daily temperatures fall below 15 °C during flowering, cause pollen sterility (Clarke et al. 2004). Pigeonpea is sensitive to low temperatures at two stages: (i) During pollen development in the flower bud, resulting in sterile pollen, and (ii) During the formation of the pollen tube (Lawlor et al. 1998). RFLP markers for chilling tolerance were identified and subsequently converted to SCAR markers for the selection of chill-tolerant progenies of Amethyst × ICCV 88516 (Millan et al. 2006).

Pigeonpea (Cajanus cajan)
During the winter season in northern India (December–January), pigeonpea experiences low temperature stress. When the minimum temperature drops below 5 °C, it negatively affects plant growth, survival, and reproductive capacity (Araújo et al. 2015). At freezing temperatures, intracellular water forms ice, causing cell contraction, wilting, and plant death (Megha and Singh 2023). Various research has shown genotypic differences in cold tolerance among pigeonpea varieties. Due to the limited knowledge about cold stress and its impact on pigeonpea, it is necessary to screen a large number of pigeonpea genotypes for low temperature tolerance under controlled conditions to generate accurate genetic information (Alagupalamuthirsolai and Vijayabharathi 2020).

Pea (Pisum sativum)

Cold stress is a major abiotic stress in pea. During frost, the temperature suddenly goes below 0 °C, which permanently damages the cell membrane system or photosynthetically active tissue (Liu et al. 2017). The occurrence of frost at the reproductive phase can damage flower buds, flowers, and pods, and seed development leads to a drop in overall grain yield (Ridge and Pye 1985). Until the vegetative stage, the field pea crop has better tolerance to frost than during the reproductive stage (Meyer and Badaruddin 2001). However, genetic variability has been observed for frost tolerance at seedling, vegetative, and reproductive stages (Shafiq et al. 2012).

Lentil (Lens culinaris)

Lentils exhibit sensitivity to frost, and at the initial stage, it leads to instant recovery from underground axillary buds. However, frost at maturity results in the ceasing of initiation of axillary buds and further restricts the plant from going further to the reproductive stage. In lentil, frost stress hinders flower initiation and pod development, injures the vegetative tissue, and damages the seed coat and leaf (Gupta et al. 2019a). Plants exposed to frost continuously become susceptible to diseases such as anthracnose botrytis, grey mould, and pest infestation (Gupta et al. 2019a). Frost tolerance in lentil has been characterised in lentil (Barrios et al. 2016). The inheritance of frost tolerance has been reported as monogenic, while winter hardiness is controlled by quantitative traits (Kahraman et al. 2004).

The details of major pulses for heat and cold tolerance and the list of genotypes of pulses are presented in Table 7.3.

7.3.3 Shade Tolerance and Response to High Light

Light quantity and quality are vital for plant development, and daily variations in light intensity and wavelengths affect growth. In nature, seeds often germinate in low-light conditions, while in agriculture, insufficient sunlight or excessive shading can affect crop yields. A three-year study on 100 high-yielding rice genotypes has revealed a 50% yield reduction under shade stress, mainly due to reduced panicle number, grains per panicle, and test weight (Ganguly et al. 2020). However, many plants have developed mechanisms to adapt to shading or low light, such as shade avoidance or shade tolerance. Generally, numerous crops can withstand a 15% decrease in solar radiation without adversely affecting their yields. A reduction of up to 30% in solar radiation is deemed advantageous for berries, fruits, and fruiting vegetables. Conversely, leafy vegetables, root crops, forages, and C3 plants exhibit less benefit in yield under such conditions. Notable yield declines have been observed in grain legumes and maize, even with minimal shade exposure (Laub et al. 2022). Plants can be classified into two categories: shade-tolerant and shade-loving (sciophilous) species. Shade tolerance refers to the minimum light necessary for a species to survive, a concept established in forest science by researchers (Zon and Graves 1911; Shirley 1943). Among shade-tolerant species, two primary strategies are employed: avoidance and tolerance. Species that adopt the avoidance

Table 7.3 Important heat-tolerant genotypes of black gram and lentil

Genotype	Screening/other	References
Heat tolerance		
Black gram		
VBG-07-001, VBG-06-010	Cellular response post-TIR	Partheeban et al. (2017)
UPUPU-85-86, IPU94-2, IPU-98/36, No.5/31, PGRU-95014, PGRU-95016, PLU-1, BGP-247,	Morphological and physiological traits	Sen Gupta et al. (2021)
PantU31, Mash114, UTTARA, IPU18-04	Physiological, reproductive and yield related traits	Chaudhary et al. (2022)
LBG-1004, TBG-129, PU-1804, LBG-995, GBG-1, TBG-104, TBG-141, LBG-1006, LBG-1010	Biochemical parameters and cluster analysis	Pavithra et al. (2023)
Lentil		
IG 3263, IG 3745, IG 2507, IG4258, and FLIP 2009	Pollen and leaf studies	Sita et al. (2017b)
ILL 7835	Screened the FIGS set for terminal heat	Rajendran et al. (2020)
Cowpea		
Prima, B89-200, TN88-63, California Blackeye No.27 (CB27), TVu 4552, IT82D889-1, IT98K-1111, and IT98K-205-8	Early maturing group	Ehlers and Hall (1998)
IAR-48, IT98K-277-2, and IT98K-1092-1	Medium maturing group	Ehlers and Hall (1998)
Cold tolerance		
Chickpea		
ILC 8262, ILC 8617 and FLIP 87-82 C	Germplasm line, mutant and breeding line, respectively	Singh et al. (1995)
ICCV 88516 (CTS 60543)	Pollen selection method	Clarke et al. (2004)
ICCV 88502, 88,503, 88,506, 88,510, 88,516	Set pods at <15 °C	ICRISAT (1994); Srinivasan et al. (1998)
BGD 112 green, BG 1100, BG 1101, PUSA 1103, BGD 1005, PUSA 1108, DG 5025, DG 5027, DG 5028, DG 5036 and DG 5042	Cold-tolerant genotypes by IARI, New Delhi	Gaur et al. (2007)
Lentil		
WA 8649041, WA 8649090, ILL 1878, ILL 669, LC 9978057, LC 9977006, Kafcas, Cifei, ILL 1878, ILL 4400, ILL 7155, LC 9977116, LC 9978013, ILL 759, ILL 8146, ILL 8611, ILL 9832, Ubek, LL 1878, ILL 662, ILL 857, ILL 975, ILL 1878, ILL 5865 and Balochistan local	Frost tolerant genotypes	Barrios et al. (2016); Kahraman et al. (2004); Hamdi and Erskine (1996); Sarker et al. (2002); Ali et al. (1999)

strategy are referred to as sun-loving, shade-intolerant, or shade-avoiding, while those that utilise the tolerance strategy are known as shade-tolerant (Valladares et al. 2016). Shade-tolerant plants typically possess broader, thinner leaves to optimise sunlight capture relative to the energy cost of leaf production. In well-lit environments, shorter plants with robust, thick leaves are preferred as they enhance light capture and utilisation. Plant growth and development are regulated through hormonal pathways (Xie et al. 2022a). Shade-intolerant species trigger shade-avoidance responses, resulting in alterations in leaf morphology, chloroplast structure, photosynthetic pigment content, light quantum yield, and other factors. These adaptations enable plants to evade canopy cover and maximise light energy absorption, facilitating early flowering and reproduction.

Recent years have seen significant progress in isolating genes related to shade avoidance and tolerance (Panigrahy et al. 2020). For instance, the Zmpif mutation in maize restricts hypocotyl elongation in dark conditions and also reduces plant elongation in response to shading. Modulating the oat PHYA gene has enhanced crop cultivars, leading to improved yield, turf quality, reduced plant height, and increased shade tolerance (Xie et al. 2022a). These findings have potential applications to other species. Developing shade-tolerant plants can be challenging, but it has significant benefits in low-light areas.

It has been suggested that improving radiation use efficiency (RUE) could potentially increase wheat yield (Molero et al. 2015). Studying the phenology of the crop is also important to better match the period of unshaded growth (Desclaux et al. 2016). Since most crop varieties are selected in full sun conditions, conducting a pre-breeding test in a greenhouse (limited area, artificial shade) to screen numerous varieties for their adaptation to shade is relevant. Using a greenhouse with various levels of artificial shading can mimic outdoor growing conditions and help in identifying high-yielding plant varieties while eliminating low-yielding ones. This method also requires less space and can lead to economic benefits. Moreover, below-ground interactions between plants and trees in the field may have a greater impact on certain genotypes than shading. By using the greenhouse shelter test, a breeder may likely dismiss certain interesting varieties or choose others that may not perform optimally in the field.

Different methods exist to create new plant traits, from traditional breeding to genetic engineering. Mutation breeding, a conventional breeding form, can effectively create new plant traits without gene flow concerns and may be useful for developing shade tolerance (Ahloowalia and Maluszynski 2001; Shu et al. 2012). While genetic engineering is powerful, there are concerns about the unintentional spread of transgenes through pollen and seeds (Hu et al. 2006; Kausch et al. 2010). In their study, Li et al. (2016) used a two-step procedure to isolate shade-tolerant mutants of perennial ryegrass. First, they screened for dominant dwarf plants among the mutants. These were then subjected to early-generation screening under varying levels of shade. The identified shade-tolerant lines were further tested under both artificial and natural shade conditions. The results indicated that the tolerant lines exhibited partial insensitivity to GA.

Qin et al. (2023) used a genome-wide association study to identify a gene, PH13, on chromosome 13 in soybeans that regulates plant height. This gene encodes a WD40 protein with three different variations in natural populations. In haplotype 3, the insertion of a Ty1/Copia-like retro-transposon results in a truncated PH13H3, leading to the accumulation of STF1/2 and reduced plant height. Deleting both PH13 and its paralogue PHP can prevent shade-induced elongation of stems (ESE) and enable high-density planting. This study sheds light on the mechanism of shade resistance and offers potential solutions for breeding high-yielding soybean cultivars suitable for high-latitude regions. Bidhari et al. (2021) measured the difference in yield between unshaded and shaded environmental conditions to evaluate average productivity in maize under normal and shaded conditions. They used the stress tolerance index (STI) and identified maize lines that were tolerant to shade. Sulistyowati et al. (2016) studied 50 tomato genotypes under 50% and 100% light intensity to find shade-tolerant varieties. The genotypes were categorised based on their response to shading. The first two principal components explained 57.19% of the variation, with the first principal component representing plant production and reproduction (37.69%) and the second principal component representing plant morphological traits (19.50%).

To address these challenges, the development of varieties that could provide practical solutions for densely planted fields, vertical farming, and indoor cultivation in urban agriculture, serving ecological, habitation, and food production needs in the future is required.

7.4 Soil Stress

In India, about 140 million hectares of land are suitable for cultivation, but around 120 million hectares face issues like soil erosion, salinity, and nutrient deficiencies, worsened by climate change. Effective management practices and the development of stress-tolerant crop varieties using traditional and modern genomic techniques are vital for food security. Understanding how these stresses impact plant responses is essential for management and breeding strategies. The progress in genetic enhancement to improve crop tolerance to salinity, acidity, heavy metal toxicity, and nutrient deficiencies is summarised here.

7.4.1 Saline and Alkaline Soil

7.4.1.1 Tolerance to Saline and Saline-Alkaline Stress in Plants

Salt stress adversely affects seed germination, root and shoot length, plant height, overall plant growth, biomass gain and fruit development (Sabagh et al. 2021; Liang et al. 2014). Soil having an electrical conductivity (EC) of $\geq$4 dS m^{-1} is referred to as saline soil (Horie et al. 2011). If the pH of the soil is greater than 8 and has a high concentration of soluble salts and an exchangeable sodium percentage (ESP) of more than 15%, then that soil is referred to as saline-alkaline. Several QTLs have been identified in many crops responsible for salinity and saline-alkaline stress tolerance, which can be further exploited in marker-assisted back-crossing (MABC) for introgressing such traits into high yielding but susceptible genotypes. For example, "SalTol" is a QTL in rice that maintains the Na:K ratio in shoots at the seedling stage (Thomson et al. 2010), imparting tolerance to salinity stress at an early stage of plant growth. Similarly, Kna1 QTL is responsible for Na^+ elimination and maintenance of the K^+/Na^+ ratio in bread wheat (*T. aestivum* L.). Some other QTLs linked to tolerance to salinity/alkalinity stresses are listed in Table 7.4.

7.4.2 Acidic Soil

7.4.2.1 Metal Toxicity in Acid Soils

As pH reduces in soil, the solubility of metal and metalloids, e.g. aluminium (Al), iron (Fe), manganese (Mn), zinc (Zn), and copper (Cu), increases, leading to toxicity affecting the growth and yield of many crop species. Among the major nutrients, P is fixed by higher concentrations of Fe^{2+} and Al^{3+}, leading to P deficiency (Suswanto et al. 2007). Among metal toxicities, Al toxicity is the most toxic and is a major constraint of crop production in 67% of the total acidic soil (Eswaran et al. 1997). Similarly, iron toxicity is also present in a wide range of acidic soils, particularly in waterlogging conditions. Genes like UGT85A111/ UGT83R1

Table 7.4 List of genes/QTLs identified to be involved in saline/alkali stress tolerance in plants

Crop	Stress	Genes	References
Rice	Salinity	*RAV2*	Takagi et al. (2015)
	Salinity	*SAPK2*	Zhang et al. (2019a)
	Salinity	*OsRR22*	Kumar et al. (2013)
	Salinity	*miR535*	Santosh Kumar et al. (2020)
Maize	Na^+ toxicity	*ZmNSA1*	Cao et al. (2020)
Leymus chinensis	Na^+ toxicity	*LcCHI2*	Liu et al. (2020)
Glycine max	Osmotic	*GmMYB3a*	He et al. (2018)
		GsSKP21	Liu et al. (2015)
Glycine soja	High pH	*GsPPCK3*	Sun et al. (2014)
Arabidopsis	High pH	*PKS5*	Fuglsang et al. (2007)
	High pH	*J3*	Yang et al. (2010b)
Soybean	Oxidative damage	*GmPKS4*	Ketehouli et al. (2021)
Soybean	Oxidative damage	*GsNAC019*	Cao et al. (2017)
Suaeda salsa	Oxidative damage	*SsMT2*	Jin et al. (2017)
Tomato	Salinity	*HyPRP1*	Lou et al. (2017)
Adzuki bean	Osmotic Oxidative damage	*VaERF3*	Li et al. (2020a)
Chickpea	Osmoregulation under salt stress	*CaLG05; CaLG07*	Ahmed et al. (2021)
Potato	Salinity	*P5CS*	Kikuchi et al. (2015)

(UDP-glycosyltransferase genes) and GmGRPL (glycine-rich protein-like gene) are responsible for Al toxicity tolerance in soybean. Several TFs (Transcription Factors) such as GmSTOP1-1 and GmSTOP1- 3 (sensitivity to proton rhizotoxicity), GmWRKYs, GmNTL (NAC TF) have also been found to be involved in Al-toxicity tolerance (Shu et al. 2022). While GmSTOP1-1 and GmSTOP1- 3 regulated cell wall modification and malate metabolism for Al and H+ tolerance, WRKY and NAC TFs like GmWRKYs, GmNTL interacted with multidrug and toxin efflux genes like GmMATE for Al tolerance. Some of the genes/QTL involved in imparting tolerance to Al-toxicity are given in Table 7.5.

7.4.3 Heavy Metal and Other Metal Tolerance

Though heavy metals and metalloids sometimes play fundamental roles in plant biology either as participants in metabolic reactions or as micronutrients (e.g. Co,

Table 7.5 Genes/QTLs involved in Al-toxicity tolerance in plants in acid soils

Crop	Stress	Genes/QTL	References
Soybean	Al/H+ toxicity	GmGSTU9/qAl06	Cai et al. (2019)
		qRRE_04/qAAC_04	Wang et al. (2019b)
		UGT85A111	Wang et al. (2021)
		GmWRKY81	Shu et al. (2022)
		GmMATE87	Zhou et al. (2019)
	P–Al interaction	*GmALMT1*	Liang et al. (2013)
Maize	Al toxicity	*MATE* (multi-drug and toxic compound extrusion)	Eswaran et al. (1997); Maron et al. (2013)
		Zea mays AltSB like (*ZmASL*)	Bian et al. (2015)
		Zea mays aluminium-activated malate transporter2 (*ALMT2*)	
		S-adenosyl-L-homocysteinase (SAHH)	
		Malic enzyme (ME)	
Rape		*BnALMT1* and *BnALMT2*	Ligaba et al. (2006)
Medicago sativa		*MsALMT1*	Chen et al. (2013a)
Holcus lanatus		*HlALMT1*	Chen et al. (2013b)
Rye		ScALMT1-M39.2, ScALMT1-M77	Collins et al. (2008)
Populus trichocarpa		*PoptrMATE54* and *PoptrALMT10*	Cardoso et al. (2020)
Triticum aestivum		*TaALMT1* and *TaMATE1*, *TaMATE1B*	Garcia-Oliveira et al. (2014); Silva et al. (2018); Liu et al. (2021)
Phaseolus vulgaris		*PvALMT* and *PvMATE*	Njobvu et al. (2020)

Mn, Zn, Cu, and Mo) (Rahman and Singh 2019), the heavy metals exceed their threshold concentrations in soil, their actions harm plant development either directly or indirectly (Blaylock 2000; Asati et al. 2016). Plants employ several physiological, biochemical, or molecular approaches for protection against heavy metal accumulation. ATPase is a heavy metal-associated (HMA) genes that are linked to accumulation, transportation, and eventual resistance to heavy metal stresses in plants. Based on their preference for specific metal substrates, two main subgroups of HMAs occur: Ag/Cu P1B-ATPase and the Pb/Zn/Cd Co P1B-ATPase (Axelsen and Palmgren 2001). HMA genes can have tissue specific expression in plants. For example, in *Populus trichocarpa,* when exposed to Ag, Cd, Cu, Zn, Pb, Mn, and Co stress conditions, HMA1 and HMA4 genes are expressed in leaves, whereas HMA 5.1 in roots. In rice, a high accumulation of copper in soil has been observed to elevate HMA5 expression (Deng et al. 2013). The H^+-ATPase gene helps *Aeluropus littoralis* in the remediation of Pb and Hg toxicity (Jam et al. 2014).

Plant-specific peroxidases (Class III) are reported to be involved in imparting tolerance to abiotic stresses and to HM, such as aluminium, zinc, cadmium, and copper, by removing ROS (Reactive Oxygen species) and RNS (Reactive Nitrogen Species) (Wu et al. 2017). Overexpression of OsPRX38 in rice transgenic has increased Arsenic (As) tolerance by regulating lignin biosynthesis in roots, leading to less absorption of As. Similarly, overexpression of a metal resistance protein 1 (LmMTP1), isolated from ryegrass, led to increased resistance to Zn, Co, and Cd in transgenic rice lines (Tong et al. 2023). Some of the genes/QTL involved in imparting tolerance to metal toxicities are given in Table 7.6.

7.4.4 Nutrient Deficiency Stress

As the majority of the cultivable soils are deficient in one or more nutrients, identification of nutrient use efficient genotypes and improving nutrient use efficiency of externally applied nutrients will be useful in enhancing nutrient availability and uptake in crop species for sustaining productivity.

7.4.4.1 Improvement in Nitrogen Use Efficiency

Nitrogen is an essential macronutrient which is fundamental in crop production to improve grain yield and quality, time of flowering and optimal root development for absorption of water and other nutrient elements from soil (Place et al. 1970). Improving nitrogen use efficiency would be pivotal in enhancing its availability and enhancing the yield of many crops. Recent advances in genomics and proteomics approaches are transforming the approach to identify the candidate genes that play a significant role in the regulation of NUE in crop plants (Liu et al. 2016b; Rose et al. 2017). Breeding for high nodulation and biological nitrogen fixation could be one of the strategies for enhancing BNF vis-à-vis nitrogen availability and improving the yield of leguminous crops. Using a doubled haploid (DH) population of barley, Saal et al. (2011) detected 82 QTL for agronomic characteristics under two N fertilisation conditions in six different environments. QTLs linked to different

Table 7.6 Some Genes/QTLs involved in metal toxicity tolerance in plants

Plant	Toxicity	QTLs/gene	References
Oryza sativa	Cadmium (Cd)	qRRDW3, qCCBR1-9, CAL 1	Shilin et al. (2021) Hu et al. (2018) Luo et al. (2018)
	Iron (Fe) and Zinc (Zn)	Fe- qFRSDW11, Zn-qZRRDW11	Liu et al. (2016a)
		Fe-qFe1.1. qFe1.2, Zn- qZn3.1, qZn7.1	Anuradha et al. (2012)
	Arsenic (As)	miR156j	Pandey et al. (2020)
Triticum aestivum	Cadmium (Cd)	Cdu1	Knox et al. (2009)
Hordeum vulgare	Manganese (Mn)	QSur.yf.3H, QLC.yf.3H	Huang et al. (2018)
Arabidopsis helleri	Zinc (Zn)	Zntol-1	Willems et al. (2007)
Zea mays	Cadmium (Cd), Copper (Cu), and Lead (Pb)	*ZmMT*	Gao et al. (2022)
Capsicum frutescens	Cadmium (Cd), Copper (Cu), Lead (Pb), Zinc (Zn), and Iron (Fe)	*CaMYB* family genes	Xie et al. (2022b)
Oryza sativa and Arabidopsis thaliana	Cadmium (Cd), Copper (Cu), Lead (Pb), and Zinc (Zn)	*OsHMP* (46 genes) and *AtHMP* (55 genes)	Li et al. (2020b)
Solanum tuberosum	Cadmium (Cd)	Mn-MTP, Zn-MTP, and Zn/Fe-MTP	Li et al. (2021)
Gossypium barbadense	Mercury (Hg)	miR164, miR171, miR390 and miR393	Noman et al. (2019)

traits like plant height, tiller number, grain yield, dry weight of shoots and roots, spikelet number, number of filled grains per panicle, 1000-grain weight, the leaf colour chart (LCC), and chlorophyll content have been identified in rice (Ali et al. 2018). Some other QTLs involved in nitrogen, phosphorus, and potassium use efficiency are listed in Table 7.7.

7.4.4.2 Improvement in Phosphorus Use Efficiency

Availability of phosphorus in soil is pH dependent as maximum availability occurs at pH >5.5 to <7.5, below or beyond which phosphorus is fixed as Fe- and Al- phosphates in acidic, and carbonate and bicarbonates in alkaline range of pH. Therefore, identification of P-efficient genotypes, or tolerant to low P in soil, and improvement of the trait further would be required to enhance crop yield. Ideally, genotypes with better P use efficiency will have longer root systems, numerous and dense root hairs, enhanced root exudation and better association with microbes like mycorrhiza and other phosphate solubilising microbes (PSM). Several key morphological and

Table 7.7 Genes/QTL involved in nutrient use efficiency of major nutrients in plants

Gene/QTL	Nutrient	Crop	References
qNGR9/DEP1	Nitrogen	Rice	Sun et al. (2014)
TOND1		Rice	Zhang et al. (2015)
NRT1.1B		Rice	Hu et al. (2015)
qNGR2/GRF4		Rice	Li et al. (2018b)
OsNRT2.1, sNRT2.2, OsNRT2.3a, OsAMT1.1, OsAMT1.2		Rice	Han et al. (2016)
OsNPF2.4		Rice	Xia et al. (2015)
OsNPF7.2		Rice	Hu et al. (2016)
TGA1/TGA4 and NTR1.1		*Arabidopsis thaliana*	Canales et al. (2017)
NRT2		*Brassica napus*	Tong et al. (2020)
TaNPF1.1, TaNPF2.1, TaNPF6.1		Wheat	Buchner and Hawkesford (2014)
OsPHT1;8	Phosphorus	Rice	Jia et al. (2011)
Pup1/OsPupK46-2/*PSTOL1*		Rice	Gamuyao et al. (2012), Heuer et al. (2009)
OsPHT1;2		Rice	Liu et al. (2010)
AtPHT1;5		*Arabidopsis thaliana*	Nagarajan et al. (2011)
PHO2		Rice	Hu et al. (2011)
AtAKT1, AtHAK5, AtKUP9, AtTPK1, AtCNGC1, AtSKOR	Potassium	*Arabidopsis thaliana*	Harada and Leigh (2006); Kanter et al. (2010)
OsHKT1;5, OsHKT8		Rice	Ren et al. (2005)
HvHKT1;5, TmHKT1;5-A, TaHKT1:5-D		Barley and wheat	Munns et al. (2012); Nguyen et al. (2013)
SlHKT1;1, SlHKT1;2		Tomato	Asins et al. (2013)

physiological traits such as plant height, number of tillers, shoot root length, relative shoot and root dry weight and leaf age and root-attributed traits such as root diameter, root hair number and number of roots have been used for screening and identifying tolerant genotypes under P deficiency conditions. Relative tiller dry weight (RTW), shoot dry weight, and plant dry weight were used as better screening criteria for identifying genotypes tolerant of low-P stress. Four subfamilies of P transporters, PHT1, PHT2, PHT3, and PHT4, regulate Pi uptake and transport in plants, and their expression can increase Pi absorption and mobilisation. Some major QTLs have been identified in crops for low P tolerance in previous studies. In maize, QTLs related to root traits (Wang et al. 2019a), biomass traits (Azevedo et al. 2015), and yield traits (Xu et al. 2018) have been detected for P use efficiency. However, the locus Pup1 for P-Starvation Tolerance 1 (PSTOL1) and biomass have been identified in rice (Gamuyao et al. 2012) and sorghum (Hufnagel et al. 2014) under P-deficient conditions. Zhang et al. (2013) identified a major QTL, qKN governing kernel numbers and Qiu et al. (2013) identified QTL AP9, responsible for acid phosphatase activity under poor soil P conditions in maize.

7.4.4.3 Breeding Strategies

To impart multiple stress tolerance in the agricultural background, multi-prong strategies are required involving conventional, genomics, proteomics and transcriptomics. Of late, genome editing tools have taken over transgenic approaches because of less stringencies in accepting genome edited plants/crops.

Conventional Breeding

Traditional breeding techniques, such as mass selection, which utilise trait-specific germplasm to introduce stress-tolerant characteristics into high-yielding yet stress-sensitive genotypes, remain among the most widely accepted approaches for enhancing stress resilience. Nevertheless, the success rates of these methods are notably low, and the processes are often time-consuming and labour-intensive. While integrating conventional breeding with "speed breeding" can accelerate generation advancement, its applicability is limited due to the photosensitive characteristics of certain crops. Furthermore, achieving tolerance to heavy metals, metalloids, aluminium toxicity, and nutrient deficiencies through conventional breeding poses significant challenges due to the polygenic nature of these traits. Fortunately, natural sources of tolerance can be found in wild accessions, and by employing molecular techniques, it is possible to identify the genes associated with multiple stress tolerances in landraces and wild relatives, which can then be introgressed into cultivated varieties.

QTL and Marker-Assisted Back-Crossing

Marker-assisted backcrossing (MABC) represents a significant advancement in traditional breeding techniques, offering a more efficient approach to the transfer of specific alleles at designated loci between donor and recipient plants (Hasan et al. 2022). This method enhances the precision with which desired quantitative trait loci (QTLs) can be incorporated into various genotypes by utilising specific markers at target loci alongside background markers within the genome (Frisch et al. 1999). The "SalTol" QTL was introduced into the rice cultivar BR11 through MABC (Gregorio et al. 2013). Despite its potential, the application of MABC for the introgression of abiotic stress tolerance has been relatively limited.

Genome Editing

The CRISPR technology has been used quite extensively in almost all crops where genome sequences are available. Zhang et al. (2019b) used a genome editing technique by engineering a Cas9-OsRR22-gRNA expressing vector for the rice OsR22 gene to develop salinity tolerant rice. These unique CRISPR/Cas9/Cas12-based genome editing tools have the potential to shorten breeder time drastically and also reduce the cost significantly, and with the application of speed breeding techniques, the time and cost will go down further.

7.5 Exploiting Germplasm Diversity

The genetic resources stored in gene banks are vital reservoirs of numerous advantageous alleles that can be leveraged for current and future crop improvement efforts. Among the 34 global biodiversity hotspots, India is home to four: the Himalayas, Western Ghats, Indo-Burma, and Sundaland (Nicobar Islands). The Indian Gene Centre (IGC) is recognised as one of the 12 mega biodiversity hotspots worldwide, rich in agro-biodiversity. It includes 811 species of cultivated plants (Singh 2017), with 167 having originated or developed diversity in this area; furthermore, it is home to 9500 species of ethnobotanical importance and over 730 species of crop wild relatives (CWR) (Pradheep et al. 2014). A concise list of traits linked to abiotic stress tolerance and possible natural habits is listed in Table 7.8.

Assessing germplasm for its resilience to various abiotic stresses must be conducted in controlled environments. This requires specialised facilities, such as phytotrons for managing light and temperature, rainout shelters to mitigate rainfall during drought stress evaluations, and micro plots with specific pH and electrical conductivity levels to investigate alkaline and saline stress conditions, respectively. The incorporation of image analysis for high-throughput plant phenotyping has opened new avenues for plant biologists studying stress tolerance. Plant phenotyping lies at the crossroads of plant biology, sensor technology, and automated engineering. Automated phenotyping employs non-invasive sensors, including spectrometry (such as hyperspectral radiometers, Fourier transform infrared spectroscopy (FTIR), infrared (IR) thermometry, and near-infrared (NIR) meters) along with various imaging techniques (such as visual imaging, hyperspectral imaging, IR thermography, NIR image analysis, chlorophyll fluorescence imaging, and bioluminescence imaging). Further, automated data processing and analysis improve the extraction of relevant phenotypic traits.

Table 7.8 Possible naturally occurring sources of abiotic stress-specific traits

Traits	Possible sources
Drought tolerance and recovery at vegetative stages	Dry seeded upland germplasm and land races grown in the hilly region under shifting (Jhum cultivation)
Cold tolerance at vegetative stage	Land races of Boro rice cultivation in Assam
Shade tolerance	Land races of traditional Sali, Asra, and Bao cultivation areas of Assam and traditional Aman varieties of Bengal and Odisha
Water logging tolerance	Coastal areas of Odisha and West Bengal, also Sali and Asra germplasm of the Brahmaputra valley of Assam
Deep water situation	Bao germplasm of Assam and land races from deep water areas of Eastern UP, Bihar, and West Bengal

7.6 Future Perspectives

The challenge of breeding for tolerance to abiotic stress is significant due to the intricate mechanisms that govern such tolerance. A comprehensive understanding of plant responses to various stressors at the molecular level is essential for the development of stress-resistant crops (Mittler et al. 2022). Key processes such as stomatal closure and the generation of reactive oxygen species represent critical convergence points in plant responses. To unravel these complex responses and pinpoint molecular targets for enhancing stress resistance in crops, advanced methodologies and multiomics analyses are required (Gong et al. 2020). In-depth studies have highlighted diverse adaptation and tolerance strategies within controlled settings; however, results from open-field studies are frequently insufficient. To address this disparity, utilizing tissue- and cell-specific phenotyping can yield valuable information on plant responses to abiotic stress, ultimately improving our ability to develop resilient crops for the future. Identifying the essential characteristics of abiotic stress, along with their physiological, biochemical, and molecular foundations—including genetic, epigenetic, transcriptomic, and metabolomic factors—will enhance our understanding and efforts in breeding crops that can withstand abiotic stress (Chang et al. 2020). Future investigations should prioritize examining the combined effects of different stress tolerance mechanisms while pinpointing critical genetic and biochemical pathways for enhancement. Emerging genomic technologies offer the potential to improve tolerance to these stresses by deepening our understanding of the underlying processes and identifying the genes involved. The application of modern phenomics tools for accurate phenotyping may significantly enhance breeding efficiency (Naik et al. 2024). High-throughput phenotyping and AI-based models will improve our capacity to forecast crop yields and comprehend climate influences on plant physiology. Collaborative efforts across disciplines are essential. Specialists from various domains—such as breeders, agronomists, geneticists, stress physiologists, ecologists, molecular biologists, engineers, and data scientists—need to unite to develop and utilise these cutting-edge tools. Data from next-generation sequencing (NGS) can be leveraged to associate and identify candidate genes linked to abiotic stress tolerance. The integration of stable quantitative trait loci that control component traits may present a viable solution. Additionally, the creation of transgenic varieties could serve as an alternative approach for developing pulse genotypes that are tolerant to heat stress (Devi et al. 2022). Despite progress, many crucial questions remain unanswered about the specific genes to target and their contributions to stress responses. There is an ongoing debate about whether all important genes are linked to particular stress factors or if a targeted selection of genes would yield better outcomes. To advance beyond traditional research, thoroughly integrated studies focusing on adaptation and tolerance mechanisms at both cellular and tissue levels are vital for creating resilient crop varieties. Tackling these issues and utilizing cutting-edge tools will be essential for refining genetic engineering approaches—such as key gene overexpression in major crops, gene editing, and the de novo creation of genes for desired traits—to enhance stress resilience and crop yield. The combination of multiomics with genetic engineering, synthetic biology, and speed breeding techniques will revolutionize the development of future crops that are resilient to stress.

References

Abiko T, Todoroki T, Pham Thi Thanh H, Nakamura T, Haraguchi Y, Tanaka T, Saisho D, Kai H (2023) Barley varieties tolerant to waterlogged reduced soil show the better root growth in hypoxia. Plant Prod Sci 26(4):429–439. https://doi.org/10.1080/1343943X.2023.2246215

Agarwal G, Garg V, Kudapa H, Doddamani D, Pazhamala LT, Khan AW, Thudi M, Lee S-H, Varshney RK (2016) Genome-wide dissection of AP2/ERF and HSP90 gene families in five legumes and expression profiles in chickpea and pigeonpea. Plant Biotechnol J 14(7):1563–1577. https://doi.org/10.1111/pbi.12520

Ahloowalia BS, Maluszynski M (2001) Induced mutations–a new paradigm in plant breeding. Euphytica 118(2):167–173. https://doi.org/10.1023/A:1004162323428

Ahmed SM, Alsamman AM, Jighly A, Mubarak MH, Al-Shamaa K, Istanbuli T, Momtaz OA, El Allali A, Hamwieh A (2021) Genome-wide association analysis of chickpea germplasms differing for salinity tolerance based on DArTseq markers. PLoS One 16(12):e0260709. https://doi.org/10.1371/journal.pone.0260709

Al-Karaki GN (2012) Phenological development-yield relationships in durum wheat cultivars under late-season high-temperature stress in a semiarid environment. ISRN Agron 2012:1–7. https://doi.org/10.5402/2012/456856

Alagupalamuthirsolai M, Vijayabharathi A (2020) Genetic diversity analysis of pigeonpea [Cajanus cajan (L.) Millsp.] under cold stress conditions. J Pharmacogn Phytochem 9(3):2048–2053. https://doi.org/10.22271/phyto.2020.v9.i3ah.11615

Ali A, Johnson DL, Stushnoff C (1999) Screening lentil (Lens culinaris) for cold hardiness under controlled conditions. J Agric Sci 133(3):313–319. https://doi.org/10.1017/S0021859699006917

Ali J, Jewel ZA, Mahender A, Anandan A, Hernandez J, Li Z (2018) Molecular genetics and breeding for nutrient use efficiency in rice. Int J Mol Sci 19(6):1762. https://doi.org/10.3390/ijms19061762

Angaji SA, Septiningsih EM, Mackill DJ, Ismail AM (2010) QTLs associated with tolerance of flooding during germination in rice (Oryza sativa L.). Euphytica 172(2):159–168. https://doi.org/10.1007/s10681-009-0014-5

Anuradha K, Agarwal S, Rao YV, Rao KV, Viraktamath BC, Sarla N (2012) Mapping QTLs and candidate genes for iron and zinc concentrations in unpolished rice of Madhukar × Swarna RILs. Gene 508(2):233–240. https://doi.org/10.1016/j.gene.2012.07.054

Araújo SS, Beebe S, Crespi M, Delbreil B, González EM, Gruber V, Lejeune-Henaut I, Link W, Monteros MJ, Prats E, Rao I, Vadez V, Patto MCV (2015) Abiotic stress responses in legumes: strategies used to cope with environmental challenges. Crit Rev Plant Sci 34(1–3):237–280. https://doi.org/10.1080/07352689.2014.898450

Asati A, Pichhode M, Nikhil K (2016) Effect of heavy metals on plants: an overview. Int J Appl Innov Eng Manag 5(3):56–66

Ashraf M, Athar HR, Harris PJC, Kwon TR (2008) Some prospective strategies for improving crop salt tolerance. Adv Agron 97:45–110. https://doi.org/10.1016/S0065-2113(07)00002-8

Asíns MJ (2002) Present and future of quantitative trait locus analysis in plant breeding. Plant Breed 121(4):281–291. https://doi.org/10.1046/j.1439-0523.2002.730285.x

Asins MJ, Villalta I, Aly MM, Olías R, Álvarez De Morales PAZ, Huertas R, Li J, Jaime-Pérez N, Haro R, Raga V, Carbonell EA, Belver A (2013) Two closely linked tomato HKT coding genes are positional candidates for the major tomato QTL involved in Na+/K+ homeostasis. Plant Cell Environ 36(6):1171–1191. https://doi.org/10.1111/pce.12051

Aski MS, Dikshit HK, Mishra GP, Yadav PS, Iquebal MA, Sarika, Bansal R, Gayacharan C, Singh A, Udupa S (2023) Current and future strategies in breeding lentil for abiotic stresses BT. In: Arjuna M, Samy P, Ramasamy A, Chinnusamy V, Kumar BS (eds) Legumes: physiology and molecular biology of abiotic stress tolerance. Springer, pp 235–257. https://doi.org/10.1007/978-981-19-5817-5_9

Aski MS, Rai N, Reddy VRP, Gayacharan DHK, Dikshit HK, Mishra GP, Singh D, Kumar A, Pandey R, Singh MP, Pratap A, Nair RM, Schafleitner R (2021) Assessment of root phenotypes in mungbean mini-core collection (MMC) from the World Vegetable Center (AVRDC) Taiwan. PLoS One 16(3):e0247810. https://doi.org/10.1371/journal.pone.0247810

Axelsen KB, Palmgren MG (2001) Inventory of the superfamily of P-type ion pumps in Arabidopsis. Plant Physiol 126(2):696–706. https://doi.org/10.1104/pp.126.2.696

Azevedo GC, Cheavegatti-Gianotto A, Negri BF, Hufnagel B, Silva LDC, Magalhaes JV, Garcia AAF, Lana UG, de Sousa SM, Guimaraes CT (2015) Multiple interval QTL mapping and searching for PSTOL1 homologs associated with root morphology, biomass accumulation and phosphorus content in maize seedlings under low-P. BMC Plant Biol 15:1–17

Bailey-Serres J, Colmer TD (2014) Plant tolerance of flooding stress–recent advances. Plant Cell Environ 37(10):2211–2215. https://doi.org/10.1111/pce.12420

Ballesta P, Mora F, Pozo AD (2019) Association mapping of drought tolerance indices in wheat: QTL-rich regions on. Sci Agric 77:e20180153

Ballesteros DC, Mason RE, Addison CK, Andrea Acuña M, Nelly Arguello M, Subramanian N, Miller RG, Sater H, Gbur EE, Miller D, Griffey CA, Barnett MD, Tucker D (2015) Tolerance of wheat to vegetative stage soil waterlogging is conditioned by both constitutive and adaptive QTL. Euphytica 201(3):329–343. https://doi.org/10.1007/s10681-014-1184-3

Banan D, Paul RE, Feldman MJ, Holmes MW, Schlake H, Baxter I, Jiang H, Leakey ADB (2018) High-fidelity detection of crop biomass quantitative trait loci from low-cost imaging in the field. Plant Direct 2(2):e00041. https://doi.org/10.1002/pld3.41

Barrios A, Aparicio T, Rodríguez MJ, Pérez de la Vega M, Caminero C (2016) Winter sowing of adapted lines as a potential yield increase strategy in lentil (Lens culinaris Medik.). Span J Agric Res 14(2):9. https://doi.org/10.5424/sjar/2016142-8092

Becklin KM, Anderson JT, Gerhart LM, Wadgymar SM, Wessinger CA, Ward JK (2016) Examining plant physiological responses to climate change through an evolutionary lens. Plant Physiol 172(2):635–649. https://doi.org/10.1104/pp.16.00793

Bernardo R (2008) Molecular markers and selection for complex traits in plants: learning from the last 20 years. Crop Sci 48(5):1649–1664. https://doi.org/10.2135/cropsci2008.03.0131

Bhalani H, Thankappan R, Mishra GP, Sarkar T, Bosamia TC, Dobaria JR (2019) Regulation of antioxidant mechanisms by AtDREB1A improves soil-moisture deficit stress tolerance in transgenic peanut (Arachis hypogaea L.). PLoS One 14(5):e0216706. https://doi.org/10.1371/journal.pone.0216706

Bhandari K, Siddique KHM, Turner NC, Kaur J, Singh S, Agrawal SK, Nayyar H (2016) Heat stress at reproductive stage disrupts leaf carbohydrate metabolism, impairs reproductive function, and severely reduces seed yield in lentil. J Crop Improv 30(2):118–151. https://doi.org/10.1080/15427528.2015.1134744

Bhandari M, Baker S, Rudd JC, Ibrahim AMH, Chang A, Xue Q, Jung J, Landivar J, Auvermann B (2021) Assessing the effect of drought on winter wheat growth using unmanned aerial system (UAS)-based phenotyping. Remote Sens 13(6):1144. https://doi.org/10.3390/rs13061144

Bhoite R, Han Y, Chaitanya AK, Varshney RK, Sharma DL (2024) Genomic approaches to enhance adaptive plasticity to cope with soil constraints amidst climate change in wheat. Plant Genome 17(1):e20358. https://doi.org/10.1002/tpg2.20358

Bian M, Jin X, Broughton S, Zhang X-Q, Zhou G, Zhou M, Zhang G, Sun D, Li C (2015) A new allele of acid soil tolerance gene from a malting barley variety. BMC Genet 16:92. https://doi.org/10.1186/s12863-015-0254-4

Bidhari LA, Effendi R, Andayani NN, Bambang S (2021) Screening of shade tolerant hybrid maize based on stress tolerance index. IOP Conf Ser Earth Environ Sci 911(1):012018. https://doi.org/10.1088/1755-1315/911/1/012018

Bidinger FR, Mahalakshmi V, Talukdar BS, Sharma RK (1995) Ann Arid Zone 34:105–110

Blaylock MJ (2000) Phytoextraction of metals. In: Phytoremediation of toxic metals, pp 53–70

Blum A (1973) Components analysis of yield responses to drought of sorghum hybrids. Exp Agric 9(2):159–167. https://doi.org/10.1017/S0014479700005627

Blum A (2000) Towards Standard assays of drought tolerance in crop plants. In: Ribaut JM, Poland D (eds) Molecular approaches for the genetic improvement of cereals for stable production in water-limited environment. A strategic planning workshop held at CIMMYT, Mexico. June 21–25, 1999, pp 29–35

Blum A (2014) Genomics for drought resistance—Getting down to earth. Funct Plant Biol 41(11):1191–1198. https://doi.org/10.1071/FP14018

Borrego-Benjumea A, Carter A, Tucker JR, Yao Z, Xu W, Badea A (2020) Genome–wide analysis of gene expression provides new insights into waterlogging responses in barley (Hordeum vulgare L.). Plants 9(2):240. https://doi.org/10.3390/plants9020240

Borrego-Benjumea A, Carter A, Zhu M, Tucker JR, Zhou MX, Badea A (2021) Genome-wide association study of waterlogging tolerance in barley (Hordeum vulgare L.) under controlled field conditions. Front Plant Sci 12:711654. https://doi.org/10.3389/fpls.2021.711654

Broughton S, Zhou GF, Teakle NL, Matsuda R, Zhou MX, Leary O (2015) Waterlogging tolerance is associated with root porosity in barley (Hordeum vulgare L.). RA, Colmer TD, Li C D. Mol Breed 35:1–15. https://doi.org/10.1007/s11032-015-0243-3

Buchner P, Hawkesford MJ (2014) Complex phylogeny and gene expression patterns of members of the nitrate transporter 1/Peptide transporter family (NPF) in wheat. J Exp Bot 65(19):5697–5710. https://doi.org/10.1093/jxb/eru231

Bueckert RA, Wagenhoffer S, Hnatowich G, Warkentin TD (2015) Effect of heat and precipitation on pea yield and reproductive performance in the field. Can J Plant Sci 95(4):629–639. https://doi.org/10.4141/cjps-2014-342

Buriro M, Oad FC, Keerio MI, Tunio S, Gandahi AW, Hassan SWU, Oad SM (2011) Wheat seed germination under the influence of temperature regimes. Sarhad J Agric 27:539–543. https://api.semanticscholar.org/CorpusID:42131739

Cai J, Zeng Z, Connor JN, Huang CY, Melino VJ, Kumar P, Miklavcic SJ (2015) RootGraph: a graphic optimization tool for automated image analysis of plant roots. J Exp Bot 66(21):6551–6562. https://doi.org/10.1093/jxb/erv359

Cai Z, Cheng Y, Xian P, Lin R, Xia Q, He X, Liang Q, Lian T, Ma Q, Nian H (2019) Fine-mapping QTLs and the validation of candidate genes for aluminum tolerance using a high-density genetic map. Plant Soil 444(1–2):119–137. https://doi.org/10.1007/s11104-019-04261-0

Campbell MT, Proctor CA, Dou Y, Schmitz AJ, Phansak P, Kruger GR, Zhang C, Walia H (2015) Genetic and molecular characterization of submergence response identifies subtol6 as a major submergence tolerance locus in maize. PLoS One 10(3):e0120385. https://doi.org/10.1371/journal.pone.0120385

Canales J, Contreras-López O, Álvarez JM, Gutiérrez RA (2017) Nitrate induction of root hair density is mediated by TGA1/TGA 4 and CPC transcription factors in Arabidopsis thaliana. Plant J 92(2):305–316. https://doi.org/10.1111/tpj.13656

Cao L, Yu Y, Ding X, Zhu D, Yang F, Liu B, Sun X, Duan X, Yin K, Zhu Y (2017) The Glycine soja NAC transcription factor GsNAC019 mediates the regulation of plant alkaline tolerance and ABA sensitivity. Plant Mol Biol 95(3):253–268. https://doi.org/10.1007/s11103-017-0643-3

Cao Y-Y, Duan H, Yang L-N, Wang Z-Q, Zhou S-C, Yang J-C (2008) Effect of heat stress during meiosis on grain yield of rice cultivars differing in heat tolerance and its physiological mechanism. Acta Agron Sin 34(12):2134–2142. https://doi.org/10.1016/S1875-2780(09)60022-5

Cao Y, Cai S (1991) Some waterlogging tolerant wheat varieties. Report of the Jiangshu Academy of Agricultural Sciences. Crop Genet Res 2:25–26

Cao Y, Zhang M, Liang X, Li F, Shi Y, Yang X, Jiang C (2020) Natural variation of an EF-hand Ca2+-binding-protein coding gene confers saline-alkaline tolerance in maize. Nat Commun 11(1):186. https://doi.org/10.1038/s41467-019-14027-y

Cao Y, Cai S, Wu Z, Zhu W, Fang X and Xiong E (1995) Studies on genetic features of waterlogging tolerance in wheat. Jiangsu J. Agric. Sci. 11(2), 11–15.

Cardoso TB, Pinto RT, Paiva LV (2020) Comprehensive characterization of the ALMT and MATE families on Populus trichocarpa and gene co-expression network analysis of its members during aluminium toxicity and phosphate starvation stresses. 3 Biotech 10(12):525. https://doi.org/10.1007/s13205-020-02528-3

Chang Y-N, Zhu C, Jiang J, Zhang H, Zhu J-K, Duan C-G (2020) Epigenetic regulation in plant abiotic stress responses. J Integr Plant Biol 62(5):563–580. https://doi.org/10.1111/jipb.12901

Chattopadhyay K, Chakraborty K, Samal P, Sarkar RK (2021) Identification of QTLs for stagnant flooding tolerance in rice employing genotyping by sequencing of a RIL population derived from Swarna × Rashpanjor. Physiol Mol Biol Plants 27(12):2893–2909. https://doi.org/10.1007/s12298-021-01107-x. Epub December 18, 2021. PubMed: 35035143, PubMed Central: PMC8720131

Chaudhary S, Jha UC, Paul PJ, Prasad PVV, Sharma KD, Kumar S, Gupta DS, Sharma P, Singh S, Siddique KHM, Nayyar H (2022) Assessing the heat sensitivity of Urdbean (Vigna mungo L. Hepper) genotypes involving physiological, reproductive and yield traits under field and controlled environment. Front Plant Sci 13:1042999. https://doi.org/10.3389/fpls.2022.1042999

Chen J, Xu W, Velten J, Xin Z, Stout J (2012) Characterization of maize inbred lines for drought and heat tolerance. J Soil Water Conserv 67(5):354–364. https://doi.org/10.2489/jswc.67.5.354

Chen Q, Wu K-H, Wang P, Yi J, Li K-Z, Yu Y-X, Chen L-M (2013a) Overexpression of MsALMT1, from the aluminum-sensitive Medicago sativa, enhances malate exudation and aluminum resistance in tobacco. Plant Mol Biol Report 31(3):769–774. https://doi.org/10.1007/s11105-012-0543-2

Chen S, Guo Y, Sirault X, Stefanova K, Saradadevi R, Turner NC, Nelson MN, Furbank RT, Siddique KHM, Cowling WA (2019) Nondestructive phenomic tools for the prediction of heat and drought tolerance at anthesis in Brassica species. Plant Phenom 2019:3264872. https://doi.org/10.34133/2019/3264872

Chen ZC, Yokosho K, Kashino M, Zhao F-J, Yamaji N, Ma JF (2013b) Adaptation to acidic soil is achieved by increased numbers of cis-acting elements regulating ALMT1 expression in Holcus lanatus. Plant J 76(1):10–23. https://doi.org/10.1111/tpj.12266

Chidambaranathan P, Jagannadham PTK, Satheesh V, Kohli D, Basavarajappa SH, Chellapilla B, Kumar J, Jain PK, Srinivasan R (2018) Genome-wide analysis identifies chickpea (Cicer arietinum) heat stress transcription factors (Hsfs) responsive to heat stress at the pod development stage. J Plant Res 131(3):525–542. https://doi.org/10.1007/s10265-017-0948-y

Choudhury DR, Tarafdar S, Das M, Kundagrami S (2012) Screening lentil (Lens culinaris Medik.) germplasms for heat tolerance. Trends Biosci 5(2):143–146. https://api.semanticscholar.org/CorpusID:87681229

Clarke HJ, Khan TN, Siddique KHM (2004) Pollen selection for chilling tolerance at hybridization leads to improved chickpea cultivars. Euphytica 139(1):65–74. https://doi.org/10.1007/s10681-004-2466-y

Collins NC, Shirley NJ, Saeed M, Pallotta M, Gustafson JP (2008) An ALMT1 gene cluster controlling aluminum tolerance at the Alt4 locus of rye (Secale cereale L.). Genetics 179(1):669–682. https://doi.org/10.1534/genetics.107.083451

Croser JS, Clarke HJ, Siddique KHM, Khan TN (2003) Low-temperature stress: Implications for chickpea (Cicer arietinum L.) improvement. Crit Rev Plant Sci 22(2):185–219. https://doi.org/10.1080/713610855

Cuellar-Ortiz SM, De La Paz Arrieta-Montiel MDLP, Acosta-Gallegos J, Covarrubias AA (2008) Relationship between carbohydrate partitioning and drought resistance in common bean. Plant Cell Environ 31(10):1399–1409. https://doi.org/10.1111/j.1365-3040.2008.01853.x

Dana I, Chatterjee S, Kumdu C (2013) Chapter 4: Twenty years of achievements of the EIRLSBN at the Rice Research Station, Chinsurah. In: EIRLSBN: Twenty years of achievements in rice breeding. IRRI International Rice Research Institute, pp 53–64

Daszkowska-Golec A, Skubacz A, Sitko K, Słota M, Kurowska M, Szarejko I (2018) Mutation in barley ERA1 (Enhanced Response to ABA1) gene confers better photosynthesis efficiency in response to drought as revealed by transcriptomic and physiological analysis. Environ Exp Bot 148:12–26. https://doi.org/10.1016/j.envexpbot.2018.01.003

De Castro J, Hill RD, Stasolla C, Badea A (2022) Waterlogging stress physiology in barley. Agronomy 12(4):780. https://doi.org/10.3390/agronomy12040780

de Lima CZ, Buzan JR, Moore FC, Baldos ULC, Huber M, Hertel TW (2021) Heat stress on agricultural workers exacerbates crop impacts of climate change. Environ Res Lett 16(4):044020. https://doi.org/10.1088/1748-9326/abeb9f

Delahunty A, Nuttall J, Nicolas M et al (2015) Genotypic heat tolerance in lentil. In: Proceedings of the 17th Asa Conference, Hobart, Australia, pp 20–24

Deng F, Yamaji N, Xia J, Ma JF (2013) A member of the heavy metal P-type ATPase OsHMA5 is involved in xylem loading of copper in rice. Plant Physiol 163(3):1353–1362. https://doi.org/10.1104/pp.113.226225

Desclaux D, Huang H, Bernazeau B, Lavène P (2016) Agroforestry: new challenge for field crop breeding? In: 3rd European Agroforestry Conference Montpellier. EURAF, p hal-02744125

Devasirvatham V, Tan D, Trethowan R, Gaur P, Mallikarjuna N (2010) Impact of high temperature on the reproductive stage of chickpea. In: Food Security from Sustainable Agriculture Proceedings of the 15th Australian Society of Agronomy Conference, pp 15–18

Devi J, Sagar V, Mishra GP, Jha PK, Gupta N, Dubey RK, Singh PM, Behera TK, Prasad PVV (2022) Heat stress tolerance in peas (Pisum sativum L.): current status and way forward. Front Plant Sci 13:1108276. https://doi.org/10.3389/fpls.2022.1108276

Dhanasekar P, Souframanien J (2023) Genetics and genomics of drought and heat tolerance in cowpea, mung bean and black gram. In: Samy MA et al (eds) Legumes: physiology and molecular biology of abiotic stress tolerance. Springer, Singapore, pp 203–233. https://doi.org/10.1007/978-981-19-5817-5_8

Dhanda SS, Munjal R (2012) Heat tolerance in relation to acquired thermotolerance for membrane lipids in bread wheat. Field Crop Res 135:30–37. https://doi.org/10.1016/j.fcr.2012.06.009

Ding FG, Tong JY, Xu R, Chen J, Xu XT, Nadeem M, Wang SP, Zhang YX, Zhu ZW, Wang FJ, Fang ZW, Hao YF (2023) Identification of stable quantitative trait loci underlying waterlogging tolerance post–anthesis in common wheat (Triticum aestivum). Crop J 11(4):1163–1170. https://doi.org/10.1016/j.cj.2023.06.004

Dixit GP, Parihar AK, Bohra A, Singh NP (2016) Achievements and prospects of grass pea (Lathyrus sativus L.) improvement for sustainable food production. Crop J 4(5):407–416. https://doi.org/10.1016/j.cj.2016.06.008

Djalovic I, Kundu S, Bahuguna RN, Pareek A, Raza A, Singla-Pareek SL, Prasad PVV, Varshney RK (2024) Maize and heat stress: physiological, genetic, and molecular insights. Plant Genom 17(1):e20378. https://doi.org/10.1002/tpg2.20378

Drew MC, Lynch JM (1980) Soil anaerobiosis, microorganisms, and root function. Annu Rev Phytopathol 18(1):37–66. https://doi.org/10.1146/annurev.py.18.090180.000345

Eathington SR, Crosbie TM, Edwards MD, Reiter RS, Bull JK (2007) Molecular markers in a commercial breeding program. Crop Sci 47(Suppl. 3):S-154–S-163. https://doi.org/10.2135/cropsci2007.04.0015IPBS

Ehlers JD, Hall AE (1998) Heat tolerance of contrasting cowpea lines in short and long days. Field Crop Res 55(1–2):11–21. https://doi.org/10.1016/S0378-4290(97)00055-5

Eswaran H, Reich P, Beinroth F (1997) Global distribution of soils with acidity. In: Moniz AC et al (eds) Plant-Soil Interactions at Low pH. Brazilian Soil Science Society, pp 159–164

Fang XW, Cao Y, Cai SB, Xiong EH, Zhu W (1997) Genetic evaluation of waterlogging tolerance in Triticummacha. Jiangsu J Agric Sci 13:73–75

Feng B, Liu P, Li G, Dong ST, Wang FH, Kong LA, Zhang JW (2014) Effect of heat stress on the photosynthetic characteristics in flag leaves at the grain-filling stage of different heat-resistant winter wheat varieties. J Agron Crop Sci 200(2):143–155. https://doi.org/10.1111/jac.12045

Francesconi S, Harfouche A, Maesano M, Balestra GM (2021) UAV-based thermal, RGB imaging and gene expression analysis allowed detection of fusarium head blight and gave new insights into the physiological responses to the disease in durum wheat. Front Plant Sci 12:628575. https://doi.org/10.3389/fpls.2021.628575

Frey FP, Presterl T, Lecoq P, Orlik A, Stich B (2016) First steps to understand heat tolerance of temperate maize at adult stage: identification of QTL across multiple environments with connected segregating populations. Theor Appl Genet 129(5):945–961. https://doi.org/10.1007/s00122-016-2674-6

Frisch M, Bohn M, Melchinger AE (1999) Comparison of selection strategies for marker-assisted backcrossing of a gene. Crop Sci 39(5):1295–1301. https://doi.org/10.2135/cropsci1999.3951295x

Fuglsang AT, Guo Y, Cuin TA, Qiu Q, Song C, Kristiansen KA, Bych K, Schulz A, Shabala S, Schumaker KS, Palmgren MG, Zhu J-K (2007) Arabidopsis protein kinase PKS5 inhibits the plasma membrane H+ -ATPase by preventing interaction with 14-3-3 protein. Plant Cell 19(5):1617–1634. https://doi.org/10.1105/tpc.105.035626

Gamuyao R, Chin JH, Pariasca-Tanaka J, Pesaresi P, Catausan S, Dalid C, Slamet-Loedin I, Tecson-Mendoza EM, Wissuwa M, Heuer S (2012) The protein kinase Pstol1 from traditional rice confers tolerance of phosphorus deficiency. Nature 488(7412):535–539. https://doi.org/10.1038/nature11346

Ganguly S, Saha S, Vangaru, Purkayastha SS, Das D, Saha AK, Roy A, Das S, Bhattacharyya PK, Mukherjee S, Bhattacharyya S (2020) Identification and analysis of low light tolerant rice genotypes in field conditions and their SSR-based diversity in various abiotic stress tolerant lines. J Genet 99:24. https://doi.org/10.1007/s12041-020-01249-z

Gao C, Gao K, Yang H, Ju T, Zhu J, Tang Z, Zhao L, Chen Q (2022) Genome-wide analysis of metallothionein gene family in maize to reveal its role in development and stress resistance to heavy metal. Biol Res 55(1):1. https://doi.org/10.1186/s40659-021-00368-w

Gao Y, Jiang W, Dai Y, Xiao N, Zhang C, Li H, Lu Y, Wu M, Tao X, Deng D, Chen J (2015) A maize phytochrome-interacting factor 3 improves drought and salt stress tolerance in rice. Plant Mol Biol 87(4–5):413–428. https://doi.org/10.1007/s11103-015-0288-z

Garcia-Oliveira AL, Martins-Lopes P, Tolrá R, Poschenrieder C, Tarquis M, Guedes-Pinto H, Benito C (2014) Molecular characterization of the citrate transporter gene TaMATE1 and expression analysis of upstream genes involved in organic acid transport under Al stress in bread wheat (Triticum aestivum). Physiol Plant 152(3):441–452. https://doi.org/10.1111/ppl.12179

Gaur PM, Gowda CLL, Knights EJ, Kumar J (2007) Breeding achievements. In: Yadav SS, Redden RJ, Chen W, Sharma B (eds) Chickpea breeding and management. CAB International, pp 391–416. https://doi.org/10.1079/9781845932138.01

Ghosal S, Quilloy FA, Casal C, Septiningsih EM, Mendioro MS, Dixit S (2020) Trait-based mapping to identify the genetic factors underlying anaerobic germination of rice: phenotyping, GXE, and QTL mapping. BMC Genet 21(1):6. https://doi.org/10.1186/s12863-020-0808-y

Gill MB, Zeng FR, Shabala L, Zhang GP, Yu M, Demidchik V, Shabala S, Zhou MX (2019) Identification of QTL related to ROS formation under hypoxia and their association with waterlogging and salt tolerance in barley. Int J Mol Sci 20(3):699. https://doi.org/10.3390/ijms20030699

Gokidi Y, Bhanu AM, Singh MN (2016) Marker assisted recurrent selection: an overview. Adv Life Sci 5(17):6493–6499. ISSN 2278-3849, 6493-6499

Gonçalves L, Rubiales D, Bronze MR, Vaz Patto MC (2022) Grass pea (Lathyrus sativus L.)—a sustainable and resilient answer to climate challenges. Agronomy 12(6):1324. https://doi.org/10.3390/agronomy12061324

Gong Z, Xiong L, Shi H, Yang S, Herrera-Estrella LR, Xu G, Chao D-Y, Li J, Wang P-Y, Qin F, Li J, Ding Y, Shi Y, Wang Y, Yang Y, Guo Y, Zhu J-K (2020) Plant abiotic stress response and nutrient use efficiency. Sci China Life Sci 63(5):635–674. https://doi.org/10.1007/s11427-020-1683-x

Gregorio G, Islam M, Vergara G, Thirumeni S (2013) Recent advances in rice science to design salinity and other abiotic stress tolerant rice varieties. SABRAO J Breed Genet 45(1):31–41

Guadarrama-Escobar LM, Hunt J, Gurung A, Zarco-Tejada PJ, Shabala S, Camino C, Hernandez P, Pourkheirandish M (2024) Back to the future for drought tolerance. New Phytol 242(2):372–383. https://doi.org/10.1111/nph.19619

Guo YY, Chen J, Kuang LH, Wang NH, Zhang GP, Jiang LX, Wu DZ (2020) Effects of waterlogging stress on early seedling development and transcriptomic responses in Brassica napus. Mol Breed 40(9):85. https://doi.org/10.1007/s11032-020-01167-z

Gupta D, Dadu RHR, Sambasivam P, Bar I, Singh M, Beera N, Biju S (2019a) Toward climate-resilient lentils: challenges and opportunities. In: Genomic designing of climate-smart pulse crops. Springer, pp 165–234

Gupta PK, Balyan HS, Gahlaut V (2017) QTL analysis for drought tolerance in wheat: present status and future possibilities. Agronomy 7(1):5. https://doi.org/10.3390/agronomy7010005

Gupta SK, Ameta VL, Pareek S, Mahala RS, Jayalekha AK, Deora VS, Verma YS, Boratkar M, Atkari D, Rai KN (2016) Genetic enhancement for flowering period heat tolerance in peart millet (Pennisetum glaucum L. (R.) Br.). In: 7th International Crop Science Congress (Beijing). Retrieved August 14–19, 2016. http://oar.icrisat.org/9757/1/Page1.pdf

Gupta SK, Rai KN, Singh P, Ameta VL, Gupta SK, Jayalekha AK, Mahala RS, Pareek S, Swami ML, Verma YS (2015) Seed set variability under high temperatures during flowering period in pearl millet (Pennisetum glaucum L. (R.) Br.). Field Crop Res 171(1):41–53. https://doi.org/10.1016/j.fcr.2014.11.005

Gupta S, Kaur S, Sehgal S, Sharma A, Chhuneja P, Bains NS (2010) Genotypic variation for cellular thermotolerance in Aegilops tauschii Coss., the D genome progenitor of wheat. Euphytica 175(3):373–381. https://doi.org/10.1007/s10681-010-0185-0

Gurgul A, Miksza-Cybulska A, Szmatoła T, Jasielczuk I, Piestrzyńska-Kajtoch A, Fornal A, Semik-Gurgul E, Bugno-Poniewierska M (2019) Genotyping-by-sequencing performance in selected livestock species. Genomics 111(2):186–195. https://doi.org/10.1016/j.ygeno.2018.02.002

Habibi F, Liu T, Shahid MA, Schaffer B, Sarkhosh A (2023) Physiological, biochemical, and molecular responses of fruit trees to root zone hypoxia. Environ Exp Bot 206:105179. https://doi.org/10.1016/j.envexpbot.2022.105179

Haldimann P, Feller URS (2005) Growth at moderately elevated temperature alters the physiological response of the photosynthetic apparatus to heat stress in pea (Pisum sativum L.) leaves. Plant Cell Environ 28(3):302–317. https://doi.org/10.1111/j.1365-3040.2005.01289.x

Hall AE (2004) Breeding for adaptation to drought and heat in cowpea. Eur J Agron 21(4):447–454. https://doi.org/10.1016/j.eja.2004.07.005

Hamdi A, Erskine W (1996) Reaction of wild species of the geneus Lens to drought. Euphytica 91(2):173–179. https://doi.org/10.1007/BF00021067

Han M, Wong J, Su T, Beatty PH, Good AG (2016) Identification of nitrogen use efficiency genes in barley: searching for QTLs controlling complex physiological traits. Front Plant Sci 7:1587. https://doi.org/10.3389/fpls.2016.01587

Harada H, Leigh RA (2006) Genetic mapping of natural variation in potassium concentrations in shoots of Arabidopsis thaliana. J Exp Bot 57(4):953–960. https://doi.org/10.1093/jxb/erj081

Hasan AN, Doubrovina E, Sottile R, Prockop S, Klatt MG, Heller G, Selvakumar A, Barnett L, Hsu KC, O'Reilly RJ (2022) Dominant epitopes presented by prevalent HLA alleles permit wide use of banked CMVpp65 T cells in adoptive therapy. Blood Adv 6(16):4859–4872. https://doi.org/10.1182/bloodadvances.2022007005

Hattori Y, Miura K, Asano K, Yamamoto E, Mori H, Kitano H, Matsuoka M, Ashikari M (2007) A major QTL confers rapid internode elongation in response to water rise in deepwater rice. Breed Sci 57(4):305–314. https://doi.org/10.1270/jsbbs.57.305

Hattori Y, Nagai K, Mori H, Kitano H, Matsuoka M, Ashikari M (2008) Mapping of three QTLs that regulate internode elongation in deepwater rice. Breed Sci 58(1):39–46. https://doi.org/10.1270/jsbbs.58.39

He Y, Yang X, Xu C, Guo D, Niu L, Wang Y, Li J, Yan F, Wang Q (2018) Overexpression of a novel transcriptional repressor GmMYB3a negatively regulates salt–alkali tolerance and stress-related genes in soybean. Biochem Biophys Res Commun 498(3):586–591. https://doi.org/10.1016/j.bbrc.2018.03.026

Hede AR, Skovmand B, Reynolds MP, Crossa J, Vilhelmsen AL, Stølen O (1999) Evaluating genetic diversity for heat tolerance traits in Mexican wheat landraces. Genet Resour Crop Evol 46(1):37–45. https://doi.org/10.1023/A:1008684615643

Heuer S, Lu X, Chin JH, Tanaka JP, Kanamori H, Matsumoto T, De Leon T, Ulat VJ, Ismail AM, Yano M, Wissuwa M (2009) Comparative sequence analyses of the major quantitative trait locus phosphorus uptake 1 (Pup1) reveal a complex genetic structure. Plant Biotechnol J 7(5):456–457. https://doi.org/10.1111/j.1467-7652.2009.00415.x

Horie T, Kaneko T, Sugimoto G, Sasano S, Panda SK, Shibasaka M, Katsuhara M (2011) Mechanisms of water transport mediated by PIP aquaporins and their regulation via phosphorylation events under salinity stress in barley roots. Plant Cell Physiol 52(4):663–675. https://doi.org/10.1093/pcp/pcr027

Hu B, Wang W, Ou S, Tang J, Li H, Che R, Zhang Z, Chai X, Wang H, Wang Y, Liang C, Liu L, Piao Z, Deng Q, Deng K, Xu C, Liang Y, Zhang L, Li L, Chu C (2015) Variation in NRT1.1B contributes to nitrate-use divergence between rice subspecies. Nat Genet 47(7):834–838. https://doi.org/10.1038/ng.3337

Hu B, Zhu C, Li F, Tang J, Wang Y, Lin A, Liu L, Che R, Chu C (2011) LEAF TIP NECROSIS1 plays a pivotal role in the regulation of multiple phosphate starvation responses in rice. Plant Physiol 156(3):1101–1115. https://doi.org/10.1104/pp.110.170209

Hu DW, Sheng ZH, Li QL, Chen W, Wei XJ, Xie LH, Jiao GA, Shao GN, Wang JL, Tang SQ, Hu PS (2018) Identification of QTLs associated with cadmium concentration in rice grains. J Integr Agric 17(7):1563–1573. https://doi.org/10.1016/S2095-3119(17)61847-1

Hu H, Xiong L (2014) Genetic engineering and breeding of drought-resistant crops. Annu Rev Plant Biol 65(1):715–741. https://doi.org/10.1146/annurev-arplant-050213-040000

Hu Q, Nelson K, Luo H (2006) FLP-mediated site-specific recombination for genome modification in turfgrass. Biotechnol Lett 28(22):1793–1804. https://doi.org/10.1007/s10529-006-9162-z

Hu R, Qiu D, Chen Y, Miller AJ, Fan X, Pan X, Zhang M (2016) Knock-down of a tonoplast localized low-affinity nitrate transporter OsNPF7.2 affects rice growth under high nitrate supply. Front Plant Sci 7:1529. https://doi.org/10.3389/fpls.2016.01529

Huang D, Sun M, Zhang A, Chen J, Zhang J, Lin C, Zhang H, Lu X, Wang X, Yan H, Tang J, Huang L (2021) Transcriptional changes in pearl millet leaves under heat stress. Genes 12(11):1716. https://doi.org/10.3390/genes12111716

Huang X, Fan Y, Shabala L, Rengel Z, Shabala S, Zhou MX (2018) A major QTL controlling the tolerance to manganese toxicity in barley (Hordeum vulgare L.). Mol Breed 38:1–9

Hufnagel B, de Sousa SM, Assis L, Guimaraes CT, Leiser W, Azevedo GC, Negri B, Larson BG, Shaff JE, Pastina MM, Barros BA, Weltzien E, Rattunde HFW, Viana JH, Clark RT, Falcão A, Gazaffi R, Garcia AAF, Schaffert RE et al (2014) Duplicate and conquer: multiple homologs of Phosphorus-Starvation Tolerance1 enhance phosphorus acquisition and sorghum performance on low-phosphorus soils. Plant Physiol 166(2):659–677. https://doi.org/10.1104/pp.114.243949

Humphreys MO, Humphreys MW (2005) Breeding for stress resistance: general principles. In: Ashraf M, Harris P (eds) Abiotic stresses: plant resistance through breeding and molecular approaches. CRC Press. https://doi.org/10.1201/9781482293609

Humplík JF, Lazár D, Husičková A, Spíchal L (2015) Automated phenotyping of plant shoots using imaging methods for analysis of plant stress responses—a review. Plant Methods 11:1–10

Hurd EA (1971) Drought injury and resistance in crops. In: Larson EL, Eastin JD (eds) Crop Science Society of America. Wiley, pp 77–78. ISBN: 978-0-891-18581-9

Inghelandt DV, Frey FP, Ries D, Stich B (2019) QTL mapping and genome-wide prediction of heat tolerance in multiple connected populations of temperate maize. Sci Rep 9(1):14418. https://doi.org/10.1038/s41598-019-50853-2

International Crops Research Institute for the Semi-Arid Tropics (1994) Cold-tolerant chickpea varieties ICCV 88503, ICCV 88506 and ICCV 88510. Plant Material Description no. 53. International Crops Research Institute for the Semi-Arid Tropics

Jagadish SVK, Cairns J, Lafitte R, Wheeler TR, Price AH, Craufurd PQ (2010) Genetic analysis of heat tolerance at anthesis in rice. Crop Sci 50(5):1633–1641. https://doi.org/10.2135/cropsci2009.09.0516

Jagadish SVK, Craufurd PQ, Wheeler TR (2008) Phenotyping parents of mapping populations of rice for heat tolerance during anthesis. Crop Sci 48(3):1140–1146. https://doi.org/10.2135/cropsci2007.10.0559

Jam M, Alemzadeh A, Tale AM, Esmaeili-Tazangi S (2014) Heavy metal regulation of plasma membrane H+-ATPase gene expression in halophyte Aeluropus littoralis. Mol Biol Res Commun 3(2):129–139

Jeffrey C, Trethowan R, Kaiser B (2021) Chickpea tolerance to temperature stress: status and opportunity for improvement. J Plant Physiol 267:153555. https://doi.org/10.1016/j.jplph.2021.153555

Jha UC, Bohra A, Parida SK, Jha R (2017) Integrated "omics" approaches to sustain global productivity of major grain legumes under heat stress. Plant Breed 136(4):437–459. https://doi.org/10.1111/pbr.12489

Jia H, Ren H, Gu M, Zhao J, Sun S, Zhang X, Chen J, Wu P, Xu G (2011) The phosphate transporter gene OsPht1;8 is involved in phosphate homeostasis in rice. Plant Physiol 156(3):1164–1175. https://doi.org/10.1104/pp.111.175240

Jin S, Xu C, Li G, Sun D, Li Y, Wang X, Liu S (2017) Functional characterization of a type 2 metallothionein gene, SsMT2, from alkaline-tolerant Suaeda salsa. Sci Rep 7(1):17914. https://doi.org/10.1038/s41598-017-18263-4

Johnson GR, Frey KJ (1967) Heritabilities of quantitative attributes of oats (Avena sp.) at varying levels of environmental Stress1. Crop Sci 7(1):43–46. https://doi.org/10.2135/cropsci1967.0011183X000700010016x

Joshi R, Wani SH, Singh B, Bohra A, Dar ZA, Lone AA, Pareek A, Singla-Pareek SL (2016) Transcription factors and plants response to drought stress: current understanding and future directions. Front Plant Sci 7:1029. https://doi.org/10.3389/fpls.2016.01029

Junker A, Muraya MM, Weigelt-Fischer K, Arana-Ceballos F, Klukas C, Melchinger AE, Meyer RC, Riewe D, Altmann T (2014) Optimizing experimental procedures for quantitative evaluation of crop plant performance in high throughput phenotyping systems. Front Plant Sci 5:770. https://doi.org/10.3389/fpls.2014.00770

Kahraman A, Kusmenoglu I, Aydin N, Aydogan A, Erskine W, Muehlbauer FJ (2004) Genetics of winter hardiness in 10 lentil recombinant inbred line populations. Crop Sci 44(1):5–12. https://doi.org/10.2135/cropsci2004.5000

Kanter U, Hauser A, Michalke B, Dräxl S, Schäffner AR (2010) Caesium and strontium accumulation in shoots of Arabidopsis thaliana: genetic and physiological aspects. J Exp Bot 61(14):3995–4009. https://doi.org/10.1093/jxb/erq213

Kausch AP, Hague J, Oliver M, Li Y, Daniell H, Mascia P, Watrud LS, Stewart CN (2010) Transgenic perennial biofuel feedstocks and strategies for bioconfinement. Biofuels 1(1):163–176. https://doi.org/10.4155/bfs.09.11

Kaushal N, Awasthi R, Gupta K, Gaur P, Siddique KHM, Nayyar H (2013) Heat-stress-induced reproductive failures in chickpea (Cicer arietinum) are associated with impaired sucrose metabolism in leaves and anthers. Funct Plant Biol 40(12):1334–1349. https://doi.org/10.1071/FP13082

Keimeso Z (2022) Review article: Plant breeding for water logging resistance. International Journal of Current Research and Academic Review 10(3):93–104. https://doi.org/10.20546/ijcrar.2022.1003.010

Ketehouli T, Zhou Y-G, Dai S-Y, Carther KFI, Sun D-Q, Li Y, Nguyen QVH, Xu H, Wang F-W, Liu W-C, Li X-W, Li H-Y (2021) A soybean calcineurin B-like protein-interacting protein kinase, GmPKS4, regulates plant responses to salt and alkali stresses. J Plant Physiol 256:153331. https://doi.org/10.1016/j.jplph.2020.153331

Khan A, Sovero V, Gemenet DC (2016a) Genome-assisted breeding for drought resistance. Curr Genomics 17(4):330–342. https://doi.org/10.2174/1389202917999160211101417

Khan MS, Khan MA, Ahmad D (2016b) Assessing utilization and environmental risks of important genes in plant abiotic stress tolerance. Front Plant Sci 7:792. https://doi.org/10.3389/fpls.2016.00792

Khan S, Anwar S, Yu S, Sun M, Yang Z, Gao Z-Q (2019) Development of drought-tolerant transgenic wheat: achievements and limitations. Int J Mol Sci 20(13):3350. https://doi.org/10.3390/ijms20133350

Kikuchi A, Huynh HD, Endo T, Watanabe K (2015) Review of recent transgenic studies on abiotic stress tolerance and future molecular breeding in potato. Breed Sci 65(1):85–102. https://doi.org/10.1270/jsbbs.65.85

Kim DM, Zhang H, Zhou H, Du T, Wu Q, Mockler TC, Berezin MY (2015) Highly sensitive image-derived indices of water-stressed plants using hyperspectral imaging in SWIR and histogram analysis. Sci Rep 5:15919. https://doi.org/10.1038/srep15919

Kim M, Lee C, Hong S, Kim SL, Baek J-H, Kim K-H (2021) High-throughput phenotyping methods for breeding drought-tolerant crops. Int J Mol Sci 22(15):8266. https://doi.org/10.3390/ijms22158266

Kirk GJD, Solivas JL, Alberto MC (2003) Effects of flooding and redox conditions on solute diffusion in soil. Eur J Soil Sci 54(3):617–624. https://doi.org/10.1046/j.1365-2389.2003.00532.x

Knox RE, Pozniak CJ, Clarke FR, Clarke JM, Houshmand S, Singh AK (2009) Chromosomal location of the cadmium uptake gene (Cdu1) in durum wheat. Genome 52(9):741–747. https://doi.org/10.1139/g09-042

Kong W, Li S, Zhang C, Qiang Y, Li Y (2022) Combination of quantitative trait locus (QTL) mapping and transcriptome analysis reveals submerged germination QTLs and candidate genes controlling coleoptile length in rice. Food Energy Secur 11(1):e354. https://doi.org/10.1002/fes3.354

Kozlowski TT (1984) Plant Responses to Flooding of Soil. Bioscience 34(3):162–167. https://doi.org/10.2307/1309751

Kumar G, Tripathi R (2008) Induced cytomictic variations through abiotic stresses in grasspea (Lathyrus sativus L.). Indian J Genet Plant Breed 68(01):58–64

Kumar J, Basu PS, Gupta S, Dubey S, Sen Gupta D, Singh NP (2018) Physiological and molecular characterisation for high temperature stress in Lens culinaris. Funct Plant Biol 45(4):474–487. https://doi.org/10.1071/FP17211

Kumar J, Sen Gupta D, Djalovic I (2020) Breeding, genetics, and genomics for tolerance against terminal heat in lentil: current status and future directions. Legum Sci 2(3):e38. https://doi.org/10.1002/leg3.38

Kumar K, Kumar M, Kim S-R, Ryu H, Cho Y-G (2013) Insights into genomics of salt stress response in rice. Rice 6(1):27. https://doi.org/10.1186/1939-8433-6-27

Kumar P, Cai J, Miklavcic SJ (2015) A complete system for 3D reconstruction of roots for phenotypic analysis. Adv Exp Med Biol 823:249–270. https://doi.org/10.1007/978-3-319-10984-8_14

Laanbroek HJ (1990) Bacterial cycling of minerals that affect plant growth in waterlogged soils: a review. Aquat Bot 38(1):109–125. https://doi.org/10.1016/0304-3770(90)90101-P

Lamichaney A, Parihar AK, Hazra KK, Dixit GP, Katiyar PK, Singh D, Singh AK, Kumar N, Singh NP (2021) Untangling the influence of heat stress on crop phenology, seed set, seed weight, and germination in field pea (Pisum sativum L.). Front Plant Sci 12:635868. https://doi.org/10.3389/fpls.2021.635868

Laub M, Pataczek L, Feuerbacher A, Zikeli S, Högy P (2022) Contrasting yield responses at varying levels of shade suggest different suitability of crops for dual land-use systems: a meta-analysis. Agron Sustain Dev 42(3):51. https://doi.org/10.1007/s13593-022-00783-7

Lawlor HJ, Siddique KHM, Sedgley RH, Thurling N (1998) Improving cold tolerance and insect resistance in chickpea and the use of AFLPs for the identification of molecular markers for these traits. In: Acta Horticulturae International symposium on Biotechnology of Tropical and Subtropical Species R. A. Drew (Ed.), *II*(461), pp 185–192. https://doi.org/10.17660/ActaHortic.1998.461.18

Li D, He G, Tian W, Saleem M, Huang Y, Meng L, Wu D, He T (2021) Comparative and systematic omics revealed low Cd accumulation of potato StMTP9 in yeast: suggesting a new mechanism for heavy metal detoxification. Int J Mol Sci 22(19):10478. https://doi.org/10.3390/ijms221910478

Li H, Vaillancourt R, Mendham N, Zhou M (2008) Comparative mapping of quantitative trait loci associated with waterlogging tolerance in barley (Hordeum vulgare L.). BMC Genomics 9(1):401. https://doi.org/10.1186/1471-2164-9-401

Li J, Zhang M, Sun J, Mao X, Wang J, Liu H, Zheng H, Li X, Zhao H, Zou D (2020b) Heavy metal stress-associated proteins in rice and Arabidopsis: genome-wide identification, phylogenetics, duplication, and expression profiles analysis. Front Genet 11:477. https://doi.org/10.3389/fgene.2020.00477

Li L, Zhang Q, Huang D (2014) A review of imaging techniques for plant phenotyping. Sensors 14(11):20078–20111. https://doi.org/10.3390/s141120078

Li S, Tian Y, Wu K, Ye Y, Yu J, Zhang J, Liu Q, Hu M, Li H, Tong Y, Harberd NP, Fu X (2018b) Modulating plant growth-metabolism coordination for sustainable agriculture. Nature 560(7720):595–600. https://doi.org/10.1038/s41586-018-0415-5

Li W-Y, Wang C, Shi H-H, Wang B, Wang J-X, Liu Y-S, Ma J-Y, Tian S-Y, Zhang Y-W (2020a) Genome-wide analysis of ethylene-response factor family in adzuki bean and functional determination of VaERF3 under saline-alkaline stress. Plant Physiol Biochem 147:215–222. https://doi.org/10.1016/j.plaphy.2019.12.019

Li W, Katin-Grazzini L, Krishnan S, Thammina C, El-Tanbouly R, Yer H, Merewitz E, Guillard K, Inguagiato J, McAvoy RJ, Liu Z, Li Y (2016) A novel two-step method for screening shade tolerant mutant plants via dwarfism. Front Plant Sci 7:1495. https://doi.org/10.3389/fpls.2016.01495

Li Z, Srivastava R, Tang J, Zheng Z, Howell SH (2018a) Cis-effects condition the induction of a major unfolded protein response factor, ZmbZIP60, in response to heat stress in maize. Front Plant Sci 9:833. https://doi.org/10.3389/fpls.2018.00833

Liang C, Piñeros MA, Tian J, Yao Z, Sun L, Liu J, Shaff J, Coluccio A, Kochian LV, Liao H (2013) Low pH, aluminum, and phosphorus coordinately regulate malate exudation through GmALMT1 to improve soybean adaptation to acid soils. Plant Physiol 161(3):1347–1361. https://doi.org/10.1104/pp.112.208934

Liang W, Cui W, Ma X, Wang G, Huang Z (2014) Function of wheat Ta-UnP gene in enhancing salt tolerance in transgenic Arabidopsis and rice. Biochem Biophys Res Commun 450(1):794–801. https://doi.org/10.1016/j.bbrc.2014.06.055

Liang Y, Zhou K, Cao L (2024) An advanced three-dimensional phenotypic measurement approach for extracting Ginkgo root structural parameters based on terrestrial laser scanning. Front Plant Sci 15:1356078. https://doi.org/10.3389/fpls.2024.1356078

Ligaba A, Katsuhara M, Ryan PR, Shibasaka M, Matsumoto H (2006) The BnALMT1 and BnALMT2 genes from rape encode aluminum-activated malate transporters that enhance the aluminum resistance of plant cells. Plant Physiol 142(3):1294–1303. https://doi.org/10.1104/pp.106.085233

Lin Y, Yang X, Liu F (1994) A study on evaluation of waterlogging tolerance in wheat varieties (Triticum aestivum L.). Acta Agric Shanghai 10(2):79–84

Liu A, Yu Y, Duan X, Sun X, Duanmu H, Zhu Y (2015) GsSKP21, a Glycine soja S-phase kinase-associated protein, mediates the regulation of plant alkaline tolerance and ABA sensitivity. Plant Mol Biol 87(1–2):111–124. https://doi.org/10.1007/s11103-014-0264-z

Liu F, Wang ZY, Ren HY, Shen CJ, Li Y, Ling H-Q, Wu CY, Lian XM, Wu P (2010) OsSPX1 suppresses the function of OsPHR2 in the regulation of expression of OsPT2 and phosphate homeostasis in shoots of rice. Plant J 62(3):508–517. https://doi.org/10.1111/j.1365-313X.2010.04170.x

Liu H, Soomro A, Zhu Y, Qiu X, Chen K, Zheng T, Yang L, Xing D, Xu J (2016a) QTL underlying iron and zinc toxicity tolerances at seedling stage revealed by two sets of reciprocal introgression populations of rice (Oryza sativa L.). Crop J 4(4):280–289. https://doi.org/10.1016/j.cj.2016.05.007

Liu L, Bai C, Chen Y, Palta JA, Delhaize E, Siddique KH (2021) Durum wheat with the introgressed TaMATE1B gene shows resistance to terminal drought by ensuring deep root growth in acidic and Al3+-toxic subsoils. Plant Soil 1–2:311–324. https://doi.org/10.1007/s11104-021-04961-6

Liu R, Fang L, Yang T, Zhang X, Hu J, Zhang H, Han W, Hua Z, Hao J, Zong X (2017) Marker-trait association analysis of frost tolerance of 672 worldwide pea (Pisum sativum L.) collections. Sci Rep 7(1):5919. https://doi.org/10.1038/s41598-017-06222-y

Liu X, Yu Y, Liu Q, Deng S, Jin X, Yin Y, Guo J, Li N, Liu Y, Han S, Wang C, Hao D (2020) A Na2CO3-responsive chitinase gene from Leymus chinensis improve pathogen resistance and saline-alkali stress tolerance in transgenic tobacco and maize. Front Plant Sci 11:504. https://doi.org/10.3389/fpls.2020.00504

Liu ZJ, Kumari S, Zhang LF, Zheng YL, Ware D (2012) Characterization of miRNAs in response to short–term waterlogging in three inbred lines of Zea mays. PLoS One 7(6):e39786. https://doi.org/10.1371/journal.pone.0039786

Liu Z, Zhu C, Jiang Y, Tian Y, Yu J, An H, Tang W, Sun J, Tang J, Chen G, Zhai H, Wang C, Wan J (2016b) Association mapping and genetic dissection of nitrogen use efficiency-related traits in rice (Oryza sativa L.). Funct Integr Genomics 16(3):323–333. https://doi.org/10.1007/s10142-016-0486-z

Lobell DB, Sibley A, Ivan Ortiz-Monasterio J (2012) Extreme heat effects on wheat senescence in India. Nat Clim Chang 2(3):186–189. https://doi.org/10.1038/nclimate1356

Longmei N, Gill GK, Zaidi PH, Kumar R, Nair SK, Hindu V, Vinayan MT, Vikal Y (2021) Genome wide association mapping for heat tolerance in sub-tropical maize. BMC Genomics 22(1):154. https://doi.org/10.1186/s12864-021-07463-y

Lou D, Wang H, Liang G, Yu D (2017) OsSAPK2 confers abscisic acid sensitivity and tolerance to drought stress in rice. Front Plant Sci 8:993. https://doi.org/10.3389/fpls.2017.00993

Luan HY, Guo BJ, Pan YH, Lv C, Shen HQ, Xu RG (2018a) Morpho-anatomical and physiological responses to waterlogging stress in different barley (Hordeum vulgare L.) genotypes. Plant Growth Regul 85(3):399–409. https://doi.org/10.1007/s10725-018-0401-9

Luan HY, Shen HQ, Pan YH, Guo BJ, Lv C, Xu RG (2018b) Elucidating the hypoxic stress response in barley (Hordeum vulgare L.) during waterlogging: a proteomics approach. Sci Rep 8(1):9655. https://doi.org/10.1038/s41598-018-27726-1

Lucas MR, Ehlers JD, Huynh B-L, Diop N-N, Roberts PA, Close TJ (2013) Markers for breeding heat-tolerant cowpea. Mol Breed 31(3):529–536. https://doi.org/10.1007/s11032-012-9810-z

Luo J-S, Huang J, Zeng D-L, Peng J-S, Zhang G-B, Ma H-L, Guan Y, Yi H-Y, Fu Y-L, Han B, Lin H-X, Qian Q, Gong J-M (2018) A defensin-like protein drives cadmium efflux and allocation in rice. Nat Commun 9(1):645. https://doi.org/10.1038/s41467-018-03088-0

Lynch PJ (1994) Recurrent selection in pearl millet for improving stand establishment at high temperature. Plant Breed 113(3):237–241. https://doi.org/10.1111/j.1439-0523.1994.tb00728.x

Ma SY, Gai PP, Wang YY, Ullah N, Zhang WJ, Fan YH, Shan YJ, Huang ZL, Hu X (2021) Carbohydrate assimilation and translocation regulate grain yield formation in wheat crops (Triticum aestivum L.) under post–flowering waterlogging. Agronomy 11(11):2209. https://doi.org/10.3390/agronomy11112209

Mamrutha HM, Sharma D, Sumanth Kumar KS, Venkatesh K, Tiwari V, Sharma I (2017) Influence of diurnal irradiance variation on chlorophyll values in wheat: a comparative study using different chlorophyll meters. Natl Acad Sci Lett 40(3):221–224. https://doi.org/10.1007/s40009-017-0544-7

Manik SMN, Quamruzzaman M, Zhao CC, Johnson P, Hunt I, Shabala S, Zhou MX (2022) Genome–wide association study reveals marker trait associations (MTA) for waterlogging-triggered adventitious roots and aerenchyma formation in barley. Int J Mol Sci 23(6):3341. https://doi.org/10.3390/ijms23063341

Mano Y, Omori F (2013) Flooding tolerance in interspecific introgression lines containing chromosome segments from teosinte (Zea nicaraguensis) in maize (Zea mays subsp. mays). Ann Bot 112(6):1125–1139. https://doi.org/10.1093/aob/mct160

Mano Y, Omori F, Takamizo T, Kindiger B, Bird RM, Loaisiga CH, Takahashi H (2007) QTL mapping of root aerenchyma formation in seedlings of a maize x rare teosinte "Zea nicaraguensis" cross. Plant Soil 295(1–2):103–113. https://doi.org/10.1007/s11104-007-9266-9

Marfo KO, Hall AE (1992) Inheritance of heat tolerance during pod set in cowpea. Crop Sci 32(4):912–918. https://doi.org/10.2135/cropsci1992.0011183X003200040015x

Maron LG, Guimarães CT, Kirst M, Albert PS, Birchler JA, Bradbury PJ, Buckler ES, Coluccio AE, Danilova TV, Kudrna D, Magalhaes JV, Piñeros MA, Schatz MC, Wing RA, Kochian LV

(2013) Aluminum tolerance in maize is associated with higher MATE1 gene copy number. Proc Natl Acad Sci USA 110(13):5241–5246. https://doi.org/10.1073/pnas.1220766110
Marschner H (1984) Mineral nutrition in higher plants. Academic Press
Mason RE, Mondal S, Beecher FW, Pacheco A, Jampala B, Ibrahim AMH, Hays DB (2010) QTL associated with heat susceptibility index in wheat (Triticum aestivum L.) under short-term reproductive stage heat stress. Euphytica 174(3):423–436. https://doi.org/10.1007/s10681-010-0151-x
Matsui T, Omasa K (2002) Rice (Oryza sativa L.) cultivars tolerant to high temperature at flowering: anther characteristics. Ann Bot 89(6):683–687. https://doi.org/10.1093/aob/mcf112
McAusland L, Davey PA, Kanwal N, Baker NR, Lawson T (2013) A novel system for spatial and temporal imaging of intrinsic plant water use efficiency. J Exp Bot 64(16):4993–5007. https://doi.org/10.1093/jxb/ert288
Megha, Singh N (2023) Perspective chapter: an insight into abiotic stresses in pigeon pea—effects and tolerance. In: Plant abiotic stress responses and tolerance mechanisms. https://doi.org/10.5772/intechopen.110368
Meyer DW, Badaruddin M (2001) Frost tolerance of ten seedling legume species at four growth stages. Crop Sci 41(6):1838–1842. https://doi.org/10.2135/cropsci2001.1838
Millan T, Clarke HJ, Siddique KHM, Buhariwalla HK, Gaur PM, Kumar J, Gil J, Kahl G, Winter P (2006) Chickpea molecular breeding: new tools and concepts. Euphytica 147(1–2):81–103. https://doi.org/10.1007/s10681-006-4261-4
Mishra GP, Singh RK, Singh R, Singh SB (2009) Molecular and physiological approaches for drought management in crops. In: Singh RK et al (eds) Molecular plant breeding: Principle, method and application. Studium Press LLC, pp 367–408. ISBN: 978-1933699493
Mishra SB, Senadhira D, Manigbas NL (1996) Genetics of submergence tolerance in rice (Oryza sativa L.). Field Crop Res 46(1–3):177–181. https://doi.org/10.1016/0378-4290(95)00088-7
Mishra S, Singh R, Kumar R, Kalia A, Panigrahy SR (2017) Impact of climate change on pigeon pea. Econ Aff 62(3):455. https://doi.org/10.5958/0976-4666.2017.00057.2
Mitra J (2001) Genetics and genetic improvement of drought resistance in crop plants. Curr Sci 80(6):758–763
Mittler R, Zandalinas SI, Fichman Y, Van Breusegem F (2022) Reactive oxygen species signalling in plant stress responses. Nat Rev Mol Cell Biol 23(10):663–679. https://doi.org/10.1038/s41580-022-00499-2
Mohammadi R (2018) Breeding for increased drought tolerance in wheat: a review. Crop Pasture Sci 69(3):223–241. https://doi.org/10.1071/CP17387
Mohapatra C, Chand R, Tiwari JK, Singh AK (2020) Effect of heat stress during flowering and pod formation in pea (Pisum sativum L.). Physiol Mol Biol Plants 26(6):1119–1125. https://doi.org/10.1007/s12298-020-00803-4
Molero G, Rivera-Amado A, Piñera-Chavez F, Negrellos E, Quintero A, González OE, Slafer G, Calderini D, Foulkes M, Berry P, Griffiths S, López-Castañeda C, Singh R, Reynolds M (2015) Dissection of yield potential related traits: what shall we focus on? In: Reynolds MP et al (eds) Proceedings of the International TRIGO (Wheat) Yield Potential Workshop, 24–26 March, 2015. CENEB, CIMMYT, Cd. Obregon
Munns R, James RA, Xu B, Athman A, Conn SJ, Jordans C, Byrt CS, Hare RA, Tyerman SD, Tester M, Plett D, Gilliham M (2012) Wheat grain yield on saline soils is improved by an ancestral Na^+ transporter gene. Nat Biotechnol 30(4):360–364. https://doi.org/10.1038/nbt.2120
Mwamahonje A, Eleblu JSY, Ofori K, Deshpande S, Feyissa T, Tongoona P (2021) Drought tolerance and application of marker-assisted selection in sorghum. Biology 10(12):1249. https://doi.org/10.3390/biology10121249
Nagarajan VK, Jain A, Poling MD, Lewis AJ, Raghothama KG, Smith AP (2011) Arabidopsis Pht1;5 mobilizes phosphate between source and sink organs and influences the interaction between phosphate homeostasis and ethylene signaling. Plant Physiol 156(3):1149–1163. https://doi.org/10.1104/pp.111.174805

Nahar K, Ahamed KU, Fujita M (2010) Phenological variation and its relation with yield in several wheat (Triticum aestivum L.) cultivars under normal and late sowing mediated heat stress condition. Not Sci Biol 2(3):51–56. https://doi.org/10.15835/nsb234723

Naik YD, Sharma VK, Aski MS, Rangari SK, Kumar R, Dikshit HK, Sahani S, Kant R, Mishra G, Mir RR, Kudapa H, Elango D, Zwart RS, Varshney RK, Thudi M (2024) Phenotypic profiling of lentil (Lens culinaris Medikus) accessions enabled identification of promising lines for use in breeding for high yield, early flowering and desirable traits. Plant Genet Resour Characterisation Util 22(2):69–77. https://doi.org/10.1017/S1479262124000042

Nature (2023) How pearl millet tolerates heat- Structural variations in its genome turn on specific genes to generate stress responses (2023) Research highlights. Nature India (Nat India). ISSN 1755-3180. Retrieved March 9, 2023. https://doi.org/10.1038/d44151-023-00029-9

Nguyen VL, Ribot SA, Dolstra O, Niks RE, Visser RGF, van der Linden CG (2013) Identification of quantitative trait loci for ion homeostasis and salt tolerance in barley (Hordeum vulgare L.). Mol Breed 31(1):137–152. https://doi.org/10.1007/s11032-012-9777-9

Njobvu J, Hamabwe SM, Munyinda K, Kelly JD, Kamfwa K (2020) Quantitative trait loci mapping of resistance to aluminum toxicity in common bean. Crop Sci 60(3):1294–1302. https://doi.org/10.1002/csc2.20043

Noman A, Sanaullah T, Khalid N, Islam W, Khan S, Irshad MK, Aqeel M (2019) Crosstalk between plant miRNA and heavy metal toxicity. In: Plant metallomics and functional omics: a system-wide perspective, pp 145–168

Ntare BR (1991) Variation in reproductive efficiency and yield of cowpea under high temperature conditions in a Sahelian environment. Euphytica 59(1):27–32. https://doi.org/10.1007/BF00025358

Nuccio ML, Wu J, Mowers R, Zhou H-P, Meghji M, Primavesi LF, Paul MJ, Chen X, Gao Y, Haque E, Basu SS, Lagrimini LM (2015) Expression of trehalose-6-phosphate phosphatase in maize ears improves yield in well-watered and drought conditions. Nat Biotechnol 33(8):862–869. https://doi.org/10.1038/nbt.3277

Pais IP, Moreira R, Semedo JN, Ramalho JC, Lidon FC, Coutinho J, Maçãs B, Scotti-Campos P (2022) Wheat crop under waterlogging: potential soil and plant effects. Plants 12(1):149. https://doi.org/10.3390/plants12010149

Panda D, Barik J (2021) Flooding tolerance in rice: focus on mechanisms and approaches. Rice Sci 28(1):43–57. https://doi.org/10.1016/j.rsci.2020.11.006

Pandey AK, Gedda MR, Verma AK (2020) Effect of arsenic stress on expression pattern of a rice specific miR156j at various developmental stages and their allied co-expression target networks. Front Plant Sci 11:752. https://doi.org/10.3389/fpls.2020.00752

Pandey GC, Sareen S, Siwach P, Tiwari R (2014) Molecular characterization of heat tolerance in bread wheat (Triticum aestivum L.) using differences in thousand-grain weights (dTGW) as a potential indirect selection criterion. Cereal Res Commun 42(1):38–46. https://doi.org/10.1556/CRC.2013.0041

Panigrahy M, Majeed N, Panigrahi KCS (2020) Low-light and its effects on crop yield: genetic and genomic implications. J Biosci 45:102. https://doi.org/10.1007/s12038-020-00070-1

Parankusam S, Bhatnagar-Mathur P, Sharma KK (2017) Heat responsive proteome changes reveal molecular mechanisms underlying heat tolerance in chickpea. Environ Exp Bot 141:132–144. https://doi.org/10.1016/j.envexpbot.2017.07.007

Parent B, Shahinnia F, Maphosa L, Berger B, Rabie H, Chalmers K, Kovalchuk A, Langridge P, Fleury D (2015) Combining field performance with controlled environment plant imaging to identify the genetic control of growth and transpiration underlying yield response to water-deficit stress in wheat. J Exp Bot 66(18):5481–5492. https://doi.org/10.1093/jxb/erv320

Parihar AK, Hazra KK, Lamichaney A, Dixit GP, Singh D, Singh AK, Singh NP (2022) Characterizing plant trait(s) for improved heat tolerance in field pea (Pisum sativum L.) under subtropical climate. Int J Biometeorol 66(6):1267–1281. https://doi.org/10.1007/s00484-022-02275-5

Parihar AK, Hazra KK, Lamichaney A, Singh AK, Dixit GP (2023) Delineating the role of plant stature towards heat stress tolerance in field pea (Pisum sativum L.). Heliyon 9(3):e14539. https://doi.org/10.1016/j.heliyon.2023.e14539

Partheeban C, Vijayaraghavan H, Boominathan P (2017) The rapid and reliable method to assess the blackgram (Vigna mungo (l.) Hepper) genotypes for thermotolerance. Int J Bioresour Stress Manag 8(1):6–11. https://doi.org/10.23910/IJBSM/2017.8.1.1710

Patel KG, Thankappan R, Mishra GP, Mandaliya VB, Kumar A, Dobaria JR (2017) Transgenic peanut (Arachis hypogaea L.) overexpressing mtlD gene showed improved photosynthetic, physio-biochemical, and yield-parameters under soil-moisture deficit stress in lysimeter system. Front Plant Sci 8:1881. https://doi.org/10.3389/fpls.2017.01881

Pavithra N, Jayalalitha K, Sujatha T, Harisatyanarayana N, Lakshmi NJ, Roja V (2023) Temperature induction response and multivariate analysis approaches to screen blackgram (Vigna mungo L.) genotypes for thermotolerance. Legum Res 1(Of):1–8. https://doi.org/10.18805/LR-5134

Peacock JM, Soman P, Jayachandran R, Rani AU, Howarth CJ, Thomas A (1993) Effects of high soil surface temperature on seedling survival in pearl millet. Exp Agric 29(2):215–225. https://doi.org/10.1017/S0014479700020664

Perata P, Alpi A (1993) Plant responses to anaerobiosis. Plant Sci 93(1–2):1–17. https://doi.org/10.1016/0168-9452(93)90029-Y

Place GA, Sims JL, Hall VL (1970) Effects of nitrogen and phosphorus fertilization on growth, yield, and cooking characteristics of rice 1. Agron J 62(2):239–243. https://doi.org/10.2134/agronj1970.00021962006200020018x

Poli Y, Basava RK, Panigrahy M, Vinukonda VP, Dokula NR, Voleti SR, Desiraju S, Neelamraju S (2013) Characterization of a Nagina22 rice mutant for heat tolerance and mapping of yield traits. Rice 6(1):36. https://doi.org/10.1186/1939-8433-6-36

Ponnamperuma FN (1972) The chemistry of submerged soils. Adv Agron 24:29–96. https://doi.org/10.1016/S0065-2113(08)60633-1

Pottorff M, Roberts PA, Close TJ, Lonardi S, Wanamaker S, Ehlers JD (2014) Identification of candidate genes and molecular markers for heat-induced brown discoloration of seed coats in cowpea [Vigna unguiculata (L.) Walp]. BMC Genomics 15(1):328. https://doi.org/10.1186/1471-2164-15-328

Pradhan GP, Prasad PVV, Fritz AK, Kirkham MB, Gill BS (2012a) High temperature tolerance in Aegilops species and its potential transfer to wheat. Crop Sci 52(1):292–304. https://doi.org/10.2135/cropsci2011.04.0186

Pradhan GP, Prasad PVV, Fritz AK, Kirkham MB, Gill BS (2012b) Effects of drought and high temperature stress on synthetic hexaploid wheat. Funct Plant Biol 39(3):190–198. https://doi.org/10.1071/FP11245

Pradheep K, Bhandari DC, Bansal KC (2014) Wild relatives of cultivated plants in India. Indian Council of Agricultural Research. ISBN: 9788191038880

Prasad PVV, Boote KJ, Allen LH, Sheehy JE, Thomas JMG (2006) Species, ecotype and cultivar differences in spikelet fertility and harvest index of rice in response to high temperature stress. Field Crop Res 95(2–3):398–411. https://doi.org/10.1016/j.fcr.2005.04.008

Pucciariello C, Perata P (2017) New insights into reactive oxygen species and nitric oxide signalling under low oxygen in plants. Plant Cell Environ 40(4):473–482. https://doi.org/10.1111/pce.12715

Punniyamoorthy D, Jegadeesan S (2023) Genetics and genomics of drought and heat tolerance in cowpea, mung bean and black gram. In: Legumes: physiology and molecular biology of abiotic stress tolerance. Springer, Singapore, pp 203–233. https://doi.org/10.1007/978-981-19-5817-5_8

Qian Y, Ren Q, Zhang J, Chen L (2019) Transcriptomic analysis of the maize (Zea mays L.) inbred line B73 response to heat stress at the seedling stage. Gene 692:68–78. https://doi.org/10.1016/j.gene.2018.12.062

Qin C, Li Y-H, Li D, Zhang X, Kong L, Zhou Y, Lyu X, Ji R, Wei X, Cheng Q, Jia Z, Li X, Wang Q, Wang Y, Huang W, Yang C, Liu L, Wang X, Xing G et al (2023) PH13 improves soybean shade traits and enhances yield for high-density planting at high latitudes. Nat Commun 14(1):6813. https://doi.org/10.1038/s41467-023-42608-5

Qin F, Kakimoto M, Sakuma Y, Maruyama K, Osakabe Y, Tran L-SP, Shinozaki K, Yamaguchi-Shinozaki K (2007) Regulation and functional analysis of ZmDREB2A in response to drought and heat stresses in Zea mays L. Plant J 50(1):54–69. https://doi.org/10.1111/j.1365-313X.2007.03034.x

Qiu FZ, Zheng YL, Zhang ZL, Xu SZ (2007) Mapping of QTL associated with waterlogging tolerance during the seedling stage in maize. Ann Bot 99(6):1067–1081. https://doi.org/10.1093/aob/mcm055

Qiu H, Mei X, Liu C, Wang J, Wang G, Wang X, Liu Z, Cai Y (2013) Fine mapping of quantitative trait loci for acid phosphatase activity in maize leaf under low phosphorus stress. Mol Breed 32(3):629–639. https://doi.org/10.1007/s11032-013-9895-z

Rahman SU, Arif M, Hussain K, Hussain S, Mukhtar T, Razaq A, Iqbal RA (2013) Evaluation of maize hybrids for tolerance to high temperature stress in Central Punjab. Am J Bioeng Biotechnol 1:30–36. https://doi.org/10.7726/ajbebt.2013.1003

Rahman Z, Singh VP (2019) The relative impact of toxic heavy metals (THMs) (arsenic (As), cadmium (Cd), chromium (Cr)(VI), mercury (Hg), and lead (Pb)) on the total environment: an overview. Environ Monit Assess 191(7):419. https://doi.org/10.1007/s10661-019-7528-7

Ramakrishna G, Kaur P, Singh A, Yadav SS, Sharma S, Singh NK, Gaikwad K (2021) Comparative transcriptome analyses revealed different heat stress responses in pigeonpea (Cajanus cajan) and its crop wild relatives. Plant Cell Rep 40(5):881–898. https://doi.org/10.1007/s00299-021-02686-5

Ray S, Satya P (2014) Next generation sequencing technologies for next generation plant breeding. Front Plant Sci 5:367. https://doi.org/10.3389/fpls.2014.00367

Rehman AIH, Ahmad N, Hussain M, Khan MA, Farooq J, Ali MA (2009) Screening wheat germplasm for heat tolerance at terminal growth stage. Plant OMICS 2:9–19

Ren Z-H, Gao J-P, Li L-G, Cai X-L, Huang W, Chao D-Y, Zhu M-Z, Wang Z-Y, Luan S, Lin H-X (2005) A rice quantitative trait locus for salt tolerance encodes a sodium transporter. Nat Genet 37(10):1141–1146. https://doi.org/10.1038/ng1643

Ridge PE, Pye DL (1985) The effects of temperature and frost at flowering on the yield of peas grown in a Mediterranean environment. Field Crop Res 12:339–346. https://doi.org/10.1016/0378-4290(85)90079-6

Ristic Z, Yang G, Martin B, Fullerton S (1998) Evidence of association between specific heat-shock protein(s) and the drought and heat tolerance phenotype in maize. J Plant Physiol 153(3–4):497–505. https://doi.org/10.1016/S0176-1617(98)80180-6

Robertson LD, Singh KB, Ocampo B (1995) A catalog of annual wild Cicer species. International Center for Agricultural Research in the Dry Areas, p 171

Rose TJ, Kretzschmar T, Waters DLE, Balindong JL, Wissuwa M (2017) Prospects for genetic improvement in internal nitrogen use efficiency in rice. Agronomy 7(4):70. https://doi.org/10.3390/agronomy7040070

Roy NN, Murty BR (1970) A selection procedure in wheat for stress environment. Euphytica 19(4):509–521. https://doi.org/10.1007/BF01902926

Rajendran K, Smouni A, Es-Safi NE et al. (2020) Screening the FIGS set of lentil (Lens Culinaris Medikus) germplasm for tolerance to terminal heat and combined drought-heat stress. Agron 10(7):1036

Saad ASI, Li X, Li H-P, Huang T, Gao C-S, Guo M-W, Cheng W, Zhao G-Y, Liao Y-C (2013) A rice stress-responsive NAC gene enhances tolerance of transgenic wheat to drought and salt stresses. Plant Sci 203–204:33–40. https://doi.org/10.1016/j.plantsci.2012.12.016

Saal B, von Korff M, Léon J, Pillen K (2011) Advanced-backcross QTL analysis in spring barley: IV. Localization of QTL × nitrogen interaction effects for yield-related traits. Euphytica 177(2):223–239. https://doi.org/10.1007/s10681-010-0252-6

Sabagh AE, Mbarki S, Hossain A, Iqbal MA, Islam MS, Raza A, Llanes A, Reginato M, Rahman MA, Mahboob W, Singhal RK, Kumari A, Rajendran K, Wasaya A, Javed T, Shabbir R, Rahim J, Barutçular C, Habib Ur Rahman M et al (2021) Potential role of plant growth regulators in administering crucial processes against abiotic stresses. Front Agron 3:648694. https://doi.org/10.3389/fagro.2021.648694

Sachs MM (1993) Molecular genetic basis of metabolic adaptation to anoxia in maize and its possible utility for improving tolerance of crops to soil waterlogging. In: Jackson MB, Black CR (eds) Interacting stresses on plants in a changing climate, NATO ASI series, vol 16. Springer, Berlin, Heidelberg, pp 375–393. https://doi.org/10.1007/978-3-642-78533-7_24

Sadras VO, Lake L, Chenu K, McMurray LS, Leonforte A (2012) Water and thermal regimes for field pea in Australia and their implications for breeding. Crop Pasture Sci 63(1):33–44. https://doi.org/10.1071/CP11321

Sallam A, Alqudah AM, Dawood MFA, Baenziger PS, Börner A (2019) Drought stress tolerance in wheat and barley: advances in physiology, breeding and genetics research. Int J Mol Sci 20(13):3137. https://doi.org/10.3390/ijms20133137

Samad A, Meisner CA, Saifuzzaman M, van Ginkel M (2001) Waterlogging tolerance. In: Reynolds MP, Ortiz-Monasterio JI, McNab A (eds) Application of Physiology in wheat breeding. CIMMYT, pp 136–144. SBN: 970-648-077-3

Sanguineti MC, Tuberosa R, Landi P, Salvi S, Maccaferri M, Casarini E, Conti S (1999) QTL analysis of drought-related traits and grain yield in relation to genetic variation for leaf abscisic acid concentration in field-grown maize. J Exp Bot 50(337):1289–1297. https://doi.org/10.1093/jxb/50.337.1289

Sanjana P (2015) Genetic improvement in pearl millet. In: Tonapi VA, Patil JV (eds) Millets-ensuring climate resilience and nutritional security. Daya Publishing House, pp 191–231

Santosh Kumar VV, Verma RK, Yadav SK, Yadav P, Watts A, Rao MV, Chinnusamy V (2020) CRISPR-Cas9 mediated genome editing of drought and salt tolerance (OsDST) gene in indica mega rice cultivar MTU1010. Physiol Mol Biol Plants 26(6):1099–1110. https://doi.org/10.1007/s12298-020-00819-w

Sarkar A, Emmrich PMF, Sarker A, Zong X, Martin C, Wang TL (2019b) Grass pea: remodeling an ancient insurance crop for climate resilience. In: Kole C (ed) Genomic designing of climate-smart pulse crops. Springer, Cham. https://doi.org/10.1007/978-3-319-96932-9_9

Sarkar T, Thankappan R, Kumar A, Mishra GP, Dobaria JR (2014) Heterologous expression of the AtDREB1A gene in transgenic peanut-conferred tolerance to drought and salinity stresses. PLoS One 9(12):e110507. https://doi.org/10.1371/journal.pone.0110507

Sarkar T, Thankappan R, Kumar A, Mishra GP, Dobaria JR (2016) Stress inducible expression of AtDREB1A transcription factor in transgenic peanut (Arachis hypogaea L.) conferred tolerance to soil-moisture deficit stress. Front Plant Sci 7:935. https://doi.org/10.3389/fpls.2016.00935

Sarkar T, Thankappan R, Mishra GP, Nawade BD (2019a) Advances in the development and use of DREB for improved abiotic stress tolerance in transgenic crop plants. Physiol Mol Biol Plants 25(6):1323–1334. https://doi.org/10.1007/s12298-019-00711-2

Sarker A, Aydin N, Aydogan A, Sabaghpour SH, Ketata H, Kusmenoglu I, Erskine W (2002) Winter lentils promise improved nutrition and income in West Asian highlands. International Center for Agricultural Research in the Dry Areas (ICARDA) Caravan 16:14–16

Sayre KD, van Ginkel M, Rajaram S, Ortiz-Monasterio I (1994) Tolerance to waterlogging losses in spring bread wheat: effect of time of onset on expression. Ann Wheat Newsl 40:165–171

Scafaro AP, Haynes PA, Atwell BJ (2010) Physiological and molecular changes in Oryza meridionalis Ng., a heat-tolerant species of wild rice. J Exp Bot 61(1):191–202. https://doi.org/10.1093/jxb/erp294

Sehgal A, Sita K, Kumar J, Kumar S, Singh S, Siddique KHM, Nayyar H (2017) Effects of drought, heat and their interaction on the growth, yield and photosynthetic function of lentil (Lens culinaris Medikus) genotypes varying in heat and drought sensitivity. Front Plant Sci 8:1776. https://doi.org/10.3389/fpls.2017.01776

Selga C, Reslow F, Pérez-Rodríguez P, Ortiz R (2022) The power of genomic estimated breeding values for selection when using a finite population size in genetic improvement of tetraploid potato. G3 Genes Genomes Genet 12(1):jkab362. https://doi.org/10.1093/g3journal/jkab362

Sen Gupta D, Basu PS, Souframanien J, Kumar J, Dhanasekar P, Gupta S, Pandiyan M, Geetha S, Shanthi P, Kumar V, Pratap Singh N (2021) Morpho-physiological traits and functional markers based molecular dissection of heat-tolerance in urdbean. Front Plant Sci 12:719381. https://doi.org/10.3389/fpls.2021.719381

Septiningsih EM, Sanchez DL, Singh N, Sendon PMD, Pamplona AM, Heuer S, Mackill DJ (2012) Identifying novel QTLs for submergence tolerance in rice cultivars IR72 and Madabaru. Theor Appl Genet 124(5):867–874. https://doi.org/10.1007/s00122-011-1751-0

Setter TL, Waters I (2003) Review of prospects for germplasm improvement for waterlogging tolerance in wheat, barley and oats. Plant Soil 253(1):1–34. https://doi.org/10.1023/A:1024573305997

Setter TL, Waters I, Sharma SK, Singh KN, Kulshreshtha N, Yaduvanshi NPS, Ram PC, Singh BN, Rane J, McDonald G, Khabaz-Saberi H, Biddulph TB, Wilson R, Barclay I, McLean R, Cakir M (2009) Review of wheat improvement for waterlogging tolerance in Australia and India: the importance of anaerobiosis and element toxicities associated with different soils. Ann Bot 103(2):221–235. https://doi.org/10.1093/aob/mcn137

Shafiq S, Mather DE, Ahmad M, Paull JG (2012) Variation in tolerance to radiant frost at reproductive stages in field pea germplasm. Euphytica 186(3):831–845. https://doi.org/10.1007/s10681-012-0625-0

Shan Q, Wang Y, Li J, Zhang Y, Chen K, Liang Z, Zhang K, Liu J, Xi JJ, Qiu J-L, Gao C (2013) Targeted genome modification of crop plants using a CRISPR-Cas system. Nat Biotechnol 31(8):686–688. https://doi.org/10.1038/nbt.2650

Sharma DK, Andersen SB, Ottosen C-O, Rosenqvist E (2015) Wheat cultivars selected for high Fv/Fm under heat stress maintain high photosynthesis, total chlorophyll, stomatal conductance, transpiration and dry matter. Physiol Plant 153(2):284–298. https://doi.org/10.1111/ppl.12245

Sharma V, Sarkar IN (2013) Bioinformatics opportunities for identification and study of medicinal plants. Brief Bioinform 14(2):238–250. https://doi.org/10.1093/bib/bbs021

Shilin D, Chaolei L, Lianguang S, Shenglong Y, Anpeng Z, Hongzhen J, Banpu R, Guonan F, Biao T, Guoyou Y, Longbiao G, Qian Q, Zhenyu G (2021) Identification of QTLs for cadmium tolerance during seedling stage and validation of qCDSL1 in rice. Rice Sci 28(1):81–88. https://doi.org/10.1016/j.rsci.2020.11.009

Shirley HL (1943) Is tolerance the capacity to endure shade? J For 41(5):339–345

Shu QY, Forster BP, Nakagawa H (2012) Plant mutation breeding and biotechnology. CABI Publishing. ISBN: 978-1-78064-085-3

Shu W, Zhou Q, Xian P, Cheng Y, Lian T, Ma Q, Zhou Y, Li H, Nian H, Cai Z (2022) GmWRKY81 encoding a WRKY transcription factor enhances aluminum tolerance in soybean. Int J Mol Sci 23(12):6518. https://doi.org/10.3390/ijms23126518

Silva CMS, Zhang C, Habermann G, Delhaize E, Ryan PR (2018) Does the major aluminium-resistance gene in wheat, TaALMT1, also confer tolerance to alkaline soils? Plant Soil 424(1–2):451–462. https://doi.org/10.1007/s11104-017-3549-6

Singh A (2015) Soil salinization and waterlogging: a threat to environment and agricultural sustainability. Ecol Indic 57:128–130. https://doi.org/10.1016/j.ecolind.2015.04.027

Singh AK (2017) Wild relatives of cultivated plants in India: A reservoir of alternative genetic resources and more. Springer, p 309p. https://doi.org/10.1007/978-981-10-5116-6

Singh KB, Malhotra RS, Saxena MC (1995) Additional sources of tolerance to cold in cultivated and wild Cicer species. Crop Sci 35(5):1491–1497. https://doi.org/10.2135/cropsci1995.0011183X003500050037x

Singh US, Dar MH, Singh S, Zaidi NW, Bari MA, Mackill DJ, Collard BCY, Singh VN, Singh JP, Reddy JN, Singh RK (2013) Field performance, dissemination, impact and tracking of submergence tolerant (Sub1) rice varieties in South Asia. SABRAO J Breed Genet 45(1):112–131

Sita K, Sehgal A, HanumanthaRao B, Nair RM, Vara Prasad PV, Kumar S, Gaur PM, Farooq M, Siddique KHM, Varshney RK, Nayyar H (2017a) Food legumes and rising temperatures: effects, adaptive functional mechanisms specific to reproductive growth stage and strategies to improve heat tolerance. Front Plant Sci 8:1658. https://doi.org/10.3389/fpls.2017.01658

Sita K, Sehgal A, Kumar J, Kumar S, Singh S, Siddique KHM, Nayyar H (2017b) Identification of high-temperature tolerant lentil (Lens culinaris Medik.) genotypes through leaf and pollen traits. Front Plant Sci 8:744. https://doi.org/10.3389/fpls.2017.00744

Soman P, Peacock JM (1985) A laboratory technique to screen seedling emergence of sorghum and pearl millet at high soil temperature. Exp Agric 21(4):335–341. https://doi.org/10.1017/S0014479700013168

Srinivasan A, Johansen C, Saxena NP (1998) Cold tolerance during early reproductive growth of chickpea (Cicer arietinum L.): characterization of stress and genetic variation in pod set. Field Crop Res 57(2):181–193. https://doi.org/10.1016/S0378-4290(97)00118-4

Sulistyowati D, Chozin MA, Syukur M, Melati M, Guntoro D (2016) Selection of shade-tolerant tomato genotypes. J Appl Hortic 18(2):154–159. https://doi.org/10.37855/jah.2016.v18i02.27

Sun H, Qian Q, Wu K, Luo J, Wang S, Zhang C, Ma Y, Liu Q, Huang X, Yuan Q, Han R, Zhao M, Dong G, Guo L, Zhu X, Gou Z, Wang W, Wu Y, Lin H, Fu X (2014) Heterotrimeric G proteins regulate nitrogen-use efficiency in rice. Nat Genet 46(6):652–656. https://doi.org/10.1038/ng.2958

Sun M, Huang D, Zhang A, Khan I, Yan H, Wang X, Zhang X, Zhang J, Huang L (2020) Transcriptome analysis of heat stress and drought stress in pearl millet based on Pacbio full-length transcriptome sequencing. BMC Plant Biol 20(1):323. https://doi.org/10.1186/s12870-020-02530-0

Sunilkumar VP, Krishna H, Devate NB, Manjunath KK, Chauhan D, Singh S, Sinha N, Singh JB, Prakasha TL, Pal D, Sivasamy M, Jain N, Singh GP, Singh PK (2023) Marker-assisted selection for transfer of QTLs to a promising line for drought tolerance in wheat (Triticum aestivum L.). Front Plant Sci 14. https://doi.org/10.3389/fpls.2023.1147200

Suswanto T, Shamshuddin J, Omar SRS, Mat P, Teh CBS (2007) Alleviating an acid sulfate soil cultivated to rice (Oryza sativa) using ground magnesium limestone and organic fertilizer. J Ilmu Tanah Lingkungan 9(1):1–9. https://doi.org/10.29244/jitl.9.1.1-9

Takagi H, Tamiru M, Abe A, Yoshida K, Uemura A, Yaegashi H, Obara T, Oikawa K, Utsushi H, Kanzaki E, Mitsuoka C, Natsume S, Kosugi S, Kanzaki H, Matsumura H, Urasaki N, Kamoun S, Terauchi R (2015) MutMap accelerates breeding of a salt-tolerant rice cultivar. Nat Biotechnol 33(5):445–449. https://doi.org/10.1038/nbt.3188

Takahashi S, Monda K, Negi J, Konishi F, Ishikawa S, Hashimoto-Sugimoto M, Gotô N, Iba K (2015) Natural variation in stomatal responses to environmental changes among Arabidopsis thaliana ecotypes. PLoS One 10(2):e0117449. https://doi.org/10.1371/journal.pone.0117449

Tang W, Zhang Z, Zou X, Zheng Y (2005) Functional genomics of maize submergence tolerance and cloning of the related gene Sicyp51. Sci China Ser C Life Sci 48:337–345. https://doi.org/10.1360/062004-27

Tenorio FA, Ye C, Redoña E, Sierra S, Laza M, Argayoso MA (2013) Screening rice genetic resources for heat tolerance. SABRAO J Breed Genet 45(3):371–381

Thi Lang N, Chi Buu B (2008) Fine mapping for drought tolerance in rice (Oryza sativa L.). OmonRice 16:9–15

Thomson MJ, de Ocampo M, Egdane J, Rahman MA, Sajise AG, Adorada DL, Tumimbang-Raiz E, Blumwald E, Seraj ZI, Singh RK, Gregorio GB, Ismail AM (2010) Characterizing the Saltol quantitative trait locus for salinity tolerance in rice. Rice 3(2–3):148–160. https://doi.org/10.1007/s12284-010-9053-8

Tian L, Li J, Bi W, Zuo S, Li L, Li W, Sun L (2019) Effects of waterlogging stress at different growth stages on the photosynthetic characteristics and grain yield of spring maize (Zea mays L.) under field conditions. Agric Water Manag 218:250–258. https://doi.org/10.1016/j.agwat.2019.03.054

Tiwari TN, Dutta D, Tewari K (2022) Performance of pigeon pea [Cajanus cajan (L.) Millsp] genotypes under cold stress. Legum Res 1(Of):1–6. https://doi.org/10.18805/LR-4887

Tong C, Hill CB, Zhou GF, Zhang X-Q, Jia Y, Li CD (2021) Opportunities for improving waterlogging tolerance in cereal crops—physiological traits and genetic mechanisms. Plants 10(8):1560. https://doi.org/10.3390/plants10081560

Tong J, Walk TC, Han P, Chen L, Shen X, Li Y, Gu C, Xie L, Hu X, Liao X, Qin L (2020) Genome-wide identification and analysis of high-affinity nitrate transporter 2 (NRT2) family genes in rapeseed (Brassica napus L.) and their responses to various stresses. BMC Plant Biol 20(1):464. https://doi.org/10.1186/s12870-020-02648-1

Tong M, Liu X, Guan J, Lin Y, Zhou A, Qiao K (2023) Novel biofortification candidate: MTP1 increases microelement contents and decreases toxic heavy metal accumulation in grains. Chemosphere 318:137967. https://doi.org/10.1016/j.chemosphere.2023.137967

Upadhyay J, Joshi R, Singh B, Bohra A, Vijayan R, Bhatt M, Bisht SPS, Wani SH (2017) Application of bioinformatics in understanding of plant stress tolerance. In: Hakeem K, Malik A, Vardar-Sukan F, Ozturk M (eds) Plant bioinformatics. Springer, Cham. https://doi.org/10.1007/978-3-319-67156-7_14

Valladares F, Laanisto L, Niinemets Ü, Zavala MA (2016) Shedding light on shade: ecological perspectives of understorey plant life. Plant Ecol Divers 9(3):237–251. https://doi.org/10.1080/17550874.2016.1210262

van Ginkel M, Rajaram S, Thijssen M (1992) Waterlogging in wheat: Germplasm evaluation and methodology development. In: The Seventh Wheat Workshop for Eastern, Central and Southern Africa; September 16–19, 1991; Nakuru, Kenya, pp 115–124

Varshney RK, Barmukh R, Roorkiwal M, Qi Y, Kholova J, Tuberosa R, Reynolds MP, Tardieu F, Siddique KHM (2021a) Breeding custom-designed crops for improved drought adaptation. Adv Genet 2(3):e202100017. https://doi.org/10.1002/ggn2.202100017

Varshney RK, Bohra A, Yu J, Graner A, Zhang Q, Sorrells ME (2021b) Designing future crops: genomics-assisted breeding comes of age. Trends Plant Sci 26(6):631–649. https://doi.org/10.1016/j.tplants.2021.03.010

Varshney RK, Roorkiwal M, Sun S, Bajaj P, Chitikineni A, Thudi M, Singh NP, Du X, Upadhyaya HD, Khan AW, Wang Y, Garg V, Fan G, Cowling WA, Crossa J, Gentzbittel L, Voss-Fels KP, Valluri VK, Sinha P et al (2021c) A chickpea genetic variation map based on the sequencing of 3,366 genomes. Nature 599(7886):622–627. https://doi.org/10.1038/s41586-021-04066-1

Vartapetian BB, Jackson MB (1997) Plant adaptations to anaerobic stress. Ann Bot 79(Suppl. 1):3–20. https://doi.org/10.1093/oxfordjournals.aob.a010303

Vijayalakshmi K, Fritz AK, Paulsen GM, Bai G, Pandravada S, Gill BS (2010) Modeling and mapping QTL for senescence-related traits in winter wheat under high temperature. Mol Breed 26(2):163–175. https://doi.org/10.1007/s11032-009-9366-8

Vinh NT, Paterson AH (2005) Genome mapping and its implication for stress resistance in plants. In: Ashraf M, Harris PJC (eds) Abiotic stresses: plant resistance through breeding and molecular approaches. CRC Press, Boca Raton, p 16. https://doi.org/10.1201/9781482293609

Voesenek LACJ, Bailey-Serres J (2013) Flooding tolerance: O2 sensing and survival strategies. Curr Opin Plant Biol 16(5):647–653. https://doi.org/10.1016/j.pbi.2013.06.008

Voesenek LACJ, Bailey-Serres J (2015) Flood adaptive traits and processes: an overview. New Phytol 206(1):57–73. https://doi.org/10.1111/nph.13209

Wahid A, Gelani S, Ashraf M, Foolad MR (2007) Heat tolerance in plants: an overview. Environ Exp Bot 61(3):199–223. https://doi.org/10.1016/j.envexpbot.2007.05.011

Wang B, Li Z, Ran Q, Li P, Peng Z, Zhang J (2018) ZmNF-YB16 overexpression improves drought resistance and yield by enhancing photosynthesis and the antioxidant capacity of maize plants. Front Plant Sci 9:709. https://doi.org/10.3389/fpls.2018.00709

Wang Q-J, Yuan Y, Liao Z, Jiang Y, Wang Q, Zhang L, Gao S, Wu F, Li M, Xie W, Liu T, Xu J, Liu Y, Feng X, Lu Y (2019a) Genome-wide association study of 13 traits in maize seedlings under low phosphorus stress. Plant Genome 12(3):1–13. https://doi.org/10.3835/plantgenome2019.06.0039

Wang R, Shakir AM, Geng M, Tian J (2024a) [Preprint] (1st version)) Dissection of QTLs underlying the genetic basis of drought resistance in wheat: a meta-analysis. Available at Research Square. https://doi.org/10.21203/rs.3.rs-4435632/v1

Wang X, Cheng Y, Yang C, Yang C, Mu Y, Xia Q, Ma Q (2019b) QTL mapping for aluminum tolerance in RIL population of soybean (Glycine max L.) by RAD sequencing. PLoS One 14(10):e0223674. https://doi.org/10.1371/journal.pone.0223674

Wang Z, Lai X, Wang C, Yang H, Liu Z, Fan Z, Li J, Zhang H, Liu M, Zhang Y (2024b) Exploring the drought tolerant quantitative trait loci in spring wheat. Plants 13(6):898. https://doi.org/10.3390/plants13060898

Wang Z, Li H, Wei Z, Sun H, He Y, Gao J, Yang Z, You J (2021) Overexpression of UDP-glycosyltransferase genes enhanced aluminum tolerance through disrupting cell wall polysaccharide components in soybean. Plant Soil 469(1–2):135–147. https://doi.org/10.1007/s11104-021-05157-8

Wilhelm EP, Mullen RE, Keeling PL, Singletary GW (1999) Heat stress during grain filling in maize: effects on kernel growth and metabolism. Crop Sci 39(6):1733–1741. https://doi.org/10.2135/cropsci1999.3961733x

Willems G, Dräger DB, Courbot M, Godé C, Verbruggen N, Saumitou-Laprade P (2007) The genetic basis of zinc tolerance in the metallophyte Arabidopsis halleri ssp. halleri (Brassicaceae): an analysis of quantitative trait loci. Genetics 176(1):659–674. https://doi.org/10.1534/genetics.106.064485

Wu G, Zhao Y, Shen R, Wang B, Xie Y, Ma X, Zheng Z, Wang H (2019) Characterization of maize phytochrome-interacting factors in light signaling and photomorphogenesis. Plant Physiol 181(2):789–803. https://doi.org/10.1104/pp.19.00239

Wu Y, Yang Z, How J, Xu H, Chen L, Li K (2017) Overexpression of a peroxidase gene (AtPrx64) of Arabidopsis thaliana in tobacco improves plant's tolerance to aluminum stress. Plant Mol Biol 95(1–2):157–168. https://doi.org/10.1007/s11103-017-0644-2

Xia X, Fan X, Wei J, Feng H, Qu H, Xie D, Miller AJ, Xu G (2015) Rice nitrate transporter OsNPF2.4 functions in low-affinity acquisition and long-distance transport. J Exp Bot 66(1):317–331. https://doi.org/10.1093/jxb/eru425

Xie X, Cheng H, Hou C, Ren M (2022a) Integration of light and auxin signaling in shade plants: from mechanisms to opportunities in urban agriculture. Int J Mol Sci 23(7):3422. https://doi.org/10.3390/ijms23073422

Xie Y-F, Zhang R-X, Qin L-J, Song LL, Zhao D-G, Xia Z-M (2022b) Genome-wide identification and genetic characterization of the CaMYB family and its response to five types of heavy metal stress in hot pepper (Capsicum annuum cv. CM334). Plant Physiol Biochem 170:98–109. https://doi.org/10.1016/j.plaphy.2021.11.024

Xu C, Zhang H, Sun J, Guo Z, Zou C, Li WX, Xie C, Huang C, Xu R, Liao H, Wang J (2018) Genome-wide association study dissects yield components associated with low-phosphorus stress tolerance in maize. Theor Appl Genet 131:1699–1714. https://doi.org/10.1007/s00122-018-3108-4

Xu KN, Mackill DJ (1996) A major locus for submergence tolerance mapped on rice chromosome 9. Mol Breed 2(3):219–224. https://doi.org/10.1007/BF00564199

Xu L, Zhao CC, Pang JY, Niu YA, Liu HQ, Zhang WY, Zhou MX (2022) Genome-wide association study reveals quantitative trait loci for waterlogging–triggered adventitious roots and aerenchyma formation in common wheat. Front Plant Sci 13:1066752. https://doi.org/10.3389/fpls.2022.1066752

Xu Y, Chu C, Yao S (2021a) The impact of high-temperature stress on rice: challenges and solutions. Crop J 9(5):963–976. https://doi.org/10.1016/j.cj.2021.02.011

Xu Y, Ma K, Zhao Y, Wang X, Zhou K, Yu G, Li C, Li P, Yang Z, Xu C, Xu S (2021b) Genomic selection: a breakthrough technology in rice breeding. Crop J 9(3):669–677. https://doi.org/10.1016/j.cj.2021.03.008

Xu Z, Ye L, Shen Q, Zhang G (2024) Advances in the study of waterlogging tolerance in plants. J Integr Agric 23(9):2877–2897. https://doi.org/10.1016/j.jia.2023.12.028

Yadav AK, Arya RK, Narwal MS (2014) Screening of pearl millet F_1 hybrids for heat tolerance at early seedling stage. Adv Agric 2014:1–17, 231301. https://doi.org/10.1155/2014/231301

Yadav AK, Narwal MS, Arya RK (2011) Genetic dissection of temperature tolerance in pearl millet (Pennisetum glaucum). Indian J Agric Sci 81:203–213

Yadava DK, Dikshit HK, Mishra GP, Tripathi S (2022) Fundamentals of field crop breeding. Springer. https://doi.org/10.1007/978-981-16-9257-4

Yamauchi T, Colmer TD, Pedersen O, Nakazono M (2018) Regulation of root traits for internal aeration and tolerance to soil waterlogging-flooding stress. Plant Physiol 176(2):1118–1130. https://doi.org/10.1104/pp.17.01157

Yan C-J, Liang G-H, Gu S-L, Yi C-D, Lu J-F, Li X, Tang S-Z, Gu M-H (2003) Molecular marker analysis and genetic basis for sterility of typical indica/japonica hybrids. Acta Genet Sin 30(3):267–276. PMID: 12812093

Yan H, Sun M, Zhang Z, Jin Y, Zhang A, Lin C, Wu B, He M, Xu B, Wang J, Qin P, Mendieta JP, Nie G, Wang J, Jones CS, Feng G, Srivastava RK, Zhang X, Bombarely A et al (2023) Pangenomic analysis identifies structural variation associated with heat tolerance in pearl millet. Nat Genet 55(3):507–518. https://doi.org/10.1038/s41588-023-01302-4

Yan K, Zhao SJ, Cui MX, Han GX, Wen P (2018) Vulnerability of photosynthesis and photosystem I in Jerusalem artichoke (Helianthus tuberosus L.) exposed to waterlogging. Plant Physiol Biochem 125:239–246. https://doi.org/10.1016/j.plaphy.2018.02.017

Yang FF, Liu T, Wang QY, Du MZ, Yang TL, Liu DZ, Li S, Liu SP (2021) Rapid determination of leaf water content for monitoring waterlogging in winter wheat based on hyperspectral parameters. J Integr Agric 20(10):2613–2626. https://doi.org/10.1016/S2095-3119(20)63306-8

Yang S, Vanderbeld B, Wan J, Huang Y (2010a) Narrowing down the targets: towards successful genetic engineering of drought-tolerant crops. Mol Plant 3(3):469–490. https://doi.org/10.1093/mp/ssq016

Yang Y, Qin Y, Xie C, Zhao F, Zhao J, Liu D, Chen S, Fuglsang AT, Palmgren MG, Schumaker KS, Deng XW, Guo Y (2010b) The Arabidopsis chaperone J3 regulates the plasma membrane H+-ATPase through interaction with the PKS5 kinase. Plant Cell 22(4):1313–1332. https://doi.org/10.1105/tpc.109.069609

Yeon J, Nguyen TTP, Kim M, Sim S-C (2024) Prediction accuracy of genomic estimated breeding values for fruit traits in cultivated tomato (Solanum lycopersicum L.). BMC Plant Biol 24(1):222. https://doi.org/10.1186/s12870-024-04934-8

Yin F, Gao J, Liu M, Qin C, Zhang W, Yang A, Xia M, Zhang Z, Shen Y, Lin H, Luo C, Pan G (2014) Genome-wide analysis of water-stress-responsive microRNA expression profile in tobacco roots. Funct Integr Genomics 14(2):319–332. https://doi.org/10.1007/s10142-014-0365-4

Yu M, Mao SL, Chen GY, Liu YX, Li W, Wei YM, Liu CJ, Zheng YL (2014) QTLs for waterlogging tolerance at germination and seedling stages in population of recombinant inbred lines derived from a cross between synthetic and cultivated wheat genotypes. J Integr Agric 13(1):31–39. https://doi.org/10.1016/S2095-3119(13)60354-8

Zaidi PH, Rashid Z, Vinayan MT, Almeida GD, Phagna RK, Babu R (2015) QTL mapping of agronomic waterlogging tolerance using recombinant inbred lines derived from tropical maize (Zea mays L) germplasm. PLoS One 10(4):e0124350. https://doi.org/10.1371/journal.pone.0124350

Zhang A, Liu Y, Wang F, Li T, Chen Z, Kong D, Bi J, Zhang F, Luo X, Wang J, Tang J, Yu X, Liu G, Luo L (2019b) Enhanced rice salinity tolerance via CRISPR/Cas9-targeted mutagenesis of the OsRR22 gene. Mol Breed 39:1–10. https://doi.org/10.1007/s11032-019-0954-y

Zhang G, Wang X, Wang B, Tian Y, Li M, Nie Y, Peng Q, Wang Z (2013) Fine mapping a major QTL for kernel number per row under different phosphorus regimes in maize (Zea mays L.). Theor Appl Genet 126(6):1545–1553. https://doi.org/10.1007/s00122-013-2072-2

Zhang J-Y, Huang S-N, Wang G, Xuan J-P, Guo Z-R (2016) Overexpression of Actinidia deliciosa pyruvate decarboxylase 1 gene enhances waterlogging stress in transgenic Arabidopsis thaliana. Plant Physiol Biochem 106:244–252. https://doi.org/10.1016/j.plaphy.2016.05.009

Zhang M, Liang X, Wang L, Cao Y, Song W, Shi J, Lai J, Jiang C (2019a) A HAK family Na+ transporter confers natural variation of salt tolerance in maize. Nat Plants 5(12):1297–1308. https://doi.org/10.1038/s41477-019-0565-y

Zhang R, Yue Z, Chen X, Huang R, Zhou Y, Cao X (2023) Effects of waterlogging at different growth stages on the photosynthetic characteristics and grain yield of sorghum (Sorghum bicolor L.). Sci Rep 13(1):7212. https://doi.org/10.1038/s41598-023-32478-8

Zhang Y, Tan L, Zhu Z, Yuan L, Xie D, Sun C (2015) TOND1 confers tolerance to nitrogen deficiency in rice. Plant J 81(3):367–376. https://doi.org/10.1111/tpj.12736

Zhang Z, Jiang H, Wei Z, Zheng Y (2003) Study on enzymology in root of maize inbred after waterlogging stress. Hubei Agric Sci 3:25–27

Zhou MX, Johnson P, Zhou GF, Li CD, Lance R (2012) Quantitative trait loci for waterlogging tolerance in a barley cross of Franklin × YuYaoXiangTian Erleng and the relationship between

waterlogging and salinity tolerance. Crop Sci 52(5):2082–2088. https://doi.org/10.2135/cropsci2012.01.0008
Zhou MX, Li HB, Mendham NJ (2007) Combining ability of waterlogging tolerance in barley. Crop Sci 47(1):278–284. https://doi.org/10.2135/cropsci2006.02.0065
Zhou X, Muhammad I, Lan H, Xia C (2022) Recent advances in the analysis of cold tolerance in maize. Front Plant Sci 13:866034. https://doi.org/10.3389/fpls.2022.866034
Zhou Y, Wang Z, Gong L, Chen A, Liu N, Li S, Sun H, Yang Z, You J (2019) Functional characterization of three MATE genes in relation to aluminum-induced citrate efflux from soybean root. Plant Soil 443:121–138. https://doi.org/10.1007/s11104-019-04192-w
Zon R, Graves HS (1911) Light in relation to tree growth (Bulletin (United States. Forest Service, 92)). United States Department of Agriculture. https://doi.org/10.5962/bhl.title.66218

Genotype × Environment Interaction: Molecular Basis and Environmental Adaptation of Demand-Led Crop Varieties

8

Muhammad Rizwan, Reena Rani, Amara Rafi, and Hussein Shimelis

Abstract

Genotype by environment interaction (G × E) influences crop ideotype selection and resilience to the growing environment conditions associated with climate change. Crop performances for economic traits have polygenic inheritance and subject to G × E, necessitating a detailed understanding of phenotypic plasticity, genetic variability, and associated impact on crop adaptation and tolerance to abiotic (e.g. drought, heat, and salinity) and biotic (e.g. insect pests and diseases) stresses. Robust statistical models and emerging molecular tools are crucial to dissect the magnitude and effects of G × E and guide genetic analysis and breeding of climate-smart crop cultivars for release and commercialization. Therefore, this chapter aims to provide an overview of the molecular basis of G × E and its associations in genotype selection for environmental adaptation. The review highlights the historical perspectives and an overview of G × E analysis and its

M. Rizwan
Plant Breeding and Genetics Division, Nuclear Institute of Agriculture (NIA), Tandojam, Pakistan

Department of Plant Sciences, University of Saskatchewan, Saskatoon, Canada

R. Rani
Agricultural Biotechnology Division, National Institute for Biotechnology and Genetic Engineering (NIBGE), Constituent College Pakistan Institute of Engineering and Applied Sciences (PIEAS), Faisalabad, Pakistan

A. Rafi
Department of Plant Protection, Faulty of Agriculture, Universiti Putra Malaysia, Serdang, Selangor, Malaysia

H. Shimelis (✉)
School of Agricultural, Earth and Environmental Sciences, African Centre for Crop Improvement, University of KwaZulu-Natal, Pietermaritzburg, South Africa
e-mail: shimelish@ukzn.ac.za

P. M. Priyadarshan, R. Ortiz (eds.), *Plant Breeding 2050*,
https://doi.org/10.1007/978-981-95-0583-8_8

applications. This is followed by the molecular basis of environmental adaptation for variable agro-ecologies; breeding for environmental resilience; integrated omics selection for adaptation, genetic modification of crops for stress adaptive traits, and participatory or demand-led plant breeding for local adaptation and marketplace. Further, the paper presents the challenges and future directions in integrating multi-environment data for G × E analysis, harnessing big data, AI, and machine learning to predict G × E effects and the future of breeding environmentally resilient crops. The review enables predictive modelling of G × E to guide crop genetics and breeding and release of demand-led and adaptive crop varieties.

Keywords

Big data · Climate resilience · Data integration · G × E interaction · Genetic modifications · Integrated omics · Multi-environment

8.1 Introduction

In crop production systems, genotype by environment interaction (G × E) affects genotype performance and ranking to changing environmental factors. Genotype-by-environment interaction impacts crop performance and yield stability. The magnitude and effect of G × E directly influence crop productivity and adaptation, making it a subject of extensive analysis in the advanced stage of plant breeding programs. When elite genotypes rank differently to the same environmental conditions for economic yield, growth rate, or stress tolerance, or when the same genotype performs differently in different environments, G × E occur. For example, a genotype tolerant to drought might perform well in dryland environments but poorly in regions with optimal rainfall conditions. A crop variety may yield better in nutrient-rich soils, while another may perform relatively better in nutrient-poor environments owing to genetic variations for nutrient-use efficiency. Plant breeders examine G × E to create and recommend crop varieties with narrow or broad adaptation to ensure that genetic advancement corresponds with yield gains under varied growing conditions.

Crop production for food security and economic returns depends on creating high-performing and stable genotypes. Best selections perform consistently across variable growing environments (e.g. sites, seasons, pests and diseases, and drought conditions), which is deciphered through G × E analyses. In cultivar development, elite genotypes are matched to their ideal growing conditions. This is achieved using advanced statistical methods, genomic tools, and environmental monitoring, which increase sustainable crop production and yield gains. Phenotyping for environmental adaptation guides new variety design and recommendations and adoption by growers and value chains. There is a need for climate-smart cultivars owing to global population expansion, climate change, and limited natural resources. By 2050, there will likely be about 9 billion people on the planet, making it difficult to produce adequate food with available resources. Resilient crop varieties and food systems are essential for maintaining food supplies. The Food and Agricultural Organisation of the United Nations (FAO) predicts climate change will affect food production in almost every region, particularly in tropical and subtropical areas (FAO 2021). Crops that are bred with better environmental adaptation would yield better and provide enough food to satisfy global food demand in the face of climate change. Furthermore, crop breeding with an emphasis on nutritional quality is essential to circumvent the burden of malnutrition. Based on local needs and environmental conditions, biofortified crops are bred with balanced levels of vital nutrients, which can help fight malnutrition (Bouis and Saltzman 2017). In this case, plant breeders harness genetic variation and meticulously assess their breeding material across representative growing sites and seasons. Computing the phenotypic variance as the total of the interactions influencing the tested genotypes' (G) performance in the various environments (E) is paramount. Estimating the G, E, and G × E interactions guarantees reliable recommendations of appropriate cultivars capable of withstanding the strain of fluctuating environmental conditions associated with climate change. Further, identification of the components of G × E aids plant breeders and plant geneticists in choosing effective selection methods to develop best-performing cultivars that are niche-specific or broadly adapted.

Breeding for environmental adaptation delivers suitable ideotypes that can withstand harsh growing conditions. For instance, as temperatures rise, heat-tolerant rice, and wheat cultivars are being created to guarantee food security (Gupta et al. 2020). Crop production on marginal agro-ecologies can be harnessed, maintained, and rejuvenated by selecting adaptive crops that can perform in poor growing conditions (Lal 2015). Likewise, cultivating crop cultivars that are efficient in soil nutrient use requires less fertilisers, reduces the cost of production, and is safe for the environment (Tilman et al. 2011). Crop diversity supports microbial diversity, which improves soil health and fosters sustainability in agricultural environments (Khoury et al. 2016). Changes in insect pests and disease patterns are another effect of climate change. Breeding crops resistant to pests and diseases will help reduce crop losses and the need for crop protection chemicals by 2050, supporting sustainable agriculture (Savary et al. 2019). Crop genotypes that have co-evolved and adapted to harsh growing environments (e.g. landraces, open-pollinated varieties, crop wild relatives) are less vulnerable to climate change, which is vital for selecting and maintaining steady crop yields (Bebber et al. 2013).

With the availability of next-generation sequencing platforms, crop genomes can be precisely tracked and modified using advanced breeding approaches such as genomic selection and clustered regularly interspaced short palindromic repeats (CRISPR) gene editing. The technologies lead to gene discovery and precision breeding to harness unique traits such as nutrient profile and drought tolerance (Simsek et al. 2024). Breeding initiatives also aim to increase resource efficiency by creating fertiliser and water-use-efficient crops, which is crucial given the world's growing population and depleting natural resources (Benitez-Alfonso et al. 2023). Genetic variation can be harnessed for crop adaptability to the changing growing environments using genetic resources, such as landraces, crop wild relatives, and mutants in breeding programmes (Gao et al. 2023).

Genotype by environment interaction influences crop ideotype selection and crop resilience to the growing environment conditions due to climate change. Crop yields and economic traits have polygenic inheritance and subject to G × E, necessitating a detailed understanding of phenotypic plasticity, genetic variability, and associated impact on crop adaptation and tolerance to abiotic (e.g. drought, heat, and salinity) and biotic stresses (e.g. insect pests and pathogen). Variable genotype performance and ranking in response to environmental conditions led to the discovery of statistical models (e.g. the additive main effects and multiplicative interaction [AMMI] and genotypic main effect plus genotype-by-environment interaction [GGE] biplots) to analyse G × E effect and distinguish adaptive genotypes and productive environments of major food security crops. Further, exploring the molecular mechanisms underlying stress responses revealed genes and epigenetic modifications and facilitated selection for environmental adaptation. Advances in genomic innovations and technologies (e.g. CRISPR-Cas9) have emerged as high throughput and transformative selection tools for precision breeding, enabling the knockout, insertion, or alteration of stress-responsive genes. The new technologies are enhancing selection response and genetic enhancement of abiotic and biotic stress tolerance in crop plants. Innovative breeding strategies, such as integrated omics selection (IOS), enable the genetic modification of crops with stress-adaptive traits. The new genotypes should be field evaluated and selected using participatory plant breeding (PPB) or demand-led breeding (DLB) approaches for local adaptation, market needs, and requirements to release high-performing commercial crop varieties with higher adoption rates.

The above opportunities to improve crop resilience and adaptation need to integrate multi-environment data for robust G × E analysis based on big data, artificial intelligence (AI), and machine learning (ML). This will enable predictive modelling of G × E to guide crop breeding and release of commercial and adaptive crop varieties. Hence, this chapter aims to provide an overview of the molecular basis of G × E and its associations in genotype selection for environmental adaptation. The review highlights historical perspectives and an overview of G × E followed by the molecular basis of environmental adaptation for variable agro-ecologies; breeding for environmental resilience; integrated omics selection for adaptation, genetic modification of crops for stress adaptive traits, and participatory or demand-led plant breeding for local adaptation and marketplace. Further, the paper presents the challenges and future directions in integrating multi-environment data for G × E analysis, harnessing big data, AI, and machine learning to predict G × E and the future of breeding environmentally resilient crops. The review enables predictive modelling of G × E to guide genetic analysis, crop breeding, and release of commercial and adaptive crop varieties.

8.2 Historical Perspectives on G × E

Early findings on variable genotype performance and ranking in response to environmental conditions marked the beginning of the study of G × E interactions from the start of the twentieth century. The G × E interactions led scholars like Wilhelm Johannsen to distinguish between phenotype and genotype. The interaction of the genetic composition (genotype) and the environment influences individuals' observed features (the phenotype). Environmental conditions such as soil type, climate, and agricultural techniques could promote or suppress traits expression of desired crop features (Al-Shammary et al. 2024). Thus, the idea of G × E interactions has substantially evolved in crop production, influencing the development of conventional and modern breeding techniques. In the past, traditional breeding concentrated on choosing crop types that were appropriate for regional conditions. Breeders routinely use designed crosses and phenotypic selection of best-performing progenies to improve yield and stability across environments. However, it is difficult to determine the genetic basis of complex traits and predict their performance across wide environmental conditions without advanced statistical methods and genetic analyses.

Our understanding of the G × E interactions is harnessed with the advent of molecular genetics in the late twentieth century. Marker-assisted selection (MAS) and subsequent genome-wide association studies (GWAS) enabled pinpointing polygenes linked to genotype performance and environmental adaptability. Furthermore, genetic databases and machine learning algorithms led to the development of genomic selection (GS), enabling forecasting of how crop genotypes will adapt in various situations (Ramirez-Villegas et al. 2020). Precision breeding, in which specific genes are altered to maximize crop resilience to environmental challenges, has also been made possible by techniques like genome editing with CRISPR. Breeders have been able to modify crops for tolerance to drought, insect pests, and extreme temperatures using molecular tools, which is crucial in the face of climate change (Cooper et al. 2016). Knowledge of G × E interactions is being harnessed and improved using precision phenotyping and sophisticated computational models, enabling sustainable crop production in harsh growing conditions. The convergence of genetics, data analytics, and environmental modelling in G × E research enables the quick development of climate-smart crop cultivars, complementing conventional breeding techniques. The advancements above highlight a paradigm shift in which data on G × E interactions are adequately analysed and used to create crops appropriate for the changing climate conditions (Chand et al. 2022).

8.3 Overview of Genotype × Environment Interaction

The G × E interaction occurs when various genotypes react differently to environmental stimuli, resulting in phenotypic expression that varies depending on the environment (Rani et al. 2023). This relationship explains why some crops perform poorly in some conditions while flourishing in others, making it essential for understanding plant performance and adaptability, particularly in cropping systems. Genetic variability and phenotypic plasticity are the critical determinants of G × E interactions. Plants vary genetically, which enables breeders to make selections based on traits that thrive in particular environments. Conversely, phenotypic plasticity refers to a genotype's capacity to modify its phenotype in response to environmental shifts without changing its genetic makeup. This adaptability is essential in breeding efforts designed to increase yield stability across a range of climates. High plasticity, however, can also be problematic because the same genotype may produce different results in different situations (Schneider and Lynch 2020).

Statistical models aid in G × E interaction analyses and make targeted genotype selection. The GGE biplots integrate both G and G × E effects to visualize genotype performance across environments, while the AMMI model distinguishes additive genetic and environmental impact. The two methods are the most commonly used in the G × E interaction analysis in breeding programmes. Breeders use these methods to find the "ideal" genotypes that balance yield stability and in varied growing environments (Arif et al. 2021). Apart from AMMI and GGE biplot, other methods including linear mixed models (LMMs), factorial designs, partial least squares (PLS) regression, and non-linear models are used to examine the G × E interaction effects. The LMMs account for random and fixed effects such as genotypes and environmental variability. LMMs can incorporate several sources of variation, such as replicates, blocks, and environmental random effects, making them versatile tools for analysing G × E data. Also, factorial designs and PLS regression are suitable to investigate the G × E interaction. The link between genotypes, environmental factors, and traits of interest can be examined using PLS regression, which minimizes the dimensionality of the data while maintaining the greatest amount of variability.

The complexities of G × E interaction analyses are captured using linear and non-linear models. Numerous traits relationship occurs, such as quadratic or exponential patterns in response to environmental changes and can be integrated by these models. Certain features that show non-linear reactions to environmental stimuli benefit from it. These statistical models offer strong frameworks for examining G × E interactions, enabling breeders and agronomists to make well-informed choices on crop management and breeding strategies. Crop performance under various climatic situations can be enhanced by comprehending and appropriately using these techniques.

G × E interactions manifest plant development, showing how various genotypes react to different environmental situations. For instance, in wheat production, G × E interaction occurs due to particular adaptations. Wheat cultivars selected for high yields in one environment frequently exhibit lower yields in other environments (Williams et al. 2008). Also, maize genotypes bred for temperate agro-ecologies are less productive in tropical regions because of variations in temperature needs and day length sensitivity (Giauffret et al. 2000). The G × E interaction may also be seen in rice breeding in Asia, where cultivars chosen for areas that flood frequently acquire characteristics like longer stems that enable them to withstand submersion. However, in drier regions, these same characteristics are detrimental (Liang et al.

2015). Furthermore, breeding programmes for millet and sorghum in sub-Saharan Africa frequently focus on particular genotypes because of their ability to withstand low-fertility soils, even though these genotypes may perform poorly in areas with high fertility (Wanga et al. 2022; Yahaya et al. 2023).

8.4 Molecular Basis of Environmental Adaptation

8.4.1 Molecular Plant Stress Responses

Numerous environmental stressors affect crop plants. Stresses can be broadly divided into two categories: biotic stresses (e.g. insects, diseases, and herbivores) and abiotic stresses (e.g. drought, salinity, and temperature extremes). Plants have developed sophisticated adaptive mechanisms at the molecular level to cope with stresses, mainly through several signalling channels and gene regulatory networks (Fig. 8.1).

Drought stress triggers several adaptation processes in plants, mainly via the abscisic acid (ABA) pathway. For the plant to save water, this pathway causes the expression of the *DREB* and *RD29A* genes, which are essential for stomatal closure

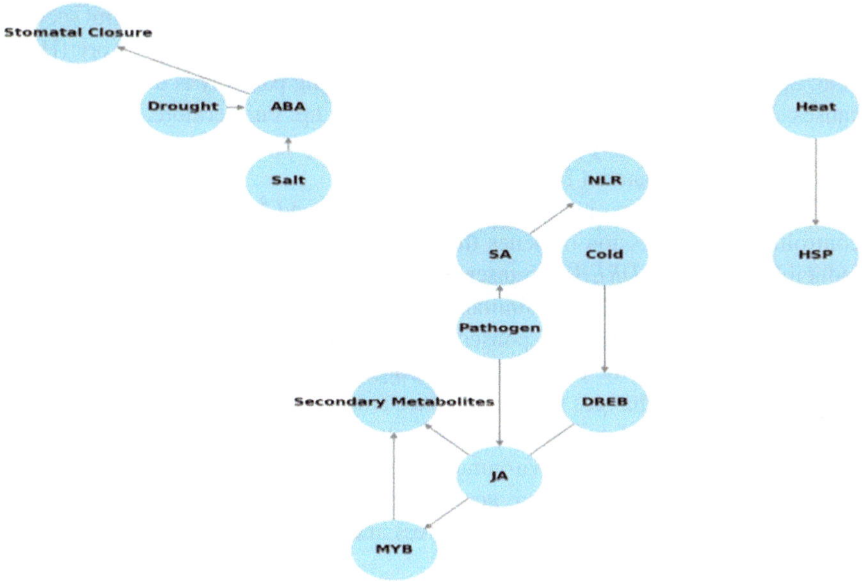

Fig. 8.1 Molecular signalling pathways in plants under abiotic (e.g. drought, salt, heat, cold) and biotic (e.g. insect pests, pathogen) stresses. Key molecules such as DREB, NLR, MYB, JA, SA, ABA, and secondary metabolites regulate stress response. The diagram illustrates how stress factors trigger hormone signalling and gene expression pathways, leading to specific adaptive responses such as stomatal closure, heat shock protein activation, and secondary metabolite production for defence

and osmotic adjustment (Yoshida et al. 2014). Plants become more tolerant of water deficits due to the subsequent upregulation of stress-responsive gene expression (Gupta et al. 2020). Likewise, ionic and osmotic stress brought on by salinity causes cellular damage and growth retardation. To prevent ion toxicity in their aerial portions, plants use sodium ion transporters such as *HKT1* and *SOS1* to maintain Na+ and K+ homeostasis (Zhu 2016). These transporters are essential for maintaining osmotic equilibrium in saline environments and eliminating excess sodium from plant cells (Yang and Guo 2018).

In plants, extreme temperatures can weaken cellular stability. Heat shock proteins (HSPs), which stabilize proteins and stop aggregation, are produced when heat stress occurs and help to preserve cellular integrity (Ohama et al. 2017). On the other hand, cold stress triggers the activation of the *CBF* pathway, which results in the production of genes that are sensitive to cold, including *COR15A*. By stabilizing cellular membranes and reducing damage from freezing temperatures, this increases the plant's tolerance (Ding et al. 2019).

Plants have evolved complex defence systems against pathogens that fall into two main categories: effector-triggered immunity (ETI) and Pathogen-associated molecular patterns (PAMPs)-triggered immunity (PTI). PAMPs are recognized by plant receptors in PTI, whereas pathogen effectors are detected by particular resistance (R) genes in ETI (Cui et al. 2015). For example, by identifying unique pathogen signals, the rice *Xa21* gene provides resistance against bacterial blight (Peng et al. 2015). Plants use the Jasmonic acid (JA) pathway to produce harmful secondary metabolites, such as terpenoids and *alkaloids*, in response to herbivory (Howe et al. 2018). When herbivores attack, the JA signalling pathway is triggered, which causes substances that discourage feeding to build up. Furthermore, as an indirect defence tactic, plants may emit volatile organic compounds (VOCs) that draw herbivores' natural predators (Turlings and Erb 2018).

Transcription factors, including DREB, WRKY, and NAC play a major role in controlling the expression of genes that respond to stress. By triggering drought-inducible genes, the *DREB2A* transcription factor is critical in improving drought tolerance (Manna et al. 2021). Similarly, during pathogen attacks, *WRKY70* is vital for regulating plant immune responses (Li et al. 2006). By modifying gene expression without affecting the DNA sequence, epigenetic processes such as DNA methylation and histone changes further influence stress responses. Certain genes can be activated or repressed by stress-induced changes in DNA methylation, which enables plants to adjust to environmental stressors (Varotto et al. 2020). Stress-responsive genes are made more accessible by histone changes like acetylation, which makes it easier for them to express themselves while under stress (Sun et al. 2022).

The coordination of plant responses to stress depends on hormonal signalling. Plant hormones such as ABA, JA, and SA play important roles in these processes. Primarily, ABA controls stomatal closure and stimulates the synthesis of osmoprotectants to regulate responses to salinity and drought (Gupta et al. 2020). On the other hand, through pathways that control the production of defence genes, JA and SA are essential for defence mechanisms against herbivores and infections, respectively (Verma et al. 2016). Plants successfully navigate the complexities of their surroundings through these varied chemical reactions, guaranteeing survival and adaptability under various stresses.

8.4.2 Genes and Pathways in Environmental Adaptation

Finding genomic areas linked to biotic and abiotic stress tolerance has been made possible by Quantitative Trait Loci (QTL) mapping. By regulating the root growth angle, Deeper Rooting 1 (*DRO1*) in rice is a reportedly crucial gene that helps plants withstand drought and reach deeper water sources (Uga et al. 2013). QTLs for salinity tolerance in maize have identified essential ionic regulatory genes, including the *HKT1* sodium transporter, necessary to preserve ion homeostasis in saline environments (Xie et al. 2019). Genome-wide association studies (GWAS) have also provided insights into the genetic underpinnings of stress tolerance by associating phenotypic features with specific genetic markers. Significant SNPs linked to heat tolerance were found in wheat, specifically in genes related to heat shock proteins and the detoxification of reactive oxygen species (ROS) (Yadav et al. 2022). SNPs linked to drought tolerance were discovered in rice genes that encode aquaporins, which control water movement through plant tissues (Shivaraj et al. 2021). The following are a few case studies:

Dehydration-responsive element-binding (DREB) transcription factors are essential for cold and drought tolerance, controlling the production of stress-responsive genes. By activating essential genes, including *RD29A* and *COR15A*, overexpression of *DREB1A* in rice and *Arabidopsis* improves tolerance to water deficiencies (Muthurajan et al. 2021). According to Peer et al. (2024), recent developments in CRISPR-Cas9 technology have made it possible to precisely modify the DREB genes in maize, increasing drought tolerance without compromising growth.

Nucleotide-Binding Leucine-Rich Repeat (NLR) proteins and Resistance (R) genes are essential for identifying and protecting against pathogen effectors. For instance, by identifying the fungal effector *Avr-Pita* and triggering immunological responses, the *Pi-ta* gene in rice provides resistance to rice blast disease (Table 8.1) (Bryan et al. 2000). Similarly, by identifying PAMPs, the *Xa21* gene protects rice from bacterial blight (Peng et al. 2015).

To protect themselves from herbivores, plants produce secondary metabolites such as phenolics, terpenoids, and *alkaloids*. The JA signalling pathway, which is activated by herbivore attack, frequently regulates these chemicals. For example, the buildup of nicotine in *Nicotiana* species offers a strong defence against insect herbivores (Steppuhn et al. 2004). Furthermore, herbivore predators are drawn to plants' volatile organic compounds (VOCs), which serve as an indirect defence mechanism (Turlings and Erb 2018).

Table 8.1 Key genes, pathways, mechanisms, and outcomes in plant stress responses

Stress type	Gene(s)	Pathway/ mechanism	Function/role	Outcome	References
Drought	*DREB1A*, *RD29A*, *COR15A*	ABA-dependent signalling, osmoprotection	Regulates osmotic adjustments and stress-responsive gene expression	Enhanced drought tolerance, reduced water loss	Muthurajan et al. (2021), Peer et al. (2024)
Salinity	*HKT1*, *SOS1*	Sodium ion transport, ion homeostasis	Regulates Na+/K+ balance to prevent ion toxicity	Improved salinity tolerance, ion homeostasis	Xie et al. (2019), Zhu (2016)
Cold stress	*CBF/DREB1*, *COR15A*	Cold acclimation, membrane stabilization	Upregulation of cold-responsive genes, ROS scavenging	Increased cold tolerance, reduced freezing damage	Ding et al. (2019), Muthurajan et al. (2021)
Heat stress	*HSP70*, *HSP101*	Heat shock response, protein stabilization	Prevents protein denaturation during heat stress	Improved heat tolerance, enhanced protein stability	Yadav et al. (2022), Ohama et al. (2017)
Nutrient deficiency	*PHO1*, *NRT1*	Phosphate and nitrate transporter regulation	Regulates nutrient uptake and metabolism	Improved growth under nutrient-limited conditions	Shivaraj et al. (2021), Uga et al. (2013)
Pathogen attack	*Pi-ta*, *Xa21*, *RPS2*	ETI, PTI, PAMP recognition	Activates immune response against fungal and bacterial pathogens	Increased resistance to pathogens, reduced disease incidence	Peng et al. (2015), Bryan et al. (2000)
Herbivory	*MYC2*, *JAZ*, *COI1*	Jasmonic acid signalling, secondary metabolite production	Induces synthesis of toxic compounds to deter herbivores	Reduced herbivore damage, attraction of natural predators	Turlings and Erb (2018), Steppuhn et al. (2004)
Heat and drought	*HSFA1*, *DREB2A*	Crosstalk between heat and drought signalling pathways	Regulates heat shock proteins and osmotic stress genes	Combined heat and drought tolerance	Muthurajan et al. (2021), Peer et al. (2024)
Combined stress	*WRKY70*, *NPR1*	Hormonal crosstalk (SA and JA antagonism)	Balances defence responses between herbivores and pathogens	Enhanced adaptability to biotic and abiotic stresses	Robert-Seilaniantz et al. (2011), Turlings and Erb (2018)

8.4.3 Epigenetics and Plant Adaptation

The control of plant gene expression in response to environmental stress is dependent mainly on epigenetic changes, including DNA methylation and histone alterations. Plants can adapt to biotic and abiotic stressors by using these changes to modify transcriptional responses dynamically. Epigenetic changes offer a versatile response mechanism that can be quickly modified in response to shifting stressors by controlling gene expression without changing the underlying DNA sequence. The addition of methyl groups to cytosine residues, known as DNA methylation, alters gene expression and chromatin accessibility. Under stress, plants can suppress or activate particular genes due to dynamic changes in DNA methylation patterns. At loci linked to stress tolerance, methylation patterns frequently alter in response to abiotic stressors such as salt or drought. These modifications enable plants to repress or activate relevant pathways to maximize ion balance, osmoprotectants, or water retention. For instance, during drought, genes involved in the manufacture of osmoprotectants, like those that encode *proline* and *glycine betaine*, are usually hypomethylated, which causes their expression (Abdulraheem et al. 2024). Because of this dynamic methylation, the plant can quickly adapt its physiology to meet environmental demands. Similar epigenetic modifications take place in response to biotic stress.

Defence-related genes are frequently demethylated in plants exposed to pathogens, enabling quick activation. As demonstrated in *Arabidopsis* following *Pseudomonas syringae* infection, this demethylation pattern at immune response loci enables a more robust and rapid response to pathogen invasion (Dowen et al. 2012). Depending on the type and degree of stress, plants may effectively balance defence and development because of the reversible nature of DNA methylation. Histone modifications work in tandem with DNA methylation by changing chromatin structure through post-translational modifications like acetylation, methylation, and ubiquitination. Because it lessens chromatin compaction and increases DNA accessibility for transcription, histone acetylation is linked to transcriptional activation. Histone acetylation is commonly seen at the locus of stress-responsive genes during abiotic stressors such as drought, which increases the expression of these genes and improves stress tolerance (Kim et al. 2010). However, depending on the location and extent of alteration, histone methylation can function as an activating or repressive signal. Histone H3 trimethylation at lysine 4 (H3K4me3), for instance, is generally linked to gene activation, whereas methylation at H3K27 (H3K27me3) is linked to gene repression, controlling the plant's defence and growth balance in stressful situations. Furthermore, plants' stress memory is influenced by epigenetic changes. By retaining molecular "memories" of previous stress exposures, this phenomenon, often referred to as epigenetic priming, allows plants to respond more effectively and quickly to similar stressors in the future. Epigenetic markers, such as particular histone modifications or methylation patterns, expedite the activation of stress-responsive genes and enable such memory effects. In *Arabidopsis*, for example, frequent exposure to drought stress causes the buildup of epigenetic markers, which facilitates the quicker reactivation of important drought-responsive genes during subsequent drought events (Ding et al. 2012). These epigenetic markers can occasionally be inherited, leading to transgenerational epigenetic inheritance, in which offspring can inherit stress-induced epigenetic modifications. Research on rice has shown that heat stress causes DNA methylation patterns to change in ways that are inheritable, improving the following generation's ability to withstand heat (Rehman et al. 2023). By combining these epigenetic processes, plants can precisely control how they react to changes in their surroundings, ensuring both short-term stress tolerance and long-term adaptability. This regulatory flexibility is crucial for plants to survive and thrive in constantly changing environments.

8.4.4 Advances in Genomic Technologies for Adaptative Crop Improvement

Enhancing abiotic and biotic stress tolerance in crops is made easier with CRISPR-Cas9, a revolutionary technique for precisely manipulating plant genomes. CRISPR-Cas9 offers a versatile platform for enhancing crop resilience by allowing the knockout, insertion, or modification of stress-responsive genes by targeting specific genes. Editing the *DREB2A* gene in wheat is a noteworthy technique that has enhanced drought tolerance by triggering osmotic regulatory pathways without affecting total plant development (Peer et al. 2024).

The ability of CRISPR-Cas9 to promote resistance to biotic stress has been established through genome editing of susceptibility genes, including *OsSWEET14* in rice. The bacterial pathogen *Xanthomonas oryzae* uses the gene *OsSWEET14* to induce infection. Targeted gene deletion boosted resistance to bacterial blight, a notorious disease affecting rice productivity (Zeng et al. 2020). This demonstrates the adaptability of genome editing in creating crop improvement by providing tolerance to biotic and abiotic challenges. Furthermore, a successful tactic for broad-spectrum stress tolerance has been modifying transcription factors, such as those belonging to the *WRKY* and *NAC* families. These transcription factors regulate complex networks of downstream genes involved in various stress responses. For example, it has been demonstrated that *WRKY70* editing increases resistance to drought and pathogens, highlighting the value of transcription factor regulation for enhancing stress resilience (Liu et al. 2017).

However, there are drawbacks to CRISPR-Cas9's effectiveness, including off-target effects and the regulatory obstacles related to genetically modified crops. More accurate genome modification is now possible without creating double-strand breaks because of recent developments such as base editing and prime editing, which reduce the number of unwanted mutations (Chen et al. 2019). Furthermore, there is potential for increasing the effectiveness and safety of CRISPR applications in plant breeding by developing more effective delivery methods, such as ribonucleoprotein complexes and nanoparticle-mediated delivery.

8.5 Adaptation to Agro-ecologies

8.5.1 Crops in Extreme Environments

Extreme climatic conditions experienced in arid and semi-arid conditions, high and low altitudes, and coastal areas present challenges to crop production. Drought-tolerant crops are essential in desert environments. Plants respond by limiting water loss through stomatal control and creating osmoprotectants such as *proline* and *trehalose*, which assist in preserving cellular turgor under water-limited situations (Mishra et al. 2016). Crops at high elevations must contend with cold temperatures and intense UV exposure. Quinoa and other plants have evolved by activating cold-responsive pathways, such as the *CBF* transcription factors, which boost freezing resistance (Bazile et al. 2016). Seawater intrusion causes crops in coastal areas to experience salinity stress. Mechanisms include ion exclusion and the compartmentalization of harmful ions like Na^+ in vacuoles, which are facilitated by transporters like *HKT1* and *NHX*, which provide salinity tolerance (Zhu 2016). In coastal regions, crops like rice have adapted to be salt-tolerant cultivars created by conventional breeding methods and molecular techniques (Gupta et al. 2020). Crops under harsh growing conditions also face biotic pressures, such as insect pests and diseases, that are particular to these areas. Aphids, for example, are pests that flourish in arid environments and are frequently encountered by desert plants. Crops like sorghum give protection from these herbivores by producing secondary metabolites like terpenoids (War et al. 2012). To guarantee sustainable harvests, resistance breeding techniques are necessary for crops susceptible to diseases like rice blast in coastal regions (Hariprasad et al. 2014). Details of crucial genes, pathways, and their function in the environmental adaptation of crop plants in different agro-ecologies are summarized in Table 8.2.

8.5.2 Impact of Climate Change on G × E

Crop growth and productivity are significantly impacted by climate change, which exacerbates abiotic stresses such as heat and drought and new pests and diseases. Crop yields have dwindled, pest dynamics have changed, and water stress is worsened by rising temperatures and altered precipitation patterns. For example, changing climates have caused insects like the corn borer to spread geographically, impacting crops in new areas (Zhou et al. 2018). Furthermore, interactions between plants and pathogens may be affected by increased CO_2. Higher CO_2 concentrations have shown to weaken plant defences, making plants more vulnerable to infections (Lake and Wade 2009). More robust crop varieties are required to withstand the combined effects of these environmental changes due to the dual impact of climate change through biotic and abiotic stresses. The intricate G × E interactions become even more noticeable when biotic and abiotic stressors are co-mingled. For instance, pathogen infections and drought frequently coexist in areas impacted by climate change. Drought stress can weaken plant immunity, increasing crops vulnerability

Table 8.2 Summary of genes and pathways involved in environmental adaptation in different agro-ecologies

Trait/agro-ecology	Gene(s)/Microbe(s)	Pathway/Mechanism	Function/Role	Outcome	References
Drought tolerance (Desert)	*DREB*, *RD29A*, *Proline*, *Trehalose*	ABA-dependent signalling, osmoprotection	Regulates stomatal closure, osmotic adjustment, *proline* synthesis	Increased drought tolerance, reduced water loss	Mishra et al. (2016), Gupta et al. (2020)
Cold tolerance (High altitude)	*CBF/DREB1*, *COR15A*	Cold acclimation, membrane stabilization	Activation of cold-responsive genes to prevent freezing damage	Enhanced cold tolerance, better survival under low temperatures	Bazile et al. (2016), Ding et al. (2019)
Salinity tolerance (Coastal)	*HKT1*, *NHX1*	Sodium ion transport, ion compartmentaliz-ation	Regulates Na+/K+ homeostasis to avoid ion toxicity	Improved salinity tolerance, better growth in saline soils	Zhu (2016), Xie et al. (2019)
Pest resistance (Desert)	*TPS* (terpenoid synthesis), *Sorghum*	Terpenoid biosynthesis, secondary metabolite production	Production of defensive metabolites to deter pests like aphids	Reduced pest damage, increased crop resilience in desert environments	War et al. (2012)
Pest resistance (Coastal)	*Pi-ta*, *Xa21*	PTI and ETI immune response, PAMP recognition	Recognizes pathogen signals and triggers immune responses	Increased resistance to pathogens like rice blast in coastal regions	Peng et al. (2015), Bryan et al. (2000)
Impact of climate Change on G × E	*R-gene-mediated resistance*, *HSP70*	Hormonal signalling, reactive oxygen species (ROS) detoxification	Mediates defence against pathogen and heat stress combination	Improved resilience to climate-induced biotic and abiotic stresses	Zhou et al. (2018)
G × E in combined stress	*DREB2A*, *R-gene*	Crosstalk between abiotic and biotic stress signalling	Mediates resistance to both drought and pathogen stress simultaneously	Enhanced combined resistance to abiotic and biotic stresses	Pandey et al. (2017), Voelckel et al. (2017)
Drought + Microbial interactions	*ACC Deaminase-producing rhizobacteria*	Ethylene regulation via ACC deaminase production	Lowers ethylene levels to reduce stress symptoms during drought	Increased drought tolerance, enhanced plant growth under water stress	Chieb and Gachomo (2023), Mohanram and Kumar (2019)

Mycorrhizal symbiosis (Drought)	*Arbuscular mycorrhizal fungi (AMF)*	Symbiotic relationship that enhances water uptake	Expands root surface area, facilitates better water absorption	Improved drought tolerance through increased water and nutrient acquisition	Tang et al. (2022)
Genotype-specific root traits	*Mycorrhizae*, *Rhizobia*	Root exudates, symbiotic relationships	Genotype-specific root exudates enhance beneficial microbe colonization	Improved drought and disease tolerance through enhanced microbe interactions	Etesami and Beattie (2017)
Microbial defence against pathogens	*Pseudomonas fluorescens*, *Bacillus* spp.	Induced systemic resistance (ISR), biocontrol	Triggers plant immune responses, protects against soil-borne pathogens	Reduced disease incidence, increased resistance to fungal pathogens	Mohanram and Kumar (2019), Hariprasad et al. (2014)

to disease attacks (Pandey et al. 2017). The necessity of breeding drought-tolerant crops resistant to multiple stresses is highlighted by the pressure of drought and insect outbreaks co-occurring, as in the case of maize (Voelckel et al. 2017).

8.5.3 Soil–Plant–Microbe Interactions

Enhancing plant resistance to biotic and abiotic stressors largely depends on soil microbes. Beneficial microbes that improve water intake and nutrient acquisition, including rhizobacteria and mycorrhizal fungi, increase drought tolerance. Mycorrhizal symbiosis, for example, increases the surface area of roots, enabling plants to obtain water from deeper soil layers (Tang et al. 2022). Rhizobacteria, especially those that generate ACC deaminase, aid plants in surviving drought by reducing stress-related ethylene levels (Table 8.2) (Chieb and Gachomo 2023). By creating systemic resistance in plants, these microorganisms also offer protection against biotic stress. For instance, rhizobacteria like *Pseudomonas fluorescens* can stimulate plant immune responses, shielding them against diseases like *Fusarium* (Mohanram and Kumar 2019). Root characteristics, which in turn impact interactions with advantageous soil bacteria, are strongly influenced by the genotype. For instance, genotype-specific root exudates can draw in helpful microorganisms like *rhizobia* and *mycorrhizae*, which can support plant development and stress tolerance (Etesami and Beattie 2017). Crops with more branching or deeper root systems are better able to develop symbiotic partnerships with these microorganisms, increasing their resistance to diseases and drought. Additionally, root exudates alter the rhizosphere microbiome's composition, improving protection against soil-borne illnesses (Lebeis et al. 2015).

8.6 Breeding for Environmental Resilience

8.6.1 Integrated Omics Selection for Adaptation

Integrating omics technologies such as transcriptomics, metabolomics, and proteomics has provided comprehensive insights into the molecular mechanisms underpinning plant adaptation to environmental stresses. These technologies facilitate the identification of stress-responsive genes, metabolites, and proteins, contributing to developing and deploying crops with enhanced resilience. Transcriptomics provides a global overview of gene expression profiles responding to stress conditions, revealing critical regulatory networks involved in stress adaptation. RNA sequencing (RNA-seq) has enabled the identification of drought-induced genes in crops such as rice, where transcripts associated with osmoprotectant synthesis, such as *proline* and *glycine betaine*, were upregulated during water deficit (Baldoni 2022). In addition, transcriptomic studies of crops subjected to heat stress have highlighted the induction of HSPs and chaperones, which play essential roles in maintaining protein homeostasis under elevated temperatures (Voelckel et al. 2017).

In biotic stress, transcriptomic analyses have elucidated the activation of defence-related pathways, including the SA and JA signalling cascades, which mediate resistance to pathogens and herbivores. For instance, a transcriptomic study of tomato plants challenged with *Pseudomonas syringae* revealed upregulation of SA-responsive genes, validating the importance of SA in bacterial resistance (Mishra et al. 2024).

Metabolomics complements transcriptomic data by quantifying the metabolites produced during stress responses. Metabolites such as osmolytes (*proline*, *trehalose*), antioxidants (*ascorbate*, *glutathione*), and secondary metabolites (*flavonoids*, *alkaloids*) are crucial for mitigating the harmful effects of stress. Metabolomic profiling of drought-tolerant maize varieties, for instance, has revealed the accumulation of osmoprotectants and antioxidants that enhance the plant's ability to withstand water deficit (Guo et al. 2018). In response to biotic stress, metabolomic analyses have identified a variety of defence-related secondary metabolites, including phenolic compounds, terpenoids, and *alkaloids*, which serve as chemical defences against herbivores and pathogens (Saito and Matsuda 2020).

Proteomics offers a complementary approach by enabling the large-scale identification and quantification of proteins involved in stress responses. Proteomic studies of plants subjected to abiotic stress, such as drought and salinity, have identified key proteins involved in reactive oxygen species detoxification, including superoxide dismutase and catalase (Kosova et al. 2011). These proteins are pivotal in mitigating oxidative stress, a common consequence of environmental challenges. Proteomics has also been instrumental in elucidating plant–pathogen interactions. In rice, proteomic analysis of plants infected with the fungal pathogen *Magnaporthe oryzae* revealed the accumulation of pathogenesis-related proteins and enzymes involved in phytoalexin biosynthesis, which are critical for limiting pathogen proliferation (Meng et al. 2019).

The integration of transcriptomics, metabolomics, and proteomics offers a holistic approach to understanding plant adaptation to environmental stresses. Combining these datasets allows for the identification of gene–metabolite–protein networks that mediate stress responses. For example, integrating transcriptomic and proteomic data can reveal the post-transcriptional regulation of stress-responsive genes, while metabolomic data can link specific metabolites to these regulatory networks (Jamil et al. 2020). Such multi-omics approaches provide a more comprehensive understanding of plant adaptation and hold the potential to identify novel targets for crop improvement. By leveraging the synergistic power of these technologies, researchers can unravel the complexity of stress responses and develop crops that are more resilient to the challenges posed by climate change and other environmental stresses.

8.6.2 Genetic Modification of Crops for Stress-Adaptive Traits

Climate change presents a significant challenge for agricultural production, highlighting the urgent need to improve climate resilience. Genetic modification, combined with other sustainable crop production technologies, can help alleviate the impacts of climate change on agriculture (Akhtar et al. 2021). Enhancing genetic diversity in crops is also a practical approach to building resilience (Garland and Curry 2022). Innovations in plant biotechnology, including genetically modified crops, have played a crucial role in increasing the genetic diversity of crops and their capacity to adapt to evolving climatic conditions, thereby protecting agricultural output in a changing climate (Fig. 8.2).

Transgenic crops, such as canola with altered oil composition, Bt potato, Bt maize, Bt cotton, bromoxynil herbicide-resistant cotton, and glyphosate-resistant soybeans, have been approved for commercial use. These commercialized transgenic crops primarily utilize microbial genes and/or genetic elements (Kumar et al. 2018). As of now, a total of 614 transgenic events across 32 crops have been commercialized (ISAAA 2024). Among these, maize has the highest number of events (290), followed by cotton (72), potato (52), soybean (50), Argentine canola (45), carnation (19), and others.

8.6.3 Participatory and Demand-led Plant Breeding for Local Adaptation

Plant breeders, farmers, value chain actors, and other stakeholders collaborate to develop new or improved crop varieties, a practice known as participatory or demand-led plant breeding (Shelton and Tracy 2016; Shimelis 2017). This collaborative and innovative effort enables sharing knowledge, making decisions, and pooling resources in breeding initiatives to design and release market-preferred varieties. Figure 8.3 illustrates the impact pathways of both conventional and participatory plant breeding, highlighting how impacts evolve as participation increases in the early stages of the breeding process. Participatory plant breeding methodologies are particularly effective in enhancing the adaptability of crops in marginal and heterogeneous environments (Ceccarelli and Grando 2019).

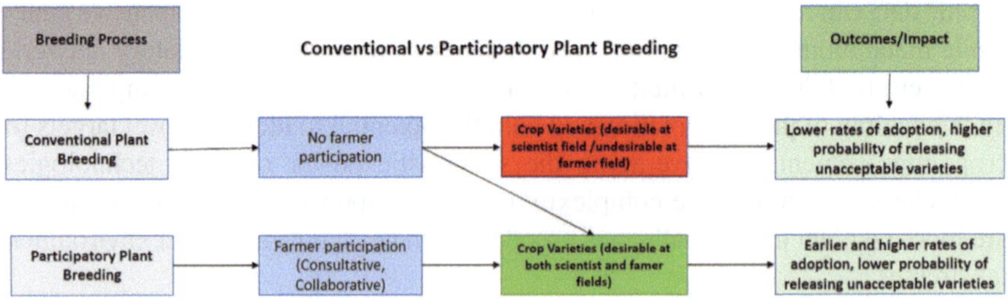

Fig. 8.2 Diagrammatic representation of genetic engineering approaches developing commercialized GM crops with enhanced resistance to insects, diseases, and abiotic stresses—cry (variants) crystal proteins, vip3a vegetative insecticidal protein, pin II proteinase inhibitor II, cpt I and II Cowpea trypsin inhibitor, pvy coat protein of potato virus Y, prsv coat protein of papaya ring spot virus, cmv coat protein of cucumber mosaic virus, zymv coat protein of zucchini yellow mosaic virus, wmv coat protein of watermelon mosaic virus, ppv coat protein of plum pox virus, Ac1 encoding viral replication protein (Rep) from bean golden mosaic virus

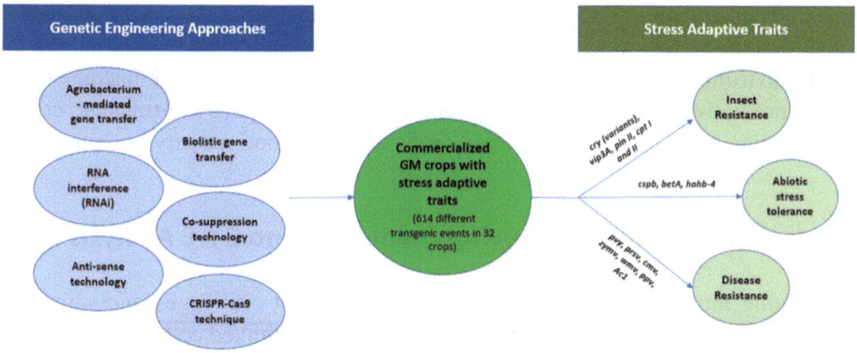

Fig. 8.3 The impact pathway of conventional and participatory or demand-led plant breeding illustrates how impacts change as participation occurs at earlier stages in the breeding process

Researchers consistently identify key motivations for initiating participatory plant breeding (PPB) or demand-led plant breeding (DLB), which include prioritizing traits for the marketplace, ensuring regional adaptation, focusing on selected breeding methods, promoting seed sovereignty, meeting culinary preferences of organic markets, diversifying underserved crops, and preserving or enhancing agrobiodiversity. Several projects have reported using participatory plant breeding to adapt grain crops to local environmental conditions, particularly due to the significant G × E interactions (Colley et al. 2021). Regional adaptation also drives breeding efforts for crops newly introduced to specific areas, such as quinoa in the United States. Additionally, sprouting broccoli—an overwintering variety with specific environmental requirements for success—has recently been introduced in northwestern France and the U.S. Pacific Northwest as part of a diversification strategy for local markets.

The importance of participatory plant breeding lies in its ability to enhance food security and resilience in agricultural systems, particularly in the face of climate change. By involving local farmers in the breeding process, this approach ensures that crop varieties are better suited to specific environmental conditions and cultural preferences, ultimately leading to increased agricultural diversity and sustainability. Furthermore, it empowers communities by fostering local knowledge and practices, making them active contributors to their own food systems.

8.7 Challenges and Future Directions

Integrating multi-environment data is key for G × E analysis and guiding genotype selection decisions. Integrative data will enable an understanding of how different varieties perform across diverse environmental conditions. By harnessing big data, artificial intelligence, and machine learning, plant breeders can develop predictive models of G × E that can modernize the breeding process and improve the accuracy of selecting climate-change-resilient crop varieties. Looking towards the future, these advancements will revolutionize agricultural practices and support sustainable food systems. However, addressing the challenges associated with integrating multi-environment data, leveraging the potential of big data, AI, and machine learning in predictive modelling of G × E, and proposing effective solutions are crucial for breeding environmentally resilient crops to ensure food security by 2050.

8.7.1 Integrating Multi-environment Data and G × E Analysis

Integrating multi-environment data presents significant challenges, including heterogeneity, scale, complexity, temporal variability, and limited collaboration. G × E studies often draw from diverse environments with varying climatic conditions, soil types, and agricultural practices, which can lead to discrepancies in measurement techniques and data quality. The vast amount of data generated from large-scale trials necessitates sophisticated statistical models that accurately capture complex interactions without oversimplifying biological processes. Additionally, changing environmental conditions requires capturing temporal variability, complicating analysis, and necessitating longitudinal data collection. Poor data sharing among research groups often hinders effective collaboration, impeding the development of comprehensive G × E models. Standardizing data collection and analysis protocols can improve comparability while encouraging open data initiatives, fostering collaboration, and allowing for more robust G × E analyses. Advanced statistical methods, such as Bayesian hierarchical models (Gunundu et al. 2023), can further elucidate the complexities of G × E interactions, providing deeper insights into genotype performance across environments.

8.7.2 Harnessing Big Data, AI, and Machine Learning for Predictive Modelling of G × E

Harnessing big data, AI, and machine learning for predictive modelling of G × E poses challenges such as data overload, model interpretability, resource intensity, and integrating diverse data sources. While big data offers opportunities, extracting meaningful patterns requires advanced computational tools and expertise. Machine learning algorithms, though powerful, often make it difficult for researchers to interpret results and apply them in breeding programs. Developing AI-driven models can also be resource-intensive, with significant computational power frequently inaccessible to many institutions. Integrating genetic, phenotypic, and environmental data from various sources also presents challenges due to differing structures and scales. Solutions like explainable AI can enhance model interpretability, while cloud-based platforms facilitate resource sharing and collaboration. Furthermore, developing dynamic models that adapt to new data can improve predictive capabilities, enabling real-time updates as conditions or genotypes change.

8.7.3 The Future of Breeding Environmentally Resilient and Demand-led Crop Cultivars for Food Security

Breeding environmentally resilient crops has multiple challenges, including climate change, biodiversity loss, regulatory barriers, and limited resources. Rapid climate shifts and changes in market demand may render current beneficial traits irrelevant, while diminishing genetic diversity restricts breeding options. Innovations like genetic modification often encounter regulatory hurdles and social acceptance. There is a need to enhance genetic diversity, adopt participatory breeding, utilize multi-omics strategies, and foster global collaboration. Conserving diverse genetic resources provides breeders with a broader array of traits, while farmer and market involvement ensures new varieties meet local needs. Integrating genomics, transcriptomics, proteomics, and metabolomics offers insights into plant responses to stress, guiding effective breeding. Establishing global partnerships can accelerate the development of resilient crop varieties, creating a coordinated response to climate challenges. Therefore, addressing the challenges in G × E analysis and leveraging the potential of big data, AI, and machine learning are essential to breeding environmentally resilient crops. By 2050, a concerted effort focusing on collaboration, innovation, and inclusivity in agricultural research and breeding can significantly contribute to global food security in the face of an ever-changing climate.

8.8 Conclusion

Comprehensive G × E data and analyses are essential for understanding how genetic variability and phenotypic plasticity impact plant adaptation to diverse environmental conditions. Early findings of crop variability in response to environmental conditions, molecular insights underlying stress responses, advances in genetic modifications, and multi-omics approaches enhance our ability to identify and select new crop ideotypes and traits that improve resilience and productivity, thereby informing future breeding strategies. However, these advancements highlight the importance of integrating genetic and environmental data to develop high-performing and demand-led crop varieties with relatively quick and higher adoption rates. A holistic approach that combines genetic research with agro-ecological considerations will be essential for sustainable agricultural practices and ensuring food security in the face of environmental challenges. Emphasizing this interconnectedness will pave the way for innovative plant breeding and adaptation solutions.

References

Abdulraheem MI, Xiong Y, Moshood AY, Cadenas-Pliego G, Zhang H, Hu J (2024) Mechanisms of plant epigenetic regulation in response to plant stress: recent discoveries and implications. Plants 13(2):163

Akhtar R, Sultana S, Masud MM, Jafrin N, Al-Mamun A (2021) Consumers' environmental ethics, willingness, and green consumerism between lower and higher income groups. Res Conserv Recycl 168:105274

Al-Shammary AAG, Al-Shihmani LSS, Fernández-Gálvez J, Caballero-Calvo A (2024) Optimizing sustainable agriculture: a comprehensive review of agronomic practices and their impacts on soil attributes. J Environ Manag 364:121487

Arif A, Parveen N, Waheed MQ, Atif RM, Waqar I, Shah TM (2021) A comparative study for assessing the drought-tolerance of chickpea under varying natural growth environments. Front Plant Sci 11:607869

Baldoni E (2022) Improving drought tolerance: can comparative transcriptomics support strategic rice breeding? Plant Stress 3:100058

Bazile D, Jacobsen SE, Verniau A (2016) The global expansion of quinoa: trends and limits. Front Plant Sci 7:622. https://doi.org/10.3389/fpls.2016.00622

Bebber DP, Ramotowski MA, Gurr SJ (2013) Crop pests and pathogens move polewards in a warming world. Nat Clim Chang 3(11):985–988

Benitez-Alfonso Y, Soanes BK, Zimba S, Sinanaj B, German L, Sharma V, Bohra A, Kolesnikova A, Dunn JA, Martin AC, Rahman MK (2023) Enhancing climate change resilience in agricultural crops. Curr Biol 33(23):R1246–R1261

Bouis HE, Saltzman A (2017) Improving nutrition through biofortification: a review of evidence from HarvestPlus, 2003 through 2016. Glob Food Secur 12:49–58

Bryan GT, Wu KS, Farrall L, Jia Y, Hershey HP, McAdams SA, Faulk KN, Donaldson GK, Tarchini R, Valent B (2000) A single amino acid difference distinguishes resistant and susceptible alleles of the rice blast resistance gene *Pi-ta*. Plant Cell 12(11):2033–2046. https://doi.org/10.1105/tpc.12.11.2033

Ceccarelli S, Grando S (2019) Participatory plant breeding: who did it, who does it and where? Exp Agric 56:1–11

Chand S, Indu B, Chauhan J, Kumar B, Kumar V, Dey P, Mishra UN, Sahu C, Singhal RK (2022) Plant–environment interaction in developing crop species resilient to climate change. In: Plant abiotic stress physiology. Apple Academic Press, pp 1–24

Chen K, Wang Y, Zhang R, Zhang H, Gao C (2019) CRISPR/Cas genome editing and precision plant breeding in agriculture. Annu Rev Plant Biol 70:667–697. https://doi.org/10.1146/annurev-arplant-050718-100049

Chieb M, Gachomo EW (2023) The role of plant growth promoting rhizobacteria in plant drought stress responses. BMC Plant Biol 23(1):407

Colley MR, Dawson JC, McCluskey C, Myers JR, Tracy WF, van Bueren EL (2021) Exploring the emergence of participatory plant breeding in countries of the Global North–a review. J Agric Sci 159(5–6):320–338

Cooper M, Technow F, Messina C, Gho C, Totir LR (2016) Use of crop growth models with whole-genome prediction: application to a maize multi-environment trial. Crop Sci 56(5):2141–2156

Cui H, Tsuda K, Parker JE (2015) Effector-triggered immunity: from pathogen perception to robust defense. Annu Rev Plant Biol 66:487–511. https://doi.org/10.1146/annurev-arplant-050213-040012

Ding Y, Fromm M, Avramova Z (2012) Multiple exposures to drought 'train' transcriptional responses in *Arabidopsis*. Nat Commun 3(1):740. https://doi.org/10.1038/ncomms1732

Ding Y, Shi Y, Yang S (2019) Advances and challenges in uncovering cold tolerance regulatory mechanisms in plants. New Phytol 222(4):1690–1704. https://doi.org/10.1111/nph.15696

Dowen RH, Pelizzola M, Schmitz RJ, Lister R, Dowen JM, Nery JR, Dixon JE, Ecker JR (2012) Widespread dynamic DNA methylation in response to biotic stress. Proc Natl Acad Sci 109(32):E2183–E2191. https://doi.org/10.1073/pnas.1209329109

Etesami H, Beattie GA (2017) Plant-microbe interactions in adaptation of agricultural crops to abiotic stress conditions. In: Kumar V, Kumar M, Sharma S, Prasad R (eds) Probiotics and plant health. Springer, Singapore. https://doi.org/10.1007/978-981-10-3473-2_7

FAO (2021) The state of food security and nutrition in the world 2021. FAO

Gao L, Kantar MB, Moxley D, Ortiz-Barrientos D, Rieseberg LH (2023) Crop adaptation to climate change: an evolutionary perspective. Mol Plant 16(10):1518–1546

Garland S, Curry HA (2022) Turning promise into practice: crop biotechnology for increasing genetic diversity and climate resilience. PLoS Biol 20(7):e3001716

Giauffret C, Lothrop J, Dorvillez D, Gouesnard B, Derieux M (2000) Genotype × environment interactions in maize hybrids from temperate or highland tropical origin. Crop Sci 40(4):1004–1012

Gunundu R, Shimelis H, Mashilo J (2023) Genomic selection and enablers for agronomic traits in maize (*Zea mays* L.): a review. Plant Breed 142(5):573–593. https://doi.org/10.1111/pbr.13127

Guo R, Shi L, Jiao Y, Li M, Zhong X, Gu F, Liu Q, Xia X, Li H (2018) Metabolic responses to drought stress in the tissues of drought-tolerant and drought-sensitive wheat genotype seedlings. AoB Plants 10(2):ply016

Gupta A, Rico-Medina A, Caño-Delgado AI (2020) The physiology of plant responses to drought. Science 368(6488):266–269. https://doi.org/10.1126/science.aaz7614

Hariprasad P, Chandrashekar S, Singh SB, Niranjana SR (2014) Mechanisms of plant growth promotion and disease suppression by *Pseudomonas aeruginosa* strain 2apa. J Basic Microbiol 54(8):792–801

Howe GA, Major IT, Koo AJ (2018) Modularity in jasmonate signaling for multistress resilience. Annu Rev Plant Biol 69:387–415. https://doi.org/10.1146/annurev-arplant-042817-040047

ISAAA database (2024) GM Approval Database retrieved on 28 Oct 2024. https://www.isaaa.org/gmapprovaldatabase/default.asp

Jamil IN, Remali J, Azizan KA, Nor Muhammad NA, Arita M, Goh HH, Aizat WM (2020) Systematic multi-omics integration (MOI) approach in plant systems biology. Front Plant Sci 11:944

Khoury CK, Achicanoy HA, Bjorkman AD, Navarro-Racines C, Guarino L, Flores-Palacios X, Engels JM, Wiersema JH, Dempewolf H, Sotelo S, Ramírez-Villegas J (2016) Origins of food crops connect countries worldwide. Proc R Soc B Biol Sci 283(1832):20160792

Kim JM, To TK, Nishioka T, Seki M (2010) Chromatin regulation functions in plant abiotic stress responses. Plant Cell Environ 38(11):2307–2321. https://doi.org/10.1111/pce.12520

Kosova K, Vítámvás P, Prášil IT, Renaut J (2011) Plant proteome changes under abiotic stress—contribution of proteomics studies to understanding plant stress response. J Proteome 74(8):1301–1322

Kumar K, Aggarwal C, Sapna B, Singh I, Yadava P (2018) Microbial genes in crop improvement. In: Crop improvement through microbial biotechnology. Elsevier, Amsterdam, pp 39–56

Lake JA, Wade RN (2009) Plant–pathogen interactions and elevated CO2: morphological changes in favour of pathogens. J Exp Bot 60(11):3123–3131

Lal R (2015) Restoring soil quality to mitigate soil degradation. Sustainability 7(5):5875–5895

Lebeis SL, Paredes SH, Lundberg DS, Breakfield N, Gehring J, McDonald M, Malfatti S, Rio TGD, Jones CD, Tringe SG, Dangl JL (2015) Salicylic acid modulates colonization of the root microbiome by specific bacterial taxa. Science 349(6250):860–864. https://doi.org/10.1126/science.aaa8764

Li J, Brader G, Kariola T, Tapio Palva E (2006) *WRKY70* modulates the selection of signaling pathways in plant defense. The Plant J 46(3):477–491

Liang S, Ren G, Liu J, Zhao X, Zhou M, McNeil D, Ye G (2015) Genotype-by-environment interaction is important for grain yield in irrigated lowland rice. Field Crop Res 180:90–99

Liu X, Xie C, Si H, Yang J (2017) CRISPR/Cas9-mediated genome editing in plants. Methods 121:94–102

Manna M, Thakur T, Chirom O, Mandlik R, Deshmukh R, Salvi P (2021) Transcription factors as key molecular target to strengthen the drought stress tolerance in plants. Physiol Plant 172(2):847–868

Meng Q, Gupta R, Min CW, Kwon SW, Wang Y, Je BI, Kim YJ, Jeon JS, Agrawal GK, Rakwal R, Kim ST (2019) Proteomics of Rice–*Magnaporthe oryzae* interaction: what have we learned so far? Front Plant Sci 10:1383

Mishra S, Kumar S, Saha B, Awasthi J, Dey M, Panda SK, Sahoo L (2016) Crosstalk between salt, drought, and cold stress in plants: toward genetic engineering for stress tolerance. In: Abiotic stress response in plants. Wiley-VCH Verlag GmbH & Co. KGaA, Weinheim, pp 57–88. https://onlinelibrary.wiley.com/doi/10.1002/9783527694570.ch4

Mishra S, Roychowdhury R, Ray S, Hada A, Kumar A, Sarker U, Aftab T, Das R (2024) Salicylic acid (SA)-mediated plant immunity against biotic stresses: an insight on molecular components and signaling mechanism. Plant Stress 11:100427

Mohanram S, Kumar P (2019) Rhizosphere microbiome: revisiting the synergy of plant-microbe interactions. Ann Microbiol 69:307–320

Muthurajan R, Ramanathan V, Bansilal Shillak A, Madhuri Pralhad S, Shankarrao CN, Rahman H, Kambale R, Nallathambi J, Tamilselvan S, Madasamy P (2021) Controlled over-expression of *AtDREB1A* enhances tolerance against drought and salinity in rice. Agronomy 11(1):159

Ohama N, Sato H, Shinozaki K, Yamaguchi-Shinozaki K (2017) Transcriptional regulatory network of plant heat stress response. Trends Plant Sci 22(1):53–65

Pandey P, Irulappan V, Bagavathiannan MV, Senthil-Kumar M (2017) Impact of combined abiotic and biotic stresses on plant growth and avenues for crop improvement by exploiting physio-morphological traits. Front Plant Sci 8:537. https://doi.org/10.3389/fpls.2017.00537

Peer LA, Bhat MY, Lone AA, Dar ZA, Mir BA (2024) Genetic, molecular and physiological crosstalk during drought tolerance in maize (*Zea mays*): pathways to resilient agriculture. Planta 260(4):81

Peng H, Chen Z, Fang Z, Zhou J, Xia Z, Gao L, Chen L, Li L, Li T, Zhai W, Zhang W (2015) Rice *Xa21* primed genes and pathways that are critical for combating bacterial blight infection. Sci Rep 5(1):12165

Ramirez-Villegas J, Molero Milan A, Alexandrov N, Asseng S, Challinor AJ, Crossa J, van Eeuwijk F, Ghanem ME, Grenier C, Heinemann AB, Wang J (2020) CGIAR modeling approaches for resource-constrained scenarios: I. Accelerating crop breeding for a changing climate. Crop Sci 60(2):547–567

Rani R, Raza G, Ashfaq H, Rizwan M, Shimelis H, Tung MH, Arif M (2023) Analysis of genotype × environment interactions for agronomic traits of soybean (Glycine max [L.] Merr.) using association mapping. Front Genet 13:1090994

Rehman S, Ahmad Z, Ramakrishnan M, Kalendar R, Zhuge Q (2023) Regulation of plant epigenetic memory in response to cold and heat stress: towards climate resilient agriculture. Funct Integr Genomics 23(4):298

Robert-Seilaniantz A, Grant M, Jones JDG (2011) Hormone crosstalk in plant disease and defense: more than just JASMONATE-SALICYLATE antagonism. Annu Rev Phytopathol 49:317–343. https://doi.org/10.1146/annurev-phyto-073009-114447

Saito K, Matsuda F (2020) Metabolomics for functional genomics, systems biology, and biotechnology. Annu Rev Plant Biol 61(1):463–489

Savary S, Willocquet L, Pethybridge SJ, Esker P, McRoberts N, Nelson A (2019) The global burden of pathogens and pests on major food crops. Nat Ecol Evol 3(3):430–439

Schneider HM, Lynch JP (2020) Should root plasticity be a crop breeding target? Front Plant Sci 11:546

Shelton AC, Tracy WF (2016) Participatory plant breeding and organic agriculture: a synergistic model for organic variety development in the United States. Elementa 4:000143

Shimelis H (2017) New variety design and product profiling. In: Persley GA, Anthony VM (eds) The business of plant breeding: market-led approaches to new variety design in Africa. CABI International, Wallington, pp 85–114

Shivaraj SM, Sharma Y, Chaudhary J, Rajora N, Sharma S, Thakral V, Ram H, Sonah H, Singla-Pareek SL, Sharma TR, Deshmukh R (2021) Dynamic role of aquaporin transport system under drought stress in plants. Environ Exp Bot 184:104367

Simsek Ö, Isak MA, Dönmez D, Dalda Şekerci A, İzgü T, Kaçar YA (2024) Advanced biotechnological interventions in mitigating drought stress in plants. Plants 13(5):717

Steppuhn A, Gase K, Krock B, Halitschke R, Baldwin IT (2004) Nicotine's defensive function in nature. PLoS Biol 2(8):e217

Sun M, Yang Z, Liu L, Duan L (2022) DNA methylation in plant responses and adaption to abiotic stresses. Int J Mol Sci 23(13):6910

Tang H, Hassan MU, Feng L, Nawaz M, Shah AN, Qari SH, Liu Y, Miao J (2022) The critical role of arbuscular mycorrhizal fungi to improve drought tolerance and nitrogen use efficiency in crops. Front Plant Sci 13:919166

Tilman D, Balzer C, Hill J, Befort BL (2011) Global food demand and the sustainable intensification of agriculture. Proc Natl Acad Sci 108(50):20260–20264

Turlings TCJ, Erb M (2018) Tritrophic interactions mediated by herbivore-induced plant volatiles: mechanisms, ecological relevance, and application potential. Annu Rev Entomol 63:433–452. https://doi.org/10.1146/annurev-ento-020117-043507

Uga Y, Sugimoto K, Ogawa S, Rane J, Ishitani M, Hara N, Kitomi Y, Inukai Y, Ono K, Kanno N, Inoue H (2013) Control of root system architecture by *Deeper Rooting 1* increases rice yield under drought conditions. Nat Genet 45(9):1097–1102. https://doi.org/10.1038/ng.2725

Varotto S, Tani E, Abraham E, Krugman T, Kapazoglou A, Melzer R, Radanović A, Miladinović D (2020) Epigenetics: possible applications in climate-smart crop breeding. J Exp Bot 71(17):5223–5236

Verma V, Ravindran P, Kumar PP (2016) Plant hormone-mediated regulation of stress responses. BMC Plant Biol 16:1–10

Voelckel C, Gruenheit N, Lockhart P (2017) Evolutionary transcriptomics and proteomics: insight into plant adaptation. Trends Plant Sci 22(6):462–471

Wanga MA, Shimelis H, Mashilo J (2022) Genotype by environment interaction of newly developed sorghum lines in Namibia. Euphytica 218(10):147

War AR, Paulraj MG, Ahmad T, Buhroo AA, Hussain B, Ignacimuthu S, Sharma HC (2012) Mechanisms of plant defense against insect herbivores. Plant Signal Behav 7(10):1306–1320

Williams RM, O'Brien L, Eagles HA, Solah VA, Jayasena V (2008) The influences of genotype, environment, and genotype× environment interaction on wheat quality. Aust J Agric Res 59(2):95–111

Xie Y, Feng Y, Chen Q, Zhao F, Zhou S, Ding Y, Song X, Li P, Wang B (2019) Genome-wide association analysis of salt tolerance QTLs with SNP markers in maize (*Zea mays* L.). Genes Genomics 41:1135–1145

Yadav MR, Choudhary M, Singh J, Lal MK, Jha PK, Udawat P, Gupta NK, Rajput VD, Garg NK, Maheshwari C, Hasan M (2022) Impacts, tolerance, adaptation, and mitigation of heat stress on wheat under changing climates. Int J Mol Sci 23(5):2838

Yahaya MA, Shimelis H, Nebie B, Mashilo J, Pop G (2023) Response of african sorghum genotypes for drought tolerance under variable environments. Agronomy 13(2):557

Yang Y, Guo Y (2018) Unraveling salt stress signaling in plants. J Integr Plant Biol 60(9):796–804. https://doi.org/10.1111/jipb.12689

Yoshida T, Mogami J, Yamguchi-Shinozaki K (2014) ABA-dependent and ABA-independent signaling in response to osmotic stress in plants. Curr Opin Plant Biol 21:133–139

Zeng X, Luo Y, Vu NTQ, Shen S, Xia K, Zhang M (2020) CRISPR/Cas9-mediated mutation of *OsSWEET14* in rice cv. Zhonghua11 confers resistance to *Xanthomonas oryzae* pv. oryzae without yield penalty. BMC Plant Biol 20:1–11

Zhou J, Xu XC, Cao JJ, Yin LL, Xia XJ, Shi K, Zhou YH, Yu JQ (2018) Heat shock factor *HsfA1a* is essential for R gene-mediated nematode resistance and triggers H2O2 production. Plant Physiol 176(3):2456–2471

Zhu JK (2016) Abiotic stress signaling and responses in plants. Cell 167(2):313–324. https://doi.org/10.1016/j.cell.2016.08.029

Part III

Omics and Next Generation Techniques

Post-genomic Era in Crop Breeding

9

J. Nikitha and Manickavelu Alagu

Abstract

The post-genomic era is considered a renaissance phase of crop breeding with technological leaps by integrating genomic tools and techniques into conventional breeding approaches at a whole new level. This paradigm shift leverages genome-wide association studies (GWAS), CRISPR-based genome editing, and high-throughput sequencing to magnify our understanding of crop genetics and facilitate the development of improved varieties. The emergence of multi-omics approaches such as transcriptomics, proteomics, and metabolomics has further enhanced our potential to dissect complex traits and alter breeding strategies. The crop breeders are now qualified to address global crises like climate change, resource scarcity, and food security with the advent of these innovations with unmatched precision. This section deals with the advancements in genomic technologies, including machine learning and artificial intelligence, and their application in crop improvement that convey breeding efficiency, trait discovery, and development of resilient crops. The importance of ethical considerations in placing these technologies is emphasized for the continued integration of the latest genomic insights and field performance data.

Keywords

Post-genomic era · Transgenics · Pan-genomes · Gene-editing · Machine learning

J. Nikitha · M. Alagu (✉)
Department of Genomic Science, Central University of Kerala, Periye, Kerala, India
e-mail: amanicks@cukerala.ac.in

P. M. Priyadarshan, R. Ortiz (eds.), *Plant Breeding 2050*,
https://doi.org/10.1007/978-981-95-0583-8_9

9.1 Introduction

During the 1800s, the human population was only one billion, whereas today, the number has reached eight billion and is growing at an unsettling rate. A significant obstacle to this expanding population is making sure that everyone has physical, social, and financial access to enough food that is sufficient, wholesome, and safe to meet their dietary needs and food preferences for an active and healthy life to achieve *food security* (The FAO 1996). Doubling the current food production by 2050 is vital for nourishing the population. Crop breeding has a significant potential for global food, nutrition, and environmental security. Adopting advances in agriculture, specifically genetic manipulation, is the primary hope for achieving the goal (Joshi et al. 2023).

Cultivating crop plants for traits of interest, mainly increased yield and nutritional quality through artificial selection, dates back into human history. Human civilization prompted the selection of the most favourable plants. The seeds and planting materials preserved were propagated in subsequent seasons, leading to continuous trait improvements that may have influenced the existence of today's food crops (Voss-Fels et al. 2019). Crop breeding as a science began backing Mendel's Laws of Heredity in 1865 and involved targeted crossing between parental genotypes essential for crop improvement. The conventional crop breeding/selective breeding methodology approaches have always involved a diligent decision-making process of integrating, testing, and mapping traits. Hence, before the advancement of genomics, the breeders were conscious of selecting superior individuals from the segregating populations phenotypically (Kaiser et al. 2020). These methods were time-consuming and less accurate for crop variety development. The initial crossing of parent plants was followed by raising successive generations to identify superior individuals that do not involve futuristic molecular breeding technologies. In addition to that, multi-year field trials across diverse locations were required to evaluate the genetic potential in varying conditions (Voss-Fels et al. 2019). To refine specific traits of crop plants, the plant breeders utilize artificial selection, often concentrating on short-duration crops (Lammerts Van Bueren et al. 2011; Kandemir and Saygili 2015).

As the prime objective of plant breeding evolved beyond the limits of increased yield, conventional breeding techniques coupled with genomics were necessary. The research on plant genomics commenced in December 2000 with the reporting of the entire genome of *Arabidopsis thaliana* (Marks et al. 2021). Strategies like genomic selection (GS), genome-wide association studies (GWAS), gene discovery, isolation, molecular markers, next-generation sequencing, and omics platforms are at the forefront of advancing breeding techniques like cis-genesis, intra-genesis, mutation breeding, and polyploidy breeding. With the advent of molecular markers during the 1900s, breeding efforts to identify and select hybrid lines became easy, time-saving, and efficient (Dreher et al. 2002; Gupta et al. 2010). The shift from a phenotypic selection of superior progenies to a genotypic selection process, strictly based on scientific research and findings, brought about revolutions in plant breeding. Molecular markers such as RFLP, AFLP, SSR, and SNP are widely used to identify QTLs in crops, including rice (Nadeem et al. 2018). Among these, the SNP markers are often used for their cost-effectiveness and ubiquity across genomes (Liakos et al. 2018; Oladosu et al. 2019). A typical case of specific genes like *Yr* for stripe rust resistance in wheat was identified using SNP markers (Farrakh et al. 2016).

Keeping the pace of the Human Genome Project (1990–2003), a paradigm shift in understanding genes and their effect on traits which are now viewed through genome expression, challenging the conventional perspectives, and fostering the rise of functional genomics has significantly contributed to the modern plant breeding techniques usually considered as the post-genomic era. The crop improvement has significantly accelerated through high-throughput crop phenotyping (HTTP) and modern accelerated breeding. The interactions among various genes/QTLs and their regulation, facilitated by high-throughput and robust phenotyping and genotyping platforms, play a crucial role in accelerating breeding programmes (Wang et al. 2007). Gene manipulation directed by genetic engineering introduced favourable traits to cultivated crops. The current intelligent breeding toolkit, which consists of genomics, bioinformatics, and advanced breeding technologies, will revolutionize agriculture for sustainable food production in the future.

9.2 Evolution of Crop Breeding: Pre-Genomic and Genomic Eras

Humans began crop improvement right from the beginning of civilizations. Initially from domestication, conventional plant breeding methods like introduction, selection, back cross-breeding, and hybridization were conducted exclusively to improve yield-related traits and biotic and abiotic stress tolerance (Adhikari et al. 2023). Launching new crop varieties with desired traits using hybridization done through pedigree, backcrossing, recurrent selection, and diallel selective mating was prevalent among plant breeders (Adhikari et al. 2023) that provided distinct purposes but faced unique challenges. While it works well for self-pollinating plants like rice and wheat, the pedigree method is also helpful in choosing superior genotypes from several generations. On the other hand, back cross-breeding eases the transfer of specific traits from donor to recipient parents (Zhang et al. 2021), which is widely used for developing stress-resistant varieties. Recurrent selection and selective mating techniques increased the adaptability of crops by diversifying genetic backgrounds.

In conventional plant breeding, the development of improved varieties of crops is limited because of the very long crop duration. It can even take one or two decades to create a new cultivar (Niazian and Niedbała 2020) as they must undergo many phases such as crossing, selection, and testing. To alleviate food scarcity and increase food security, developing new and improved varieties suited for the then-existing abiotic and biotic stresses and increased yield is crucial. The significant milestones in crop breeding are represented in Fig. 9.1 (Naqvi et al. 2022). The genetic variability, being the soul of plant breeding, has gradually decreased by solely adopting traditional methods (Prabhu et al. 2023). Mutation breeding

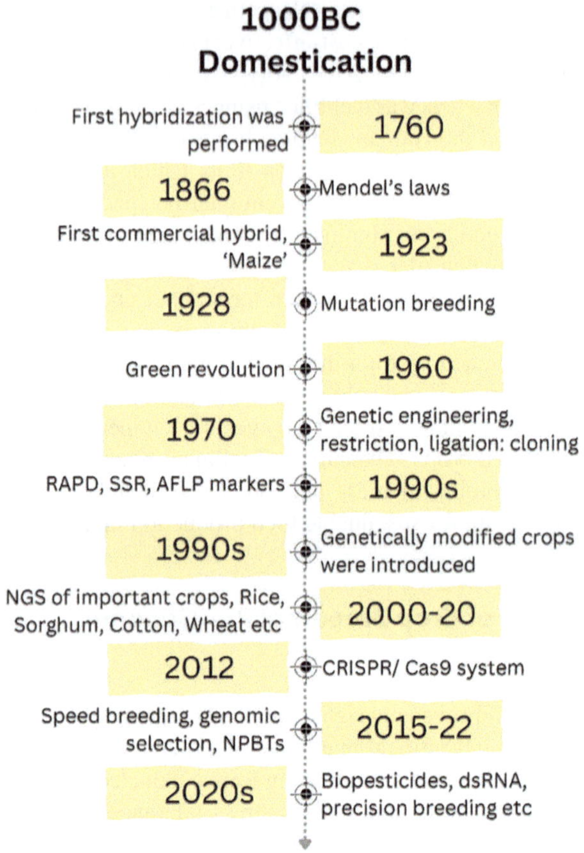

Fig. 9.1 Chronology of significant crop breeding discoveries over the years. (Source: Adapted from Naqvi et al. (2022), licensed under CC BY 4.0)

administers various mutagens to induce genetic variations even though it is labour-intensive and offers unpredictable and untargeted results. The various conventional breeding techniques, their merits and demerits are listed in Table 9.1.

The genomics era revolutionized crop breeding through precise identification and manipulation of traits at the molecular level, thereby understanding the sequences, gene function, and regulatory mechanisms that play crucial roles in enhancing plant stress tolerance and yield (Pourkheirandish et al. 2020). Quantitative trait loci (QTL) analysis, which identifies genomic areas associated with specific traits, was used in marker-assisted selection (MAS) to alter breeding techniques significantly (Table 9.2).

Genome-wide association studies (GWAS) were made possible by the correlation between desired qualities and genetic variants that were found utilizing technologies such as genotyping by sequencing (GBS) and RADSeq. Similarly, the breeding efficiency in crops was enhanced by predicting breeding values using

Table 9.1 Conventional plant breeding approaches

Breeding method	Technique adopted	Merits	Demerits	Achievements
Introduction	A group or individual genotype is taken to a new environment where they were not grown previously.	Providing entirely new crops to an area will help reduce disease incidence, germplasm collection and maintenance, and the quickest method for genetic variability.	The introduction of alien diseases, pests, and weeds can be associated with the process.	Sioux, Roma, Marglobe, Best of All and La Bonita of tomato, dwarf wheat varieties like Sonora 64 and Lerma Rojo, rice dwarf varieties TN-1, IR-8
Pure line Selection (individual plant selection)	The selective breeding of a superior organism continuously for several generations to develop a new progeny with only the desired traits established (pure line)	Extremely uniform varieties achieve the maximum possible improvement over the original variety	Comparatively, less adaptability and stability, more time, space, and expensive yield trials are required, and trials are limited to self-pollinated species only.	Mungbean varieties T1 and B1; rai variety L18; cotton varieties Cambodia Coimbatore 2, Gadag 1; Rice varieties Mtul, Mtu2, Mtu7, T29; barley varieties C251, C50; tobacco varieties NP28, NP63, NP7; cotton—C7, CO2, K 1, IIHR tomato—Arka Abha
Hybridization	Production of progeny by crossing two individuals of unlike genetical constitution (same species—intraspecific or inter-varietal & different species—interspecific)	Genetic variation can be introduced to a desirable plant, and hybrids better adapted to the environment are developed, also developed by asexual means or protoplast fusion in tissue culture media	Infertile offspring—farmers need to buy seeds year after year, and the high cost and expertise required for hybridization	Rice—IR36, Sona, Swarna; Wheat—Sonalika, Kalyan Sona, PBW343, HD2967; Maize—Deccan, Jawahar, Navjyot; Millets—HHB67 Improved, Hybrids of Pearl Millet; Pulses—Pusa Komal (Pigeon Pea), Vishal (Chickpea); Tomato—Arka Vardhan

(continued)

Table 9.1 (continued)

Breeding method	Technique adopted	Merits	Demerits	Achievements
Pedigree	Individual plants from segregating generations, their offsprings are grown, and a record of all parents–offspring relationships is maintained	Selection is based on phenotype and genotype, a high degree of genetic purity is produced, and breeding value is ascertained through progeny test	Tedious, time-consuming (10–12 years or more), expensive, suitable only for specific crops like rice, wheat, cotton pigeon pea, chickpea, etc	Wheat varieties—NP52, NP80–5, and NP120. Rice varieties—Bala, Cauveri, Karuna, Krishna, Ratna, Sabarmati
Back cross-breeding	Crossing of hybrid to one of its parents/closely related individual for transferring superior traits from the donor parent to the genetic background of the recurrent parent	An extensive yield trial is not needed, a smaller population is needed compared to the pedigree method	Problem of linkage drag exists, hybridization of each generation to be done	Wheat variety—Transfer (rust resistance), cotton varieties—170-Co2 and 134-Co2M
Recurrent selection	Reselection generation after generation with inbreeding of selected ones to provide for genetic recombination (Hull)	Developing synthetic varieties improves desirable genes in a population, and isolation of inbred lines improves population	Much effort and many recurrent cycles required, not directly used in variety development, continuous selfing leads to loss of genetic variability	Improvement of oil content in maize, fibre strength in cotton, and sugar content in sugarbeet and sugarcane
Mutation breeding	Deliberate induction of mutation and development of mutant lines for crop improvement	Cheap and rapid method for new variety development, desirable mutant allele not present in the germplasm, suitable for oligogenic characters, production of haploids by irradiating pollens	Generally random and unpredictable, strong negative pleiotropic effect, health risks for handling, most mutants are rare and predominantly recessive	Mustard- Rai 5, groundnut-TG mutants, TAP 7 in mungbean, TT-6 in pigeon pea, TRC77-4, TC-901 in cowpea, TCDM-1, Vikram-TCR in rice

(continued)

Table 9.1 (continued)

Breeding method	Technique adopted	Merits	Demerits	Achievements
Polyploidy breeding	The intentional manipulation and utilization of polyploid plants, those with more than two sets of chromosomes	Creation of new species is possible, widening the base of existing polyploids, induction of variation, and hybrid vigour is maintained	Effects cannot be predicted, low fertility, cytogenetic and genetic instability in many cases.	Watermelon varieties—Pusa Bedana, Shonima, Swarna, Sugarbaby, Arka Madhura. Palak variety Pusa Jyothi, Tapioca varieties SreeHarsha, SreeApoorva

genome-wide markers rather than depending on extensive phenotyping through genomic selection (GS). Time consumption, being the major obstacle in the conventional breeding approaches, was circumvented with the application of speed breeding techniques, where the environmental conditions were manipulated to achieve multiple generations per year by incorporating single seed descent (SSD) and MAS for rapid selection of traits. Sequenced genomic data of significant crops are utilized for breeding (Table 9.3).

9.3 Post-Genomic Era in Crop Breeding

Sufficient food to feed the increasing population and food security have been significant concerns in crop breeding strategies. Despite adopting conventional and genomic techniques, the goal still needs to be fully achieved. In order to advance the traditional time-consuming cross-breeding methods, brand-new techniques have been added to the plant breeding toolkit that can assist breeders in developing crops capable of adapting to climate changes through a genomic-led approach for predicting the breeding values, targeted mutagenesis through gene editing, trait evaluation through high-throughput phenomics, machine learning for managing massive datasets, and speed breeding approaches that promote early flowering and seed production (Ortiz 2023).

Genome editing and pangenomics refinement happen by unlocking complex genetic traits and enhancing environmental adaptability. Similarly, transgenic plants, made possible by developing recombinant DNA technology, are essential for finding potential genes and verifying their activities in laboratory settings. To maintain sustainable agriculture, addressing several issues brought on by pests, diseases, and climate change, and raise output to meet rising food demand with nutritional quality is a pressing need. improved crops need to be developed with limited time and efficiency. The newer crop breeding techniques during the post-genomic era are detailed below.

Table 9.2 Crop varieties released using molecular markers

Crop	Varieties	Trait	QTL/ molecular markers	Released year	Released from
Wheat	PBW 723 (Unnat PBW 343)	Stripe and leaf rust resistance	SSR & CAPS	2017	PAU, Ludhiana (Punjab) & ICAR-AICRP on Wheat & Barley
	PBW 761 (Unnat PBW 550), PBW 752, PBW 757	Stripe rust resistance	SSR	2019	
	PBW 771	Stripe and leaf rust resistance	SSR & CAPS	2020	
Rice	Improved Pusa Basmati 1	Bacterial blight resistance	CAPS and STS	2007	ICAR-IARI and ICAR-NIPB, New Delhi
	Improved Samba Mahsuri	High yield and grain quality similar to Samba Mahsuri	SSR	2008	ICAR-IIRR & CSIR-CCMB, Hyderabad (Telangana)
	Improved Lalat, Improved Tapaswini	Bacterial blight resistance	STS	2013	ICAR-NRRI, Cuttack (Odisha)
	PR 121, PR 122, PR 123, PR 124	Bacterial blight resistance	SSR	2015	PAU, Ludhiana (Punjab) & ICAR-AICRP on Rice
	Pusa 1592	Bacterial blight resistance	SSR & STS	2015	ICAR-IARI, New Delhi
	DRR Dhan 51	Blast resistance	SSR	2018	ICAR-IIRR Hyderabad (Telangana)
	CR Dhan 802 (Subhash), CR Dhan 801	Submergence & drought tolerance	SSR	2019	ICAR-NRRI, Cuttack (Odisha) & IRRI, Philippines
	DRR Dhan 59	Bacterial blight resistance	SSR	2021	ICAR-IIRR Hyderabad (Telangana)
	DRR Dhan 62	Bacterial blight and blast resistance	SSR	2021	
	Pusa Basmati 1985	Herbicide tolerance (Imazethapyr)	SSR	2021	ICAR-IARI, New Delhi

(continued)

Table 9.2 (continued)

Crop	Varieties	Trait	QTL/ molecular markers	Released year	Released from
Maize	Vivek QPM 9 (quality protein maize)	High lysine and tryptophan	SSR	2008	ICAR-VPKAS, Almora (Uttarakhand)
	Pusa HM4 improved, Pusa HM8 improved, Pusa HM9 improved	High lysine and tryptophan	SSR	2017	ICAR-IARI, New Delhi & CCS-HAU, Hisar (Haryana)
	Pusa Vivek QPM9 Improved	Provitamin-A	InDel	2017	ICAR-IARI, New Delhi & ICAR-VPKAS, Almora (Uttarakhand)
	Pusa Vivek Hybrid-27 Improved	Provitamin-A	InDel	2020	
	Pusa HQPM-5 Improved, Pusa HQPM-7 Improved	Provitamin-A	InDel	2020	ICAR-IARI, New Delhi & CCS-HAU, Hisar (Haryana)
	Pusa HQPM-1 Improved	Provitamin-A	InDel	2021	
Pearl millet	HHB 67 Improved	Downy mildew resistance	RFLP	2005	CCS-HAU, Hisar (Haryana) & ICAR-AICRP on pearlmillet with ICRISAT, Telangana
	HHB 67 Improved 2	Downy mildew resistance	SSR	2021	
Chickpea	Super Annigeri-1	Fusarium wilt resistance	SSR	2020	UAS, Raichur & ICAR AICRP on chickpea with ICRISAT, Telangana
	Pusa Chickpea 20211	Fusarium wilt resistance	SSR	2021	ICAR-IARI New Delhi
	IPCMB 19–3 (Samriddhi), IPCL4–14	Fusarium wilt resistance	SSR	2021	ICAR-IIPR, Kanpur (UP)
	Pusa Chickpea 10216	Drought tolerance	SSR	2020	ICAR-IARI New Delhi & ICRISAT, Telangana
	Pusa Chickpea 4005	Drought tolerance	SSR	2021	ICAR-IARI New Delhi

(continued)

Table 9.2 (continued)

Crop	Varieties	Trait	QTL/ molecular markers	Released year	Released from
Soybean	NRC 127	Kunitz trypsin inhibitor free	STS & SSR	2018	ICAR-IISR, Indore (MP)
	MACSNRC 1667	Kunitz trypsin inhibitor free	STS & SSR	2021	
	NRC 132	Less beany flavour	STS & SSR	2021	
	NRC 142	Kunitz trypsin inhibitor free & less beany flavour	STS & SSR	2021	
	NRCSL 1	YMV resistance	SSR	2021	
	NRC 138	Early maturity	STS & SSR	2021	
Groundnut	Girnar 4, Girnar 5	Oleic acid	AS-PCR and CAPS	2020	ICAR-DGR, Junagadh (Gujrat)

Source: Kumar et al. (2022)

9.3.1 Transgenics

Transgenics was coined in 1981 to cite genetically altered creatures (animals, plants, and microorganisms) (Gordon and Ruddle 1981). The genetic enhancement of organisms by biotechnology is termed transgenic breeding. It is a genetic transformation process that involves recombinant DNA technology. The existing genetic makeup is edited by genetic engineering to introduce a new trait that is not present in that species. The first transgenic plant was created by Fraley et al. (1983) for antibiotic resistance in tobacco (Ciro et al. 2018). The possibilities of plant breeding through transgenic breeding extend by (1) the introduction of novel genetic variation that is absent in the gene pool and (2) the production of desired phenotypes from known genes (Kundu 2024).

Transgenic technology is considered a second "green revolution" worldwide for being a reliable medium for developing designer crops. Many transgenic lines can be created using candidate genes found by molecular techniques such as QTL analysis, GWAS, genomic selection, and genome editing (Noman et al. 2017). The critical focus of crop breeding is increased yield, improved shelf life, nutritional quality, and tolerance to various biotic and abiotic stresses; transgenic breeding (Obembe et al. 2022) is even utilized for expressing proteins with pharmaceutical and industrial values. Similarly, they are also utilized in bioremediation of contaminated soils. For instance, transgenic crops successfully remove toxic pollutants like mercury, selenium, and organic pollutants from soils. Plants are distinct in producing vaccines or antibiotics, namely "plantibodies", that help reduce screening costs for viruses and bacterial toxins as they are devoid of human diseases (Salyaev and Rekoslavskaya 2017). Several crops like tobacco, tomato, corn, potato, alfalfa,

Table 9.3 Features of major sequenced crop genome

Sl. No.	Scientific name	Common name	Haploid chromosome number	Estimated genome size (Mb)	Assembly size (Mb)	Number of gene predictions	Repeat (%)	References
1	*Actinidia eriantha*	Kiwi fruit	29	690.6	615.95	35,530	56.03	Liao et al. (2023)
2	*Allium sativum*	Garlic	8	16,900	16,243	57,561	91.3	Li et al. (2022), Sun et al. (2020)
3	*Asparagus officinalis*	Garden Asparagus	10	1308	1200	32,237	74.08	Tsugama and Fujino (2020).
4	*Azadirachta indica*	Neem	12	364	–	20,000	13.03	Krishnan et al. (2012)
5	*Beta vulgaris*	Sugar beet	9	714.00–758.00	567	27,421	63	Dohm et al. (2014)
6	*Brassica napus*	Rapeseed	19	1130	892.7	101,040	34.8	Chalhoub et al. (2014)
7	*Brassica oleracea var. capitata*	Cabbage	9	630	535.5	45,758	38.8	Liu et al. (2014)
8	*Brassica juncea*	Brown mustard	18	1056.53	933	92,878	33	Kang et al. (2021)
9	*Brassica rapa*	Chinese cabbage	10	529	283.8	41,174	39.5	Wang et al. (2011)
10	*Cajanus cajan*	Pigeon pea	11	833.07	605.78	48,680	51.67	Varshney et al. (2011)
11	*Cametina sativa*	Camelina	20	785	641.45	89,418	28	Kagale et al. (2014)
12	*Carica papaya*	Papaya	9	372	271	24,746	52	Ming et al. (2008)
13	*Cannabis sativa*	Marijuana	10	820	534.7	30,000	–	Van Bakel et al. (2011)
14		Hemp	10	818	220.8	–	–	
15	*Capsicum annum*	Hot pepper	12	3480	3060.00	34,903	76.4	Kim et al. (2014)
16	*Chenopodium quinoa*	Quinoa	18	1300	1390.00	44,776	64	Jarvis et al. (2017)

(continued)

Table 9.3 (continued)

Sl. No.	Scientific name	Common name	Haploid chromosome number	Estimated genome size (Mb)	Assembly size (Mb)	Number of gene predictions	Repeat (%)	References
17	*Cicer arietinum*	Chickpea	8	~738.00	532.29	28,269	49.41	Varshney et al. (2013)
18	*Citrullus lanatus*	Water melon	11	~425.00	353.5	23,440	45.2	Guo et al. (2013)
19	*Citrus clementina*	Clementine mandarin	9	367	301.4	24,533	45	Wu et al. (2013)
20	*Citrus sinensis*	Sweet orange	9	367	320.5	29,445	20.5	Xu et al. (2013)
21	*Coffea canephora*	Robusta coffee	11	710	568.6	25,574	50	Denoeud et al. (2014)
22	*Coriandrum sativum*	Coriander/ Cilantro	11	2130	2118	40,747	80.58	Song et al. (2020)
23	*Cucumis melo*	Melon	12	450	375	27,427	19.7	González et al. (2010)
24	*Cucumis sativus*	Cucumber	7	367	243.5	26,682	24	Huang et al. (2009)
25	*Dioscorea alata*	Greater yam	20	2238	480	76,255	73	Bredeson et al. (2022)
26	*Elaeis guineensis*	Oil palm	16	1800	1535.00	34,802	57	Thottathil et al. (2016)
27	*Eragrostis tef*	Tef	20	772	672	–	14	Cannarozzi et al. (2014)
28	*Eucalyptus, grandis*	Eucalyptus	11	640	605	36,796	50	Myburg et al. (2011)
29	*Fragaria vesca*	Strawberry	7	240	209.8	34,809	16	Shulaev et al. (2011)
30	*Glycine max*	Soybean	20	1115	950	46.43	57	Schmutz et al. (2010)
31	*Gossypium hirsutum*	Cotton	26	2430	2300	70,478	67.2	Chen et al. (2020)
32	*Helianthus annuus*	Sunflower	17	3600	3000	61,205	62	Badou et al. (2017)
33	*Hevea brasiliensis*	Rubber tree	18	2150	1900	68,955	78	Rahman et al. (2013)

34	*Hordeum vulgare*	Barley	7	4790	4200	58,438	–	Beier et al. (2017)
35	*Mangifera indica*	Mango	20	439	492	43,247	77.14	Singh (2016)
36	*Manihot esculenta*	Cassava	18	772	637.2	32,858	–	Daniell et al. (2008)
37	*Musa acuminata*	Banana	11	523	472.2	36,542	43.72	Hont et al. (2012)
38	*Nicotiana tabacum*	Tobacco	12	4500	3700.00	90,000	72.00–78.00	Sierro et al. (2014)
39	*Oryza sativa-spp indica*	Rice	12	430	466	46,022–55,615	42.2	Yu et al. (2002)
40	*Oryza sativa-spp japonica*	Rice	12	420	389.8	37,544	35	Goff et al. (2002)
41	*Phaseolus vulgaris*	Common bean	11	587	473	27,197	45.37	Schmutz et al. (2014)
42	*Phoenix dactylifera*	Date palm	18	671	605.4	41,660	21.99	Al-Dous et al. (2011), Al-Mssallem et al. (2013)
43	*Phyllostachys heterocycla*	Moso bamboo	24	2075	2050.00	31,987	59	Peng et al. (2013)
44	*Populus trichocarpa*	Poplar	19	485	410	45.555	44	Tuskan et al. (2006)
45	*Prunus mume*	Chinese plum/Mei	8	280	237	31,390	45	Zhang et al. (2012)
46	*Pyruss bretschneideri*	Pear	8	265	226.6	27,852	29.6	Verde et al. (2013)
47	*Pyruss bretschneideri*	Pear	17	527	512	42,812	53.1	Wu et al. (2013)
48	*Ricinus communis*	Castor bean	10	320	350	31,237	50.33	Chan et al. (2010)

(continued)

Table 9.3 (continued)

Sl. No.	Scientific name	Common name	Haploid chromosome number	Estimated genome size (Mb)	Assembly size (Mb)	Number of gene predictions	Repeat (%)	References
49	*Setaria italica*	Foxtail millet	9	490	423	38,801	46	Zhang et al. (2012), Bennetzen et al. (2012)
50	*Solanum lycopersicum*	Tomato	12	900	760	34,727	63.28	The Tomato Genome Consortium (2012)
51	*Solanum melongena*	Eggplant	12	1126	833.1	85,446	70.4	Hirakawa et al. (2014)
52	*Solanum tuberosum*	Potato	12	844	727	39,031	62.2	The Potato Genome Sequencing Consortium (2011)
53	*Sorghum bicolour*	Sorghum	10	~730.00	698	27,640	62	Paterson et al. (2009)
54	*Theobroma cacao*	Cocoa	10	430	326.9	28,798	25.7	Argout et al. (2011)
55	*Triticum aestivum*	Bread wheat	21	17,000	3800.33	94,000–90,000	80	Brenchley et al. (2012)
56	*Vaccinium macrocarpon*	Cranberry	12	470	420	36,364	5.6	Polashock et al. (2014)
57	*Vigna radiata*	Mungbean	11	579	431	22,427	43	Kang et al. (2014)
58	*Vitis vinifera*	Grape	19	475	487	30,434	41.4	The French–Italian Public Consortium for Grapevine Genome Characterization (2007)
59	*Zea mays*	Maize	10	2300	2048.00	32,540	85	Schnable et al. (2009)
60	*Ziziphus jujuba*	Jujube	12	444	437.65	32,808	49.49	Liu et al. (2014)

banana, and canola were transgenically modified for expressing proteins. Among them, tobacco, tomatoes, and bananas were widely used.

The genetic makeup of crop plants is modified mainly through three transformation techniques: *Agrobacterium tumefaciens* mediated, biolistic and protoplast transformations. The former method, also known as the stable transformation, is mediated through *Agrobacterium tumefaciens*, a *Rhizobium* bacterium that can control host cell metabolism and cause crown gall disease. It transfers a single-stranded form of transfer DNA (T-DNA), having the transgene or the gene of interest incorporated into the host genome and will get expressed permanently. This method is conveniently chosen over other transgenic techniques in many plant species as the efficiency is greater regardless of the transgene size. This technique is mainly adopted for potatoes, tomatoes, and tobacco plants.

Biolistic transformation or particle bombardment is a physical process that requires specific equipment or a gene gun to discharge DNA-coated metal particles into the target cell at high velocity. The DNA particle can be delivered to any material through this technique with a relatively high frequency of transient expression rather than stable transformation. The method is versatile and practical, especially in monocot plants like rice, maize, and wheat. Even though the technique is fast and straightforward, if the DNA is integrated into the reproductive cell line or tissue, these lines must be refined to generate the transgenic line.

In the protoplast transformation, protoplasts of desired cells are generated by enzymatic digestion of plant cell wall and the vectors are integrated by transfection using electroporation or chemical methods. The more extended culture period reduces the regeneration rate, leading to unexpected somatic mutation in most cases. Hence, this method is rarely used for stable transformations. Even then, protoplast transformation is being re-evaluated to improve efficiency, as it can directly take up proteins and RNA.

Around the world, several transgenic crop varieties have been commercialized despite continuous discussions about their usage for human consumption. 90% of canola, soybean, and castor production are transgenic crops in North America. Some of the crop transgenes developed include (1) rice—golden rice developed to overcome Vitamin A deficiency in children by incorporating phytoene synthase gene through the gene gun method, which in turn led to an increase in Vitamin A content in rice grains (2) cotton—Bt cotton which is resistant to bollworms, (3) tomato—Flavr Savr tomato with longer shelf life due to delayed ripening.

Transgenic breeding offers various advantages over the conventional breeding methods. The transfer of genes from microbes and animals is possible with this method. As transgenics bypass the sexual process, chances of linkage drag are absent. Direct manipulation of one or more genes is possible, whereas indirect manipulation occurs in the conventional breeding approaches. Similarly, less time is required for new varietal release. GM technology has revolutionized crop production for farmers as they develop crops of desired traits, such as tolerance to herbicides and various biotic and abiotic stresses. About 525 distinct transgenic modifications were carried out and approved for commercial cultivation around the globe in 32 crops (Kumar et al. 2020). Some of the significant milestones in crop breeding through transgenic technology are listed in Table 9.4. That represents its wide application in crop breeding.

Consumers worldwide perceive the technology rather philosophically. The release of transgenic food crops often encountered moral and ethical issues, such as consuming products from different species of organisms that raise concern among the end users for allergies, even though there is uncertainty regarding the fact that the transgenes are responsible for the reaction. Similarly, antibiotic-resistant crops

Table 9.4 Overview of some key transgenically modified crops

Genetic modification	Character	Crops	Transgene	From organism
Herbicide resistance	Glyphosate	Tomato, potato, tobacco, cotton, etc.	aro A	*Salmonella typhimurium* and *E. coli*
	Sulphonylurea	Tobacco	mutant ALS (acetolactate synthase) gene	*Arabidopsis thaliana*
	Atrazine	Tobacco	QB protein of photosystem II	*Amaranthus*
Insect resistance	Insect pests of order Lepidoptera, Coleoptera, Hymenoptera, Diptera, and Nematoda.	Corn, cotton, potato, tobacco	*Cry1Aa, Cry1Ab, Cry1Ac, Cry2Aa, Cry3Aa*	*Bacillus thuringiensis*
Virus resistance	*Tobacco Mosaic Virus*	Tobacco, tomato	*N, NN*	*Nicotiana glutinosa*
	Potato virus	Potato	*CP* (Coat Protein) gene	*Potato Virus Y (PVY)*
	RSV (*Rice Stripe Virus*)	Rice	*CP* (Coat Protein) gene	*Rice Stripe Virus*
	YMV (*Yellow Mosaic Virus*)	Black gram, green gram	*Ty-1* gene	Tomato (*Solanum lycopersicum*)
Nutrient improvement	Vitamin A	Rice (golden rice)	*Psy* (phytoene synthase) gene & C*rtI* (carotene desaturase) gene	Daffodil (*Narcissus pseudonarcissus*) or Maize (*Zea mays*) & *Erwinia uredovora*, respectively
Abiotic stress	Salt tolerance	Maize, tomato	*AtNHX1* gene	*Arabidopsis thaliana Avicennia marina* (Mangroves)
Molecular farming	Vaccines, antibodies, and other therapeutic proteins, e.g. edible vaccine against Hepatitis B	Potato, banana	*HBsAg* (Hepatitis B surface antigen) gene	*Hepatitis B virus*

with resistance genes can develop pest resurgence, liberating superbugs that cannot be treated using these antibiotics in the future. Likewise, another concern is for the environment; for instance, the pollen from transgenic corn can disperse up to 60 m and kill monarch butterfly larvae, potentially threatening non-targeted organisms. Also, a transgenic breeding programme requires technical skills and sophisticated, well-equipped labs.

The proposed disadvantages of the technique are a subject for research and discussion, which can be examined over time. Meanwhile, the technique solely addresses the food demands of the hiking populations but carries hidden costs of environmental and health concerns. The future of GM crops remains uncertain as the issue continues to be controversial. So far, cotton is the only transgenic crop commercially being cultivated in India. The government approved the cultivation of Bt cotton in 2002 to resist crop loss due to the American pink bollworm. Other edible crops like brinjal, tomato, maize, and chickpea are still in various trial stages. The Genetic Engineering Appraisal Committee (GEAC) approved the release of mustard hybrid DMH-11 (Obembe et al. 2022), a step closer to complete commercial cultivation in 2022, but it has not yet been commercially released. However, the Supreme Court of India still seeks permission to cultivate transgenic food crops commercially.

9.3.2 Pangenomics

Reference genomes exist as representative nucleic acid sequences of an idealized individual of a particular species and have been the basis for understanding the genetic variation in organisms since the beginning of the genomic era. With the advancement of genomic science, new insights into biological genetic diversity, population history, and genome-based breeding can be obtained (Gong et al. 2023) through newer techniques. The fragment variants like single nucleotide polymorphisms (SNPs), small insertions and deletions (indels), long-fragment structural variants (SVs) like copy number variations (CNVs), rearrangements and presence/absence variants (PAVs) can be identified by aligning sequencing reads with a reference genome.

Previous studies reveal that when there is a significant difference between the reference genome and large segments of variants longer than 50 bp (Gong et al. 2023), alignment of a single reference genome and sequencing techniques with reduced detection efficiency yield various limitations. Similarly, the complexity of structural variant computational models, the mixing repeats, and errors created by sequencing make it the most challenging type of variant to capture by short-read sequencing (Gong et al. 2023). The individual reference sequence can poorly provide complete information regarding the real-world blueprint of that species. Hence, the pressing priority was to develop a composite reference from multiple genomes with a wide range of variability and diversity at a particular chromosome site, termed a pangenome, a collection of all nonredundant DNA sequences in a population or species.

Multiple genomes of closely related species are now accessible by modern technologies, which was the major constraint of earlier reference genomes. This tool was earlier established for microorganisms and plants, whereas it was introduced to the vertebrate world. The first pangenome was developed for the single-celled microbe *Streptococcus agalactiae* in 2005. The first assembly of eukaryotic pangenomes started in 2014 in soybeans, followed by wheat, maize, and rice. Crop improvement utilizes the technique for various biotic and abiotic stress tolerance traits (Eisenstein 2023).

Over the last 20 years, genome diversity and evolution fields were transformed significantly by the advancements in high-throughput sequencing. Previously prevalent low-throughput Sanger sequencing methods utilized markers like simple sequence repeats (SSRs) for studying large populations (Schmid et al. 2005). Next-generation sequencing (NGS) encircled more focused studies around multiple genomes and populations (Redon et al. 2006). Pangenomes are also applicable in evolutionary and ecological studies as they provide evidence about the diversity of plant species and how they adapt to new selection pressures by humans or changing environments.

Due to the genomic processes such as gene duplication, genome rearrangements, transposon activity, deletions and recombination within the populations, the comparative studies are centred around SNPs of various plant species populations (Bennetzen 2000; Chen et al. 2015; Gabur et al. 2019; Yu et al. 2014). Even though the SNPs were the focus of genomic diversity studies earlier, the structural variations in genomes are gaining more significance now (Zhao et al. 2018). The genetic characterization of agronomic traits is widely identified from crops by the structural variants such as the presence–absence variants (PAVs) and copy number variants (CNVs) (Li et al. 2014; Lu et al. 2015; Springer et al. 2009).

The pangenome concept is again categorized into the core genome, which comprises genes universally present across all individuals, and variable genomes partially shared or unique to specific individuals (Tettelin et al. 2005). The variable genome, which includes genes associated with adaptation to biotic and abiotic stresses such as phosphorus deficiency in rice (Gamuyao et al. 2012), head smut resistance in maize (Zuo et al. 2015), and extreme temperatures (Tao et al. 2019), holds great potential for unravelling the genetic basis of these critical agronomic traits and driving advancements in crop improvement (Danilevicz et al. 2020).

To date, various methodologies have been employed to construct pangenomes in crop species, including iterative assembly approaches (Golicz et al. 2016; Hurgobin et al. 2018; Montenegro et al. 2017), map-to-pan approach (Lu et al. 2015), and comparative de novo approaches (Gordon et al. 2017; Li et al. 2014; Schatz et al. 2014; Zhao et al. 2018). In the iterative mapping and assembly, NGS reads are initially mapped to reference genomes and the sequences that are not aligned are extracted, which are used to supplement the reference genome and thereby utilized to construct the pangenome (Gong et al. 2023). This is applied to building pangenomes for species such as humans, red jungle fowl (*Gallus gallus*), and *Sorghum bicolour* L. (Li et al. 2021a, b; Ruperao et al. 2021; Wang et al. 2021).

According to Gong et al. (2023), in the map-to-pan method, the de novo assembled contigs are generated for each sample and aligned to the reference genome to identify non-reference sequences. These sequences added to the reference genome together form the species' pangenome. This strategy is used in the pangenome making of *Solanum lycopersicum* (Gao et al. 2019) and cotton (Li et al. 2021a, b). These two methods rely primarily on the previously generated short-read data. However, detection efficiency, accuracy, and the power to apprehend specific structural variants beyond the reference genome are the limitations of both methods.

In the de novo assembly approach, the genomes of each individual are assembled and annotated from the beginning, and a comparative analysis is performed to identify different fragments or genes (Gong et al. 2023). These will be defined as nonessential genes (such as in the pangenomes of *Sesamum indicum* (Yu et al. 2019) and *Sus scrofa* (Tian et al. 2020). This method can accurately classify sequence variants but involves high costs and assembly and annotation errors as the assemblies must be obtained from scratch for comparison.

In addition to these methods, a graph genome was recently constructed using bidirected variation graphs and Graph genome toolkits (Lei et al. 2021; Paten et al. 2017). The graphical pangenome is built by taking into consideration the genomic location of variable sequences from haploid/inbred organisms or completely haplotype-phased contigs based on de novo assemblies and adding them to the reference genome progressively (Garrison et al. 2018; Rakocevic et al. 2019). The potential of using genome maps includes improving the accuracy of read mapping and variant calling, reducing mapping bias and identifying ATAC-seq peaks and variants which cannot be identified by linear genomes, e.g. *Zea mays* (Lu et al. 2015), *Solanum pennelli* (Schmidt et al. 2017), *Solanum verrucosum* (Paajanen et al. 2019), *Glycine max* (Liu et al. 2020a, b), *Oryza sativa* (Qin et al. 2021), *Sorghum bicolour* (Tao et al. 2021), *Cicer arietinum* (Khan et al. 2024), *Phyllostachys edulis* (Hou et al. 2024), and *Citrullus* species (Zhang et al. 2024).

In developing the pangenome concept, the primary significance is for the core or dispensable genome, new sequences and SVs, including GWAS and transcriptomics, for examining the kinship between genes, their function, biology, and adaptation. Hence, among the four approaches for analysing short-read datasets, the first two methods are more reliable, which satisfies the need to use this method for large-scale genomic data analysis. The latter approach has the advantage of accurately mapping genes controlling traits and SVs, emphasizing the quantity and quality of de novo genome assembly. Graph-based pangenomics illustrates the heritability of various complex traits in crop breeding (Zhou et al. 2022). For example, the graph-based pangenome approach was used to identify missing heritability in diverse tomato accessions, which resulted in a 24% increase in heritability compared to previously measured methods (Gong et al. 2023). The graphical representation of pangenome has been widely used recently for its ability to precisely collect and present spatial information of genetic variation in the genome (Gong et al. 2023).

Pangenomics significantly contributes to crop breeding, mainly for understanding the genetic diversity of both wild and cultivated species. Recently, the world has witnessed various publications of crop pangenomes for soybean (Li et al. 2014), brassica (Lin et al. 2014), maize (Hirsch et al. 2014), and rice (Schatz et al. 2014). In molecular breeding and evolutionary studies, replacing single-sample reference genomes with crop pangenomes aims to limit sampling errors by providing a more comprehensive representation of genetic diversity (Golicz et al. 2016). For example, mapping sequenced data of 354 sorghum varieties of diverse genetic backgrounds to sorghum pangenome produced approximately two million SNPs, among which 398 were associated with agronomic traits and 1788 with drought stress response diversity (Golicz et al. 2016). Another significant advancement is the development of a super pangenome for plants. This involves analysing inter-individual gene exchange and genome evolution post polyploidy, which provides a theoretical basis for the rational use and conservation of wild germplasm resource diversity (Golicz et al. 2016). These findings about pangenomes highlight their potential to enhance genome-wide association studies (GWAS) and provide a stage for using SVs in developing molecular markers.

Even though advances in various research on pangenomes in domestic animals are projected, research in crops and humans highlights the significance of structural variants (SVs) in understanding species domestication, migration, and adaptation. The progress in next-generation sequencing (NGS), third-generation sequencing (TGS), and improved genomic analysis pave the way for future research in analysing genomic variation efficiently and accurately. Future crop pan-genomic research must include noncoding regions to address broader scientific questions.

9.3.3 Genome Editing

The creation of genetic variation through mutation breeding is usually random and tedious, whereas transgenic techniques face environmental and ethical challenges. This has paved the way for a newer genome editing technique with utmost precision and wide acceptance. With the increased availability of pangenomes and whole genome DNA sequences for a wide range of crops, desired variation could be created in crops with a new level of accuracy and predictability through genome editing previously unavailable for crop breeding purposes.

There are mainly three methods of genome editing by site-directed nuclease (SDN) (de Yassitepe et al. 2021). The SDN1 creates double-stranded DNA breaks, which are repaired by non-homologous end joining, leading to random nucleotide changes, finally resulting in frameshift mutations (Soppe and Lebbink 2017). SDN2 uses homologous recombination with a programmed DNA template to add, delete, or replace specific nucleotides. In addition, SDN3 adds gene segments or entire genes at specific sites through homologous recombination that results in transgenic products.

Hence, genome editing has become a creative technology in crop breeding that utilizes site-directed nucleases (SDNs) like meganucleases (MN), Zinc-finger nucleases (ZFN), transcription activators like effector nucleases (TALEN), and the most widely accepted clustered regularly interspaced short palindromic repeats CRISPR/CRISPR-associated (Cas) system (Gosal and Wani 2020).

Meganucleases (MN)/homing endonucleases were the first sequence-specific nucleases for targeted double-stranded breaks found in prokaryotes, archaebacteria, and unicellular eukaryotes (Wani and Aftab 2022). They can cleave dsDNA at specific recognition sites ranging from 14 to 40 bp, which is why it is categorized among the most specific naturally occurring restriction enzymes. They can recognize more extensive DNA sequences and modify their recognition sequences by enabling targeted replacement, elimination, or modification of any DNA sequence. Homing is a process of meganuclease that facilitates horizontal gene transfer by encoding with mobile genetic elements within introns or inteins. It has DNA binding and cleavage domains and double-strand breaks repaired by homologous recombination (HR), where the meganucleases are inserted into the gene of interest, allowing targeted genome engineering. This method has been applied since the 1990s for modifying plant genomes and understanding plant genome integrity. In 1999, SceI meganuclease was introduced into tobacco cells following the above-mentioned procedure. Similarly, in 2008, *BAR* gene expression was induced in maize that render phosphinothricin (PPT) resistance.

On the other hand, zinc finger nucleases (ZFNs) are artificially designed restriction enzymes containing non-specific cleavage domains linked to sequence-specific DNA binding domains (Plant Symposia 2014). DSBs created through ZFNs are repaired through error-prone non-homologous end joining (NHEJ), leading to InDels (Ciro et al. 2018). Likewise, specific DNA modifications can be made by supplying a donor DNA template homologous to the sequence, flanking the cleavage site through homology-directed repair (HDR) (Wang et al. 2017). Using ZFNs, targeted deletions of intervening DNA sequences can also be obtained by which concurrent DSBs are created. Similarly, site-specific transgene integration into ZFN-induced double-stranded breaks is possible through NHEJ or HDR (Plant Symposia 2014).

Following the footsteps of ZFNs came the TALE nucleases (TALEN) that revolutionized genome editing. These were the first-ever methods that could be designed and built relatively quickly to target any specific genomic locus with high precision and efficiency. TALEN was rapidly applied to crops, livestock, and other model and non-model organisms. This method was applied in curing cancer in 2015, and the first genome-edited crop (Calyno, a high oleic soybean oil) was produced in the market in 2019 (Gosal and Wani 2020). Later, in 2010, a modified TALEN was developed with a fusion of transcription activator-like effector (TALE) and the catalytic domain of the restriction endonuclease FokI (Becker and Boch 2021). These were more accessible and led to various applications in the general scientific community.

The CRISPR/Cas system is widely adopted for its rapid, accurate, and coherent method for creating targeted genetic variations like epigenetic changes, InDels and gene replacements across multiple loci simultaneously (Tang et al. 2017; Kushwaha and Mohan 2021). This technique depends on two main components: a site-specific nuclease-Cas nuclease (Cas9 or Cpf1) and a guide RNA (gRNA) (Ricroch et al. 2023).The gRNA is programmed to bind to specific target DNA sequences, thereby directing Cas nuclease to induce double-strand breaks (DSBs) at these sites. These

DSBs in plants are (Anuar et al. 2017) naturally repaired through the error-prone non-homologous end joining (NHEJ) pathway or homologous recombination (HR) that can lead to targeted modifications, including gene knock-ins, knock-outs, and structural alterations in crops. The comparison of all four methods is listed in Table 9.5.

Table 9.5 Comparison of various genome editing methods

Characteristic feature	Meganucleases	ZFNs	TALENs	CRISPR
Origin	Derived from microbial genetic elements	Derived from eukaryotic gene regulators	Derived from the bacterium *Xanthomonas*	Derived from adaptive immune system of bacteria and archae
Core components	DNA binding domain and DNA cleaving domain	Zinc finger domain with *FokI* nuclease	TALE-DNA binding domain with *FokI* nuclease	CrRNA, Cas9 protein
Dimerization requirement	Dimeric	Dimeric	Dimeric	Monomeric
Target sequence length	14–40 bp target sequence	24–36 bp target sequence	24–59 bp target sequence	20–22 bp target sequence
Mode of action	Single/chimeric work mode per target	Paired work mode per target	Paired work mode per target	Single work mode per target
Catalytic domain	DNA binding site contains catalytic domain	Catalytic domain	*FokI* catalytic domain	RUVC and HNH catalytic domain
Protein engineering criteria	Protein engineering is required	Protein engineering is required	Protein engineering is required	Not difficult to test gRNA
Cloning criteria	Cloning is essential	Cloning is essential	Cloning is essential	Cloning is not essential
gRNA requirement	Required	Required	Required	Not required and is easy to produce
Genome altering	Generates DSB in target DNA	Generates DSB in target DNA	Generates DSB in target DNA	Generates single-strand nicks or DSB in target DNA
Delivery ease	Easy in vivo delivery; small size allows use in a variety of viral vectors	Easy in vivo delivery; small size allows use in a variety of viral vectors	Difficult in vivo delivery; large size and repetitive nature causes unwanted recombination events	Moderately easy in vivo delivery; small size allows use in a variety of viral vectors (Cas9 involves packaging problems in some vectors)
Target recognition efficiency	Low to moderate	Moderate	Moderate	High

(continued)

Table 9.5 (continued)

Characteristic feature	Meganucleases	ZFNs	TALENs	CRISPR
Mutation rate	High	Medium	Medium	Low
Multiplexing	Not possible	Difficult	Difficult	Easier
Methylation sensitivity	High	High	High	Low
Cytotoxic effect	Variable to high	Variable to high	Low	Low
Vector packaging	Easy	Difficult	Difficult	Moderate
Cost effectiveness	Not cost effective	Not cost effective	Moderately cost effective	Highly cost effective

All the tools in genome editing allow crop breeders to alter target genes to improve crop yield, quality, stress tolerance, and herbicide resistance, which makes the technique the next-generation breeding strategy. The CRISPR/Cas methods transformed plant breeding by contributing specific modifications to single gene traits without compromising the agronomic qualities. The significant constraints of cereal crops include mineral accumulation, stress tolerance, quality improvement, and domestication of wild species using these technologies (Chandra et al. 2020). However, new developments in the method, such as creating a multi-genome editing toolkit, facilitate high-throughput genome editing in various plant species. In addition, CRISPR/Cas technology advancements must be made to overcome environmental limitations and improve agricultural yield under challenging circumstances.

Examples of these breakthroughs include high-amylopectin waxy maize and browning-resistant mushrooms. To address issues like disease and drought, CRISPR-Cas9 technology has aided in the quick production of cereal crops that are both climate-resilient and more nutritious. By precisely targeting genes for enhanced features, merging CRISPR with genetic diversity research and transcriptomics promises to improve breeding efforts for potatoes, which face several problems connected to the environment and pests. Scientists are planting a CRISPR/Cas9-modified rice variety in test fields across India after a recent relaxation of regulations on gene-edited crops, mainly altering the plant traits through SDN1 and SDN2 gene editing processes without adding any foreign DNA. CRISPR/Cas9-edited mustard with less glucosinolate content, drought and salt-tolerant rice, and Vitamin A-fortified bananas that showcase the potential of gene editing in crop improvement.

Regulation and proper legislation are critical for marketing genetically edited crops as in the case of transgenic breeding. Even though genome editing involves small InDels, base-pair changes, and specific short sequence changes through homologous recombination, which are indistinguishable from the natural genome variants, many countries do not count these types of mutants as genetically modified organisms (GMO) and are exempted from GMO regulations (Kim and Kim 2016; Turnbull et al. 2021). Gene editing is widespread in countries like the United States,

Japan, and China, where they are approved for release for commercial cultivation. A stable regulatory framework is essential to ensure the safety and efficacy of these innovations while promoting scientific progress.

9.3.4 Phenomics and Artificial Intelligence

During the 1990s, a new discipline complements to genomics emerged in complex human diseases to bridge the gap between genes and clinical endpoints (Zavafer et al. 2023). In 2020, Yang et al. defined phenomics as acquiring high-dimensional phenotypic data on an organism-wide scale. In plant science, Fiorani and Schurr (2013) referred to plant phenotyping as the methodologies and protocols used to accurately measure plant growth, architecture, and composition at different scales. Similarly, we can define crop phenomics as the multidisciplinary study of high-throughput, accurate acquisition, and analysis of multidimensional phenotypes on an organism-wide scale through crop development (Yang et al. 2020). The advanced tools of next-generation sequencing, like quantitative trait loci (QTL) mapping and genome-wide association studies (GWASs), have identified many genes controlling major agronomic traits. For example, 3000 rice genes were well characterized in 2017 compared to 130 genes in 2000. In addition to that, the acquisition of phenotypic data in traditional crop breeding remains labour-intensive, time-consuming, and many times destructive to plants. Even though the omics approaches progressed rapidly, the phenotyping technologies have lagged and often need to be completed. This has led to the evolution of an omics approach to phenotyping data.

Since the last decade, plant phenomics has developed into a more comprehensive research area by including multidimensional phenotypic data from the cell to the population level. The phenotypic expression of plants is the outcome of the interaction of genotype with the environmental parameters that affect the growth and development of the crops measured by both structural traits and physiological traits (Etingoff 2014). The internal phenotypes correlate with external phenotypes like morphology, biomass, and yield. Hence, to obtain complete information on crop phenomics, integrating agronomy, information technology, mathematics, life science, and engineering with high-performance computing and artificial intelligence is pivotal for crops grown in a complex environment. Hence, the goal is to develop an effective technical system for phenotyping crops in a high-throughput, multidimensional, automated measuring integrating large datasets with intelligence (Zhao et al. 2019).

High-throughput (HT) phenotyping uses cutting-edge technologies like aerial imagery and remote sensing to quickly and reliably gather large amounts of data in agriculture. These techniques record spatiotemporal fluctuations within fields, which is essential for narrowing down huge population genotypes with desirable features (Fahlgren et al. 2015). Unmanned aerial vehicles (UAVs) equipped with multispectral and hyperspectral cameras provide remote sensing, which detects reflectance over a range of electromagnetic spectrum wavelengths to evaluate the health and growth of plants (Kim et al. 2020). Furthermore, vegetation colour is analysed using RGB colour space models, which offer essential insights into crop phenotypes (Kushwaha and Mohan 2021). To support efforts to enhance crops, these spectral data are transformed into indices that can forecast characteristics like leaf chlorophyll concentration and biomass under stress (Condorelli et al. 2018). Extensive phenotyping techniques for Crop morphology, structure, and physiology across various scales are discussed below.

9.3.4.1 Addressing the Limitations of Micro-Phenotyping

Compared to whole-plant phenotyping, tissue and cellular-level phenotyping involves significant challenges due to the destructive nature of pre-treatment procedures and the inefficiency of high-resolution imaging at the micron level (Hall et al. 2016). These limitations are overcome by various algorithms and tools like computer-aided phenotypic extraction methods, as in the case of wheat roots (Wu et al. 2011), high-throughput 3D imaging for root anatomy (Burton et al. 2012), and automated software for root phenotyping (Chopin et al. 2015). The efficiency and accuracy for measuring vascular bundles were significantly improved by the innovations in X-ray micro-CT in maize root tissue analysis (Pan et al. 2017). Similarly, automated and semi-automated methods were developed for data analysis to reduce the complexity of crop stalk microstructures (Legland et al. 2017; Zhang et al. 2013). The maize stalks vascular bundles analysis by Du et al. (2016) was advanced using tools such as VesselParser 1.0 algorithm, and the phenotyping capacity is aimed for further enhancement by X-ray contrast techniques and image processing workflows (Zhang et al. 2018).

9.3.4.2 Three-Dimensional Phenotyping Techniques for Organ-Level Analysis

The phenotyping accuracy at the organ level is usually compromised in high-throughput plant phenotyping (HTPPs). In traditional methods and smartphone apps, the leaf angles and lengths can be captured with 2D imaging but lack critical 3D data that can be addressed by multi-view stereo (MVS) for organ-level analysis (Duan et al. 2016). High-resolution 3D scanners and LiDAR systems produce comprehensive phenotypic data but are expensive and limited to tall plants (Hu et al. 2018; Rist et al. 2018). For extensive plant phenotypic data gathering efficiently, advanced automated HTPPs in controlled environments with a variety of imaging techniques are used, e.g. RGB, chlorophyll fluorescence, hyperspectral, thermal, and LiDAR (Di Gennaro et al. 2018). Being less accessible and less affordable especially in developing regions, new efforts are made to create small-scale phenotyping technologies that can normalize its usage for analysing complex traits (Pereyra-Irujo et al. 2012).

9.3.4.3 Automated Phenotyping Methods in Controlled Environments

In crop improvement programmes, cellular, tissue, organ, and plant-level phenotyping are crucial; at the same time, field-based phenotyping (FBP) is essential for capturing the effects of genetics and environmental parameters on primary traits like yield and stress tolerance. According to White et al. (2012) and Zhang and Kovacs (2012), traditional ground-based platforms with motorized gadgets and rigid stages with cameras and sensors become inefficient for larger plots due to high maintenance costs, scalability, and high construction. Recent advancements, including FIELD SCANALYZERS and cable-suspended platforms, are convenient for more diminutive crops and enable continuous and high-throughput monitoring for more extensive areas but remain costly. Unmanned aerial vehicles (UAVs) and manned aircraft were promoted as adaptable tools for field-level phenotyping that offer high spatial, spectral, and temporal resolutions with sensors unique for various crop characters. However, these methods exhibit challenges in data processing speed, airspace regulations, and weather responsiveness (Berni et al. 2009; Yang et al. 2017). Hence, a comprehensive approach combining ground-based and aerial platforms can provide flexibility, such as tractor-based systems paired with UAVs for detailed crop trait analysis (Furbank et al. 2019; Potgieter et al. 2018).

After collecting comprehensive data through various techniques mentioned so far, the second phase is the phenotypic data extraction, standardization, storage, and analysis. In addition to image-based phenotyping, machine learning techniques, especially deep learning, can improve trait detection and analysis accuracy and automation. Large volumes of phenotypic data need to be standardized and stored based on principles such as the minimum information approach and cloud-based storage solutions. The phenotypic and genotypic association studies favoured by the genomic data with phenotypic variation can help understand the genetic foundation for complex crop traits that can support QTL mapping and GWAS for crop improvement. These techniques open our understanding of plant genetics to improve crop breeding programmes by overcoming challenges like data integration and the continued development of artificial intelligence technologies to manage complex phenomic data.

9.3.5 Machine Learning

Recent advancements in high-throughput techniques have generated vast datasets of plant genotypes and phenotypes, necessitating sophisticated methods to extract meaningful insights and integrate diverse datasets. Concurrently, machine learning has rapidly progressed and is now extensively utilized in plant genotyping and phenotyping (van Dijk et al. 2021). Genomics not only involves acquiring molecular phenotypes but also employs robust data mining tools to predict and interpret them effectively (Wang et al. 2020). Machine learning, a burgeoning interdisciplinary field, offers computational models that analyse large, heterogeneous, and unstructured datasets on a Big Data scale, proving invaluable in biological research (Jordan and Mitchell 2015; Ma et al. 2014). Machine learning algorithms learn from data patterns without explicit programming rules to efficiently uncover core structures, similarities, and dissimilarities to classify new data experiences accurately (Singh et al. 2016). These algorithms are adept at handling massive datasets plagued by high noise levels, dimensionality, or incompleteness (Liu et al. 2020a, b; Mahood et al. 2020).

Machine learning encompasses three fundamental approaches: supervised and unsupervised, linear and non-linear algorithms, and shallow and deep learning models. Supervised learning categorizes each example in a dataset and aims to build models that map predictors (e.g. DNA sequences) to target variables, e.g. histone marks (Liu et al. 2020a, b). Applications of supervised learning in genomics include predicting regulatory regions in maize genomes (Mejía-Guerra and Buckler 2019), mRNA expression levels (Washburn et al. 2019), and stress responses in plants, such as identifying polyadenylation sites in *Arabidopsis* (Gao et al. 2018) and nutrient deficiencies in tomato (Tran et al. 2019). On the other hand, unsupervised learning utilizes algorithms like clustering to analyse data without predefined labels (Libbrecht and Noble 2015), while semi-supervised (that comes in between supervised and unsupervised) learning utilizes labelled and unlabelled data examples (Libbrecht and Noble 2015). Additionally, machine learning-driven digital imaging has successfully detected diseases in various crop plants, such as bacterial blight in rice (Lu et al. 2017), maize (DeChant et al. 2017), soybean (Ghosal et al. 2018), and tomato (Prabhakar et al. 2020). Techniques like hyperspectral imaging have also been instrumental in detecting diseases like yellow rust in wheat (Zhang et al. 2019) and *potato virus Y* in potatoes (Polder et al. 2019). Non-linear, non-parametric methods of machine learning algorithms include artificial neural networks (ANNs), deep neural networks (DNNs), convolutional neural networks (CNNs), random forest (RF), and support vector machines (SVMs) (Niazian and Niedbała 2020) (Table 9.6).

Machine learning offers more comprehensive applications in crop breeding that include:

Table 9.6 The merits and demerits of non-linear machine learning algorithms

Learning algorithm	Merits	Demerits
ANNs	Good learning capabilities	Overfitting and local minima in a small number of training data Lack of interpretation capability Implementing a very small number of hidden neurons
CNNs	Ability of automatic feature extraction	Absence of interpretation capability Large amount of data is required for training Skill and experience are required to select suitable hyperparameters
SVMs	Use great number of hidden units For formulation of learning problems, quadratic optimization task is used	Shallow architecture
RF	Noise handling ability It prevents overfitting Large number of features can be managed	Shallow architecture

9.3.5.1 Classifying and Analysing Genetic Diversity in Plants

Genetic diversity, the crucial factor for selecting and developing valuable accessions for future use, was determined using molecular markers such as AFLP, RFLP, RAPD, SSR, ISSR, and SNP that provided more precise assessment. With the advent of deep learning, especially CNNs, processing morphological parameters, and DNA sequences with high accuracy is possible, which is an efficient alternative to traditional methods. For example, CNNs outmatched traditional methods in classifying morphologically similar species and have attained high classification accuracy in genetic diversity studies of soybeans and *Cinnamomum osmophloeum*. The neural networks make applying genomic predictions and selection and improve performance in predicting phenotypic traits from genomic data easier than other methods.

9.3.5.2 Yield Component Analysis and Predictive Selection

Yield is a complex trait directed by many genes and environmental interactions that lead to low heritability and affect direct selection. However, increased economic yield, such as seed, oil content, and biomass, is the primary goal of any plant breeding programme. Hence, to address this issue, breeders usually select highly correlated traits as indirect selection criteria for yield. Even though traditional statistical approaches like correlation coefficient analysis, phenotypic combining ability (PCA), path analysis, and multiple regression are often used for this purpose, the non-linear relationships cannot be measured using these methods. Machine learning algorithms, primarily artificial neural networks (ANNs), provide advanced solutions by productively handling non-linearities to predict yield with high accuracy compared to linear models.

9.3.5.3 Yield Stability and GxE Interactions

The genotype-environment (GxE) interaction presents annual yield fluctuations and phenotypic changes that challenge genotypic selection. The conventional methods involve Finlay and Wilkinson's regression, Eberhart and Russell's regression coefficient analysis, and non-parametric approaches used for stability analysis. However, these methods require more simplicity and specific data assumptions. Similarly, less accessible and complex multivariate analysis techniques like principal component analysis, Additive Main Effects and Multiplicative Interaction (AMMI), and Genotype + Genotype × Environment (GGE) biplots were used for more precise GxE analysis. At the same time, artificial neural networks (ANNs) provide a more efficient alternative and accurate and rapid assessment of GxE by taking both linear and non-linear relationships between variables, thereby improving the evaluation of genetic stability and plant–environment interactions.

9.3.5.4 Assessment of Biotic and Abiotic Stresses

Numerous methods are adopted to develop crop plants that are resistant and tolerant to various biotic and abiotic stresses. The morphological data obtained under stressful and normal conditions were collected to develop various indices like tolerance (TOL), mean productivity (MP), and stress tolerance index (STI) to evaluate drought tolerance. Similarly, various physiological and biochemical responses and the production of secondary metabolites and cellular antioxidants also play crucial roles in tolerance levels. In machine learning, deep convolutional neural networks (CNNs) provide advanced solutions for high-throughput stress phenotyping by combining phenomic, metabolomic, and genomic data, improving prediction accuracy and selecting stress-resistant genotypes.

9.3.5.5 Conventional Mating Designs and Hybrid Breeding Approaches

Applying statistics in crop breeding started way beyond classical mating designs—mean generation analysis, line-tester analysis, diallel crosses, North Carolina designs and triple test crosses that helped study gene actions and estimate heritability. The outcome of cross-breeding methods can be predicted, and the genetic control of critical traits could be understood using these methods. For the selection of superior homozygous lines in F1-hybrid, the prediction of the performance of parental combination is critical, but regulation of numerous cross combinations is necessary and costly. This is made possible by artificial neural networks (ANNs) that can predict yield performances and efficiently calculate general and specific combining abilities (GCA and SCA). Khaki et al. (2020) demonstrated that a combination of matrix factorization with the ANN model outperformed methods like deep factorization and generalized matrix factorization for yield prediction of maize cross combinations based on historical data.

Crop breeding is not just the hybridization of two superior plants nowadays. Science and technology have outspoken that it is used for broader applications. The handling of more extensive data was the limitation addressed by the collaboration of artificial intelligence in analysing complex phenotypic information.

The challenges the high-throughput, multi-dimensional, and multi-scale data still face, emphasize the need for comprehensive phenomics at various levels. Advancement in AI techniques like machine learning for managing and interpreting numerous digital data. Modelling tools must be validated and refined to better use and understand genotype–environment interactions. Detailed analysis of the genotype–phenotype–environment interaction must be developed with intelligent data mining analytics for a better and more resilient superior crop.

9.4 Conclusion

The post-genomic era has revolutionized crop breeding by offering extraordinary insights into various crop traits' genetic and molecular basis. Next-generation sequencing and advanced bioinformatics tools coming under high-throughput genomic technologies have significantly enhanced the ability to identify and manipulate genes associated with particular traits of interest, and this period yielded improved crop varieties with enhanced productivity, climate resilience, nutritional superiority, and various complex challenges during the genomic era. Developing crops of choice in the future will transcend our current expectations. Integrating newer techniques will unlock possibilities far beyond our imagination in developing designer crops as illustrated in Fig. 9.2 (Naqvi et al. 2022).

However, the transition from the genomic era to the present state has represented various challenges, including the need for robust data integration, the development

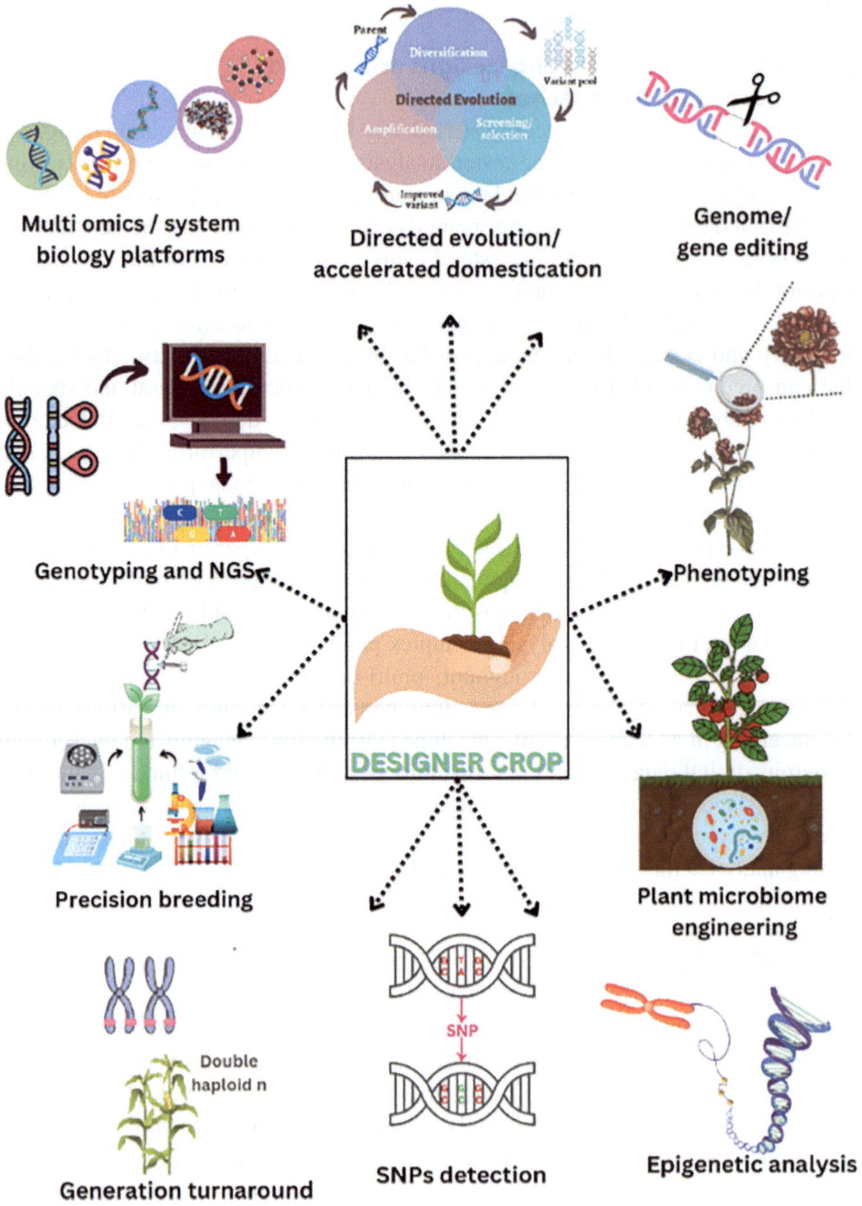

Fig. 9.2 Futuristic approach for developing crops of choice. (Source: Adapted from Naqvi et al. (2022), licensed under CC BY 4.0)

of new breeding strategies that include genomic data, and the addressing of ethical and regulatory concerns regarding the commercialization of genetic modifications. The succeeding challenge for these methods will depend on how well the breeders can adopt the techniques effectively by ensuring an equitable approach and sustainability without compromising environmental concerns.

For the prospects, interdisciplinary collaboration, continuous innovation, and regular monitoring are crucial in accelerating genomic advances to attain food security that is on par with environmental sustainability goals. The kinship of genomic research and practical breeding applications offers a new edge in crop improvement, indicating a period where agriculture can adapt to changing conditions and demands.

References

Adhikari S, Joshi A, Chandra AK, Bharati A, Sarkar S, Dinkar V, Kumar A, Singh AK (2023) SMART plant breeding from pre-genomic to post-genomic era for developing climate-resilient cereals. In: Smart plant breeding for field crops in post-genomics era. Springer Nature Singapore, pp 41–97. https://doi.org/10.1007/978-981-19-8218-7_2

Al-Dous EK, George B, Al-Mahmoud ME, Al-Jaber MY, Wang H, Salameh YM, Malek JA (2011) De novo genome sequencing and comparative genomics of date palm (*Phoenix dactylifera*). Nat Biotechnol 29(6):521–527. https://doi.org/10.1038/nbt.1860

Al-Mssallem, I., Hu, S., Zhang, X. et al. Genome sequence of the date palm *Phoenix dactylifera* L. Nat Commun 4, 2274 (2013). https://doi.org/10.1038/ncomms3274

Anuar ND, Field M, Kurscheid S, Zhang L, Rebar E, Gregory P, Soboleva T (2017) Gene editing of the multi-copy H2A.B gene family by a single pair of TALENS. bioRxiv:233379. https://doi.org/10.1101/233379

Argout X, Salse J, Aury J-M, Guiltinan MJ, Droc G, Gouzy J, Lanaud C (2011) The genome of *Theobroma cacao*. Nat Genet 43(2):101–108. https://doi.org/10.1038/ng.736

Badouin H, Gouzy J, Grassa CJ, Murat F, Staton SE, Cottret L, Langlade NB (2017) The sunflower genome provides insights into oil metabolism, flowering and Asterid evolution. Nature 546(7656):148–152. https://doi.org/10.1038/nature22380

van Bakel H, Stout JM, Cote AG, Tallon CM, Sharpe AG, Hughes TR, Page JE (2011) The draft genome and transcriptome of *Cannabis sativa*. Genome Biol 12(10):R102. https://doi.org/10.1186/gb-2011-12-10-r102

Becker S, Boch J (2021) TALE and TALEN genome editing technologies. Gene Genome Edit 2:100007. https://doi.org/10.1016/j.ggedit.2021.100007

Beier S, Himmelbach A, Colmsee C, Zhang X-Q, Barrero RA, Zhang Q, Mascher M (2017) Construction of a map-based reference genome sequence for barley, Hordeum vulgare L. Sci Data 4(1):170044. https://doi.org/10.1038/sdata.2017.44

Bennetzen JL (2000) Transposable element contributions to plant gene and genome evolution. Plant Mol Biol 42(1). https://doi.org/10.1023/A:1006344508454

Bennetzen JL, Schmutz J, Wang H, Percifield R, Hawkins J, Pontaroli AC, Devos KM (2012) Reference genome sequence of the model plant Setaria. Nat Biotechnol 30(6):555–561. https://doi.org/10.1038/nbt.2196

Berni J, Zarco-Tejada PJ, Suarez L, Fereres E (2009) Thermal and narrowband multispectral remote sensing for vegetation monitoring from an unmanned aerial vehicle. IEEE Trans Geosci Remote Sens 47(3):722–738. https://doi.org/10.1109/TGRS.2008.2010457

Bredeson JV, Lyons JB, Oniyinde IO, Okereke NR, Kolade O, Nnabue I, Rokhsar DS (2022) Chromosome evolution and the genetic basis of agronomically important traits in greater yam. Nat Commun 13(1):2001. https://doi.org/10.1038/s41467-022-29114-w

Brenchley R, Spannagl M, Pfeifer M, Barker GLA, D'Amore R, Allen AM, Hall N (2012) Analysis of the bread wheat genome using whole-genome shotgun sequencing. Nature 491(7426):705–710. https://doi.org/10.1038/nature11650

Burton AL, Williams M, Lynch JP, Brown KM (2012) RootScan: software for high-throughput analysis of root anatomical traits. Plant Soil 357(1–2):189–203. https://doi.org/10.1007/s11104-012-1138-2

Cannarozzi G, Plaza-Wüthrich S, Esfeld K, Larti S, Wilson YS, Girma D, Tadele Z (2014) Genome and transcriptome sequencing identifies breeding targets in the orphan crop tef (*Eragrostis tef*). BMC Genomics 15(1):581. https://doi.org/10.1186/1471-2164-15-581

Chalhoub B, Denoeud F, Liu S, Parkin IAP, Tang H, Wang X, Wincker P (2014) Plant genetics. Early allopolyploid evolution in the post-Neolithic *Brassica napus* oilseed genome. Science (New York, N.Y.) 345(6199):950–953. https://doi.org/10.1126/science.1253435

Chan AP, Crabtree J, Zhao Q, Lorenzi H, Orvis J, Puiu D, Rabinowicz PD (2010) Draft genome sequence of the oilseed species *Ricinus communis*. Nat Biotechnol 28(9):951–956. https://doi.org/10.1038/nbt.1674

Chandra AK, Kumar A, Bharati A, Joshi R, Agrawal A, Kumar S (2020) Microbial-assisted and genomic-assisted breeding: a two way approach for the improvement of nutritional quality traits in agricultural crops. 3 Biotech 10(1). https://doi.org/10.1007/s13205-019-1994-z

Chen JY, Huang JQ, Li NY, Ma XF, Wang JL, Liu C, Liu YF, Liang Y, Bao YM, Dai XF (2015) Genome-wide analysis of the gene families of resistance gene analogues in cotton and their response to Verticillium wilt. BMC Plant Biol 15(1). https://doi.org/10.1186/s12870-015-0508-3

Chen ZJ, Sreedasyam A, Ando A, Song Q, De Santiago LM, Hulse-Kemp AM, Schmutz J (2020) Genomic diversifications of five Gossypium allopolyploid species and their impact on cotton improvement. Nat Genet 52(5):525–533. https://doi.org/10.1038/s41588-020-0614-5

Chopin J, Laga H, Huang CY, Heuer S, Miklavcic SJ (2015) RootAnalyzer: a cross-section image analysis tool for automated characterization of root cells and tissues. PLoS One 10(9):e0137655. https://doi.org/10.1371/journal.pone.0137655

Ciro Gianmaria Amoroso Maria Raffaella Ercolano Dott Gelsomina Formisano, D (2018) Genomic approaches to improve tomato response to multiple stress

Condorelli GE, Maccaferri M, Newcomb M, Andrade-Sanchez P, White JW, French AN, Sciara G, Ward R, Tuberosa R (2018) Corrigendum: comparative aerial and ground based high throughput phenotyping for the genetic dissection of NDVI as a proxy for drought adaptive traits in durum wheat. Front Plant Sci 9:893. https://doi.org/10.3389/fpls.2018.00893

Daniell H, Wurdack KJ, Kanagaraj A, Lee SB, Saski C, Jansen RK (2008) The complete nucleotide sequence of the cassava (*Manihot esculenta*) chloroplast genome and the evolution of atpF in Malpighiales: RNA editing and multiple losses of a group II intron. Theor Appl Genet 116(5):723–737. https://doi.org/10.1007/s00122-007-0706-y

Danilevicz MF, Tay Fernandez CG, Marsh JI, Bayer PE, Edwards D (2020) Plant pangenomics: approaches, applications and advancements. Curr Opin Plant Biol 54. https://doi.org/10.1016/j.pbi.2019.12.005

DeChant C, Wiesner-Hanks T, Chen S, Stewart EL, Yosinski J, Gore MA, Nelson RJ, Lipson H (2017) Automated identification of northern leaf blight-infected maize plants from field imagery using deep learning. Phytopathology 107(11). https://doi.org/10.1094/PHYTO-11-16-0417-R

Denoeud F, Carretero-Paulet L, Dereeper A, Droc G, Guyot R, Pietrella M, Lashermes P (2014) The coffee genome provides insight into the convergent evolution of caffeine biosynthesis. Science (New York, N.Y.) 345(6201):1181–1184. https://doi.org/10.1126/science.1255274

Di Gennaro SF, Rizza F, Badeck FW, Berton A, Delbono S, Gioli B, Toscano P, Zaldei A, Matese A (2018) UAV-based high-throughput phenotyping to discriminate barley vigour with visible and near-infrared vegetation indices. Int J Remote Sens 39(15–16):5330–5344. https://doi.org/10.1080/01431161.2017.1395974

van Dijk ADJ, Kootstra G, Kruijer W, de Ridder D (2021) Machine learning in plant science and plant breeding. Science 24(1). https://doi.org/10.1016/j.isci.2020.101890

Dohm JC, Minoche AE, Holtgräwe D, Capella-Gutiérrez S, Zakrzewski F, Tafer H, Himmelbauer H (2014) The genome of the recently domesticated crop plant sugar beet (*Beta vulgaris*). Nature 505(7484):546–549. https://doi.org/10.1038/nature12817

Dreher K, Morris ML, Khairallah M, Ribaut JM, Pandey S, Srinivasan G, Zilberman D (2002) Is marker-assisted selection cost-effective compared with conventional plant breeding methods? The case of quality protein maize. In: Economic and social issues in agricultural biotechnology. CABI publishing, pp 203–236. https://doi.org/10.1079/9780851996189.0203

Du J, Zhang Y, Guo X, Ma L, Shao M, Pan X, Zhao C (2016) Micron-scale phenotyping quantification and three-dimensional microstructure reconstruction of vascular bundles within maize stalks based on micro-CT scanning. Funct Plant Biol: FPB 44(1):10–22. https://doi.org/10.1071/FP16117

Duan T, Chapman SC, Holland E, Rebetzke GJ, Guo Y, Zheng B (2016) Dynamic quantification of canopy structure to characterize early plant vigour in wheat genotypes. J Exp Bot 67(15):4523–4534. https://doi.org/10.1093/jxb/erw227

Eisenstein M (2023) Every base everywhere all at once: pangenomics comes of age. Nature 616(7957):618–620. https://doi.org/10.1038/d41586-023-01300-w

Etingoff K (ed) (2014) Organic agricultural practices. Apple Academic Press. https://doi.org/10.1201/b17368

Fahlgren N, Gehan MA, Baxter I (2015) Lights, camera, action: high-throughput plant phenotyping is ready for a close-up. Curr Opin Plant Biol 24. https://doi.org/10.1016/j.pbi.2015.02.006

Farrakh S, Khalid S, Rafique A, Riaz N, Mujeeb-Kazi A (2016) Identification of stripe rust resistant genes in resistant synthetic hexaploid wheat accessions using linked markers. Plant Genet Resour: Characterisation and Utilisation 14(3). https://doi.org/10.1017/S1479262115000283

Fiorani F, Schurr U (2013) Future scenarios for plant phenotyping. Annu Rev Plant Biol 64(1):267–291. https://doi.org/10.1146/annurev-arplant-050312-120137

FAO (1996) Food and Agriculture Organization of the United Nations; Rome, Italy: 1996. Report of the World Food Summit, 13-17 November 1996

Fraley RT, Rogers SG, Horsch RB, Sanders PR, Flick JS, Adams SP, Bittner ML, Brand LA, Fink CL, Fry JS, Galluppi GR, Goldberg SB, Hoffmann NL, Woo SC (1983) Expression of bacterial genes in plant cells (plant protoplasts/transformation/foreign DNA/antibiotic resistance/selectable markers). Genetics 80

Furbank RT, Jimenez-Berni JA, George-Jaeggli B, Potgieter AB, Deery DM (2019) Field crop phenomics: enabling breeding for radiation use efficiency and biomass in cereal crops. New Phytol 223(4):1714–1727. https://doi.org/10.1111/nph.15817

Gabur I, Chawla HS, Snowdon RJ, Parkin IAP (2019) Connecting genome structural variation with complex traits in crop plants. Theor Appl Genet 132(3). https://doi.org/10.1007/s00122-018-3233-0

Gamuyao R, Chin JH, Pariasca-Tanaka J, Pesaresi P, Catausan S, Dalid C, Slamet-Loedin I, Tecson-Mendoza EM, Wissuwa M, Heuer S (2012) The protein kinase Pstol1 from traditional rice confers tolerance of phosphorus deficiency. Nature 488(7412). https://doi.org/10.1038/nature11346

Gao X, Zhang J, Wei Z, Hakonarson H (2018) DeepPolyA: a convolutional neural network approach for polyadenylation site prediction. IEEE Access 6. https://doi.org/10.1109/ACCESS.2018.2825996

Gao L, Gonda I, Sun H, Ma Q, Bao K, Tieman DM, Burzynski-Chang EA, Fish TL, Stromberg KA, Sacks GL, Thannhauser TW, Foolad MR, Diez MJ, Blanca J, Canizares J, Xu Y, van der Knaap E, Huang S, Klee HJ, Fei Z (2019) The tomato pan-genome uncovers new genes and a rare allele regulating fruit flavor. Nat Genet 51(6):1044–1051. https://doi.org/10.1038/s41588-019-0410-2

Garrison E, Sirén J, Novak AM, Hickey G, Eizenga JM, Dawson ET, Jones W, Garg S, Markello C, Lin MF, Paten B, Durbin R (2018) Variation graph toolkit improves read mapping by representing genetic variation in the reference. Nat Biotechnol 36(9):875–879. https://doi.org/10.1038/nbt.4227

Ghosal S, Blystone D, Singh AK, Ganapathysubramanian B, Singh A, Sarkar S (2018) An explainable deep machine vision framework for plant stress phenotyping. Proc Natl Acad Sci USA 115(18). https://doi.org/10.1073/pnas.1716999115

Goff SA, Ricke D, Lan T, Presting G, Wang R, Dunn M, Glazebrook J (2002) A draft sequence of the rice genome *(Oryza sativa L). Oryza Sativa L. Ssp. Japonica*. Science 296(5565):92–100

Golicz AA, Bayer PE, Barker GC, Edger PP, Kim HR, Martinez PA, Chan CKK, Severn-Ellis A, McCombie WR, Parkin IAP, Paterson AH, Pires JC, Sharpe AG, Tang H, Teakle GR, Town CD, Batley J, Edwards D (2016) The pangenome of an agronomically important crop plant *Brassica oleracea*. Nat Commun 7. https://doi.org/10.1038/ncomms13390

Gong Y, Li Y, Liu X, Ma Y, Jiang L (2023) A review of the pangenome: how it affects our understanding of genomic variation, selection and breeding in domestic animals? J Anim Sci Biotechnol 14(1). BioMed Central Ltd. https://doi.org/10.1186/s40104-023-00860-1

González VM, Benjak A, Hénaff EM, Mir G, Casacuberta JM, Garcia-Mas J, Puigdomènech P (2010) Sequencing of 6.7 Mb of the melon genome using a BAC pooling strategy. BMC Plant Biol 10(1):246

Gordon JW, Ruddle FH (1981) Integration and stable germ line transmission of genes injected into mouse pronuclei. New Series 214(4526)

Gordon SP, Contreras-Moreira B, Woods DP, Des Marais DL, Burgess D, Shu S, Stritt C, Roulin AC, Schackwitz W, Tyler L, Martin J, Lipzen A, Dochy N, Phillips J, Barry K, Geuten K, Budak H, Juenger TE, Amasino R, Vogel JP (2017) Extensive gene content variation in the *Brachypodium distachyon* pan-genome correlates with population structure. Nat Commun 8(1). https://doi.org/10.1038/s41467-017-02292-8

Gosal SS, Wani SH (eds) (2020) Accelerated plant breeding, vol 1, 1st edn. https://doi.org/10.1007/978-3-030-41866-3

Guo S, Zhang J, Sun H, Salse J, Lucas WJ, Zhang H, Zheng Y et al (2013) The draft genome of watermelon (*Citrullus lanatus*) and resequencing of 20 diverse accessions. Nat Genet 45(1):51–58

Gupta PK, Kumar J, Mir RR, Kumar A (2010) Marker-assisted selection as a component of conventional plant breeding. Plant Breed Rev 33. https://doi.org/10.1002/9780470535486.ch4

Hall HC, Fakhrzadeh A, Luengo Hendriks CL, Fischer U (2016) Precision automation of cell type classification and sub-cellular fluorescence quantification from laser scanning confocal images. Front Plant Sci 7. https://doi.org/10.3389/fpls.2016.00119

Hirakawa H, Shirasawa K, Miyatake K, Nunome T, Negoro S, Ohyama A, Yamaguchi H et al (2014) Draft genome sequence of eggplant (Solanum melongena L.): the representative solanum species indigenous to the old world. DNA Res 21(6):649–660

Hirsch CN, Foerster JM, Johnson JM, Sekhon RS, Muttoni G, Vaillancourt B, Peñagaricano F, Lindquist E, Pedraza MA, Barry K, de Leon N, Kaeppler SM, Robin Buell C (2014) Insights into the maize pan-genome and pan-transcriptome. Plant Cell 26(1):121–135. https://doi.org/10.1105/tpc.113.119982

Hont D, Denoeud A, Aury F, Baurens J-M, Carreel F-C, Garsmeur F, Noel O (2012) The banana *(Musa acuminata*) genome and the evolution of monocotyledonous plants. Nature 488(7410):213–217

Hou Y, Gan J, Fan Z, Sun L, Garg V, Wang Y, Zhao H (2024) Haplotype-based pangenomes reveal genetic variations and climate adaptations in moso bamboo populations. Nat Commun 15(1):8085. https://doi.org/10.1038/s41467-024-52376-5

Hu Y, Wang L, Xiang L, Wu Q, Jiang H (2018) Automatic non-destructive growth measurement of leafy vegetables based on Kinect. Sensors 18(3):806. https://doi.org/10.3390/s18030806

Huang S, Li R, Zhang Z, Li L, Gu X, Fan W, Lucas WJ et al (2009) The genome of the cucumber, *Cucumis sativus L.* Nat Genet 41(12):1275–1281

Hurgobin B, Golicz AA, Bayer PE, Chan CKK, Tirnaz S, Dolatabadian A, Schiessl SV, Samans B, Montenegro JD, Parkin IAP, Pires JC, Chalhoub B, King GJ, Snowdon R, Batley J, Edwards D (2018) Homoeologous exchange is a major cause of gene presence/absence variation in the amphidiploid *Brassica napus*. Plant Biotechnol J 16(7). https://doi.org/10.1111/pbi.12867

Jarvis DE, Ho YS, Lightfoot DJ, Schmöckel SM, Li B, Borm TJA, Tester M (2017) The genome of *Chenopodium quinoa*. Nature 542(7641):307–312. https://doi.org/10.1038/nature21370

Jordan MI, Mitchell TM (2015) Machine learning: trends, perspectives, and prospects. Science 349(6245). https://doi.org/10.1126/science.aaa8415

Joshi BK, Shrestha HK, Ayer DK (2023) Plant breeding strategies and methods for food security: review on the technology. In: Ghosh S, Kumari Panda A, Jung C, Singh Bisht S (eds) Emerging solutions in sustainable food and nutrition security. Springer International Publishing, pp 213–254. https://doi.org/10.1007/978-3-031-40908-0_9

Kagale S, Koh C, Nixon J, Bollina V, Clarke WE, Tuteja R, Spillane C et al (2014) The emerging biofuel crop Camelina sativa retains a highly undifferentiated hexaploid genome structure. Nat Commun 5:3706. https://doi.org/10.1038/ncomms4706

Kaiser N, Douches D, Dhingra A, Glenn KC, Herzig PR, Stowe EC, Swarup S (2020) The role of conventional plant breeding in ensuring safe levels of naturally occurring toxins in food crops. Trends Food Sci Technol 100. https://doi.org/10.1016/j.tifs.2020.03.042

Kandemir N, Saygili İ (2015) Apomixis: new horizons in plant breeding. Turk J Agric For 39(4). https://doi.org/10.3906/tar-1409-74

Kang YJ, Kim SK, Kim MY, Lestari P, Kim KH, Ha B-K, Lee SH (2014) Genome sequence of mungbean and insights into evolution within Vigna species. Nat Commun 5(1):5443. https://doi.org/10.1038/ncomms6443

Kang L, Qian L, Zheng M, Chen L, Chen H, Yang L, Liu Z (2021) Genomic insights into the origin, domestication and diversification of *Brassica juncea*. Nat Genet 53(9):1392–1402. https://doi.org/10.1038/s41588-021-00922-y

Khaki S, Zahra K, Lizhi W (2020) Predicting yield performance of parents in plant breeding: a neural collaborative filtering approach. PLoS One 15:e0233382

Khan AW, Garg V, Sun S, Gupta S, Dudchenko O, Roorkiwal M, Varshney RK (2024) Cicer super-pangenome provides insights into species evolution and agronomic trait loci for crop improvement in chickpea. Nat Genet 56(6):1225–1234. https://doi.org/10.1038/s41588-024-01760-4

Kim J, Kim JS (2016) Bypassing GMO regulations with CRISPR gene editing. Nat Biotechnol 34(10):1014–1015. https://doi.org/10.1038/nbt.3680

Kim S, Park M, Yeom SI, Kim YM, Lee JM, Lee HA, Seo A et al (2014) Genome sequence of the hot pepper provides insights into the evolution of pungency in *Capsicum* species. Nat Genet 46(3):270–278

Kim J, Kim KS, Kim Y, Chung YS (2020) A short review: comparisons of high-throughput phenotyping methods for detecting drought tolerance. Sci Agric 78(4). https://doi.org/10.1590/1678-992x-2019-0300

Krishnan NM, Pattnaik S, Jain P, Gaur P, Choudhary R, Vaidyanathan S, Deepak S et al (2012) A draft of the genome and four transcriptomes of a medicinal and pesticidal angiosperm *Azadirachta indica*. BMC Genomics 13(1):464

Kumar K, Gambhir G, Dass A, Tripathi AK, Singh A, Jha AK, Yadava P, Choudhary M, Rakshit S (2020) Genetically modified crops: current status and future prospects. Planta 251(4). Springer. https://doi.org/10.1007/s00425-020-03372-8

Kumar D, Firoz Y, Partha H, Choudhury R, Kumar D, Kumar A, Tilak S, Sharma R, Mohapatra T (2022) Crop cultivars developed through molecular breeding, 2nd edn. Indian Council of Agricultural Research, New Delhi

Kundu V (2024) Transgenic breeding. In: Elements of plant breeding. Iterative International Publishers, Selfypage Developers Pvt Ltd, pp 14–34. https://doi.org/10.58532/nbennurch218

Kushwaha AK, Mohan N (2021) Emerging tools and techniques in crop improvement for higher productivity and multiple stress tolerance. In: Agricultural research, technology and policy: innovations and advances. ICAR-NAARM. https://www.naarm.org.in

Lammerts Van Bueren ET, Jones SS, Tamm L, Murphy KM, Myers JR, Leifert C, Messmer MM (2011) The need to breed crop varieties suitable for organic farming, using wheat, tomato and broccoli as examples: a review. NJAS - Wageningen J Life Sci 58(3–4). https://doi.org/10.1016/j.njas.2010.04.001

Legland D, El-Hage F, Méchin V, Reymond M (2017) Histological quantification of maize stem sections from FASGA-stained images. Plant Methods 13(1):84. https://doi.org/10.1186/s13007-017-0225-z

Lei L, Goltsman E, Goodstein D, Wu GA, Rokhsar DS, Vogel JP (2021) Plant pan-genomics comes of age. Annu Rev Plant Biol 72(1):411–435. https://doi.org/10.1146/annurev-arplant-080720-105454

Li YH, Zhou G, Ma J, Jiang W, Jin LG, Zhang Z, Guo Y, Zhang J, Sui Y, Zheng L, Zhang SS, Zuo Q, Shi XH, Li YF, Zhang WK, Hu Y, Kong G, Hong HL, Tan B, Qiu LJ (2014) De novo assembly of soybean wild relatives for pan-genome analysis of diversity and agronomic traits. Nat Biotechnol 32(10). https://doi.org/10.1038/nbt.2979

Li J, Yuan D, Wang P, Wang Q, Sun M, Liu Z, Si H, Xu Z, Ma Y, Zhang B, Pei L, Tu L, Zhu L, Chen LL, Lindsey K, Zhang X, Jin S, Wang M (2021a) Cotton pan-genome retrieves the lost sequences and genes during domestication and selection. Genome Biol 22(1):119. https://doi.org/10.1186/s13059-021-02351-w

Li Q, Tian S, Yan B, Liu CM, Lam T-W, Li R, Luo R (2021b) Building a Chinese pan-genome of 486 individuals. Commun Biol 4(1):1016. https://doi.org/10.1038/s42003-021-02556-6

Li N, Zhang X, Sun X, Zhu S, Cheng Y, Liu M, Liu T (2022) Genomic insights into the evolutionary history and diversification of bulb traits in garlic. Genome Biol 23(1):188. https://doi.org/10.1186/s13059-022-02756-1

Liakos KG, Busato P, Moshou D, Pearson S, Bochtis D (2018) Machine learning in agriculture: a review. Sensors (Switzerland) 18(8). https://doi.org/10.3390/s18082674

Liao GL, Huang CH, Jia DF, Zhong M, Tao JJ, Qu XY, Xu XB (2023) A high-quality genome of *Actinidia eriantha* provides new insight into ascorbic acid regulation. J Integr Agric 22(11):3244–3255. https://doi.org/10.1016/j.jia.2023.07.018

Libbrecht MW, Noble WS (2015) Machine learning applications in genetics and genomics. Nat Rev Genet 16(6). https://doi.org/10.1038/nrg3920

Lin K, Zhang N, Severing EI, Nijveen H, Cheng F, Visser RGF, Wang X, de Ridder D, Bonnema G (2014) Beyond genomic variation - comparison and functional annotation of three *Brassica rapa* genomes: a turnip, a rapid cycling and a Chinese cabbage. BMC Genomics 15(1). https://doi.org/10.1186/1471-2164-15-250

Liu MJ, Zhao J, Cai QL, Liu GC, Wang JR, Zhao ZH, Liu P et al (2014) The complex jujube genome provides insights into fruit tree biology. Nat Commun 5:5315. https://doi.org/10.1038/ncomms6315

Liu S, Liu Y, Yang X, Tong C, Edwards D, Parkin IA, Zhao M et al (2014b) The *Brassica oleracea* genome reveals the asymmetrical evolution of polyploid genomes. Nat Commun 5:3930. https://doi.org/10.1038/ncomms4930

Liu Y, Du H, Li P, Shen Y, Peng H, Liu S, Zhou GA, Zhang H, Liu Z, Shi M, Huang X, Li Y, Zhang M, Wang Z, Zhu B, Han B, Liang C, Tian Z (2020a) Pan-genome of wild and cultivated soybeans. Cell 182(1):162–176.e13. https://doi.org/10.1016/j.cell.2020.05.023

Liu J, Li J, Wang H, Yan J (2020b) Application of deep learning in genomics. Sci China Life Sci 63(12). https://doi.org/10.1007/s11427-020-1804-5

Lu F, Romay MC, Glaubitz JC, Bradbury PJ, Elshire RJ, Wang T, Li Y, Li Y, Semagn K, Zhang X, Hernandez AG, Mikel MA, Soifer I, Barad O, Buckler ES (2015) High-resolution genetic mapping of maize pan-genome sequence anchors. Nat Commun 6. https://doi.org/10.1038/ncomms7914

Lu Y, Yi S, Zeng N, Liu Y, Zhang Y (2017) Identification of rice diseases using deep convolutional neural networks. Neurocomputing 267. https://doi.org/10.1016/j.neucom.2017.06.023

Ma C, Xin M, Feldmann KA, Wang X (2014) Machine learning-based differential network analysis: a study of stress-responsive transcriptomes in *Arabidopsis*. Plant Cell 26(2). https://doi.org/10.1105/tpc.113.121913

Mahood EH, Kruse LH, Moghe GD (2020) Machine learning: a powerful tool for gene function prediction in plants. Appl Plant Sci 8(7). https://doi.org/10.1002/aps3.11376

Marks RA, Hotaling S, Frandsen PB, VanBuren R (2021) Representation and participation across 20 years of plant genome sequencing. Nat Plants 7(12):1571–1578. https://doi.org/10.1038/s41477-021-01031-8

Mejía-Guerra MK, Buckler ES (2019) A k-mer grammar analysis to uncover maize regulatory architecture. BMC Plant Biol 19(1). https://doi.org/10.1186/s12870-019-1693-2

Ming R, Hou S, Feng Y, Yu Q, Dionne-Laporte A, Saw JH, Senin P et al (2008) The draft genome of the transgenic tropical fruit tree papaya (*Carica papaya Linnaeus*). Nature 452(7190):991–996

Montenegro JD, Golicz AA, Bayer PE, Hurgobin B, Lee HT, Chan CKK, Visendi P, Lai K, Doležel J, Batley J, Edwards D (2017) The pangenome of hexaploid bread wheat. Plant J 90(5). https://doi.org/10.1111/tpj.13515

Myburg A, Grattapaglia D, Tuskan G, Jenkins J, Schmutz J, Mizrachi E, Hefer C, et al. (2011) The Eucalyptus grandis Genome Project: genome and transcriptome resources for comparative analysis of woody plant biology. In: *BMC Proceedings;* IUFRO Tree Biotechnology Conference 2011: from genomes to integration and delivery; Arraial d'Ajuda, Bahia, Brazil. 26 June – 2 July 2011. p I20

Nadeem MA, Nawaz MA, Shahid MQ, Doğan Y, Comertpay G, Yıldız M, Hatipoğlu R, Ahmad F, Alsaleh A, Labhane N, Özkan H, Chung G, Baloch FS (2018) DNA molecular markers in plant breeding: current status and recent advancements in genomic selection and genome editing. Biotechnol Biotechnol Equip 32(2):261–285. Taylor and Francis Ltd. https://doi.org/10.1080/13102818.2017.1400401

Naqvi RZ, Siddiqui HA, Mahmood MA, Najeebullah S, Ehsan A, Azhar M, Farooq M, Amin I, Asad S, Mukhtar Z, Mansoor S, Asif M (2022) Smart breeding approaches in post-genomics era for developing climate-resilient food crops. Front Plant Sci 13. Frontiers Media S.A. https://doi.org/10.3389/fpls.2022.972164

Niazian M, Niedbała G (2020) Machine learning for plant breeding and biotechnology. Agriculture (Switzerland) 10(10):1–23. MDPI AG. https://doi.org/10.3390/agriculture10100436

Noman A, Aqeel M, Deng J, Khalid N, Sanaullah T, Shuilin H (2017) Biotechnological advancements for improving floral attributes in ornamental plants. Front Plant Sci 8. https://doi.org/10.3389/fpls.2017.00530

Obembe OO, Ekundayo EO, Okoli AS, Gidado A, Adetunji CO, Ibrahim AB, Ubi BE (2022) Agricultural biotechnology, biodiversity and bioresources conservation and utilization. CRC Press. https://doi.org/10.1201/9781003178880

Oladosu Y, Rafii MY, Samuel C, Fatai A, Magaji U, Kareem I, Kamarudin ZS, Muhammad I, Kolapo K (2019) Drought resistance in rice from conventional to molecular breeding: a review. Int J Mol Sci 20(14). https://doi.org/10.3390/ijms20143519

Ortiz R (2023) Challenges for crop improvement. Emerg Topics Life Sci 7(2):197–205. https://doi.org/10.1042/ETLS20230106

Paajanen P, Kettleborough G, López-Girona E, Giolai M, Heavens D, Baker D, Clark MD (2019) A critical comparison of technologies for a plant genome sequencing project. GigaScience 8(3):giy163. https://doi.org/10.1093/gigascience/giy163

Pan X, Ma L, Zhang Y, Wang J, Du J, Guo X (2017) Three-dimensional reconstruction of maize roots and quantitative analysis of Metaxylem vessels based on X-ray micro-computed tomography. Can J Plant Sci. https://doi.org/10.1139/cjps-2017-0179

Paten B, Novak AM, Eizenga JM, Garrison E (2017) Genome graphs and the evolution of genome inference. Genome Res 27(5):665–676. https://doi.org/10.1101/gr.214155.116

Paterson AH, Bowers JE, Bruggmann R, Dubchak I, Grimwood J, Gundlach H, Haberer G et al (2009) The *Sorghum bicolor* genome and the diversification of grasses. Nature 457(7229):551–556

Peng Z, Lu Y, Li L, Zhao Q, Feng Q, Gao Z, Lu H et al (2013) The draft genome of the fast-growing non-timber forest species moso bamboo (*Phyllostachys heterocycla*). Nat Genet 45(4):456–461

Pereyra-Irujo GA, Gasco ED, Peirone LS, Aguirrezábal LAN (2012) GlyPh: a low-cost platform for phenotyping plant growth and water use. Funct Plant Biol 39(11):905. https://doi.org/10.1071/fp12052

Plant Symposia (2014) In vitro cellular & developmental biology - animal. Plant Symposia 50(S1):13–25. https://doi.org/10.1007/s11626-014-9772-z

Polashock J, Zelzion E, Fajardo D, Zalapa J, Georgi L, Bhattacharya D, Vorsa N (2014) The American cranberry: first insights into the whole genome of a species adapted to bog habitat. BMC Plant Biol 14(1):165–181

Polder G, Blok PM, de Villiers HAC, van der Wolf JM, Kamp J (2019) Potato virus Y detection in seed potatoes using deep learning on hyperspectral images. Front Plant Sci 10. https://doi.org/10.3389/fpls.2019.00209

Potgieter AB, Watson J, Eldridge M, Laws K, George-Jaeggli B, Hunt C, Borrell A, Mace E, Chapman SC, Jordan DR, Hammer GL (2018) Determining crop growth dynamics in sorghum breeding trials through remote and proximal sensing technologies. In: 38th IEEE International Geoscience and Remote Sensing Symposium (IGARSS), Valencia, Spain, Jul 22–27, 2018. IEEE, New York. https://doi.org/10.1109/IGARSS.2018.8519296

Pourkheirandish M, Golicz AA, Bhalla PL, Singh MB (2020) Global role of crop genomics in the face of climate change. Front Plant Sci 11. https://doi.org/10.3389/fpls.2020.00922

Prabhakar M, Purushothaman R, Awasthi DP (2020) Deep learning based assessment of disease severity for early blight in tomato crop. Multimed Tools Appl 79(39–40). https://doi.org/10.1007/s11042-020-09461-w

Prabhu KR, Kumar A, Yumkhaibam RS, Janeja HS, Krishna B, Talekar N (2023) A review on conventional and modern breeding approaches for developing climate resilient crop varieties. J Appl Nat Sci 15(3):987–997. Applied and Natural Science Foundation. https://doi.org/10.31018/jans.v15i3.4653

Qin P, Lu H, Du H, Wang H, Chen W, Chen Z, Li S (2021) Pan-genome analysis of 33 genetically diverse rice accessions reveals hidden genomic variations. Cell 184(13):3542–3558.e16. https://doi.org/10.1016/j.cell.2021.04.046

Rahman AYA, Usharraj AO, Misra BB, Thottathil GP, Jayasekaran K, Feng Y, Alam M (2013) Draft genome sequence of the rubber tree *Hevea brasiliensis*. BMC Genomics 14(1):75. https://doi.org/10.1186/1471-2164-14-75

Rakocevic G, Semenyuk V, Lee W-P, Spencer J, Browning J, Johnson IJ, Arsenijevic V, Nadj J, Ghose K, Suciu MC, Ji S-G, Demir G, Li L, Toptaş BÇ, Dolgoborodov A, Pollex B, Spulber I, Glotova I, Kómár P, Kural D (2019) Fast and accurate genomic anal's using genome graphs. Nat Genet 51(2):354–362. https://doi.org/10.1038/s41588-018-0316-4

Redon R, Ishikawa S, Fitch KR, Feuk L, Perry GH, Andrews TD, Fiegler H, Shapero MH, Carson AR, Chen W, Cho EK, Dallaire S, Freeman JL, González JR, Gratacòs M, Huang J, Kalaitzopoulos D, Komura D, MacDonald JR, Hurles ME (2006) Global variation in copy number in the human genome. Nature 444(7118). https://doi.org/10.1038/nature05329

Ricroch A, Eriksson D, Miladinović D, Sweet J, Van Laere K, Woźniak-Gientka E (2023) A roadmap for plant genome editing. Springer

Rist F, Herzog K, Mack J, Richter R, Steinhage V, Töpfer R (2018) High-precision phenotyping of grape bunch architecture using fast 3D sensor and automation. Sensors 18(3):763. https://doi.org/10.3390/s18030763

Ruperao P, Thirunavukkarasu N, Gandham P, Selvanayagam S, Govindaraj M, Nebie B, Rathore A (2021) Sorghum pan-genome explores the functional utility for genomic-assisted breeding to accelerate the genetic gain. Front Plant Sci 12:666342. https://doi.org/10.3389/fpls.2021.666342

Salyaev RK, Rekoslavskaya NI (2017) Plant-based peroral vaccines. In: Multifunctional systems for combined delivery, biosensing and diagnostics. Elsevier, pp 193–210. https://doi.org/10.1016/B978-0-323-52725-5.00010-1

Schatz MC, Maron LG, Stein JC, Hernandez Wences A, Gurtowski J, Biggers E, Lee H, Kramer M, Antoniou E, Ghiban E, Wright MH, Chia JM, Ware D, McCouch SR, McCombie WR (2014) Whole genome de novo assemblies of three divergent strains of rice, *Oryza sativa*, document novel gene space of aus and indica. Genome Biol 15(11). https://doi.org/10.1186/s13059-014-0506-z

Schmid KJ, Ramos-Onsins S, Ringys-Beckstein H, Weisshaar B, Mitchell-Olds T (2005) A multilocus sequence survey in *Arabidopsis thaliana* reveals a genome-wide departure from a neutral model of DNA sequence polymorphism. Genetics 169(3). https://doi.org/10.1534/genetics.104.033795

Schmidt MH-W, Vogel A, Denton AK, Istace B, Wormit A, van de Geest H, Usadel B (2017) De Novo assembly of a new *Solanum pennellii* accession using nanopore sequencing. Plant Cell 29(10):2336–2348. https://doi.org/10.1105/tpc.17.00521

Schmutz J, Cannon SB, Schlueter J, Ma J, Mitros T, Nelson W, Jackson SA (2010) Erratum: genome sequence of the palaeopolyploid soybean. Nature 465(7294):120–120. https://doi.org/10.1038/nature08957

Schmutz J, McClean PE, Mamidi S, Wu GA, Cannon SB, Grimwood J, Jackson SA (2014) A reference genome for common bean and genome-wide analysis of dual domestications. Nat Genet 46(7):707–713. https://doi.org/10.1038/ng.3008

Schnable PS, Ware D, Fulton RS, Stein JC, Wei F, Pasternak S et al (2009) The B73 maize genome: complexity, diversity, and dynamics. Science (New York, N.Y.) 326(5956):1112–1115. https://doi.org/10.1126/science.1178534

Shulaev V, Sargent DJ, Crowhurst RN, Mockler TC, Folkerts O, Delcher AL, Folta KM (2011) The genome of woodland strawberry (*Fragaria vesca*). Nat Genet 43(2):109–116. https://doi.org/10.1038/ng.740

Sierro N, Battey JN, Ouadi S, Bakaher N, Bovet L, Willig A, Geopfert S, Peitsch MC, Ivanov NV (2014) The tobacco genome sequence and its comparison with those of tomato and potato. Nat Commun 5:3833. https://doi.org/10.1038/ncomms4833

Singh NK (2016) Origin, diversity and genome sequence of mango (*Mangifera indica* L.). Indian J Hist Sci 51(2.2). https://doi.org/10.16943/ijhs/2016/v51i2.2/48449

Singh A, Ganapathysubramanian B, Singh AK, Sarkar S (2016) Machine learning for high-throughput stress phenotyping in plants. Trends Plant Sci 21(2). https://doi.org/10.1016/j.tplants.2015.10.015

Song X, Nie F, Chen W, Ma X, Gong K, Yang Q, Wang X (2020) Coriander genomics database: a genomic, transcriptomic, and metabolic database for coriander. Hortic Res 7(1):55. https://doi.org/10.1038/s41438-020-0261-0

Soppe JA, Lebbink RJ (2017) Antiviral goes viral: harnessing CRISPR/Cas9 to combat viruses in humans. Trends Microbiol 25(10):833–850. https://doi.org/10.1016/j.tim.2017.04.005

Springer NM, Ying K, Fu Y, Ji T, Yeh CT, Jia Y, Wu W, Richmond T, Kitzman J, Rosenbaum H, Iniguez AL, Barbazuk WB, Jeddeloh JA, Nettleton D, Schnable PS (2009) Maize inbreds exhibit high levels of copy number variation (CNV) and presence/absence variation (PAV) in genome content. PLoS Genet 5(11). https://doi.org/10.1371/journal.pgen.1000734

Sun X, Zhu S, Li N, Cheng Y, Zhao J, Qiao X, Liu T (2020) A chromosome-level genome assembly of garlic (*Allium sativum*) provides insights into genome evolution and allicin biosynthesis. Mol Plant 13(9):1328–1339. https://doi.org/10.1016/j.molp.2020.07.019

Tang X, Lowder LG, Zhang T, Malzahn AA, Zheng X, Voytas DF, Zhong Z, Chen Y, Ren Q, Li Q, Kirkland ER, Zhang Y, Qi Y (2017) A CRISPR-Cpf1 system for efficient genome editing and transcriptional repression in plants. Nat Plants 3. https://doi.org/10.1038/nplants.2017.18

Tao Y, Zhao X, Mace E, Henry R, Jordan D (2019) Exploring and exploiting pan-genomics for crop improvement. Mol Plant 12(2). https://doi.org/10.1016/j.molp.2018.12.016

Tao Y, Luo H, Xu J, Cruickshank A, Zhao X, Teng F, Hathorn A, Wu X, Liu Y, Shatte T, Jordan D, Jing H, Mace E (2021) Extensive variation within the pan-genome of cultivated and wild sorghum. Nat Plants 7(6):766–773. https://doi.org/10.1038/s41477-021-00925-x

Tettelin H, Masignani V, Cieslewicz MJ, Donati C, Medini D, Ward NL, Angiuoli SV, Crabtree J, Jones AL, Durkin AS, DeBoy RT, Davidsen TM, Mora M, Scarselli M, Margarit Y, Ros I, Peterson JD, Hauser CR, Sundaram JP, Nelson WC, Fraser CM (2005) Genome analysis of multiple pathogenic isolates of *Streptococcus agalactiae*: Implications for the microbial "pan-genome.". Proc Natl Acad Sci USA 102(39). https://doi.org/10.1073/pnas.0506758102

The French–Italian Public Consortium for Grapevine Genome Characterization (2007) The grapevine genome sequence suggests ancestral hexaploidization in major angiosperm phyla. Nature 449(7161):463–467

The Potato Genome Sequencing Consortium (2011) Genome sequence and analysis of the tuber crop potato. Nature 475(7355):189–195

The Tomato Genome Consortium (2012) The tomato genome sequence provides insights into fleshy fruit evolution. Nature 485(7400):635–641

Thottathil GP, Jayasekaran K, Othman AS (2016) Sequencing crop genomes: a gateway to improve tropical agriculture. Trop Life Sci Res 27(1):93–114

Tian X, Li R, Fu W, Li Y, Wang X, Li M, Du D, Tang Q, Cai Y, Long Y, Zhao Y, Li M, Jiang Y (2020) Building a sequence map of the pig pan-genome from multiple de novo assemblies and Hi-C data. Sci China Life Sci 63(5):750–763. https://doi.org/10.1007/s11427-019-9551-7

Tran TT, Choi JW, Le TTH, Kim JW (2019) A comparative study of deep CNN in forecasting and classifying the macronutrient deficiencies on development of tomato plant. Appl Sci (Switzerland) 9(8). https://doi.org/10.3390/app9081601

Tsugama D, Fujino K (2020) Data of whole genome sequencing of five garden asparagus (*Asparagus officinalis*) individuals with the MinION nanopore sequencer. Data Brief 28:104838. https://doi.org/10.1016/j.dib.2019.104838

Turnbull C, Lillemo M, Hvoslef-Eide TAK (2021) Global regulation of genetically modified crops amid the gene edited crop boom—A review. Front Plant Sci 12. https://doi.org/10.3389/fpls.2021.630396

Tuskan GA, Difazio S, Jansson S, Bohlmann J, Grigoriev I, Hellsten U, Putnam N et al (2006) The genome of black cottonwood, *Populus trichocarpa* (Torr. & Gray). Science 313(5793):1596–1604

Varshney RK, Chen W, Li Y, Bharti AK, Saxena RK, Schlueter JA, Donoghue MTA et al (2011) Draft genome sequence of pigeonpea (*Cajanus cajan*), an orphan legume crop of resource-poor farmers. Nat Biotechnol 30(1):83–90

Varshney RK, Song C, Saxena RK, Azam S, Yu S, Sharpe AG, Cannon S et al (2013) Draft genome sequence of chickpea (*Cicer arietinum*) provides a resource for trait improvement. Nat Biotechnol 31(3):240–248

Verde I, Abbott AG, Scalabrin S, Jung S, Shu S, Marroni F, Zhebentyayeva T, Dettori MT, Grimwood J, Cattonaro F, Zuccolo A, Rossini L, Jenkins J, Vendramin E, Meisel LA, Decroocq V, Sosinski B, Prochnik S, Mitros T, Rokhsar DS (2013) The high-quality draft genome of peach (*Prunus persica*) identifies unique patterns of genetic diversity, domestication and genome evolution. Nat Genet 45(5):487–494. https://doi.org/10.1038/ng.2586

Voss-Fels KP, Stahl A, Hickey LT (2019) Q&A: modern crop breeding for future food security. BMC Biol 17(1). https://doi.org/10.1186/s12915-019-0638-4

Wang JK, Wolfgang H, P. (2007) Simulation modeling in plant breeding: principles and applications. Agric Sci China 6(8). https://doi.org/10.1016/S1671-2927(07)60129-1

Wang X, Wang H, Wang J, Sun R, Wu J, Liu S, Bai Y, Mun J-H, Bancroft I, Cheng F, Huang S, Li X, Hua W, Wang J, Wang X, Freeling M, Pires JC, Paterson AH, Chalhoub B, Zhang Z (2011) The genome of the mesopolyploid crop species *Brassica rapa*. Nat Genet 43(10):1035–1039. https://doi.org/10.1038/ng.919

Wang L, Chen L, Li R, Zhao R, Yang M, Sheng J, Shen L (2017) Reduced drought tolerance by CRISPR/Cas9-Mediated SlMAPK3 mutagenesis in tomato plants. J Agric Food Chem 65(39):8674–8682. https://doi.org/10.1021/acs.jafc.7b02745

Wang H, Cimen E, Singh N, Buckler E (2020) Deep learning for plant genomics and crop improvement. Curr Opin Plant Biol 54. https://doi.org/10.1016/j.pbi.2019.12.010

Wang K, Hu H, Tian Y, Li J, Scheben A, Zhang C, Li Y, Wu J, Yang L, Fan X, Sun G, Li D, Zhang Y, Han R, Jiang R, Huang H, Yan F, Wang Y, Li Z, Kang X (2021) The chicken Pan-genome reveals gene content variation and a promoter region deletion in IGF2BP1 affecting body size. Mol Biol Evol 38(11):5066–5081. https://doi.org/10.1093/molbev/msab231

Wani KI, Aftab T (2022) Tools and techniques used in plant molecular farming. Springer, pp 11–30. https://doi.org/10.1007/978-3-031-12794-6_2

Washburn JD, Mejia-Guerra MK, Ramstein G, Kremling KA, Valluru R, Buckler ES, Wang H (2019) Evolutionarily informed deep learning methods for predicting relative transcript abundance from DNA sequence. Proc Natl Acad Sci USA 116(12). https://doi.org/10.1073/pnas.1814551116

White JW, Andrade-Sanchez P, Gore MA, Bronson KF, Coffelt TA, Conley MM, Feldmann KA, French AN, Heun JT, Hunsaker DJ, Jenks MA, Kimball BA, Roth RL, Strand RJ, Thorp KR, Wall GW, Wang G (2012) Field-based phenomics for plant genetics research. Field Crop Res 133:101–112. https://doi.org/10.1016/j.fcr.2012.04.003

Wu H, Jaeger M, Wang M, Li B, Zhang BG (2011) Three-dimensional distribution of vessels, passage cells and lateral roots along the root axis of winter wheat (*Triticum aestivum*). Ann Bot 107(5):843–853. https://doi.org/10.1093/aob/mcr005

Wu J, Wang Z, Shi Z, Zhang S, Ming R, Zhu S, Khan MA et al (2013) The genome of the pear (*Pyrus bretschneideri Rehd.*). Genome Res 23(2):396–408

Xu Q, Chen LL, Ruan X, Chen D, Zhu A, Chen C, Bertrand D et al (2013) The draft genome of sweet orange (*Citrus sinensis*). Nat Genet 45(1):59–66

Yang G, Liu J, Zhao C, Li Z, Huang Y, Yu H, Xu B, Yang X, Zhu D, Zhang X, Zhang R, Feng H, Zhao X, Li Z, Li H, Yang H (2017) Unmanned aerial vehicle remote sensing for field-based crop phenotyping: current status and perspectives. Front Plant Sci 8. https://doi.org/10.3389/fpls.2017.01111

Yang W, Feng H, Zhang X, Zhang J, Doonan JH, Batchelor WD, Xiong L, Yan J (2020) Crop phenomics and high-throughput phenotyping: past decades, current challenges, and future perspectives. Mol Plant 13(2):187–214. https://doi.org/10.1016/j.molp.2020.01.008

de Yassitepe JECT, da Silva VCH, Hernandes-Lopes J, Dante RA, Gerhardt IR, Fernandes FR, da Silva PA, Vieira LR, Bonatti V, Arruda P (2021) Maize transformation: from plant material to the release of genetically modified and edited varieties. Front Plant Sci 12. https://doi.org/10.3389/fpls.2021.766702

Yu J, Hu S, Wang J, Wong KS, Li S, Liu B, Deng Y et al (2002) A draft sequence of the rice genome (*Oryza sativa L. ssp. indica*). Science 296(5565):79–92

Yu J, Tehrim S, Zhang F, Tong C, Huang J, Cheng X, Dong C, Zhou Y, Qin R, Hua W, Liu S (2014) Genome-wide comparative analysis of NBS-encoding genes between *Brassica* species and *Arabidopsis thaliana*. BMC Genomics 15(1). https://doi.org/10.1186/1471-2164-15-3

Yu J, Golicz AA, Lu K, Dossa K, Zhang Y, Chen J, Wang L, You J, Fan D, Edwards D, Zhang X (2019) Insight into the evolution and functional characteristics of the pan-genome assembly from sesame landraces and modern cultivars. Plant Biotechnol J 17(5):881–892. https://doi.org/10.1111/pbi.13022

Zavafer A, Bates H, Mancilla C, Ralph PJ (2023) Phenomics: conceptualization and importance for plant physiology. Trends Plant Sci 28(9):1004–1013. https://doi.org/10.1016/j.tplants.2023.03.023

Zhang C, Kovacs JM (2012) The application of small unmanned aerial systems for precision agriculture: a review. Precis Agric 13(6):693–712. https://doi.org/10.1007/s11119-012-9274-5

Zhang G, Liu X, Quan Z, Cheng S, Xu X, Pan S, Xie M et al (2012) Genome sequence of foxtail millet (*Setaria italica*) provides insights into grass evolution and biofuel potential. Nat Biotechnol 30(6):549–554

Zhang Q, Chen W, Sun L, Zhao F, Huang B, Yang W, Tao Y et al (2012b) The genome of *Prunus mume*. Nat Commun 3:1318

Zhang Y, Legay S, Barrière Y, Méchin V, Legland D (2013) Color quantification of stained maize stem section describes lignin spatial distribution within the whole stem. J Agric Food Chem 61(13):3186–3192. https://doi.org/10.1021/jf400912s

Zhang Y, Ma L, Pan X, Wang J, Guo X, Du J (2018) Micron-scale phenotyping techniques of maize vascular bundles based on X-ray microcomputed tomography. J Vis Exp 140. https://doi.org/10.3791/58501

Zhang X, Han L, Dong Y, Shi Y, Huang W, Han L, González-Moreno P, Ma H, Ye H, Sobeih T (2019) A deep learning-based approach for automated yellow rust disease detection from high-resolution hyperspectral UAV images. Remote Sens 11(13). https://doi.org/10.3390/rs11131554

Zhang F, Shi Y, Ali J, Xu J, Li Z (2021) Breeding by selective introgression: theory, practices, and lessons learned from rice. Crop J 9(3):646–657. Institute of Crop Sciences. https://doi.org/10.1016/j.cj.2021.03.006

Zhang Y, Zhao M, Tan J, Huang M, Chu X, Li Y, He H (2024) Telomere-to-telomere *Citrullus* super-pangenome provides direction for watermelon breeding. Nat Genet 56(8):1750–1761. https://doi.org/10.1038/s41588-024-01823-6

Zhao Q, Feng Q, Lu H, Li Y, Wang A, Tian Q, Zhan Q, Lu Y, Zhang L, Huang T, Wang Y, Fan D, Zhao Y, Wang Z, Zhou C, Chen J, Zhu C, Li W, Weng Q, Huang X (2018) Pan-genome analysis highlights the extent of genomic variation in cultivated and wild rice. Nat Genet 50(2). https://doi.org/10.1038/s41588-018-0041-z

Zhao C, Zhang Y, Du J, Guo X, Wen W, Gu S, Wang J, Fan J (2019) Crop phenomics: current status and perspectives. Front Plant Sci 10). Frontiers Media S.A. https://doi.org/10.3389/fpls.2019.00714

Zhou Y, Zhang Z, Bao Z, Li H, Lyu Y, Zan Y, Wu Y, Cheng L, Fang Y, Wu K, Zhang J, Lyu H, Lin T, Gao Q, Saha S, Mueller L, Fei Z, Städler T, Xu S, Huang S (2022) Graph pangenome captures missing heritability and empowers tomato breeding. Nature 606(7914). https://doi.org/10.1038/s41586-022-04808-9

Zuo W, Chao Q, Zhang N, Ye J, Tan G, Li B, Xing Y, Zhang B, Liu H, Fengler KA, Zhao J, Zhao X, Chen Y, Lai J, Yan J, Xu M (2015) A maize wall-associated kinase confers quantitative resistance to head smut. Nat Genet 47(2). https://doi.org/10.1038/ng.3170

Phenomics and Next-Generation Phenotyping to Increase Genetic Gains in Crop Breeding

10

Sneha Priya Pappula Reddy, Neeraj Kumar, Minhui Bi, Sudhir Kumar, and C. Bharadwaj

Abstract

Unprecedented difficulties have been brought forth by climate change. Many plant breeders hate phenotyping, the traditional method of measuring plant attributes, which is labor-intensive and time-consuming. It is still difficult and restricts plant breeding for complex characteristics like heat and drought. High-throughput, accurate, and non-destructive measurements of intricate plant characteristics are possible with next-generation phenotyping (NGP) technology. These sophisticated solutions overcome the disadvantages of traditional phenotyping techniques by combining automation, robots, state-of-the-art sensors, and data analytics. Employing sophisticated imaging systems, sensor technologies, and automated platforms, next-generation phenotyping (NGP) approaches allow for the accurate and high-throughput extraction of a variety of plant features. These methods are intended to accurately and consistently evaluate the morphological, physiological, and biochemical traits of plants in a non-invasive way

S. P. P. Reddy
The UWA Institute of Agriculture, School of Natural and Agricultural Sciences, The University of Western Australia, Perth, WA, Australia

School of Biological Sciences, The University of Western Australia, Perth, WA, Australia

N. Kumar · S. Kumar
ICAR-Indian Agricultural Research Institute, New Delhi, India

M. Bi
State Key Laboratory of Herbage Improvement and Grassland AgroEcosystems, College of Ecology, Lanzhou University, Lanzhou, Gansu, China

C. Bharadwaj (✉)
The UWA Institute of Agriculture, School of Natural and Agricultural Sciences, The University of Western Australia, Perth, WA, Australia

ICAR-Indian Agricultural Research Institute, New Delhi, India

P. M. Priyadarshan, R. Ortiz (eds.), *Plant Breeding 2050*,
https://doi.org/10.1007/978-981-95-0583-8_10

across sizable populations. Strong data management storage, analysis and retrieval systems are essential for data produced by various phenotyping platforms. Computational networks make it easier to combine phenotypic data from several sources, such as field sensors, spectroscopy, and photos, to produce a single framework for assessing traits. Researchers may access and evaluate phenotypic data in real time by connecting these networks to cloud-based systems, which improves breeding program scalability and collaboration.

Keywords

Next-generation phenomics · High-throughput phenotyping · Genetic gains · Crop breeding · Deep neural networks

10.1 Introduction

Climate change has introduced unprecedented challenges, including increased droughts, heat waves, and soil salinity, significantly threatening crop yields and food security. Simultaneously, the growing global demand for food has heightened the urgency of developing resilient and high-yielding crop varieties (Saleem et al. 2024). However, phenotyping—the measurement of plant traits—has historically been a bottleneck in breeding programs due to its labor-intensive and time-consuming nature (Anand et al. 2023). Conventional breeding approaches, though effective in the past, are limited in their capacity to assess complex traits like drought tolerance, nutrient use efficiency, and disease resistance. Additionally, traditional methods struggle to capture and utilize complex, quantitative traits governed by multiple genes, such as yield stability and stress tolerance (Hafeez et al. 2023).

High-throughput phenotyping (HTP) and next-generation phenotyping (NGP) can replace conventional phenotyping to overcome these challenges and accelerate genetic gains in breeding programs. The efficiency of HTP and NGP is critical for addressing complex traits influenced by multiple genes, which traditionally requires extensive field trials across diverse environments (Munjal et al. 2023). Leveraging imaging systems, spectroscopy, and sensor-based methods, HTP and NGP capture multidimensional datasets to provide detailed insights into plant morphology, physiology, and biochemical characteristics and enable breeders to evaluate genotypes with less time and resources (D'Agostino and Tripodi 2017; Fig. 10.1). Integration of HTP data with genomic tools such as genome-wide association studies (GWAS) and genomic selection (GS), provides a better understanding of genotype-by-environment interactions and identifies quantitative trait loci (QTLs) associated with desirable traits. This integration bridges the gap between genotype and phenotype, enabling predictive modeling to identify superior genotypes before field trials (Xu et al. 2022). Consequently, breeders accelerate the delivery of improved crop varieties to farmers in less time and resources. By integrating robotics, imaging systems, and sensors, HTP and NGP provide detailed, multidimensional data on plant traits that can be linked with genomic information, enhancing selection

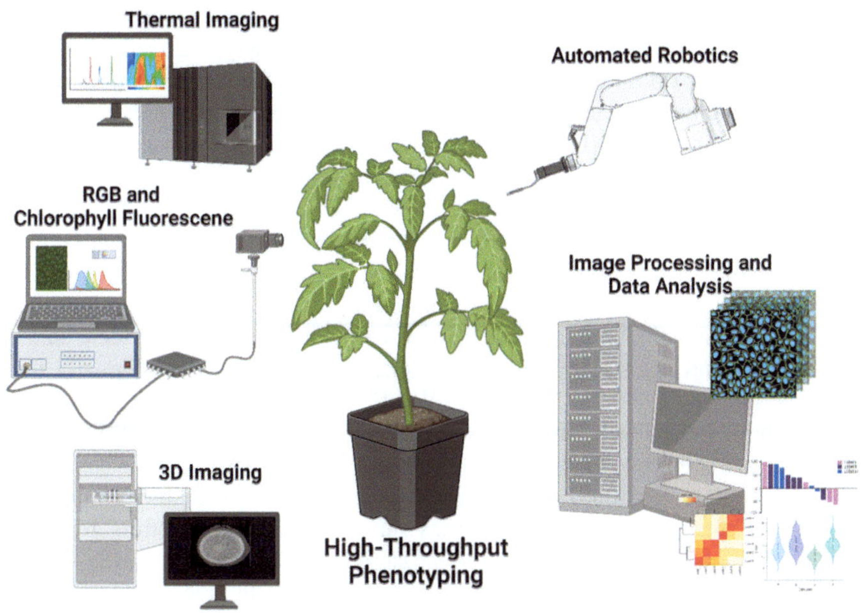

Fig. 10.1 Various next-generation phenotyping techniques applied in crop breeding

efficiency, and shortening breeding cycles (Zhang et al. 2024a, b). The adoption of HTP and NGP contributes to sustainable agriculture by enabling the development of crops with higher resource-use efficiency, such as improved nitrogen and water-use efficiency, thereby reducing reliance on fertilizers and irrigation (Geng and Yufeng 2023). The integration of phenomics with machine learning and artificial intelligence will pave the way for predictive breeding, where crop improvement decisions are data-driven and more accurate (Xu et al. 2022). By accelerating genetic gains and enhancing crop resilience to environmental stresses, HTP and NGP align with the United Nations' Sustainable Development Goals, particularly those related to zero hunger and climate action (Kundu et al. 2024).

10.2 Understanding Phenomics and the Next-Generation Phenotyping Platforms

Phenomics, the comprehensive study of plant phenotypes, integrates advanced technologies, data analytics, and biological understanding to unravel genotype–phenotype relationships and transform plant breeding. Its components can be categorized into next-generation phenotyping, phenotyping platforms, sensing and imaging technologies, data management systems, and data analysis and integration (Tripodi et al. 2022). Next-generation phenotyping (NGP) technologies offer high-throughput, precise, and non-destructive measurement of complex plant traits.

These advanced tools integrate automation, robotics, cutting-edge sensors, and data analytics to address the limitations of conventional phenotyping methods (Marsh et al. 2021; Fig. 10.1). Visible (RGB) imaging is a cost-effective method for capturing morphological traits such as leaf area and plant height, useful in both field and controlled environments (Li et al. 2014). Multispectral imaging extends this capability by measuring reflectance across discrete wavelengths, providing insights into chlorophyll content, photosynthetic activity, and stress responses (Serbin et al. 2012). Hyperspectral imaging takes this further by capturing a continuous light spectrum, enabling precise biochemical trait assessment such as nitrogen content and water status (Mahlein 2016; Pappula-Reddy et al. 2024). Thermal imaging detects infrared radiation to monitor canopy temperature, providing critical data on transpiration and heat stress responses (Prashar and Jones 2014). Additionally, fluorescence imaging evaluates photosynthetic efficiency and detects early stress signals at the cellular level, while 3D imaging constructs detailed plant models for structural and morphological analysis (Pérez-Bueno et al. 2019). Sensor-based technologies complement imaging by capturing real-time data on biochemical and structural traits. Spectroscopy quantifies nutrient content and stress responses through light absorption and reflection measurements (Kalaji et al. 2018). LIDAR (Light Detection and Ranging) uses laser pulses to create precise 3D models of canopy architecture and field heterogeneity (Jin et al. 2021). Below-ground traits, often overlooked in traditional phenotyping, can now be accessed with root phenotyping sensors such as MRI, X-ray CT, and ground-penetrating radar, enabling comprehensive analysis of root systems and their interactions with soil (Tracy et al. 2010; Metzner et al. 2015). Automation and robotics enhance phenotyping efficiency and precision, reducing human error and enabling large-scale assessments. Conveyor systems automate plant movement and imaging in controlled environments, while field robots equipped with advanced sensors provide real-time data under natural conditions (Atefi et al. 2021). Environmental monitoring systems and remote sensing technologies, including drones and UAVs equipped with diverse cameras, allow large-scale monitoring of crop health, stress responses, and yield predictions (Awais et al. 2022). Satellites further extend these capabilities to macro-scale phenotyping, assessing crop performance across vast agricultural regions. IoT-based sensors measure soil moisture, temperature, humidity, and light intensity to understand genotype × environment (G × E) interactions (Ali et al. 2023).

10.3 Traits of Interest

Next-generation phenotyping (NGP) techniques enable the precise and high-throughput extraction of various plant traits by utilizing advanced imaging systems, sensor technologies, and automated platforms. These techniques are designed to assess morphological, physiological, and biochemical characteristics of plants in a non-invasive manner, ensuring accuracy and consistency across large populations (Fahlgren et al. 2015). The process typically involves three main stages: data acquisition, data processing, and trait extraction. NGP platforms employ diverse technologies such as hyperspectral imaging, thermal imaging, LiDAR (light detection and ranging), and fluorescence imaging to collect high-resolution data on plant attributes. For instance, hyperspectral imaging captures detailed reflectance spectra that reveal information about leaf pigment composition, water content, and nutrient status (Zhang et al. 2024a, b). Thermal imaging measures canopy temperature, which is indicative of stomatal conductance and transpiration rates, key traits for drought tolerance (Smigaj et al. 2024). LiDAR generates three-dimensional models of plant architecture, aiding in analyzing structural traits like plant height, leaf area, and biomass. These technologies are often integrated with robotics or drones to enable rapid and efficient data collection in controlled and field environments (Paturkar et al. 2021). Raw data collected from NGP tools undergo preprocessing to remove noise and standardize measurements. Advanced computational tools, including machine learning algorithms, are then applied to extract meaningful patterns and features (Dhanya et al. 2024; Table 10.1). For example, hyperspectral data are analyzed to identify specific wavelengths associated with biochemical compounds such as chlorophyll or anthocyanins (Zhang et al. 2024a, b). Thermal data are processed to quantify temperature differentials across plant tissues, providing insights into stress responses (Smigaj et al. 2024). Similarly, image segmentation techniques are used to isolate plant structures from the background, enabling accurate measurements of traits like canopy coverage or root length (Buckner et al. 2021). Once the data are processed, specific traits are quantified and linked to underlying biological processes. Morphological characteristics such as leaf size, plant height, and branching patterns are extracted using three-dimensional imaging and geometric modeling (Noshita et al. 2022). These extracted traits are compiled into multidimensional datasets that comprehensively describe the plant's performance under specific environmental conditions.

Data generated from phenotyping and environmental sensing requires robust data management systems. Integrating phenotypic and genomic data enables GWAS and GS, to refine breeding strategies and accelerate the development of superior crop varieties (Robert et al. 2022a, b, c). Efficient data handling is essential for NGP, with cloud-based platforms enabling the storage, processing, and analysis of large datasets generated by phenotyping platforms (Munjal et al. 2023; Table 10.1). Raw data from imaging and sensors often require preprocessing to remove noise, correct for environmental variations, and standardize measurements. Preprocessing involves steps like image segmentation, feature extraction, and normalization (Buckner et al. 2021). Integration of data from multiple sensing modalities, such as combining thermal and hyperspectral data, provides a comprehensive view of plant performance across diverse traits. NGP platforms generate vast and complex datasets, particularly from high-resolution imaging systems like hyperspectral, thermal, and multispectral cameras (Munjal et al. 2023). The analysis and interpretation of these images are facilitated by advanced computational tools, including machine learning (ML), artificial intelligence (AI), modeling, and computational networks. These technologies enable feature extraction from images and spectral data, predictive modeling of trait performance, and improve the accuracy and speed of phenotypic predictions (Crossa et al. 2024). Machine learning algorithms detect patterns and relationships within high-dimensional datasets, making them ideal for analyzing phenotypic images. Supervised learning algorithms, such as support vector machines (SVMs) and random forests, are often used to classify images and quantify specific traits like leaf area, plant height, or disease symptoms (Kok et al. 2021). Unsupervised learning techniques, such as clustering and principal component analysis (PCA), help identify underlying trends and groupings in phenotypic data without prior labels, enabling the discovery of novel trait correlations. Deep learning, a subset of ML, leverages convolutional neural networks (CNNs) to process complex image data. CNNs are particularly effective in recognizing intricate patterns in plant morphology, stress responses, and growth dynamics, making them indispensable for next-generation phenotyping (Mansoor et al. 2024; Table 10.1). Reinforcement learning algorithms, for example, can simulate and optimize breeding scenarios by predicting the outcomes of specific crosses or environmental interventions. Quantified traits are derived from processed data, including morphological (plant height, leaf area), physiological (chlorophyll content, water-use efficiency), and biochemical (secondary metabolite concentrations) characteristics (Yoosefzadeh Najafabadi et al. 2023). Statistical and computational models are then used to identify relationships between these traits and environmental variables. Real-time acquisition, wireless transfer, and storage on cloud-based platforms ensure data security, accessibility, and organization. Standardizing data using metadata frameworks such as MIAPPE (Minimum Information About a Plant Phenotyping Experiment) allows seamless integration and comparison across datasets (Tripodi et al. 2022). Image processing tools like OpenCV and ImageJ extract key traits such as canopy cover and growth dynamics (Ye et al. 2023). Statistical software such as R, SAS, and Python libraries identify significant phenotypic relationships, while machine learning tools like TensorFlow and PyTorch analyze complex datasets to predict genetic performance or uncover patterns (Jayasinghe et al. 2023). Structural-functional plant models (SFPs) use phenotypic data from imaging platforms to recreate three-dimensional representations of plant architecture. These models provide a comprehensive understanding of how traits interact with each other and the environment (Soualiou et al. 2021). Additionally, process-based models integrate phenotypic and environmental data to simulate physiological processes such as photosynthesis, water uptake, and nutrient use, offering valuable insights into trait performance (Dubbert et al. 2023). Deep neural networks (DNNs) process layers of data, identifying complex, non-linear relationships between traits and environmental variables. Recurrent neural networks (RNNs) and long short-term memory (LSTM) networks are particularly useful for temporal data analysis, allowing researchers to track plant growth and stress responses over time (Paramasivan 2021; Table 10.1). Computational networks facilitate the integration of phenotypic data from diverse sources, such as images, spectroscopy, and field sensors, creating a unified framework for trait evaluation. By linking these networks to cloud-based platforms, researchers can access and process phenotypic data in real time, enhancing collaboration and scalability in breeding programs (Sheikh et al. 2024). These models guide breeders in selecting candidate lines for further evaluation and field trials, to advance promising genotypes breeding programs and to develop predictive breeding pipelines. Using

Table 10.1 Overview of phenotyping systems and methods

		Applications	Tools and technologies	Advantages
Phenotyping platforms	Field-based phenotyping platforms	Large-scale phenotyping under natural conditions (Oliver et al. 2013)	Drones and UAVs (Unmanned Aerial Vehicles) equipped with multispectral, hyperspectral, or thermal cameras (Zhang and Kovacs 2012; Holman et al. 2016) Tractor-mounted sensors for canopy reflectance, NDVI (Normalized Difference Vegetation Index), and growth measurements (Oliver et al. 2013)	High-throughput, real-time data collection (Araus and Cairns 2014) Captures genotype × environment (G × E) interactions effectively (Millet et al. 2019)
	Controlled environment platforms	Precision phenotyping in greenhouses or growth chambers under controlled conditions.	Automated systems with robotic arms for imaging and data collection (Humplík et al. 2015) Environmental control systems to manipulate temperature, light, humidity, and CO_2	Precise environmental manipulation (Granier et al. 2006) Reproducible results for specific trait evaluations (e.g., drought tolerance) (Wasaya et al. 2018)
	High-throughput screening platforms	Large-scale screening of populations	Conveyor-based imaging systems for rapid trait assessment (Fahlgren et al. 2015) Automated plant movement and monitoring systems (Tardieu et al. 2017)	Accelerates trait evaluations (Fahlgren et al. 2015) Supports early-stage breeding programs (Tardieu et al. 2017)

Table 10.1 (continued)

		Applications	Tools and technologies	Advantages
Sensing and imaging technologies	Visible imaging	Captures morphological traits (e.g., plant height, leaf area, and growth patterns).	RGB cameras for high-resolution imaging (Fahlgren et al. 2015)	Non-destructive, high-resolution data for detailed morphological analysis (Li et al. 2014)
	Multispectral and hyperspectral imaging	Assess plant health, chlorophyll content, and stress responses by measuring reflectance across multiple wavelengths (Serbin et al. 2012)	Spectral cameras and handheld spectrometers	Detailed spectral information and stress (Zhang and Kovacs 2012) Early detection of physiological changes (Mahlein 2016)
	Thermal imaging	Monitors transpiration, water-use efficiency, and heat stress responses (Prashar and Jones 2014)	Infrared cameras for canopy temperature mapping (Prashar and Jones 2014)	Early detection stress responses before visible symptoms appear (Prashar and Jones 2014) Provides spatial mapping of temperature variations across plant canopies for large-scale screening (Pérez-Bueno et al. 2019)
	Fluorescence imaging	Measures photosynthetic efficiency and stress responses at the cellular level	Chlorophyll fluorescence cameras (Pérez-Bueno et al. 2019)	Early detection of abiotic and biotic stress responses at the cellular and sub-cellular levels (Kalaji et al. 2018)
	Root imaging and soil sensors	Evaluates root architecture, growth, and interactions with soil	MRI, X-ray CT scanners, and rhizotrons	Non-destructive monitoring of root systems (Tracy et al. 2010) Three-dimensional visualization of root architecture (Metzner et al. 2015)

(continued)

Table 10.1 (continued)

		Applications	Tools and technologies	Advantages
Environmental sensing		Monitors environmental factors affecting plant performance, including soil moisture, light intensity, temperature, and humidity	IoT-based weather stations (Metzner et al. 2015) Soil moisture sensors and evapotranspiration monitors (Jones 2007)	Provides critical context for interpreting phenotypic data in field trials
Data management systems	Data acquisition		Real-time collection of data from sensors and imaging devices Wireless data transfer and integration into centralized databases	Enables automated, continuous data collection, reducing manual labor and increasing data precision (Oliver et al. 2013) Improves reproducibility and standardization of phenotyping experiments (Krajewski et al. 2015)
	Data storage and curation		Cloud-based platforms, high-performance computing systems, and relational databases	Ensures secure, scalable storage solutions (Arend et al. 2014) Ensures data accessibility and organization for efficient retrieval and analysis (Arend et al. 2014)
	Data standardization		Use of metadata standards like MIAPPE (Minimum Information About a Plant Phenotyping Experiment) (Krajewski et al. 2015)	Harmonizes diverse datasets to allow seamless integration and comparison (Krajewski et al. 2015)

(continued)

Table 10.1 (continued)

		Applications	Tools and technologies	Advantages
Data analysis and interpretation	Image processing and feature extraction	Extracts traits such as canopy cover, leaf area index, and growth dynamics from images	OpenCV, ImageJ, and custom machine learning models	Enables high-throughput analysis, significantly reducing manual effort and time (Fahlgren et al. 2015) Improves accuracy and consistency in trait measurement by minimizing human error
	Statistical analysis	Identifies significant phenotypic differences and relationships	R, SAS, or python-based statistical libraries (Wickham and Grolemund 2017)	Supports advanced modeling techniques Enables reproducible research through scripting and automation of statistical workflows (VanderPlas 2016)
	Machine learning and artificial intelligence	Analyzes complex phenotypic datasets to predict genetic performance or identify patterns (Hafeez et al. 2023; Shinde and Shah 2018)	TensorFlow, PyTorch, or specialized phenomics software	Identifies complex relationships and patterns in phenotypic datasets that traditional statistical methods may miss (Shinde and Shah 2018)
	Genotype–phenotype integration	Combines phenotypic data with genomic data to perform GWAS or genomic selection	Genomic prediction models and bioinformatics pipelines	Enables GWAS and QTL mapping for discovering genes controlling complex traits (Dehghan 2018)

next-generation phenotyping data, researchers can link phenotypic traits to underlying genetic markers through genome-wide association studies (GWAS) or quantitative trait locus (QTL) mapping (Munjal et al. 2023).

10.4 Phenomics in Crop Breeding

Phenomics is the study of the phenome, encompassing all observable traits expressed by a genotype due to interactions between environmental factors and its developmental program (Zavafer et al. 2023). The term "phenome," introduced by Davis in 1949, originates from the Greek words φαίνω ("to bring to light") and τύπος ("shape" or "pattern"). A phenotype comprises measurable traits—physical (e.g., color, shape), chemical (e.g., metabolites, pH), and biological (e.g., growth, physiology)—that vary across developmental stages and environmental conditions. Like other "omics" fields, phenomics investigates this diverse trait spectrum, with High-Throughput Phenotyping (HTP) improving efficiency through automation and imaging. As noted by Zavafer et al. (2023), phenomics examines the full range of phenotypes a genotype can express, shaped by both environmental influences and developmental programming.

Non-invasive, image-based high-throughput phenotyping (HTP), combined with advanced data analytics, plays a crucial role in bridging the genotype–phenotype gap and accelerating genetic progress in crop breeding (Houle et al. 2010; Großkinsky et al. 2018). Deep phenotyping allows for identifying genetic factors influencing traits such as stress tolerance and yield. By focusing on key traits, HTP helps establish correlations between phenotypes and genetic markers, aiding in the identification of quantitative trait loci (QTLs) or genes. Traditional phenotyping methods, being manual and subjective, lack robustness, repeatability, and long-term reliability, leading to a "phenotyping bottleneck" that limits the understanding of genotype–phenotype relationships and environmental interactions (Furbank and Tester 2011). HTP typically involves capturing images to quantify traits throughout the crop life cycle (Ampatzidis and Partel 2019). A key advancement in HTP is its ability to non-destructively monitor plant traits over time, enabling time-series analyses of growth and stress responses in individual plants. High-throughput image-based phenotyping can process hundreds of plants daily, facilitating mutant population analysis, QTL mapping, and gene–environment interaction studies (Fahlgren et al. 2015b). By leveraging automated data extraction from sensor inputs, HTP, integrated with machine learning (ML) and deep learning (DL), offers a transformative approach for genetic improvement. It enhances selection accuracy and intensity, playing a vital role in advancing plant science (Araus et al. 2018; van Dijk et al. 2021). The integration of high-dimensional sensor data—collected via RGB, near-infrared (NIR), infrared (IR), fluorescent, and hyperspectral cameras—with systems biology and multi-omics approaches has significant potential to improve crop resilience to climate challenges (Großkinsky et al. 2018). Efficient data pipelines in HTP convert large image and sensor datasets into actionable insights through advanced analysis and modeling, accelerating the selection of climate-resilient crop varieties (Tardieu et al. 2017). Addressing big data and image processing challenges is crucial for accurate phenotyping, as rapid crop improvement relies on cutting-edge tools such as computer vision, artificial intelligence (AI), ML, and DL (Hickey et al. 2019).

10.5 High-Throughput Phenotyping Platforms (HTPP)

HTPPs have replaced traditional low-throughput, destructive methods for assessing abiotic stress tolerance. These platforms integrate automated robotics, various sensors (RGB, thermal, chlorophyll fluorescence, VNIR, SWIR, 3D imaging like CT, MRI, PET), image processing, and trait extraction for precise, non-destructive phenotyping. Advancements in digital and information technology have led to the development of HTPPs such as WPScanConveyor, TraitMill, Germination Scanalyzer, Lab Scanalyzer HTS, Greenhouse Scanalyzer, PhenoCenter, KeyGene Digital Phenotyping, Plant Screen, PlantEye F500, DroughtSpotter, PHENOSPEX, Rice Automatic Phenotyping Platform (RAP), SCREEN House, and PHENOPSIS. Major global facilities hosting these technologies include the Australian Plant Phenomics Facility (Australia), LemnaTec (Germany), Phenospex (Netherlands), and Nanaji Deshmukh Plant Phenomics Facility (India) (Kumari et al. 2024).

10.6 High-Throughput Phenotyping at NDPPC

The Indian Council of Agricultural Research (ICAR) in New Delhi has initiated the establishment of phenomics facilities across India. These facilities are located at ICAR-Indian Agricultural Research Institute in New Delhi, ICAR-Central Research Institute for Dryland Agriculture in Hyderabad, ICAR-Indian Institute of Horticultural Research in Bengaluru, and the National Institute of Abiotic Stress Management in Baramati, India.

Specifically, ICAR-IARI in New Delhi has developed an advanced automated high-throughput plant phenomics facility, NDPPC, funded by the National Agricultural Science Fund (NASF), ICAR, New Delhi. The Nanaji Deshmukh Plant Phenomics Centre (NDPPC) exemplifies a state-of-the-art platform for advanced plant phenotyping and genomics research, addressing critical challenges in global food security. Equipped with eight imaging sensors (Fig. 10.2), the facility provides a robust multidimensional phenotyping platform. High-resolution imaging capabilities, including hyperspectral VNIR (Visible Near-Infrared) and SWIR (Shortwave Infrared) sensors, enable detailed analysis of physiological and morphological traits, contributing to precise, non-invasive plant monitoring. The center's expertise in monitoring plant growth dynamics across entire crop cycles using high-throughput imaging systems allows for the identification of genotype-specific growth patterns and stress responses under controlled environmental conditions.

NDPPC has achieved significant milestones in phenotypic trait detection, particularly in wheat. The development of algorithms for ear or spike detection using RGB sensors and machine learning techniques has advanced phenological trait monitoring and yield prediction. Similarly, cutting-edge tools for high-resolution spikelet detection have facilitated detailed floral trait analysis, critical for yield enhancement. These methods, integrated with genetic analysis, provide valuable insights into spikelet development under diverse environmental conditions. The

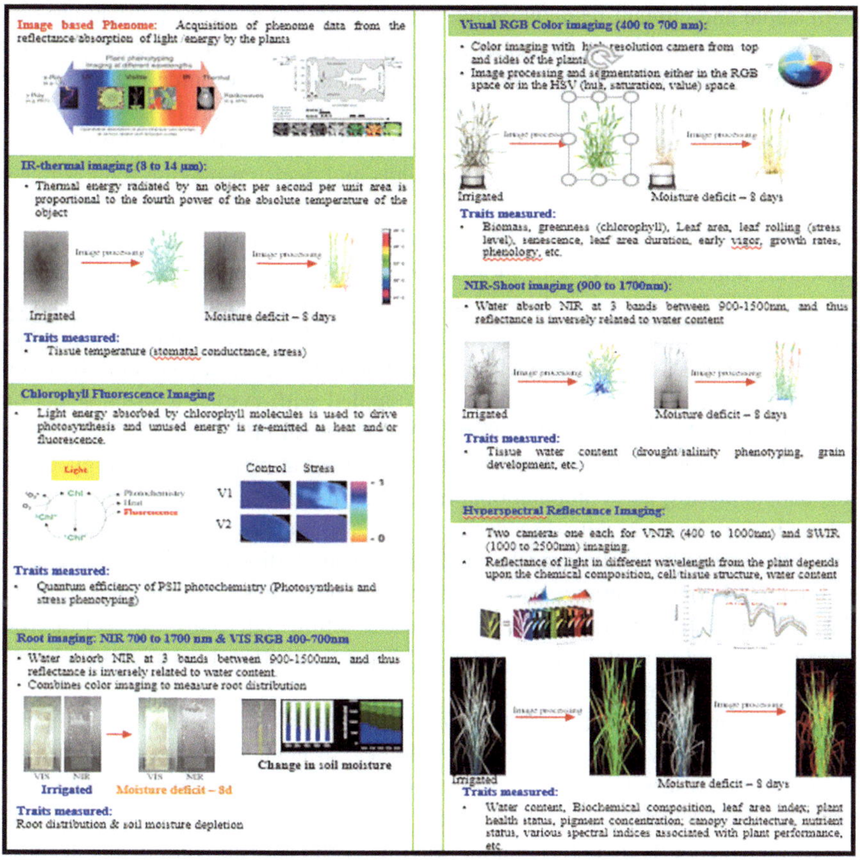

Fig. 10.2 Analysis of physiological traits from images captured using sensors (Kumar et al. 2016)

center's innovative approaches to biomass estimation using RGB imaging ensure real-time, non-destructive, and high-throughput assessments, with significant implications for precision agriculture and resource-use efficiency.

The facility is equipped with a precise weighing and watering station that measures water uptake and evapotranspiration, enabling critical studies on water use efficiency and drought stress responses. Automation in weighing and watering ensures consistent experimental conditions, supporting high-quality data generation. NDPPC also leads targeted research on abiotic stresses, particularly heat and drought, and nutrient-use efficiency, addressing key productivity challenges in the context of climate change. The integration of abiotic stress studies with genetic and phenotypic analyses enhances understanding of plant resilience and adaptive mechanisms.

The center's research spans a diverse range of strategic crops, including wheat, rice, *Brassica* (mustard), and pearl millet, which are vital for food security and nutrition in India. This diversity enables translational research with applications in both staple and climate-resilient crops. Precision phenotyping for nutrient use efficiency has also positioned the center as a pioneer in sustainable agriculture, reducing dependency on chemical fertilizers and enhancing resource-use efficiency. Advanced imaging and data acquisition systems further support large-scale phenotyping experiments. The integration of spectral imaging with environmental data enables the development of predictive models and decision-making tools for crop improvement.

A key strength of the center lies in its focus on climate-resilient agriculture. Research outcomes align with global priorities of developing crops resilient to changing climatic scenarios, directly supporting national and international goals for sustainable agriculture. Pioneering studies in wheat have identified novel quantitative trait loci (QTLs) associated with heat and drought tolerance, particularly in root and stem architecture. These discoveries, coupled with marker-assisted selection, are instrumental in breeding climate-resilient wheat varieties.

NDPPC fosters interdisciplinary research by integrating plant physiology, genetics, data analytics, and computational modelling to decode complex traits. Its collaborative framework with breeders, geneticists, and agronomists ensures that research outputs have high translational value. The facility's groundbreaking work in integrated stress tolerance studies combines phenomics and genomics to elucidate wheat's adaptability to heat and drought. Multi-trait analyses, such as root morphology and stem elongation under stress, provide actionable insights for breeding resilient cultivars.

In summary, the Nanaji Deshmukh Plant Phenomics Centre is a hub of innovation, leveraging advanced imaging technologies, strategic crop focus, and targeted abiotic stress research to redefine plant science. Its contributions in monitoring growth rates, detecting phenological traits, estimating biomass, and discovering QTLs underscore its pivotal role in advancing sustainable agriculture and ensuring food security in a changing climate.

10.7 Role of Phenomics in Crop Improvement

10.7.1 Genomic Selection

The current rate of genetic gain in crops is insufficient to keep pace with the 2% annual population growth, projected to reach 9.8 billion by 2050 (Ray et al. 2012, 2013). Breeders must develop high-yield, nutrient-rich, climate-resilient, and pest-resistant crops by integrating advanced techniques such as next-generation sequencing, GWAS, genomic selection (GS), high-throughput phenotyping (HTP), and speed breeding with traditional methods (Varshney et al. 2021). Genomic selection (GS), first applied in maize in 2007 (Bernardo and Yu 2007), has since been adopted in various crops (Crossa et al. 2014; Sun et al. 2017). It uses dense molecular markers to predict breeding values (BVs), estimating both parental averages and Mendelian sampling effects without extensive phenotyping, thereby accelerating breeding cycles and enhancing genetic gain (Meuwissen et al. 2001; Crossa et al.

2017). Encouraged by success in cereals, efforts are underway to apply GS in legumes such as pea, soybean, chickpea, groundnut, and pigeon pea to boost genetic gains (Varshney 2016). The integration of phenomics with modern genomics has significantly advanced breeding programs (Masuka et al. 2017a, 2017b; Araus et al. 2018). Multi-trait genomic selection (MTGS) models further enhance predictions by leveraging correlated traits from HTP, as demonstrated in maize (Anche et al. 2020) and wheat (Kaushal et al. 2024). Details on GS models, their characteristics, and source code are available in Sandhu et al. (2022).

10.7.2 Phenomic Selection

Phenomic selection (PS), introduced by Rincent et al. (2018), is a breeding method like genomic selection (GS) but utilizes near-infrared (NIR) spectroscopy instead of genotyping. This cost-effective, high-throughput approach captures non-additive effects and genetic similarities between genotypes, enabling accurate predictions of polygenic traits. Traditionally, NIRS is used to assess the chemical composition of tissues. In PS, NIR spectra and other phenomic data serve as proxies for complex traits such as grain yield, water stress, and chlorophyll content, often using vegetation indices (VIs) like NDVI. These indices, correlated with photosynthetic activity and sometimes yield, facilitate indirect trait selection, reducing the need for expensive phenotyping before harvest (Robert et al. 2022a, b, c). Studies have validated the effectiveness of PS using NIRS data in wheat (Cuevas et al. 2019; Robert et al. 2022a, 2022b), as well as in maize (Lane et al. 2020; Weiß et al. 2022), grapevine (Brault et al. 2022), triticale (Zhu et al. 2022), soybean (Zhu et al. 2021), and rapeseed (Roscher-Ehrig et al. 2024). A comparative analysis of grain yield prediction in hard winter wheat using phenomic versus genomic approaches was conducted by Winn et al. (2023).

10.7.3 PheWAS

Phenome-wide association study (PheWAS) complements genome-wide association studies (GWAS) by analyzing the relationship between genetic variants and a broad spectrum of phenotypic traits, whereas GWAS typically focuses on a limited number of traits. PheWAS utilizes comprehensive phenotypic datasets to uncover genotype–phenotype associations (Denny et al. 2010; Pendergrass et al. 2013; Bhat et al. 2015).

10.7.4 GWAS

Genomics-assisted breeding (GAB) approaches, such as genome-wide association studies (GWAS), have become powerful tools for dissecting complex quantitative traits like abiotic stress tolerance in crops, providing high-resolution insights and allelic diversity (Varshney et al. 2021). The integration of high-throughput phenotyping (HTP), GWAS, and machine learning/deep learning (ML/DL) models has helped uncover the genetic basis of plant traits, such as maize plant height (Wang et al. 2023). HTP data has also been used for predictive modeling of plant growth and drought response in barley (Chen et al. 2014, 2018) and has identified 470 loci associated with drought response in rice (Guo et al. 2018). Additionally, HTP sensors have predicted key traits such as plant color, water content, and temperature, while assessing photosynthetic efficiency and water use in drought-tolerant rice (Kim et al. 2020). The combination of HTP and image analysis has also facilitated the study of maize growth dynamics under stress conditions (Dodig et al. 2019, 2021).

10.8 Challenges and Limitations

The integration of High-Throughput Phenotyping (HTP) and sensor technologies results in the generation of extensive datasets, necessitating robust data analysis and management solutions. Traditional statistical methods fall short when addressing the complexities inherent in these large phenotypic datasets. Consequently, there is an urgent requirement for advanced analytical tools and data management platforms capable of effectively organizing, sharing, and retrieving phenotypic information. Key components of these systems include databases and cloud-based storage solutions that facilitate comprehensive data management (Arend et al. 2014). Furthermore, data transparency and interoperability with other research initiatives are crucial, achievable through adherence to standardized protocols and metadata profiles (Krajewski et al. 2015). This study aims to systematically assess the limitations of HTP and sensor-based phenotyping technologies within the context of their application in commercial breeding programs.

The Challenge of Big Data Analysis in Phenomics

- Image analysis methodologies in phenomics have traditionally relied on predetermined sequences of image processing and measurement operations tailored explicitly by their developers for distinct applications. Consequently, adapting a particular tool to address a slightly varied problem or operational context often necessitates a near-total overhaul of the software infrastructure.
- One of the fundamental requirements for maximizing the efficacy of deep learning tools in this domain is the availability of large-scale training datasets that span diverse environmental conditions. The challenges inherent in procuring such datasets are significant.
- Volume of Data: While technological advancements have enabled the collection of vast amounts of data, they do not inherently reduce the time spent on data acquisition. Instead, this capability provides breeders access to more extensive datasets, facilitating more informed decision-making processes.
- Cost Considerations: The financial implications of using UAVs can be considerable, depending on the model and sensors selected. For instance, the DJI Phantom

4—with integrated sensors—represents a more economical option. In contrast, the DJI Matrice 300, which does not come with sensors, is approximately six times more expensive than the Phantom 4. The additional costs for commonly used RGB and multispectral sensors to equip the Matrice can double the overall price. Moreover, supplementary expenses for accessories such as batteries and specialized data processing software must also be considered.
- Software Development: The underlying software for these technologies is still evolving, leading to various operational inefficiencies. For example, glitches during drone initialization can extend the startup time beyond simple activation, which poses a challenge for researchers constrained by weather conditions and the developmental stages of crops.
- Data Storage Requirements: High-throughput phenotyping (HTP) generates vast data, necessitating substantial data storage solutions. For context, storing RGB and multispectral images from a field approximately 2800 feet (equivalent to 1/20 of an acre) could require around 60 gigabytes of storage capacity.
- Indirect Measurements: Direct measurement of specific phenotypic traits is often impractical, leading to reliance on proxy measurements that correlate with the characteristics of interest. For instance, while UAVs cannot directly identify crop diseases, sensors can assess plant health through the normalized difference vegetation index (NDVI). This health metric can serve as a proxy for disease presence, yet it is critical to acknowledge that drought and nutrient deficiencies can also influence plant health. Field presence for visual assessments remains essential for accurate evaluations.

Infrastructure and Cost Implications

- Phenomics technologies require significant initial investments in high-throughput platforms, imaging systems, and sophisticated data analytics. However, these upfront costs are mitigated by the long-term financial benefits inherent in the policy framework. Phenomics enhances breeding efficiency and accelerates the development of superior germplasm. This approach yields considerable economic advantages through increased crop yields and reduced input requirements. For example, cultivating varieties with lower irrigation and nutrient needs significantly cuts expenses on water management and fertilizers, thus benefitting growers economically. Similarly, deploying disease-resistant plants diminishes the reliance on chemical pesticides (Mahlein et al. 2019), offering both cost savings and environmental sustainability.
- A comprehensive cost–benefit analysis of phenomics must encompass both direct and indirect costs. These technologies' immediate benefits include enhanced harvest volumes and lowered production expenses.
- Functional phenomics strategies will provide critical insights for precision agriculture and targeted breeding efforts, enabling rapid local adaptation to changing environmental conditions. Farming practices tailored to climate variability will likely achieve greater efficiency. Phenomics plays a crucial role in promoting sustainable agricultural practices by optimizing resource utilization and boosting

crop productivity, thereby fostering resilient agricultural systems to environmental stresses. Additionally, phenomics identifies essential germplasm for enhancing stress tolerance and promoting agro-diversity, which is vital for sustainable agriculture.

Nonetheless, integrating phenotypic data into decision support systems continues to pose challenges for advancing Precision Agriculture, although the potential of IoT-based networks to improve modeling and forecasting capabilities holds promise (Rodrigues et al. 2023).

10.9 Future Perspectives

10.9.1 Integration with IoT

The integration of plant phenomics with advanced technologies like the Internet of Things (IoT) and blockchain provides a substantial enhancement to holistic methodologies for monitoring and regulating plant growth within controlled environmental chambers. IoT facilitates the interconnectedness of a multitude of sensors and devices, enabling autonomous collection and analysis of in situ phenotypic data, which yields precise insights into plant–environment interactions. Supported by contemporary research, IoT frameworks empower plant scientists and agricultural engineers to manipulate environmental parameters in real time, informed by plant physiological models. This dynamic adaptability is facilitated through an open application programming interface (API), which allows for the development and deployment of algorithms aimed at optimizing plant growth control within experimental settings (Yuan et al. 2022).

10.9.2 Blokchain in Phenomics

Moreover, blockchain technology introduces a decentralized, immutable database for the management of phenotypic data, bolstering both traceability and the integrity of data. The combination of IoT and deep learning within this system fortifies data integrity, enhancing the reliability of agricultural practices (Frikha et al. 2023). This framework ensures the accuracy of phenotypic records, validating the efficacy of various growth strategies, and promotes secure inter-organizational data sharing for research or commercial objectives (Frikha et al. 2023). By decentralizing access, blockchain not only grants full data ownership to the creator but also enhances data sharing capabilities, thereby improving security and privacy in plant phenotyping initiatives (Samaniego et al. 2019). Thus, with the synergy of IoT and blockchain, phenomics lays a robust foundation for regulating plant developmental processes. Real-time phenotyping benefits significantly from IoT, which facilitates remote data collection that can be adjusted dynamically, while blockchain safeguards the integrity and authenticity of the accumulated datasets.

10.9.3 The Role of Artificial Intelligence (AI) and Machine Learning (ML)

The role of AI/ML is pivotal in plant phenomics, particularly in the development of predictive models for plant behavior and the fine-tuning of growth conditions. AI leverages real-time data to make dynamic adjustments to environmental parameters, ensuring optimal conditions for enhanced plant growth and yield. These technologies excel at analyzing and quantifying data, identifying critical patterns and correlations between plant traits and environmental variables. By integrating diverse environmental datasets and utilizing sophisticated algorithms, predictive models can be effectively trained to forecast plant responses across varied conditions (Gill et al. 2022).

The utilization of both supervised and unsupervised learning techniques allows for the decoding of complex phenotypic traits that are often obscured and not easily addressed through traditional analytical methods (Omta et al. 2020). Prominent supervised learning models, such as support vector machines (SVM) and random forests, are extensively employed for classifying phenotypic traits and predicting plant performance based on historical training datasets (Kok et al. 2021; Danilevicz et al. 2022). In contrast, clustering techniques, such as principal component analysis (PCA), facilitate the grouping of similar phenotypes and assist in data dimensionality reduction (Xu and Wu 2020). Additionally, subfields of ML, particularly deep learning, leverage multi-layered neural networks, demonstrating significant promise in phenotyping applications, especially image-based ones, which enable the extraction of traits from composite images (Ubbens and Stavness 2017; Murphy et al. 2024).

AI and machine learning are pivotal in modeling plant growth through the development of predictive models that leverage extensive datasets detailing plant behavior across varying environmental conditions. By analyzing the interplay and correlations between environmental variables and phenotypic traits, these models enhance the accuracy of forecasting plant responses. They encompass the integration of phenotypic characteristics, environmental factors, and genomic data to assess plant performance under diverse conditions such as light, temperature, water availability, and nutrient levels (Yasrab et al. 2021; Tripodi et al. 2022).

10.9.4 Artificial Neural Networks (ANNs)

These have become a standard approach for modeling plant growth, applicable to both steady-state and dynamic models of plant physiology. This methodology is especially relevant for optimizing food, air, and water production in controlled environments, essential for scenarios such as space travel, as discussed by Zee and Bubenheim (1997) and Kujawa and Niedbała (2021). Additionally, generative adversarial networks (GANs) significantly advance plant phenotyping and growth prediction by employing segmentation masks of root and shoot systems, improving data annotation and predictive accuracy (Yasrab et al. 2021; Lu et al. 2015). The use of convolutional neural networks (CNNs) for analyzing leaf images enables the detection of stress conditions and the automation of stress response assessments (Jiang and Li 2020).

The future of plant phenomics is exceptionally promising, driven by anticipated technological advancements and methodological innovations. Upcoming technologies are poised to enhance the speed and precision of phenotyping, thereby improving the resolution of plant traits. This enhancement will capitalize on advanced sensor technologies and imaging systems, alongside automation to facilitate high-throughput phenotyping (HTP) and deepen our understanding of plant biology. Furthermore, integrating phenomic data with various omics datasets—including genomics, transcriptomics, proteomics, and metabolomics—will deepen our understanding of the molecular underpinnings driving phenotypic variability. Such integrative strategies will facilitate a targeted and efficient breeding paradigm by harnessing multiple omics methodologies. The application of cutting-edge AI and ML algorithms to phenotypic data analysis will augment predictive capabilities, unearth novel patterns, and support real-time decision-making in crop management. Advancements in AI technology will foster the creation of optimized algorithms capable of programming environmental conditions for maximal crop output and resource efficiency. Phenomics will also play a crucial role in developing climate-smart crops tailored for cultivation in changing climates. By elucidating traits associated with stress tolerance and resource-use efficiency, phenomics will enhance sustainable agricultural practices, mitigating the environmental impact of crop production.

10.10 Conclusion

The integration of phenomic technologies marks a significant advancement in enhancing reproducibility and reducing bias in both basic and applied research. Achieving success in this domain necessitates a robust amalgamation of sensor technologies alongside sophisticated wavelength and image acquisition methods to accurately identify the subjects of analysis. While considerable advancements have been made in controlled indoor environments that facilitate high-quality imaging, outdoor scenarios pose substantial challenges due to inherent limitations in imaging devices and the uncontrolled conditions that can adversely impact image integrity. Emerging technologies such as high-definition LIDAR and multi-hyperspectral cameras hold considerable promise for overcoming these outdoor challenges and are expected to evolve rapidly. Effective preprocessing and segmentation of captured data are critical aspects of data management that require meticulous design to prevent distortions and ensure reproducibility. Given that images originate from machine-based processes, the variability in imaging types and processing techniques necessitates standardization in image capture, preprocessing, and segmentation protocols. It is essential to recognize that no single analysis procedure is universally superior; instead, researchers must evaluate various algorithms to tailor an optimized approach for their specific research setups. The imminent establishment of databases housing raw images—akin to NCBI for genomic data or Uniprot for proteomics—can significantly enhance the field of phenomics. As the cost of hyperspectral imaging instruments continues to decline, it enables new experimental designs that yield expansive datasets. These larger datasets will require sophisticated artificial intelligence-based analytical methods to translate data into interpretable results for researchers.

In the context of plant phenomics, which serves as a transformative paradigm in agricultural advancement, the integration of high-throughput phenotyping (HTP), artificial intelligence (AI), and machine learning (ML) facilitates precise quantification of phenotypic traits, optimization of environmental conditions, and the development of climate-resilient crops. These innovations are pivotal in enhancing crop yields, optimizing resource use, and promoting sustainability, directly addressing pressing issues such as food security, resource limitations, and climate change challenges. The interplay of advanced imaging technologies, sensor networks, and predictive analytics has redefined early disease detection, stress tolerance assessments, and precision agriculture practices. Techniques such as hyperspectral imaging, fluorescence analysis, and CRISPR gene editing further empower researchers to fine-tune crop performance and adapt plant phenotypes to specific environmental contexts. Moreover, the deployment of data-driven models in phenomics is enabling reduced resource inputs while maximizing agricultural productivity.

As a multidisciplinary field, plant phenomics effectively bridges genomics, physiology, and environmental science, offering comprehensive strategies to understand genotype–environment interactions. This integrative approach not only accelerates breeding cycles but also sets a foundation for the development of climate-smart crops, sustainable food systems, and eco-friendly agricultural practices. With ongoing advancements in automation, Internet of Things (IoT) technologies, and blockchain integration, the future of plant phenomics is poised to redefine agriculture as a sustainable, efficient, and resilient enterprise. Continued investment in phenomics research and infrastructure will be crucial for unlocking its full potential in addressing global agricultural challenges.

References

Abebe AM, Kim Y, Kim J, Kim SL, Baek J (2023) Image-based high-throughput phenotyping in horticultural crops. Plants 12(10):2061

Ali A, Hussain T, Tantashutikun N, Hussain N, Cocetta G (2023) Application of smart techniques, internet of things and data mining for resource use efficient and sustainable crop production. Agriculture 13(2):397

Ampatzidis Y, Partel V (2019) UAV-based high throughput phenotyping in citrus utilizing multispectral imaging and artificial intelligence. Remote Sens 11(4):410

Anand KJ, Nagre SP, Shrivastava MK, Amrate PK, Patel T, Katara VK (2023) Enhancing crop improvement through synergistic integration of advanced plant breeding and proximal remote sensing techniques: a review. Int J Plant Soil Sci 35(19):121–138

Anche MT, Kaczmar NS, Morales N, Clohessy JW, Ilut DC, Gore MA, Robbins KR (2020) Temporal covariance structure of multi-spectral phenotypes and their predictive ability for end-of-season traits in maize. Theor Appl Genet 133:2853–2868

Araus JL, Cairns JE (2014) Field high-throughput phenotyping: the new crop breeding frontier. Trends Plant Sci 19(1):52–61

Araus JL, Kefauver SC, Zaman-Allah M, Olsen MS, Cairns JE (2018) Translating high-throughput phenotyping into genetic gain. Trends Plant Sci 23(5):451–466

Arend D, Lange M, Chen J, Colmsee C, Flemming S, Hecht D, Scholz U (2014) e!DAL–a framework to store, share and publish research data. BMC Bioinf 15:214

Atefi A, Ge Y, Pitla S, Schnable J (2021) Robotic technologies for high-throughput plant phenotyping: contemporary reviews and future perspectives. Front Plant Sci 12:611940

Awais M, Li W, Cheema MJM, Zaman QU, Shaheen A, Aslam B et al (2022) UAV-based remote sensing in plant stress imagine using high-resolution thermal sensor for digital agriculture practices: A meta-review. Int J Environ Sci Technol:1–18

Bernardo R, Yu J (2007) Prospects for genomewide selection for quantitative traits in maize. Crop Sci 47(3):1082–1090

Bhat JA, Salgotra RK, Dar MY (2015) Phenomics: a challenge for crop improvement in genomic era. Mol Plant Breed 6(22):1–8

Brault C, Lazerges J, Doligez A, Thomas M, Ecarnot M, Roumet P et al (2022) Interest of phenomic prediction as an alternative to genomic prediction in grapevine. Plant Methods 18(1):108

Buckner E, Tong H, Ottley C, Williams C (2021) High-throughput image segmentation and machine learning approaches in the plant sciences across multiple scales. Emerg Top Life Sci 5(2):239–248

Chen D, Neumann K, Friedel S, Kilian B, Chen M, Altmann T, Klukas C (2014) Dissecting the phenotypic components of crop plant growth and drought responses based on high-throughput image analysis. Plant Cell 26(12):4636–4655

Chen D, Shi R, Pape JM, Neumann K, Arend D, Graner A et al (2018) Predicting plant biomass accumulation from image-derived parameters. GigaScience 7(2):giy001

Crossa J, Perez P, Hickey J, Burgueno J, Ornella L, Cerón-Rojas J et al (2014) Genomic prediction in CIMMYT maize and wheat breeding programs. Heredity 112(1):48–60

Crossa J, Pérez-Rodríguez P, Cuevas J, Montesinos-López O, Jarquín D, De Los Campos G et al (2017) Genomic selection in plant breeding: methods, models, and perspectives. Trends Plant Sci 22(11):961–975

Crossa J, Martini JW, Gianola D, Pérez-Rodríguez P, Jarquin D, Juliana P et al (2019) Deep kernel and deep learning for genome-based prediction of single traits in multienvironment breeding trials. Front Genet 10:1168

Crossa J, Montesinos-Lopez OA, Costa-Neto G, Vitale P, Martini JW, Runcie D et al (2024) Machine learning algorithms translate big data into predictive breeding accuracy. Trends Plant Sci 30:167

Cuevas J, Montesinos-López O, Juliana P, Guzmán C, Pérez-Rodríguez P, González-Bucio J et al (2019) Deep kernel for genomic and near infrared predictions in multi-environment breeding trials. G3: Genes Genomes Genetics 9(9):2913–2924

D'Agostino N, Tripodi P (2017) NGS-based genotyping, high-throughput phenotyping and genome-wide association studies laid the foundations for next-generation breeding in horticultural crops. Diversity 9(3):38

Danilevicz MF, M Gill R Anderson et al (2022) "Plant Genotype to Phenotype Prediction Using Machine Learning." Frontiers in Genetics 13:822173

Dallinger HG, Löschenberger F, Bistrich H, Ametz C, Hetzendorfer H, Morales L et al (2023) Predictor bias in genomic and phenomic selection. Theor Appl Genet 136(11):235

Dehghan A (2018) Genome-wide association studies. Methods Mol Biol (Clifton, N.J.) 1793:37–49

Denny JC, Ritchie MD, Basford MA, Pulley JM, Bastarache L, Brown-Gentry K et al (2010) PheWAS: demonstrating the feasibility of a phenome-wide scan to discover gene–disease associations. Bioinformatics 26(9):1205–1210

Dhanya VG, Subeesh A, Susmita C, Amaresh, Saji SJ, Dilsha C et al (2024) High throughput phenotyping using hyperspectral imaging for seed quality assurance coupled with machine learning methods: principles and way forward. Plant Physiol Rep 29:1–20

Dodig D, Božinović S, Nikolić A, Zorić M, Vančetović J, Ignjatović-Micić D et al (2019) Image-derived traits related to mid-season growth performance of maize under nitrogen and water stress. Front Plant Sci 10:814

Dodig D, Božinović S, Nikolić A, Zorić M, Vančetović J, Ignjatović-Micić D et al (2021) Dynamics of maize vegetative growth and drought adaptability using image-based phenotyping under controlled conditions. Front Plant Sci 571

Dubbert M, Couvreur V, Kübert A, Werner C (2023) Plant water uptake modelling: added value of cross-disciplinary approaches. Plant Biol 25(1):32–42

Fahlgren N, Feldman M, Gehan MA, Wilson MS, Shyu C, Bryant DW, Hill ST, McEntee CJ, Warnasooriya SN, Kumar I, Ficor T, Turnipseed S, Gilbert KB, Brutnell TP, Carrington JC, Mockler TC, Baxter I (2015) A versatile phenotyping system and analytics platform reveals diverse temporal responses to water availability in Setaria. Mol Plant 8(10):1520–1535

Fahlgren N, Gehan MA, Baxter I (2015b) Lights, camera, action: high-throughput plant phenotyping is ready for a close-up. Curr Opin Plant Biol 24:93–99

Frikha T, J Ktari B Zalila O Ghorbel and NB Amor (2023) "Integrating Blockchain and Deep Learning for Intelligent Greenhouse Control and Traceability." Alexandria Engineering Journal 79:259–273

Furbank RT, Tester M (2011) Phenomics–technologies to relieve the phenotyping bottleneck. Trends Plant Sci 16(12):635–644

Geng BAI, Yufeng GE (2023) Crop stress sensing and plant phenotyping systems: a review. Smart Agric 5(1):66

Gill T, SK Gill DK Saini Y Chopra JP de Koff and KS Sandhu (2022) "A Comprehensive Review of High Throughput Phenotyping and Machine Learning for Plant Stress Phenotyping." Phenomics 2, no. 3:156–183

Granier C, Aguirrezabal L, Chenu K, Cookson SJ, Dauzat M, Hamard P, Thioux JJ, Rolland G, Bouchier-Combaud S, Lebaudy A, Muller B, Simonneau T, Tardieu F (2006) PHENOPSIS, an automated platform for reproducible phenotyping of plant responses to soil water deficit in Arabidopsis thaliana permitted the identification of an accession with low sensitivity to soil water deficit. New Phytol 169(3):623–635

Großkinsky DK, Syaifullah SJ, Roitsch T (2018) Integration of multi-omics techniques and physiological phenotyping within a holistic phenomics approach to study senescence in model and crop plants. J Exp Bot 69(4):825–844

Guo Z, Yang W, Chang Y, Ma X, Tu H, Xiong F et al (2018) Genome-wide association studies of image traits reveal genetic architecture of drought resistance in rice. Mol Plant 11(6):789–805

Hafeez A, Ali B, Javed MA, Saleem A, Fatima M, Fathi A et al (2023) Plant breeding for harmony between sustainable agriculture, the environment, and global food security: an era of genomics-assisted breeding. Planta 258(5):97

Hickey LT, Hafeez AN, Robinson H, Jackson SA, Leal-Bertioli S, Tester M et al (2019) Breeding crops to feed 10 billion. Nat Biotechnol 37(7):744–754

Holman FH, Riche AB, Michalski A, Castle M, Wooster MJ, Hawkesford MJ (2016) High throughput field phenotyping of wheat plant height and growth rate in field plot trials using UAV based remote sensing. Remote Sensing 8(12):1031

Houle D, Govindaraju DR, Omholt S (2010) Phenomics: the next challenge. Nat Rev Genet 11(12):855–866. https://doi.org/10.1038/nrg2897

Humplík JF, Lazár D, Husičková A, Spíchal L (2015) Automated phenotyping of plant shoots using imaging methods for analysis of plant stress responses—a review. Plant Methods 11:29

Jackson R, Buntjer JB, Bentley AR, Lage J, Byrne E, Burt C, Jack P, Berry S, Flatman E, Poupard B, Smith S, Hayes C, Barber T, Love B, Gaynor RC, Gorjanc G, Howell P, Mackay IJ, Hickey JM, Ober ES (2023) Phenomic and genomic prediction of yield on multiple locations in winter wheat. Front Genet 14:1164935

Jannink JL, Lorenz AJ, Iwata H (2010) Genomic selection in plant breeding: from theory to practice. Brief Funct Genomics 9(2):166–177

Jayasinghe SL, Thomas DT, Anderson JP, Chen C, Macdonald BC (2023) Global application of regenerative agriculture: A review of definitions and assessment approaches. Sustainability 15(22):15941

Jiang Y, and C Li (2020) "Convolutional Neural Networks for Image-Based High-Throughput Plant Phenotyping: A Review." Plant Phenomics 2020:4152816

Jin S, Sun X, Wu F, Su Y, Li Y, Song S et al (2021) Lidar sheds new light on plant phenomics for plant breeding and management: recent advances and future prospects. ISPRS J Photogramm Remote Sens 171:202–223

Jones HG (2007) Monitoring plant and soil water status: established and novel methods revisited and their relevance to studies of drought tolerance. J Exp Bot 58(2):119–130

Kalaji HM, Bąba W, Gediga K, Goltsev V, Samborska IA, Cetner MD, Dimitrova S, Piszcz U, Bielecki K, Karmowska K, Dankov K, Kompała-Bąba A (2018) Chlorophyll fluorescence as a tool for nutrient status identification in rapeseed plants. Photosynth Res 136(3):329–343

Kaushal S, Gill HS, Billah MM, Khan SN, Halder J, Bernardo A et al (2024) Enhancing the potential of phenomic and genomic prediction in winter wheat breeding using high-throughput phenotyping and deep learning. Front Plant Sci 15:1410249

Kim SL, Kim N, Lee H, Lee E, Cheon KS, Kim M et al (2020) High-throughput phenotyping platform for analyzing drought tolerance in rice. Planta 252(3):1–18

Kok ZH, Shariff ARM, Alfatni MSM, Khairunniza-Bejo S (2021) Support vector machine in precision agriculture: a review. Comput Electron Agric 191:106546

Krajewski P, Chen D, Ćwiek H, van Dijk AD, Fiorani F, Kersey P, Klukas C, Lange M, Markiewicz A, Nap JP, van Oeveren J, Pommier C, Scholz U, van Schriek M, Usadel B, Weise S (2015) Towards recommendations for metadata and data handling in plant phenotyping. J Exp Bot 66(18):5417–5427

Kujawa S, and G Niedbała (2021) "Artificial Neural Networks in Agriculture." Agriculture 11, no. 6:497

Kumar S, Raju D, Sahoo RN, Chinnusamy V (2016) Phenomics: unlocking the hidden genetic variation for breaking the barriers in yield and stress tolerance. Indian J Plant Physiol. https://doi.org/10.1007/s40502-016-0261-0

Kumari P, Bhatt A, Meena VK, Adhikari S, Dhar N, Chawda H et al (2024) Plant Phenomics: the force behind tomorrow's crop phenotyping tools. J Plant Growth Regul:1–19

Kundu S, Saini DK, Meena RK, Bahuguna RN, Jagadish SK (2024) High-throughput phenotyping and AI technologies for deciphering crop resilience to heat stress. Plant Physiol Rep 29:1–17

Lane HM, Murray SC, Montesinos-López OA, Montesinos-López A, Crossa J, Rooney DK et al (2020) Phenomic selection and prediction of maize grain yield from near-infrared reflectance spectroscopy of kernels. Plant Phenome J 3(1):e20002

Li L, Zhang Q, Huang D (2014) A review of imaging techniques for plant phenotyping. Sensors (Basel, Switzerland) 14(11):20078–20111

Lu Y, Liu Y, Niu X, Yang Q, Hu X, Zhang HY, Xia J (2015) Systems genetic validation of the SNP-metabolite association in rice via metabolite-pathway-based phenome-wide association scans. Front Plant Sci 6:1027

Mahlein AK (2016) Plant disease detection by imaging sensors – parallels and specific demands for precision agriculture and plant phenotyping. Plant Dis 100(2):241–251

Mansoor S, Karunathilake EMBM, Tuan TT, Chung YS (2024) Genomics, Phenomics, and machine learning in transforming plant research: advancements and challenges. Horticultural. Plant J 11:486

Marsh JI, Hu H, Gill M, Batley J, Edwards D (2021) Crop breeding for a changing climate: integrating phenomics and genomics with bioinformatics. Theor Appl Genet 134:1677–1690

Masuka B, Atlin GN, Olsen M, Magorokosho C, Labuschagne M, Crossa J et al (2017a) Gains in maize genetic improvement in Eastern and Southern Africa: I. CIMMYT hybrid breeding pipeline. Crop Sci 57(1):168–179

Masuka B, Magorokosho C, Olsen M, Atlin GN, Bänziger M, Pixley KV et al (2017b) Gains in maize genetic improvement in Eastern and Southern Africa: II. CIMMYT open-pollinated variety breeding pipeline. Crop Sci 57(1):180–191

Metzner R, Eggert A, van Dusschoten D, Pflugfelder D, Gerth S, Schurr U, Uhlmann N, Jahnke S (2015) Direct comparison of MRI and X-ray CT technologies for 3D imaging of root systems in soil: potential and challenges for root trait quantification. Plant Methods 11:17

Meuwissen TH, Hayes BJ, Goddard M (2001) Prediction of total genetic value using genome-wide dense marker maps. Genetics 157(4):1819–1829

Millet EJ, Kruijer W, Coupel-Ledru A, Alvarez Prado S, Cabrera-Bosquet L, Lacube S, Charcosset A, Welcker C, van Eeuwijk F, Tardieu F (2019) Genomic prediction of maize yield across European environmental conditions. Nat Genet 51(6):952–956

Munjal R, Benıwal J, Dhundwal A, Goyal A, Kumarı A, Behl RK (2023) Accelerating crop breeding in the 21st century: A comprehensive review of next generation phenotyping techniques and strategies. Ekin J Crop Breed Genet 9(2):160–171

Murphy KM, E Ludwig J Gutierrez and MA Gehan (2024) "Deep Learning in Image-Based Plant Phenotyping." Annual Review of Plant Biology 75:771–795

Noshita K, Murata H, Kirie S (2022) Model-based plant phenomics on morphological traits using morphometric descriptors. Breed Sci 72:19–30

Oliver M, Bishop T, Marchant B (eds) (2013) Precision agriculture for sustainability and environmental protection. Routledge

Omta WA, RG van Heesbeen I Shen et al (2020) "Combining Supervised and Unsupervised Machine Learning Methods for Phenotypic Functional Genomics Screening." SLAS Discovery: Advancing the Science of Drug Discovery 25, no. 6:655–664

Pappula-Reddy SP, Kumar S, Pang J, Chellapilla B, Pal M, Millar AH, Siddique KH (2024) High-throughput phenotyping for terminal drought stress in chickpea (Cicer arietinum L.). Plant Stress 11:100386

Paramasivan SK (2021) Deep learning based recurrent neural networks to enhance the performance of wind energy forecasting: a review. Revue d'Intelligence Artificielle 35(1):1

Paturkar A, Sen Gupta G, Bailey D (2021) Making use of 3D models for plant physiognomic analysis: a review. Remote Sens 13(11):2232

Pendergrass SA, Brown-Gentry K, Dudek S, Frase A, Torstenson ES, Goodloe R et al (2013) Phenome-wide association study (PheWAS) for detection of pleiotropy within the Population Architecture using Genomics and Epidemiology (PAGE) Network. PLoS Genet 9(1):e1003087

Pérez-Bueno ML, Pineda M, Barón M (2019) Phenotyping plant responses to biotic stress by chlorophyll fluorescence imaging. Front Plant Sci 10:1135

Prashar A, Jones HG (2014) Infra-red thermography as a high-throughput tool for field phenotyping. Agronomy 4(3):397–417

Ray DK, Ramankutty N, Mueller ND, West PC, Foley JA (2012) Recent patterns of crop yield growth and stagnation. Nat Commun 3(1):1293

Ray DK, Mueller ND, West PC, Foley JA (2013) Yield trends are insufficient to double global crop production by 2050. PLoS One 8(6):e66428

Rincent R, Charpentier JP, Faivre-Rampant P, Paux E, Le Gouis J, Bastien C, Segura V (2018) Phenomic selection is a low-cost and high-throughput method based on indirect predictions: proof of concept on wheat and poplar. G3: Genes GenomesGenetics 8(12):3961–3972

Robert P, Auzanneau J, Goudemand E, Oury FX, Rolland B, Heumez E et al (2022a) Phenomic selection in wheat breeding: identification and optimisation of factors influencing prediction accuracy and comparison to genomic selection. Theor Appl Genet 135:1–20

Robert P, Brault C, Rincent R, Segura V (2022b) Phenomic selection: a new and efficient alternative to genomic selection. In: Genomic prediction of complex traits: methods and protocols, pp 397–420

Robert P, Goudemand E, Auzanneau J, Oury FX, Rolland B, Heumez E et al (2022c) Phenomic selection in wheat breeding: prediction of the genotype-by-environment interaction in multi-environment breeding trials. Theor Appl Genet 135(10):3337–3356

Rodrigues L, P Moura F Terra et al (2023) "Synergizing Crop Growth Models and Digital Phenotyping: The Design of a Cost-Effective Internet of Things-Based Sensing Network." In Biology and Life Sciences Forum, Vol. 27, No. 1, p. 41. Basel, Switzerland: MDPI

Roscher-Ehrig L, Weber SE, Abbadi A, Malenica M, Abel S, Hemker R et al (2024) Phenomic selection for hybrid rapeseed breeding. Plant Phenomics 6:0215

Saleem A, Anwar S, Nawaz T, Fahad S, Saud S, Ur Rahman T et al (2024) Securing a sustainable future: the climate change threat to agriculture, food security, and sustainable development goals. J Umm Al-Qura Univ Appl Sci:1–17

Samaniego M, C Espana and R Deters (2019) "Access Control Management for Plant Phenotyping Using Integrated Blockchain." In Proceedings of the 2019 ACM International Symposium on Blockchain and Secure Critical Infrastructure, 39–46

Sandhu KS, Merrick LF, Sankaran S, Zhang Z, Carter AH (2022) Prospectus of genomic selection and phenomics in cereal, legume and oilseed breeding programs. Front Genet 12:829131

Serbin SP, Dillaway DN, Kruger EL, Townsend PA (2012) Leaf optical properties reflect variation in photosynthetic metabolism and its sensitivity to temperature. J Exp Bot 63(1):489–502

Sheikh M, Iqra F, Ambreen H, Pravin KA, Ikra M, Chung YS (2024) Integrating artificial intelligence and high-throughput phenotyping for crop improvement. J Integr Agric 23(6):1787–1802

Shinde PP, Shah S (2018) A review of machine learning and deep learning applications. In: 2018 Fourth international conference on computing communication control and automation (ICCUBEA), Pune, India, pp 1–6

Smigaj M, Agarwal A, Bartholomeus H, Decuyper M, Elsherif A, de Jonge A, Kooistra L (2024) Thermal infrared remote sensing of stress responses in forest environments: a review of developments, challenges, and opportunities. Curr For Rep 10(1):56–76

Soualiou S, Wang Z, Sun W, de Reffye P, Collins B, Louarn G, Song Y (2021) Functional–structural plant models mission in advancing crop science: opportunities and prospects. Front Plant Sci 12:747142

Sun J, Rutkoski JE, Poland JA, Crossa J, Jannink JL, Sorrells ME (2017) Multitrait, random regression, or simple repeatability model in high-throughput phenotyping data improve genomic prediction for wheat grain yield. Plant Genome 10(2):plantgenome2016-11

Tardieu F, Cabrera-Bosquet L, Pridmore T, Bennett M (2017) Plant phenomics, from sensors to knowledge. Curr Biol 27(15):R770–R783

Tracy SR, Roberts JA, Black CR, McNeill A, Davidson R, Mooney SJ (2010) The X-factor: visualizing undisturbed root architecture in soils using X-ray computed tomography. J Exp Bot 61(2):311–313

Tripodi P, Nicastro N, Pane C, Cammarano D (2022) Digital applications and artificial intelligence in agriculture toward next-generation plant phenotyping. Crop Pasture Sci 74:597

Ubbens JR, and I Stavness (2017) "Deep Plant Phenomics: A Deep Learning Platform for Complex Plant Phenotyping Tasks." Frontiers in Plant Science 8:1190

van Dijk ADJ, Kootstra G, Kruijer W, de Ridder D (2021) Machine learning in plant science and plant breeding. IScience 24(1):101890

van Klompenburg T, Kassahun A, Catal C (2020) Crop yield prediction using machine learning: A systematic review. Comput Electron Agric 177:105709

VanderPlas J (2016) Python data science handbook. O'Reilly Media

Varshney RK (2016) Exciting journey of 10 years from genomes to fields and markets: some success stories of genomics-assisted breeding in chickpea, pigeonpea and groundnut. Plant Sci 242:98–107

Varshney RK, Bohra A, Roorkiwal M, Barmukh R, Cowling WA, Chitikineni A et al (2021) Fast-forward breeding for a food-secure world. Trends Genet 37(12):1124–1136

Wang W, Guo W, Le L, Yu J, Wu Y, Li D et al (2023) Integration of high-throughput phenotyping, GWAS, and predictive models reveals the genetic architecture of plant height in maize. Mol Plant 16(2):354–373

Wasaya A, Zhang X, Fang Q, Yan Z (2018) Root phenotyping for drought tolerance: a review. Agronomy 8(11):241

Weiß TM, Zhu X, Leiser WL, Li D, Liu W, Schipprack W et al (2022) Unraveling the potential of phenomic selection within and among diverse breeding material of maize (Zea mays L.). G3 12(3):jkab445

Wickham H, Grolemund G (2017) R for data science. O'Reilly Media

Winn ZJ, Amsberry AL, Haley SD, DeWitt ND, Mason RE (2023) Phenomic versus genomic prediction—a comparison of prediction accuracies for grain yield in hard winter wheat lines. Plant Phenome J 6(1):e20084

Xu Z, and C Wu (2020) "Machine Learning and Statistical Approaches for Plant Phenotyping." In Intelligent Image Analysis for Plant Phenotyping, 195–220. Boca Raton, FL: CRC Press

Xu Y, Zhang X, Li H, Zheng H, Zhang J, Olsen MS et al (2022) Smart breeding driven by big data, artificial intelligence, and integrated genomic-enviromic prediction. Mol Plant 15(11):1664–1695

Yasrab R, J Zhang P Smyth and MP Pound (2021) "Predicting Plant Growth From Time-Series Data Using Deep Learning." Remote Sensing 13, no. 3:331

Ye D, Wu L, Li X, Atoba TO, Wu W, Weng H (2023) A synthetic review of various dimensions of non-destructive plant stress phenotyping. Plants (Basel, Switz) 12(8):1698

Yoosefzadeh Najafabadi M, Hesami M, Eskandari M (2023) Machine learning-assisted approaches in modernized plant breeding programs. Genes 14(4):777

Yuan S, Tang H, Fu Lj, Tan Jl, Govindjee G, Guo Y (2022) An open Internet of Things (IoT)-based framework for feedback controlof photosynthetic activities Photosynthetica 60(SPECIAL ISSUE 2022) 79–87. https://doi.org/10.32615/ps.2021.066

Zavafer A, Bates H, Mancilla C, Ralph PJ (2023) Phenomics: conceptualization and importance for plant physiology. Trends Plant Sci 28:1004

Zee F, and D Bubenheim (1997) "Plant Growth Model Using Artifical Neural Networks." SAE Transactions 106:1020–1028

Zhang C, Kovacs JM (2012) The application of small unmanned aerial systems for precision agriculture: A review. Precis Agric 13(6):693–712

Zhang Y, Huang G, Zhao Y, Lu X, Wang Y, Wang C et al (2024a) Revolutionizing crop breeding: next-generation artificial intelligence and big data-driven intelligent design. Engineering 44:245

Zhang Q, Luan R, Wang M, Zhang J, Yu F, Ping Y, Qiu L (2024b) Research Progress of spectral imaging techniques in plant phenotype studies. Plants 13(21):3088

Zhu X, Leiser WL, Hahn V, Würschum T (2021) Phenomic selection is competitive with genomic selection for breeding of complex traits. Plant Phenome J 4(1):e20027

Zhu X, Maurer HP, Jenz M, Hahn V, Ruckelshausen A, Leiser WL, Würschum T (2022) The performance of phenomic selection depends on the genetic architecture of the target trait. Theor Appl Genet 135:1–13

Zia S, Romano G, Spreer W, Sanchez C, Cairns J, Araus JL, Müller J (2013) Infrared thermal imaging as a rapid tool for identifying water-stress tolerant maize genotypes of different phenology. J Agron Crop Sci 199(2):75–84

11 Phenomics to Improve Genebank Value and Utility

Andrey Vega Alfaro, Liam Dixon Scully, and Irwin L. Goldman

Abstract

One of the most pressing problems in agricultural science is understanding global stored seed bank variation and how it might be successfully deployed to feed a growing population. Less than 1% of accessions held by Plant Genetic Resources for Food and Agriculture are utilized for crop improvement. Phenomics, a science that integrates genotype, environment, and time, has the potential to characterize stored genetic variation in genebanks. Phenomic approaches that make use of digital imaging and machine learning may provide high-quality data that can be used for germplasm assessment. However, much of these data at present are neither centralized nor attached to germplasm repositories. An enormous opportunity exists to integrate physical collections of seeds with phenomic information from those accessions. A global approach to centralizing phenomic data on crop germplasm may require the cooperation of world governments and require substantial resources, but the potential to more fully characterize our crop genetic resources using these techniques should foster improved opportunities for plant breeders to increase food production.

Keywords

Plant genetic resources · Seed bank · Phenome · Phenotyping · Germplasm

---

A. V. Alfaro · L. D. Scully · I. L. Goldman (✉)
Department of Plant and Agroecosystem Sciences, University of Wisconsin-Madison, Madison, WI, USA
e-mail: ilgoldma@wisc.edu

P. M. Priyadarshan, R. Ortiz (eds.), *Plant Breeding 2050*,
https://doi.org/10.1007/978-981-95-0583-8_11

11.1 Introduction

Of some 380,000 known plant species on earth, humans rely on about 100 of these for their survival. Of these 100 species, about 15 provide for roughly 80% of our plant-based calories (McCouch et al. 2013), and only three—maize, wheat, and rice—deliver as much as 60% of these calories (Listman and Ordonez 2019). By contrast, thousands of species are curated in seed banks throughout the world, represented by more than seven million distinct accessions (Anglin et al. 2018). One is tempted to consider the mismatch between seed bank inventory and in-field diversity as a hedge against an uncertain future for humanity's efforts to feed itself. However, phenotypic characterization of that seed bank inventory is woefully inadequate. One of the most pressing problems in agricultural science is understanding our stored seed bank variation and how it might be successfully deployed to feed a growing population. Currently, less than 1% of accessions held by Plant Genetic Resources for Food and Agriculture are utilized for crop improvement (Sharma et al. 2013). Nevertheless, in the USA, the National Plant Germplasm System (NPGS) processes on average 250,000 requests a year (Richards 2021).

Seed banks, sometimes known as genebanks, are often funded and managed by world governments. Long considered a "public good," global crop germplasm may be one of the most critically important and poorly understood natural resources upon which our future depends. Management of the planet's global germplasm resources faces a number of critical challenges, including the difficulty of maintaining sufficient genetic diversity through multiplication of genebank accessions, finding consistent sources of public funding to support maintenance of the collections, and rapid habitat destruction leading to a scarcity of natural landscapes for crop wild relatives to survive (Khoury et al. 2020). Among the most pressing challenges facing the users of genebanks is a general lack of information about the characteristics and potential value of any given germplasm accession.

Since it is often far beyond the available resources and scope for genebanks to provide comprehensive characterization of their collections for all traits of potential value, the responsibility for screening and evaluation has traditionally rested with the user community. To that end, modern technologies that allow for large-scale trait assessment and characterization (i.e., phenotyping) have tremendous potential to enhance the utility of genebanks. What traditionally may have been referred to as "phenotyping" has, in a modern "omics" context, morphed into the term "phenomics." This word encompasses the full characterization of the phenome, which has been defined by Zavafer et al. (2023) as the set of phenotypes that a given genotype may express in relation to the factors that caused their expression. Those factors include the genotype (G), time or developmental stage (t), and environment (E). As such, the discipline of phenomics aims to characterize the *hypervolume* of phenotypes that result from the $G \times t \times E$ interaction (Fig. 11.1).

Thus, while key morphological characteristics such as flower color, growth habit, or cultivar type may be recorded in a genebank's database, phenomic information is often lacking. Such data consider environmental interactions that are critical to

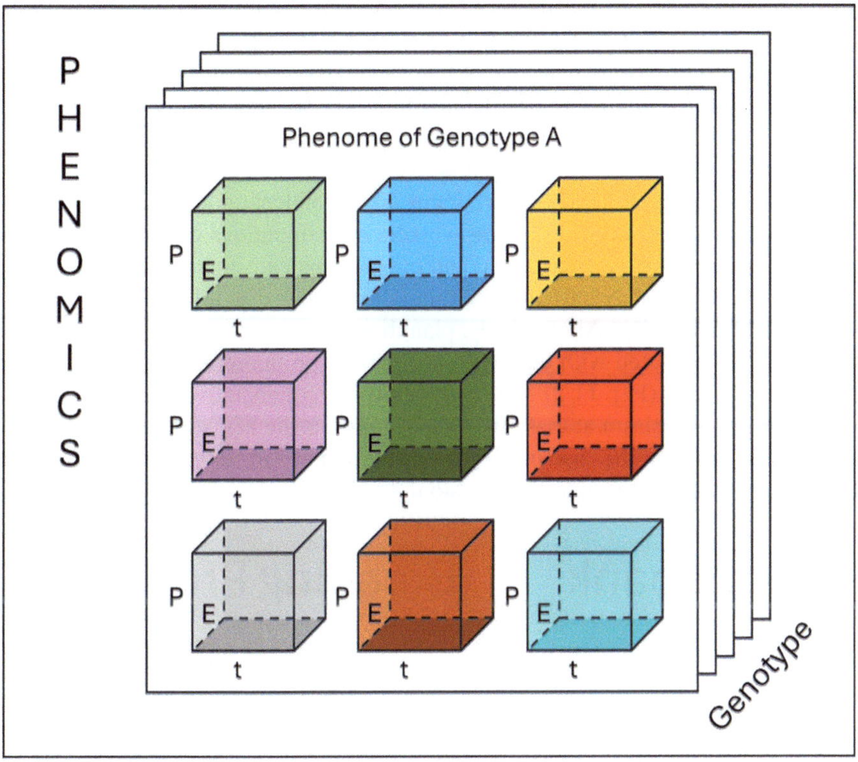

Fig. 11.1 Visual description of the phenomics discipline. Phenomics may be defined as the study of the phenome. The phenome, in turn, may be described as the hypervolume of phenotypes that a given genotype can express due to the interaction between genetic and non-genetic factors, namely environmental conditions and developmental stage. The phenome of genotype A is represented by a collection of nine traits, each assigned a different color, and each displayed as a plot with axes for phenotype (P), environment (E), and developmental stage (t). Although nine traits are shown for convenience, in truth, the number of traits, be them physical, chemical, or biological, that comprise the phenome of a genotype is virtually infinite. Moreover, full trait characterization involves measuring the phenotype in all possible environments and across all developmental stages

understanding more complex traits. Traditionally, studies designed to understand complex traits have been carried out over years and in multiple cropping locations by researchers who invest substantial resources in trait measurement and the statistical analysis of results from a series of environments. Such data are rarely found accompanying lists of accessions held by genebanks, in part because each experiment designed to collect such data is unique to the objectives and constraints of a particular research group. However, beginning in 2022, government policy changes in the USA will influence how those data are communicated and shared. The Office of Science and Technology Policy has required that all federal agencies supporting scientific research develop plans by the end of 2024 for requiring data to be freely available to the public by the end of 2025. It is therefore more plausible that publicly accessible genebanks could accumulate phenomic information on genebank accessions in part to fulfill this mandate. While such data would still be tied to specific objectives and research locations, the possibility of accumulating evaluative information on genebank accessions in a single location is a compelling step in increasing the value of banks for future plant breeding efforts.

Genebanks and plant breeding programs are dependent upon germplasm assessment and characterization, and often workers at both entities collaborate on such efforts. Once germplasm is collected and passport data are entered into a database, germplasm curators may have an opportunity to provide trait information on accessions in their collection. Alternatively, plant breeders may request germplasm from genebanks and proceed with their own assessments in the screening phase of their breeding programs (Fig. 11.2). The opportunity for collaborative phenomic assessment of crop germplasm at these two stages presents the best opportunity for each entity to describe (in the case of the genebank) and utilize (in the case of the plant breeding program) a germplasm accession.

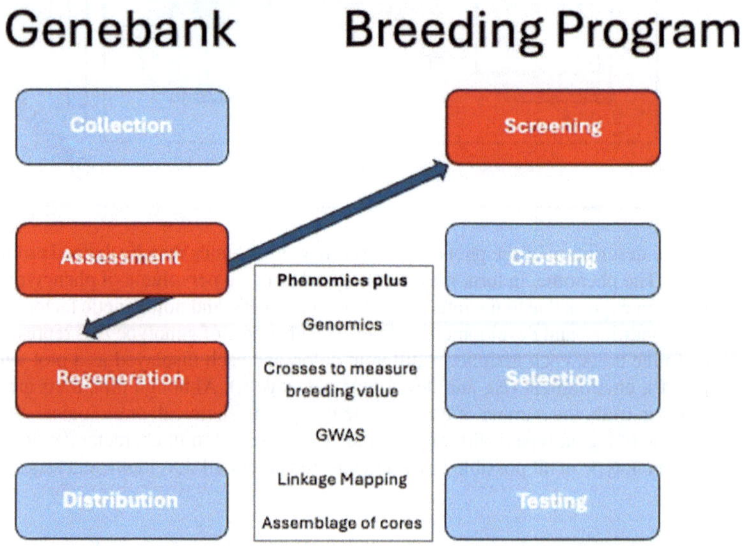

Fig. 11.2 Workflow diagram for genebanks and plant breeding programs. Rectangles indicate the four major steps for each entity, and red rectangles indicate the steps in which phenomic characterization of germplasm accessions may occur. Loci of potential collaboration for application of phenomics to characterize germplasm accessions exist between genebanks and breeding programs at the steps of regeneration and screening, as indicated by the double arrow. Phenomic approaches may also include supporting data from crosses, association and linkage studies, genomics, and core collection subsets

11.2 Genebanks

Genebanks have two central functions—to conserve genetic material and to make conserved materials available to the user community. Fulfillment of these functions requires a complex and interdependent set of operations involving germplasm collection, assessment, regeneration, and distribution. New samples may be evaluated alongside already registered accessions by morphological, biochemical, and molecular approaches to limit conservation of genetic duplicates (Rao et al. 2006). During regeneration, there may be opportunities to contribute further to accession passport information by reporting on specific traits of interest to the user community (Nguyen and Norton 2020). There may also not be opportunities to conduct such work owing to limited resources, or regeneration procedures that complicate phenotypic data collection including lack of appropriate replication. In such cases as well as in general, genebanks may benefit from collaboration with plant breeding programs that regularly collect phenomic data on several traits during the screening phase (Fig. 11.2). In addition, further phenomic characterization of genebank accessions may be possible by expanding data collection to the citizen scientist community. Doing so may considerably increase the number of environments in which genebank accessions are evaluated. Digital applications, such as SeedLinked, offer a streamlined and user-friendly data collection experience that may be leveraged to allow citizen scientist participation in this work while maintaining high data quality. In a 2024 pilot study between the USDA and SeedLinked, 200 growers in the USA were recruited to contribute notes on plant maturity, height, and flavor profiles of 20 common bean genebank accessions. The growers volunteered within 2 hours, demonstrating high enthusiasm for germplasm characterization within SeedLinked's 10,000-member user community. Incorporating crowdsourced data, either from the research or citizen scientist communities, to genebank accession profiles has the potential to improve genebank value and utility among plant breeders.

It is also possible that the post-genomic and phenomic era will usher in a change in the way information and seed are both collected and distributed by genebanks. In the past, it was common to think of variation in relation to a single reference genome for a crop. This gave way to the development of pangenomes, which expanded the collection of genomic information to an entire clade. In today's post-genomic era, the information most valuable to the user community may be the variation in relation to all relatives, based on evolutionary relationships. Aubry (2023) and Mascher et al. (2019) hypothesized that genebank activities will shift from management of genetic resources to distribution of digital information. In this scenario, farmers may be empowered as primary contributors to a global database that describes the world's genebank accessions. Digitization of genomic and phenomic information held by genebanks could have a democratizing effect on agricultural science, potentially leveling the playing field for the many users of these data. The promise of large-scale digitization of genebank information represents an exciting opportunity for crop breeding in the twenty-first century, and however, substantial barriers to wide-scale adoption, including unequal access to digital technologies, unresolved intellectual property issues, and imbalance with respect to the digitization of economically valuable crop species, may make this a global challenge in the coming decades. In addition, recent efforts have expanded on the digitization of genebanks and genetic information. "Genebanking" thus, not only represents a technological advancement but also reflects deeper changes regarding the future of food systems. The blurred line between physical (seed, accessions) and informational genetic resources (DNA sequences, genes, alleles, genomes) could affect future decisions about access, ownership, and control of plant genetic resources (Aubry 2023).

11.3 Global Genebank Landscape

Plant genetic resources (PGR) are central to sustainable crop improvement. The loss and unavailability of genetic diversity hinders the ability to improve crops sustainably (Fernando et al. 2023). Genebanks preserve genetic diversity over the long term and ensure its availability to breeders and professionals, helping to advance genetic gain and food security. Unfortunately, many genebanks operate independently, disconnected from one another and using different protocols and standards for data management, storage, and conservation. This lack of uniformity leads to inefficiencies and difficulties in sharing and accessing genetic resources globally.

There is often limited communication and collaboration between genebanks, leading to redundant efforts and gaps in the global conservation network. However, there are two current large-scale efforts with the mission of providing scalable versions of germplasm resources: GRIN-Global (https://www.grin-global.org/) and Genesys PGR (https://www.genesys-pgr.org/). The Genesys system is a centralized portal to information about crop diversity conserved in member genebanks that connects users, data providers, and genebanks to work efficiently in the complex landscape of genetic conservation. Genesys PGR aims to standardize data management practices and improve the visibility and access to plant genetic resources. On the other hand, GRIN-Global provides a scalable database designed to be implemented by virtually any genebank. The mission of GRIN-Global is to facilitate the sharing of germplasm information while promoting collaboration and efficiency. Genesys PGR facilitates institutes with collections of crop diversity to make data publicly available.

Genesys PGR aggregates data from two main categories of institutions: (1) Research and Development Institutions and (2) Public and Private Collections. The system publicly records about 530 public and private institutions with accessions registered in the system. Many accessions with data recorded in Genesys PGR are available upon request from their holding genebank. Breeders and other professionals can use the system to relay requests for plant materials to the holding genebank. Most institutions distribute a small sample of the plant material for research and breeding purposes, and access is subject to legal and phytosanitary conditions.

Genesys PGR and GRIN-Global are examples of centralized genebank governance of information and resources which flows in two directions. Plant breeders can request material and information from Genesys PGR and GRIN-Global, and breeders or other data providers can contribute by uploading characterization and evaluation data to the portal and make it publicly available after a review process. In Genesys PGR, for example, this can be done using an independent application designed for handling accession-level data. It enables users to map their Excel files, CSVs, or database SQL queries using a unified format and then upload the mapped data for publication.

The Genesys PGR database is accessible through the R package Genesysr (https://cran.r-project.org/web/packages/genesysr/index.html) and a webpage (https://www.genesys-pgr.org/a/overview). Although this system offers a convenient way to obtain data about the number of accessions and institutions that conserve germplasm, the picture is likely incomplete. Other accounts suggest that the estimated number of accessions is 7.4–7.5 million in 1750 genebanks worldwide (Fernando et al. 2023; FAO 2014), while Genesys reports 4.4 million accessions in 530 institutions. Nevertheless, the initiatives stand out for its coordinated efforts to consolidate data on millions of accessions from genebanks worldwide into a dynamic portal with unified data formats which will be valuable in the long term.

11.3.1 Types of Genebanks

There are three overall types of genebanks according to the *Genebank standards for plant genetic resources* (FAO 2014). These include: seed banks, field genebanks, and in vitro genebanks. Seed banks are designed to conserve seeds which can be dried to a low humidity content and stored at low temperatures to prolong seed longevity. Examples of species include maize, wheat, rice, chickpea, cotton, and sunflower.

Field genebanks, in contrast, serve the purpose of conserving plants that are propagated vegetatively or require a long-life cycle to generate breeding materials. Examples include species with long breeding or flowering cycles and trees. Field genebanks are usually ascribed to research institutions or botanical gardens. For example, CATIE (The Tropical Agricultural Research and Higher Education Center, in Turrialba, Costa Rica) maintains a genebank field collection of over 14 hectares of coffee (*Coffea* spp) with more than 1800 accessions in their tropical climate campus.

Last, in vitro genebanks are useful for the conservation of plant genetic resources through slow growth (in vitro) for medium-term storage or cryopreservation for long-term conservation. In vitro culture involves maintaining cultures under growth-limiting conditions on artificial media, while cryopreservation involves storing biological materials at ultra-low temperatures, usually in liquid nitrogen at −196 °C, to halt biochemical and physiological processes.

11.3.2 Institutional Nature of Genebanks

Most genebanks are owned by world governments; however, some have non-governmental, parastatal, private or regional denominations (Fig. 11.3).

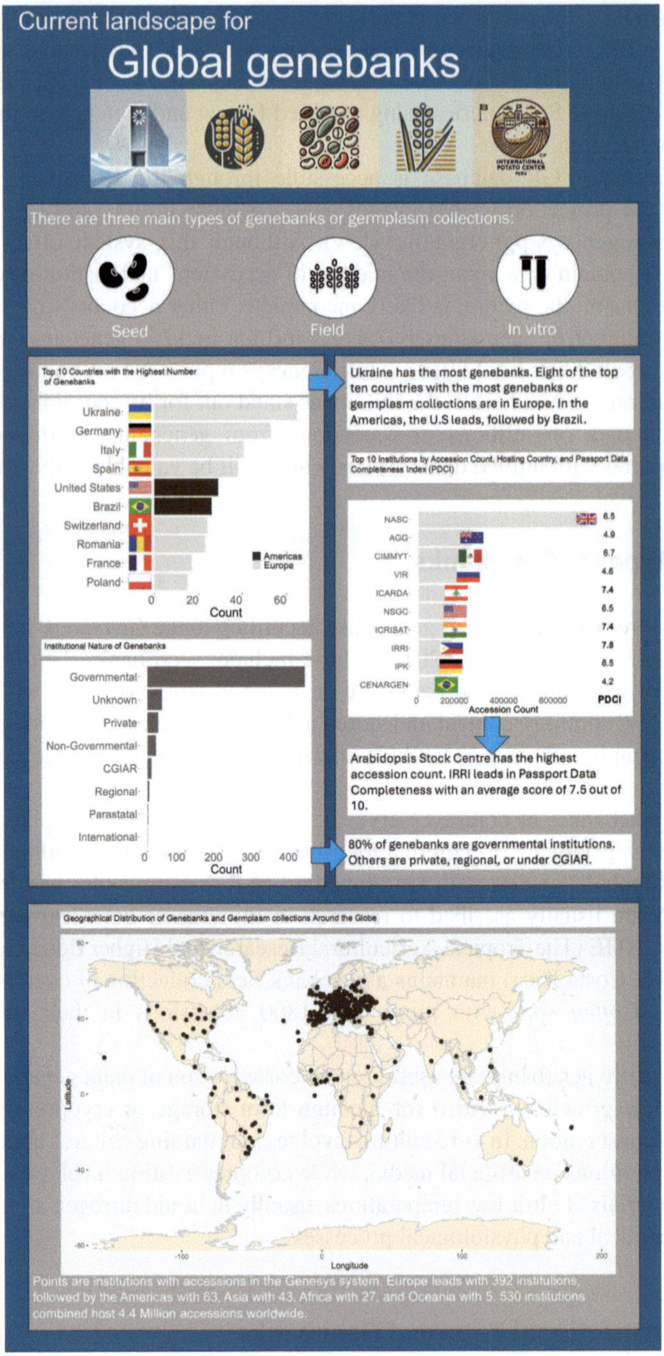

Fig. 11.3 Infographic of the current global landscape of global genebanks. Icons in the first box of infographic are by The Noun Project. Test tubes by Alone forever, Agriculture by Justin Blake. Seeds by Rafi Al Hakim. All icons (CC BY 3.0). Data for infographic was obtained through the R package Genesys PGR Client (*genesysr*) following the tutorial by Obreza and Castañeda (https://cran.r-project.org/web/packages/genesysr/vignettes/tutorial.html)

Non-governmental institutions include non-profit organizations like Seed Savers Exchange in the USA (https://seedsavers.org/) and the World Vegetable Center in Taiwan (https://avrdc.org/) which are founded by private individuals or public institutions that reside apart from the government. Other genebanks are categorized as parastatal which work with governments under some capacity, for example, The Plant Genetic Resources Centre (PGRC) in Uganda (https://www.pgrc.go.ug/) but are part of larger initiatives of national or regional resource conservation. The last two categories are private and regional institutions. Private genebanks are usually breeding companies that host plant genetic resources. Last, regional genebanks have an impact on the host genetic diversity of a region, for example, the Centre for Pacific Crops and Trees in Fiji (https://www.spc.int/resource-centre/centre-for-pacific-crops-and-trees-cepact) or CATIE in Central America (https://www.catie.ac.cr/en/). The infographic of Fig. 11.3 also provides information on the countries with the highest number of genebanks, top institutions with the highest number of stored accessions, institutions with the most passport data completeness, and the institutional nature of genebanks.

11.3.3 Species Diversity

To explore the diversity of species, we sampled more than 200,000 accessions from three genebanks (Table 11.1) using the Genesys PGR web page and R package Genesys PGR Client (*genesysr*; https://github.com/cran/genesysr). The three genebanks sampled were Centro Internacional de la Papa (CIP, Lima, Peru), Centro Internacional de Agricultura Tropical (CIAT, Cali, Colombia), and National Small Grains Collection (NSGC, Aberdeen, Idaho, USA).

The three genebanks—CIP, CIAT, and NSGC—collectively hold 232,738 accessions, each specializing in different crops. CIP focuses on tuber crops such as potatoes and sweet potatoes. CIAT emphasizes legumes, forages, and tropical crops, with a significant focus on beans and cassava. NSGC, the largest of the three, primarily specializes in small grains, with extensive collections of common wheat, barley, and rice. The data in Table 11.1 highlight a strong focus on diversity within major crops of interest, rather than diversity across multiple crops. By concentrating on a limited number of crop species, genebanks are able to use resources efficiently.

It is possible to imagine that the specialization already in place by genebanks enables the systematic expansion of phenotypic data collection during the regeneration phase. Since regeneration is an opportunity to collect comprehensive phenotypic data, focusing on a smaller subset of species may allow for deeper and more consistent trait characterization, thereby improving the quantity and quality of information available to breeders and researchers. This approach enhances genebanks' role in improving accession characterization, but additional financial resources and talent may be needed to support increased responsibilities like phenotyping.

Table 11.1 Diversity of species in three genebanks: Centro Internacional de la Papa (CIP, Lima, Peru), Centro Internacional de Agricultura Tropical (CIAT, Cali, Colombia), and National Small Grains Collection (NSGC, Aberdeen, Idaho, USA)

Genebanks	Total accessions	Genus number	Species number	Top 3 Crop	Species	Accession count
CIP	16,354	11	249	Sweet potato	*Ipomoea batatas*	5311
				Potatoes	*Solanum tuberosum*	3546
				Oca	*Oxalis tuberosa*	881
CIAT	66,507	122	679	Beans	*Phaseolus vulgaris*	32,330
				Cassava	*Manihot esculenta*	5746
				Lima bean	*Phaseolus lunatus*	3301
NSGC	149,880	87	14,823	Wheat	*Triticum aestivum*	53,505
				Barley	*Hordeum vulgare*	36,249
				Rice	*Oryza sativa*	18,703

Data retrieved on September 25, 2024, from: https://www.genesys-pgr.org/a/overview

11.4 Phenomic Characterization

Characterizing the plant phenome presents a number of technical challenges, chiefly the way the plant genome interacts with the growing environment to produce divergent phenotypes due to photoperiod, temperature, fertility, irrigation, and many other variables. Tardieu et al. (2017) described three platforms for phenomic characterization: high precision, whole-plant multi-environment platforms, and field multi-environment networks. High precision platforms are intended to assess variation at the level of the plant organ and might involve 4D imaging to better understand metabolism or signaling. Whole-plant multi-environment platforms focus on plants grown in either field or controlled environments, where variables such as light, water, or nutrient transport, and whole-plant signaling might be assessed. Field multi-environment platforms encompass field-based measurements of plant canopies across a range of environments and may focus on traits like yield and biomass. This latter category is likely to include the use of drone and robotic technologies for large-scale assessments of field-grown crop germplasm. Each of these platforms may generate large amounts of data that must be processed prior to utilization. Convolutional Neural Networks, which take advantage of machine learning capabilities, have been harnessed to establish phenomic databases in crop species (Wang and Su 2022).

Many view a primary opportunity for the success of phenomics to be based on HTP, or high-throughput phenotyping, approaches (Nguyen and Norton 2020). The use of light spectra between 400 and 2500 nm directed at plant material has the potential to provide information about nutrient status, stress levels, yield potential, and various growth parameters. Three-dimensional imaging platforms have long been available that make use of this technology for controlled-environment situations. Both ground and aerially based sensor technologies are in widespread use in both research and farming to monitor crop phenotypes. These approaches make use of both passive and active sensors, each of which yields different types of data. Passive sensors, such as those that make use of RGB and hyperspectral imaging technologies, reflect natural light and are therefore influenced by environmental conditions. Active sensors, such as LiDAR, provide their own light source and are more stable with respect to environmental conditions. The NDVI, or normalized difference vegetation index, is an example of the way sensors can be harnessed to provide useful phenomic data. This index can be developed from RGB or hyperspectral imaging inputs that are then translated into traits of interest, hundreds of which have been explored in a variety of crops (Nguyen and Norton 2020). Presumably, these strategies would be deployed on crop genebank accessions during the regeneration phase; that is, when seed or propagules are being produced for future storage. It is at this stage when the accession is being grown in an environment most suitable for exposure to HTP approaches, which might include replicated plots.

For many years, sensors have been employed to collect near-infrared and red spectra to develop indices that estimate biomass, nitrogen status, yield, and other parameters. One of the most common is the normalized difference vegetation index (NDVI). Spectral data can also be used to make structural models that can predict the values for various agronomic traits. Rutkowski et al. (2016) were able to show that NDVI measurements were helpful in formulating better yield predictions of wheat accessions in a breeding program. Adak et al. (2024) performed field-based HTP on vegetation indices in maize collected from drone images in an effort to predict agronomic traits. Various models that incorporated phenomic-only, or genomic-only information, were compared to examine both tested and untested genotypes in both observed and unobserved environments. The authors found that phenomic data were comparable to, and in some cases exceeded, genomic data for prediction accuracy.

Artificial intelligence (AI) and machine learning (ML) approaches are now being used for the characterization of crop germplasm. Recently, Xue et al. (2024) have developed an AI program known as Cucumber AI, which begins with a digital image of the fruit. Various models are employed to identify 11 primary cucumber fruit shapes and 27 additional characteristics to deliver a thorough description of the fruits. Manual measurements and photographs of cucumber fruits were used to train this model, which can more fully characterize cucumber germplasm. Evaluation of crop root systems is extremely difficult and labor-intensive, and thus, the characterization of root system traits in crop germplasm has been limited. Methods such as shovelomics (Trachsel et al. 2011), which are based on excavating and imaging root crowns within a 25 cm cylinder using a combination of digging and washing, show promise in advancing our ability to characterize crop root systems. Others have deployed PET, MRI, and CT sensors, in some cases with the use of AI and ML, to assess root systems nondestructively. Weihs et al. (2024) reviewed the use of AI in studying and characterizing crop root systems. More recently, approaches that feature both root excavation and both AI and ML techniques have been employed to characterize root systems in cotton, maize, soybean, and alfalfa (Xu et al. 2022; Falk et al. 2020).

11.5 Examples of Phenomics-Assisted Characterization

Menard et al. (2021) described how direct-seeded rice (DSR) is considered more sustainable than puddled transplanted rice (PTR), but suffers from challenges with emergence, mesocotyl length, and stand establishment—particularly when seeds are sown deeply. The length of the mesocotyl is therefore a valuable trait for plant breeding programs. Numerous GWAS projects have identified QTL for mesocotyl length, but because these studies were conducted in different environments, including field and controlled environments with various types of soil and media, there has been little overlap of the identified QTL. Menard et al. (2021) reported the development of a laboratory-based phenomics approach that allows for analysis of multiple seedling traits in rice. Their phenotyping platform involves a customized device that allows for seed sterilization and imbibition, a sowing tray with barcoding capability, plates that are arranged at 80 degrees upright in a growth chamber, and a customized digital imaging station. They screened seedling traits in 684 accessions from the 3000 Rice Genomes population in their laboratory system and in the field at three different planting depths. Laboratory-based measurements of mesocotyl length were good predictors of field performance. Furthermore, QTL for three of the key desired traits: mesocotyl length, seedling emergence, and shoot biomass, co-locate on the same arm of chromosome seven. Using this information, haplotypes from the indica group can be used to predict seedling vigor of accessions in the collection. This example brings together genomic and phenomic information to improve our ability to preserve and utilize genetic diversity. Thus, a laboratory-based phenomics-assisted screening method might be a very valuable tool for rice breeders to identify useful germplasm in global seed banks.

High-throughput phenotyping has been used to develop deep learning platforms for the assessment of crop seed bank accessions. Kienbaum et al. (2021) trained a model on nearly 20,000 maize cobs obtained from thousands of images of a large number of genebank accessions at the Universidad Nacional Agraria in La Molina, Peru. Such machine learning approaches are facilitated by tools like the Videometer (https://videometer.com). This tool, in particular, has also been useful for conducting non-destructive seed phenotype measurements, including seed size, shape, and color (Lee et al. 2020). De Carvalho et al. (2022) used imaging systems to assess carotenoid levels in cassava roots. Carotenoids are notoriously laborious to measure using spectrophotometric techniques, and imaging systems offer a more streamlined approach. In this study, a neural network prediction model exhibited strong predictive ability of total carotenoids in cassava roots ($R^2 = 0.94$) and is therefore potentially useful in concert with a breeding program.

Koc et al. (2022) sought to use spectral vegetation indices and machine learning methods to assess rust sensitivity in wheat genotypes. In particular, detection of rust was made possible through vegetation indices that were sensitive to changes in leaf pigment concentration. Detection can be accomplished with a hyperspectral imaging sensor on an unmanned aerial vehicle. They used a random forest approach to connect phenomic data with disease scores recorded from visual observations.

Schulthess et al. (2022) performed genotyping by sequencing for more than 7700 wheat accessions from a European genebank as well as more than 300 modern cultivars. Most of these accessions and cultivars were also phenotypically characterized in a series of field trials for resistance to yellow rust. To mitigate the challenge of characterizing unadapted germplasm resources, breeding values for yield for a subset of these accessions were calculated from crosses made between the accessions and several dozen cultivars. The crossing step allowed for the identification of accessions that had agronomic potential in addition to their disease reaction. This study demonstrates how both genotypic and phenotypic data on genebank accessions can be leveraged to identify the most useful accessions, even in very large collections.

The USDA's National Plant Germplasm System Germplasm Resources Information Network (GRIN) allows public access to hundreds of thousands of crop germplasm accessions from hundreds of species. The curatorial efforts behind this collection represent one of the most substantial programs of germplasm collection and distribution of any world government. However, information about individual accessions in the collection may be limited, requiring the user community to perform its own evaluations.

To illustrate the potential for a phenomics-assisted genebank strategy, we display information on two table beet (*Beta vulgaris*) accessions, Long Smoothe Dark Red (Ames 22164), and Winter Keeper (NSL 6320) (Table 11.2). Both are open-pollinated cultivars that have been in the public domain for many decades, qualifying them as heirlooms. A small amount of information about these accessions is present in the GRIN database, including hypocotyl color and donor.

Since phenomics encompasses the hypervolume of genotype × time × environment interactions for particular traits, the greatest leveraging of the genebank might be accomplished through the inclusion of data summarizing these complex interactions. Over a 5-year period, our laboratory has evaluated these two accessions for a multitude of traits, including three important disease resistances (Rhizoctonia root and crown rot) (Wigg and Goldman 2020), Cercospora leaf spot (Dixon and Goldman 2024), and Bacterial leaf spot (Morrison and Goldman 2024), soluble solids (Pelikan and Goldman 2024), and dozens of shape parameters (Vega et al. 2024). In addition, one of these accessions has been evaluated for fungal storage rots by another laboratory (Dorn 2023). Each of these studies represents a separate set of environmental conditions and years, and in each case, these multi-environment trials included replicated plots and detailed phenotypic characterizations. Disease reactions were assessed following inoculation with causal pathogens under both

Table 11.2 Phenomic assessment of USDA-NPGS GRIN table beet (Beta vulgaris) accessions Ames 22164 and Winter Keeper evaluated in six separate studies, five of which utilized multiple years and environments

	Study								
GRIN Accession	Rhizoctonia root rot reaction (% disease), averaged over environments	BLUE* for bacterial leaf spot (based on a 1-5 visual rating), averaged over environments, based on visual rating	BLUE for *Cercospora* leaf spot (% diseased leaf area), averaged over years, based on visual rating	BLUE for total dissolved solids (Degrees Brix), averaged over years and environments	Root rot in storage (% root coverage by fungal pathogens**)	Root Length (mm), assessed by digital imaging, averaged over years	Root Width (mm), assessed by digital imaging, averaged over years	Root Length-width-ratio, assessed by digital imaging, averaged over years	Root Digital Biomass (mm2), assessed by digital imaging, averaged over years
Reference	Wigg and Goldman, 2020	Morrison and Goldman, 2024	Dixon and Goldman, 2024	Pelikan and Goldman, 2024	Dorn, 2023	Vega, Oravec, and Goldman, 2024	Vega, Oravec, and Goldman, 2024	Vega, Oravec, and Goldman, 2024	Vega, Oravec, and Goldman, 2024
Ames 22164, Long Smoothe Dark Red	1.9	1.9	134	12.02	29	95.35	58.09	1.5	4622.03
Winter Keeper	20.5	1.93	95	13.02	-	105.25	44.2	1.83	4196.91
Experiment Wise Mean	21.4	2.25	571	11.5	58.4	78.24	65.85	1.21	4182.69

Gray boxes indicate statistically significant differences from the experiment-wise mean

[a]Best linear unbiased estimate

[b]Root storage pathogens included Athelia-like basidiomycetes, Penicillium *expansum*, and *Penicillium cellarum*

controlled environments and field conditions. Imaging studies were performed using custom-built algorithms that digitally quantify root contours (Brainard et al. 2021).

Taken together, the performance data accumulated for these accessions provides a fuller phenomic picture that should be useful to the user community in assessing these accessions. Without these data, a plant breeder perusing the GRIN database might identify these accessions but have little ability to make decisions about their inclusion in a breeding program. With these data, a plant breeder would be able to identify that these accessions offer promise for increasing resistance to all three diseases, increasing soluble solids, and reducing storage fungal rots. In addition, shape characteristics, which define market class and, to some degree consumer acceptance, could be easily assessed.

Here, we present an additional example from our laboratory showing how phenomics data can enhance our understanding of complex traits and add value to genebank collections. Once a phenotyping platform is established, digital imaging captures a range of carrot (*Daucus carota* var *sativus*) root traits across different growth stages and environments. While the phenome captured is incomplete, the datasets generated can aid breeders' decision-making. Phenomics, seen as the interaction of genotype, time, and environments, can highlight the importance of thoroughly evaluating accessions to fully understand their value (Fig. 11.4).

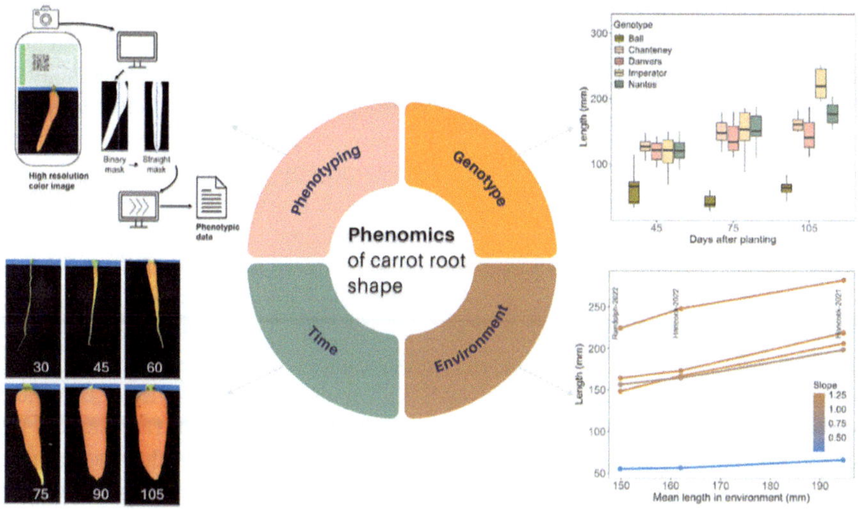

Fig. 11.4 Example of carrot root phenomics. Integration of Genotype, Environment, and Time in the phenomics of carrot (*Daucus carota* var *sativus*) root shape. Clockwise, a phenotyping platform that aids the automation of phenotypic data collection. Boxplots show the root length of distinct carrot genotypes including genetic characterization of each genotype. The performance of genotypes in three Wisconsin environments (Randolph 2022; Hancock 2022) is depicted, with slopes representing each genotype's performance relative to the mean trait values in each environment (GxE). A sample of carrot roots was evaluated at 6 different physiological stages or times from 30 to 105 days after planting

11.6 Utilization of Both Phenomic and Genomic Tools in Germplasm Characterization

While phenomic approaches to characterizing germplasm accessions in genebanks represent a powerful strategy for the future of plant breeding, other approaches to predicting the value of germplasm accessions have been developed. Hanafi et al. (2023) sought to evaluate genome-wide prediction approaches to assessing the value of seed bank accessions across the world, which is an alternative to a strict phenotyping approach. They chose more than 10,000 accessions from two seed banks, the German Federal Ex Situ Genebank for Agricultural and Horticultural Crops (IPK) and the International Center for Agriculture Research in the Dry Areas (ICARDA) genebank, along with historical data on flowering/heading date, plant height, and kernel weight. They considered three different scenarios for each of these traits: the first of which was where the test and training sets for genomic prediction were composed entirely of ICARDA accessions; the second of which was to use the IPK accessions as the training set and the ICARDA accessions as the test set; and the third of which was to use the IPK accessions along with 30% of the ICARDA accessions to predict the rest of the ICARDA accessions. Their results indicate that for certain traits, genome-wide prediction might be useful as an alternative to launching large phenomics-based projects to characterize seed bank accessions. However, genotype × environment interactions may limit the extent to which the use of test and training sets from different seed bank collections might be employed. The authors further suggest that the grouping of accessions into mega-environments with respect to adaptation may aid in increasing the predictive ability of these approaches. A similar scheme was attempted by Yu et al. (2016), who chose a set of 962 sorghum accessions for genotyping. A training set of 299 accessions was chosen to represent the overall diversity and characterized for performance traits. High prediction accuracy was measured for biomass yield, and a validation set chosen from the reference accessions revealed high prediction accuracy in experiments. To identify useful variation in mungbean, Fumia et al. (2023) collected genotypic data of a mungbean collection and compared single-environment trials, where yield components were measured, with multi-environment trials that measured overall yield. They also used a core subset of 292 accessions representing the global diversity of mungbean. Notably, they found that the same three accessions from the core collection were consistently identified using either phenotypic data alone or through a combined genomic-phenotypic approach.

Despite the availability of high-quality phenotypic data for many accessions as the examples above demonstrate, it is not stored in a centralized, easily accessible database that breeders can use for decision-making.

A challenging aspect of genomics-based approaches for characterizing genebank accessions is the inherent heterogeneity of certain accessions. Heterogeneous accessions may be represented by multiple alleles at certain loci, and any single individual sampled from these accessions may not fully represent true allele frequencies or capture the presence of rare alleles. In such cases, pooled genotyping strategies may be deployed (Volk et al. 2021). Such strategies offer the potential of capturing a pan-genome-like signature of accessions, but may make identification of specific alleles in the collection difficult. In contrast, phenomics-based characterization of heterogeneous accessions provides an opportunity for a global assessment of the inherent variability of the accession, especially if the accession is measured in multiple environments and measures of variance for key traits are provided in the assessment.

An important development assisting phenomics has been the adoption of machine-readable ontologies for crop characterization (Volk et al. 2021; Arnaud et al. 2020). Vocabularies have been developed that include trait information and relationship terms, allowing complex phenotypes to be recorded through machine processes. The Crop Ontology Curation Tool (Shrestha et al. 2012) and other tools for collecting this information are assisting curators and breeders with meeting FAIR (findability, accessibility, interoperability, and reusability) data standards (Volk et al. 2021).

11.7 Future Perspectives

Genetic diversity within a species is *heritable* variation and provides plant breeders with the ability to react, adapt, and be resilient against challenges in their breeding programs (O'Brien et al. 2022). With the continuous advancement in long-read sequencing, telomere-to-telomere genome assemblies (Garg et al. 2024), pangenomes (Schraiber et al. 2024), and super-pangenomes (Li et al. 2023), genebanks will likely include even more comprehensive genomic databases alongside plant material collections, effectively moving genomic advances into conservation (Hogg 2024).

How will genebanks transform in the next 25 years? We believe that with the push toward digitization and the use of comprehensive phenomics data in genebanks, rather than large, centralized institutions, genebanks might adopt a decentralized model, with regionally specialized repositories connected via a global network (Aubry 2023). This would enhance access, minimize risks by focusing on crops and traits relevant to specific environments.

By 2050, there may be a greater emphasis on supplementary ex-situ conservation efforts involving farmers, home gardeners, and citizen scientists. These actors will be contributing more directly to (and benefiting from) genebanks (Engels and Ebert 2024). The participatory approach could be enhanced by mobile apps and open-access platforms that facilitate the sharing and tracking of plant genetics across regions and institutions such as the previously mentioned SeedLinked App. In addition, with the growing dimensionality of phenomics data generated, genebanks' policy will likely be shaped anew by more strict global agreements around intellectual property and access to genetic resources. In addition, ownership and fair use of plant materials and phenomics data may also be put to the test (Thanopoulos et al. 2024) as genebanks strike a balance between commercial interests and global food security.

As genebanks continue to collect, organize, and process phenomics data on their accessions, genebanks will gain a profound understanding of genetic diversity within their collections. Phenomics will help identify unique traits and avoid redundancy across different genebanks. We believe this will lead to a specialization of genebanks in one or a few related crops, focusing their resources and expertise on maintaining and improving collections. In part, specialization can boost collaboration among genebanks. Each institution can develop into a center of excellence for a crop or groups of related crops (e.g., vegetables, tubers, or grains) leading to effective digitization and sharing of knowledge, ultimately benefiting global plant breeding efforts.

The phenomics definition offered here broadens the traditional view of plant phenomics, shifting focus from the development of high-throughput tools to a more conceptual framework. We predict that integrating phenomics as a discipline of plant science may help push toward the digitization of genebanks. Nevertheless, the trend toward the digitization of genebanks may have long-term potential risks. A transformation toward digital biological centers may offer a more efficient alternative than the current model but a new concern is that digitalization could undermine the importance of maintaining physical genetic resources (Aubry 2023).

In addition, the approach of digitization does not align easily with the historical understanding of genetic resources conservation, which is characterized by years of preservation through scientific practices. This paradigm shift will demand organizational and governance changes to accommodate the evolving role of digital phenomics information and its management, not forgetting that the main goals are preserving biodiversity and promoting equitable access to genetic resources. For the evolution from genebanks to bio-digital centers (*i-genebanks*) (Aubry 2023) to occur, a strong focus on capacity building and standardization will be needed.

Currently, genebank curators serve as providers, offering accession diversity (physical specimens) to researchers and breeders for evaluation, gene discovery, and plant improvement. However, the flow of information is mostly one-way, as phenomics data generated by researchers and the plant science community is rarely linked back to the original genebank accessions that were trialed (Richards 2021). Unfortunately, this represents a missed opportunity, as genebanks could enhance curation efficiency by accessing and integrating information about their materials, while researchers and breeders would benefit from having more data on genebank collections. Unfortunately, there are few formal channels for researchers to easily share information with genebanks, though platforms which facilitate this like Genesys PGR exist. In addition, if researchers genuinely share information with genebanks, it will require integrating copious amounts of data, increasing the responsibilities of both genebank curators and researchers (Richards 2021).

We are aware of two current large-scale efforts with the mission of providing scalable versions of germplasm resources: GRIN-Global (https://www.grin-global.org/) and Genesys PGR (https://www.genesys-pgr.org/). Both initiatives are database platforms that have been implemented by several genebanks worldwide. A third effort was the DivSeek International Network (https://divseekintl.org/). DivSeek did influence the interaction among the community of genebanks around the world, but as of June 2024, DivSeek, unfortunately, has formally dissolved due to the absence of ongoing funding support.

Intuitively, large language models and AI could potentially accelerate the process of tracing and integrating accession information from the literature into genebank databases.

One good example is NARO, the National Agriculture and Food Research Organization. Japan's central coordinating institute for the conservation of genetic resources related to agriculture. NARO provides 15 or more traits of each accession evaluated in several locations. For example, see (https://www.gene.affrc.go.jp/databases-plant_search_detail_en.php?jp=32592). However, from our perspective, it is difficult to determine whether the data characterizing accessions originates from NARO as a genebank or from independent researchers who generated phenomics data later integrated into the accession description.

Although there is little incentive for researchers or plant breeders to share data with genebanks, we believe many would do so if formal, user-friendly channels were in place. Given that genebanks are a public good, a streamlined, standardized process that requires minimal time could encourage more researchers to link their data back to these repositories, ultimately enhancing the value of genebanks through phenomics.

11.8 Conclusions

The definition of phenomics presented (Fig. 11.1) emphasizes integrating genotype, environment, and time. By applying advanced phenotyping, genomic characterization, and collecting traits across environments and growth stages, genebank collections can be better characterized, potentially transforming underutilized genebanks (Diez et al. 2018). A thorough understanding of each genebank accession may reduce duplication and transform genebanks into vibrant digital centers of information and genetic variability (*i-genebanks*).

We have shown examples from both our research laboratory and other researchers producing and publishing high-quality data on accessions evaluated across multiple environments and physiological stages for various traits. However, much of this data is not centralized and often fails to return to genebanks to enhance accession descriptions.

We believe that genebanks will hold physical collections in addition to a treasure trove of phenomics information all connected through a centralized database similar to The National Center for Biotechnology Information (NCBI) which hosts databases relevant to biotechnology and biomedicine. However, the success of genebank evolution depends on a commitment from several actors including genebanks and researchers, funding availability, and supporting government policies.

References

Adak A, DeSalvio AJ, Arik MA, Murray SC (2024) Field-based high-throughput phenotyping enhances phenomic and genomic predictions for grain yield and plant height across years in maize. G3. https://doi.org/10.1093/g3journal/jkae092

Anglin NL, Amri A, Kehel Z, Ellis D (2018) A case of need: linking traits to genebank accessions. Biopreserv Biobank 16:337–349. https://doi.org/10.1089/bio.2018.0033

Arnaud E, Laporte M, Kim S, Aubert C, Leonelli S, Miro B et al (2020) The ontologies community of practice: a CGIAR initiative for big data in agrifood systems. Patterns 1(7). https://doi.org/10.1016/j.patter.2020.100105

Aubry S (2023) Genebanking plant genetic resources in the post genomic era. Agric Human Values 40:961–971. https://doi.org/10.1007/s10460-023-10417-7

Brainard SH, Bustamante JA, Dawson JC, Spalding EP, Goldman IL (2021) A digital image-based phenotyping platform for analyzing root shape attributes in carrot. Front Plant Sci 12:690031. https://doi.org/10.3389/fpls.2021.690031

de Carvalho RRB, Marmolejo Cortes DF, Bandeira e Sousa M, de Oliveira LA, de Oliveira EJ (2022) Image-based phenotyping of cassava roots for diversity studies and carotenoids prediction. PLoS One 17:e0263326. https://doi.org/10.1371/journal.pone.0263326

Díez MJ, De la Rosa L, Martín I, Guasch L, Cartea ME, Mallor C et al (2018) Plant genebanks: present situation and proposals for their improvement. The case of the spanish network. Front Plant Sci 9:1794. https://doi.org/10.3389/fpls.2018.01794

Dixon LS, Goldman IL (2024) Genetic architecture of Cercospora leaf spot response in table beet with implications for other Beta vulgaris crop types. Crop Sci. https://doi.org/10.1002/csc2.21396

Dorn K (2023) USDA-ARS Plant Introduction lines evaluated for rhizomania and storage rot resistance in Idaho, 2022. Plant Disease Management Reports 17:V114:1-2.

Engels JM, Ebert AW (2024) How can we strengthen the global genetic resources' conservation and use system? Plants 13(5):702. https://doi.org/10.3390/plants13050702

Falk KG, Jubery TZ, Mirnezami SV, Parmley KA, Sarkar S, Singh A, Ganapathysubramanian B, Singh AK (2020) Computer vision and machine learning enabled soybean root phenotyping pipeline. Plant Methods 16:5

FAO (2014) Genebank standards for plant genetic resources for food and agriculture. https://www.fao.org/agriculture/crops/thematic-sitemap/theme/seeds-pgr/gbs/en/

Fernando EAJ, Selvaraj M, Ghamkhar K (2023) The power of phenomics: improving genebank value and utility. Mol Plant 16(7):1099–1101. https://doi.org/10.1016/j.molp.2023.05.011

Food and Agriculture Organization (2009) International treaty on plant genetic resources for food and agriculture. Available at http://go.nature.com/ikflqh

Fumia N, Nair R, Lin Y-P, Lee C-R, Chen H-W, von Wettberg EB, Kantar M, Schafleitner R (2023) Leveraging genomics and phenomics to accelerate improvement in mungbean: a case study in how to go from GWAS to selection. Plant Phenome J 6:e20088. https://doi.org/10.1002/ppj2.20088

Garg V, Bohra A, Mascher M, Spannagl M, Xu X, Bevan MW et al (2024) Unlocking plant genetics with telomere-to-telomere genome assemblies. Nat Genet 1–12. https://doi.org/10.1038/s41588-024-01830-7

Hanafi ELS, Jiang Y, Kehel Z, Schulthess AW, Zhao Y, Mascher M et al (2023) Genomic predictions to leverage phenotypic data across genebanks. Front Plant Sci 14:1227656. https://doi.org/10.3389/fpls.2023.1227656

Hogg CJ (2024) Translating genomic advances into biodiversity conservation. Nat Rev Genet 25(5):362–373. https://doi.org/10.1038/s41576-023-00671-0

Johal GS, Balint-Kurti P, Weil CF (2008) Mining and harnessing natural variation: a little MAGIC. Crop Sci 48(6):2066–2073. https://doi.org/10.2135/cropsci2008.03.0150

Khoury CK, Carver D, Greene SL, Williams KA, Achicanoy HA, Schori M, León B, Wiersema JH, Frances A (2020) Crop wild relatives of the United States require urgent conservation action. Proc Natl Acad Sci U S A. 117:33351–33357. https://doi.org/10.1073/pnas.2007029117

Kienbaum L, Correa Abondano M, Blas R, Schmid K (2021) DeepCob: precise and high-throughput analysis of maize cob geometry using deep learning with an application in genebank phenomics. Plant Methods 17:91. https://doi.org/10.1186/s13007-021-00787-6

Koc A, Odilbekov F, Alamrani M, Henriksson T, Chawade A (2022) Predicting yellow rust in wheat breeding trials by proximal phenotyping and machine learning. Plant Methods 18:30. https://doi.org/10.1186/s13007-022-00868-0

Kurup S (2021) Laboratory phenomics predicts field performance and identifies superior indica haplotypes for early season vigour in dry direct-seeded rice. Genomics. https://doi.org/10.1016/j.ygeno.2021.11.006

Lee JS, Chebotarov D, Platten JD, McNally K, Kohli A (2020) Advanced strategic research to promote the use of rice genetic resources. Agronomy 10:1629. https://doi.org/10.3390/agronomy10111629

Li N, He Q, Wang J, Wang B, Zhao J, Huang S et al (2023) Super-pangenome analyses highlight genomic diversity and structural variation across wild and cultivated tomato species. Nat Genet 55(5):852–860. https://doi.org/10.1038/s41588-023-01340-y

Listman M, Ordonez R (2019) Ten things you should know about maize and wheat. CIMMYT. https://www.cimmyt.org/news/ten-things-you-should-know-about-maize-and-wheat/. Accessed 29 Sept 2024

Mascher M, Schreiber M, Scholz U, Graner A, Reif JC, Stein N (2019) Genebank genomics bridges the gap between the conservation of crop diversity and plant breeding. Nat Genet 51:1076–1081. https://doi.org/10.1038/s41588-019-0443-6

McCouch S, Baute GJ, Bradeen J, Bramel P, Bretting PK, Buckler E et al (2013) Feeding the future. Nature 499(7456):23–24. https://doi.org/10.1038/499023a

Menard G, Sandhu N, Anderson D, Catalos M, Hassall KI, Eastmond PJ, Kumar A, Kurup S (2021) Laboratory phenomics predicts field performance and identifies superior indica haplotypes for early seedling vigour in dry direct-seeded rice. Genomics 113(6):4227–4236. https://doi.org/10.1016/j.ygeno.2021.11.006

Morrison AK, Goldman IL (2024) Chromosome one QTLs associated with response to bacterial leaf spot in Beta vulgaris. Crop Sci 65(1):e21448

Nguyen GN, Norton SL (2020) Genebank phenomics: a strategic approach to enhance value and utilization of crop germplasm. Plants 9:817. https://doi.org/10.3390/plants9070817

O'Brien D, Laikre L, Hoban S, Bruford MW, Ekblom R, Fischer MC et al (2022) Bringing together approaches to reporting on within species genetic diversity. J Appl Ecol 59(9):2227–2233. https://doi.org/10.1111/1365-2664.14225

Pelikan A, Goldman IL (2024) Variation for QTL alleles associated with total dissolved solids among crop types in a GWAS of a Beta vulgaris diversity panel. Plant Genome 18(1):e70014

Rao NK, Hanson J, Dulloo EM, Ghosh K, Nowell D, Larinde M (2006) Manual of seed handling in genebanks. In: Handbooks for genebanks No. 8. Bioversity International, Rome. https://hdl.handle.net/10568/2945

Richards C (2021) Genomics of plant genebanks: prospects for managing and delivering diversity in the digital age. In: Rajora OP (ed) Population genomics. New York, Springer Nature, pp 1–33

Rutkoski J, Poland J, Mondal S, Autrique E, Pérez LG, Crossa J, Reynolds M, Singh R (2016) Canopy temperature and vegetation indices from high-throughput phenotyping improve accuracy of pedigree and genomic selection for grain yield in wheat. G3 (Bethesda) 6(9):2799–2808. https://doi.org/10.1534/g3.116.032888

Schreiber M, Jayakodi M, Stein N, Mascher M (2024) Plant pangenomes for crop improvement, biodiversity and evolution. Nat Rev Genet 25:1–15. https://doi.org/10.1038/s41576-024-00691-4

Schulthess AW, Kale SM, Zhao Y et al (2022) Large-scale genotyping and phenotyping of a worldwide winter wheat genebank for its use in pre-breeding. Sci Data 9:784. https://doi.org/10.1038/s41597-022-01891-5

Sharma S, Upadhyaya HD, Varshney RK, Gowda CLL (2013) Pre-breeding for diversification of primary gene pool and genetic enhancement of grain legumes. Front Plant Sci 4:309. https://doi.org/10.3389/fpls.2013.00309

Shrestha R, Matteis L, Skofic M, Portugal A, McLaren G, Hyman G et al (2012) Bridging the phenotypic and genetic data useful for integrated breeding through a data annotation using the crop ontology developed by the crop communities of practice. Front Physiol 3:326. https://doi.org/10.3389/fphys.2012.00326

Tardieu F, Cabrera-Bosquet L, Pridmore T, Bennett M (2017) Plant phenomics, from sensors to knowledge. Curr Biol 27(15):R770–R783. https://doi.org/10.1016/j.cub.2017.05.055

Thanopoulos R, Negri V, Pinheiro de Carvalho MAA, Petrova S, Chatzigeorgiou T, Terzopoulos P et al (2024) Landrace legislation in the world: status and perspectives with emphasis in EU system. Genet Resour Crop Evol 71(3):957–997. https://doi.org/10.1007/s10722-023-01824-0

Trachsel S, Kaeppler SM, Brown KM et al (2011) Shovelomics: high throughput phenotyping of maize (*Zea mays* L.) root architecture in the field. Plant Soil:75–87. https://doi.org/10.1007/s11104-010-0623-8

Vega A, Oravec M, Goldman IL (2024) Population structure and genome-wide association offer insights into the genetic architecture of root shape in *Beta vulgaris* crop complex. In preparation for Horticulture Research

Volk GM, Byrne PF, Coyne CJ, Flint-Garcia S, Reeves PA, Richards C (2021) Integrating genomic and phenomic approaches to support plant genetic resources conservation and use. Plants (Basel) 10(11):2260. https://doi.org/10.3390/plants10112260. PMID: 34834625

Wang Y-H, Su W-H (2022) Convolutional neural networks in computer vision for grain crop phenotyping: a review. Agronomy 12(11):2659. https://doi.org/10.3390/agronomy12112659

Weihs BJ, Heuschele DJ, Tang Z, York LM, Zhang Z, Xu Z (2024) The state of the art in root system architecture image analysis using artificial intelligence: a review. Plant Phenom 6:0178. https://doi.org/10.34133/plantphenomics.0178

Wigg, KS, and Goldman, IL (2020) Variability in reaction to root and crown rot caused by Rhizoctonia Solani among table beet cultivars, breeding lines, and plant introductions in controlled environment conditions. HortScience. 55:1482-94. https://doi.org/10.21273/HORTSCI15011-20

Wigg KS, Brainard SH, Metz N, Dorn KM, Goldman IL (2023) Novel QTL associated with Rhizoctonia solani Kühn resistance identified in two table beet × sugar beet F2:3 populations using a new table beet reference genome. Crop Sci 63:535–555. https://doi.org/10.1002/csc2.20865

Xu Z, York LM, Seethepalli A, Bucciarelli B, Cheng H, Samac DA (2022) Objective phenotyping of root system architecture using image augmentation and machine learning in alfalfa (*Medicago sativa* L.). Plant Phenom 2022:9879610

Xue W, Ding H, Jin T, Meng J, Wang S, Liu Z et al (2024) CucumberAI: cucumber fruit morphology identification system based on artificial intelligence. Plant Phenom 6:0193. https://doi.org/10.34133/plantphenomics.0193

Yu X, Li X, Guo T et al (2016) Genomic prediction contributing to a promising global strategy to turbocharge genebanks. Nature Plants 2:16150. https://doi.org/10.1038/nplants.2016.150

Zavafer A, Bates H, Mancilla C, Ralph PJ (2023) Phenomics: conceptualization and importance for plant physiology. Trends Plant Sci 28(9):1004–1013. https://doi.org/10.1016/j.tplants.2023.03.023

Doubled Haploids: Accelerating the Breeding of Resilient Crops for the Future

12

Riniya Najeeb, Abhinav Uniyal, Ramesh Bondada, and Ravi Maruthachalam

Abstract

Environmental challenges, including rising temperatures, extreme weather events, soil degradation, and the spread of pests and diseases, pose increasing threats to global food security. These pressures underscore the need to breed crops with enhanced resilience to such biotic and abiotic stresses. While traditional breeding methods are effective, they are often too slow to quickly develop varieties that can thrive under these stresses. Doubled haploid (DH) technology offers a valuable approach by enabling the rapid creation of fully homozygous lines, facilitating the swift incorporation of traits such as stress tolerance and disease resistance within a single generation. This chapter discusses the advantages of DH technology in boosting genetic gain, improving precision in quantitative trait locus (QTL) mapping, and advancing marker-assisted breeding. Despite its potential, DH technology remains underutilized in resilience breeding, primarily due to challenges in producing large haploid populations for comprehensive genetic screening. Advances in haploid generation techniques, alongside recent innovations in plant breeding, now allow for large-scale production of genetically uniform plants. Incorporating DH technology into breeding programs can thus expedite the development of environmentally resilient crops,

R. Najeeb · A. Uniyal · R. Maruthachalam (✉)
School of Biology, Indian Institute of Science Education and Research, Thiruvananthapuram (IISER TVM), Maruthamala, Vithura, Kerala, India
e-mail: ravi@iisertvm.ac.in

R. Bondada
School of Biology, Indian Institute of Science Education and Research, Thiruvananthapuram (IISER TVM), Kerala, India

Max-Planck-Institute for Biology, Tübingen, Germany

P. M. Priyadarshan, R. Ortiz (eds.), *Plant Breeding 2050*,
https://doi.org/10.1007/978-981-95-0583-8_12

meeting the urgent need for sustainable agriculture in the face of mounting environmental challenges.

Keywords

Haploid · DH · Homozygous · Inbred · QTL · Climate · Breeding

12.1 Authors Riniya Najeeb, Abhinav Uniyal and Ramesh Bondada have equally contributed to this chapter.Introduction

By 2050, the global population is projected to grow to nearly ten billion, creating unprecedented demands on food production systems. This growth alone requires food production to increase by approximately 60% compared to current levels to meet the nutritional needs of an expanded global population. However, producing this additional food supply is complicated by several overlapping challenges, notably climate change, environmental degradation, and limitations in arable land availability (Food and Agriculture Organization 2020). These factors make it essential for agricultural systems to become more resilient, adaptable, and efficient to secure food for future generations.

Climate change, in particular, poses serious threats to global food security. Rising temperatures, unpredictable weather patterns, and an increase in pests and diseases are already impacting crop yields and are expected to intensify. The global average temperature is projected to increase by 1.5 °C by 2050 and certain regions could experience temperature rises of up to 3–5 °C by the late twenty-first century (Intergovernmental Panel on Climate Change 2018; Fatima et al. 2020). Studies show global maize and wheat production dropped by 3.8% and 5.5%, respectively (Lobell et al. 2011), and 1 °C rise in temperature could lead to a decline in wheat yield by 6% (Asseng et al. 2015). High temperature has been shown to reduce grain yield and kernel number by 23.8% and 25.1%, respectively, due to meiotic defects, while also shortening pollen shedding duration and decreasing pollen viability (Wang et al. 2020; Draeger and Moore 2017). Plant disease outbreaks threaten global food security and environmental sustainability by reducing productivity and biodiversity, impacting the ecosystems (Singh et al. 2023). Anthropogenic climate changes have reduced global agricultural total factor productivity by approximately 21% since 1961 (Ortiz-Bobea et al. 2021).

Given the profound impacts of climate change on crop productivity and the growing challenge of increasing yields, developing crop varieties resilient to these stresses is essential (Saleem et al. 2024; Shivanna 2022; Food and Products n.d.). Rapid, smart, and advanced breeding techniques are needed to meet these demands. Biotechnological innovations, including genome editing, synthetic biology, high-throughput genomics, artificial intelligence (AI), and machine learning (ML), offer transformative potential for developing climate-resilient crop varieties (Naqvi et al. 2022; Ahmar et al. 2020).

The conventional breeding process to develop any superior cultivar typically involves crossing two distinct parent lines with desirable traits, resulting in hybrid seeds that exhibit hybrid vigor (heterosis) (Schnable and Springer 2013). This hybrid vigor often leads to higher productivity and improved traits like disease resistance, drought tolerance, or faster growth rates (Hochholdinger and Hoecker 2007). However, heterosis diminishes in subsequent generations due to genetic segregation and inbreeding depression, which causes a decline in performance. Because of this, farmers and breeders need to produce new hybrid seeds each season, a laborious and costly process. In addition, conventional breeding methods require multiple generations—at least six—to stabilize desirable traits in pure lines, where all genetic loci become homozygous (Rutkoski et al. 2022). This makes the process time-consuming, limiting the ability to quickly respond to emerging agricultural challenges, such as climate change. However, modern techniques like speed breeding, marker-assisted selection, and DH can dramatically reduce the time to develop pure lines. These advancements accelerate the creation of homozygous lines, where all desired traits are fixed in fewer generations, improving the efficiency of crop productivity enhancement by accelerated fixing of the hybrid vigor (Forster et al. 2007; Asif 2013; Yan et al. 2017).

Here, we focus on the pivotal role of DH technology in breeding climate-resilient cultivars. DHs are plants derived from haploid cells that undergo chromosome doubling to reach the diploid chromosome number ($2n$) (Kasha and Maluszynski 2003). This technique significantly accelerates the development of homozygous lines, reducing the time required compared to traditional breeding methods to achieve genetic uniformity. These DH plants allow for clearer interpretation of phenotypic effects and are invaluable in gene function studies. Their genetic uniformity minimizes variability, allowing breeders to attribute trait differences to environmental factors, improving the precision and reliability of trait selection. Despite its potential, DH technology remains underutilized in climate resilience breeding due to the lack of suitable methods for generating large populations of haploids for screening traits of interest. By integrating DH technology into breeding programs, we can significantly accelerate the development of climate-resilient crops with enhanced stress tolerance and adaptability. In this chapter, we discuss the advantages of DH in boosting genetic gain, improving the precision of QTL mapping, and supporting marker-assisted breeding. We highlight the potential of DHs to enhance stress tolerance and adaptability across various environments, emphasizing how genetic uniformity aids in selecting traits related to environmental resilience. Additionally, we review recent developments in methods for generating haploids for crop improvement.

12.2 DH in Breeding for Environmental Resilience

One of the challenges in breeding for environmental resilience is the fixation of desired phenotypes, a process complicated by various genetic and environmental factors. The phenotype is influenced by three types of variances: additive variance, dominance variance, and environmental variance. Additive variance which is also called fixable variance is the cumulative effect of inter-allelic interactions inherited from parents, playing a crucial role in trait inheritance across generations (Byers 2008). Dominance variance, non-fixable variance, arises from intra-allelic interactions at the same genetic locus (Class and Brommer 2020) and minimizing this is essential for developing pure lines with consistent traits. Environmental variance reflects changes in phenotypic expression due to external conditions (Pham and McConnaughay 2014). Together, these variances influence the phenotype and pose challenges in stabilizing desired traits, particularly hindering the efficient development of environment-resilient crops, as the resultant phenotype can be the result of any of these three factors.

In traditional breeding, selfing and backcrossing are routinely employed to develop nearly homozygous individuals over multiple generations, gradually increasing additive genetic variance (V_A) while reducing dominance genetic variance (V_D). Over time, as the inbreeding coefficient approaches one (indicating complete homozygosity), any remaining dominance variance diminishes significantly. Although selfing accelerates the fixation of additive genetic effects, it takes considerable time to achieve these genetic states. However, the generation of haploids and subsequent DH offers a solution to this problem. The DH system mitigates both the time constraints and dominance variance challenges inherent in conventional breeding methods by producing fully homozygous lines within a single generation. With DH lines, all genetic variation is additive, eliminating dominance variation and simplifying the interpretation of phenotypic traits (Fig. 12.1). This reduction in genetic noise enhances the ability to detect the influence of environmental factors on traits, leading to more accurate selection of traits that truly respond to these conditions. Furthermore, the environmental variation observed within this generation can be leveraged to evaluate and screen plants for stress tolerance and adaptability, optimizing the selection process for traits critical to crop resilience. As a result, breeders can identify lines that show superior performance and stability across various environments with greater precision. Moreover, selecting for environment-resilient crops among DH populations offers an easy and quick means to identify the genetic factors governing important traits. Utilizing DHs for quantitative trait locus (QTL) mapping accelerates the mapping of genes associated with stress tolerance and adaptability. This approach simplifies the genetic background, making it easier to detect significant QTLs, and is discussed in detail elsewhere in this chapter. Thus, the DH system offers significant advantages in the development of environment-resilient crops, allowing breeders to more efficiently select for traits that improve stress resistance, adaptability, and yield stability in response to changing environmental conditions. Genetic gain, which refers to the annual improvement in performance achieved through artificial selection (Xu et al. 2017), is closely tied to genetic variance. DH populations benefit from elevated additive genetic variance, leading to enhanced genetic gain and shorter breeding cycles compared to conventional breeding progenies. However, these populations undergo only a single recombination cycle, which can limit the development of improved germplasm and hybrid cultivars (Al-Khayri et al. 2016). Additionally, the limited recombination may preserve

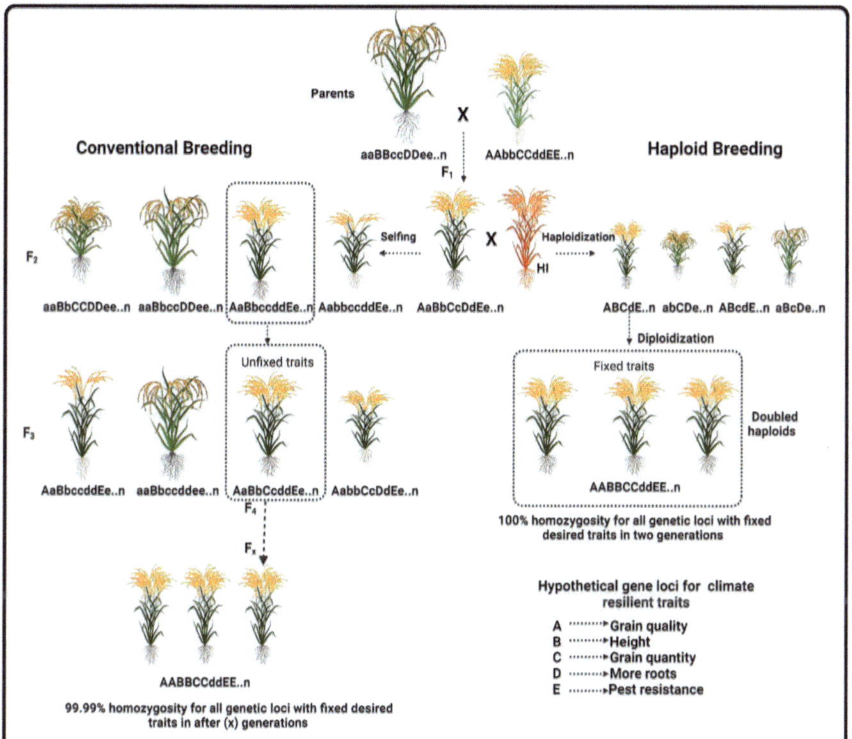

Comparison of Traditional Breeding and Doubled Haploid (DH) Systems in terms of Homozygosity and Genetic Variance

Generations	Inbreeding coefficient	Additive variance (V_A)	Dominance variance (V_D)
F_2	0	1	1
F_3	0.5	1.5	0.75
F_4	0.75	1.75	0.44
F_5	0.875	1.875	0.23
F_6	0.9375	1.9375	0.121
Fx	1	2	0

Generations	Inbreeding coefficient	Additive variance (V_A)	Dominance variance (V_D)
F_2	0	2	0

Fig. 12.1 Comparison of the traditional breeding approach with the DH system showing hypothetical gene loci for environment-resilient traits in terms of time to homozygosity, genetic variance, environmental influence, and selection efficiency. In traditional breeding, multiple generations of selfing are needed to increase homozygosity for the genetic loci and reduce dominance variance. The DH system achieves complete homozygosity in a single generation, eliminating dominance variance. The DH system allows breeders to more effectively assess and select for traits influenced by environmental factors, improving the efficiency of developing high-performing, environment-resilient cultivars. (Created using BioRender)

undesirable genetic linkages. To address these limitations, genomic selection—which integrates both phenotypic and genotypic data—enhances genetic gain by predicting the performance of new germplasm without the need for detailed phenotypic characterization. This method has become a valuable tool in modern breeding programs, enabling more accurate and efficient selection.

Furthermore, Daetwyler et al. 2015, proposed an extension of genomic selection called optimal haploid value (OHV) selection, which predicts the most advantageous DH that can be derived from the haploid gametes produced by a heterozygous parent. By optimizing haplotypes, OHV selection aligns the breeding process with the goal of developing elite fixed lines (Daetwyler et al. 2015). Importantly, this method preserves a greater degree of genetic diversity within the breeding population, ensuring long-term genetic progress and mitigating the risks associated with reduced genetic variability. In contrast, conventional breeding introduces and compounds environmental variances across generations, often reducing selection efficiency due to high dominance and low additive variance. This causes the parent–offspring correlation value to be less than one, as differences between parents and offspring become more pronounced (Snape 1989). DH lines, by comparison, reduce this genetic variation, making it easier to detect significant environmental influences on traits, thereby increasing the statistical power of analyses. The clear data derived from DH lines' responses to various environments not only allows breeders to make informed decisions on which lines to advance but also lays the groundwork for applying advanced breeding strategies such as QTL mapping, marker-assisted selection, and addressing biotic and abiotic stressors, ultimately driving the development of robust and high-performing cultivars.

12.2.1 DH and QTL Mapping

A QTL is a genomic region that governs specific traits, controlled by multiple genes, and mapping these loci is an effective method for elucidating the genetic basis of phenotypic variation (Lynch 1997). In plants, QTLs play a pivotal role in determining essential agronomic traits for environmental resilience breeding, necessitating their mapping to specific chromosomal locations for successful crop improvement. Among the various mapping populations employed for QTL analysis in plants, DH populations are noteworthy due to their unique attributes. These permanent mapping populations, characterized by their complete homozygosity, are optimal for associating genetic markers with traits of interest. DH lines can be used to study the changes in the genotype due to environmental interaction (Filiault et al. 2017). DH populations are among the few true-breeding lines that facilitate precise QTL mapping and can be generated in a single generation, offering a rapid alternative compared to other non-segregating true-breeding populations, such as near-isogenic lines (NILs) and recombinant inbred lines (RILs). Consequently, DH populations represent an efficient and effective resource for gene and QTL mapping (Forster and Thomas 2003; Forster et al. 2007; Ali and Wani 2021). Traditional mapping populations like RILs and backcross inbred lines (BILs) require a minimum of 6–7 generations to achieve the homozygosity necessary for effective QTL mapping. In contrast, DHs can reach full homozygosity within just one or two generations, which makes them highly advantageous for QTL and genome-wide association studies (GWAS) (Seymour et al. 2012). The complete homozygosity of DHs reduces genetic background variation, thereby minimizing confounding effects from gene–environment interactions. Nevertheless, a significant limitation of DHs is that they undergo only a single recombination cycle, resulting in fewer opportunities for novel recombinations or the disruption of undesirable linkages compared to conventional self- or sib-pollination methods, which involve multiple generations of recombination (Yan et al. 2017). RILs obtained by selfing have been shown with larger variations, increased QTL x Environment interaction, higher yield, and transgressive segregation for agronomic traits than in the DH population (Zhao et al. 2010). DH can fix rare recombinants in a single generation and reduce the occurrence of false-positive phenotypes compared to lines developed by selfing, which may be heterozygous and remain advantageous for rapid and stable genetic analysis. DH technology can be combined with mutation breeding to achieve much quicker genetic gains than conventional mutagenesis programs (Sharma et al. 2022). DH saves time and resources due to increased selection efficiency of $(1/2)^n$ compared to $(1/4)^n$ for conventional breeding (Snape 1989). It can also be used for reverse breeding techniques where inbreds are developed from successful hybrids by inhibiting meiotic crossovers in F1.

12.2.2 DH and Marker-Assisted Breeding (MAB)

MAB is a vital molecular breeding approach that uses molecular markers, like DNA sequences, to establish connections between desirable traits and their associated genetic markers. This technique enhances the selection of favorable traits in crops while effectively minimizing linkage drag (Elsharawy and Refat 2023). Thus, it is essential for environmental resilience breeding, enabling the selection of traits that enhance a crop's ability to withstand environmental stresses. MAB includes methods such as Marker-Assisted Selection (MAS) and Marker-Assisted Backcrossing (MABC) and involves key steps like selecting parent plants, creating a mapping population, and conducting phenotyping and genotyping (Jiang 2015). By focusing on markers closely linked to target genes or quantitative trait loci (QTLs), MAB allows for precise trait selection. The introgression of complex traits from crop wild relatives is a promising approach to enhance cultivated crops. In a study focused on rapeseeds, which exhibit low tolerance to Verticillium wilt, researchers phenotyped DH lines under various conditions and identified two major QTLs associated with resistance to Verticillium, using molecular markers (Rygulla et al. 2008). Another study aimed at mapping the genetic factors influencing green plant regeneration in anther culture, researchers employed bulked segregant analysis along with AFLP markers and composite interval mapping techniques to identify QTLs associated with the percentage of green plants produced (Orp et al. 2001; Torp and Andersen 2008). Marker-assisted selection (MAS) can be exploited to enhance haploid induction rate and their identification. For example, by utilizing both the *qhir1* and *qhir8* loci, MAS facilitates the breeding of haploid inducers with high HIR while accurately identifying true haploids using *R1-nj* marker (Khammona et al. 2024). The use of MAS for qhir1 has been shown in breeding high-oil inducers as well (Dong et al. 2014).

Sclerotinia sclerotiorum resistance has been improved in the elite rapeseed line by introgressing resistance alleles from a partially resistant cultivar through back-crossing, phenotypic selection, and marker-assisted background selection (Yu et al. 2010). Four resistant double haploid restorer lines were developed and are now used to produce commercial F1 hybrids, demonstrating the effectiveness of these methods. Integrating DH technology with marker-assisted breeding enhances efficiency and allows for precise customization of the DH population to target specific traits. This combination improves both qualitative and quantitative traits by stabilizing the incidence of a gene variant in a population (gene frequencies) (Humphreys and Knox 2015). Utilizing molecular markers in DH breeding further refines the genetic profile by allowing direct screening at the haploid stage. Since haploid plants have only one set of chromosomes and thus one allele per gene, this enables accurate selection for traits such as disease resistance, abiotic stress tolerance, or specific quality attributes (Liu et al. 2014; Somers and Humphreys 2009). MAB significantly enhances the selection for recessively inherited traits, particularly at the haploid stage of DH production. By working at this stage, breeders can directly express and select for recessive traits, leading to a population enriched for the desired characteristic, especially when using perfect markers linked to the trait. However, there is a risk that focusing on specific traits, like disease resistance, could reduce genetic variability for other important attributes such as yield and quality. To mitigate this, generating larger DH populations can help maintain a broad genetic base, while utilizing elite parents in crosses can ensure that desirable traits are present in the resulting lines (Humphreys and Knox 2015).

12.2.3 DH and Biotic and Abiotic Stress Tolerance/Resistance Breeding

DH has significantly advanced the screening and selection of stress-tolerant genes, offering a powerful approach to improve crop resilience under adverse conditions. The pollen grain culture offers a better system for treatment and selection for variants as it allows for rapid screening since millions of pollen grains, with diverse genetic backgrounds, can be cultured simultaneously, allowing for high-throughput screening and selection, thereby increasing the likelihood of finding rare beneficial variants. As recessive mutations are not masked by dominant alleles in haploids, expression traits can be analyzed in the double haploids, making it easier to identify and select desired traits or mutations. DH technology has been utilized to generate cultivars which are tolerant to salt, drought, toxic, and other abiotic and biotic stresses. In a study conducted on developing salt-tolerant plants of *Brassica napus cv. Topas* (Hafizur Rahman et al. 1995), researchers selected microspore-derived embryos that could survive mutagen treatment and lethal doses of NaCl. One of the lines derived had better salt tolerance, maintaining a favorable Na^{+}-K^{+} ratio by accumulating less sodium and retaining more potassium than controls. This approach was extended to investigating the high-frequency selection of salt-tolerant DHs in rice using anther culture (Lee et al. 2003). Climate change-induced submergence stress threatens rice production, but developing DH rice varieties offers a sustainable solution. A systematic screening of DH lines for submergence tolerance found lines that showed strong adaptability and stability (Anshori et al. 2023). Using anther culture, DH lines of upland rice were developed that exhibit tolerance to aluminum (Al) toxicity and shade, as well as resistance to blast fungus (Purwoko et al. 2010). These traits are crucial for expanding rice cultivation to stress-prone areas and adapting to intercropping systems.

QTL analysis of three DH canola populations identified shared *Sclerotinia* stem rot resistance loci on specific linkage groups. This study is the first to report common SR resistance QTLs in DH lines of *Brassica napus* (Behla et al. 2017). In wheat, DH lines have been successfully developed to incorporate the eyespot resistance gene *pch1* using marker-assisted selection (Wiśniewska et al. 2019). Furthermore, a mapping population of DH lines was utilized to identify resistance genes against stripe rust, providing QTLs that aid in marker-assisted selection. This facilitates the identification of the best DH lines for high yield, quality, and stripe rust resistance in future breeding efforts (Tehseen et al. 2022). DH in *Brassica napus* has assisted in identifying QTL associated with several agronomic traits such as disease resistance, seed traits, and plant architecture (Starosta et al. 2023). The evaluation of DH lines from tropical backcross populations in Kenya showed a yield increase of 0.6% to 32.7% in well-watered conditions and 11.8% to 40.9% under drought stress (Yoseph et al. 2017). A population of 1090 DH lines was developed from Polish F1 materials to enhance drought and freezing tolerance, with ten extreme lines identified for further analysis of photosynthetic function, antioxidative activity, and proteomic profiles under stress conditions (Gołębiowska-Pikania et al. 2017). Additionally, DH lines have been employed to investigate stress-specific defense reactions, focusing on the role of phytohormones and their interactions with the antioxidative system in winter barley's defense against cold and drought stress (Malaga et al. 2020).

DH technology in maize has been successfully utilized to enhance genetic homozygosity, genetic enrichment, gamete selection, gene mutation, and molecular mapping (Chang and Coe Jr. 2009). Interestingly, doubled haploids (DHs), by itself, have proven valuable in developing better maternal haploid inducers by enabling breeders to effectively select recombinant pollen parents with optimized alleles for higher haploid induction rates (Chen et al. 2023a, b). Several improved protocols have been developed to produce DH lines via in vivo maternal haploid induction in maize (Grüning et al. 2024). DH technology has been widely adopted by the International Maize and Wheat Improvement Center (CIMMYT) maize breeding programs replacing traditional breeding methods (Chaikam et al. 2019a, b). DH technology is also employed by commercial maize breeding programs in Europe, North America, and China (Molenaar et al. 2019). Several improved maize hybrids featuring DH lines have been released in Africa, demonstrating enhanced grain yield and drought tolerance, leading to hybrids recommended for commercial cultivation in East and Southern Africa (Yoseph et al. 2017; Chaikam et al. 2018). Furthermore, DH technology has been instrumental in identifying genomic regions associated with stress in maize (Jiao et al. 2020), as well as in other crops like rice (Calayugan et al. 2020; Park et al. 2014; Xu et al. 2011), wheat (Gahlaut et al. 2017; Liu et al. 2013; Asif et al. 2018), and other crops. A study by (Sserumaga et al. 2018) analyzed genotype × environment interactions for grain yield and agronomic traits in maize across East African agro-ecologies. Hybrids from DH inbred lines outperformed commercial hybrids by 23% and 43%, respectively (Sserumaga et al. 2018). DH technology offers significant advantages in crop breeding by facilitating the transfer of favorable traits to offspring without the confounding effects of dominant trait or gene epistasis (Prasanna et al. 2013). This makes it a powerful tool for developing superior maize cultivars with enhanced stress tolerance and adaptability, contributing to global food security.

12.3 Advancing Plant Breeding by Synergizing Haploid Induction and New Breeding Innovations

To expedite the development of environment-resilient DH lines, we can integrate several advanced breeding techniques with the HI system, including speed breeding, genomic selection (GS), genome editing, and machine learning. Recent research has also showcased the transformative potential of CRISPR/Cas editing systems in accelerating crop breeding. Further, minichromosome (mChr) technology, cytoplasmic male sterility, and exploiting unreduced gametes to produce clonal seeds can be used in synergy with DH technology to fast-track breeding. Therefore, in the following sections, we will discuss the applications of CRISPR-mediated HI editing, mChr manipulation, cytoplasmic male sterility (CMS), synthetic apomixis, and other modern innovations in comparison with traditional breeding (Fig. 12.2a), highlighting their role in enhancing DH technology and its impact on crop improvement (Fig. 12.2).

12.3.1 Haploid-Inducer Mediated Genome Editing (HI-Edit): Overcoming CRISPR Limitations in Crop Breeding

CRISPR (clustered regularly interspaced short palindromic repeats)–Cas (CRISPR-associated protein) utilizes DNA–RNA recognition system for sequence-specific nucleic acid cleavage to introduce double-strand breaks at the target site (Zhu et al. 2020). Recent advancements in CRISPR-Cas9 genome editing have enabled precise targeting of desired loci within crop genomes, holding immense potential to revolutionize agriculture (Shi et al. 2017; Nekrasov et al. 2017; Pramanik et al. 2021). However, significant limitations persist, such as off-target mutations and the necessity to segregate out the CRISPR transgene in the F_2 generation, adding complexity

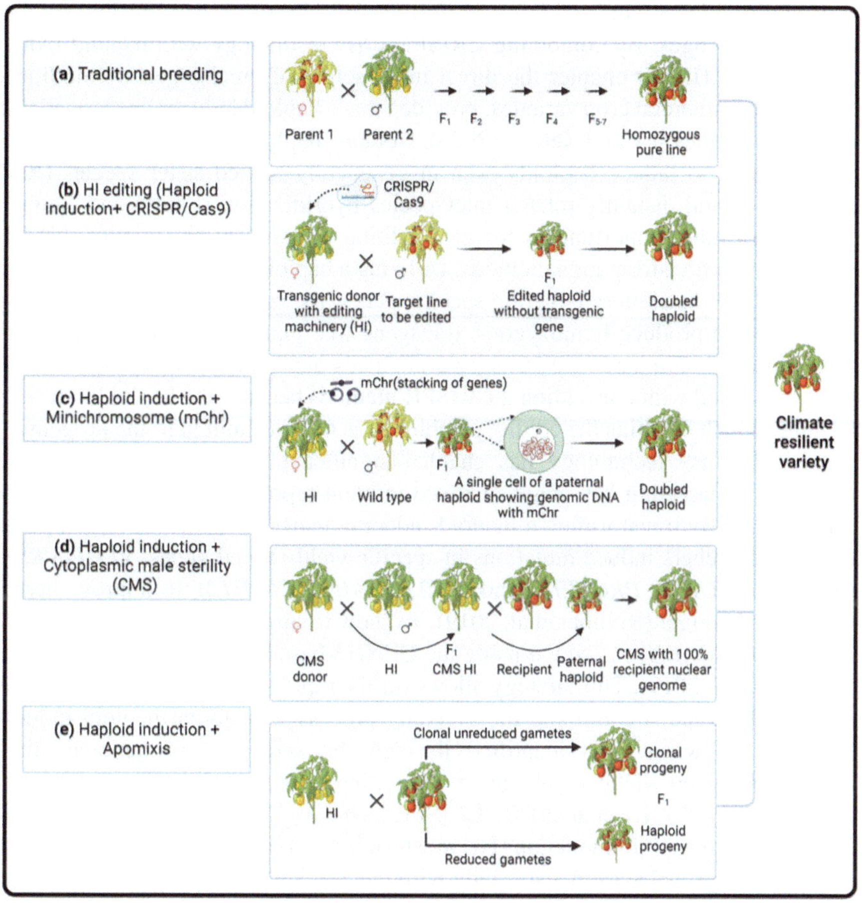

Fig. 12.2 Synergistic approaches in plant breeding: This figure illustrates the (**a**) traditional breeding; and integration of haploid induction with: (**b**) CRISPR-mediated genome editing: HI-edit, (**c**) mChr manipulation, (**d**) CMS, and (**e**) Synthetic clonal reproduction. These innovations collectively enhance DH technology, leading to significant improvements in crop genetic gain and resilience. (Created using BioRender)

to the process of generating mutants (Li et al. 2023). Genome editing is limited by the poor transformation efficiency for many commercial crop varieties. Key traits like yield and drought tolerance require testing in breeding lines, but most crops are recalcitrant to CRISPR-Cas9 delivery via agrobacterium-mediated transformation or biolistics, constraining practical applications (Son and Park 2022). This limitation restricts breeders to a few transformable varieties for initial testing, requiring laborious introgression of successful edits into elite germplasm-a process that can take several years and incur substantial costs—thereby complicating and delaying breeding advancements.

Haploid-Inducer mediated Genome Editing (HI-edit) offers a promising solution to these challenges. By combining CRISPR-Cas9 technology with haploid induction systems, HI-edit enables the direct introduction of precise genetic modifications into commercial crop varieties, provided that a haploid inducer line is available within the species (Lv and Kelliher 2023). Notably, haploid inducer lines have been demonstrated to induce haploids even in genetically related sister species (Ravi et al. 2014) and distantly related interspecies hybridization (Laurie and Bennett 1988), presenting opportunities for gene editing in cultivars or varieties that are recalcitrant to transformation methods. Utilizing a haploid inducer line carrying the CRISPR-Cas9 machinery targeting specific genes, the edited haploid genomes can be doubled to produce homozygous, transgene-free plants in a single generation. This approach eliminates the transgenic elements and minimizes off-target mutations associated with conventional CRISPR methods because the haploid inducer's genome, which contains the editing machinery, is not transmitted to the progeny.

CRISPR/Cas9 technology has enabled significant advancements in haploid induction by targeting key genes involved in plant reproduction and development (Fig. 12.2b). By transforming a haploid-inducing maize line with Cas9 and guide RNAs, researchers induce mutations in specific yield-related genes, *VLHP(VRS1-LIKE HOMEOBOX PROTEIN)* and *GW2 (GRAIN WEIGHT2),* to improve kernel number and weight (Kelliher et al. 2019). To show its applicability in dicots, the *gl1* gene in *Arabidopsis* was edited using the CENH3-based haploid induction system (Kelliher et al. 2019). This strategy allows for the rapid selection of desirable traits in haploid plants, streamlining the breeding process. A paternal haploid inducer (HI) line in wheat was identified through the screening of genome-edited *TaCENH3α*-heteroallelic combinations, resulting in a haploid induction rate of approximately 7% (Lv et al. 2020). CRISPR technology has been shown to significantly enhance haploid induction (HI) in wheat by enabling precise gene editing. In a study, the SpCas9 system, especially with the TaU3 promoter and dual sgRNAs, achieved high editing efficiency (80.5%) and a haploid induction rate of 18.9% (Liu et al. 2020b). Further, the CRISPR/Cas9 system has been successfully employed to edit genes associated with haploid induction in several crop species, as well as in the model plant *Arabidopsis thaliana*. Notable examples include, *CENH3* in *Arabidopsis thaliana* (Kuppu et al. 2020), *Triticum aestivum* (Lv et al. 2020), and *Zea mays* (Wang et al. 2021), *DMP (DOMAIN OF UNKNOWN FUNCTION)* in *Arabidopsis thaliana* (Zhong et al. 2020), *Zea mays* (Zhong et al. 2019), *ig1* (*indeterminate gametophyte1*) in *Oryza sativa* (Yao et al. 2018), and *MTL/PLA1 (MATRILINEAL/PHOSPHOLIPASE A1)* in *Triticum aestivum* (Liu et al. 2020a). One innovative approach, Haploid-Inducer Mediated Genome Editing (IMGE), utilizes a CRISPR/Cas9 cassette in maize to swiftly generate homozygous DH lines with desired traits within just two generations (Wang et al. 2019). Using in silico analysis, putative orthologs of genes involved in haploid induction have been identified in rice and wheat as potential targets for gene editing to develop a haploid inducer stock with high haploid induction frequency (Goyal et al. 2024). Additionally, a novel strategy that combines multiplex gene editing with in vivo haploid induction in maize has resulted in genetically uniform plants with multiple homozygous edits, leading to notable improvements in traits like leaf size (Impens et al. 2023). DH technology rapidly produces genetically uniform plants, while CRISPR–Cas9 enables precise editing of genes responsible for traits like drought tolerance, disease resistance, and yield. This synergy can allow breeders to quickly stabilize and propagate desirable traits, creating crop varieties better suited to withstand environment-related stresses.

12.3.2 Integrating Haploids with Minichromosome Technology for Resilience Breeding

Minichromosomes (mChr) are small, engineered chromosomes that can carry multiple gene constructs, enabling concatenation of several gene aggregates into a single autonomous element, unlike conventional genome manipulations that integrate a few genes randomly. Minichromosomes can be generated by top-down or bottom-up approaches (Houben et al. 2008; Houben and Schubert 2007). This technology is particularly valuable for engineering entire metabolic pathways, as it allows the stacking of multiple genes (Fig. 12.2c). Engineered minichromosomes enable the precise insertion of transgenes at defined loci on independent chromosomes, ensuring stable and predictable expression while avoiding the linkage drag and mutagenic effects associated with random integration into the host genome (Birchler et al. 2010). It has been demonstrated that foreign genes integrated into both normal and truncated miniB chromosomes in maize can be transmitted faithfully and stably expressed (Yu et al. 2007).

A recent study has demonstrated a reliable method for generating in vivo mChr in haploids through CENH3-mediated genome elimination, which induces genome instability and karyotypic changes, specifically on one parental genome. This method has been shown to produce heritable, rearranged mChrs that are stably inherited across generations. For example, mChr derived from centromeric regions of specific chromosomes was consistently transmitted in haploid lines over multiple generations (Tan et al. 2023). In breeding applications, incorporation of mChr stacked with desired genes in haploid-inducing lines which are known for their high haploid induction rates, can be particularly useful. A study by Zhao et al. 2013, found that a truncated chromosome from the haploid inducer line persists during haploid formation, suggesting it may function independently as a mChr. This fragment, mapping near the centromere of chromosome 4, could remain stable if the broken DNA ends are repaired, potentially carrying centromeric repeats for maintenance. This indicates the potential for transferring engineered mChr to haploids (Zhao et al. 2013). Minichromosomes can be introduced into a haploid-inducing line, crossed with target lines, and the resulting haploids that carry the minichromosome can be doubled to create homozygous lines with the transgene, eliminating the need for repeated introgression (Birchler 2015). These DH lines are completely homozygous and can carry the engineered mChr with multiple genes conferring tolerance to abiotic stresses such as drought, heat, salinity, and cold. This enables the simultaneous introduction of complex traits essential for crop survival in variable climates circumventing the lengthy process of traditional backcrossing, enabling faster development and testing of new varieties (Birchler and Swyers 2020).

12.3.3 Integrating Haploids for Synthetic Apomixis and Easy Transfer of Male Sterility

Haploid induction plays a crucial role in synthetic clonal reproduction through seeds, or apomixis, a groundbreaking advancement in plant breeding designed to enable clonal propagation of hybrid cultivars through seeds. Apomixis combines the benefits of both vegetative and seed propagation, involving the generation of diploid asexual gametes (apomeiosis), embryo development without fertilization (parthenogenesis), and the formation of endosperm to nourish the embryo (Ronceret and Vielle-Calzada 2015). *Arabidopsis* mutants like MiMe and *dyad* demonstrate key aspects of these processes, but a fully self-reproducing clonal system remains a challenge (Ravi et al. 2008; d'Erfurth et al. 2009; Mieulet et al. 2016). Notably, when MiMe and *dyad* are crossed with haploid-inducing strains, clonal progeny are produced, showing significant potential for practical application in achieving apomixis (Marimuthu et al. 2011) (Fig. 12.2e). Recently, *BBM (BABY BOOM)* (Boutilier et al. 2002; Conner et al. 2015, 2017; Khanday et al. 2019) and *PAR* (*PARTHENOGENESIS*) (Underwood et al. 2022) have shown promising advancements in inducing parthenogenesis and apomictic reproduction in plants and are pivotal in potentially overcoming the current challenges of establishing a fully functional clonal reproduction system (Mahlandt et al. 2023).

Uniparental Genome Elimination (UGE) can advance breeding efficiency by facilitating rapid transfer of cytoplasmic male sterility (CMS) using a haploid-inducing strains, allowing for a swift and effective CMS transfer in a single generation (Fig. 12.2d). Recent studies have illustrated that paternal haploids can streamline CMS transfer in maize and broccoli, eliminating the need for lengthy backcross generations (Bortiri et al. 2024; Han et al. 2024). Utilizing CRISPR–Cas to edit *CENH3* genes, this method creates CMS–haploid inducer lines that significantly improve hybrid seed production efficiency (Maruthachalam 2024). Environmental resilience can be enhanced through these technologies by rapidly developing crops that can withstand environmental stresses.

Additionally, recessive mutations, which are often screened in the M2 generation, can be identified more quickly in the M1 generation by using haploids. By crossing mutant pollen with a haploid-inducing parent, phenotypically distinct haploid M1 progeny was obtained, attributed to an indel causing a variegated phenotype. This method, demonstrated in *Arabidopsis* (Ravi et al. 2014), accelerates the process by direct screening of mutation, but only works for viable and fertile mutants. Further, to ease genetic analysis and selection, the ploidy of polyploids can be reduced for analytical breeding using haploid genetics (Bondada et al. 2020). DH technology can enable the efficient stacking of multiple beneficial traits, including those for climate adaptability (Fulcher and Riha 2015). By accelerating the breeding cycle and incorporating traits for drought tolerance, disease resistance, and other climate-related factors, these methods contribute to the development of crops better suited to changing environmental conditions, thus supporting sustainable agriculture and enhancing food security.

12.3.4 DH and Next-Generation Breeding Approaches (Speed Breeding, Genomic Selection, and Machine Learning)

Speed breeding has been shown to accelerate the growth cycle of plants, allowing multiple generations to be produced within a single year. It is a crop improvement technique that accelerates plant development by controlling environmental conditions and extending light exposure, and allowing up to 6 generations per year for crops like wheat and barley, compared to 2–3 under standard conditions. This method supports rapid phenotyping, mutant studies, and inbreeding through single-seed descent (Watson et al. 2018; Hickey et al. 2019). This rapid generation turnover is crucial for testing and selecting DH lines that can withstand environmental stressors, thereby shortening the time required to develop resilient varieties.

As discussed earlier, MAS uses linked DNA markers to track key genes or traits in crop breeding that are influenced by major mutations. Meanwhile, Genomic selection utilizes genome-wide DNA markers to predict the genetic potential of plants for complex traits, such as yield, which are influenced by numerous genes with small individual effects (Bhat et al. 2016). This approach allows for the estimation of genomic estimated breeding values (GEBVs), which facilitate the selection of superior genotypes, and accelerates the breeding cycle, and reduces the costs associated with breeding line development (Crossa et al. 2017; Meuwissen et al. 2001). Combining genomic selection with speed breeding can significantly enhance the development of DH lines for environmental resilience as the rapid generation turnover allows for the application of GS in each cycle to identify optimal parents based on GEBVs (Watson et al. 2018). This integration allows for efficient stacking of beneficial haplotypes—gene combinations that confer desirable traits—thereby improving overall performance in environment-resilient crops (Kemper et al. 2012). Further, ML can be used as an application in genomic selection to predict important agronomic traits (Tong and Nikoloski 2021) and facilitate high-throughput stress phenotyping in plants (Singh et al. 2016). Inducing in vitro meiosis to bypass flowering offers a promising strategy for accelerating the development of improved crop varieties (Cook et al. 2023, 2024). This approach takes inspiration from successful demonstration in mammals where somatic cells are converted to germline cells. Unlike animals, where the germline is specified early, plant germline cells are determined much later in development. This difference opens unique possibilities for inducing meiosis in somatic cells, potentially enabling the rapid generation of haploids or other genetic modifications without relying on the flowering process. Altogether, by leveraging these advanced techniques, we can rapidly develop DH lines that are better adapted to environmental stresses, contributing to more resilient agricultural systems.

12.4 Advances in Methods for Generating Haploids

Haploids are typically generated using two primary approaches: *in vitro* culturing of gametes or gametophytic cells harvested from reproductive organs and tissues such as pistils, ovules, anthers, or immature pollen (Mujeeb-Kazi et al. 2006; Bernardo 2009; Germanà 2011; Tripathy et al. 2019). The *in vivo* method involves creating haploids through hybridization, where one set of chromosomes is selectively lost during early embryo development. In addition, haploids can be generated through parthenogenesis and pollen treated with radiation (Sestili and Ficcadenti 1996), chemicals (Roldán-Arjona and Ariza 2009; Jiang et al. 2022a), or high temperatures (Kalinowska et al. 2019).

The production of *in-planta* haploids has been recognized since the early twentieth century as a genetic consequence of post-fertilization genome conflict in artificial distant or wide hybridization crosses (Blakeslee et al. 1922; Clausen and Mann 1924). However, the importance of haploids as a crop improvement tool became apparent only after the discovery of tissue culture (gametophyte)-mediated in vitro haploid induction (Guha and Maheshwari 1964, 1966). Since then, in vitro haploid production has been successfully applied across numerous agriculturally important crops (Dunwell 2010). This section provides a concise review of current haploid induction methods, with particular focus on advancements in in vivo haploid induction.

12.4.1 In Vitro Methods to Generate Haploids

In vitro methods for haploid induction include *androgenesis and gynogenesis* (Fig. 12.3a). Androgenesis involves culturing immature anthers or microspores to quickly produce haploids (Guha and Maheshwari 1964, 1966), which are effective for crops like barley (Clapham 1973; Szarejko 2003) and wheat (García-Llamas et al. 2004; Kim and Stephen Baenziger 2005; Mujeeb-Kazi et al. 2006). Anther culture is a crucial method in plant tissue culture that accelerates breeding cycles by generating haploids (Kasha and Maluszynski 2003; El-Hennawy et al. 2011). The success of this technique is dependent on the genotype of the plant, the growth conditions, and the developmental stage of the microspores. Additionally, pretreatment of the plant tissues, the physiological state of the donor plants, the composition of the culture medium, and the overall cultural conditions, such as temperature and light, all play significant roles in determining the efficiency of haploid induction (Lazar et al. 1984; Andersen et al. 1987; Zhou and Konzak 1989; Ekiz and Konzak 1991; Zheng and Konzak 1999). Further, anther culture is also limited by high albino rates, and high cost. Microspore culture, which utilizes immature, uninucleate pollen grains, offers key advantages over anther culture. It facilitates efficient genetic manipulation, avoids interference from diploid tissues (e.g., anther walls),

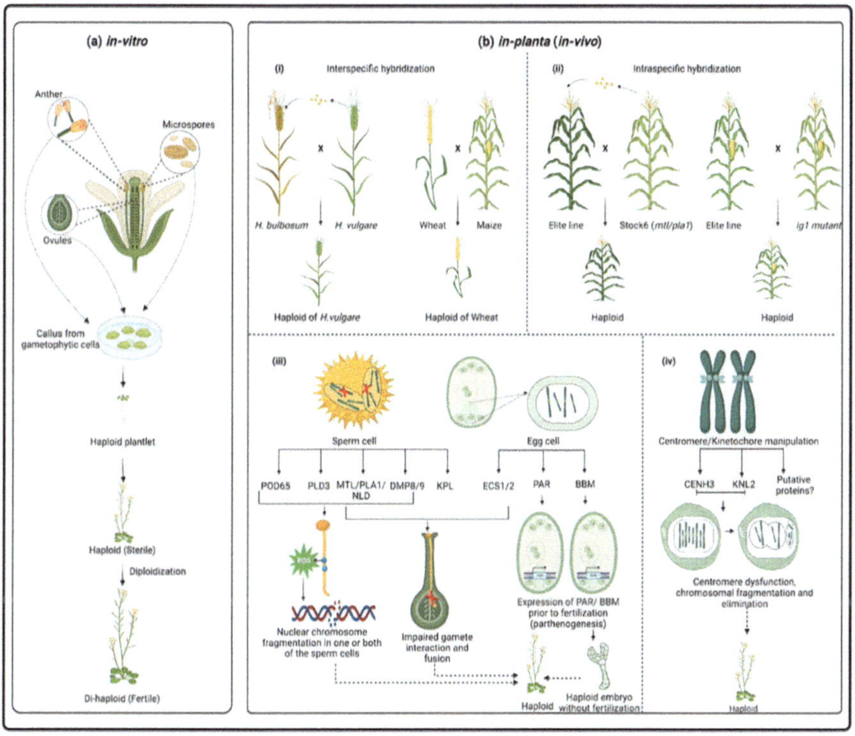

Fig. 12.3 Overview of haploid induction methods: (**a**) In vitro methods include androgenesis (microspore/anther culture), gynogenesis (ovule/ovary culture). (**b**) *In-planta* (in vivo) methods: (i) Interspecific hybridization and wide hybridization techniques (e.g., bulbosum method and wheat-maize system) that use chromosome elimination for efficient haploid production. (ii) In maize, intraspecific hybridization produces haploids using the *ig1* system for paternal haploids and the Stock6 system to generate haploids. (iii) Gene mutations in sperm cells (*MTL, ZmPOD65, ZmPLD3)* increase ROS levels in sperm nuclei, leading to chromosomal fragmentation; mutations in *DMP* and *KPL* impair gamete interaction. Mutation in egg cell-related genes, *ECS1/ECS2* mutants result in semigamous embryos, parthenogenesis-related genes (*BBM* and *PAR*) also generate haploids. (iv) Centromere and kinetochore proteins such as *CENH3* and *KNL2* manipulation. (Created using BioRender)

and increases the likelihood of spontaneous chromosome doubling, promoting stable homozygous line development (Castillo et al. 2008; Ferrie and Caswell 2011; Shariatpanahi and Ahmadi 2016). Limitation of this method includes genotype dependency (Murovec and Bohanec 2012) and difficulties in identifying the optimal microspore stage for haploid induction. Gynogenesis entails culturing female gametes to generate genetically stable haploids with fewer albinos, but it is less commonly used due to its limited effectiveness, restricted to certain cultivars, and the limited number of available embryo sacs (Rakha et al. 2012).

12.4.2 In Vivo Methods to Generate Haploids

While haploids can be produced in vitro via anther/microspore or ovule/ovary culture (Dunwell 2010), this method demands controlled, aseptic conditions for callus regeneration. Further, in vitro haploid induction is genotype dependent, and many agronomic crops exhibit recalcitrance to it. In contrast, the in vivo *(in-planta)* systems require simple crosses to produce haploids within a single generation. Several distinct in vivo methods are known for *in-planta* haploid induction.

Many distant crosses (interspecific and intergeneric) typically lead to the loss of one set of chromosomes which may be due to conflicts arising from genomic and epigenomic differences between the different parental genomes (Riddle and Birchler 2003; Marimuthu et al. 2021). A well-studied approach is the bulbosum method, which uses hybridization between diploid *Hordeum vulgare* × tetraploid *H. bulbosum* to produce diploid barley (*Hordeum vulgare*)-like plants (Davies 1956; Kasha and Kao 1970) (Fig. 12.3b, (i)). The wheat-maize system (Zenkteler and Nitzsche 1984), another effective technique, induces haploid wheat embryos by eliminating maize chromosomes after a few mitotic divisions (Kisana et al. 1993; Inagaki and Mujeeb-Kazi 1995; Pratap et al. 2006; Jing-Lin et al. 1991). Similarly, the wheat–*Imperata cylindrica* system provides benefits like synchronization with wheat flowering, without the need for a greenhouse, and elimination of crossability inhibitor genes and enhanced embryo formation efficiency (Chaudhary et al. 2005; Kishore et al. 2011). Intraspecific hybridization can also produce haploids through crosses within the same species (Fig. 12.3b, (ii)). In maize, for example, the *ig1* system induces paternal haploids with a haploid induction rate (HIR) of around 3% (Kermicle 1969). Similarly, the Stock6-derived system, an inbred line in maize, produces maternal haploids with an HIR of 2.3–3.2% (Coe 1959). Ravi and Chan (2010) discovered that intraspecific hybridization between plants polymorphic for centromere-specific histones can also induce haploids. This finding has sparked considerable interest within the plant science community, prompting further exploration of in-planta (in vivo) haploid induction across various crops, as discussed below.

Centromere-mediated UGE approach involves crossing of an engineered haploid inducer line (HI) in *Arabidopsis thaliana,* which has mutation in the centromere-specific histone protein (CENH3), complemented by an altered CENH3 variant, GFP-tailswap to a wild-type plant to produce haploids (Ravi and Chan 2010; Ravi and Bondada 2016) (Fig. 12.3b, (iv)). This outcome is attributed to competition between the haploid inducer and wild-type centromeres, leading to reduced spindle attachment and subsequent chromosome fragmentation and loss (Thondehaalmath et al. 2021; Marimuthu et al. 2021). This method addresses the limitations of in vitro methods such as genotype dependency and the recalcitrance of plant cells for haploid induction. CENH3 mediated haploid induction has been successfully translated into several crop species with mixed success, such as maize (Kelliher et al. 2016; Wang et al. 2021), wheat (Lv et al. 2020), rice (WO2017200386/KEYGENE 2017; Kalinowska et al. 2019), cucumber (Van Dun et al. 2018), melon (WO2017081011A1/RIJK ZWAAN 2017), tomato (WO2017081011A1/RIJK ZWAAN 2017), onion

(Manape et al. 2024), and cotton (Gao et al. 2020). Further, targeting kinetochore protein, KNL2 has also been shown to induce haploids in *Arabidopsis* (Fig. 12.3b, (iv)) (Ahmadli et al. 2023).

Apart from the CENH3-mediated haploid induction, mutations in several other genes are known to cause in-planta haploid induction in various plant systems. Reactive oxygen species (ROS) are known to have a significant role in inducing haploids in maize and wheat, primarily through their capacity to cause chromosomal damage (Li et al. 2017; Jiang et al. 2022a; Sun et al. 2022). Mutant lines with disrupted *MTL /PLA1/NLD(NOT LIKE DAD)* (Kelliher et al. 2017; Liu et al. 2017; Gilles et al. 2017), *PEROXIDASE (ZmPOD65)* (Jiang et al. 2022b), and *PHOSPHOLIPASE D3 (ZmPLD3)* (Li et al. 2021) exhibit elevated ROS levels in sperm nuclei, resulting in chromosomal degradation (Fig. 12.3b, (iii)). When *mtl*, *pod65*, and *pld3* mutant lines are used as pollen donors, increased ROS levels can lead to nuclear chromosome fragmentation in the sperm cells, thus resulting in haploid offspring. Furthermore, mutant lines such as *dmp* (Zhong et al. 2019; Cyprys et al. 2019; Wang et al. 2022b), *KOKOPELLI* (*kpl*) (Zhang et al. 2023b; Jacquier et al. 2023), *mtl*, or *pld3* crossed with wild-type (WT) females, or *ecs1/2* double mutants crossed with WT males, exhibit significant impairment in gamete interaction and fusion (Fig. 12.3b, (iii)). If the central cell fails to fuse with the sperm cell, seed abortion occurs; conversely, if the egg cell and sperm cell do not fuse, it results in HI. *ECS1* and *ECS2* are *EC-SPECIFIC ASPARTIC ENDOPEPTIDASES* (ECSs) in *Arabidopsis* (Yu et al. 2021) and *OsECS* in rice (Zhang et al. 2023a), which are critical for the effective interaction and fusion of egg and sperm cell nuclei. When *ecs1/2* mutant lines are fertilized with WT pollen, it can lead to sperm fusion with the egg cell without concurrent nuclear fusion, resulting in the formation of semigamous embryos (Jiang et al. 2022b; Mao et al. 2023; Zhang et al. 2023a) (Fig. 12.3b, (iii)). Mutation of these genes in crops has shown promising results for enhancing production through in vivo haploids. Notably, mutations in the in *dmp8/9* has been translated into rapeseed (Zhong et al. 2022b), tobacco (Zhang et al. 2022b), tomato (Zhong et al. 2022a), *Arabidopsis* (Zhong et al. 2020), *Medicago truncatula* (Wang et al. 2022a), *Brassica oleracea* (Zhao et al. 2022), potato (Zhang et al. 2022a), and watermelon (Tian et al. 2023; Chen et al. 2023a, b). Similarly, *mtl* mutants in wheat (Liu et al. 2020a), maize (Gilles et al. 2017; Kelliher et al. 2017; Liu et al. 2017), rice (Yao et al. 2018; Jang et al. 2023), foxtail millet (Cheng et al. 2021), *pod65* (Jiang et al. 2022b), and *pld3 mutants* in maize (Li et al. 2021) have been generated to induce haploids in these respective crops.

Another method involves inducing parthenogenesis, where haploids are produced from unfertilized ovules, exemplified by the *PAR* and *BBM*. Expressing the *PAR* gene, from apomictic dandelion in sexual lettuce under the control of an egg-cell-specific promoter, induces the formation of haploid embryos, without fertilization (Underwood et al. 2022). The key gene linked to parthenogenesis has been identified in the apomictic grass *Pennisetum squamulatum* (pearl millet), known as *PsASGR-BABY BOOM-like* (*PsASGR-BBML*). This gene is expressed in egg cells before fertilization and has been shown to induce parthenogenesis in sexual pearl millet (Conner et al. 2015) and rice (Khanday et al. 2019; Vernet et al. 2022).

Haploid technology is increasingly recognized for its potential benefits in crop improvement programs across various crops, including cereals, legumes, fruits, and more. The diverse in vivo methods for haploid induction, including interspecific hybridization, centromere-mediated genome elimination, and parthenogenesis, highlight the potential for innovative biotechnological approaches to enhance crop breeding and improve the efficiency of haploid production across various plant species.

12.5 Conclusion

Meeting the food demands of 2050 will require a multi-faceted approach that combines traditional agricultural knowledge, cutting-edge biotechnology, and sustainable practices. While conventional breeding methods have been crucial to agricultural advancements, they are often too slow to address these urgent demands. To keep pace with rapid population growth, climate change, biotic stresses, and shifting dietary preferences, accelerated crop breeding is essential. Adopting innovative breeding strategies will be crucial to ensuring global food security in the decades to come. Developing resilient crop varieties that can withstand heat, drought, and pest pressures will be crucial for food security in the face of climate change. DH technology emerges as one of the powerful solutions to address these limitations, with the potential to significantly accelerate breeding processes and improve crop resilience. Its ability to quickly eliminate heterozygosity enhances trait selection precision, making it invaluable in stress tolerance studies and mapping. Integrating DH technology with other modern breeding innovations can further accelerate genetic gains and develop environment-resilient crops efficiently. However, to fully integrate DH technology into large-scale breeding programs, certain challenges need to be addressed. For instance, current haploid induction methods and chromosome doubling protocols need improvement to increase the efficiency and scalability of DH systems. Producing large populations of haploid plants and reliably doubling their chromosomes to form fertile, stable lines are essential steps that, if optimized, could enhance DH technology's effectiveness. Improving these protocols will make it more feasible for breeding programs to produce large numbers of environment-resilient varieties swiftly, a capability that is increasingly important as climate change intensifies. In summary, by harnessing DH technology, breeders can accelerate the development of crops with superior traits, including stress tolerance, adaptability, and yield stability. This approach not only addresses the immediate demands of environmental adaptation in agriculture but also ensures long-term sustainability by producing varieties capable of withstanding a rapidly changing environment. The integration of DH technology into breeding programs represents a pivotal step toward achieving global food security in an era marked by environmental uncertainty.

References

Ahmadli U, Kalidass M, Khaitova LC, Fuchs J, Cuacos M, Demidov D, Zuo S et al (2023) High temperature increases centromere-mediated genome elimination frequency and enhances Haploid induction in Arabidopsis. Plant Commun 4(3):100507

Ali J, Wani SH (eds) (2021) Rice Improvement. Springer Nature, Cham

Al-Khayri JM, Jain SM, Johnson DV (eds) (2016) Advances in plant breeding strategies: breeding, biotechnology and molecular tools, 1st edn. Springer International Publishing, Cham

Andersen SB, Due IK, Olesen A (1987) The response of anther culture in a genetically wide material of winter wheat (Triticum Aestivum L.). Plant Breeding = Zeitschrift Fur Pflanzenzuchtung 99(3):181–186

Anshori MF, Purwoko BS, Dewi IS, Suwarno WB, Ardie SW (2023) Systematic selection to adaptive doubled Haploid Rice lines under different environments of submergence screening methods. J Agric Food Res 14(100775):100775

Asif M (2013) Progress and opportunities of doubled Haploid production, SpringerBriefs in plant science 6. Springer International Publishing, Cham

Asif MA, Schilling RK, Tilbrook J, Brien C, Dowling K, Rabie H, Pearson. (2018) Mapping of Novel Salt Tolerance QTL in an Excaliburx Kukri Doubled Haploid Wheat Population. Theor Appl Genet 131:2179–2196

Asseng S, Ewert F, Martre P, Rötter RP, Lobell DB, Cammarano D, Kimball BA et al (2015) Rising temperatures reduce global wheat production. Nat Clim Chang 5(2):143–147

Behla R, Hirani AH, Zelmer CD, Fengqun Y, Dilantha Fernando WG, McVetty P, Li G (2017) Identification of common QTL for resistance to sclerotinia sclerotiorum in three doubled haploid populations of brassica Napus (L.). Euphytica/Netherlands J Plant Breed 213(11). https://doi.org/10.1007/s10681-017-2047-5

Bernardo R (2009) Should maize doubled haploids be induced among F 1 or F 2 plants? Theor Appl Genet 119:255–262

Bhat JA, Ali S, Salgotra RK, Mir ZA, Dutta S, Jadon V, Tyagi A et al (2016) Genomic selection in the era of next generation sequencing for complex traits in plant breeding. Front Genet 7(December):221

Birchler JA (2015) Engineered Minichromosomes in plants. Chromosome Res 23(1):77–85

Birchler JA, Swyers NC (2020) Engineered Minichromosomes in plants. Exp Cell Res 388(2):111852

Blakeslee AF, Belling J, Farnham ME, Bergner AD (1922) A Haploid mutant in the jimson weed, 'Datura Stramonium. Science 55:646–647

Bondada R, Somasundaram S, Marimuthu MP, Badarudeen MA, Puthiyaveedu VK, Maruthachalam R (2020) Natural Epialleles of Arabidopsis SUPERMAN display superwoman phenotypes. Commun Biol 3(1):772

Bortiri E, Selby R, Egger R, Tolhurst L, Dong S, Beam K, Meier K et al (2024) Cyto-swapping in maize by Haploid induction with a cenh3 mutant. Nat Plants 10(4):567–571

Boutilier K, Offringa R, Sharma VK, Kieft H, Ouellet T, Zhang L, Hattori J et al (2002) Ectopic expression of BABY BOOM triggers a conversion from vegetative to embryonic growth. Plant Cell 14(8):1737–1749

Byers D (2008) Components of phenotypic variance. Nat Educ 1(1):161

Calayugan MIC, Formantes AK, Amparado A, Descalsota-Empleo GI, Nha CT, Inabangan-Asilo MA, Swamy. (2020) Genetic analysis of agronomic traits and grain iron and zinc concentrations in a doubled Haploid population of Rice (Oryza Sativa L). Sci Rep 10(1):2283

Castillo AM, Cistué L, Vallés MP, Soriano M (2008) Chromosome doubling in monocots. In: Advances in Haploid production in higher plants. Springer Netherlands, Dordrecht, pp 329–338

Chaikam V, Nair SK, Martinez L, Lopez LA, Friedrich Utz H, Melchinger AE, Boddupalli PM (2018) Marker-assisted breeding of improved maternal Haploid inducers in maize for the tropical/subtropical regions. Front Plant Sci 9(October):1527

Chaikam V, Molenaar W, Melchinger AE, Boddupalli PM (2019a) Doubled haploid technology for line development in maize: technical advances and prospects. Theor Appl Genet 132(12):3227–3243

Chaikam V, Gowda M, Nair SK, Melchinger AE, Boddupalli PM (2019b) Genome-wide association study to identify genomic regions influencing spontaneous fertility in maize haploids. Euphytica 215(8):138

Chang M-T, Coe EH Jr (2009) Doubled haploids. In: Genetic M (ed) Approaches to maize improvement. Springer Berlin Heidelberg, Berlin/Heidelberg, pp 127–142

Chaudhary HK, Sethi GS, Singh S, Pratap A, Sharma S (2005) Efficient Haploid induction in wheat by using pollen of Imperata Cylindrica. Plant Breeding = Zeitschrift Fur Pflanzenzuchtung 124(1):96–98

Chen X, Li Y, Ai G, Chen J, Guo D, Zhu Z, Zhu X et al (2023a) Creation of a watermelon Haploid inducer line via ClDMP3-mediated single fertilization of the central cell. Hortic Res 10(6):uhad081

Chen Y-R, Lübberstedt T, Frei UK (2023b) Development of doubled Haploid inducer lines facilitates selection of superior Haploid inducers in maize. Front Plant Sci 14:1320660

Cheng Z, Sun Y, Yang S, Zhi H, Yin T, Ma X, Zhang H et al (2021) Establishing in planta Haploid inducer line by edited SiMTL in foxtail millet (Setaria Italica). Plant Biotechnol J 19(6):1089–1091

Clapham D (1973) Haploid Hordeum plants from anthers in vitro. Z Pflanzenzücht 69:142–155

Class B, Brommer JE (2020) Can dominance Genetic variance be ignored in evolutionary quantitative Genetic analyses of wild populations? Evol Int J Org Evol 74(7):1540–1550

Clausen RE, Mann MC (1924) Inheritance in Nicotiana Tabacum: V. The occurrence of Haploid plants in interspecific progenies. Proc Natl Acad Sci USA 10(4):121–124

Coe EH (1959) A line of maize with high Haploid frequency. Am Nat 93(873):381–382

Conner JA, Mookkan M, Huo H, Chae K, Ozias-Akins P (2015) A parthenogenesis gene of apomict origin elicits embryo formation from unfertilized eggs in a sexual plant. Proc Natl Acad Sci USA 112(36):11205–11210

Conner JA, Podio M, Ozias-Akins P (2017) Haploid embryo production in Rice and maize induced by PsASGR-BBML transgenes. Plant Reproduct 30(1):41–52

Cook TM, Isenegger D, Dutta S, Sahab S, Kay P, Aboobucker SI, Biswas E et al (2023) Overcoming roadblocks for in vitro nurseries in plants: induction of meiosis. Front Plant Sci 14(June):1204813

Cook TM, Biswas E, Aboobucker SI, Dutta S, Lübberstedt T (2024) A cell-based fluorescent system and statistical framework to detect meiosis-like induction in plants. Front Plant Sci 15(July):1386274

Crossa J, Pérez-Rodríguez P, Cuevas J, Montesinos-López O, Jarquín D, de los Campos G, Burgueño J et al (2017) Genomic selection in plant breeding: methods, models, and perspectives. Trends Plant Sci 22(11):961–975

Cyprys P, Lindemeier M, Sprunck S (2019) Gamete fusion is facilitated by two sperm cell-expressed DUF679 membrane proteins. Nat Plants 5(3):253–257

d'Erfurth I, Jolivet S, Froger N, Catrice O, Novatchkova M, Mercier R (2009) Turning Meiosis into Mitosis. PLoS Biol 7(6):e1000124

Daetwyler HD, Hayden MJ, Spangenberg GC, Hayes BJ (2015) Selection on optimal Haploid value increases Genetic gain and preserves more Genetic diversity relative to genomic selection. Genetics 200(4):1341–1348

Davies DR (1956) Cytogenetic studies in wild and cultivated species of Hordeum. Doctoral dissertation

Dong X, Xiaowei X, Li L, Liu C, Tian X, Li W, Chen S (2014) Marker-assisted selection and evaluation of high oil in vivo Haploid inducers in maize. Mol Breed New Strateg Plant Improv 34(3):1147–1158

Draeger T, Moore G (2017) Short periods of high temperature during meiosis prevent Normal meiotic progression and reduce grain number in Hexaploid wheat (Triticum Aestivum L.). Theor Appl Genet 130(9):1785–1800

Dunwell JM (2010) Haploids in flowering plants: origins and exploitation. Plant Biotechnol J 8(4):377–424

Ekiz H, Konzak CF (1991) Nuclear and cytoplasmic control of anther culture response in wheat: III. Common wheat crosses. Crop Sci 31(6):1432–1436

El-Hennawy MA, Abdalla AF, Shafey SA, Al-Ashkar IM (2011) Production of doubled Haploid wheat lines (Triticum Aestivum L.) using anther culture technique. Ann Agric Sci 56(2):63–72

Elsharawy H, Refat M (2023) CRISPR/Cas9 genome editing in wheat: enhancing quality and productivity for global Food security-a review. Funct Integr Genomics 23(3):265

Fatima Z, Ahmed M, Hussain M, Abbas G, Ul-Allah S, Ahmad S, Ahmed N et al (2020) The fingerprints of climate warming on cereal crops phenology and adaptation options. Sci Rep 10(1):18013

Ferrie AMR, Caswell KL (2011) Isolated microspore culture techniques and recent Progress for Haploid and doubled Haploid plant production. Plant Cell Tissue Organ Cult 104(3):301–309

Filiault DL, Seymour DK, Maruthachalam R, Maloof JN (2017) The generation of doubled Haploid lines for QTL mapping. Methods Mol Biol 1610:39–57

Food and Agriculture Organization (2020) The state of Food security and nutrition in the world 2020. The state of Food security and nutrition in the world (SOFI). Food & Agriculture Organization of the United Nations (FAO), Rome

Food, Fibre, and Other Ecosystem Products (n.d.) Climate change 2022: impacts, adaptation and vulnerability. In: Pörtner H-O, Roberts DC, Tignor M, Poloczanska ES, Mintenbeck K, Alegría A, Craig M et al (eds) Contribution of working group II to the sixth assessment report of the intergovernmental panel on climate change. Cambridge University Press, Cambridge/New York

Forster BP, Thomas W (2003) Doubled haploids in Genetic mapping and genomics. In: Doubled Haploid production in crop plants. Springer, Dordrecht

Forster BP, Heberle-Bors E, Kasha KJ, Touraev A (2007) The resurgence of haploids in higher plants. Trends Plant Sci 12(8):368–375

Fulcher N, Riha K (2015) Using centromere mediated genome elimination to elucidate the functional redundancy of candidate telomere binding proteins in Arabidopsis Thaliana. Front Genet 6:349

Gahlaut V, Jaiswal V, Tyagi BS, Singh G, Sareen S, Balyan HS, Gupta PK (2017) QTL mapping for nine drought-responsive agronomic traits in bread wheat under irrigated and rain-fed environments. PLoS One 12(8):e0182857

Gao X, Guo H, Jianfei W, Fan Y, Zhang L, Guo H, Lian X et al (2020) Haploid bio-induction in plant through mock sexual reproduction. iScience 23(7):101279

García-Llamas C, Martín A, Ballesteros J (2004) Differences among auxin treatments on Haploid production in durum wheatx maize crosses. Plant Cell Rep 23:46–49

Germanà MA (2011) Gametic embryogenesis and Haploid technology as valuable support to plant breeding. Plant Cell Rep 30(5):839–857

Gilles LM, Khaled A, Laffaire JB, Chaignon S, Gendrot G, Laplaige J, Berges H, Beydon G, Bayle V, Barret P (2017) Loss of pollen-specific phospholipaseNOT LIKE DAD triggers Gynogenesis in maize. EMBO J 36:707–717

Gołębiowska-Pikania G, Kopeć P, Surówka E, Janowiak F, Krzewska M, Dubas E, Nowicka A et al (2017) Changes in protein abundance and activity induced by drought during generative development of winter barley (Hordeum Vulgare L.). J Proteome 169(October):73–86

Goyal L, Kaur M, Mandal M, Panda D, Karmakar S, Molla KA, Bhatia D (2024) Potential gene editing targets for developing haploid inducer stocks in rice and wheat with high haploid induction frequency. 3 Biotech 14(1):14

Grüning VK, Lübberstedt T, Frei UK (2024) Doubled Haploid technology: generation of doubled Haploid maize lines using Haploid inducers. Cold Spring Harbor Protocols. https://doi.org/10.1101/pdb.prot108624

Guha S, Maheshwari SC (1964) In vitro production of embryos from anthers of Datura. Nature 204(4957):497–497

Guha S, Maheshwari SC (1966) Cell division and differentiation of embryos in the pollen grains of Datura in vitro. Nature 212(5057):97–98

Hafizur Rahman M, Krishnaraj S, Thorpe TA (1995) Selection for salt tolerance in vitro using microspore-derived embryos of brassica Napus Cv. Topas, and the characterization of putative tolerant plants. In Vitro Cell Dev Biol Plant 31:116–121

Han F, Zhang X, Liu Y, Liu Y, Zhao H, Li Z (2024) One-step creation of CMS lines using a BoCENH3-based Haploid induction system in brassica crop. Nat Plants 10(4):581–586

Hickey LT, Hafeez AN, Robinson H, Jackson SA, Leal-Bertioli SCM, Tester M, Gao C, Godwin ID, Hayes BJ, Wulff BBH (2019) Breeding crops to feed 10 billion. Nat Biotechnol 37(7):744–754

Hochholdinger F, Hoecker N (2007) Towards the molecular basis of Heterosis. Trends Plant Sci 12(9):427–432

Houben A, Schubert I (2007) Engineered plant Minichromosomes: a resurrection of B chromosomes? Plant Cell 19(8):2323–2327

Houben A, Kelly Dawe R, Jiang J, Schubert I (2008) Engineered plant Minichromosomes: a bottom-up success? Plant Cell 20(1):8–10

Humphreys DG, Knox RE (2015) Doubled Haploid breeding in cereals. In: Advances in plant breeding strategies: breeding, biotechnology and molecular tools. Springer International Publishing, Cham, pp 241–290

Impens L, Lorenzo CD, Vandeputte W, Wytynck P, Debray K, Haeghebaert J, Herwegh D et al (2023) Combining multiplex gene editing and doubled Haploid Technology in Maize. New Phytol 239(4):1521–1532

Inagaki M, Mujeeb-Kazi A (1995) Comparison of Polyhaploid production frequencies in crosses of Hexaploid wheat with maize, pearl millet and Sorghuml. Ikushugaku Zasshi 45(2):157–161

Intergovernmental Panel on Climate Change (2018) Global warming of 1.5° C: An IPCC special report on the impacts of global warming of 1.5° C above pre-industrial levels and related global greenhouse gas emission pathways, in the context of strengthening the global response to the threat of climate change, sustainable development, and efforts to eradicate poverty. Intergovernmental Panel on Climate Change

Jacquier NMA, Calhau ARM, Fierlej Y, Martinant J-P, Rogowsky PM, Gilles LM, Widiez T (2023) In planta Haploid induction by Kokopelli mutants. Plant Physiol 193(1):182–185

Jang JH, Noh G, Seo HS, Jung K-H, Kim Y-J, Lee OR (2023) Loss of function of pollen-expressed Phospholipase OsMATL2 triggers Haploid induction in japonica Rice. Plant Physiol 193(3):1749–1752

Jiang G-L (2015) Molecular marker-assisted breeding: a plant breeder's review. In: Advances in plant breeding strategies: breeding, biotechnology and molecular tools. Springer International Publishing, Cham, pp 431–472

Jiang C, Sun J, Li R, Yan S, Chen W, Guo L, Qin G et al (2022a) A reactive oxygen species burst causes Haploid induction in maize. Mol Plant 15(6):943–955

Jiang J, Stührwohldt N, Liu T, Huang Q, Li L, Zhang L, Hongya G et al (2022b) Egg cell-secreted Aspartic proteases ECS1/2 promote gamete attachment to prioritize the fertilization of egg cells over central cells in Arabidopsis. J Integr Plant Biol 64(11):2047–2059

Jiao Y, Li J, Li W, Chen M, Li M, Liu W, Chen. (2020) QTL mapping and prediction of Haploid male fertility traits in maize (Zea Mays L). Plants 9(7):836

Jing-Lin W, Jing-San S, Tie-Gang L, Ren F, Hai-Rui C, Shu-Zeng C, Cai Y (1991) Fertilization and embryo development in wheat X maize crosses. J Integr Plant Biol 33(9)

Kalinowska K, Chamas S, Unkel K, Demidov D, Lermontova I, Dresselhaus T, Kumlehn J, Dunemann F, Houben A (2019) State-of-the-art and novel developments of in vivo Haploid technologies. Theor Appl Genet 132(3):593–605

Kasha KJ, Kao KN (1970) High frequency Haploid production in barley (Hordeum Vulgare L.). Nature 225(5235):874–876

Kasha KJ, Maluszynski M (2003) Production of doubled haploids in crop plants. An introduction. In: Doubled haploid Production in crop plants. Springer Netherlands, Dordrecht, pp 1–4

Kelliher T, Starr D, Wang W, McCuiston J, Zhong H, Nuccio ML, Martin B (2016) Maternal haploids are preferentially induced by CENH3-Tailswap transgenic complementation in maize. Front Plant Sci 7(March):414

Kelliher T, Starr D, Richbourg L, Chintamanani S, Delzer B, Nuccio ML, Green J et al (2017) MATRILINEAL, a sperm-specific Phospholipase, triggers maize Haploid induction. Nature 542(7639):105–109

Kelliher T, Starr D, Xiujuan S, Tang G, Chen Z, Carter J, Wittich PE et al (2019) One-step genome editing of elite crop germplasm during Haploid induction. Nat Biotechnol 37(3):287–292

Kemper KE, Bowman PJ, Pryce JE, Hayes BJ, Goddard ME (2012) Long-term selection strategies for complex traits using high-density Genetic markers. J Dairy Sci 95(8):4646–4656

Kermicle JL (1969) Androgenesis conditioned by a mutation in maize. Science (New York, NY) 166(3911):1422–1424

Khammona K, Dermail A, Suriharn K, Lübberstedt T, Wanchana S, Thunnom B, Poncheewin W, Toojinda T, Ruanjaichon V, Arikit S (2024) Accelerating Haploid induction rate and Haploid validation through marker-assisted selection for qhir1 and qhir8 in maize. Front Plant Sci 15. https://doi.org/10.3389/fpls.2024.133746

Khanday I, Skinner D, Yang B, Mercier R, Sundaresan V (2019) A male-expressed Rice embryogenic trigger redirected for asexual propagation through seeds. Nature 565(7737):91–95

Kim K-M, Stephen Baenziger P (2005) A simple wheat haploid and doubled haploid production system using anther culture. In Vitro Cell Dev Biol Plant 41(1):22–27

Kisana NS, Nkongolo KK, Quick JS, Johnson DL (1993) Production of doubled haploids by anther culture and wheat X maize method in a wheat breeding Programme. Plant Breed = Zeitschrift Fur Pflanzenzuchtung 110(2):96–102

Kishore N, Chaudhary HK, Chahota RK, Kumar V, Sood SP, Jeberson S, Tayeng T (2011) Relative efficiency of the maize-and Imperata Cylindrica-mediated chromosome elimination approaches for induction of haploids of wheat-Rye derivatives. Plant Breed = Zeitschrift Fur Pflanzenzuchtung 130(2):192–194

Kuppu S, Ron M, Marimuthu MPA, Li G, Huddleson A, Siddeek MH, Terry J et al (2020) A variety of changes, including CRISPR/Cas9-mediated deletions, in CENH3 Lead to Haploid induction on outcrossing. Plant Biotechnol J 18(10):2068–2080

Laurie DA, Bennett MD (1988) The production of Haploid wheat plants from wheat X maize crosses. Theor Appl Genet 76(3):393–397

Lazar MD, Schaeffer GW, Baenziger PS (1984) Cultivar and cultivar X environment effects on the development of callus and Polyhaploid plants from anther cultures of wheat. Theor Appl Genet 67(2–3):273–277

Lee Y, Ho Lee S, Oh Kwon J (2003) Selection of salt-tolerant doubled haploids in Rice anther culture. Plant Cell Tissue Organ Cult 74:143–149

Li X, Meng D, Chen S, Luo H, Zhang Q, Jin W, Yan J (2017) Single nucleus sequencing reveals spermatid chromosome fragmentation as a possible cause of maize haploid induction. Nat Commun 8(1). https://doi.org/10.1038/s41467-017-00969-8

Li Y, Lin Z, Yue Y, Zhao H, Fei X, Lizhu E, Liu C, Chen S, Lai J, Song W (2021) Loss-of-function alleles of ZmPLD3 cause Haploid induction in maize. Nat Plants 7(12):1579–1588

Li T, Yang Y, Qi H, Cui W, Zhang L, Xiuxiu F, He X, Liu M, Li P-F, Tao Y (2023) CRISPR/Cas9 therapeutics: progress and prospects. Signal Transduct Target Ther 8(1):36

Liu X, Li R, Chang X, Jing R (2013) Mapping QTLs for seedling root traits in a doubled Haploid wheat population under different water regimes. Euphytica/Netherlands J Plant Breed 189(1):51–66

Liu S, Rudd JC, Bai G, Haley SD, Ibrahim AMH, Xue Q, Hays DB, Graybosch RA, Devkota RN, St. Amand P (2014) Molecular markers linked to important genes in hard winter wheat. Crop Sci 54(4):1304–1321

Liu C, Li X, Dexuan Meng Y, Zhong C, Dong X, Xiaowei X et al (2017) A 4-Bp insertion at ZmPLA1 encoding a putative Phospholipase a generates Haploid induction in maize. Mol Plant 10(3):520–522

Liu C, Zhong Y, Qi X, Chen M, Liu Z, Chen XT et al (2020a) Extension of the in vivo Haploid induction system from diploid maize to Hexaploid wheat. Plant Biotechnol J 18(2):316–318

Liu H, Wang K, Jia Z, Gong Q, Lin Z, Lipu D, Pei X, Ye X (2020b) Efficient induction of Haploid plants in wheat by editing of TaMTL using an optimized agrobacterium-mediated CRISPR system. J Exp Bot 71(4):1337–1349

Lobell DB, Schlenker W, Costa-Roberts J (2011) Climate trends and global crop production since 1980. Science 333(6042):616–620

Lv J, Kelliher T (2023) Recent advances in engineering of in vivo Haploid induction systems. Methods Mol Biol (Clifton, NJ) 2653:365–383

Lv J, Kun Y, Wei J, Gui H, Liu C, Liang D, Wang Y et al (2020) Generation of paternal haploids in wheat by genome editing of the Centromeric histone CENH3. Nat Biotechnol 38(12):1397–1401

Lynch M (1997) Mapping and characterizing QTLs: inbred line crosses. In: Genetics and analysis of quantitative traits, pp 431–489

Mahlandt A, Singh DK, Mercier R (2023) Engineering Apomixis in crops. Theor Appl Genet 136(6):131

Malaga S, Janeczko A, Janowiak F, Waligórski P, Oklestkova J, Dubas E, Krzewska M et al (2020) Involvement of Homocastasterone, salicylic and Abscisic acids in the regulation of drought and freezing tolerance in doubled Haploid lines of winter barley. Plant Growth Regul 90(1):173–188

Manape TK, Satheesh V, Somasundaram S, Soumia PS, Khade YP, Mainkar P, Mahajan V, Singh M, Anandhan S (2024) RNAi-mediated downregulation of AcCENH3 can induce in vivo haploids in onion (Allium Cepa L.). Sci Rep 14(1):14481

Mao Y, Nakel T, Serbes IE, Joshi S, Tekleyohans DG, Baum T, Groß-Hardt R (2023) ECS1 and ECS2 suppress polyspermy and the formation of Haploid plants by promoting double fertilization. eLife 12(July). https://doi.org/10.7554/eLife.85832

Marimuthu MPA, Jolivet S, Ravi M, Pereira L, Davda JN, Cromer L, Wang L et al (2011) Synthetic clonal reproduction through seeds. Science (New York, NY) 331(6019):876

Marimuthu MPA, Maruthachalam R, Bondada R, Kuppu S, Tan EH, Britt A, Chan SWL, Comai L (2021) Epigenetically mismatched parental centromeres trigger genome elimination in Hybrids. Sci Adv 7(47):eabk1151

Maruthachalam R (2024) Haploids fast-track hybrid plant breeding. Nat Plants 10(4):530–532

Meuwissen TH, Hayes BJ, Goddard ME (2001) Prediction of total genetic value using genome-wide dense marker maps. Genetics 157(4):1819–1829

Mieulet D, Jolivet S, Rivard M, Cromer L, Vernet A, Mayonove P, Pereira L et al (2016) Turning Rice meiosis into mitosis. Cell Res 26(11):1242–1254

Molenaar WS, University of Hohenheim, Germany, Melchinger AE (2019) Production of doubled Haploid lines for hybrid breeding in maize. In: Advances in breeding techniques for cereal crops. Burleigh Dodds Science Publishing, pp 143–172

Mujeeb-Kazi A, Gul A, Ahmed J, Mirza JI (2006) A simplified and effective protocol for production of bread wheat haploids (n= 3x= 21, ABD) with some application areas in wheat improvement. Pak J Bot 38(2):393

Murovec J, Bohanec B (2012) Haploids and doubled haploids in plant breeding. In: Plant breeding. InTech

Naqvi RZ, Siddiqui HA, Mahmood MA, Najeebullah S, Ehsan A, Azhar M, Farooq M et al (2022) Smart breeding approaches in post-genomics era for developing climate-resilient food crops. Front Plant Sci 13(September):972164

Nekrasov V, Wang C, Win J, Lanz C, Weigel D (2017) Kamoun rapid generation of a transgene-free powdery mildew resistant tomato by genome deletion Sci. Sci Rep 7:482

Orp AM, Hansen AL, Andersen SB (2001) Chromosomal regions associated with Green plant regeneration in wheat (Triticum Aestivum L.) anther culture. Euphytica/Netherlands J Plant Breed 119:377–387

Ortiz-Bobea A, Ault TR, Carrillo CM, Chambers RG, Lobell DB (2021) Anthropogenic climate change has slowed global agricultural productivity growth. Nat Clim Chang 11(4):306–312

Park GH, Kim J-H, Kim K-M (2014) QTL analysis of yield components in Rice using a Cheongcheong/Nagdong doubled Haploid Genetic map. Am J Plant Sci 05(09):1174–1180

Pham B, McConnaughay K (2014) Plant phenotypic expression in variable environments. In: Ecology and the environment. Springer New York, New York, pp 119–141

Pramanik D, Shelake RM, Kim MJ, Kim J-Y (2021) CRISPR-mediated engineering across the central dogma in plant biology for basic research and crop improvement. Mol Plant 14(1):127–150

Prasanna BM, Cairns J, Xu Y (2013) Genomic tools and strategies for breeding climate resilient cereals. In: Genomics and breeding for climate-resilient crops. Springer Berlin Heidelberg, Berlin/Heidelberg, pp 213–239

Pratap A, Sethi GS, Chaudhary HK (2006) Relative efficiency of anther culture and chromosome elimination techniques for Haploid induction in triticale × wheat and triticale × triticale hybrids. Euphytica 150(3):339–345

Purwoko BS, Dewi IS, Khumaida N (2010) Rice anther culture to obtain doubled-haploids with multiple tolerances. Asia Pac J Mol Biol Biotechnol 18(1):55–57

Rakha MT, Metwally EI, Moustafa SA, Etman AA, Dewir YH (2012) Evaluation of regenerated strains from six Cucurbita interspecific hybrids obtained through anther and ovule in vitro cultures. Aust J Crop Sci 6(1):23–30

Ravi M, Bondada R (2016) Genome Elimination by Tailswap CenH3: in vivo haploid production in arabidopsis thaliana. Methods Mol Biol (Clifton, N.J.) 1469:77–99

Ravi M, Chan SWL (2010) Haploid plants produced by centromere-mediated genome elimination. Nature 464(7288):615–618

Ravi M, Marimuthu MPA, Siddiqi I (2008) Gamete formation without meiosis in Arabidopsis. Nature 451(7182):1121–1124

Ravi M, Marimuthu MPA, Tan EH, Maheshwari S, Henry IM, Marin-Rodriguez B, Urtecho G et al (2014) A Haploid genetics toolbox for Arabidopsis Thaliana. Nat Commun 5(1):5334

Riddle NC, Birchler JA (2003) Effects of reunited diverged regulatory hierarchies in allopolyploids and species hybrids. Trends Genet: TIG 19(11):597–600

Roldán-Arjona T, Ariza RR (2009) Repair and tolerance of oxidative DNA damage in plants. Mutat Res 681(2–3):169–179

Ronceret A, Vielle-Calzada J-P (2015) Meiosis, unreduced gametes, and parthenogenesis: implications for engineering clonal seed formation in crops. Plant Reprod 28(2):91–102

Rutkoski JE, Krause MR, Sorrells ME (2022) Breeding methods: line development. In: Wheat improvement. Springer International Publishing, Cham, pp 69–82

Rygulla W, Snowdon RJ, Friedt W, Happstadius I, Cheung WY, Chen D (2008) Identification of quantitative trait loci for resistance against Verticillium Longisporum in oilseed rape (Brassica Napus). Phytopathology 98(2):215–221

Saleem A, Anwar S, Nawaz T, Fahad S, Saud S, Rahman TU, Khan MNR, Nawaz T (2024) Securing a sustainable future: the climate change threat to agriculture, food security, and sustainable development goals. J Umm Al-Qura Univ Appl Sci. https://doi.org/10.1007/s43994-024-00177-3

Schnable PS, Springer NM (2013) Progress toward understanding Heterosis in crop plants. Annu Rev Plant Biol 64(1):71–88

Sestili S, Ficcadenti N (1996) Irradiated pollen for Haploid production. In: In in vitro Haploid production in higher plants, Current plant science and biotechnology in agriculture. Springer Netherlands, Dordrecht, pp 263–274

Seymour DK, Filiault DL, Henry IM, Monson-Miller J, Ravi M, Pang A, Comai L, Chan SWL, Maloof JN (2012) Rapid creation of Arabidopsis doubled Haploid lines for quantitative trait locus mapping. Proc Natl Acad Sci USA 109(11):4227–4232

Shariatpanahi ME, Ahmadi B (2016) Isolated microspore culture and its applications in plant breeding and genetics. In: Plant tissue culture: propagation, conservation and crop improvement, pp 487–507

Sharma D, Gahtyari NC, Parul Sharma RK, Khulbe RS, Pal, Kant L (2022) Doubled Haploidy: an accelerated breeding tool for stress resilience breeding in cereals. In: Next-generation plant breeding approaches for stress resilience in cereal crops. Springer Nature Singapore, Singapore, pp 199–240

Shi J, Gao H, Hongyu Wang H, Lafitte R, Archibald RL, Yang M, Hakimi SM, Mo H, Habben JE (2017) ARGOS8 variants generated by CRISPR-Cas9 improve maize grain yield under field drought stress conditions. Plant Biotechnol J 15(2):207–216

Shivanna KR (2022) Climate change and its impact on biodiversity and human welfare. Proc Indian Natl Sci Acad 88(2):160–171

Singh A, Ganapathysubramanian B, Singh AK, Sarkar S (2016) Machine learning for high-throughput stress phenotyping in plants. Trends Plant Sci 21(2):110–124

Singh BK, Delgado-Baquerizo M, Egidi E, Guirado E, Leach JE, Liu H, Trivedi P (2023) Climate change impacts on plant pathogens, food security and paths forward. Nat Rev Microbiol 21(10):640–656

Snape J (1989) Doubled Haploid breeding: theoretical basis and practical applications. In: Review of advances in plant biotechnology. CIMMYT and IRRI

Somers DJ, Humphreys G (2009) State of QTL detection and marker-assisted selection in wheat improvement. In: Wheat science and trade. Wiley-Blackwell, Oxford, pp 309–326

Son S, Park SR (2022) Challenges facing CRISPR/Cas9-based genome editing in plants. Front Plant Sci 13(May):902413

Sserumaga JP, Beyene Y, Pillay K, Kullaya A, Oikeh SO, Mugo S, Machida L et al (2018) Grain-yield stability among tropical maize hybrids derived from doubled-Haploid inbred lines under random drought stress and optimum moisture conditions. Crop Pasture Sci 69(11):691–712

Starosta E, Szwarc J, Niemann J, Szewczyk K, Weigt D (2023) Brassica Napus Haploid and double Haploid production and its latest applications. Curr Issues Mol Biol 45(5):4431–4450

Sun G, Geng S, Zhang H, Jia M, Wang Z, Deng Z, Tao S et al (2022) Matrilineal empowers wheat pollen with Haploid induction potency by triggering Postmitosis reactive oxygen species activity. New Phytol 233(6):2405–2414

Szarejko I (2003) Anther culture for doubled Haploid production in barley (Hordeum Vulgare L.). In: Haploid D (ed) Production in crop plants. Springer Netherlands, Dordrecht, pp 35–42

Tan EH, Ordoñez B, Thondehaalmath T, Seymour DK, Maloof JN, Maruthachalam R, Comai L (2023) Establishment and inheritance of Minichromosomes from Arabidopsis Haploid induction. Chromosoma 132(2):105–115

Tehseen MM, Tonk FA, Tosun M, Randhawa HS, Kurtulus E, Ozseven I, Akin B, Zulfuagaoglu ON, Nazari K (2022) QTL mapping of adult plant resistance to stripe rust in a doubled Haploid wheat population. Front Genet 13(May):900558

Thondehaalmath T, Kulaar DS, Bondada R, Maruthachalam R (2021) Understanding and exploiting uniparental genome elimination in plants: insights from Arabidopsis Thaliana. J Exp Bot 72(13):4646–4662

Tian S, Zhang J, Zhao H, Zong M, Li M, Gong G, Wang J et al (2023) Production of double Haploid watermelon via maternal Haploid induction. Plant Biotechnol J 21(7):1308–1310

Tong H, Nikoloski Z (2021) Machine learning approaches for crop improvement: leveraging phenotypic and genotypic big data. J Plant Physiol 257(153354):153354

Torp AM, Andersen SB (2008) Albinism in Microspore Culture. In: Advances in Haploid production in higher plants. Springer Netherlands, Dordrecht, pp 155–160

Tripathy SK, Swain D, Mohapatra PM, Prusti AM, Sahoo B, Panda S, Behera. (2019) Exploring factors affecting anther culture in Rice (Oryza Sativa L). J Appl Biol Biotechnol 7(02):87–92

Underwood CJ, Vijverberg K, Rigola D, Okamoto S, Oplaat C, Rik HM, den Camp O, Radoeva T et al (2022) A PARTHENOGENESIS allele from apomictic dandelion can induce egg cell division without fertilization in lettuce. Nat Genet 54(1):84–93

Van Dun CMP, Lelivelt CLC, Movahedi S (2018) U.S. Patent Application No. 15/774,779

Vernet A, Meynard D, Lian Q, Mieulet D, Gibert O, Bissah M, Rivallan R et al (2022) High-frequency synthetic Apomixis in hybrid Rice. Nat Commun 13(1):7963

Wang B, Zhu L, Zhao B, Zhao Y, Xie Y, Zheng Z, Li Y, Sun J, Wang H (2019) Development of a Haploid-inducer mediated genome editing system for accelerating maize breeding. Mol Plant 12(4):597–602

Wang Y, Tao H, Zhang P, Hou X, Sheng D, Beijing Tian P, Wang, and Shoubing Huang. (2020) Reduction in seed set upon exposure to high night temperature during flowering in maize. Physiol Plant 169(1):73–82

Wang N, Gent JI, Kelly Dawe R (2021) Haploid induction by a maize cenh3 null mutant. Sci Adv 7(4):eabe2299

Wang N, Xia X, Jiang T, Li L, Zhang P, Niu L, Cheng H, Wang K, Lin H (2022a) In planta Haploid induction by genome editing of DMP in the model legume Medicago Truncatula. Plant Biotechnol J 20(1):22–24

Wang W, Xiong H, Zhao P, Peng X, Sun M-X (2022b) DMP8 and 9 regulate HAP2/GCS1 trafficking for the timely Acquisition of Sperm Fusion Competence. Proc Natl Acad Sci USA 119(45):e2207608119

Watson A, Ghosh S, Williams MJ, Cuddy WS, Simmonds J, María-Dolores Rey M, Hatta AM et al (2018) Speed breeding is a powerful tool to accelerate crop research and breeding. Nat Plants 4(1):23–29

Wiśniewska H, Majka M, Kwiatek M, Gawłowska M, Surma M, Adamski T, Kaczmarek Z et al (2019) Production of wheat-doubled haploids resistant to eyespot supported by marker-assisted selection. Electron J Biotechnol 37(January):11–17

WO2017200386A1/KEYGENE (2017) Method for the production of haploid and subsequent doubled haploid plants

Xu Q, Yuan X, Hanyong Y, Wang Y, Tang S, Wei X (2011) Mapping quantitative trait loci for sheath blight resistance in rice using double Haploid population. Plant Breed = Zeitschrift Fur Pflanzenzuchtung 130(3):404–406

Xu Y, Li P, Zou C, Yanli L, Xie C, Zhang X, Prasanna BM, Olsen MS (2017) Enhancing Genetic gain in the era of molecular breeding. J Exp Bot 68(11):2641–2666

Yan G, Liu H, Wang H, Lu Z, Wang Y, Mullan D, Hamblin J, Liu C (2017) Accelerated generation of Selfed pure line plants for gene identification and crop breeding. Front Plant Sci 8(October). https://doi.org/10.3389/fpls.2017.01786

Yao L, Zhang Y, Liu C, Liu Y, Wang Y, Liang D, Liu J, Sahoo G, Kelliher T (2018) OsMATL mutation induces Haploid seed formation in Indica Rice. Nat Plants 4(8):530–533

Yoseph B, Mugo Stephen O, Sylvester O, Collins J, Mike O, M. Prasanna B. (2017) Hybrids performance of doubled Haploid lines derived from 10 tropical bi-parental maize populations evaluated in contrasting environments in Kenya. Afr J Biotechnol 16(8):371–379

Yu B, Liu P, Hong D, He Q, Yang G (2010) Improvement ofSclerotiniaresistance of a Polima CMS restorer line of rapeseed via phenotypic selection, marker-assisted background selection and microspore culture. Plant Breed = Zeitschrift Fur Pflanzenzuchtung 129(1):39–44

Yu X, Zhang X, Zhao P, Peng X, Chen H, Bleckmann A, Bazhenova A, Shi C, Dresselhaus T, Sun M-X (2021) Fertilized egg cells secrete Endopeptidases to avoid Polytubey. Nature 592(7854):433–437

Zenkteler M, Nitzsche W (1984) Wide hybridization experiments in cereals. Theor Appl Genet 68(4):311–315

Zhang J, Yin J, Luo J, Tang D, Zhu X, Wang J, Liu Z et al (2022a) Construction of homozygous diploid potato through maternal haploid induction. aBIOTECH 3(3):163–168

Zhang X, Zhang L, Zhang J, Jia M, Cao L, Jing Y, Zhao D (2022b) Haploid induction in Allotetraploid tobacco using DMPs mutation. Planta 255(5):98

Zhang X, Shi C, Li S, Zhang B, Luo P, Peng X, Zhao P, Dresselhaus T, Sun MX (2023a) A female in vivo Haploid-induction system via mutagenesis of egg cell specific peptidases. Mol Plant 16:471–480

Zhang Y, Maruyama D, Toda E, Kinoshita A, Okamoto T, Mitsuda N, Takasaki H, Ohme-Takagi M (2023b) Transcriptome analyses uncover reliance of endosperm gene expression on Arabidopsis embryonic development. FEBS Lett 597(3):407–417

Zhao X, Qin Y, Jia B, Kim S-M, Lee H-S, Eun M-Y, Kim K-M, Sohn J-K (2010) Comparison and analysis of Main effects, Epistatic effects, and QTL × environment interactions of QTLs for agronomic traits using DH and RILs populations in Rice. J Crop Sci Biotechnol 13(4):235–241

Zhao X, Xiaowei X, Xie H, Chen S, Jin W (2013) Fertilization and uniparental chromosome elimination during crosses with maize Haploid inducers. Plant Physiol 163(2):721–731

Zhao X, Yuan K, Liu Y, Zhang N, Yang L, Zhang Y, Wang Y et al (2022) In vivo maternal Haploid induction based on genome editing of DMP in brassica Oleracea. Plant Biotechnol J 20(12):2242–2244

Zheng MY, Konzak CF (1999) Effect of 2, 4-Dichlorophenoxyacetic acid on callus induction and plant regeneration in anther culture of wheat (Triticum Aestivum L.). Plant Cell Rep 19:69–73

Zhong Y, Liu C, Qi X, Jiao Y, Wang D, Wang Y, Liu Z et al (2019) Mutation of ZmDMP enhances Haploid induction in maize. Nat Plants 5(6):575–580

Zhong Y, Chen B, Li M, Wang D, Jiao Y, Qi X, Wang M et al (2020) A DMP-triggered in vivo maternal Haploid induction system in the dicotyledonous Arabidopsis. Nat Plants 6(5):466–472

Zhong Y, Chen B, Wang D, Zhu X, Li M, Zhang J, Chen M et al (2022a) In vivo maternal Haploid induction in tomato. Plant Biotechnol J 20(2):250–252

Zhong Y, Wang Y, Chen B, Liu J, Wang D, Li M, Qi X, Liu C, Boutilier K, Chen S (2022b) Establishment of a Dmp based maternal Haploid induction system for Polyploid brassica Napus and Nicotiana Tabacum. J Integr Plant Biol 64(6):1281–1294

Zhou H, Konzak CF (1989) Improvement of anther culture methods for Haploid production in wheat. Crop Sci 29(3):817–821

Zhu H, Li C, Gao C (2020) Applications of CRISPR-Cas in agriculture and plant biotechnology. Nat Rev Mol Cell Biol 21(11):661–677

Speed Breeding in Cereals: A Case for Africa

13

Kwame W. Shamuyarira and Maryke T. Labuschagne

Abstract

Plant breeding can play an important role in buffering crop production systems against the vagaries of climate by producing climate smart varieties that are more resilient to harsh environments while using limited resources efficiently. However, conventional breeding takes ~10–12 years to complete the entire breeding cycle (from the selection of parental genotypes to the registration of new varieties). These long generation times impede the ability of breeders to develop new adapted cultivars swiftly in response to farmers' immediate needs. The advent of speed breeding (SB) has opened opportunities to improve crops faster, especially in Africa, where there is a disproportionate burden to achieve food security. SB is a plant breeding tool that significantly reduces generation time by manipulating environmental conditions for the rapid advancement of the crop cycle. Growth is accelerated from the vegetative to the reproductive stage, allowing several plant generations to be grown in a single year. This allows rapid evaluation and selection of desirable traits. However, SB protocols have mainly been developed for long day (LD) crops with only a few protocols developed for short day (SD) crops. SD crops are the predominant food crops in Africa and SB can increase the efficiency of plant breeding pipelines and shorten the breeding cycle to ensure early variety release and more flexibility in responding to farmers' needs in Africa. Although clear benefits are achievable, integration of SB in Africa will require infrastructure development as well as skills training of plant breeders to translate them to tangible benefits that will contribute to sustainable food production.

K. W. Shamuyarira · M. T. Labuschagne (✉)
Department of Plant Sciences (Plant Breeding), University of the Free State, Bloemfontein, South Africa
e-mail: labuscm@ufs.ac.za

P. M. Priyadarshan, R. Ortiz (eds.), *Plant Breeding 2050*,
https://doi.org/10.1007/978-981-95-0583-8_13

Keywords

Accelerated breeding · Climate resilience · Crop-cycle · Short-day crops

13.1 Introduction

Crop improvement has been very successful and been a major driver of increasing food production across the globe. With environmental conditions becoming more unfavourable for crop production because of climate change, plant breeding will continue to play a role in increasing productivity. It is predicted that crops will grow under extreme temperatures and under unpredictable rainfall patterns, which will increase the vulnerability of most cropping systems. To overcome these challenges, developing new varieties offers a sustainable method to ensure food security. Unfortunately, genetics progress in breeding is hindered by the long generation times and the high cost of advancing genetics materials over years (Sharma et al. 2022). The conventional plant breeding process takes ~10–12 years (from the selection of parental genotypes to the registration of new varieties), which limits the ability of breeders to develop new adapted cultivars swiftly to respond to the changing climate.

Modern breeding technologies such as genetically modified organisms (GMOs) and gene editing have been proposed as possible solutions to overcome the long generation times of conventional breeding programs. However, GMOs and gene editing face regulatory resistance and negative views from most African consumers. They are viewed to have adverse health and environmental impacts that reduce the rate of their adoption (Qaim 2020). There are also concerns that genetically modified (GM) products are largely beneficial for large multinational seed companies that own the sole rights for producing these crops, which excludes participation in the seed industry from local players (Qaim 2020). Therefore, their value in the African context is diminished and there is need for a technology that is acceptable to the consumers. Speed breeding (SB), a novel and new method proposed by Watson et al. (2018), gives breeders the ability to reduce the length of the breeding cycle without the negative undertones that are associated with GMOs and gene editing.

13.2 Speed Breeding and Its Applications

Speed breeding is a plant breeding tool that significantly reduces generation time by manipulating environmental conditions for the rapid advancement of a crops' life cycle (Fig. 13.1). Growth is accelerated from the vegetative to the reproductive stage, allowing several plant generations to be grown in a single year. This allows rapid evaluation and selection of desirable traits (Wanga et al. 2021; Begna 2022). Researchers at the University of Queensland were the first to coin the term "speed breeding" in 2003, where they applied the technique in wheat breeding without any specialized facilities (Houston et al. 2020). Speed breeding uses external stimuli,

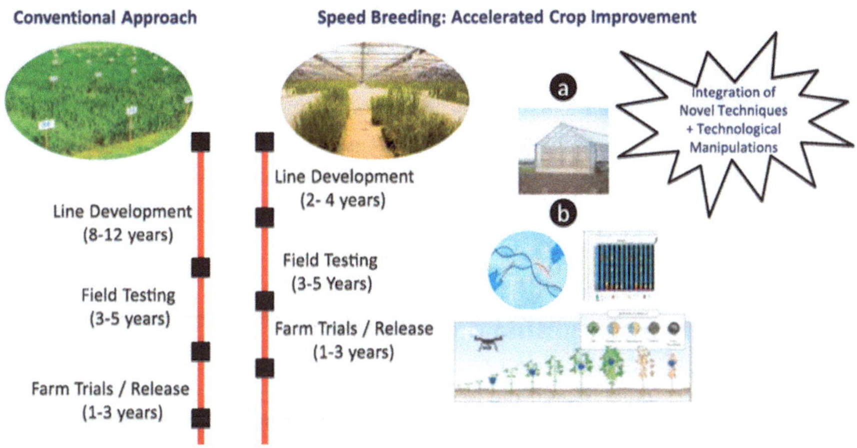

Fig. 13.1 The advantages of speed breeding vs conventional breeding in crops. a. Plants grown under controlled conditions. b. Use of high-throughput genotyping platforms (from Shanmugavel et al. 2023)

including mainly photoperiod and temperature manipulation, to hasten plant development and reduce generation times (Sharma et al. 2022). For successful SB, growing conditions must be manipulated, which include lighting, temperature, plant density, and the composition of the soil (Ghosh et al. 2018). Other plant breeding methods such as single-seed descent (SSD) was developed earlier for rapid inbreeding of populations before selection, thereby accelerating the breeding process (Cazzola et al. 2020). Likewise doubled haploid (DH) technology has been used extensively to accelerate breeding. This technology allows for the rapid development of homozygous lines (Liu et al. 2022). Both these techniques have limitations, for example the DH technology is very time consuming and requires a high level of skill, and the technology is strongly genotype dependent. Crops also often respond poorly to tissue culture procedures.

A further advantage of SB is that it allows the integration of other technologies to further decrease the breeding cycle. For instance, it is particularly well aligned with the principles of SSD and DH technology and can be applied in programs using these technologies. Combining DH and SB can significantly accelerate the breeding process (Potts et al. 2023). Other technologies such as marker-assisted selection (MAS) and genomics selection (GS) can also be integrated in SB and can be used to make accurate predictions and increase genetics gains through selection of parents early for the next breeding cycle (Ahmar et al. 2020). Gene pyramiding has been done in wheat for diseases like rust and *Fusarium* head blight, as well as for salinity tolerance and preharvest sprouting by combining SB and MAS (Hickey et al. 2019).

In countries where gene editing tools are embraced, SB can be combined with CRISPR-Cas9 (clustered regularly interspaced short palindromic repeat and their associated protein, Cas-9), which will allow integration of desirable traits at a faster pace (Varshney et al. 2021). In hybrid breeding, it offers more efficiency by hastening inbred line development (Raju and Sagar 2020) and can rapidly generate multiple generations allowing the creation and evaluation of multi-trait combinations in inbred lines. Speed breeding also has other applications such as the development of mapping populations, which has always been a tedious process (Tan and Gören 2023), and for the development of genetically modified (GM) crops. SB can speed up the GM process from gene insertion to ultimate field trials and commercial release (Jähne et al. 2020).

13.3 Influence of Lighting on Flowering

The light intensity under which the plants grow has an impact on the efficiency of photoperiod on plants (Adams and Langton 2005; Paradiso and Proietti 2022). Light affects the growth and development of crops by influencing the rate of photosynthesis and accumulation of assimilates. According to Paradiso and Proietti (2022), small variations in plant exposure to light can have a significant impact on the time of flowering. For SB protocols, extending the photoperiod by illuminating the plants with continuous light for 24 hours can significantly reduce the vegetative period and initiate early transition to the reproductive phase of LD plants (Sysoeva et al. 2010). However, it was found that the continuous presence of light was extremely strenuous and affected the normal development of plants. In the process of developing SB protocols, Watson et al. (2018) proposed a 2 h rest period to improve the overall health and development of crops.

Light quality that refers to the spectral composition of the light has an influence on the plant's growth and physiological responses (Sharma et al. 2022). Light does not only affect the photosynthetic efficiency of plants but also trigger specific signals in plants depending on the light spectrum and the light intensity (Paradiso and Proietti 2022). Controlling light spectrum was previously impossible until the use of LED (light emitting diode) lights on horticultural crops that are grown under controlled conditions was initiated. LED lights can produce a specific light spectrum that other light sources such as sodium vapour lamps (SVLs) cannot produce (Hitz et al. 2019). This has made it possible for researchers to manipulate the light intensity and spectrum to meet the requirements of both LD and SD plants to flower by using different combinations of LED lighting (SharathKumar et al. 2024). This is particularly useful for SD plants that require a shorter daylength to flower and where manipulation of the light spectrum will be more effective for flowering than the simple extension of the daylength.

LED lights are more efficient than other light sources like SVLs and are increasingly becoming cheaper, making them a cost-effective and environmentally friendly source of lighting. LED lights produce monochromatic light that allows scientists to study the effect of different wavelengths of light and their effect on photoperiodism in plants. For instance, the flowering time of many SD crops is influenced by far red, red, and blue light, while the ratio of red to far red light is effective in triggering flowering in some crops such as *Arabidopsis* (Shibuya and Kanayama 2014). For

LD plants, Ghosh et al. (2018) also stressed the importance of considering the different wavelengths in the light spectrum with a particular focus on far red, red, and blue wavelength as well as their respective ratios when initiating SB trials. Most LED modules are designed by manufacturers and can be regulated to fit the breeders' needs, but some modules can be made to specifications of the breeder (Jähne et al. 2020).

13.4 Influence of Other Environmental Factors on Speed Breeding

Environmental factors can influence floral induction individually, but each factor can change the flowering threshold for other factors due to strong interactions among them (Bernier et al. 1993). So, there are different pathways that can lead to the transition of plants to flowering. Manipulation of temperature, light quality and nutrition can cause some SD and LD plants of certain species to flower even under unfavourable photoperiods (Bernier et al. 1993). Among these factors, temperature has the greatest impact on flowering on most plants. Some temperate crops such as winter wheat, and some *Arabidopsis* genotypes do not flower until the vernalization requirement has been reached. Such ecotypes grown at high latitudes can be induced to flower earlier if they are grown under low temperature for extended periods (Reeves and Coupland 2000). For other plants to transition from one developmental stage to another, a threshold of heat units must be exceeded (Jagadish et al. 2016). Therefore, such plants will flower earlier under warmer growing conditions due to an acceleration of crop development (Craufurd and Wheeler 2009). Historical data on flowering plants show plants flowering up to 6 days earlier for every 1 °C increment in average ambient temperature (Jagadish et al. 2016).

Stress conditions such as heat, cold, and drought can modify gene expression by causing epigenetic changes in plants and cause early transition to flowering to ensure reproductive success (Ausin et al. 2005; Yaish et al. 2011). Production of viable seed when plants are induced to flower due to stress is a critical plant adaptation strategy by plants to ensure species survival (Takeno 2016). This form of adaptation has been observed under different experimental conditions in many plant species. For instance, in a glasshouse experiment consisting of two water regimes, Franks (2011) observed that *Brassica rapa* families generally flowered earlier under drought conditions as a drought escape mechanism. Also, SD flowering plants, *Pharbitis nil* and *Perilla frutescens* var. *crispa*, flower earlier if exposed to nutritional deficiencies and poor light intensity under controlled conditions (Wada et al. 2010). However, plants that are forced into floral induction by stress often produce fewer and shrivelled seeds but in most cases the progeny will germinate and develop normally until maturity (Watson et al. 2018). Therefore, developing SB protocols requires running experiments using different environmental factors applied at different levels to determine the optimum environment that will induce early flowering and maturity for rapid generation advance.

13.5 Current Status of Speed Breeding in the World

Current speed breeding protocols that have been developed rely mainly on manipulation of light and temperature with minimal variation of the other factors (Ghosh et al. 2018). A single protocol for all plant species is not possible as there is variation in response to photoperiod between and within species. Plants have been categorized into different groups based on the critical daylength needed to initiate flowering. There are long-day, short-day, and day-neutral plants. By manipulating the critical daylength period in the long- and short-day plants through SB, the flowering and subsequent breeding can be accelerated (Bhatta et al. 2021).

13.5.1 Speed Breeding for Long Day Crops

Extensive research has been done on LD plants (such as bread and durum wheat, barley, chickpea, and quinoa) with many protocols developed for the different crops. Ghosh et al. (2018) developed protocols that involved extending the photoperiod to 22 hours of continuous lighting, which stimulated crops to hasten plant development and reduce the growing cycle substantially. The authors also suggested including a 2 h dark period that would allow the plants to recover from rapid physiological development caused by the continued exposure to light. Most of the crops grew normally, which allowed phenotyping of some important traits, such as days to flowering. Based on the protocols developed for LD crops, both LEDs, SVLs, and other light sources can be used to induce early flowering. What is critical is to ensure that the light source produces light within the photosynthetically active radiation that is between 400 and 700 nm (Watson et al. 2018).

The photoperiod had the strongest influence on flowering, while other environmental conditions such as temperature and humidity had a smaller impact on flowering of plants. A 22 °C maximum temperature and a 17 °C minimum temperature were proven optimum for LD crops with the time allocated to the maximum and minimum having a small impact on plant development. Similarly, a humidity between 60 and 70% was proposed, with slight variations to be considered depending on crop adaptations. However, what is critical is to provide the growing plants with sufficient nutrients for development with suggestions to use slow-release fertilizers to avoid nutrient deficiencies during the early vegetative growth phases (Ghosh et al. 2018).

13.5.2 Speed Breeding of Short Day Crops

The continuous exposure to light shortens the time to reproduction for LD plants but for SD plants such as peanut and millets, only enhanced vegetative growth is observed (Sysoeva et al. 2010). SD plants require a light period that is less than the critical daylength to induce flowering (Sharma et al. 2022). However, some SD crops that are phytochrome-deficient initiate flowering earlier when daylength exceeds 12 h (Tsubokura et al. 2014; Jähne et al. 2020). Most tropical and subtropical crops are SD crops and require a photoperiod of less than 12 hours to induce flowering. These crops include rice, sorghum, soybean, cassava, peanut. and millets, although the level of daylength sensitivity may vary among crops and individual genotypes (SharathKumar et al. 2024).

When breeding for SD plants, the light quality and intensity becomes of great importance to reduce the time to flowering. SB protocols for SD plants have been developed by Jähne et al. (2020) for soybean, rice, and amaranth. The SB protocols are based on LED modules with different light quality and intensity, which were effective in reducing the flowering time of the three crops. However, the effect of different wavelengths of light on plant growth and development is genotype- and species-sensitive and optimization of SB protocols requires many experiments of long duration that take large amounts of space (Hitz et al. 2019). As an example, for soybean, Jähne et al. (2020) used small pots that had a volume of 392 cm^3 and in some instances used cell plates that had a volume of 32 cm^3. The small pots that were used allowed high plant densities and the constrained rooting area may have also contributed to early flowering and maturity. These protocols demonstrated that similar protocols could also be developed for SD crops such as sorghum and millet that are important for Africa. The manipulation of the light spectrum may be critical to find a breakthrough that will lead to the development of SB protocols for these crops. These crops are of local importance and are often referred to as orphan crops, since they do not receive research funding that is comparable to other SD crops that have high global commercial value, such as soybean. Interestingly, focus is beginning to grow on these crops with new funding programs recognizing the importance of these crops for food security and their unique capacity to offer resilience against climate vagaries.

13.6 Speed Breeding Status in Africa

13.6.1 Funding and Infrastructural Challenges in Africa

In Africa, the unavailability of sufficient funding to support infrastructure development to increase research capacity is a major drawback when trying to implement SB. There is limited spending on agricultural research in Africa. In 2014, over half of the US$2.5 billion spent on agricultural research in Africa were driven by only three countries namely South Africa, Nigeria, and Kenya (Beintema and Stads 2017). This leaves the rest of the African continent with very little funds directed to agricultural infrastructural development and capacity building to drive agricultural innovation. As a result, most African countries do not have the infrastructural capacity to carry out long term agricultural projects that require a constant source of funding to achieve project goals.

Several SB protocols have been developed and optimized for different crops, particularly temperate crops such as wheat, canola, and barley and a few subtropical crops such as chickpea (Ghosh et al. 2018; Watson et al. 2018). These protocols require complete control of environmental conditions in either growth chambers or green houses to manipulate light and temperature to induce plant transition from vegetative growth to flowering (Sharma et al. 2022). All these activities require a readily available source of power. However, successful implementation of SB may be reduced due to high energy needs that are required for continuous lighting and temperature control under greenhouse conditions (Sysoeva et al. 2010). Energy provision in most African countries is not dependable, including in the biggest economies such as South Africa and Nigeria, where there are significant electricity interruptions (Adenikinju 2005; Akpeji et al. 2020; Alemzero et al. 2021). Furthermore, the cost of electricity continues to increase and can reduce adoption of SB in institutions that would otherwise have the infrastructural capacity to implement SB. Investments in alternative forms of energy such as solar can help in ensuring continued availability of light for green houses or growth chambers. However, the initial cost of setting up solar panels and purchasing batteries may be prohibitively high and be beyond the reach of many institutions.

The total component cost (£707.39) of constructing a SB benchtop growth cabinet that was proposed by Ghosh et al. (2018) was relatively cheap and would be affordable for many breeding programs in Africa. Using LED lighting is more energy efficient and much cheaper compared to SVL and will thus be more ideal for cost saving. Use of alternative sources of electricity to supplement the national grid can circumvent the situation but has the added disadvantage to increase the cost of the implementation of SB. However, there are some institutions that have greater financial capacity such as universities, international research centers, and private seed companies. At such institutions, harvesting premature seeds for embryo rescue can be used to further shorten the breeding cycle. Where resources or infrastructure is limited, some universities can work jointly with other research institutions to tap into this infrastructural capacity and synergize capacity and expertise.

Most of the crops that have been evaluated to develop SB protocols elsewhere in the world are generally short growing compared to important cereal crops in Africa, such as sorghum and millets. The size of the plants may complicate the ability of growing these crops under controlled conditions, particularly in growth chambers and other confined environments. Jähne et al. (2020) noted that reducing the far-red light wavelengths in LED modules contributed to reducing plant height in soybean, which allowed SB protocols to be compatible to compact systems that can handle large plant numbers. Assessing the impact of far-red light on the elongation of sorghum and millet plants may open an opportunity to grow these crops in confined environments, which would facilitate multiple experimental combinations to optimize SB of these crops.

Ghosh et al. (2018) suggested the use of cheaper ways to construct SB cabinets that allow for light and humidity control that can be used for smaller projects. These cabinets will allow researchers to begin experiments that can run in parallel while imposing different treatments of light quality, daylength, and other environmental conditions. In most cases, however, it may be cheaper for African institutions to adopt developed protocols at present. However, limited experiments to optimize the developed protocols can be done in cases where resources and infrastructure are constrained.

13.6.2 Partners Who Could Be Involved

There should be an increase in the collaboration between universities, National Agricultural Research and Extension Services (NARES) and the Consortium of International Agricultural Research Centres (CGIARs) for speed breeding to be implemented effectively in Africa. In most African countries, NARES and local universities may not have the capacity to implement SB due to challenges in funding and infrastructural capacity. However, some richer universities (particularly in South Africa) and CGIARs have good infrastructure and often have better access to donor and government funding that can support the establishment and management of SB infrastructure. Furthermore, new funding opportunities on orphan crops such as the Vision for Adapted Crops and Soils (VACSs), which involve the United States Agency for International Development (USAID) and the International Maize and Wheat Improvement Center (CIMMYT) is an example of an opportunity to increase research capacity on orphan crops.

13.7 Conclusions

Conventional plant breeding is a tedious process requiring many years before commercial crop release. Speed breeding is a relatively new breeding technique where generation time is significantly reduced by manipulating environmental conditions such as light, for the rapid advancement of the crop cycle. With this technique several generations can be propagated in one year, making it easier for breeders to respond to farmers' immediate needs. It requires less resources and skills than other breeding techniques, which are used to accelerate breeding, such as double haploid breeding. Most speed breeding protocols developed so far have focused on long day crops. Speed breeding holds significant potential for accelerated crop development in Africa where lack of food security is a consistent threat. Most African staple crops are, however, short day crops, such as the millets. A concerted effort is needed to develop protocols for short day crops and to develop and fund infrastructure needed for speed breeding. With regional cooperation between universities, CGIAR centers, and national agricultural research centers, and adequate funding, this is possible.

References

Adams SR, Langton FA (2005) Photoperiod and plant growth: a review. J Hortic Sci Biotechnol 80:2–10. https://doi.org/10.1080/14620316.2005.11511882

Adenikinju A (2005, February) Analysis of the cost of infrastructure failures in a developing economy: the case of the electricity sector in Nigeria. AERC Research Paper 148 African Economic Research Consortium, Nairobi

Ahmar S, Gill RA, Jung KH, Faheem A, Qasim MU, Mubeen M, Zhou W (2020) Conventional and molecular techniques from simple breeding to speed breeding in crop plants: recent advances and future outlook. Int J Mol Sci 21:2590. https://doi.org/10.3390/ijms21072590

Akpeji KO, Olasoji AO, Gaunt CT, Oyedokun DT, Awodele KO, Folly KA (2020) Economic impact of electricity supply interruptions in South Africa. SAIEE Afr Res J 111:73–87

Alemzero DA, Sun H, Mohsin M, Iqbal N, Nadeem M, Vo XV (2021) Assessing energy security in Africa based on multi-dimensional approach of principal composite analysis. Environ Sci Pollut Res 28:2158–2171. https://doi.org/10.1007/s11356-020-10554-0

Ausin I, Alonso-Blanco C, Martinez-Zapater JM (2005) Environmental regulation of flowering. Int J Dev Biol 49:689–705. https://doi.org/10.1387/ijdb.052022ia

Begna T (2022) Speed breeding to accelerate crop improvement. Int J Agric Sci Food Technol 8:178–186. https://doi.org/10.17352/2455-815X.000161

Beintema N, Stads GJ (2017, April) A comprehensive overview of investments and human resource capacity in African agricultural research. Agricultural Science and Technology Indicators (ASTI) Synthesis Report, International Food Policy Research Institute (IFPRI), Washington, DC. www.asti.cgiar.org/sites/default/files/pdf/SSA-Synthesis-Report-2017.pdf

Bernier G, Havelange A, Houssa C, Petitjean A, Lejeune P (1993) Physiological signals that induce flowering. Plant Cell 5:1147. https://doi.org/10.1105/tpc.5.10.1147

Bhatta M, Sandro P, Smith MR, Delaney O, Voss-Fels KP, Gutierrez L et al (2021) Need for speed: manipulating plant growth to accelerate breeding cycles. Curr Opin Plant Biol 60:101986. https://doi.org/10.1016/j.pbi.2020.101986

Cazzola F, Bermejo CJ, Guindon MF, Cointry E (2020) Speed breeding in pea (*Pisum sativum* L.), an efficient and simple system to accelerate breeding programs. Euphytica 216:178–185. https://doi.org/10.1007/s10681-020-02715-6

Craufurd PQ, Wheeler TR (2009) Climate change and the flowering time of annual crops. J Exp Bot 60:2529–2539. https://doi.org/10.1093/jxb/erp196

Franks SJ (2011) Plasticity and evolution in drought avoidance and escape in the annual plant *Brassica rapa*. New Phytol 190:249–257. https://doi.org/10.1111/j.1469-8137.2010.03603.x

Ghosh S, Watson A, Gonzalez-Navarro OE, Ramirez-Gonzalez RH, Yanes L, Mendoza-Suárez M, Simmonds J, Wells R, Rayner T, Green P, Hafeez A (2018) Speed breeding in growth chambers and glasshouses for crop breeding and model plant research. Nat Protoc 13:2944–2963. https://doi.org/10.1038/s41596-018-0072-z

Hickey LT, Hafeez AN, Robinson H, Jackson SA, Leal-Bertioli SCM, Tester M et al (2019) Breeding crops to feed 10 billion. Nat Biotechnol 37:744–754. https://doi.org/10.1038/s41587-019-0152-9

Hitz T, Henke M, Graef-Hönninger S, Munz S (2019) Three-dimensional simulation of light spectrum and intensity within an LED growth chamber. Comput Electron Agric 156:540–548. https://doi.org/10.1016/j.compag.2018.11.043

Houston RD, Bean TP, Macqueen DJ, Gundappa MK, Jin YH, Jenkins TL, Selly SLC, Martin SAM, Stevens JR, Santos EM et al (2020) Harnessing genomics to fast-track genetic improvement in aquaculture. Nat Rev Genet 21:389–409. https://doi.org/10.1038/s41576-020-0227-y

Jagadish SK, Bahuguna RN, Djanaguiraman M, Gamuyao R, Prasad PV, Craufurd PQ (2016) Implications of high temperature and elevated CO_2 on flowering time in plants. Front Plant Sci 7:913. https://doi.org/10.3389/fpls.2016.00913

Jähne F, Hahn V, Würschum T, Leiser WL (2020) Speed breeding short-day crops by LED-controlled light schemes. Theor Appl Genet 133:2335–2342. https://doi.org/10.1007/s00122-020-03601-4

Liu K, He R, He X, Tan J, Chen Y, Li Y et al (2022) Speed breeding scheme of hot pepper through light environment modification. Sustainability 14:12225. https://doi.org/10.3390/su141912225

Paradiso R, Proietti S (2022) Light-quality manipulation to control plant growth and photomorphogenesis in greenhouse horticulture: the state of the art and the opportunities of modern LED systems. J Plant Growth Regul 41:742–780. https://doi.org/10.1007/s00344-021-10337-y

Potts J, Jangra S, Michael VN, Wu X (2023) Speed breeding for crop improvement and food security. Crops 3:276–291. https://doi.org/10.3390/crops3040025

Qaim M (2020) Role of new plant breeding technologies for food security and sustainable agricultural development. Appl Econ Perspect Policy 42:129–150. https://doi.org/10.1002/aepp.13044

Raju CSN, Sagar CK (2020) Speed breeding in agriculture future prospects. Int J Curr Microbiol Appl Sci 9:1059–1076. https://doi.org/10.20546/ijcmas.2020.912.128

Reeves PH, Coupland G (2000) Response of plant development to environment: control of flowering by daylength and temperature. Curr Opin Plant Biol 3:37–42. https://doi.org/10.1016/S1369-5266(99)00041-2

Shanmugavel P, Ramasamy G, Vellingiri G et al (2023) Speed breeding: a propitious technique for accelerated crop improvement. In: Case studies of breeding strategies in major plant species. IntechOpen. https://doi.org/10.5772/intechopen.105533

SharathKumar M, Luo J, Xi Y, van Ieperen W, Marcelis LF, Heuvelink E (2024) Sci Hortic 325:112657. https://doi.org/10.1016/j.scienta.2023.112657

Sharma S, Kumar A, Dhakte P, Raturi G, Vishwakarma G, Barbadikar KM, Das BK, Shivaraj SM, Sonah H, Deshmukh R (2022) Speed breeding opportunities and challenges for crop improvement. J Plant Growth Regul 42:46–59. https://doi.org/10.1007/s00344-021-10551-8

Shibuya T, Kanayama Y (2014) Flowering response to blue light and its molecular mechanisms in Arabidopsis and horticultural plants. Adv Hortic Sci 28:179–183. https://www.jstor.org/stable/24586832

Sysoeva MI, Markovskaya EF, Shibaeva TG (2010) Plants under continuous light: a review. Plant Stress 4:5–17

Takeno K (2016) Stress-induced flowering: the third category of flowering response. J Exp Bot 67:4925–4934. https://doi.org/10.1093/jxb/erw272

Tan U, Gören HK (2023) Speed breeding: an innovative approach for accelerated genetic improvement in agricultural crops and integrating with molecular-based approaches. In: Bobat A (ed) Pioneer and contemporary studies in agriculture, forest and water issues. Duvar Publishing, Izmir, pp 241–264

Tsubokura Y, Watanabe S, Xia Z, Kanamori H, Yamagata H, Kaga A, Katayose Y, Abe J, Ishimoto M, Harada K (2014) Natural variation in the genes responsible for maturity loci E1, E2, E3 and E4 in soybean. Ann Bot 113:429–441. https://doi.org/10.1093/aob/mct269

Varshney RK, Bohra A, Yu J, Graner A, Zhang Q, Sorrells ME (2021) Designing future crops: genomics-assisted breeding comes of age. Trends Plant Sci 26:631–649. https://doi.org/10.1016/j.tplants.2021.03.010

Wada KC, Kondo H, Takeno K (2010) Obligatory short-day plant, *Perilla frutescens* var. Crispa can flower in response to low-intensity light stress under long-day conditions. Physiol Plant 138:339–345. https://doi.org/10.1111/j.1399-3054.2009.01337.x

Wanga MA, Shimelis H, Mashilo J, Laing MD (2021) Opportunities and challenges of speed breeding: a review. Plant Breed 140:185–194. https://doi.org/10.1111/pbr.12909

Watson A, Ghosh S, Williams MJ, Cuddy WS, Simmonds J, Rey MD, Asyraf Md Hatta M, Hinchliffe A, Steed A, Reynolds D, Adamski NM (2018) Speed breeding is a powerful tool to accelerate crop research and breeding. Nat Plants 4:23–29. https://doi.org/10.1038/s41477-017-0083-8

Yaish MW, Colassanti J, Rothstein SJ (2011) The role of epigenetic processes in controlling flowering time in plants exposed to stress. J Exp Bot 62:3727–3735. https://doi.org/10.1093/jxb/err177

Computational Biology and Bioinformatics Tools and Databases for Next-Generation Plant Breeding

14

Saurav Kumar Mishra, Sneha Roy, Tabsum Chhetri, Anagha Balakrishnan, Kusum Gurung, and John J. Georrge

Abstract

Plant breeding is the study of creating new plant types with desired characteristics such as amplified yields, resistance, and adaptation to ecological stresses. Modern plant breeding employs computer tools and bioinformatics to facilitate the effective management, analysis, and integration of large-scale genetic and phenotypic data. Developments in genomics, transcriptomics, and technologies provide the computational frameworks within bioinformatics to organize and analyze this data, allowing breeders to look for genetic markers, predict outcomes of traits, and optimize breeding strategies. Hence, key resources include databases for storing and sharing genomics information and breeding management systems that may integrate private and public datasets. These provide insight into genetic diversity, trait inheritance, the environment to support marker-assisted selection, genomic prediction, and many more with the help of computational tools. Advanced algorithm-driven tools enable predictive modelling and support data-driven decisions, substantially increasing breeding strategies' accuracy. Therefore, this study aims to provide brief knowledge of computational tools and databases in plant breeding research and discuss the methods and advancements in this research, along with other essential information to get complete insight.

Keywords

Database · Crop · Genomics · Plant breeding · Tools · Plant

S. K. Mishra · S. Roy · T. Chhetri · A. Balakrishnan · K. Gurung · J. J. Georrge (✉)
Department of Bioinformatics, University of North Bengal, District-Darjeeling, West Bengal, India

P. M. Priyadarshan, R. Ortiz (eds.), *Plant Breeding 2050*,
https://doi.org/10.1007/978-981-95-0583-8_14

14.1 Introduction

This will be a significant stride for agricultural science in integrating bioinformatics into plant breeding, fundamentally changing the methods used in crop development and improvement. Modern plant breeding no longer relies on traditional historical methods such as the selection of plants merely based on observable features such as yield, resistance to diseases, and adaptability to ecological conditions. This kind of phenotypic selection worked in the long run, but it was slow in operation and dependent on natural variations within crop populations. In the latter part of the twentieth century, the whole method has been changed with the advent of molecular markers like RFLPs (Restriction Fragment Length Polymorphisms), SSRs (Simple Sequence Repeats), and now single-nucleotide polymorphism (SNP). These molecular approaches would enable breeders to select a plant based on genetic information rather than just by observed traits, thereby achieving much-increased speed and precision in the breeding of new plant varieties (Jung et al. 2021; Mu et al. 2022; Nair and Pandey 2024). Although genomics, transcriptomics, and proteomics technologies have further amplified what was developed in plant breeding, these technologies help determine specific genes associated with key agronomic traits. This capability opens new ways toward more targeted, efficient plant breeding strategies designed to conquer critical climate and food security challenges. CRISPR-Cas9 and many other genome-editing tools have emerged recently, enabling direct editing of genetic material in the plants being studied. A critical dimension of this capability is speeding the enhancement of desired traits within unprecedented accuracy scales and scopes, opening new channels for improving crop performance responses to global agricultural demands (Chen et al. 2019; Shen et al. 2022). Furthermore, the inescapable utility of bioinformatics, integrating biology with computer science and information technology to manage this sea of data that high-throughput genomic technologies are generating. The computational tools to properly analyze, store, and interpret such data have proven indispensable (Fig. 14.1). It helps researchers extract meaningful information from complex biological datasets by providing them with the right fundamental frameworks that would have been complex to analyze through normal methods. The three core areas in which bioinformatics is very important include genome assembly, gene annotation, and the discovery of genetic markers linked to significant agricultural characters. More sequence data can be obtained by developing new next-generation sequencing (NGS) technologies at lesser costs and much shorter periods than any earlier approach (Ibrahim et al. 2021; Roychoudhury 2020). Moreover, genomic selection (GS) would be the most effective use of bioinformatics in plant breeding, which aids in making the selection processes more efficient and accurate. Bioinformatics tools integrated with traditional breeding techniques would help breeders solve the most pressing global problems, such as agricultural sustainability and feeding the planet's increasing population during environmental changes (Sinha et al. 2023). Comprehensive databases have been established to host genomic sequences, gene annotations, and phenotypic data; this allows easy access to critical information to researchers while facilitating intercommunicative efforts across the scientific community (Fig. 14.1) (Chao et al. 2024;

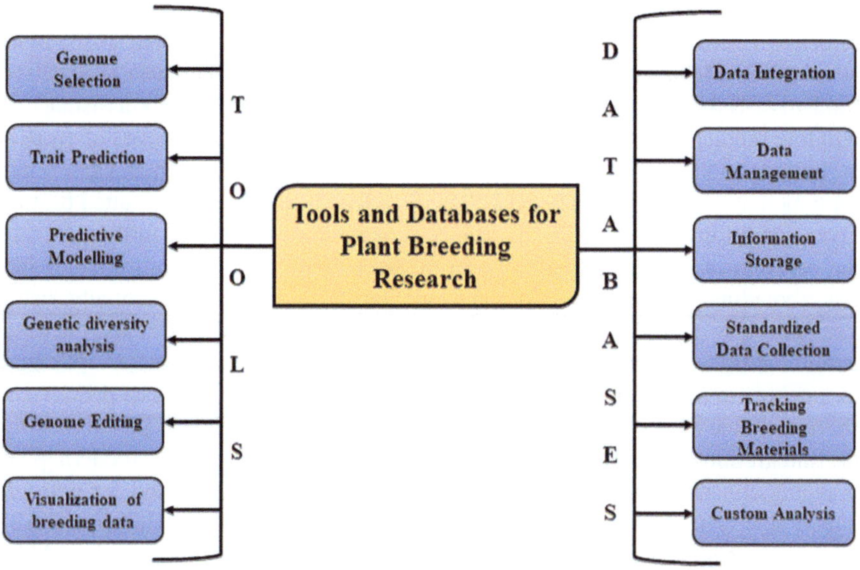

Fig. 14.1 Involvement of tools and databases in plant breeding

Mishra et al. 2024; Yoosefzadeh Najafabadi et al. 2023). Moreover, advanced algorithms and analytical techniques using computational biology accelerate the breeding process and improve its accuracy. Computational biology relies on big amounts of data from genomics, phenomics, and other environmental sources, equipping breeders with smart decisions through data analysis without relying purely on conventional methods. This helps identify the genetic markers of desirable traits and thus allows for marker-assisted selection and genomic prediction. One of the applications is genomic selection; it uses statistical models to probably compute their genetic information. With the combination of genomic and phenotypic data, the breeders can increase efficiency and accuracy in selection. It allows faster development cycles of new varieties, while making sure that the new varieties are equipped with the traits to enhance yield and resistance (Anilkumar et al. 2023; Degen and Müller 2023; Jeon et al. 2023; Vinjoda et al. 2024). Technologies involving artificial intelligence (AI) and machine learning (ML) can analyze big, complex data that are human-incomputable, discovering patterns hidden here and there. The AI-driven methods in conventional breeding practices improve selection and enhance the efficiency of breeding programs. With bioinformatics advancements, plant breeding is at the start of new challenges and opportunities. Multi-omics data that link genomics with transcriptomics and metabolomics give greater depth in understanding plant biology but demand advanced analytical instruments capable of large-scale dataset management. Emergent trends such as pan-genomics and metagenomics also open up the path of new possibilities for plant breeding, where genetic diversity and microbial interplay affect crop performance (Schreiber et al. 2024; Yoosefzadeh Najafabadi et al. 2023). Apart from that, the involvement of the application in examining the plants' associated data also leads to the potential for therapeutic selection (Detroja et al. 2024; Jogani et al. 2024; Johny and Mishra 2023; Maru et al. 2023; Mehta et al. 2023; Mishra and Georrge 2023; Vaghasia et al. 2024; Vakhariya Sakina et al. 2023).Moreover, another is bioinformatics assistance in building predictive models that can forecast the activity of new varieties of crops at various surroundings of the environment. Combining environmental factors into such models with omics data could allow breeders to select varieties suited best for given conditions of enormous importance for agriculture. This will further alleviate the problem of data handling and interpretation in integrating bioinformatics into plant breeding. Because genomics information is expanding exponentially, better strategies for storing, retrieving, and analyzing these data are also needed. Bioinformatics provides frameworks that streamline such processes, allowing researchers to access the relevant datasets more quickly. Although data privacy and sharing ethical issues are gaining relevance in bioinformatics-driven plant breeding, the integration of bioinformatics and computational biology into modern plant breeding enhances traditional practices while opening new pathways for research aimed at various channels of plant breeding (Heslot et al. 2012; Hu et al. 2018). Moreover, this chapter helps us explore the role of computational biology and bioinformatics in plant breeding and discusses its relevance in research and other related areas.

14.2 Databases Used for Plant Breeding Research

Plant breeding research relies on multiple databases, making it easy to manage, integrate, and analyze extensive genetic and phenotypic data. It is the critical resource from which breeders get data to enhance crop traits such as quality of nutrients, and many more. The complexity of plant breeding requires the use of complex data management systems that can manage large volumes of information generated through modern breeding techniques, including genomics, transcriptomics, and metabolomics analyzes (Price et al. 2020; Tan et al. 2022) One primary type of databases used in plant breeding is the breeding management system that allows breeders to save and analyze their own data with publically available datasets. Such systems enable the integration of various types of information, including pedigree records, performance data from field trials, and genetic marker information, into decision-making on which plants to cross or select for additional development. For example, breeders can use these databases to trace the pedigree of specific cultivars, monitor their performance in different environments and years, and trace breeding materials during the breeding process (Han et al. 2020; Jung et al. 2021). Furthermore, omics databases have emerged as essential tools in plant breeding, providing access to large-scale datasets that include genomic, transcriptomic, proteomic, and metabolomic information. Integrating several datasets provides a comprehensive view of the biological mechanisms that produce desirable features in plants. By analyzing multi-omics data, breeders find key genes associated with particular traits and devise predictive models for forecasting the performance of novel kinds under different environmental circumstances (Chao et al. 2024; Mahmood et al. 2022). Open-source platforms also had a substantial effect on research, and these platforms allow researchers from many different institutions to collaborate free of charge because they share data and tools. They also enable data formats and ontologies in a standard form, the most important need for successful integration and analysis of heterogeneous data from many sources. This generally enhances breeding program efficiency because collaboration means that knowledge and resources would be shared, and custom analysis would be offered for needs required in a given study. Moreover, specialized databases on specific crops or families will provide more targeted information when working with specific species for breeders (Luby et al. 2015; Morales et al. 2020).

For instance, databases of major crops- rice, maize, or wheat-hold a goldmine of phenotypic and genotypic material that can be exploited under marker-assisted selection breeding. Such crop-specific information usually provides a detailed listing of genetic diversity in its germplasm collections of the respective crop, allowing plant breeders to choose desired parental lines for crossing in the breeding program. Besides, modern breeding databases generally address the issue of interoperability of data. Most of these systems are meant to be compatible with other bioinformatics tools and frameworks, hence allowing for smooth data sharing between different platforms (Peng et al. 2020; Woodhouse et al. 2024). This is important to ensure that new research findings are quickly integrated into practical breeding applications and that efficient data storage is generated during breeding trials. Further, ML and AI are gradually being incorporated into plant breeding databases. These technologies help provide complex analyzes revealing hidden patterns in huge datasets and making better predictions about traits' inheritance and performance. A researcher can explore the interaction between genes and environmental influences on plant growth through algorithms from machine learning (Niazian and Niedbała 2020; Xu et al. 2022). Additionally, community-driven databases will ensure the democratization of access to breeding data by assuring that the industry community will publish freely accessible and open data. These platforms (Table 14.1) invite researchers from any part of the world and create a great repository where anyone interested in plant breeding can use the data generated by the researchers (Yu et al. 2021).

14.3 Tools Used for Plant Breeding Research

Plant breeding researchers employ innovative computational tools and approaches to accomplish, analyze, and understand massive datasets created throughout the breeding process. These technologies allow researchers to get valuable insights into complicated biological data related to crop attributes like yield, disease resistance, and stress tolerance. One of these computing resources' most critical foundational elements is the capability to process genomics data. Such high-throughput sequencing technologies have opened avenues through which scientists can produce enormous amounts of genomics information and comprehend the genetics nature of different plant species (Kim 2020; Medina-Lozano and Díaz 2022). Among this

Table 14.1 List of useful databases used in plant breeding research

Sl. No.	Name	URL	Reference
1.	AgBioData	https://www.agbiodata.org/	Harper et al. (2018)
2.	Metacrop	https://metacrop.ipk-gatersleben.de/apex/f?p=269:111	Schreiber et al. (2012)
3.	Breedbase	https://breedbase.org/	Morales et al. (2022)
4.	PlantCellMaker	https://www.tobaccodb.org/pcmdb/homePage	Jin et al. (2022)
5.	PlantEXP	https://biotec.njau.edu.cn/plantExp/	Liu et al. (2023a)
6.	PPRD	https://plantrnadb.com/	Yu et al. (2022)
7.	PMN	https://plantcyc.org/	Hawkins et al. (2021)
8.	Milletdb	http://milletdb.novogene.com/home	Sun et al. (2023)
9.	BnIR	https://yanglab.hzau.edu.cn/BnIR	Yang et al. (2023)
10.	AMIR	https://yanglab.hzau.edu.cn/AMIR/home/	Liu et al. (2024)

genomics information, vital information must be garnered about SNPs, insertions and deletions (INDELs), and structural dissimilarities that would describe how traits are inherited. Using leading algorithms in computation, scientists can infer genetic markers attributed to specific traits for more focused breeding and genome selection. Another important role these computational tools perform is integrating data from various types. Therefore, genotype–phenotype relationships and trait prediction are used to create models for selecting superior plants in breeding projects. High-throughput phenotyping technologies are also part of modern plant breeding efforts. These technologies involve automated systems and imaging techniques to measure a wide array of agronomic traits over large populations (Kim et al. 2020; Yan and Wang 2023). For instance, sensors can quickly and accurately capture data on plant height, area, and other responses. The resulting phenotypic data can be utilized with genomics data to enhance the precision of selection processes. Machine learning plays a great role in decoding breeding data. Sophisticated ML algorithms can recognize outlines and associations in huge datasets that are often not readily apparent to traditional statistical analysis. Using such historical data, model training can help researchers better predict how certain genotypes perform under given conditions. This enables easy prediction of outcomes involving complex traits controlled by many genes and environmental interactions (Crossa et al. 2021; Yan and Wang 2023). In addition to predictive modelling, these computational tools also help in marker-assisted selection by allowing the design and optimization of SNP panels. These SNP panels are very important for hereditary variety analysis in germplasm assemblies and for tracing alleles related with desirable traits from the breeding process. Using these markers wisely allows breeders to decide which plants to cross or select for further breeding. Most of the methodologies applied in plant breeding involve the usage of both forward and reverse genetics approaches. Forward genetics involves screening large populations to identify an individual with desirable characteristics based on prior knowledge of a genetic basis. Reverse genetics focuses on understanding gene function based on the manipulation of certain genes. Both approaches are supplemented by computational tools to examine the data, accelerating the identification of valuable genetics targets in crop improvement. Furthermore, some other technologies that have recently come into the forefront of plant breeding research are the emerging genome editing technologies. These will allow for precise changes in the specific genes in plant genomes, introducing or enhancing desired traits in the plant without the crossing method. Computational tools also aid in designing to guide RNAs for genome editing applications to ensure that targeted modifications are made efficiently. Moreover, good data management systems are essential for handling data generated throughout breeding programs. These systems provide a structured environment for storing, managing, and analyzing diverse datasets, including genomics sequences, phenotypic measurements, environmental conditions, and historical breeding records (Ahmar et al. 2020; Aklilu 2021; Guo et al. 2021). Integrating external genomics and phenotypes into breeding programs can help researchers understand genetic diversity and trait associations across different populations. This kind of collaboration enriches individual breeding programs and fosters innovation through shared knowledge and resources. In recent years, this has developed with integrative multi-omics approaches. Crop improvement strategies have come up strongly with genetic engineering and modification methods due to the integrated input coming from genomic transcriptomics, proteomic, and metabolomic data at every level of investigation; as a result, integration through holistic understanding has created possible more informed choices during traits selection and subsequent production. This implies that with technological advancement, computational tools (Table 14.2) will be significantly important in plant breeding (Hasan et al. 2021; Thumadath Palayullaparambil A Krishna et al. 2023).

14.4 High-Throughput Sequencing and Data Management in Plant Breeding

The population growth, coupled with the world's changing climate in recent years, has created soaring demand for food whilst putting tremendous pressure on the agricultural systems to invent more high-yielding and sustainable crops with good nutrition value (van Dijk et al. 2021). Plant breeding is one of the major factors that help achieve the demands and ensure food security. In the early days, plant breeding relied on the traditional methods wherein phenotype-based selection was used, which was not cost-efficient, often resulted in lots of trial and error, and was time-consuming. The process was effective; however, there were challenges, such as the plant genomes were diverse and had long breeding cycles accompanied by limited knowledge of the former (Nizamani et al. 2023). Modern-day plant breeding, however, is accompanied with high-throughput sequencing (HTS) technologies, which has various advantages and hold significant power over traditional plant breeding. These technologies have revolutionized the industry by making the process of deciphering plant genomes quick. This has reduced the cost of the process and shortened

Table 14.2 List of useful tools used in plant breeding research

Sl. No.	Name	URL	Reference
1.	MAKER	https://www.yandell-lab.org/software/maker.html	Cantarel et al. (2008)
2.	PLINK	https://zzz.bwh.harvard.edu/plink/	Purcell et al. (2007)
3.	TASSEL	https://www.maizegenetics.net/tassel	Bradbury et al. (2007)
4.	Phytozome	https://phytozome-next.jgi.doe.gov/	Goodstein et al. (2012)
5.	R/qtl	https://rqtl.org/	Broman et al. (2003)
6.	BrAPI	https://brapi.org/	Selby et al. (2019)
7.	ChromaX	https://chromax.readthedocs.io/en/latest/	Younis et al. (2023)
8.	MAGE	https://www.maizegdb.org/effect/maize/	Zhuo et al. (2024)
9.	PedigreeNet	https://www.maizegdb.org/breeders_toolbox	Braun et al. (2019)
10.	iPat	https://zzlab.net/iPat/	Chen and Zhang (2018)
11.	PERSUS	https://bioinformatics.cragenomica.es/perseus	Pradas et al. (2024)
12.	BSASeq	https://www.sciapps.org/?page_id=bsa	Wang et al. (2021)
13.	Tomato expression atlas	https://tea.solgenomics.net/	Fernandez-Pozo et al. (2017)
14.	MaizeGDB	https://www.maizegdb.org/	Lawrence et al. (2004)
15.	AlphaMPSim	https://vet.ed.ac.uk/roslin	Fernandez-Pozo et al. (2017)

the breeding period. HTS has aided in identifying genetics variations, linked them to phenotypic traits, and later incorporated this knowledge into breeding programs. HTS has led the process of plant breeding into a more developed route as it helps to elucidate the genetic information of the plant and help sequence the entire genome of the plant, identify genetic variants, and develop molecular markers that can be used to enhance the breeding process further (Kim et al. 2024; Krishna et al. 2023).

High-throughput sequencing, called next-generation sequencing (NGS), aids in rapidly sequencing bulk amount of fragments in a single run, allowing an inclusive view of the plant genome (Alekseyev et al. 2018). HTS has had its own evolution, and at present, HTS is a vital fundamental approach used for the plant breeding process (Sun et al. 2022). Plants are known to have complex genomes, which makes it difficult for them to map; however, with the HTS technique, such as whole genome sequencing (WGS), it is now easy to gather complete information about the plant's genome. WGS elucidates the genetic variation at the nucleotide level, including single-nucleotide polymorphism (SNP), insertions and deletions (INDELs), and many more, which enables a knowledge of the role of genetic factors (Schreiber et al. 2024; Xie et al. 2024). HTS can be utilized in GWAS data to recognize the association between various genetic variations of the known variation and helps identify alleles and loci connected to vital agricultural traits such as disease and climate resistance, enhanced nutrient value, etc. (Khan et al. 2021). Further, HTS has aided in genomic selection (GS) (Fig. 14.2), where genome-wide markers are identified and used for the prediction of the genetic value of a gene and the breeding value of the plant carrying that gene. Initially, it took several years to grow the plant and evaluate the trials in the field GS, making it easier for the genes to be screened and assessed based on the genotype. This has led not only to the reduction of breeding time but has allowed the rapid development of crops with the desired features (T. P. A. Krishna et al. 2023). Another important advancement is the plant's transcriptomic analysis, which evaluates its gene expression under different conditions, making it easier to understand a plant's response to various stresses, such as temperature fluctuations or attacks from some invasive species. It helps the selection of favorable genes and aids further in breeding as plants can be grown with the gene (Afzal et al. 2020). Plant epigenetics can be studied with HTS, emphasizing DNA methylation and other alterations that control gene expression. Methods such as bisulfite sequencing can be used to gain insight into how environmental conditions affect plant growth and stress adaptations. Compared to traditional methods, plant breeding can benefit from HTS since it can analyze and screen bulk data in a single day and reduce sequencing time. It also makes it possible to precisely identify genes

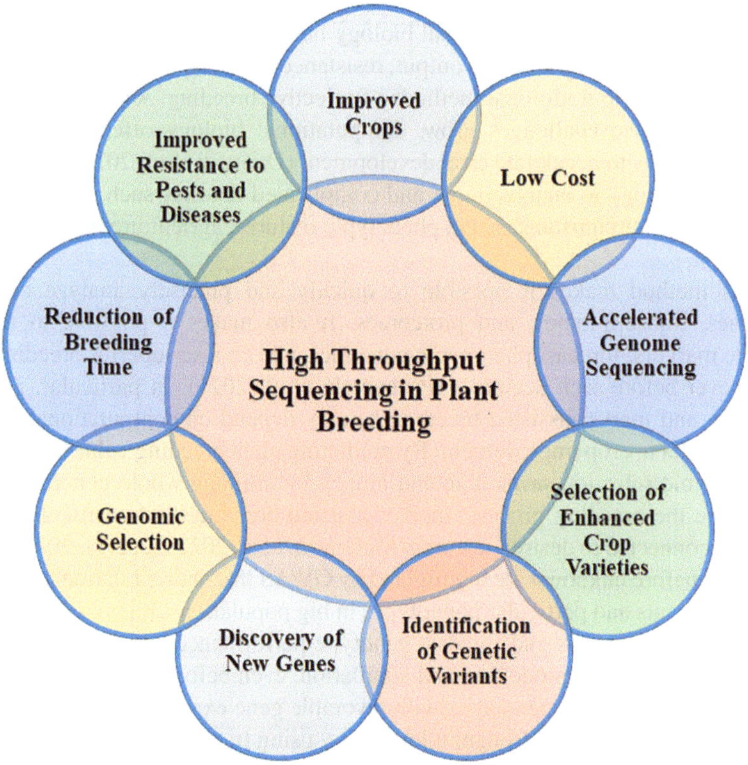

Fig. 14.2 Advantages of incorporating high throughput sequencing in plant breeding

and genetic variants, which results in finding novel genes, regulatory elements, and molecular markers (Fig. 14.2) (Dalakouras and Vlachostergios 2021; Lucibelli et al. 2022; Turcotte et al. 2022).

HTS has made data screening and processing rapid, and with this increase in data processing comes challenges such as data storage. The sheer volume of data is very high, and storing them has been challenging. Another important thing to consider is the data quality (Kim et al. 2020). HTS data usually have errors, which include low-quality reads and incorrect base calls. Therefore, it is essential to filter and correct the error before, which requires expertise in the former (Kayihan et al. 2023). Another major challenge is the integration and visualization of genomics information with the phenotypic data, such as the traits of the plant and environmental conditions, followed by the data sharing among the various platforms, making it accessible for all and standardization of the data for the different databases and tools. It is also important to maintain the security and privacy of the data and protect it from unauthorized access (Joshi and Kaushik 2021; Kim 2020).

14.5 Applications of Computational Biology in Crop Improvement

The advancements in computational biology have introduced numerous innovative methods to improve agricultural output, resistance, and adaptability, and they are no longer restricted to traditional methods of selective breeding. As food security and climate resilience challenges grow, computational biology offers precise, data-driven solutions to accelerate crop development (Dwivedi et al. 2020). Using computational biology to analyze large and complicated datasets such as genetic data, environmental circumstances, and phenotypic features, agriculture has undergone a revolution.

This method makes it possible to quickly and precisely analyze complete genomes, transcriptomes, and proteomes. It also makes it possible to identify genetic markers, forecast phenotypic outcomes, and choose superior breeding lines with never-before-seen accuracy (Pazhamala et al. 2021). In particular, genomic selection and marker-assisted breeding heavily depend on computational biology, which is vital to crop improvement. By predicting plant breeding value before planting, genomic selection saves time and money by utilizing whole-genome data. To accelerate the breeding process, marker-assisted breeding finds particular genetic markers connected to desired features (McGowan et al. 2021; Xu et al. 2021). Genes that were before unknown are highlighted by GWAS that shows relationships within genetic variants and particular phenotypes in big populations. It is possible to simulate breeding scenarios and forecast genotype performance under different conditions using predictive modelling and simulation, even before sowing a single seed. It is possible to choose genotypes with favorable gene expression profiles for traits like disease resistance or drought tolerance by using transcriptomics and functional genomics, which offer insights into gene expression under various situations. These applications are, especially helpful as genetically modified organisms (GMOs) are developed (Majumder 2024; Tibbs Cortes et al. 2021). Computational biology rapidly combines AI and ML methods to find patterns in large, complicated datasets. Artificial intelligence is revolutionizing genotype–phenotype mapping by accurately predicting phenotypic outcomes and decreasing the need for field experiments. Drones and high-throughput platforms are used in ML-based image processing for phenotyping to examine plant growth, disease symptoms, and nutrient deficits. The time needed for manual evaluations can be decreased by quickly selecting superior genotypes using this automated, high-precision procedure (Fig. 14.3). This strategy makes it possible to create fresh approaches to improving crop resilience, production, and nutritional value. It opens new avenues for crop science and plant breeding using data analytics and sequencing technologies (Khan et al. 2022; Xu et al. 2022; Yoosefzadeh Najafabadi et al. 2023). Integrating data from genomics, proteomics, transcriptomics, and metabolomics, systems biology makes it possible to uncover crucial genes and pathways involved in desirable traits using network-based techniques for trait discovery. Additionally, it permits multi-omics integration, combining information from several levels to understand how genetic variants affect phenotypic features. This enhances crop breeding efficiency and precision by enabling more precise predictions of complex traits (Yan and Wang 2023; Yang et al. 2021).

In agricultural development, comprehending how plants manufacture and control vital substances that affect growth, and other factors, by identifying the metabolic pathways that produce vital nutrients, metabolic profiling aids in identifying targets for biofortification, is important. Metabolic fingerprinting identifies biomarkers linked to stress tolerance to help breeding programs create resilient crops. When metabolic pathways are under stress, bottlenecks can be found using computational methods for metabolic flow analysis (Fernandez et al. 2021; Raza 2022; Razzaq et al. 2022). Protein–protein interaction (PPI) networks that show how

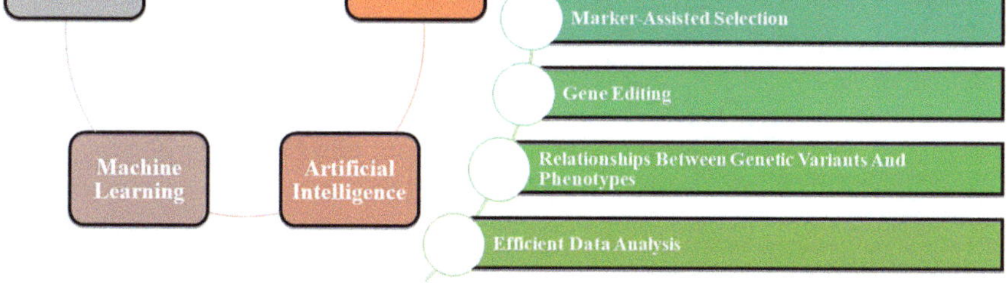

Fig. 14.3 An overview of the application of computational biology in crop development

plants react to environmental stressors and govern complex features are studied computationally in biology. With sophisticated models, it is possible to discover hub proteins essential for stress responses and can help farmers create more resilient crops (Thanasomboon et al. 2020). The field of computational biology is transforming agricultural development by offering instruments that expand, accelerate, and fine-tune plant breeding initiatives. However, because of the massive volumes of genomics data produced by contemporary breeding operations, data management is a major difficulty. To guarantee that data are readily shared and reusable throughout research organizations, initiatives such as the findable, accessible, interoperable, and reusable (FAIR) data principles are being embraced (Papoutsoglou et al. 2023). To fully utilizes computational biology in crop development, however, a significant roadblock is still the requirement for enhanced computational infrastructure knowledge. Despite these obstacles, computational biology is revolutionizing crop development by giving breeders an effective toolkit to address contemporary issues in agriculture.

14.6 Genomics Selection and Predictive Breeding Models

Genomic selection (GS) is a new predictive breeding strategy that employs genome-wide genetic data to determine the potential of individuals in breeding using their genomics estimated breeding values (GEBVs) (Yuan et al. 2018). The GS does not depend on phenotypic selection as in traditional breeding; instead, it uses dense genetic markers, such as SNPs, to predict complicated qualities like yield, disease resistance, and drought tolerance (Budhlakoti et al. 2022). Bioinformatics is an integral part of GS in plants, enabling breeders to handle bulk genetic data, identify important genetic markers, and make accurate forecasts about plant performance.

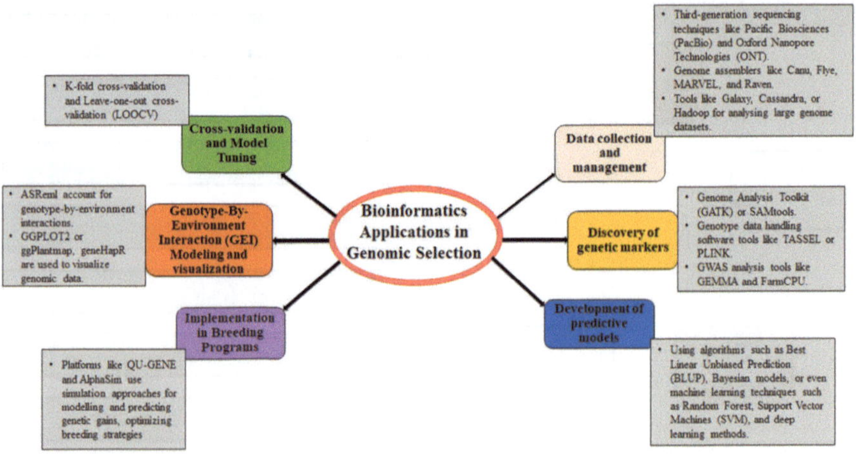

Fig. 14.4 Applications of bioinformatics in genomic selection (GS)

Figure 14.4 shows the applications of bioinformatics in genomic selection. Recently, accelerated enhancement in ML and computational modelling have greatly improved the accuracy of GEBV prediction, particularly based on multi-omics data, including genomes and transcriptomics, and accounting for genotype-environment relationships. Different predictive breeding models, including linear mixed models (LMMs), Bayesian models, and others are used to train datasets and forecast the best prospects for breeding (Chen et al. 2023; Chung and Liao 2022). This enables quicker breeding cycles, increased genetic gain, and a potent tool for managing crop and livestock traits in response to environmental difficulties or to produce increasingly needed food crops. We provide a variety of bioinformatics apps for GS.

14.6.1 Data Collection and Management

Genome sequencing and assembly aided in evaluating intraspecific polymorphism and identifying major genes and molecular markers involved in developing desired traits (Dmitriev et al. 2022). High-throughput sequencing technologies like third-generation sequencing techniques generate raw sequencing data. Pacific Biosciences (PacBio) and Oxford Nanopore Technologies (ONT) are two prominent sequencing techniques for providing long reads and are important for resolving complex genomic regions. PacBio employs single molecule real-time (SMRT) sequencing that can produce highly accurate long reads with high consensus accuracy, but it is usually more expensive and takes longer to run. In ONT, a nanopore technology is used, which allows the DNA strands to pass through nanopores and detect changes in electrical current, providing real-time sequencing with longer reads than PacBio and is portable and more affordable (Athanasopoulou et al. 2021; De Coster et al. 2021; van Dijk et al. 2018). Assembling the sequenced genomes allows a complete representation of a plant's genetic material. This is critical in understanding genetic diversity and the functional roles of various genes. Currently, various genome assemblers are employed, including Canu, Flye, MARVEL, Raven, etc. to assemble plant genomes with better results (Kolmogorov et al. 2019; Koren et al. 2017; Nowoshilow et al. 2018; Vaser and Sikic 2021). Thus, third-generation sequencing is critical for producing high-quality genotypic data, which drives genomic selection.

High-throughput sequencing and genotyping generate big data, which implies that the storing, managing, and processing demands are sophisticated. Because of this, bioinformatics tools like Galaxy, Cassandra, or Hadoop frameworks allows breeders to handle and analyze large genomic datasets (Ko et al. 2020; Lakshman and Malik 2010; Ngo et al. 2018; Ostrovsky et al. 2021). Modern bioinformatics enables different types of omics data, which must be merged to understand how features have such complexity. For example Sariel et al. have used RNA sequencing data to identify drought- and stress-associated genes in barley from different habitats (Hubner et al. 2015). Various bioinformatics tools handle massive amounts of phenotypic data generated in field trials (Araus and Cairns 2014). Minguo et al. developed an automated platform called MAIZTRO for measuring ear phenotypes in maize. This platform aids in identifying new traits associated with stable yield (Liu et al. 2023b).

14.6.2 Discovery of Genetic Markers

SNPs are the commonly used kinds of markers for genomic selection. High throughput sequencing data necessitates SNP calling, which is accomplished using bioinformatics software such as Genome Analysis Toolkit (GATK) or SAMtool. These technologies establish genetic variation in a large population and may be used to determine the SNPs connected with the attributes of interest (Altmann et al. 2012; P. J. Bradbury et al. 2007; Li et al. 2009; Pereira et al. 2020; Zhou et al. 2024). Further, genotyping-by-sequencing (GBS) techniques generate large datasets that are subsequently processed by bioinformatics pipelines (Tripodi 2023). Genome data handling, filtering, and quality control are often done using genotype data handling software tools like TASSEL or PLINK (P. J. Bradbury et al. 2007; Chang et al. 2015). Bioinformatics pipelines annotate genetic variants and describe which SNPs contribute to the phenotypic changes in plants, including SnpEff or ANNOVAR (Cingolani et al. 2012; Wang et al. 2010). GWAS uncover associations among genetic dissimilarities and phenotypic features like disease susceptibility, crop productivity, drought tolerance, and so on (Wang and Qin 2017). Bioinformatics tools, such as GEMMA and FarmCPU are used to analyze GWAS and identify SNPs statistically correlated with traits (Clauw et al. 2024; Tibbs Cortes et al. 2021).

14.6.3 Developing Predictive Models

Genomics estimated breeding values (GEBVs) are the predictions of potential genetic makeup for qualities such as yield, disease resistance, and so on based on genome-wide genetic marker data. This estimates how well an individual will likely pass on desirable traits (e.g. yield, disease resistance) to the next generation (Alemu et al. 2024; Miedaner et al. 2020). The GEBVs are computed with predictive models that depend on a training population with known genotypes and phenotypes. A training population is established using the genotypic and phenotypic records for the qualities under consideration. The markers were then used to create predictive models for their association with the phenotypic traits. The association between these genetic markers and attributes can be predicted using algorithms like Best Linear Unbiased Prediction (BLUP), Bayesian models, ML techniques like RF, SVM, and DL methods. (Alves et al. 2019; Gianola et al. 2018; Ogutu et al. 2011; Ren et al. 2021; Xu et al. 2024). With these, once these models are fitted and trained, they could make predictions for the GEBVs on the breeding population having genotypic but no phenotypic information for ranking the best individual with genetic potential. This remarkably reduces the demand for extensive phenotypic tests, accelerates cycles in breeding programs, and improves the precision of selection for complicated traits regulated by multiple genes. Altogether, this will direct more efficient and targeted breeding strategies.

14.6.4 Cross-Validation and Model Tuning

Cross-validation (CV) determines how well a model generalizes toward unseen data, allowing the breeder to evaluate a model's stability and predictive ability. Common methods of CV include K-fold cross-validation and Leave-one-out cross-validation (LOOCV). In K-fold, the data will be separated into folds, with one as a test others as a training set. This process is performed k times, and the average estimate accuracy is computed (Kristensen et al. 2018). This approach ensures the model is validated on multiple datasets, preventing overfitting. LOOCV is an instance of k-fold, where k equals the total number of samples; each individual is left out once and used as the test set. This method is computationally expensive but can provide more accurate results for small datasets (Cericola et al. 2017). Thus, cross-validation techniques are used in bioinformatics pipelines to assess the activity of genomic selection models. The most common one is the k-fold cross-validation technique, aiding in assessing the accuracy of prediction and developing models toward robust predictions for unseen populations (Yoosefzadeh Najafabadi et al. 2023). Bioinformatics tools such as GridSearchCV from machine learning libraries are used to search for the best model configuration (Sandhu et al. 2022).

14.6.5 Genotype-by-Environment Interaction (GEI) Modelling and Visualization Tools

Genotype-by-environment interaction modelling aids in identifying genotypes with wider stability and yield under different environmental conditions. GEI accounts for the differential performance of a genotype to different ecological conditions (Bhartiya et al. 2018). Two main methodological approaches facilitate the elucidation of GEI patterns, that is the regression-based models and advanced multivariate stability prediction approaches, including AMMI and GGE biplot analysis (Gupta et al. 2023; Hudson et al. 2022; Khare et al. 2024). Tools like ASReml account for genotype-by-environment interactions, allowing breeders to select plants that will perform well across diverse environments (Rebollo et al. 2023). Tools like R, Python, and specialized bioinformatics packages like GGPLOT2 or ggPlantmap, geneHapR are used to visualize genomics data, marker distributions, etc. (Cao et al. 2023; Jo and Kajala 2023; Zhang et al. 2023). Breeders must make informed decisions based on the graphical outputs of genetic data and predictions. Bioinformatics tools also draw Manhattan plots and heatmaps depicting genomic relationships and breeding values if SNP-trait associations are significant in GWAS (Narawatthana et al. 2023).

14.6.6 Implementation of Breeding Programs

Decision support systems (DSSs) are key tools in modern plant breeding programs that make data-based decisions, integrating genomics and phenomics levels with the environment (Gunawan et al. 2024). DSSs can provide a detailed assessment of potential crops by considering any criteria, such as suitability for any climate, market demand, soil fertility, and sustainability measures. It goes much beyond the conventional single-factor approaches because the system considers a range of criteria at any one time, providing a more holistic view of complex issues to stakeholders (Barons et al. 2024; Chang-Brahim et al. 2024). Platforms like QU-GENE and AlphaSim use simulation approaches to model and predict genetic gains, optimize breeding strategies, and evaluate different selection methods (Benčič et al. 2024; Podlich and Cooper 1998). QU-GENE is an application platform that simulates genetic architecture and breeding processes. It helps breeders develop models for inheriting various traits and test their breeding strategies in the computer before implementation. AlphaSim supplements this by simulating the whole process of the breeding pipeline from the generation of founder populations to the final selection of superior varieties. The model can simulate complex breeding scenarios, including genome-wide selection, marker-assisted selection, and environmental effects (Gaynor et al. 2021; Merrick et al. 2022). These tools, in tandem with genomic selection platforms, such as genomic prediction models, can refine the breeder's programs to be more efficient, more accurate, and quicker at achieving genetic gain, all while saving time and money when developing new plant varieties.

14.7 Challenges and Limitations

Bioinformatics has improved plant breeding by analysing genetic data to discover critical markers for traits, hence minimising the need for many field experiments. However, various problems prevent its full implementation. Plant genomes are vast, repetitive, and complicated, making data handling challenging. Additionally, diverse genomics datasets (e.g. environmental, transcriptomics, and phenomics) require advanced computational methods for reliable integration and analysis. Poor genome annotation limits the discovery of biological functions, restricting bioinformatics' potential. Data standardization is also lacking, with research groups storing data inconsistently, complicating sharing and integration. Many commercial breeders keep key datasets proprietary, slowing research progress. Public genome databases often provide limited, low-quality data, restricting researchers' access to vital information. Bioinformatics demands high-performance computing and cloud infrastructure that are expensive to maintain and operate. While sequencing costs have reduced, they remain exorbitant for large-scale research, especially for complex plant genomes. Developing countries struggle with limited infrastructure and funding, creating inequalities in access to bioinformatics tools and technologies. A shortage of skilled professionals is another challenge, as traditional plant breeders often lack the expertise to work with advanced computational tools. Intellectual property concerns also arise, with genetic discoveries requiring a balance between protection and fair access to data. Ultimately, bioinformatics in plant breeding holds great potential, but infrastructure gaps, funding limitations, and skill shortages, particularly in developing regions, must be addressed to ensure equitable progress and food security.

14.8 Future Perspectives

In the future, advancements in bioinformatics, computational approaches, and emerging trends like ML and AI will significantly enhance plant breeding. ML algorithms analyze large omics datasets to identify novel genes or biomarkers linked to disease resistance, higher yields, and climate resilience. AI is key in processing complex data, predicting plant performance, selecting ideal breeding candidates, and reducing experimental time, costs, and labour. The integration of precision agriculture techniques will further transform the industry. When merged with bioinformatics, tools and sensors providing real-time data on soil, plant growth, and environmental conditions can guide genetic modifications tailored to specific environments, resulting in better-quality, climate-adaptive crops. AI-based breeding tools can use real-world farm data to develop sustainable crop varieties optimized for specific regions. Techniques like soil mapping and genetic insights will enable precise crop improvement. Genome editing technologies like CRISPR allow breeders to modify or remove genes associated with vulnerabilities like drought susceptibility, pests, or diseases. Bioinformatics enables targeted gene editing, boosting crop resilience and productivity to support sustainable agriculture and global food security.

14.9 Conclusions

Computational biology and bioinformatics have accelerated plant breeding quickly and accurately. Features like genomic selection, multi-omics analysis, and AI modelling enable breeders to breed crops with higher yield, disease resistance, and climate change tolerance traits. This becomes more important with growing food demand and limited farming land availability. Bioinformatics will help develop stronger and more productive crops by organizing and analyzing large amounts of genetic data, thus supporting sustainable agriculture. Bioinformatics will play an even bigger role in securing global food supplies and supporting innovative farming solutions as it evolves.

Acknowledgement The authors acknowledge the Department of Bioinformatics, University of North Bengal, India, for providing the facility to complete this work.

Author Contribution *SKM* and *JJG* conceptualized and designed the study. *SKM* collected data and wrote the manuscript. *SR*, *TC*, *AB*, and *KG* conducted formal analysis and helped prepare the original draft. *JJG* performed critical analysis, reviewed, and revised the final manuscript.

Consent for Publication Not applicable.

Conflicts of Interest Not applicable.

Funding None.

References

Afzal M, Alghamdi SS, Migdadi HH, Khan MA, Nurmansyah, Mirza SB, El-Harty E (2020) Legume genomics and transcriptomics: from classic breeding to modern technologies. Saudi J Biol Sci 27(1):543–555. https://doi.org/10.1016/j.sjbs.2019.11.018

Ahmar S, Saeed S, Khan MHU, Ullah Khan S, Mora-Poblete F, Kamran M, Faheem A, Maqsood A, Rauf M, Saleem S (2020) A revolution toward gene-editing technology and its application to crop improvement. Int J Mol Sci 21(16):5665

Aklilu E (2021) Review on forward and reverse genetics in plant breeding. All Life 14(1):127–135

Alekseyev YO, Fazeli R, Yang S, Basran R, Maher T, Miller NS, Remick D (2018) A next-generation sequencing primer-how does it work and what can it do? Acad Pathol 5:2374289518766521. https://doi.org/10.1177/2374289518766521

Alemu A, Astrand J, Montesinos-Lopez OA, Isidro YSJ, Fernandez-Gonzalez J, Tadesse W, Vetukuri RR, Carlsson AS, Ceplitis A, Crossa J, Ortiz R, Chawade A (2024) Genomic selection in plant breeding: key factors shaping two decades of progress. Mol Plant 17(4):552–578. https://doi.org/10.1016/j.molp.2024.03.007

Altmann A, Weber P, Bader D, Preuß M, Binder EB, Müller-Myhsok B (2012) A beginners guide to SNP calling from high-throughput DNA-sequencing data. Hum Genet 131:1541–1554

Alves FC, Granato ISC, Galli G, Lyra DH, Fritsche-Neto R, de Los Campos G (2019) Bayesian analysis and prediction of hybrid performance. Plant Methods 15:14. https://doi.org/10.1186/s13007-019-0388-x

Anilkumar C, Muhammed Azharudheen T, Sah RP, Sunitha NC, Devanna BN, Marndi BC, Patra BC (2023) Gene based markers improve precision of genome-wide association studies and accuracy of genomic predictions in rice breeding. Heredity 130(5):335–345

Araus JL, Cairns JE (2014) Field high-throughput phenotyping: the new crop breeding frontier. Trends Plant Sci 19(1):52–61. https://doi.org/10.1016/j.tplants.2013.09.008

Athanasopoulou K, Boti MA, Adamopoulos PG, Skourou PC, Scorilas A (2021) Third-generation sequencing: the spearhead towards the radical transformation of modern genomics. Life (Basel) 12(1). https://doi.org/10.3390/life12010030

Bančič J, Greenspoon P, Gaynor RC, Gorjanc G (2024) Plant breeding simulations with AlphaSimR. Crop Sci

Barons MJ, Walsh LE, Salakpi EE, Nichols L (2024) A decision support system for sustainable agriculture and food loss reduction under uncertain agricultural policy frameworks. Agriculture 14(3):458

Bhartiya A, Aditya J, Kumari V, Kishore N, Purwar J, Agrawal A, Kant L, Pattanayak A (2018) Stability analysis of soybean [Glycine max (L.) Merrill] genotypes under multi-environments rainfed condition of North Western Himalayan hills. Indian J Genet Plant Breed 78(03):342–347

Bradbury PJ, Zhang Z, Kroon DE, Casstevens TM, Ramdoss Y, Buckler ES (2007) TASSEL: software for association mapping of complex traits in diverse samples. Bioinformatics 23(19):2633–2635. https://doi.org/10.1093/bioinformatics/btm308

Braun BL, Schott DA, Portwood JL, Andorf CM, Sen TZ (2019) PedigreeNet: a web-based pedigree viewer for biological databases. Bioinformatics 35(20):4184–4186

Broman KW, Wu H, Sen S, Churchill GA (2003) R/qtl: QTL mapping in experimental crosses. Bioinformatics 19(7):889–890. https://doi.org/10.1093/bioinformatics/btg112

Budhlakoti N, Kushwaha AK, Rai A, Chaturvedi KK, Kumar A, Pradhan AK, Kumar U, Kumar RR, Juliana P, Mishra DC, Kumar S (2022) Genomic selection: a tool for accelerating the efficiency of molecular breeding for development of climate-resilient crops. Front Genet 13:832153. https://doi.org/10.3389/fgene.2022.832153

Cantarel BL, Korf I, Robb SM, Parra G, Ross E, Moore B, Holt C, Alvarado AS, Yandell M (2008) MAKER: an easy-to-use annotation pipeline designed for emerging model organism genomes. Genome Res 18(1):188–196

Cao T, Li Q, Huang Y, Li A (2023) plotnineSeqSuite: a python package for visualizing sequence data using ggplot2 style. BMC Genomics 24(1):585. https://doi.org/10.1186/s12864-023-09677-8

Cericola F, Jahoor A, Orabi J, Andersen JR, Janss LL, Jensen J (2017) Optimizing training population size and genotyping strategy for genomic prediction using association study results and pedigree information. A case of study in advanced wheat breeding lines. PLoS One 12(1):e0169606

Chang CC, Chow CC, Tellier LC, Vattikuti S, Purcell SM, Lee JJ (2015) Second-generation PLINK: rising to the challenge of larger and richer datasets. Gigascience 4:7. https://doi.org/10.1186/s13742-015-0047-8

Chang-Brahim I, Koppensteiner LJ, Beltrame L, Bodner G, Saranti A, Salzinger J, Fanta-Jende P, Sulzbachner C, Bruckmüller F, Trognitz F (2024) Reviewing the essential roles of remote phenotyping, GWAS and explainable AI in practical marker-assisted selection for drought-tolerant winter wheat breeding. Front Plant Sci 15:1319938

Chao H, Zhang S, Hu Y, Ni Q, Xin S, Zhao L, Ivanisenko VA, Orlov YL, Chen M (2024) Integrating omics databases for enhanced crop breeding. J Integr Bioinform 20(4):20230012

Chen CJ, Zhang Z (2018) iPat: intelligent prediction and association tool for genomic research. Bioinformatics 34(11):1925–1927

Chen K, Wang Y, Zhang R, Zhang H, Gao C (2019) CRISPR/Cas genome editing and precision plant breeding in agriculture. Annu Rev Plant Biol 70(1):667–697

Chen S-P, Tung C-W, Wang P-H, Liao C-T (2023) A statistical package for evaluation of hybrid performance in plant breeding via genomic selection. Sci Rep 13(1):12204

Chung P-Y, Liao C-T (2022) Selection of parental lines for plant breeding via genomic prediction. Front Plant Sci 13:934767

Cingolani P, Platts A, Wang LL, Coon M, Nguyen T, Wang L, Land SJ, Lu X, Ruden DM (2012) A program for annotating and predicting the effects of single nucleotide polymorphisms, SnpEff: SNPs in the genome of Drosophila melanogaster strain w1118; iso-2; iso-3. fly 6(2):80–92

Clauw P, Ellis TJ, Liu HJ, Sasaki E (2024) Beyond the standard GWAS - a guide for plant biologists. Plant Cell Physiol 66. https://doi.org/10.1093/pcp/pcae079

Crossa J, Fritsche-Neto R, Montesinos-Lopez OA, Costa-Neto G, Dreisigacker S, Montesinos-Lopez A, Bentley AR (2021) The modern plant breeding triangle: optimizing the use of genomics, phenomics, and enviromics data. Front Plant Sci 12:651480

Dalakouras A, Vlachostergios D (2021) Epigenetic approaches to crop breeding: current status and perspectives. J Exp Bot 72(15):5356–5371

De Coster W, Weissensteiner MH, Sedlazeck FJ (2021) Towards population-scale long-read sequencing. Nat Rev Genet 22(9):572–587. https://doi.org/10.1038/s41576-021-00367-3

Degen B, Müller NA (2023) A simulation study comparing advanced marker-assisted selection with genomic selection in tree breeding programs. G3 13(10):jkad164

Detroja K, Sharma K, Mishra SK, Georrge JJ (2024) Phytochemical profiling of Tamarindus indica: a medicinal plant for snakebite. In: Herbal formulations, phytochemistry and pharmacognosy. Elsevier, pp 413–421

Dmitriev AA, Pushkova EN, Melnikova NV (2022) Plant genome sequencing: modern technologies and novel opportunities for breeding. Mol Biol (Mosk) 56(4):531–545. https://doi.org/10.31857/S0026898422040048

Dwivedi SL, Goldman I, Ceccarelli S, Ortiz R (2020) Advanced analytics, phenomics and biotechnology approaches to enhance genetic gains in plant breeding. Adv Agron 162:89–142

Fernandez O, Millet EJ, Rincent R, Prigent S, Pétriacq P, Gibon Y (2021) Plant metabolomics and breeding. In: Advances in botanical research, vol 98. Elsevier, pp 207–235

Fernandez-Pozo N, Zheng Y, Snyder SI, Nicolas P, Shinozaki Y, Fei Z, Catala C, Giovannoni JJ, Rose JK, Mueller LA (2017) The tomato expression atlas. Bioinformatics 33(15):2397–2398

Gaynor RC, Gorjanc G, Hickey JM (2021) AlphaSimR: an R package for breeding program simulations. G3 11(2):jkaa017

Gianola D, Cecchinato A, Naya H, Schon CC (2018) Prediction of complex traits: robust alternatives to best linear unbiased prediction. Front Genet 9:195. https://doi.org/10.3389/fgene.2018.00195

Goodstein DM, Shu S, Howson R, Neupane R, Hayes RD, Fazo J, Mitros T, Dirks W, Hellsten U, Putnam N (2012) Phytozome: a comparative platform for green plant genomics. Nucleic Acids Res 40(D1):D1178–D1186

Gunawan MI, Sitopu JW, Sechan G, Gunawan I (2024) Optimizing crop selection: a multi-criteria decision support system for sustainable agriculture. Int J Enterprise Model 18(3):113–123

Guo Z, Yang Q, Huang F, Zheng H, Sang Z, Xu Y, Zhang C, Wu K, Tao J, Prasanna BM (2021) Development of high-resolution multiple-SNP arrays for genetic analyses and molecular breeding through genotyping by target sequencing and liquid chip. Plant Commun 2(6):100230

Gupta V, Mehta G, Kumar S, Ramadas S, Tiwari R, Singh GP, Sharma P (2023) AMMI and GGE biplot analysis of yield under terminal heat tolerance in wheat. Mol Biol Rep 50(4):3459–3467. https://doi.org/10.1007/s11033-023-08298-4

Han Y-Y, Wang K-Y, Liu Z-Q, Pan S-H, Zhao X-Y, Zhang Q, Wang S-F (2020) Research on hybrid crop breeding information management system based on combining ability analysis. Sustainability 12(12):4938

Harper L, Campbell J, Cannon EK, Jung S, Poelchau M, Walls R, Andorf C, Arnaud E, Berardini TZ, Birkett C (2018) AgBioData consortium recommendations for sustainable genomics and genetics databases for agriculture. Database 2018:bay088

Hasan N, Choudhary S, Naaz N, Sharma N, Laskar RA (2021) Recent advancements in molecular marker-assisted selection and applications in plant breeding programmes. J Genet Eng Biotechnol 19(1):128

Hawkins C, Ginzburg D, Zhao K, Dwyer W, Xue B, Xu A, Rice S, Cole B, Paley S, Karp P (2021) Plant Metabolic Network 15: a resource of genome-wide metabolism databases for 126 plants and algae. J Integr Plant Biol 63(11):1888–1905

Heslot N, Yang HP, Sorrells ME, Jannink JL (2012) Genomic selection in plant breeding: a comparison of models. Crop Sci 52(1):146–160

Hu H, Scheben A, Edwards D (2018) Advances in integrating genomics and bioinformatics in the plant breeding pipeline. Agriculture 8(6):75

Hubner S, Korol AB, Schmid KJ (2015) RNA-Seq analysis identifies genes associated with differential reproductive success under drought-stress in accessions of wild barley Hordeum spontaneum. BMC Plant Biol 15:134. https://doi.org/10.1186/s12870-015-0528-z

Hudson AI, Odell SG, Dubreuil P, Tixier MH, Praud S, Runcie DE, Ross-Ibarra J (2022) Analysis of genotype-by-environment interactions in a maize mapping population. G3 (Bethesda) 12(3). https://doi.org/10.1093/g3journal/jkac013

Ibrahim MA, El-Sherbieny GA, Ismail E, Ali SM, Ahmed NK, Ratib NR, Aziz MM, Bishay SK, Fayez MS, Nasry NN (2021) Bioinformatics approaches toward plant breeding programs. Asian J Res Rev Agric:107–116

Jeon D, Kang Y, Lee S, Choi S, Sung Y, Lee T-H, Kim C (2023) Digitalizing breeding in plants: a new trend of next-generation breeding based on genomic prediction. Front Plant Sci 14:1092584

Jin J, Lu P, Xu Y, Tao J, Li Z, Wang S, Yu S, Wang C, Xie X, Gao J (2022) PCMDB: a curated and comprehensive resource of plant cell markers. Nucleic Acids Res 50(D1):D1448–D1455

Jo L, Kajala K (2023) ggPlantmap: an R package for creation of informative and quantitative ggplot maps derived from plant images. bioRxiv:2023.2011. 2030.569429

Jogani R, Mishra SK, Sharma K, Georrge JJ (2024) Metabolite profiling of Rauvolfia serpentina: an antivenom plant. In: Herbal formulations, Phytochemistry and Pharmacognosy. Elsevier, pp 255–260

Johny A, Mishra SK (2023) Pharmacological effects of bioactive compounds from Allium sativum. In: Pharmacological benefits of natural agents. IGI Global, pp 13–30

Joshi A, Kaushik V (2021) Big data and its analytics in agriculture. In: Bioinformatics for agriculture: high-throughput approaches, pp 71–83

Jung S, Lee T, Gasic K, Campbell BT, Yu J, Humann J, Ru S, Edge-Garza D, Hough H, Main D (2021) The Breeding Information Management System (BIMS): an online resource for crop breeding. Database 2021:baab054

Kayihan C, Yilmaz H, Özden Çiftçi Y (2023) Next-generation sequencing in plant breeding: challenges and possibilities. In: Advanced crop improvement, volume 1: Theory and practice, pp 507–535

Khan SU, Saeed S, Khan MHU, Fan C, Ahmar S, Arriagada O, Shahzad R, Branca F, Mora-Poblete F (2021) Advances and challenges for QTL analysis and GWAS in the plant-breeding of high-yielding: a focus on rapeseed. Biomolecules 11(10). https://doi.org/10.3390/biom11101516

Khan MHU, Wang S, Wang J, Ahmar S, Saeed S, Khan SU, Xu X, Chen H, Bhat JA, Feng X (2022) Applications of artificial intelligence in climate-resilient smart-crop breeding. Int J Mol Sci 23(19). https://doi.org/10.3390/ijms231911156

Khare V, Shukla RS, Pandey S, Singh SK, Singh C (2024) Exploring the genotype-environment interaction of bread wheat in ambient and high-temperature planting conditions: a rigorous investigation. Sci Rep 14(1):2402. https://doi.org/10.1038/s41598-024-53052-w

Kim JY (2020) Roadmap to high throughput phenotyping for plant breeding. J Biosyst Eng 45:43–55

Kim KD, Kang Y, Kim C (2020) Application of genomic big data in plant breeding: past, present, and future. Plants (Basel) 9(11):1454. https://doi.org/10.3390/plants9111454

Kim S-H, Choi I, Kim J-B (2024) Advancing plant breeding with next-generation technologies: insights from recent research. Plants 13(20):2877

Ko G, Kim PG, Cho Y, Jeong S, Kim JY, Kim KH, Lee HY, Han J, Yu N, Ham S, Jang I, Kang B, Shin S, Kim L, Lee SW, Nam D, Kim JF, Kim N, Kim SY, Lee S, Roh TY, Lee B (2020) Bioinformatics services for analyzing massive genomic datasets. Genomics Inform 18(1):e8. https://doi.org/10.5808/GI.2020.18.1.e8

Kolmogorov M, Yuan J, Lin Y, Pevzner PA (2019) Assembly of long, error-prone reads using repeat graphs. Nat Biotechnol 37(5):540–546. https://doi.org/10.1038/s41587-019-0072-8

Koren S, Walenz BP, Berlin K, Miller JR, Bergman NH, Phillippy AM (2017) Canu: scalable and accurate long-read assembly via adaptive k-mer weighting and repeat separation. Genome Res 27(5):722–736. https://doi.org/10.1101/gr.215087.116

Krishna TPA, Veeramuthu D, Maharajan T, Soosaimanickam M (2023) The era of plant breeding: conventional breeding to genomics-assisted breeding for crop improvement. Curr Genomics 24(1):24–35. https://doi.org/10.2174/1389202924666230517115912

Kristensen PS, Jahoor A, Andersen JR, Cericola F, Orabi J, Janss LL, Jensen J (2018) Genome-wide association studies and comparison of models and cross-validation strategies for genomic prediction of quality traits in advanced winter wheat breeding lines. Front Plant Sci 9:69. https://doi.org/10.3389/fpls.2018.00069

Lakshman A, Malik P (2010) Cassandra: a decentralized structured storage system. ACM SIGOPS 44(2):35–40

Lawrence CJ, Dong Q, Polacco ML, Seigfried TE, Brendel V (2004) MaizeGDB, the community database for maize genetics and genomics. Nucleic Acids Res 32(suppl_1):D393–D397

Li H, Handsaker B, Wysoker A, Fennell T, Ruan J, Homer N, Marth G, Abecasis G, Durbin R, Genome Project Data Processing S (2009) The Sequence Alignment/Map format and SAMtools. Bioinformatics 25(16):2078–2079. https://doi.org/10.1093/bioinformatics/btp352

Liu J, Zhang Y, Zheng Y, Zhu Y, Shi Y, Guan Z, Lang K, Shen D, Huang W, Dou D (2023a) PlantExp: a platform for exploration of gene expression and alternative splicing based on public plant RNA-seq samples. Nucleic Acids Res 51(D1):D1483–D1491

Liu M, Zhang S, Li W, Zhao X, Wang XQ (2023b) Identifying yield-related genes in maize based on ear trait plasticity. Genome Biol 24(1):94. https://doi.org/10.1186/s13059-023-02937-6
Liu D, Luo C, Dai R, Huang X, Chen X, He L, Mao H, Li J, Zhang L, Yang Q-Y (2024) AMIR: a multi-omics data platform for Asteraceae plants genetics and breeding research. Nucleic Acids Res:gkae833
Luby CH, Kloppenburg J, Michaels TE, Goldman IL (2015) Enhancing freedom to operate for plant breeders and farmers through open source plant breeding. Crop Sci 55(6):2481–2488
Lucibelli F, Valoroso MC, Aceto S (2022) Plant DNA methylation: An epigenetic mark in development, environmental interactions, and evolution. Int J Mol Sci 23(15). https://doi.org/10.3390/ijms23158299
Mahmood U, Li X, Fan Y, Chang W, Niu Y, Li J, Qu C, Lu K (2022) Multi-omics revolution to promote plant breeding efficiency. Front Plant Sci 13:1062952
Majumder P (2024) Golden opportunities: harnessing bioinformatics to revolutionize plant research and unleash the power of Golden Rice in crop breeding. In: Bioinformatics for plant research and crop breeding, pp 505–537
Maru H, Sharma K, Mishra SK (2023) A comprehensive review of ethnomedical uses, phytochemical studies, and properties of holy basil (Ocimum Sanctum Linn.). In: Pharmacological benefits of natural agents, pp 49–68
McGowan M, Wang J, Dong H, Liu X, Jia Y, Wang X, Iwata H, Li Y, Lipka AE, Zhang Z (2021) Ideas in genomic selection with the potential to transform plant molecular breeding: a review. Plant Breed Rev 45:273–319
Medina-Lozano I, Díaz A (2022) Applications of genomic tools in plant breeding: crop biofortification. Int J Mol Sci 23(6):3086
Mehta HJ, Mishra SK, Sharma K (2023) Phytochemical studies of Piper nigrum L: a comprehensive review. In: Pharmacological benefits of natural agents, pp 31–48
Merrick LF, Herr AW, Sandhu KS, Lozada DN, Carter AH (2022) Utilizing genomic selection for wheat population development and improvement. Agronomy 12(2):522
Miedaner T, Boeven ALG, Gaikpa DS, Kistner MB, Grote CP (2020) Genomics-assisted breeding for quantitative disease resistances in small-grain cereals and maize. Int J Mol Sci 21(24). https://doi.org/10.3390/ijms21249717
Mishra SK, Georrge JJ (2023) In Silico investigation of Epigallocatechin-3-Gallate and its derivatives against the VEGF-A cancerous protein through molecular docking and MD simulation studies. Recent Trends in Science and Technology, pp 22–38. https://doi.org/10.5281/zenodo.10696756
Mishra SK, Jeba Praba J, Georrge JJ (2024) An emerging trends of bioinformatics and big data analytics in healthcare. In: Digital Transformation in Healthcare 5.0: Volume 2: Metaverse, nanorobots and machine learning, p 159
Morales N, Kaczmar NS, Santantonio N, Gore MA, Mueller LA, Robbins KR (2020) ImageBreed: open-access plant breeding web–database for image-based phenotyping. Plant Phenome J 3(1):e20004
Morales N, Ogbonna AC, Ellerbrock BJ, Bauchet GJ, Tantikanjana T, Tecle IY, Powell AF, Lyon D, Menda N, Simoes CC (2022) Breedbase: a digital ecosystem for modern plant breeding. G3 12(7):jkac078
Mu H, Wang B, Yuan F (2022) Bioinformatics in plant breeding and research on disease resistance. Plants 11(22):3118
Nair RJ, Pandey MK (2024) Role of molecular markers in crop breeding: a review. Agric Rev 45(1)
Narawatthana S, Phansenee Y, Thammasamisorn BO, Vejchasarn P (2023) Multi-model genome-wide association studies of leaf anatomical traits and vein architecture in rice. Front Plant Sci 14:1107718. https://doi.org/10.3389/fpls.2023.1107718
Ngo VM, Le-Khac N-A, Kechadi M (2018) An efficient data warehouse for crop yield prediction. arXiv preprint arXiv:1807.00035
Niazian M, Niedbała G (2020) Machine learning for plant breeding and biotechnology. Agriculture 10(10):436

Nizamani MM, Zhang Q, Muhae-Ud-Din G, Wang Y (2023) High-throughput sequencing in plant disease management: a comprehensive review of benefits, challenges, and future perspectives. Phytopathol Res 5(1):1–17

Nowoshilow S, Schloissnig S, Fei JF, Dahl A, Pang AWC, Pippel M, Winkler S, Hastie AR, Young G, Roscito JG, Falcon F, Knapp D, Powell S, Cruz A, Cao H, Habermann B, Hiller M, Tanaka EM, Myers EW (2018) The axolotl genome and the evolution of key tissue formation regulators. Nature 554(7690):50–55. https://doi.org/10.1038/nature25458

Ogutu JO, Piepho HP, Schulz-Streeck T (2011) A comparison of random forests, boosting and support vector machines for genomic selection. BMC Proc 5 Suppl 3(Suppl 3):S11. https://doi.org/10.1186/1753-6561-5-S3-S11

Ostrovsky A, Hillman-Jackson J, Bouvier D, Clements D, Afgan E, Blankenberg D, Schatz MC, Nekrutenko A, Taylor J, Team TG (2021) Using galaxy to perform large-scale interactive data analyses—an update. Curr Protoc 1(2):e31

Papoutsoglou EA, Athanasiadis IN, Visser RG, Finkers R (2023) The benefits and struggles of FAIR data: the case of reusing plant phenotyping data. Sci Data 10(1):457

Pazhamala LT, Kudapa H, Weckwerth W, Millar AH, Varshney RK (2021) Systems biology for crop improvement. Plant Genome 14(2):e20098

Peng H, Wang K, Chen Z, Cao Y, Gao Q, Li Y, Li X, Lu H, Du H, Lu M (2020) MBKbase for rice: an integrated omics knowledgebase for molecular breeding in rice. Nucleic Acids Res 48(D1):D1085–D1092

Pereira R, Oliveira J, Sousa M (2020) Bioinformatics and computational tools for next-generation sequencing analysis in clinical genetics. J Clin Med 9(1):132

Podlich DW, Cooper M (1998) QU-GENE: a simulation platform for quantitative analysis of genetic models. Bioinformatics 14(7):632–653. https://doi.org/10.1093/bioinformatics/14.7.632

Pradas N, Jurado-Ruiz F, Onielfa C, Arús P, Aranzana MJ (2024) PERSEUS: an interactive and intuitive web-based tool for pedigree visualization. Bioinformatics 40(2):btae060

Price EJ, Drapal M, Perez-Fons L, Amah D, Bhattacharjee R, Heider B, Rouard M, Swennen R, Becerra Lopez-Lavalle LA, Fraser PD (2020) Metabolite database for root, tuber, and banana crops to facilitate modern breeding in understudied crops. Plant J 101(6):1258–1268

Purcell S, Neale B, Todd-Brown K, Thomas L, Ferreira MA, Bender D, Maller J, Sklar P, De Bakker PI, Daly MJ (2007) PLINK: a tool set for whole-genome association and population-based linkage analyses. Am J Hum Genet 81(3):559–575

Raza A (2022) Metabolomics: a systems biology approach for enhancing heat stress tolerance in plants. Plant Cell Rep 41(3):741–763

Razzaq A, Wishart DS, Wani SH, Hameed MK, Mubin M, Saleem F (2022) Advances in metabolomics-driven diagnostic breeding and crop improvement. Metabolites 12(6):511

Rebollo I, Aguilar I, Pérez de Vida F, Molina F, Gutiérrez L, Rosas JE (2023) Genotype by environment interaction characterization and its modeling with random regression to climatic variables in two rice breeding populations. Crop Sci 63(4):2220–2240

Ren D, An L, Li B, Qiao L, Liu W (2021) Efficient weighting methods for genomic best linear-unbiased prediction (BLUP) adapted to the genetic architectures of quantitative traits. Heredity (Edinb) 126(2):320–334. https://doi.org/10.1038/s41437-020-00372-y

Roychoudhury A (2020) Next generation sequencing: prospects in plant breeding and crop improvement. SF J Agric Crop Manag 1(1):1004

Sandhu KS, Patil SS, Aoun M, Carter AH (2022) Multi-trait multi-environment genomic prediction for end-use quality traits in winter wheat. Front Genet 13:831020

Schreiber F, Colmsee C, Czauderna T, Grafahrend-Belau E, Hartmann A, Junker A, Junker BH, Klapperstück M, Scholz U, Weise S (2012) MetaCrop 2.0: managing and exploring information about crop plant metabolism. Nucleic Acids Res 40(D1):D1173–D1177

Schreiber M, Jayakodi M, Stein N, Mascher M (2024) Plant pangenomes for crop improvement, biodiversity and evolution. Nat Rev Genet 25(8):563–577. https://doi.org/10.1038/s41576-024-00691-4

Selby P, Abbeloos R, Backlund JE, Basterrechea Salido M, Bauchet G, Benites-Alfaro OE, Birkett C, Calaminos VC, Carceller P, Cornut G (2019) BrAPI—an application programming interface for plant breeding applications. Bioinformatics 35(20):4147–4155

Shen Y, Zhou G, Liang C, Tian Z (2022) Omics-based interdisciplinarity is accelerating plant breeding. Curr Opin Plant Biol 66:102167

Sinha D, Maurya AK, Abdi G, Majeed M, Agarwal R, Mukherjee R, Ganguly S, Aziz R, Bhatia M, Majgaonkar A (2023) Integrated genomic selection for accelerating breeding programs of climate-smart cereals. Genes 14(7):1484

Sun Y, Shang L, Zhu Q-H, Fan L, Guo L (2022) Twenty years of plant genome sequencing: achievements and challenges. Trends Plant Sci 27(4):391–401

Sun M, Yan H, Zhang A, Jin Y, Lin C, Luo L, Wu B, Fan Y, Tian S, Cao X (2023) Milletdb: a multi-omics database to accelerate the research of functional genomics and molecular breeding of millets. Plant Biotechnol J 21(11):2348–2357

Tan YC, Kumar AU, Wong YP, Ling APK (2022) Bioinformatics approaches and applications in plant biotechnology. J Genet Eng Biotechnol 20(1):106

Thanasomboon R, Kalapanulak S, Netrphan S, Saithong T (2020) Exploring dynamic protein-protein interactions in cassava through the integrative interactome network. Sci Rep 10(1):6510

Tibbs Cortes L, Zhang Z, Yu J (2021) Status and prospects of genome-wide association studies in plants. Plant Genome 14(1):e20077

Turcotte H, Hooker J, Samanfar B, Parent J-S (2022) Can epigenetics guide the production of better adapted cultivars? Agronomy 12(4):838

Tripodi, P. (2023). The Evolution of Molecular Genotyping in Plant Breeding. Agronomy 13(10), 2569. https://doi.org/10.3390/agronomy13102569

Vaghasia VV, Sharma K, Mishra SK, Georrge JJ (2024) In silico identification of natural product inhibitor for multidrug resistance proteins from selected gram-positive bacteria. In: Nanotechnology and in silico tools. Elsevier, pp 309–317

Vakhariya Sakina S, Mishra SK, Sharma K, Georrge JJ (2023) Designing of a novel curcumin analogue to inhibit mitogen-activated protein kinase: a cheminformatics approach. J Phytonanotechnol Pharma Sci 3(1):37–47

van Dijk EL, Jaszczyszyn Y, Naquin D, Thermes C (2018) The third revolution in sequencing technology. Trends Genet 34(9):666–681. https://doi.org/10.1016/j.tig.2018.05.008

van Dijk M, Morley T, Rau ML, Saghai Y (2021) A meta-analysis of projected global food demand and population at risk of hunger for the period 2010-2050. Nat Food 2(7):494–501. https://doi.org/10.1038/s43016-021-00322-9

Vaser R, Sikic M (2021) Time- and memory-efficient genome assembly with Raven. Nat Comput Sci 1(5):332–336. https://doi.org/10.1038/s43588-021-00073-4

Vinjoda P, Mishra SK, Sharma K, Georrge JJ (2024) In silico identification of novel drug target and its natural product inhibitors for herpes simplex virus. In: Nanotechnology and in Silico tools. Elsevier, pp 377–383

Wang K, Li M, Hakonarson H (2010) ANNOVAR: functional annotation of genetic variants from high-throughput sequencing data. Nucleic Acids Res 38(16):e164. https://doi.org/10.1093/nar/gkq603

Wang L, Lu Z, Regulski M, Jiao Y, Chen J, Ware D, Xin Z (2021) BSAseq: an interactive and integrated web-based workflow for identification of causal mutations in bulked F2 populations. Bioinformatics 37(3):382–387

Wang H, & Qin F (2017) Genome-Wide Association Study Reveals Natural Variations Contributing to Drought Resistance in Crops. Front Plant Sci, 8, 1110. https://doi.org/10.3389/fpls.2017.01110

Woodhouse MR, Cannon EK, Portwood JL, Gardiner JM, Hayford RK, Haley O, Andorf CM (2024) Tools and resources at the maize genetics and genomics database (MaizeGDB). Cold Spring Harbor Protocols

Xie L, Gong X, Yang K, Huang Y, Zhang S, Shen L, Sun Y, Wu D, Ye C, Zhu QH, Fan L (2024) Technology-enabled great leap in deciphering plant genomes. Nat Plants 10(4):551–566. https://doi.org/10.1038/s41477-024-01655-6

Xu Y, Ma K, Zhao Y, Wang X, Zhou K, Yu G, Li C, Li P, Yang Z, Xu C (2021) Genomic selection: a breakthrough technology in rice breeding. Crop J 9(3):669–677

Xu Y, Zhang X, Li H, Zheng H, Zhang J, Olsen MS, Varshney RK, Prasanna BM, Qian Q (2022) Smart breeding driven by big data, artificial intelligence, and integrated genomic-enviromic prediction. Mol Plant 15(11):1664–1695

Xu Y, Zhang Y, Cui Y, Zhou K, Yu G, Yang W, Wang X, Li F, Guan X, Zhang X (2024) GA-GBLUP: leveraging the genetic algorithm to improve the predictability of genomic selection. Brief Bioinform 25(5)

Yan J, Wang X (2023) Machine learning bridges omics sciences and plant breeding. Trends Plant Sci 28(2):199–210

Yang Y, Saand MA, Huang L, Abdelaal WB, Zhang J, Wu Y, Li J, Sirohi MH, Wang F (2021) Applications of multi-omics technologies for crop improvement. Front Plant Sci 12:563953

Yang Z, Wang S, Wei L, Huang Y, Liu D, Jia Y, Luo C, Lin Y, Liang C, Hu Y (2023) BnIR: a multi-omics database with various tools for Brassica napus research and breeding. Mol Plant 16(4):775–789

Yoosefzadeh Najafabadi M, Hesami M, Eskandari M (2023) Machine learning-assisted approaches in modernized plant breeding programs. Genes 14(4):777

Younis OG, Turchetta M, Ariza Suarez D, Yates S, Studer B, Athanasiadis IN, Krause A, Buhmann JM, Corinzia L (2023) ChromaX: a fast and scalable breeding program simulator. Bioinformatics 39(12):btad691

Yu J, Jung S, Cheng C-H, Lee T, Zheng P, Buble K, Crabb J, Humann J, Hough H, Jones D (2021) CottonGen: the community database for cotton genomics, genetics, and breeding research. Plants 10(12):2805

Yu Y, Zhang H, Long Y, Shu Y, Zhai J (2022) Plant public RNA-seq database: a comprehensive online database for expression analysis of~ 45 000 plant public RNA-seq libraries. Plant Biotechnol J 20(5):806

Yuan Y, Cairns JE, Babu R, Gowda M, Makumbi D, Magorokosho C, Zhang A, Liu Y, Wang N, Hao Z, San Vicente F, Olsen MS, Prasanna BM, Lu Y, Zhang X (2018) Genome-wide association mapping and genomic prediction analyses reveal the genetic architecture of grain yield and flowering time under drought and heat stress conditions in maize. Front Plant Sci 9:1919. https://doi.org/10.3389/fpls.2018.01919

Zhang R, Jia G, Diao X (2023) geneHapR: an R package for gene haplotypic statistics and visualization. BMC Bioinform 24(1):199

Zhou Y, Kathiresan N, Yu Z, Rivera LF, Yang Y, Thimma M, Manickam K, Chebotarov D, Mauleon R, Chougule K (2024) A high-performance computational workflow to accelerate GATK SNP detection across a 25-genome dataset. BMC Biol 22(1):13

Zhuo Y, Du H, Diao C, Li W, Zhou L, Jiang L, Jiang J, Liu J (2024) MAGE: metafounders-assisted genomic estimation of breeding value, a novel additive-dominance single-step model in crossbreeding systems. Bioinformatics 40(2):btae044

Metabolic Engineering of Essential Micronutrients

15

Roy Chen and Jianjun Chen

Abstract

Essential micronutrients are minerals and vitamins that are crucial in human growth, development, disease prevention, and well-being. The human body cannot synthesize them; they must come from diets. Although various sources, such as fruits, vegetables, eggs, meats, and seafood, are rich in some essential micronutrients, most people live on a diet based on staple crops, such as rice, corn, wheat, millet, cassava, and potatoes. These crops contain low levels of essential micronutrients. An insufficient intake of them can cause either malnutrition or hidden hunger, leading to detrimental effects on human health. It is estimated that about two billion people worldwide suffer from deficiencies of essential micronutrients, which signifies a serious public health concern worldwide. Biofortification of staple crops, such as conventional breeding and agronomic approaches, has made progress in increasing the contents of essential micronutrients, but such methods are relatively less effective. Metabolic engineering is another strategy for enhancing essential micronutrients. This chapter is intended to briefly introduce priority essential micronutrients, their roles and recommended dietary allowance, and conventional breeding efforts made to improve their contents in major staple crops. Emphasis is placed on metabolic engineering of priority essential micronutrients, approaches for their engineering, and current progress. The United Nations Sustainable Development Goal 2 (UN-SDG2) has set forth a goal to end all forms of hunger by 2030. Metabolic engineering should make a significant contribution to the UN-SDG2 goal.

R. Chen
Department of Molecular Biosciences, University of South Florida, Tampa, FL, USA

J. Chen (✉)
Mid-Florida Research and Education Center, Department of Environmental Horticulture, Institute of Food and Agricultural Sciences, University of Florida, Apopka, FL, USA
e-mail: jjchen@ufl.edu

P. M. Priyadarshan, R. Ortiz (eds.), *Plant Breeding 2050*,
https://doi.org/10.1007/978-981-95-0583-8_15

Keywords

Essential micronutrients · Genetic transformation · Metabolic engineering · Staple crops · Vitamins

15.1 Introduction

Essential micronutrients are minerals and vitamins that humans require for normal metabolism and development. Essential mineral elements are cobalt (Co), chromium (Cr), copper (Cu), iodine (I), iron (Fe), manganese (Mn), selenium (Se), and zinc (Zn) (Islam et al. 2023). Essential vitamins include vitamins A, B1 (thiamine), B2 (riboflavin), B3 (niacin), B5 (pantothenic acid), B6 (pyridoxine), B9 (folate), B12 (cobalamin), C (ascorbic acid), D, E, K, and H (biotin) (Tiwari et al. 2024). These micronutrients have the following characteristics: (a) the human body cannot synthesize them and must come from diets; (b) they are present in small amounts and not used for energy but play a vital role in human metabolism and preventing chronic diseases; and (c) due to their concentrations at minimum levels, their deficiency and/or hidden hunger occur in a significant number of people, especially those in developing countries. Among the eight mineral elements and 13 vitamins, deficiencies in Fe, Zn, Se, and I microelements and vitamins A, B9, and E are more prevalent than the others (Tulchinsky 2010; Islam et al. 2023). Herein, they are defined as priority essential micronutrients.

15.1.1 Roles, Allowance, and Deficiencies of Priority Essential Microelements

Iron is an essential constituent of myoglobin and hemoglobin and a component of heme and nonheme enzymes that catalyze oxidation–reduction reactions (La Frano et al. 2014). The recommended dietary allowance (RDA) ranges from 8 to 18 mg daily, depending on age, gender, and health status (Table 15.1). It was estimated that about one-third of the population in the world suffers from anemia (Chaparro and Suchdev 2019) that is mainly caused by Fe deficiency known as iron deficiency anemia (IDA). IDA also causes cognitive and physical impairment in children while affecting the physical performance of adults. As a result, Fe deficiency is the most common nutrition disorder worldwide (Lopez et al. 2016).

Zinc is the second most abundantly distributed micronutrient in the body after Fe (Saper and Rash 2009). It is necessary for the activity of more than 300 enzymes (Maret 2013) with an RDA ranging from 8 to 11 mg daily (Table 15.1), but Zn deficiency in humans has become an emerging health issue in the world by affecting two billion people (Gibson and Ferguson 1998; Hussain et al. 2022). Zn deficiency includes pneumonia in children, diarrhea, alopecia, and glossitis (Sezer et al. 2011; Lassi et al. 2016).

Selenium is essential to various metabolic pathways, such as thyroid hormone metabolism, immune function, and antioxidant defense systems (Brown and Arthur

Table 15.1 Allowance and deficiency of priority essential micronutrients

Category	Nutrient	RDA[a]	Deficiency symptom	References
Mineral	Iron (Fe)	8–18 mg	Anemia, impaired immunity	Institute of Medicine (2001)
	Zinc (Zn)	8–11 mg	Pneumonia in children, diarrhea, and diabetes	Institute of Medicine (2001)
	Selenium (Se)	55 μg (0.7 μmol)	Risk of cardiovascular disease, risk of cancer, and Keshan disease	Institute of Medicine (2000)
	Iodine (I)	90–250 μg	Goiter, hypothyroidism	Zimmermann et al. (2008)
Vitamin	A	700–900 μg of RAE[b]	Risk of blindness and death from infections such as measles and diarrhea	Institute of Medicine (2001)
	B9 (Folate)	400 μg of DFE[c]	Neural tube defects, cancer, depression, and heart disease and stroke	Bailey and Gregory III (1999)
	E	15 mg (35 μmol) of α-tocopherol	Impaired reflexes and coordination	Institute of Medicine (2000)

[a]*RDA* Recommended dietary allowance
[b]*RAE* Retinol activity equivalent
[c]*DFE* Dietary folate equivalent

2001). Its RDA is 55 μg daily, while the maximum tolerable adult intake without side effects is set at 400 μg a day (Radomska et al. 2021). A typical disease associated with Se is known as Keshan and Kashin-Beck disease. The characteristics of this disease are short stature, joint pain, deformity, and even disability. Se deficiency affects about 0.5 to 1 billion people worldwide (Radomska et al. 2021).

Iodine is involved in the biosynthesis of thyroid hormones in the human body, and its RDA ranges from 90 to 250 μg a day (Zimmermann and Andersson 2012) (Table 15.1). Inadequate intake of iodine can lead to clinical and social issues known as "iodine deficiency disorder" (IDD). The classic signs of IDDs are goiter and the enlargement of the thyroid gland, which happen in all age groups. This can lead to adverse health outcomes, including pregnancy loss and infant mortality. Currently, IDDs still affect about one-third of the human population (Gonzali et al. 2017). It was estimated that about two billion people are at risk of iodine deficiency (Tulchinsky 2010).

15.1.2 Roles, Allowance, and Deficiencies of Priority Essential Vitamins

Human deficiencies in vitamins A, B9, and E are generally more common than the other vitamins. Vitamin A is a compound that maintains the structural and functional integrity of epithelial tissue and the immune system. It is essential for normal vision, reproductive health, and fetal growth, with an RDA varying from 700 to 900 μg RAE (retinol activity equivalent). Deficiency in vitamin A leads to skin integrity loss, poor eyesight, night blindness, and an increased risk of infection. It was reported that about 254 million preschool children were at risk of vitamin A deficiency (Tulchinsky 2010).

The most commonly occurring vitamin B deficiency is B9. B9 is required for DNA synthesis and cell division and plays an important role during pregnancy. The RDA for B9 is 400 μg DFE (dietary folate equivalent). B9 deficiency can cause neural tube defects, cancers, and depression (Table 15.1).

Vitamin E is a major component in the cell antioxidant defense system and is exclusively obtained from diet. It is critical in scavenging free radicals, forming red blood cells, and widening blood vessels to prevent blood clots. It also helps maintain the immune system's defense against viruses and bacteria (Shahidi et al. 2021). The daily RDA of vitamin E is 15 mg (35 μmol) of α-tocopherol (Table 15.1). It was estimated that over 90% of Americans fail to consume sufficient vitamin E as α-tocopherol to meet the RDA (Traber 2014).

15.2 Occurrence in Essential Micronutrient Deficiencies and Strategies for Intervention

15.2.1 Occurrence in Essential Micronutrient Deficiencies

Essential micronutrients exist in various food sources, particularly in vegetables, fruits, grains, meats, seafood, and fortified foods (Beal and Ortenzi 2022; Chungchunlam and Moughan 2024). However, over two billion people still suffer essential micronutrient deficiencies (WHO 2006). The deficiencies are related to multiple factors (Keen et al. 2003; Bailey et al. 2015). (a) Socioeconomic limitations: either inability to afford or lack of access to nutrient-rich foods, largely attributed to poverty, particularly in developing countries. (b) Health issues: chronic diseases and malabsorption that limit the body from effectively absorbing nutrients from food. (c) Dietary patterns include vegetarianism, overconsumption of processed foods, or limited food diversity. For example rice and wheat are dominant foods for more than half of the world population (Fitzgerald et al. 2009), but they contain low contents of essential micronutrients (Kiely et al. 2021). (d) Environmental conditions: geographic limitations for accessing nutrient-rich foods and soil depletion of some minerals, resulting in low contents of essential microelements in the edible portion of crops.

15.2.2 Strategies for the Intervention of Essential Micronutrient Deficiencies

Strategies for intervention of essential micronutrient deficiencies include dietary modifications and education, supplementation, and biofortification. Biofortification represents an effective nutritional intervention, (Li et al. 2024) including agronomic and genetic fortification. Agronomic approaches involve production practices, mainly soil or foliar application of micronutrient fertilizers or seed priming, to increase micronutrients in edible parts of crops (Saeid et al. 2019). Genetic fortification is to improve plant absorption of microelements from soils and transport them to the edible portion of crops and/or enhance the biosynthesis of essential micronutrients, such as vitamins, in edible organs or tissue through conventional breeding and genetic engineering (Zulfiqar et al. 2024). This effort has been largely focused on cereals, such as rice, corn, wheat, sorghum, and starchy food, such as potatoes and cassava.

15.3 Improvement Through Conventional Breeding

Conventional breeding is referred to as the hybridization of parental plants with complementary traits, selection of offspring with combined desirable characteristics using different methods, evaluation of the selections, and release of developed cultivars. The methods include crossbreeding and mutation breeding. The past breeding effort has been primarily focused on crop yield with little attention to essential micronutrients. With the realization of malnutrition and hidden hunger, breeding for improving micronutrient contents has been initiated.

15.3.1 Genetic Resources and Progress of Improvement

Breeding crops with high contents of essential micronutrients first requires identifying genetic resources of interest. Several international organizations, including the International Rice Research Institute (IRRI), International Maize and Wheat Improvement Center (CIMMYT), and International Center for Tropical Agriculture (CIAT), have made substantial efforts to identify valuable genetic resources with high micronutrient contents. CIMMYT screened 132 wheat accessions and reported that Fe contents varied from 28.8 to 56.5 μg/g, and Zn differed from 25.2 to 53.3 μg/g (Graham et al. 1999). IRRI evaluated 939 rice accessions and found that Fe and Zn contents ranged from 7.5 to 24.2 μg/g and 13.5 to 58.4 μg/g, respectively. In both evaluated wheat and rice accessions, those with higher Fe contents were also higher in Zn contents. More than 1000 accessions of common beans were evaluated at CIAT, and results showed that Fe contents ranged from 34 to 89 μg/g, and Zn contents were 21 to 54 μg/g. Some accessions collected from Peru contained Fe higher than 100 μg/g, suggesting that significant genetic variation existed among collected germplasm (Graham et al. 1999).

Significant progress has been made in biofortification of essential micronutrients through breeding. Quantitative trait loci (QTL) related to essential micronutrients have been established using molecular markers. For example QTLs of crtRB1–3'TE-1, crtRB1–5'TE-2, crtRB13'TE, LCYE, PSY1, and crtRb1 were identified in maize for pro-vitamin A. An SSR (simple sequence repeats or microsatellites) marker for β-carotene was established in maize recombinant inbred line populations (Gaikwad et al. 2020). A summary of QTLs related to the contents of

Fe, Zn, pro-vitamin A, and β-carotene as well as other compounds in wheat, barley, rice, maize, common bean, soybean, and other crops were reported by Gaikwad et al. (2020). A notable organization known as HarvestPlus was established in 2003, which is a program within the International Food Policy Research Institute (IFPRI). As a nonprofit organization, HarvestPlus has been addressing essential micronutrient deficiency through the biofortification of staple crops, including agronomic practices, conventional breeding, and genetic modification (Pfeiffer and McClafferty 2007). A series of vitamin A enriched sweet potato (13–35.4 μg/g), cassava (4.12–30 μg/g), and maize (2.2–11.3 μg/g); Fe enriched beans (62–113 μg/g), cowpea (43–100 μg/g), and pearl millet (59–78 μg/g); Zn enriched wheat (30–48.5 μg/g), rice (22–34 μg/g), maize (60–35.1 μg/g), and sorghum (32 μg/g) were released (https://bcr.harvestplus.org/).

15.3.2 Challenges with Conventional Breeding Methods

The content of crop's essential micronutrients is largely polygenic, quantitatively controlled, and highly influenced by the environment. Thus, improving such traits through conventional breeding is generally difficult. One challenge in maize breeding is that grain iron and zinc concentrations negatively correlate with grain yield, and breeding for high Fe and Zn contents may decrease crop yield. Increasing aleurone layers has been considered a fast track to overcome the problem as Fe and Zn in maize kernels are localized in a single cell layer known as aleurone (Bänziger and Long 2000). Plant foods (especially seeds and grains) contain various antinutrients in amounts depending on both genetic and environmental factors that can reduce the bioavailability of dietary nonheme Fe, Zn, and other nutrients to humans (Welch and House 1984).

15.4 Metabolic Engineering of Essential Micronutrients

The term metabolic engineering was coined by James Bailey in 1991, which is defined as "the improvement of cellular activities by manipulating enzymatic, transport, and regulatory functions of the cell with the use of recombinant DNA technology" (Bailey 1991). It started with the recombinant DNA technology in microbes in the early 1970s (Cohen et al. 1973). With the arrival of the omics era in the early 2000s and the quick development of analytical tools, synthetic biology, and gene editing, metabolic engineering has become more sophisticated and applicable to crop plants. The objective of metabolic engineering is to enhance the production and accumulation of desirable target products in naturally low organisms that cannot meet the demands of humans. It represents a promising alternative to conventional plant breeding to increase the production of the commonly deficient essential micronutrients (Zhu et al. 2020; Jiang et al. 2021).

15.4.1 Methods of Metabolic Engineering

Metabolic engineering aims to increase a target compound in particular crops. The first step requires a complete understanding of metabolic pathways. The metabolic pathway may be a linear pathway that is either unidirectional or reversible, or it may be branched with conversion points. Some enzymes engaged in the biosynthesis may come from a gene family. For the engineering process, genes catalyzing the reaction should be specified, and rate-limiting enzymes should be identified. The biosynthesis pathway may involve several organelles: nucleus, cytosol, mitochondria, and chloroplast. Thus, the flow of molecules through the metabolic pathway and the accumulation of target products in organelles should be clear. Several computational tools, including genome-scale metabolic models, metabolic flux analysis, and related algorithms, like OptKnock, have been developed for predicting genetic modifications, which may help identify target genes more precisely (Volk et al. 2022).

The next step is to design and construct vectors for genes of interest. Several strategies can be used to enhance the biosynthesis of target products: (a) Enhanced expression of genes that encode rate-limiting enzymes or several key enzymes in the pathway of the target products. For example, the over-expression of the gene encoding γ-tocopherol methyltransferase *(γ-TMT)* in maize increased α-tocopherol content by 6.5-fold in transgenic maize kernels (Zhang et al. 2020). Expression of rice family genes encoding nicotianamine synthase (*NAS*), *OsNAS1*, *OsNAS2*, and *OsNAS2* increased Fe and Zn contents in rice endosperm (Johnson et al. 2011). (b) Overexpression of transcription factors that control the expression of genes encoding rate-limiting enzymes or related key enzymes. Expression of a member of transcription factors, such as *ATAF* (arabidopsis transcription activation factor), *CUC* (cup-shaped cotyledon), and *NAM-B1* (no apical meristem-B1) resulted in high levels of Fe and Zn in wheat grain (Ricachenevsky et al. 2013). (c) Suppression or knockout genes via antisense, siRNAs (small interfering RNAs), or shRNAs (short hairpin RNAs) in the branch pathways that compete with the path of interest to avoid the intermediates being diverted to the other pathways. (d) Combined application of the above strategies to optimize the production of target products. (e) The use of *CRISP/dCase9* technologies to activate or repress genes encoding key enzymes.

Step three is to select crop species and varieties with great potential for success in metabolic engineering. This requires understanding the crops of interest, including the genetic variation of species and even cultivars within a species in adaptation to abiotic and biotic stress, target compounds' contents, and genetic transformation's feasibility. For example wheat and soybean are generally difficult to regenerate and transform, but cultivar differences exist. Metabolic engineering should select cultivars that have a wide adaptation to abiotic and biotic stress, relatively higher overall nutrient contents, and are easy to transform for initiating genetic manipulation.

The fourth step is to develop regeneration systems. Based on the above selection, a regeneration system should be established through either shoot organogenesis or somatic embryogenesis for the selected cultivars. The systems must be fine-tuned to ensure that plantlets can be successfully regenerated from explants cultured on selected media supplemented with growth regulators at defined concentrations. Cultivars and sources of explants differ in response to *in vitro* induction. Different cultivars and different explants per cultivar should be evaluated to ensure the reliability of the established regeneration systems.

The fifth step is to test methods for transformation. Current methods include *Agrobacterium tumefaciens*-mediated transformation, microparticle bombardment, direct DNA transfer system, chemically mediated introduction into protoplasts and regeneration, pollen tube pathway, electroporation, aerosol bean injection, and whiskers-mediated transformation (Su et al. 2023). Among these, *Agrobacterium tumefaciens*-mediated transformation is the dominant method, followed by plant cells or tissue microparticle bombardment. The number of released transgenic crops transformed through the former was 195 compared to 45 with the latter method (Anjanappa and Gruissem 2021). Recently, developmental regulators have been used in plant's transformation (Lian et al. 2022); it represents a new approach to genetic transformation, but its applicability and dependability require extensive investigation.

Step six is to evaluate plants derived from the transformation. This includes overall plant performance, i.e. if plant growth rates, tolerance or resistance to abiotic and biotic stresses, and grain yield are similar to or greater than the non-transformed control plants. The more important parameter is if the target compounds occur in the expected organs or tissues. The target compounds in barley, corn, rice, sorghum, and wheat should be significantly high in grains. The target compounds should be high in tubers for tuber crops, such as cassava and potato. Because of the differences in copy numbers and location of gene insertion, transformed individuals vary in the mentioned parameters. Desirable individual plants should be identified, and their growth needs to be appropriately managed. For grain crops, genes will be segregated in the next generation. Homozygous lines should be identified and maintained for commercial production. As to the tuber crops, they can be propagated vegetatively. Propagation through vegetative means to increase the starting materials can be used directly for commercial production.

There is an additional step for incorporating two or more transgenes into individual lines to engineer crops with increased contents of multiple essential micronutrients. For example, to increase both Fe and vitamin A contents, transgenic rice lines with high Fe content can be hybridized with a transgenic rice line with high vitamin A content by monitoring segregated populations and selecting individuals with higher levels of Fe and vitamin A. Homozygous lines with higher contents of Fe and vitamin A can be developed for commercial production.

15.4.2 Engineering of Priority Essential Microelements

15.4.2.1 Engineering for Increasing Fe Contents

As mentioned above, Fe is the most commonly deficient essential microelement, which is likely related to the consumption of staple grains since rice, wheat, and maize account for more than 60% of the daily energy intake of the world population

(Stanton et al. 2022). These gains, however, are classified as poor dietary sources of essential microelements. The baseline concentrations of Fe in whole grain of wheat, maize, pearl millet, and beans are 30, 30, 50, and 47 mg/kg, which are substantially lower than the recommended levels of 59, 60, 88, and 107 mg/kg, respectively (Vasconcelos et al. 2017). Rice that is consumed by nearly 3.5 billion people contains only about 2 mg/kg Fe in polished grain (Van Der Straeten et al. 2020; Kong et al. 2022). On the other hand, the target content of Fe in polished rice grain is expected to be 15 mg/kg (Van Der Straeten et al. 2020). Thus, polished rice grain needs to increase an additional 13 mg/kg to reach the target level.

Plants have developed two strategies for absorbing Fe from soils. The reduction for non-graminaceous species as first strategy and the chelation for graminaceous plants as second strategy (Marschner et al. 1986). The first strategy involves the acidification of the rhizosphere by root release of proton through H^+-ATPase 2 (*HA*2), the reduction of Fe^{3+} to Fe^{2+} by a membrane-localized reductase (FRO2), and root uptake of Fe^{2+} by a membrane-localized high-affinity Fe transporter (IRT1) (Liang 2022). The second strategy involves the release of phytosiderophores, compounds primarily belonging to the muriatic acid family, from roots to solubilize and chelate Fe^{3+} in soils, and the chelated Fe^{3+} is transported into roots by yellow stripe 1-like (YSL) proteins. Mugineic acids are synthesized using a precursor known as nicotinamide (NA), and NA is synthesized by NA synthase (NAS) (Liang 2022). Because iron in paddy soil occurs as Fe^{2+}, rice has developed mechanisms to absorb both Fe^{3+} and Fe^{2+} through the action of OsIRT1 and OsIRT2. In addition to the two strategies, plants have a ubiquitous protein known as ferritin for storage of Fe, and it can store about 4000 Fe atoms within its complex structure (Theil 2003).

Several genes have been manipulated for elevating Fe contents in seeds (Table 15.2). The overexpression of *ferritin* under the control of different promoters, primarily endosperm-specific promoters, increased Fe content in rice, maize, and wheat. Additionally, the enhanced expression of *OsNAS2* and *OsYSL2* elevated Fe contents in rice, and the expression of *FEA1* (algae iron assimilatory gene) increased Fe in cassava tubers. It is important to be aware that consumable products of wheat, maize, pearl millet, beans, and soybean are whole grains; edible organs of potato and cassava are tubers and roots, but the consumable tissue of rice is polished grains or white rice. Fe concentration in brown rice (unpolished grains) generally ranged from 10 to 20 mg/kg. But polishing removed 25%–84% of Fe from the brown rice, resulting in particularly lower Fe contents in polished grains (Prom-u-thai et al. 2007). As a result, A major effort has been focused on rice, mainly Fe contents in polished grains. Masuda et al. (2012) introduced multiple genes: *ferritin* under the control of endosperm-specific promotes, *NAS* under the control of *OsActin1*, and *OsYSL2* under the control of *OsSUT1* promoter into rice. Fe contents in polished seeds of paddy soil-grown T_3 transgenic plants were 4.4-fold higher than those of non-transgenic plants. However, Fe contents still failed to reach the 15 mg/kg target in polished rice grain. The authors explained that the Fe content in the rice variety used for the transformation was only 0.9 mg/kg. Using a variety that endogenously had high Fe content would achieve the target Fe content. Subsequently, Trijatmiko et al. (2016) transformed both rice nicotianamine synthase (*OsNAS2*)

Table 15.2 Metabolic engineering of increasing Fe, Zn, Se, and I in staple crops

Micronutrients	Genes	Gene sources	Manipulation methods	Traits improved	References
Iron (Fe)	*Ferritin*	*Ferritin* from *Glycine max*	*Agrobacterium tumefaciens*	Increased Fe content in rice seeds up to 38.1 μg/g DW.	Goto et al. (1999)
	Ferritin	*Ferritin* from *Phaseolus vulgaris*	*Agrobacterium tumefaciens*	Increased Fe content in rice seeds up to 22.1 μg/g DW.	Lucca et al. (2001)
	PhyA (phytoene synthase) and *ferritin*	*PhyA* from *Aspergillus niger* and *ferritin* from *G. max*	Particle bombardment	Increased Fe content in maize seeds up to 35 μg/g DW and its bioavailability.	Drakakaki et al. (2005)
	OsYSL2 (yellow stripe 1 (*YS1*)-like)	*OsYSL2* from *Oryza sativa*	*Agrobacterium tumefaciens*	Fe content increased 4.4 times in polished rice compared to the control.	Ishimaru et al. (2010)
	Ferritin	*Ferritin* from *G. max*	Particle bombardment	Increased Fe content in maize seeds to 70 μg/g DW.	Aluru et al. (2011)
	Ferritin	*TaFer1-A* from *Triticum aestivum*	Microprojectile bombardment	Increased Fe content in wheat seeds up to 95 μg/g DW and its bioavailability.	Borg et al. (2012)
	OsNAS2 (nicotianamine synthase 2)	*OsNAS2* from *O. sativa*	*Agrobacterium tumefaciens*	Increased Fe content in rice seeds to about 26 μg/g DW.	Lee et al. (2012)
	FEA1 (algae iron assimilatory gene)	*FEA1* from *Chlamydomonas*	*Agrobacterium tumefaciens*	Fe in mature cassava roots increased from 10 to 36 μg/g DW.	Ihemere et al. (2012)
Zinc (Zn)	*OsZIP1* (Zn transporter)	*OsZIP1* from *O. sativa*	*Agrobacterium tumefaciens*	Zn in finger millet seeds increased 10 to 20 μg/g DW from the control level (about 68 μg/g DW).	Ramegowda et al. (2013)
	PhyA (phytase)	*PhyA* from *Aspergillus japonicus*	*Agrobacterium tumefaciens*	Fe and Zn contents in wheat flour were 14.28% and 4% higher than non-transgenic control.	Abid et al. (2017)

	HvMTP1 (metal tolerance protein)	*HvMTP1* from *Hordeum vulgare*	*Agrobacterium tumefaciens*	Zn content in barley endosperm increased 31% compared to the control.	Menguer et al. (2018)
	OsZIP9 (Zn transporter9)	*OsZIP9* from *O. sativa*	*Agrobacterium tumefaciens*	Zn in brown rice grain increased by 25% to 35% compared with the wild type.	Tan et al. (2020)
Fe and Zn	*OsNAS3* (nicotianamine synthase 3)	*OsNAS2* from *O. sativa*	*Agrobacterium tumefaciens*	Fe and Zn contents in milled rice seeds were about 12 and 35 μg/g DW, respectively.	Lee et al. (2009)
	OsNAS2 (nicotianamine synthase 2)	*OsNAS2* from *O. sativa*	*Agrobacterium tumefaciens*	Fe and Zn contents in unpolished rice seeds were 64 and 91 μg/g DW, respectively.	Johnson et al. (2011)
	Gmferritin H2, *HvNAS1*, and *OsYSL2*	*Ferritin* from *G. max*, *HvNAS1* from *Hordeum vulgare, and YSL2 from O. sativa*	*Agrobacterium tumefaciens*	Fe and Zn in polished rice seeds were 4.4 and 1.6 times higher than the control, respectively.	Masuda et al. (2012)
	OsNAS2 and *SferH-1*	*SferH-1* (soybean *ferritin*)	*Agrobacterium tumefaciens*	Fe and Zn contents in polished rice grains were 15 μg/g and 45.7 μg/g.	Trijatmiko et al. (2016)
	phyA (phytase)	*PhyA* from *Aspergillus japonicus*	*Agrobacterium tumefaciens*	Fe and Zn contents in wheat flour were 14.28% and 4% higher than non-transgenic control.	Abid et al. (2017)
	OsNAS2	*OsNAS2* from *O. sativa*	Particle bombardment	Increased Fe and Zn contents in wheat white flour to 22 and 31 μg/g DW, respectively.	Beasley et al. (2019)

(continued)

Table 15.2 (continued)

Micronutrients	Genes	Gene sources	Manipulation methods	Traits improved	References
	OsIPK5/6 K1 (inositol 1,3,4,5,6-pentakisphosphate 2-kinase)	*OsIPK5/6 K1* from *O. sativa*	*Agrobacterium tumefaciens*	RNAi-mediated down-regulation of *OsIPK5/6 K1* decreased PA by 46.2% and increased Fe and Zn contents in polished rice seeds by 1.3 and 1.6 folds, respectively.	Karmakar et al. (2020)
	HsSL (selenocysteine lyase), *AbSMT* (selenocysteine methyltransferase)	*HsSL* from *Homo sapiens, AbSMT* from *Astragalus bisulcatus*, and	*Agrobacterium tumefaciens*	When treated with selenate and selenite, respectively, transgenic rice seedlings accumulated 38.5% and 128.6% more Se than the wild-type plants.	Li et al. (2022)
Selenium (Se)	*APS* (ATP sulfurylase)	*APS from Arabidopsis thaliana*	*Agrobacterium tumefaciens*	Shoot Se levels were 2-fold higher than wild type treated with 50 μM selenate.	Pilon-Smits et al. (1999)
	SMT (SeCys methyltransferase)	*SMT* from *Astragalus bisulcatus*	*Agrobacterium tumefaciens*	Mature transgenic Indian mustard accumulated 2-fold higher levels of Se than wild type when treated with 50 μM selenite.	LeDuc et al. (2004)
	APS and *SMT*	*APS* from *Arabidopsis thaliana* and *SMT* from *Astragalus bisulcatus*	Crossing APS and SMT transgenic plants	Homozygous *APS* and *SMT* Indian mustard plants accumulated 4000 μg/g DW Se when treated with 500 μM selenate.	LeDuc et al. (2006)
Iodine (I)	*NIS* (human sodium-iodide symporter), *HOL* (harmless to ozone layer or halide methylatransferase)	*NIS* from human and *Arabidopsis hol-1* mutants	*Agrobacterium*-mediated transgenic NIS plants were crossed with *hol-1* mutants	Transgenic *NIS Arabidopsis* plants with defect *HOL-1* gene accumulated more than 50 and 170 μg/g FW I when applied with 3.5 and 35 μM of iodide, respectively.	Landini et al. (2012)

and soybean ferritin (*SferH-1*) genes into an indica rice variety IR64, which endogenously contained 2 mg/kg Fe in polished grain. Fe and Zn contents in polished grain of transgenic plants of IR64 were 15 mg/k and 45.7 mg/kg, respectively, indicating that Fe content in polished transgenic grain reached the targeted 15 mg/kg. This result documented the importance of selecting suitable varieties for genetic transformation.

Genetic engineering of key regulators should be another approach for increasing Fe contents in staple crops. Several transcription factors, such as *OsbHLH156*, *OsIRO2/OsbHLH56*, *OsPRI1/OsbHLH60*, *OsPRI2/OsbHLH58*, *OsPRI3/OsbHLH59*, and *OsIRO3/OsbHLH63* have been identified in rice (Kong et al. 2022; Liang 2022). OsbHLH156 and OsIRO2 have been shown to form a functional transcription activation complex to regulate Fe uptake (Wang et al. 2020). Rice has additional regulators, such as *OsHRZ1* and *OsHRZ2*, two iron-binding hemerythrin motif-containing E3 ubiquitin ligases . Knocking down *OsHRZ* increased Fe accumulation in shoots and gains (Kobayashi et al. 2013). Thus far, no report has been available on manipulating these regulators in improving Fe contents in polished rice grains.

15.4.2.2 Engineering for Increasing Zn Contents

Zinc in grains of staple food is low (Rashid and Fox 1992). The average Zn content is 16 mg/kg in polished rice and 25 mg/kg in wheat (Bouis et al. 2011), but the current target Zn contents in rice and wheat are 28 and 40 mg/kg, respectively (Bouis and Saltzman 2017). Increasing Zn accumulation depends on root uptake of Zn from soils, transport of absorbed Zn from roots to shoots, and remobilization into grains. Several zinc-regulated (ZR) transporters, iron-regulated transporter (ZR-IRT)-like protein (ZIP) transporters play important roles in Zn uptake and intracellular transport. In rice, *OsZIP1, OsZIP2, OsZIP3, OsZIP15, OsZIP18,* and *OsZIP9* have been extensively studied (Kong et al. 2022). As shown in Table 15.2, the transformation of *OsZIP1* into finger millet led to a 10 to 20 mg/kg increase of Zn in seeds compared to the non-transgenic control (Ramegowda et al. 2013). Overexpression of *OsZIP9* resulted in a 25% to 35% increase in Zn in brown rice (Tan et al. 2020). *OsZIP5* and *OsZIP8* also contribute to Zn uptake, but *OsZIP2* and *OsZIP3* involve Zn translocation by unloading Zn from the xylem to developing rice tissues (Kong et al. 2022).

Other proteins also engage in Zn uptake and transport. Endosperm-specific expression of *HvMTP1* (*Hordeum vulgare* metal tolerance protein) resulted in a 31% increase in Zn content in endosperm (Menguer et al. 2018). *OsHMA2* (heavy metal *ATPase*) plays an important role in the transport of Zn from roots to shoots (Yamaji et al. 2013), while OsHMA3 mediates Zn sequestration into the vacuoles (Cai et al. 2019). Since Zn and Fe share similar properties (ionic charge and size) in relation to plant uptake, proteins that facilitate Fe uptake also enhance plant uptake of Zn. Table 15.2 shows that transgenic rice and wheat plants overexpressing *OsNAS2* increased both Fe and Zn contents in seeds. Overexpression of *OsNAS3* enhanced Fe and Zn contents in polished rice. Transformation of multiple genes, such as *Gmferritin H2, HvNAS1,* and *OsYSL2,* as well as *HsSL* and *AbSMT,* elevated Fe and Zn in rice seeds. More notably, overexpression of both *OsNAS2* and *SferH-1* increased Fe and Zn in polished rice to 15 mg/kg and 45.7 mg/kg, respectively, which met the target Fe and Zn contents of 15 and 28 mg/kg in polished rice.

15.4.2.3 Manipulating Phytic Acid to Increase Fe and Zn Contents

Phytic acid (PA) occurs naturally in high concentrations in cereal grains and legumes as the primary storage form of phosphorus (P). On the other hand, PA is a chelator and can form insoluble complexes with metal ions, including Fe and Zn, in the small intestine, resulting in impaired absorption of Fe and Zn (Raboy 2020). As a result, high concentrations of PA in dietary are associated with Fe and Zn deficiencies. Furthermore, PA is mainly present in the aleurone layer of seeds, which can restrict the translocation of Fe or Zn into the endosperm, resulting in low concentrations of Fe and Zn in the endosperm. Meanwhile, PA is the only source of P for seed germination. Thus, establishing appropriate levels of PA in seeds, known as nonlethal low PA, is important for different grains and legumes. Traditional breeding has successfully developed nonlethal low PA in barley, maize, and rice (Stanton et al. 2022). Transgenic approaches for reducing PA have also been conducted. Overexpression of a phytase gene (*phyA*) from *Aspergillus japonicus* resulted in a 12–76% reduction of PA and significantly increased Fe and Zn contents in wheat seeds (Abid et al. 2017). RNA interference (RNAi) technology was used for reducing PA as well. Transgenic rice with down-regulated expression of a major *ITPK* homolog (*OsITP5/6 K-1*) showed a 46.2% reduction in PA and 1.3 and 1.6-fold increase in Fe and Zn contents in polished rice grains, respectively (Karmakar et al. 2020).

15.4.2.4 Engineering for Increasing Se Content

Selenium is not essential but beneficial to plant growth and development. Roots absorb Se using the sulfate transport system due to their chemical similarity (Schiavon and Pilon-Smits 2017). Plants can absorb both inorganic Se as selenate (SeO_4) and selenite (SeO_3) and organic forms of Se, such as selenocysteine (SeCys) and selenomethionine (SeMet). When grown in non-seleniferous soils, Se contents in most plants range from 10 to 100 mg/kg dry weight (White 2017) and exist in both inorganic and organic forms. Inorganic forms of Se are usually complexed with sodium to form Na_2SeO_3 and Na_2SeO_4. The most common organic form of Se is selenomethionine (SeMet), followed by selenocysteine (Sec). Organic forms are more bioavailable to humans than inorganic forms (Radomska et al. 2021). Currently, the biofortification of Se in maize, rice, and wheat is largely through agronomic means, such as the application of Se fertilizers (Sarwar et al. 2020; Ghorai et al. 2022). The genetic engineering is focused on (a) reducing the conversion of selenate to selenite by manipulating ATP sulfurylase (APS) because the bioavailability of selenate to humans is higher than selenite; (b) enhancing the conversion of SeCys to SeMet by SeCys methyltransferase (SMT) as SeCys is less bioavailable than SeMet; (c) restraining SeCys into proteins by regulating SMT; and (d) Se volatilization (Zhu et al. 2009; Malagoli et al. 2015). These approaches, primarily the first two, are limited to some non-cereal plants. Overexpression of *APS* or *SMT* resulted in a 2-fold increase of Se in shoots of Indian mustard compared to non-transgenic plants after all plants were fertilized with 50 μM selenate (Table 15.2).

To further increase Se contents, *APS* and *SMT* transgenic lines were hybridized, and homozygous lines carrying both *APS* and *SMT* in Indian mustard accumulated 4000 μg/g DW Se when they were fertilized with 500 μM selenate (LeDuc et al. 2004, 2006). However, these genes have not been manipulated in staple crops.

15.4.2.5 Engineering for Increasing Iodine Contents

Like Se, iodine is not essential to plants but is beneficial to plant growth at low concentrations (Zhang et al. 2023). Iodine in soils exists in inorganic as IO^{-}_{3} (iodate) and I^{-} (iodide) and organic forms with an average concentration of 5.1 mg/kg (Gonzali et al. 2017). Plants can take up iodine from soils and transport it to shoots. Additionally, plants can emit volatile methyl iodine through a reaction catalyzed by a S-adenosyl-L-methionine (SAM)-dependent halide/thiol methyltransferase (HTMT) (Gonzali et al. 2017). Applying iodine increases plant uptake of iodine, but iodine transporter has not been identified. Landini et al. (2012) studied iodine accumulation in *Arabidopsis* and reported that the expression of a human sodium-symporter (*NIS*) or knocking down the *HOL-1* gene increased iodine content. Subsequently, Carlessi et al. (2021) knocked out rice *OsHOL1* and *OsHOL2* using *CRISPR/Cas9* technology and found that the emission of CH_3I from mutated *OsHOL1* T1 lines was substantially reduced, but from mutated *OsHOL2* T_1 lines was variable. A new development with rice is sea rice that has been increasingly cultivated in China (Chen et al. 2017); iodine uptake, transport, and accumulation in sea rice grain could be appropriately investigated. Iodine in seawater ranges from 40 to 60 mg/L; sea rice may evolve a mechanism for the absorption and transport of iodine. It is likely that iodine in seeds of sea rice could be highly elevated.

15.4.3 Engineering of Vitamins

Whole grain cereals generally contain relatively higher vitamins A, B1–B3, B5, B6, B9, E, and K than B12, C, and D (Acosta-Estrada et al. 2019). Seed coats are richer in these vitamins than the endosperm; processing or milling of whole grains causes a significant loss of the vitamins (Garg et al. 2021). Starchy root and tuber crops vary substantially in these vitamins. For example raw sweet potatoes, cassava, and yam have higher vitamin A concentrations than raw potatoes. Folate in raw potatoes, cassava, and yam is much higher than in sweet potatoes (Chandrasekara and Josheph Kumar 2016). The metabolic engineering of vitamins in staple crops has been largely focused on vitamins A, B9, and E.

15.4.3.1 Engineering of Vitamin A

Humans obtain vitamin A from two sources: preformed vitamin A, including retinol, retinal, and retinyl esters from animal tissue, and provitamin A carotenoids, including α-carotene, β-carotenes, and β-cryptoxanthin from plants. To use both forms of vitamin A, our body must convert them to retinal and retinoic acid, the active forms of the vitamin (Garg et al. 2021). Other carotenoids, like lycopene, lutein, and zeaxanthin, benefit our health but cannot be converted to vitamin A. The major provitamin A in cereal grains are β-carotenes and β-cryptoxanthin (Trono 2019). It was reported that 74%–86% of carotenoids are in the corneous endosperm, 9–23% in the endosperm, 2%–4% in the germ, and 1% in the pericarp (Garg et al. 2021).

Carotenoid biosynthesis in higher plants occurs in plastids through the methylerythritol phosphate (MEP) pathway, of which deoxy-D-xylulose 5-phosphate synthase (DXS) is a key rate-limiting enzyme for providing precursors. Phytoene synthase (PSY) is the first-rate limiting enzyme catalyzing phytoene formation, a colorless carotenoid acting as a precursor to other carotenoids. Metabolic engineering started initially with rice (Table 15.3). The *psy* and *β-lcy* (lycopene β-cyclase), originating from *Narcissus pseudonarcissus* plants under the control of the endosperm-specific glutelin promoter and *crtI* (phytoene desaturase) from *Erwinia uredovora* under the control of CaMV 35S promoter were transformed into rice through *Agrobacterium tumefaciens*. This metabolic engineering process detected 1.6 mg/kg carotenoid in the endosperm (Ye et al. 2000). The success in enriching vitamin A in rice endosperm led to the term: golden rice (Beyer et al. 2002) that is known as the original or first generation of golden rice. The same genes were transformed through particle bombardment, which increased carotenoids from 0.297 to 1.65 mg/kg (Datta et al. 2003).

The transformation of *ZmPsy* derived from maize along with the bacterial *crtI* increased carotenoids in rice endosperm by up to 37 mg/kg that was 23-fold higher than the carotenoid concentration in the original golden rice (Paine et al. 2005); this transformation was known as the second generation of golden rice. Additionally, the expression of *ZmPsy* and *crtI* along with *Arabidopsis AtDXS* and *AtOP* (an orange gene to create a metabolic sink) resulted in 31.78 mg/kg total carotenoids in rice grain (Bai et al. 2016). The concept was subsequently used in other crops. The expression of *Erwinia herbicolar CrtB* and *CrtI* genes in maize increased β-carotene by 7- to 13-fold in maize seeds compared to non-transgenic yellow maize (Aluru et al. 2008). Additionally, *ZmPsy* from *Z. may, PaCrtI* from *Pantoea ananatis, Gllych* (β-carotene hydroxylase), *Glbch* (lycopene β-cyclase) from *Gentiana lutea,* and *ParacrtW* (β-carotene ketolase) from *Paracoccus* ssp were transformed into maize (Zhu et al. 2008). β-carotene contents at 48.87 mg/kg in the endosperm of transgenic lines expressing *Zmpsy1* + *PacrtI* + *Gllycb*, and 57.35 μg/kg DW in transgenic lines carrying *Zmpsy1* and *PacrtI* were achieved.

The total carotenoid in the endosperm of transgenic wheat plants expressing *zmY1* (phytoene synthase from maize) and *cetI* reached 4.96 mg/kg (Cong et al. 2009). Another work on the transformation of *CrtB* and *CrtI* led to 4.76 mg/kg total carotenoids and 3.21 mg/kg β-carotene in wheat endosperm (Wang et al. 2014).

Sorghum expressing *zmPSY1*, *CrtI,* and *LPA-a* (low phytic acid 1) elevated provitamin carotenoids up to 14.0 mg/kk in grain (Lipkie et al. 2013). The expression of *ZmPSY1*, *CrtI*, A*tDXS* (arogenate dehydratases), and *HvHGGT* (homogentisate geranylgeranyl transferase) resulted in β-carotene in sorghum endosperm ranging from 2.5 to 9.2 mg/kg (Storozhenko et al. 2007).

15.4.3.2 Engineering of Folates

Folates comprise three key moieties: a pterin ring whose precursors are synthesized in the 0.cytosol; a *para*-aminobenzoate produced in plastids, and a glutamate tail. The first two are imported into the mitochondria in which folates are synthesized. Overexpression of *GTPCHI* encoding the first enzyme GTP cyclohydrolase I in pterin biosynthesis and *ADCS* encoding aminodeoxychorismate synthase in the *para*-aminodeoxychorismate branch from *Arabidopsis* increased folate content in polished rice grains (Storozhenko et al. 2007). However, vitamins are sensitive to heat-induced and photooxidative degradation during storage. Folate levels decrease by 50% after four months of storage (Blancquaert et al. 2015). To increase the folate stability of rice grains of GA (*GTPCHI* and *ADCS*) transgenic lines, two approaches were implemented: one was the polyglutamylation of folate by folylpolyglutamate synthetase (FPGS), and the other was to complex with folate binding proteins (FBP). Two transgenic lines: GA-CAFBP and GA-ctF-CAFBP resulted from the transformation of *CAFBP* (*sFBP* fused with a gene encoding Arabidopsis β-carbonic anhydrase 2 to GA and *ctF* (cytosolic FPGS) and *CAFBP* to GA, respectively, that had significantly higher levels of folate than GA lines, folate was stable for four months at 28 °C (Blancquaert et al. 2015). Potato has low concentrations of folate, about 0.18 mg/kg fresh weight. However, the transformation of the GA genes did not significantly increase folate content (Blancquaert et al. 2013). De Lepeleire et al. (2018) transformed *HPPK/DHPS* and/or *FPGS* (folylpolyglutamate synthetase, extending glutamate tail) into GA potato lines, resulting in a total folate content of 19.25 mg/kg based on dry weight. Moreover, folates in tubers were stable upon long-term storage. The *HPPK/DHPS* encodes a bifunctional enzyme that catalyzes activities of both 6-hydroxymethyl-7,8-dihydropterin pyrophosphokinase (HPPK) and dihydropteroate synthase (DHPS) in mitochondrial folate biosynthesis.

Cereals, in general, are low in folates but contain high levels of MeFox (a pyrazine-s-triazine derivative of 4α-hydroxy-5-methyltetrahydrofolate), a biologically inactive, oxidized folate. Liang et al. (2024) identified a rice gene *OsGFT1* (glutamate formiminotransferase 1) that plays an important role in MeFox formation. Using *CRISPR/Cas9* technology, the authors knocked out this gene. Knocked-out mutants substantially reduced MeFox levels. Subsequently, a binary vector harboring *AtGTPCHI* and *AtADCS* was transformed into a wild type of knocked-out mutant. Transgenic knocked-out mutant lines were crossed with transgenic GA lines, progenies resulting from the cross had high concentrations of folate, up to 4.59 mg/kg based on fresh weight. Thus, increasing folates in cereals involves not only overexpression of the genes in the biosynthesis of folates but also knocking out genes in the production of the inactive forms of folates. Polyglutamylation of folate and/or complexing folate with folate binding proteins are effective approaches for stabilizing folates.

15.4.3.3 Engineering of Vitamin E

Vitamin E in plants consists of tocotrienols and tocopherols. Based on the number and position of methyl groups on the aromatic ring, each has four members, α, β, γ, and δ. Tocotrienols occur mainly in seeds of most monocots and a few dicots. Seeds

Table 15.3 Metabolic engineering for increasing vitamins A, B9, and E in major staple crops

Vitamins	Genes	Gene sources	Manipulation methods	Traits improved	References
A	*psy* (phytoene synthase), *β, β-lcy* (lycopene β-cyclase), and *crtI* (phytoene desaturase)	Both *psy* and *β-lcy* from *Narcissus pseudonarcissus* and *crtI* from *Erwinia uredovora*	*Agrobacterium*-mediated	Rice endosperm contained carotenoid at 1.6 μg/g DW	Ye et al. (2000)
	psy, *lcy*, and *crtI*	The same sources as mentioned above	Particle bombardment	Total carotenoid levels varied from 0.297 to 1.05 1.6 μg/g in rice endosperm.	Datta et al. (2003)
	psy, *crtI*	*Psy* from maize and *crtI* from *Erwinia uredovora*	*Agrobacterium tumefaciens*	The highest level of total carotenoids was 37 μg/g that was 23-fold higher than original Golden Rice.	Paine et al. (2005)
	psy (phytoene synthase) and *crtI* (phytoene desaturase)	Erwinia herbicola	Particle bombardment	Increased β-carotene to 7–13-fold compared to non-transgenic yellow maize, B73.	Aluru et al. (2008)
	ZmPsy 1 (phytoene synthase), *PaCrtI* (phytoene desaturase), *Gllycb* (lycopene β-cyclase), *Glbch* (β-carotene hydroxylase), and *ParacrtW* (β-carotene ketolase)	*ZmPsy* from *Z. may, PaCrtI* from *Pantoea ananatis*, *Gllych* and *Glbch* from *Gentiana lutea,* and *ParacrtW* from *Paracoccus* ssp.	Particle bombardment	Maize endosperm β-carotene contents were 48.87 μg/g DW in transgenic lines harboring *Zmpsy1*+ *PacrtI* + *Gllycb* and 57.35 μg/g DW in transgenic lines carrying *Zmpsy1* and *PacrtI*.	Zhu et al. (2008)

	ZmY1 (phytoene synthase) and *crtI* (phytoene desaturase)	*zmY1* from *Z. mays* and *crtI* from *E. uredovora*	Particle bombardment	Total carotenoids were up to 4.96 μg/g DW in wheat endosperm.	Cong et al. (2009)
	ZmPSY1 (phytoene synthase), *Crt1* (phytoene desaturase), and *LPA-1* (low phytic acid 1)	*PSY1* from *Z. mays*, *Crt1* from *E. uredovora*, and *LPA-1* from Sorghum bicolor	*Agrobacterium tumefaciens*	Provitamin A carotenoids were up to 14.0 μg/g DW increase in sorghum grain.	Lipkie et al. (2013)
	CrtB (phytoene synthase) and *CrtI* (phytoene desaturase)	*Erwinia uredovora*	Particle bombardment	Total carotenoids were up to 4.76 μg/g DW and β-carotene 3.21 μg/g DW in wheat endosperm.	Wang et al. (2014)
	ZmPSY1, *Crt1*, *AtDXS* (1-deoxy-D-xylulose 5-phosphate synthase) or *AtOR* (an ORANGE gene)	*AtDXS* from *Arabidopsis thaliana*,	Particle bombardment	Total carotenoids in rice grain were up to 31.78 mg/kg.	Bai et al. (2016)
	ZmPSY1, *Crt1*, *AtDXS*, *HvHGGT* (homogentisate geranylgeranyl transferase)	*HyHGGT* from *Hordeum vulgare*	*Agrobacterium tumefaciens*	Top five transgenic lines accumulated β-carotene ranging from 2.5 to 9.168 μg/g DW in sorghum endosperm.	Chen et al. (2002)
	GTPCHI (GTP-cyclohydrolase I) and *ADCS* (aminodeoxychorismate synthase)	*Arabidopsis thaliana*	*Agrobacterium tumefaciens*	Transgenic lines overexpressing both genes accumulated 15–100 times more folate than the control lines.	Storozhenko et al. (2007)

(continued)

Table 15.3 (continued)

Vitamins	Genes	Gene sources	Manipulation methods	Traits improved	References
B9 (Folate)	*GTPCHI*, *ADCS*, *mtFPGS* (mitochondrial folylpolyglutamate synthetase), *ctFPGS* (cytosolic FPGS), *CAFBP* [β- carbonic anhydrase 2 with *fsFBP* (synthetic folate binding protein)]	*Arabidopsis thaliana*	*Agrobacterium tumefaciens*	The greatest folate content in transgenic rice seeds was 25.29 μg/g FW, which was 150-fold higher than that of wild type.	Blancquaert et al. (2015)
	FPGS (folylpolyglutamate synthetase), *HPPK/ DHPS* (bifunctional hydroxymethyldihydropterin pyrophosphokinase/ dihydropteroate synthase)	*FPGS* from *A. thaliana*), *HPPK/ DHPS* from *O. sativa*	*Agrobacterium tumefaciens*	Total folate levels in potato tubers were up to 19.25 μg/g DW, which was 12-fold higher than the control.	De Lepeleire et al. (2018)
	γ-TMT (γ-tocopherol methyltransferase)	*A. thaliana*	*Agrobacterium mediated*	The α-tocotrienol content increased 1.8-fold in transgenic rice seeds compared to the control at about 10 μg/g DW.	Zhang et al. (2013a)
	GmTMTa (γ-tocopherol methyltransferase)	*Glycine max*	*Agrobacterium tumefaciens*	The α-tocotrienol contents in seeds of transgenic maize lines were 3–4.5-fold higher than wild type.	Zhang et al. (2013b)
	HGGT (homogentisate geranylgeranyl transferase), *γ-TMT*	*HGGT* from *Hordeum vulgare* and *γ-TMT from soybean*	Particle bombardment	Transgenic soybean seeds had 8- to 10-fold increases in tocotrienols in δ and γ forms, which were largely shifted to α and γ forms with co-expression of *γ-TMT*.	Konda et al. (2020)
	ZmTMT (γ-tocopherol methyltransferase)	*Z. mays*	*Agrobacterium tumefaciens*	The α-tocopherol content in the kernels of transgenic maize increased 6.5-fold compared to the control.	Zhang et al. (2020)

	AtHPT (homogentisate phytyltransferase), *Atγ-TMT*	*Arabidopsis thaliana*	*Agrobacterium tumefaciens*	The α-tocopherol contents in tubers of transgenic potato ranged from 32.5 to 51.0 mg/kg compared to 18.4 mg/kg in control tubers.	Upadhyaya et al. (2021)
	HaHPT and *HaTMT*	*Helianthus annuus*	*Agrobacterium tumefaciens*	The α-tocopherol contents in rice grain increased 2.3 times in transgenic *HaTMT* rice plants compared to wild-type plants.	Song et al. (2024)

of staple grains such as rice, oats, and wheat contain more than 50% tocotrienols. Tocopherols occur more widely than tocotrienols, presenting in the leaves and seeds of most dicots. (Cahoon et al. 2003). Tocotrienols have a higher ability than tocopherols to scavenge free radicals but are not as easily absorbed by humans as α-tocopherol (Packer et al. 2001). On the other hand, all tocopherols are equally absorbed during digestion, but the human body preferentially retains and absorbs α-tocopherol. As a result, α-tocopherol is considered to be the most important to human health due to its highest vitamin E activity.

Dietary intake of vitamin E is largely from edible oils, but γ-tocopherol contents in those oils vary greatly. Soybean oil accounts for 80% and 25% of the edible oil consumed in the U.S. and the world, respectively, but it contains 7% of α-tocopherol and 70% of γ-tocopherol (Shintani and DellaPenna 1998). The other oilseed crops, such as canola, corn, cottonseed, and palm oils, have similarly low α- to γ-tocopherol ratios. Thus, substantial effort on the increase of α-tocopherol content has been made through metabolic engineering.

One approach is engineering γ-tocopherol methyltransferase (γ-TMT), an enzyme catalyzing the final step for converting γ-tocopherol to α-tocopherol. In general, γ-tocopherol, the precursor to α-tocopherol, is much higher than α-tocopherol in many oilseeds, suggesting that the final reaction for converting γ-tocopherol to α-tocopherol catalyzed by γ-TMT is the limiting step. Shintani and DellaPenna (1998) produced transgenic *Arabidopsis* lines by overexpressing a *γ-TMT* driven by a seed-specific carrot DC3 promoter. Seeds of the transgenic lines had 95% of their total tocopherol pool as α-tocopherol. This suggests that overexpression of *γ-TMT* shifted seed oil compositions in favor of α-tocopherol. This concept was applied to enhance vitamin E contents in other crops. The *Arabidopsis* *γ-TMT* was transformed into rice, and α-tocopherol in rice seeds increased 1.8-fold compared to seeds of non-transgenic control (Zhang et al. 2013a). Transformation of a *GmTMTa* isolated from soybean into maize resulted in a 3–4.5-fold increase of α-tocotrienol in maize seeds compared to non-transgenic plants (Zhang et al. 2013b). Moreover, a γ-TMT was isolated from *Zea mays*, *ZmTMT*, and its overexpression increased α-tocopherol content by 6.5-fold in transgenic maize kernels (Zhang et al. 2020).

Another approach involves the manipulation of either homogentisate geranylgeranyl transferase (HGGT) or homogentisic phytyltransferase (HPT). Tocotrienols arise from the condensation of geranylgeranyl diphosphate (GGDP and homogentisic acid (HGA) catalyzed by HGGT, and tocopherols initially start with the condensation of HGA and phytyl diphosphate (PDP) catalyzed by HPT. Overexpressing barley *HGGT* in corn seeds increased tocotrienol and tocopherol contents as much as 6-fold (Cahoon et al. 2003). Subsequently, Konda et al. (2020) transformed the barley *HGGT* into soybean, and seeds of transgenic lines expressing *HGGT* had 8- to 10-fold increases in total vitamin E, primarily tocotrienols in δ and γ forms, but they were largely shifted to α and γ forms with the co-expression of soybean *γ-TMT*. On the other hand, Upadhyaya et al. (2021) co-transformed *HPT* and *γ-TMT* isolated from *Arabidopsis* to potato. The transgenic potato lines had significantly higher α-tocopherol contents ranging from 32.5 to 51.0 mg/kg compared to the non-transformed control plants of 18.4 mg/kg. Similarly, the transformation of *HPT* and *γ-TMT* derived from sunflower increased α-tocopherol 2.3 times in rice grain compared to the non-transformed control (Song et al. 2024).

15.5 Prospects

Deficiencies in essential micronutrients have become a global health issue as more than two billion people in the world are deficient in key vitamins and minerals (WHO 2006). Biofortification through agronomic means and conventional breeding have made significant progress in fighting against the deficiencies. Metabolic engineering, as documented in this chapter, represents a viable alternative to accelerate the progress.

Metabolic engineering is a rapidly evolving field. With the arrival of omics technologies and whole genome sequencing of more economically important crops, biosynthesis pathways, and key genes involved in the biosynthesis of vitamins will become clearer, and mechanisms underlying plant uptake, transport, and accumulation of essential microelements will be revealed. More organ- and tissue-specific promoters enhancing target gene expression will be identified and used. Gene editing will facilitate gene manipulation, and in-plant transformation will speed up the transformation process. The current effort has been mainly focused on manipulating structure genes. The engineering of transcription factors could be a future emphasis, which will lead to the simultaneous expression of multiple genes of interest and the improvement of different essential micronutrients in staple crops. Breakthrough could be made in rice and corn to increase the content of Fe, Zn, and vitamins.

Molecular biologists and plant breeders should work together to increase the efficiency of metabolic engineering. Plant breeders can provide valuable genetic resources for molecular biologists to jump-start the project. Plant species or cultivars that possess desirable agronomic traits, contain high levels of target compounds, and are relatively easy for transformation should be recommended to molecular biologists. Breeders can develop effective methods for selecting transgenic lines, hybridizing to incorporate different transgenes into single plants, and implementing strategies for maintaining stable transgenic lines for commercial production.

The current regulatory policy for genetic engineered crops, including genetic transformation and gene editing, should be reviewed and revised based on the decade's scientific data regarding their potential adverse effects on living organisms and the environment. Genetic engineered crops without detrimental effects should be released and produced for commercial production. The United Nations Sustainable Development Goal 2 (UN-SDG2 2016) has been proposed to end all forms of hunger by 2030. This daunting test requires metabolic engineering to make concrete contributions to biofortification.

References

Abid N, Khatoon A, Maqbool A, Irfan M, Bashir A, Asif I, Shahid M, Saeed A, Brinch-Pedersen H, Malik KA (2017) Transgenic expression of phytase in wheat endosperm increases bioavailability of iron and zinc in grains. Transgenic Res 26:109–122

Acosta-Estrada BA, Gutiérrez-Uribe JA, Serna-Saldivar SO. (2019) Minor constituents and phytochemicals of the Kernel. In: Serna-Saldivar O, editor. Corn. AACC International Press, p 369–403. https://doi.org/10.1016/B978-0-12-811971-6.00014-0

Aluru M, Xu Y, Guo R, Wang Z, Li S, White W, Wang K, Rodermel S (2008) Generation of transgenic maize with enhanced provitamin A content. J Exp Bot 59:3551–3562

Aluru MR, Rodermel SR, Reddy MB (2011) Genetic modification of low phytic acid 1-1 maize to enhance iron content and bioavailability. J Agric Food Chem 59:12954–12962. https://doi.org/10.1021/jf203485a. Pubmed: 22088162

Anjanappa RB, Gruissem W (2021) Current progress and challenges in crop genetic transformation. J Plant Physiol 261:153411

Bai C, Capell T, Berman J, Medina V, Sandmann G, Christou P, Zhu C (2016) Bottlenecks in carotenoid biosynthesis and accumulation in rice endosperm are influenced by the precursor-product balance. Plant Biotechnol J 14:195–205

Bailey JE (1991) Toward a science of metabolic engineering. Science 252:1668–1675

Bailey LB, Gregory JF III (1999) Folate metabolism and requirements. J Nutr 129:779–782

Bailey RL, West KP Jr, Black RE (2015) The epidemiology of global micronutrient deficiencies. Ann Nutr Metabol 66(Suppl. 2):22–33

Bänziger M, Long J (2000) The potential for increasing the iron and zinc density of maize through plant-breeding. Food Nutr Bull 21:397–400

Beal T, Ortenzi F (2022) Priority micronutrient density in foods. Front Nutr 9:806566

Beasley JT, Bonneau JP, Sánchez-Palacios JT, Moreno-Moyano LT, Callahan DL, Tako E, Glahn RP, Lombi E, Johnson AA (2019) Metabolic engineering of bread wheat improves grain iron concentration and bioavailability. Plant Biotechnol J 17:1514–1526

Beyer P, Al-Babili S, Ye X, Lucca P, Schaub P, Welsch R, Potrykus I (2002) Golden rice: Introducing the β-caroteen biosynthesis pathway into rice endosperm by genetic engineering to defeat vitamin A deficiency. J Nutr 132:506S–510S

Blancquaert D, Storozhenko S, Van Daele J, Stove C, Visser RGF, Lambert W, Van Der Straeten D (2013) Enhancing pterin and para-aminobenzoate content is not sufficient to successfully biofortify potato tubers and Arabidopsis thaliana plants with folate. J Exp Bot 64:3899–3909

Blancquaert D, Van Daele J, Strobbe S, Kiekens F, Storozhenko S, De Steur H, Gellynck X, Lambert W, Stove C, Van Der Straeten D (2015) Improving folate (vitamin B9) stability in biofortified rice through metabolic engineering. Nat Biotechnol 33:1076–1078

Borg S, Brinch-Pedersen H, Tauris B, Madsen LH, Darbani B, Noeparvar S, Holm PB (2012) Wheat ferritins, improving the iron content of the wheat grain. J Cereal Sci 56:204–213

Bouis HE, Saltzman A (2017) Improving nutrition through biofortification: a review of evidence from HarvestPlus, 2003 through 2016. Glob Food Sec 12:49–58

Bouis HE, Hotz C, McClafferty B, Meenakshi JV, Pfeiffer WH (2011) Biofortification: A new tool to reduce micronutrient malnutrition. Food Nutr Bull 32:S31–S40

Brown KM, Arthur JR (2001) Selenium, selenoproteins and human health: a review. Public Health Nutr 4(2b):593–599

Cahoon EB, Hall SE, Ripp KG, Ganzke TS, Hitz WD, Coughlan SJ (2003) Metabolic redesign of vitamin E biosynthesis in plants for tocotrienol production and increased antioxidant content. Nat Biotechnol 21:1082–1087

Cai H, Huang S, Che J, Yamaji N, Ma JF (2019) The tonoplast-localized transporter OsHMA3 plays an important role in maintaining Zn homeostasis in rice. J Exp Bot 70:2717–2725

Carlessi M, Mariotti L, Giaume F, Fornara F, Perata P, Gonzali S (2021) Targeted knockout of the gene OsHOL1 removes methyl iodide emissions from rice plants. Sci Rep 11:17010

Chandrasekara A, Josheph Kumar T (2016) Roots and tuber crops as functional foods: a review on phytochemical constituents and their potential health benefits. Int J Food Sci 3631647:1
Chaparro CM, Suchdev PS (2019) Anemia epidemiology, pathophysiology, and etiology in low- and middle-income countries. Ann N Y Acad Sci 1450:15–31
Chen L, Yang F, Xu J, Hu Y, Hu Q, Zhang Y, Pan G (2002) Determination of selenium concentration of rice in China and effect of fertilization of selenite and selenate on selenium content of rice. J Agric Food Chem 50:5128–5130
Chen R, Cheng Y, Han S, Van Handel B, Dong L, Li X, Xie X (2017) Whole genome sequencing and comparative transcriptome analysis of a novel seawater adapted, salt-resistant rice cultivar–sea rice 86. BMC Genomics 18:1–11
Chungchunlam SM, Moughan PJ (2024) Comparative bioavailability of vitamins in human foods sourced from animals and plants. Crit Rev Food Sci Nutr 64:11590–11625
Cohen SN, Chang ACY, Boyer HW, Helling RB (1973) Construction of biologically functional bacterial plasmids in vitro. Proc Natl Acad Sci USA 70:3240–3244
Cong L, Wang C, Chen L, Liu H, Yang G, He G (2009) Expression of phytoene synthase1 and carotene desaturase crtI genes result in an increase in the total carotenoids content in transgenic elite wheat (*Triticum aestivum* L.). J Agric Food Chem 57:8652–8660
Datta K, Baisakh N, Oliva N, Torrizo L, Abrigo E, Tan J, Rai M, Rehana S, Al-Babili S, Beyer P, Potrykus I (2003) Bioengineered 'golden'indica rice cultivars with β-carotene metabolism in the endosperm with hygromycin and mannose selection systems. Plant Biotechnol J 1:81–90
De Lepeleire J, Strobbe S, Verstraete J, Blancquaert D, Ambach L, Visser R.GF, Stove C, Van Der Straeten D (2018) Folate biofortification of potato by tuber-specific expression of four folate biosynthesis genes. Mol Plant 11:175–188
Drakakaki G, Marcel S, Glahn RP, Lund EK, Pariagh S, Fischer R, Christou P, Stoger E (2005) Endosperm-specific co-expression of recombinant soybean ferritin and Aspergillus phytase in maize results in significant increases in the levels of bioavailable iron. Plant Mol Biol 59:869–880
Fitzgerald MA, McCouch SR, Hall RD (2009) Not just a grain of rice: The quest for quality. Trend Plant Sci 14:133–139
Gaikwad KB, Rani S, Kumar M, Gupta V, Babu PH, Bainsla NK, Yadav R (2020) Enhancing the nutritional quality of major food crops through conventional and genomics-assisted breeding. Front Nutr 7:533453
Garg M, Sharma A, Vats S, Tiwari V, Kumari A, Mishra V, Krishania M (2021) Vitamins in cereals: a critical review of content, health effects, processing losses, bioaccessibility, fortification, and biofortification strategies for their improvement. Front Nutr 8:586815
Ghorai M, Kumar V, Kumar V, Al-Tawaha AR, Shekhawat MS, Pandey DK, Batiha GES, Bursal E, Jha NK, Gadekar VS, Radha (2022) Beneficial role of selenium (Se) biofortification in developing resilience against potentially toxic metal and metalloid Stress in crops: recent trends in genetic engineering and omics approaches. J Soil Sci Plant Nutr 22:2347–2377
Gibson RS, Ferguson EL (1998) Nutrition intervention strategies to combat zinc deficiency in developing countries. Nutr Res Rev 11(1):115–131
Gonzali S, Kiferle C, Perata P (2017) Iodine biofortification of crops: agronomic biofortification, metabolic engineering and iodine bioavailability. Cur Opin Biotechnol 44:16–26
Goto F, Yoshihara T, Shigemoto N, Toki S, Takaiwa F (1999) Iron fortification of rice seed by the soybean ferritin gene. Nat Biotechnol 17:282–286
Graham RD, Senadhira C, Beebe SE, Iglesias C, Monasterio I (1999) Breeding for micronutrient density in edible portions of staple food crops: conventional approaches. Field Crop Res 60:57–80
Hussain A, Jiang W, Wang X, Shahid S, Saba N, Ahmad M, Dar A, Masood SU, Imran M, Mustafa A (2022) Mechanistic impact of zinc deficiency in human development. Front Nutr 9:717064
Ihemere UE, Narayanan NN, Sayre RT (2012) Iron biofortification and homeostasis in transgenic cassava roots expressing the algal iron assimilatory gene, FEA1. Front Plant Sci 3:171
Institute of Medicine (2000) Dietary reference intakes for vitamin C, vitamin E, selenium, and carotenoids. The National Academies Press, Washington, DC, p 528

Institute of Medicine (2001) Dietary reference intakes for vitamin A, vitamin K, arsenic, boron, chromium, copper, iodine, iron, manganese, molybdenum, nickel, silicon, vanadium, and zinc. National Academies Press (US), Washington, DC. https://doi.org/10.17226/10026

Ishimaru Y, Masuda H, Bashir K, Inoue H, Tsukamoto T, Takahashi M, Nakanishi H, Aoki N, Hirose T, Ohsugi R et al (2010) Rice metal-nicotianamine transporter, OsYSL2, is required for the long-distance transport of iron and manganese. Plant J 62:379–390

Islam MR, Akash S, Jony MH, Alam MN, Nowrin FT, Rahman MM, Rauf A, Thiruvengadam M (2023) Exploring the potential function of trace elements in human health: a therapeutic perspective. Mol Cell Biochem 478:2141–2171

Jiang L, Strobbe S, Van Der Straeten D, Zhang C (2021) Regulation of plant vitamin metabolism: backbone of biofortification for the alleviation of hidden hunger. Mol Plant 14:40–60

Johnson AAT, Kyriacou B, Callahan DL, Carruthers L, Stangoulis J, Lombi E, Tester M (2011) Constitutive overexpression of the OsNAS gene family reveals single-gene strategies for effective iron and zinc-biofortification of rice endosperm. PLoS ONE 6:e24476

Karmakar A, Bhattacharya S, Sengupta S, Ali N, Sarkar SN, Datta K, Datta SK (2020) RNAi-mediated silencing of ITPK gene reduces phytic acid content, alters transcripts of phytic acid biosynthetic genes, and modulates mineral distribution in rice seeds. Rice Sci 27:315–328

Keen CL, Clegg MS, Hanna LA, Lanoue L, Rogers JM, Daston GP, Oteiza P, Uriu-Adams JY (2003) The plausibility of micronutrient deficiencies being a significant contributing factor to the occurrence of pregnancy complications. J Nutr 133:1597S–1605S

Kiely ME, McCarthy EK, Hennessy A (2021) Iron, iodine and vitamin D deficiencies during pregnancy: epidemiology, risk factors and developmental impacts. Proc Nutr Soc 80:290–302

Kobayashi T, Nagasaka S, Senoura T, Itai RN, Nakanishi H, Nishizawa NK (2013) Iron-binding haemerythrin RING ubiquitin ligases regulate plant iron responses and accumulation. Nat Commun 4:2792

Konda AR, Nazarenus TJ, Nguyen H, Yang J, Gelli M, Swenson S, Shipp JM, Schmidt MA, Cahoon RE, Ciftci ON (2020) Metabolic engineering of soybean seeds for enhanced vitamin E tocochromanol content and effects on oil antioxidant properties in polyunsaturated fatty acid-rich germplasm. Metab Eng 57:63–73

Kong D, Khan SA, Wu H, Liu Y, Ling HQ (2022) Biofortification of iron and zinc in rice and wheat. J Integ Plant Bio 64:1157–1167

La Frano MR, de Moura FF, Boy E, Lönnerdal B, Burri BJ (2014) Bioavailability of iron, zinc, and provitamin A carotenoids in biofortified staple crops. Nutr Rev 72:289–307

Landini M, Gonzali S, Kiferle C, Tonacchera M, Agretti P, Dimida A, Vitti P, Alpi A, Pinchera A, Perata P (2012) Metabolic engineering of the iodine content in *Arabidopsis*. Sci Rep 2:338

Lassi ZS, Moin A, Bhutta ZA (2016) Zinc supplementation for the prevention of pneumonia in children aged 2 months to 59 months. Cochrane Database Syst Rev 12:CD005978

LeDuc DL, Tarun AS, Montes-Bayon M, Meija J, Malit MF, Wu CP, AbdelSamie M, Chiang CY, Tagmount A, deSouza M, Neuhierl B (2004) Overexpression of selenocysteine methyltransferase in Arabidopsis and Indian mustard increases selenium tolerance and accumulation. Plant Physiol 135:377–383

LeDuc DL, AbdelSamie M, Móntes-Bayon M, Wu CP, Reisinger SJ, Terry N (2006) Overexpressing both ATP sulfurylase and selenocysteine methyltransferase enhances selenium phytoremediation traits in Indian mustard. Environ Poll 144:70–76

Lee S, Jeon US, Lee SJ, Kim YK, Persson DP, Husted S et al (2009) Iron fortification of rice seeds through activation of the nicotianamine synthase gene. Proc Natl Acad Sci USA 106:22014–22019

Lee S, Kim YS, Jeon US, Kim YK, Schjoerring JK, An G (2012) Activation of rice nicotinamine synthase 2 (OsNAS2) enhances iron availability for biofortification. Mol Cell 33:269–275

Li J, Martin C, Fernie A (2024) Biofortification's contribution to mitigating micronutrient deficiencies. Nat Food 5:19–27

Lian Z, Nguyen CD, Liu L, Wang G, Chen J, Wang S, Yi G, Wilson S, Ozias-Akins P, Gong H, Huo H (2022) Application of developmental regulators to improve in planta or in vitro transformation in plants. Plant Biotechnol J 20:1622–1635

Liang G (2022) Iron uptake, signaling, and sensing in plants. Plant Commun 14:100349
Liang Q, Zhang W, Pang J, Zhang S, Hou X, Zhang C (2024) Creation of folate-biofortified rice by simultaneously enhancing biosynthetic flux and blocking folate oxidation. Mol Plant 17:1487–1489
Lipkie TE, De Moura FF, Zhao ZY, Albertsen MC, Che P, Glassman K, Ferruzzi MG (2013) Bioaccessibility of carotenoids from transgenic provitamin a sorghum. J Agri Food Chem 61:5764–5771
Lopez A, Cacoub P, Macdougall IC, Peyrin-Biroulet L (2016) Iron deficiency anaemia. Lancet 387:907–916
Lucca P, Hurrell R, Potrykus I (2001) Genetic engineering approaches to improve the bioavailability and the level of iron in rice grains. Theor Appl Genet 102:392–397
Malagoli M, Schiavon M, Pilon-Smits EA (2015) Effects of selenium biofortification on crop nutritional quality. Front Plant Sci 6:280
Maret W (2013) Zinc biochemistry: from a single zinc enzyme to a key element of life. Adv Nutr 4:82–91
Marschner H, Römheld V, Kissel M (1986) Different strategies in higher plants in mobilization and uptake of iron. J Plant Nutr 9:695–713
Masuda H, Ishimaru Y, Aung MS, Kobayashi T, Kakei Y, Takahashi M, Higuchi K, Nakanishi H, Nishizawa NK (2012) Iron biofortification in rice by the introduction of multiple genes involved in iron nutrition. Sci Rep 2(1):543
Menguer PK, Vincent T, Miller AJ, Brown JK, Vincze E, Borg S, Holm PB, Sanders D, Podar D (2018) Improving zinc accumulation in cereal endosperm using Hv MTP 1, a transition metal transporter. Plant Biotechnol J 16:63–71
Packer L, Weber SU, Rimbach G (2001) Molecular aspects of α-tocotrienol antioxidant action and cell signalling. J Nutr 131:369S–373S
Paine JA, Shipton CA, Chaggar S, Howells RM, Kennedy MJ, Vernon G, Wright S, Hinchliffe E, Adams JL, Silverstone AL, Drake R (2005) Improving the nutritional value of Golden Rice through increased pro-vitamin A content. Nat Biotechnol 23:482–487
Pfeiffer WH, McClafferty B (2007) HarvestPlus: breeding crops for better nutrition. Crop Sci 47:S–88
Pilon-Smits E, Hwang S, Lytle CM, Zhu Y, Tai JC, Bravo RC, Chen Y, Leustek T, Terry N (1999) Over-expression of ATP sulfurylase in Brassica juncea leads to increased selenate uptake, reduction and tolerance. Plant Physiol 119:123–132
Prom-u-thai C, Fukai S, Godwin ID, Huang L (2007) Genotypic variation of iron partitioning in rice grain. J Sci Food Agri 87:2049–2054
Raboy V (2020) Low phytic acid crops: observations based on four decades of research. Plants 9:140
Radomska D, Czarnomysy R, Radomski D, Bielawska A, Bielawski K (2021) Selenium as a bioactive micronutrient in the human diet and its cancer chemopreventive activity. Nutrients 13:1649
Ramegowda Y, Venkategowda R, Jagadish P, Govind G, Hanumanthareddy RR, Makarla U, Guligowda SA (2013) Expression of a rice Zn transporter, OsZIP1, increases Zn concentration in tobacco and finger millet transgenic plants. Plant Biotechnol Rep 7:309–319
Rashid A, Fox RL (1992) Evaluating internal zinc requirements of grain crops by seed analysis. Agron J 84:469–474
Ricachenevsky FK, Menguer PK, Sperotto RA (2013) kNACking on heaven's door: how important are NAC transcription factors for leaf senescence and Fe/Zn remobilization to seeds? Front Plant Sci 4:226
Saeid A, Patel A, Jastrzębska M, Korczyński M (2019) Food biofortification. J Chem 2019:5718426
Saper RB, Rash R (2009) Zinc: an essential micronutrient. Amer Family Phys 79(9):768
Sarwar N, Akhtar M, Kamran MA, Imran M, Riaz MA, Kamran K, Hussain S (2020) Selenium biofortification in food crops: Key mechanisms and future perspectives. J Food Comp Anal 93:103615
Schiavon M, Pilon-Smits EA (2017) The fascinating facets of plant selenium accumulation–biochemistry, physiology, evolution and ecology. New Phytol 213:1582–1596

Sezer RG, Ipek IO, Bozaykut A, Bayoglu DS (2011) Zinc deficiency in acute gastroenteritis and upper respiratory tract infections. ICAN 3:49–52

Shahidi F, Pinaffi-Langley ACC, Fuentes J, Speisky H, de Camargo AC (2021) Vitamin E as an essential micronutrient for human health: common, novel, and unexplored dietary sources. Free Rad Biol Med 176:312–321

Shintani D, DellaPenna D (1998) Elevating the vitamin E content of plants through metabolic engineering. Science 282:2098–2100

Song S, Li H, Lin S, Dong X, Tian R, Wu Z, Li Q, Li M, Zhang K, Liu X, Wan J (2024) Heterologous expression of sunflower HaHPT and HaTMT genes enhances rice-grain vitamin E content. Plants 13(17):2392

Stanton C, Sanders D, Krämer U, Podar D (2022) Zinc in plants: Integrating homeostasis and biofortification. Mol Plant 15(1):65–85

Storozhenko S, De Brouwer V, Volckaert M, Navarrete O, Blancquaert D, Zhang GF et al (2007) Folate fortification of rice by metabolic engineering. Nat Biotechnol 25(11):1277–1279. https://doi.org/10.1038/nbt1351x. Pubmed: 22088162

Su W, Xu M, Radani Y, Yang L (2023) Technological development and application of plant genetic transformation. Int J Mol Sci 24(13):10646

Tan L, Qu M, Zhu Y, Peng C, Wang J, Gao D, Chen C (2020) ZINC TRANSPORTER5 and ZINC TRANSPORTER9 function synergistically in zinc/cadmium uptake. Plant Physiol 183(3):1235–1249

Theil EC (2003) Ferritin: at the crossroads of iron and oxygen metabolism. J Nutr 133:1549–1553

Tiwari N, Mehrotra N, Tripathi P, Sharma S (2024) Vitamins and their impact on human health. In: Singh RL, Singh P, Pathak N (eds) Clinical applications of biomolecules in disease diagnosis. Springer, Singapore. https://doi.org/10.1007/978-981-97-4723-8_6

Traber MG (2014) Vitamin E inadequacy in humans: causes and consequences. Adv Nutr 5(5):503–514

Trijatmiko KR, Dueñas C, Tsakirpaloglou N, Torrizo L, Arines FM, Adeva C, Balindong J, Oliva N, Sapasap MV, Borrero J, Rey J (2016) Biofortified indica rice attains iron and zinc nutrition dietary targets in the field. Sci Rep 6(1):19792

Trono D (2019) Carotenoids in cereal food Crops: composition and retention throughout grain storage and food processing. Plants 8:551. https://doi.org/10.3390/plants8120551

Tulchinsky TH (2010) Micronutrient deficiency conditions: global health issues. Public Health Rev 32:243–255

UN SDG (United Nations Sustainable Development Goals) (2016). Retrieved from https://sustainabledevelopment.un.org/?menu=1300

Upadhyaya DC, Bagri DS, Upadhyaya CP, Kumar A, Thiruvengadam M, Jain SK (2021) Genetic engineering of potato (Solanum tuberosum L.) for enhanced α-tocopherols and abiotic stress tolerance. Physiol Plant 173(1):116–128

Van Der Straeten D, Bhullar NK, De Steur H, Gruissem W, MacKenzie D, Pfeiffer W, Qaim M, Slamet-Loedin I, Strobbe S, Tohme J, Trijatmiko KR (2020) Multiplying the efficiency and impact of biofortification through metabolic engineering. Nat Commun 11(1):5203–5210

Vasconcelos MW, Gruissem W, Bhullar NK (2017) Iron biofortification in the 21st century: setting realistic targets, overcoming obstacles, and new strategies for healthy nutrition. Curr Opin Biotechnol 44:8–15

Volk MJ, Tran VG, Tan S, Mishra S, Fatma Z, Boob A, Li H, Xue P, Martin TA, Zhao H (2022) Metabolic engineering: methodologies and applications. Chem Rev 123(9):5521–5570

Wang C, Zeng J, Li Y, Hu W, Chen L, Miao Y, Deng P, Yuan C, Ma C, Chen X, Zang M (2014) Enrichment of provitamin A content in wheat (Triticum aestivum L.) by introduction of the bacterial carotenoid biosynthetic genes CrtB and CrtI. J Exp Bot 65(9):2545–2556

Wang S, Li L, Ying Y, Wang J, Shao JF, Yamaji N, Whelan J, Ma JF, Shou H (2020) A transcription factor OsbHLH156 regulates Strategy II iron acquisition through localising IRO2 to the nucleus in rice. New Phytol 225:1247–1260

Welch RM, House WA (1984) Factors affecting the bioavailability of mineral nutrients in plant foods. In: Welch RM, Gabelman WH (eds) Crops as sources of nutrients for humans. American Society of Agronomy, Madison, pp 37–54

White PJ (2017) The genetics of selenium accumulation in plants. In: PilonSmits EAH, Winkel LHE, Lin ZQ (eds) Selenium in plants: molecular, physiology ecology evolutionary aspects. Springer, Cham, pp 143–163

WHO (World Health Organization) (2006) Preventing and controlling micronutrient deficiencies in populations affected by an emergency. In: Preventing and controlling micronutrient deficiencies in populations affected by an emergency, p 2

Yamaji N, Xia J, Mitani-Ueno N, Yokosho K, Ma JF (2013) Preferential delivery of zinc to developing tissues in rice is mediated by P- type heavy metal ATPase OsHMA2. Plant Physiol 162:927–939

Ye X, Al-Babili S, Klöti A, Zhang J, Lucca P, Beyer P, Potrykus I (2000) Engineering the provitamin a (β-carotene) biosynthetic pathway into (carotenoids-free) rice endosperm. Science 287:303–305

Zhang GY, Liu RR, Xu G, Zhang P, Li Y, Tang KX, Liang GH, Liu QQ (2013a) Increased α-tocotrienol content in seeds of transgenic rice overexpressing *Arabidopsis* γ-tocopherol methyltransferase. Transgenic Res 22:89–99

Zhang L, Luo Y, Zhu Y, Zhang L, Zhang W, Chen W, Xu M, Fan Y, Wang L (2013b) GmTMT2a from soybean elevates the α-tocopherol content in corn and *Arabidopsis*. Transgenic Res 22:1021–1028

Zhang L, Luo Y, Liu B, Zhang L, Zhang W, Chen R, Wang L (2020) Overexpression of the maize γ-tocopherol methyltransferase gene (ZmTMT) increases α-tocopherol content in transgenic Arabidopsis and maize seeds. Transgenic Res 29:95–104

Zhang Y, Cao H, Wang M, Zou Z, Zhou P, Wang X, Jin J (2023) A review of iodine in plants with biofortification: uptake, accumulation, transportation, function, and toxicity. Sci Total Environ 878:163203

Zhu C, Naqvi S, Breitenbach J, Sandmann G, Christou P, Capell T (2008) Combinatorial genetic transformation generates a library of metabolic phenotypes for the carotenoid pathway in maize. Proc Natl Acad Sci USA 105:18232–18237

Zhu YG, Pilon-Smits EAH, Zhao FJ, Williams PN, Meharg AA (2009) Selenium in higher plants: understanding mechanisms for biofortification and phytoremediation. Trend Plant Sci 14:436–442

Zhu Q, Wang B, Tan J, Liu T, Li L, Liu YG (2020) Plant synthetic metabolic engineering for enhancing crop nutritional quality. Plant Commun 1:100017

Zimmermann MB, Andersson M (2012) Assessment of iodine nutrition in populations: past, present, and future. Nutr Rev 70:553–570

Zimmermann MB, Jooste PL, Pandav CS (2008) Iodine-deficiency disorders. Lancet 372:1251–1262

Zulfiqar U, Khokhar A, Maqsood MF, Shahbaz M, Naz N, Sara M, Maqsood S, Sahar S, Hussain S, Ahmad M (2024) Genetic biofortification: advancing crop nutrition to tackle hidden hunger. Funct Integr Genomics 24(2):34

Part IV

Artificial Intelligence and Digital Devices

Machine Learning and Pangenomics: Revolutionizing Plant Breeding for a Sustainable Future

16

Felipe López-Hernández, Federico Roda, and Andrés J. Cortés

Abstract

In a world where the global population continues to grow and natural resources are increasingly limited, ensuring food security has become one of the greatest challenges of our time. Climate change, soil degradation, and agro-biodiversity loss have intensified threats to agricultural systems, which are already under significant pressure to produce more food in less space and with a reduced environmental impact. Considering these difficulties, achieving the Sustainable Development Goal of Zero Hunger is an urgent necessity to secure the health and stability of future generations. In this context, crop genetic breeding plays a fundamental role, not only by increasing yields but also by developing varieties resilient to changing conditions. Modern biotechnology tools, such as genomics, and more recently, machine learning (ML), a branch of artificial intelligence (AI), have paved an innovative path in this endeavor. ML enables systems to learn and improve automatically from data, without explicit programming, which has revolutionized multiple fields, including agriculture, by enabling the analysis of large

F. López-Hernández
Corporación Colombiana de Investigación Agropecuaria—AGROSAVIA, CI La Selva, Rionegro, Colombia
e-mail: llopez@agrosavia.co

F. Roda
Max Planck Tandem Group in Evolutionary Genomics of Specialized Metabolism, Facultad de Ciencias, Universidad Nacional de Colombia, Bogotá, Colombia
e-mail: frodaf@unal.edu.co

A. J. Cortés (✉)
Department of Plant Breeding, Swedish University of Agricultural Sciences, Alnarp, Sweden

Facultad de Ciencias Agrarias—Departamento de Ciencias Forestales, Universidad Nacional de Colombia—Sede Medellín, Medellín, Colombia
e-mail: andres.cortes@slu.se

P. M. Priyadarshan, R. Ortiz (eds.), *Plant Breeding 2050*,
https://doi.org/10.1007/978-981-95-0583-8_16

volumes of genomic data to identify complex patterns that predict phenotypic traits, such as disease resistance or productivity under stress, or optimize demographic clustering. Specifically, supervised ML algorithms may enhance genomic prediction accuracy while unsupervised methods reveal deeper population structure and genetic diversity for pre-breeding. Meanwhile, pangenomics is addressing single-reference genome biases and missing heritability by incorporating structural variants (SV), copy number variation (CNV), and transposable element (TE) diversity into genomic analyses. Genome graph models, boosted by long-read sequencing, improve pangenomic variant detection, mapping accuracy, and heritability estimates for complex traits. Nowadays, the development of "super-pangenomes", which merge cultivated and wild relatives, is further expanding the reservoir of adaptive and consumer-preferred traits. Together, ML and pangenomics offer a powerful framework to speed up crop improvement, prioritize germplasm conservation, and harness genomic diversity for climate-resilient and sustainable agriculture. The integration of comparative genomics through pangenomics and ML will ultimately accelerate the identification of improved varieties and optimize decision-making in agriculture, driving a crucial shift toward more precise and resilient agricultural practices.

Keywords

Supervised machine learning · Unsupervised machine learning · Genomic prediction · Population stratification · Genomic diversity · Structural variants (SVs) · Presence-absence variants (PAVs) · Variable transposable elements (TEs)

16.1 Introduction

Genomics has revolutionized the field of plant breeding by providing a detailed and precise view of the genetic bases that underlies all observable traits, or phenotypes, in a plant. Before the genomic era, breeding relied on the direct observation of these phenotypes across multiple generations. This process was lengthy and costly, as it required extensive field trials to assess characteristics like pest resistance, productivity, or adaptation to extreme climatic conditions (Bernardo 2008; Peláez et al. 2022). With the advent of genomic sequencing, scientists can now look directly at a plant's DNA and use this vast amount of information to guide the selection process far more efficiently (Varshney et al. 2014). In plant breeding, genomics has enabled the development of various strategies for selecting key traits. Marker-assisted selection (MAS) focuses on identifying specific genetic markers associated with desirable traits, often using Genome-Wide Association Studies (GWAS—López-Hernández et al. 2023; Wu et al. 2020, 2024) or Environmental Association Analysis (GEA—Cortés and Blair 2018; López-Hernández and Cortés 2019; Cortés et al. 2022, 2025) to pinpoint genes linked to key characteristics including adaptation (Huang and Han 2014). These approaches allow breeders to predict traits without needing to evaluate all characteristics in the field. For example, in maize breeding, MAS based on GWAS' associated markers has been used to identify genes associated with resistance to late blight, accelerating the development of resistant varieties (Collard and

Mackill 2008; Guevara-Escudero et al. 2021). In contrast, genomic selection uses data from thousands of markers across the genome to predict plants' genetic value, even without prior knowledge of the specific genes. This strategy is particularly valuable for improving complex traits, such as yield (Meuwissen et al. 2001; Xu et al. 2017) and abiotic stresses tolerance (Arenas et al. 2021; Cortés 2024).

Advances in sequencing technologies have enabled the rapid and cost-effective acquisition of complete genomes for multiple plant species, driving the development of bioinformatic tools that link the genome to observed traits (Varshney et al. 2014; Michael and VanBuren 2015). However, analyzing this vast amount of data remains a significant challenge. Here, machine learning (ML) plays a central role, as it allows the processing of large volumes of genomic data and the discovery of complex patterns that may go unnoticed by traditional summary statistics (Libbrecht and Noble 2015; Montesinos-López et al. 2021). Thus, ML has become an essential tool for extracting maximum value from genomic data and advancing plant breeding, optimizing both marker-assisted selection (MAS) and genomic selection (Crossa et al. 2011, 2017; Xu et al. 2017).

16.2 Machine Learning as a Catalyst in Genomic Breeding

As genomic data grow exponentially, the need for advanced tools that can analyze and extract value from this information has become crucial. This is where machine learning (ML) has proven to be an effective solution for handling the analysis of large volumes of genetic data and its application in crop breeding (Libbrecht and Noble 2015; Montesinos-López et al. 2021). ML involves the development of algorithms that can learn and improve from experience without being explicitly programmed for each specific task. This is an ideal tool to explore complex patterns in genomic data and making predictions based on those patterns.

Given the complexity of genomic data, various machine learning techniques have been developed to harness its potential in practical applications. Among these, supervised machine learning is one of the most widely used approaches in genomic analysis. This approach involves training algorithms on a known dataset where correct responses are already identified (for example, genes that are already known to be associated with disease resistance) (Crossa et al. 2011, 2017; Xu et al. 2017). The model learns from this training data and is then applied to new data to predict the behavior of other genes or identify genetic traits of interest. In the context of plant breeding, this predictive capability is invaluable, allowing researchers to foresee how certain genetic variations will impact important phenotypic traits, such as yield or disease resistance. Such predictive power has laid the groundwork for more advanced applications in plant genetics (Cortés and Du 2023), including genomic prediction, which leverages large-scale data to enhance selection accuracy and accelerate the breeding process (Meuwissen et al. 2001).

16.3 Speed Crop Breeding with Supervised Machine Learning: Predictive Models for Efficient Genomic Selection

Genomic prediction has emerged as one of the most powerful applications of ML in plant breeding, providing a more complete representation of polygenic quantitative traits compared to traditional GWAS-based genetic mapping (Crossa et al. 2011; Mir et al. 2012; Cortés et al. 2020; Cortés and López-Hernández 2021). Through genomic prediction, phenotypic data are used to calibrate marker-based predictive models where additive effects are adjusted by considering a large number of SNPs (Meuwissen et al. 2001; de los Campos et al. 2013). A significant statistical challenge in genomic prediction is estimating these effects in situations where the number of individuals is much smaller than the number of SNPs (Habier et al. 2009). Advanced methods such as BayesA, BayesB, and BayesC (Meuwissen et al. 2001; Habier et al. 2011), along with regularization techniques like LASSO (Tibshirani 1996) and Ridge Regression (Hoerl and Kennard 1970), have been developed to manage this complexity, minimizing overfitting and optimizing predictive accuracy.

Genomic selection using machine learning has been applied in essential crops like wheat, rice, beans, and sorghum. For example, in wheat, models based on Random Forest have been used to predict grain yield under drought conditions. These models have been trained with genomic and phenotypic data collected across various locations, demonstrating superior accuracy in yield prediction (Crossa et al. 2017; Farooq et al. 2023). On the other hand, Support Vector Machines (SVM) algorithms have proven useful for genotype classification. In sorghum, SVM has enabled the classification of varieties based on their tolerance to extreme drought, optimizing the selection of more robust varieties for areas with severe water stress (Zhao et al. 2020). In rice, the use of Ridge regression has facilitated the prediction of resistance to blast disease, allowing a rapid selection of the most resistant varieties without the need for prolonged field trials (Spindel et al. 2016). Also, one of the greatest advantages of supervised machine learning is its ability to handle incomplete or noisy data, which is common in genomic trials due to environmental variability and potential measurement errors. Algorithms can filter out noise, focus on relevant genetic patterns, and thereby improve prediction accuracy without requiring perfect datasets. This capability is crucial in plant breeding, where field conditions can vary significantly across different locations, affecting data consistency (Libbrecht and Noble 2015; Montesinos-López et al. 2018).

16.4 Unsupervised Learning in Genomics: Unveiling Population Structure and Genomic Diversity for Crop Pre-breeding

Unsupervised learning, complementary to supervised learning, plays a crucial role as a descriptive analysis tool for genomic data and serves as a valuable source of covariates in genotype prediction models. Unlike supervised learning, which relies on labeled training data, unsupervised learning explores hidden patterns and relationships within the data without prior knowledge of target outcomes. In plant genomics, this approach is indispensable for understanding population structure and characterizing genetic diversity, both of which are essential for crop breeding programs and managing genetic resources in germplasm banks (Falush et al. 2003; Evanno et al. 2005). Principal Component Analysis (PCA), a commonly used dimensionality reduction method, reveals population structure by simplifying large genomic datasets into fewer components, helping researchers infer genetic clustering and underlying patterns across populations (Jolliffe and Cadima 2016).

ML-based clustering techniques, both partition-based methods such as k-means and PAM (Partitioning Around Medoids) and hierarchical methods, including divisive and agglomerative clustering, are essential in identifying genetically similar subpopulations. These clustering methods allow researchers to divide populations into genetically homogeneous groups, supporting the characterization of genetic diversity and facilitating the creation of core collections in germplasm banks. For example, k-means has been applied in maize to group varieties based on drought tolerance, enabling breeding programs to focus on developing more resilient crops for regions impacted by climate change (Romero Navarro et al. 2017). Likewise, hierarchical methods, used to explore genetic relationships among individuals or populations, are particularly useful for germplasm bank management, where the formation of core collections preserves a representative subset of genetic diversity. By combining unsupervised learning with dimensionality reduction techniques like PCA, researchers can better understand population structure (Burbano et al. 2021; López-Hernández and Cortés 2022; Berdugo et al. 2023), enabling optimized conservation and providing critical insights for breeding programs aimed at developing crops better adapted to climate stress (Pritchard et al. 2000).

16.5 Missing Heritability and the Reference Genome Approach

Pangenomic methodologies use the genomes of multiple related organisms to build a framework for comparing other genomes. Pangenomics appeared as an alternative to the prevalent practice of genotyping, which Has to align sequence reads from different taxa to a single high-quality reference genome. The so-called "reference genome" approach leads to multiple problems (Eizenga et al. 2020; Lei et al. 2021). First, there is a mapping bias toward variants matching the reference sequence. Second, aligning sequence reads to a reference genome misses sequences not found in the reference genome or highly divergent from this reference. These include large structural variants (SVs) like insertions, duplications, and deletions.

Finally, the reference genome approach can lead to missing heritability, an important issue in genome-wide association studies (GWAS) and breeding programs. Missing heritability is the discrepancy between heritability estimates from family-based genetic studies and the variance explained by GWAS (Reyes et al. 2020; Cañas et al. 2022). Missing heritability can be caused by incomplete detection of causal genomic variants, particularly SVs, leading to inaccurate estimates of linkage disequilibrium between genetic markers and causal variants (Zhou et al. 2022). The greatest advantage of the pangenomic approach is its ability to incorporate SV variation and therefore has been particularly productive in plants because of their genomic dynamics (Lei et al. 2021; Naithani et al. 2023).

The adaptive diversification of land plants has been made possible by unparalleled high rates of gene duplications followed by partitioning of ancestral functions (subfunctionalization) or gain of new functions (neofunctionalization). For instance, plants interact with their environment largely via the production of thousands of specialized metabolites, which originated through the duplication and subsequent divergence of enzyme-encoding genes (Fernie et al. 2024). Additionally, plant receptor genes that recognize herbivores and pathogens are members of large gene families existing as vast groups of tandem duplications. Genomic regions containing tandem gene duplications are remarkably variable, presenting frequent presence/absence and copy number variants within and between species (Macho and Zipfel 2014; Yuan et al. 2021). This variation is responsible for diversity in many agronomically important traits like stress resistance, yield, flavor, aroma, and nutritional quality. Large SVs containing multiple genes have played a crucial role in the adaptative divergence of plant species and crop varieties (Cortés and Wheeler 2018), where large inversions and deletions often harbor genes that determine the ability of plants to survive in specific environments or their mating preferences (Wolf and Ellegren 2017).

Noncoding sequences, including transposable elements (TEs), constitute a majority of plant genomes and are the most variable fraction. There is increasing evidence that TE variation is important for gene regulation, reproductive isolation, and the generation of SVs like inversions and translocations. For instance, TEs are involved in the adaptation to cold and dry environments in cultivated and wild potato species (Bozan et al. 2023). Unfortunately, TE variation is usually not included in genetic analyses due to limitations related with the determination of genotypes in these genomic regions. One of the greatest promises of pangenomic methods is the possibility of incorporating TE-related variants in association studies and breeding programs.

Most reference genomes were created from highly homozygous plants whose genomes were impacted by inbreeding depression or the generation of double haploids. However, polyploidy is frequent across the plant kingdom, and high ploidy and heterozygosity levels are important for the adaptation of crops to extreme environments. Additionally, many crops are hybrid allopolyploids propagated asexually, whose genomes contain a high prevalence of SVs and a high genetic load (Manrique-Carpintero et al. 2018). Given that polysomic inheritance and genetic load can hamper agronomic trait performance, it is necessary to incorporate SV variation in studies. Recent works indicate that a pangenomic approach can help identifying deleterious mutations to mitigate the effect of genetic load (Bao et al. 2022; Bozan et al. 2023).

16.6 Pangenomic Methods in Plant Breeding

The most basic pangenomic analyses focus on the presence or absence of genes across different organisms to determine a core (commonly present), accessory (frequently absent), and private or cloud (present only in a sample or a few samples) pangenome (Eizenga et al. 2020). A pangenomic model combines a set of genomic sequences into a smaller data structure representing the full set (REF). Genome graphs are the most used type of pangenomic model and serve to compress sequences into graphs composed of nodes, edges, and paths that represent the recombinations of the genomes included in the model (Eizenga et al. 2020; Usadel 2022). An advantage of genome graphs is that they represent structural variation using the same components (edges, nodes, and paths) used to represent single-base changes, allowing more accurate read mapping. Pangenomic graphs can be used to genotype different types of SVs, including Copy Number Variants (CNVs) and Prescence-Absence Variants (PAVs). Many different programs have been devised to create and use genome graphs. Models differ on the need for a unidirectionally ordered reference and the amount of computational resources that they require (Hickey et al. 2020; Usadel 2022; Sibbesen et al. 2023). The most popular pangenomic software is the vg-tools package, which is very computationally efficient because it uses succinct indexes and prioritizes alignments consistent with known sequences (Naithani et al. 2023).

Pangenomic pipelines largely rely on the use of long-read sequencing technologies, to create and compare telomere-to-telomere genomic sequences from multiple related organisms. This allows the identification of SVs and variable TEs, which can be incorporated into pangenome graphs and used to genotype a broader array of taxa using short-read technologies and low sequencing coverage (Hickey et al. 2020; Bozan et al. 2023). Genotypic information can then be used for genome-wide association studies based on PAVs (PAV-GWAS), CNVs (CNV-GWAS), and TEs (TE-GWAS). The use of these methodologies in a growing number of crops has led to an increase in estimates of heritability for agronomically important traits due to the identification of causal variants that were not detected by studies based on a reference genome single-nucleotide polymorphisms (SNPs). Pangenomic references can also be used to analyze transcriptomic data, allowing better estimates of gene expression, particularly for genes involved in duplications or deletions (Sibbesen et al. 2023).

The first pangenomic studies in plants focused on intraspecific variation across varieties and cultivars from crop species. Later studies that included the closest wild relatives of these crops have been key in understanding the genomic and phenotypic impacts of domestication. However, a pan-genome that focuses on domesticated forms and their close progenitor species may not capture the full spectrum of relevant variations. The assembly of "super-pangenomes" containing a diverse repertoire of related wild species and crop cultivars has proven to be of great potential to harness novel genetic variation associated to stress tolerance, as well as consumer-preferred traits (Li et al. 2023). This deeper knowledge of the genomic basis of traits promises to speed up genetic improvement through marker-assisted selection and genomic selection.

16.7 The Future of Pangenomics: Harnessing Genomic Diversity

Pangenomic methods have drastically improved our understanding of the genomic causes of plant evolution and diversification beyond crops and model species. This deeper understanding can be used to face some of the modern challenges of crop breeding. First, pangenomics promises to revolutionize the study of crops with high levels of hybridization and introgression. For instance, a study in different species of peppers (*Capsicum* spp.) showed that the same phenotypes were achieved through selection in distinct genomic regions and that interspecific introgression was crucial to provide stress tolerance during domestication (Liu et al. 2023). Pangenomics has also been used for the identification of segregating inversions or deletions that can affect crossing success and trigger linkage drag during breeding, but can also contain agronomically important genes. For instance, genomic inversions present in Barley impact reproductive success but contain genes that expanded the geographical range of this crop (Jayakodi et al. 2020). Finally, pangenomic methods can be particularly useful in the precision breeding of clonally propagated polyploid crops like potatoes, by helping to recognize and purge dysfunctional and deleterious alleles that increase the genetic load (Bao et al. 2022; Achakkagari et al. 2024).

Pangenomic methods can also be helpful in mitigating the risks related to food insecurity and poor agricultural resilience derived from the lack of agro-biodiversity. Modern food production is based on a handful of crop species from only three plant families, but humans have historically used hundreds of indigenous crops. These indigenous crops can be used in crop improvement since they often belong to the same families as major crops and have interesting traits absent in the major crops. In this context, super-pangenomes including wild species as well as major and indigenous crops can be used for "de-domestication" or "de-novo" domestication programs. De-domestication involves the transfer to major crops of important genes lost during domestication (Wu et al. 2021). De novo domestication involves the creation of new crops by manipulating major genes in wild species or indigenous crops using genome editing technologies (Caestecker 2012; Fernie and Yan 2019). For example, pangenomic analyses in the megadiverse and economically important genus *Solanum* identified genes that, when disrupted with CRISPR, result in prickle loss in multiple species without off-target phenotypic effects (Satterlee et al. 2024).

One of the most promising applications of the combined analyses of the genomes from multiple plant species has been metabolic engineering of medicinal and nutraceutical substances. The comparison of the genomes of plants with different metabolic profiles has allowed the identification of biosynthetic genes that can then be transferred to high biomass plants or microorganisms to assemble de novo the pathways of interest (Courdavault and Papon 2024). This approach can be facilitated by the identification of biosynthetic gene clusters, genomic regions containing multiple genes from the same pathway that can be efficiently transferred between species (Bharadwaj et al. 2021). The power of pangenomics is greater when used in combination with traditional genetics, molecular biology, and genome editing. For

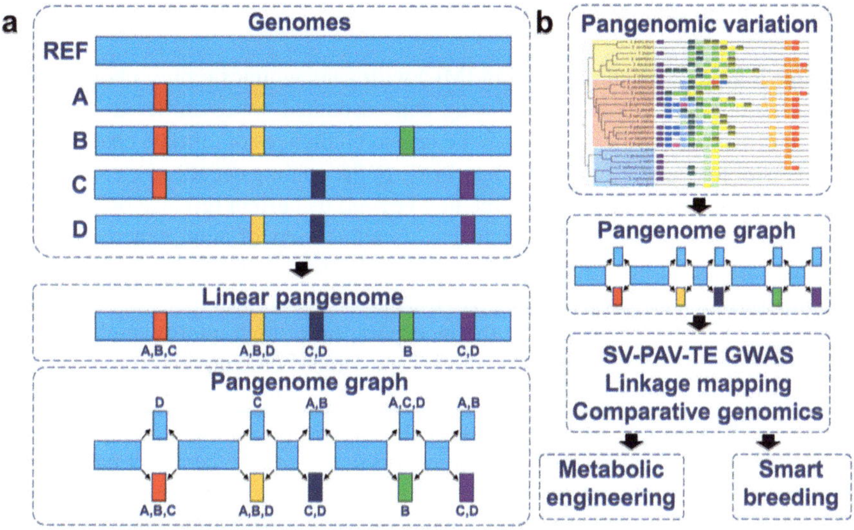

Fig. 16.1 Pangenomic approach: (**a**) The pangenomic approach is based on the comparison of multiple genomes (blue bars labeled with letters) to identify genetic variants (colored squares) that are used to create a pangenomic reference, which can be linear or take the form of a graph pangenome. (**b**) Pangenomic references are used to detect genetic variants in a broader array of samples, including single nucleotide polymorphisms as well as structural variants (SVs), presence-absence variants (PAVs), and variable transposable elements (TEs). These genotypes can then be used to study the genetic basis of important traits through genome-wide association studies (GWAS), linkage mapping, and comparative genomics. Finally, candidate genes and pathways can be manipulated via smart breeding or metabolic engineering.

instance, a study of the late-blight-resistant species *Solanum americanum*, combined a pangenomic analysis of immune receptor genes with the screening for recognition of pathogen effectors in order to clone genes that can be used for engineering late-blight-resistant in susceptible plants (Lin et al. 2023).

Additionally, besides merging various transdisciplinary paradigms (Fig. 16.1), a deeper understanding of paralog histories and functions is necessary to increase the predictability of phenotypic effects based on the pangenomic approach. Gene duplication and paralog diversification are major obstacles to genotype-phenotype predictability given that genes often acquire new functions. Therefore, it is also necessary to complement pangenomic studies with functional analyses aimed at testing functional conservation between species (Benoit et al. 2025). Finally, it is indispensable to consider additive and epistatic effects from background genetic modifiers that complicate the transfer of new genetic variants into a different background in breeding programs. The creation of AI-based computational tools to integrate genomic and phenotypic data across a large number of taxa promises to these issues.

16.8 Perspectives

The integration of machine learning (ML) with pangenomics represents a transformative shift in plant breeding, offering unprecedented opportunities to accelerate crop improvement while addressing the urgent challenges of climate change, agrobiodiversity loss, and global food insecurity. In the coming decade, the convergence of these fields promises to move breeding beyond its current reliance on reference-based genomics toward a more inclusive, dynamic view of plant genetic diversity, one that captures structural variants, transposable element dynamics, and presence–absence variation as drivers of adaptive traits.

An imminent next step will be the incorporation of high-quality, multi-genome references and pangenomic graphs into ML pipelines, allowing algorithms to operate directly on the full spectrum of genomic diversity rather than filtered, reference-biased SNP datasets. As computational frameworks mature, deep learning architectures and graph-based neural networks could enable the direct modeling of complex genotype–phenotype–environment relationships, capturing epistatic interactions and genotype-specific environmental responses with higher precision.

However, achieving this will require coordinated progress on multiple fronts. First, the generation of diverse, representative, and open-access pangenomes for major and underutilized crops must become a global priority, supported by standardized metadata and phenotyping protocols. Second, scalable ML models must be trained on datasets that integrate genomic, transcriptomic, metabolomic, and environmental information, bridging the gap between genomic prediction and systems-biology crop design. Third, ethical considerations, including equitable access to genomic resources, benefit-sharing with indigenous and smallholder farming communities, and the mitigation of biases accessing ML tools, must be embedded in research proposals from the beginning. Merging the data-richness pangenomics with the analytical power of ML will enable plant molecular breeders to design resilient, high-yielding, and resource-efficient cultivars at the pace demanded by a rapidly changing world. Ultimately, this synergy promises redefining breeding strategies by somehow transforming them from reactive responses to emerging threats into proactive, predictive systems that safeguard food security for generations to come.

16.9 Conclusions

Machine learning and pangenomics are redefining the scope of modern plant breeding. Machine learning enables the analysis of vast, complex genomic datasets, improving the prediction of phenotypic traits and accelerating selection cycles, even under incomplete or noisy data conditions. Supervised approaches have enhanced genomic prediction accuracy, while unsupervised methods have deepened our understanding of population structure and genetic diversity, critical for targeted breeding and germplasm conservation strategies. Meanwhile, pangenomics complements these advances by overcoming the limitations of single-reference genome analyses, capturing structural variants, presence–absence variation, and transposable element diversity that underly many agronomically important traits. Incorporating this genomic richness in structural variation will enable breeders to effectively address the challenge of missing heritability and unlock adaptive potential hidden in noncoding regions and gene duplications. Together, these innovations provide the foundation for next generation breeding programs capable of developing high-yielding, stress-resilient, and resource-efficient crop varieties. As climate change and agro-biodiversity loss intensify global food security challenges, the combined application of machine learning and pangenomics will be essential for delivering sustainable, data-driven solutions in agriculture.

Acknowledgments F.L.-H., A.J.C., and F.R. thank the Capsicum project consortium for insightful networking and discussions, which ultimately led the authors to conceive this chapter. A.J.C. also appreciates M.J. Torres' care while editing this chapter. F.L.-H., A.J.C., and F.R. are in debt with the editors P.M. Priyadarshan and R. Ortíz for envisioning, making possible, and encouraging submissions to the cutting-edge compilation on "Plant Breeding 2050: Next-Gen Crops".

Author Contribution F.L.-H., F.R., and A.J.C. envisioned this chapter. F.L.-H. and F.R. drafted a first version with further insights from A.J.C. F.L.-H., F.R., and A.J.C. edited and amended posterior versions of the chapter. All authors contributed to the chapter and endorsed its final publication.

Disclaimer The authors declare no conflict of interest.

Funding F.L.-H. recognizes support from MinCiencias and Ministerio de Agricultura y Desarrollo Rural (MADR). Additionally, A.J.C. acknowledges Kungliga Vetenskapsakademien, the British Council, and Vetenskapsrådet (VR) for funding throughout BS20170036, 527023146, and 2016–00418/2022–04411 grants, respectively.

References

Achakkagari SR, Bozan I, Camargo-Tavares JC, McCoy HJ, Portal L, Soto J et al (2024) The phased Solanum okadae genome and Petota pangenome analysis of 23 other potato wild relatives and hybrids. Sci Data 11:1–17. https://doi.org/10.1038/s41597-024-03300-5

Arenas S, Cortés AJ, Mastretta A, Jaramillo JP (2021) Evaluating the accuracy of genomic prediction for the management and conservation of relictual natural tree populations. Tree Genet Genom 17:12. https://doi.org/10.1007/s11295-020-01489-1

Bao Z, Li C, Li G, Wang P, Peng Z, Cheng L et al (2022) Genome architecture and tetrasomic inheritance of autotetraploid potato. Mol Plant 15:1211–1226. https://doi.org/10.1016/j.molp.2022.06.009

Benoit M, Jenike KM, Satterlee JW, Ramakrishnan S, Gentile I, Hendelman A et al (2025) *Solanum* pan-genetics reveals paralogues as contingencies in crop engineering. Nature 640(8057):135–145. https://doi.org/10.1038/s41586-025-08619-6

Berdugo-Cely JA, Cortés AJ, López-Hernández F, Delgadillo-Durán P, Cerón-Souza I, Reyes-Herrera P, Navas-Arboleda AA, Yockteng R (2023) Pleistocene-dated genomic divergence of avocado trees supports cryptic diversity in the Colombian germplasm. Tree Genet. Genom. 19:42. https://doi.org/10.1007/s11295-023-01616-8

Bernardo R (2008) Molecular markers and selection for complex traits in plants: learning from the last 20 years. Crop Sci 48:1649–1664. https://doi.org/10.2135/cropsci2008.03.0131

Bharadwaj R, Kumar SR, Sharma A, Sathishkumar R (2021) Plant metabolic gene clusters: evolution, organization, and their applications in synthetic biology. Front Plant Sci 12. https://doi.org/10.3389/fpls.2021.697318

Bozan I, Achakkagari SR, Anglin NL, Ellis D, Tai HH, Strömvik MV (2023) Pangenome analyses reveal impact of transposable elements and ploidy on the evolution of potato species. Proc Natl Acad Sci 120. https://doi.org/10.1073/pnas.2211117120

Burbano-Erazo E, León-Pacheco RI, Cordero-Cordero CC, López-Hernández LF, Tofiño-Rivera AP, Cortés AJ (2021) Multi-environment yield components in advanced common bean (*Phaseolus vulgaris* L.) × Tepary bean (*P. acutifolius* A. Gray) interspecific lines for heat and drought tolerance. Agronomy 11:1978. https://doi.org/10.3390/agronomy11101978

Caestecker E (2012) De novo transcriptome assembly and annotation in heterozygous crop species guided by comparative genomics

Cañas GP, Sepulveda-Ortega S, López-Hernández F, Navas-Arboleda AA, Cortés AJ (2022) Inheritance of yield components and morphological traits in avocado cv. Hass from "criollo" 'elite trees' via half-sib seedling rootstocks. Front Plant Sci 13:843099. https://doi.org/10.3389/fpls.2022.843099

Collard BC, Mackill DJ (2008) Marker-assisted selection: an approach for precision plant breeding in the twenty-first century. Philos Trans R Soc B Biol Sci 363:557–572. https://doi.org/10.1098/rstb.2007.2170

Cortés AJ (2024) Abiotic stress tolerance boosted by genetic diversity in plants. Int J Mol Sci 25(10):5367. https://doi.org/10.3390/ijms25105367

Cortés AJ (2025) Unlocking genebanks for climate adaptation, vol 15. Nat. Clim. Change

Cortés AJ, Blair MW (2018) Genotyping by sequencing and genome—environment associations in wild common bean predict widespread divergent adaptation to drought. Front Plant Sci 9:128. https://doi.org/10.3389/fgene.2022.910386

Cortés AJ, Du D (2023) Molecular genetics modernizes plant breeding. Int J Mol Sci 24(12):9977. https://doi.org/10.3390/ijms24129977

Cortés AJ, López-Hernández F (2021) Harnessing crop wild diversity for climate change adaptation. Genes 12. https://doi.org/10.3390/genes12050783

Cortés AJ, Wheeler JA (2018) The environmental heterogeneity of mountains at a fine scale in a changing world. In: Hoorn C, Perrigo A, Antonelli A (eds) Mountains, climate, and biodiversity. NY

Cortés AJ, Restrepo-Montoya M, Bedoya-Canas LE (2020) Modern strategies to assess and breed forest tree adaptation to changing climate. Front Plant Sci 11:583323. https://doi.org/10.3389/fpls.2020.583323

Cortés AJ, López-Hernández F, Blair MW (2022) Genome–environment associations, an innovative tool for studying heritable evolutionary adaptation in orphan crops and wild relatives. Front Genet 13:910386. https://doi.org/10.3389/fgene.2022.910386

Courdavault V, Papon N (2024) Metabolic engineering of plant medicines. Nat Rev Genet 25:307–307. https://doi.org/10.1038/s41576-024-00705-1

Crossa J, Pérez P, de los Campos G, Mahuku G, Dreisigacker S, Magorokosho C (2011) Genomic selection and prediction in plant breeding. J Crop Improv 25:239–261. https://doi.org/10.1080/15427528.2011.558767

Crossa J, Pérez P, Hickey J, Burgueño J, Ornella L, Cerón-Rojas J et al (2014) Genomic prediction in CIMMYT maize and wheat breeding programs. Heredity (Edinb). 112:48–60. https://doi.org/10.1038/hdy.2013.16

Crossa J, Pérez-Rodríguez P, Cuevas J, Montesinos-López O, Jarquín D, de los Campos G et al (2017) Genomic selection in plant breeding: methods, models, and perspectives. Trends Plant Sci 22:961–975. https://doi.org/10.1016/j.tplants.2017.08.011

de los Campos G, Hickey JM, Pong-Wong R, Daetwyler HD, Calus MPL (2013) Whole-genome regression and prediction methods applied to plant and animal breeding. Genetics 193:327–345. https://doi.org/10.1534/genetics.112.143313

Eizenga JM, Novak AM, Sibbesen JA, Heumos S, Ghaffaari A, Hickey G et al (2020) Pangenome graphs. Annu Rev Genomics Hum Genet 21:139–162. https://doi.org/10.1146/annurev-genom-120219-080406
Evanno G, Regnaut S, Goudet J (2005) Detecting the number of clusters of individuals using the software STRUCTURE: a simulation study. Mol Ecol 14:2611–2620. https://doi.org/10.1111/j.1365-294X.2005.02553.x
Falush D, Stephens M, Pritchard JK (2003) Inference of population structure using multilocus genotype data: linked loci and correlated allele frequencies. Genetics 164:1567–1587. https://doi.org/10.1093/genetics/164.4.1567
FAO (2021) Sustainable development goal 2: zero hunger. Food and Agriculture Organization of the United Nations. https://www.fao.org/sustainable-development-goals/goals/goal-2/en/
Farooq M, van Dijk ADJ, Nijveen H, Mansoor S, de Ridder D (2023) Genomic prediction in plants: opportunities for ensemble machine learning based approaches. F1000Res 11:802. https://doi.org/10.12688/f1000research.122437.2
Fernie AR, Yan J (2019) De novo domestication: an alternative route toward new crops for the future. Mol Plant 12:615–631. https://doi.org/10.1016/j.molp.2019.03.016
Fernie, A. R., Vries, S. De, Vries, J. De (2024). Evolution of plant metabolism: the state- of-the-art
Guevara-Escudero M, Osorio AN, Cortés AJ (2021) Integrative pre-breeding for biotic resistance in forest trees. Plants 10. https://doi.org/10.3390/plants10102022
Habier D, Fernando RL, Dekkers JCM (2009) Genomic selection using low-density marker panels. Genetics 182:343–353. https://doi.org/10.1534/genetics.108.100289
Habier D, Fernando RL, Kizilkaya K, Garrick DJ (2011) Extension of the bayesian alphabet for genomic selection. BMC Bioinformatics 12. https://doi.org/10.1186/1471-2105-12-186
Hickey G, Heller D, Monlong J, Sibbesen JA, Sirén J, Eizenga J et al (2020) Genotyping structural variants in pangenome graphs using the vg toolkit. Genome Biol 21:1–17. https://doi.org/10.1186/s13059-020-1941-7
Hoerl AE, Kennard RW (1970) Ridge regression: biased estimation for nonorthogonal problems. Technometrics 12:55–67. https://doi.org/10.1080/00401706.1970.10488634
Huang X, Han B (2014) Natural variations and genome-wide association studies in crop plants. Annu Rev Plant Biol 65:531–551. https://doi.org/10.1146/annurev-arplant-050213-035715
IPCC (2022) Climate change 2022: impacts, Adaptation and Vulnerability. Intergovernmental Panel on Climate Change. https://www.ipcc.ch/reports/
Jayakodi M, Padmarasu S, Haberer G, Bonthala VS, Gundlach H, Monat C et al (2020) The barley pan-genome reveals the hidden legacy of mutation breeding. Nature 588:284–289. https://doi.org/10.1038/s41586-020-2947-8
Jolliffe IT, Cadima J (2016) Principal component analysis: a review and recent developments. Philos Trans R Soc A Math Phys Eng Sci 374:20150202. https://doi.org/10.1098/rsta.2015.0202
Lei L, Goltsman E, Goodstein D, Wu GA, Rokhsar DS, Vogel JP (2021) Plant pan-genomics comes of age. 411–435
Li N, He Q, Wang J, Wang B, Zhao J, Huang S et al (2023) Super-pangenome analyses highlight genomic diversity and structural variation across wild and cultivated tomato species. Nat Genet 55:852–860. https://doi.org/10.1038/s41588-023-01340-y
Libbrecht MW, Noble WS (2015) Machine learning applications in genetics and genomics. Nat Rev Genet 16:321–332. https://doi.org/10.1038/nrg3920
Lin X, Jia Y, Heal R, Prokchorchik M, Sindalovskaya M, Olave-Achury A et al (2023) Solanum americanum genome-assisted discovery of immune receptors that detect potato late blight pathogen effectors. Nat Genet 55:1579–1588. https://doi.org/10.1038/s41588-023-01486-9
Liu F, Zhao J, Sun H, Xiong C, Sun X, Wang X et al (2023) Genomes of cultivated and wild capsicum species provide insights into pepper domestication and population differentiation. Nat Commun 14. https://doi.org/10.1038/s41467-023-41251-4
López-Hernández LF, Cortés AJ (2019) Last generation genome—environment associations reveal the genetic basis of heat tolerance in common bean (Phaseolus vulgaris L.). Front Genet 10:954. https://doi.org/10.3389/fgene.2019.00954

López-Hernández F, Cortés AJ (2022) Whole transcriptome sequencing unveils the genomic determinants of putative somaclonal variation in mint (*Mentha* L.). Int J Mol Sci 23(10):5291. https://doi.org/10.3390/ijms23105291

López-Hernández F, Burbano-Erazo E, León-Pacheco RI, Cordero-Cordero CC, Villanueva-Mejía DF, Tofiño-Rivera AP, Cortés AJ (2023) Multi-environment genome wide association studies of yield traits in common bean (*Phaseolus vulgaris* L.) × Tepary bean (*P. acutifolius* A. gray) interspecific advanced lines at the humid and dry Colombian Caribbean sub-regions. Agronomy 13(5):1396. https://doi.org/10.3390/agronomy13051396

Macho AP, Zipfel C (2014) Plant PRRs and the activation of innate immune signaling. Mol Cell 54:263–272. https://doi.org/10.1016/j.molcel.2014.03.028

Manrique-Carpintero NC, Coombs JJ, Pham GM, Laimbeer FPE, Braz GT, Jiang J et al (2018) Genome reduction in tetraploid potato reveals genetic load, haplotype variation, and loci associated with agronomic traits. Front Plant Sci 9:1–15. https://doi.org/10.3389/fpls.2018.00944

Meuwissen T, Hayes B, Goddard ME (2001) Prediction of total genetic value using genome-wide dense marker maps. Genetics 157:1819–1829

Michael TP, VanBuren R (2015) Progress, challenges and the future of crop genomes. Curr Opin Plant Biol 24:71–81. https://doi.org/10.1016/j.pbi.2015.02.002

Mir RR, Zaman-Allah M, Sreenivasulu N, Trethowan R, Varshney RK (2012) Integrated genomics, physiology and breeding approaches for improving drought tolerance in crops. Theor Appl Genet 125:625–645. https://doi.org/10.1007/s00122-012-1904-9

Montesinos-López A, Montesinos-López OA, Gianola D, Crossa J, Hernández-Suárez CM (2018) Multi-environment genomic prediction of plant traits using deep learners with dense architecture. G3 genes. Genomes, Genet 8:3813–3828. https://doi.org/10.1534/g3.118.200740

Montesinos-López OA, Montesinos-López A, Pérez-Rodríguez P, Barrón-López JA, Martini JWR, Fajardo-Flores SB et al (2021) A review of deep learning applications for genomic selection. BMC Genomics 22:1–23. https://doi.org/10.1186/s12864-020-07319-x

Naithani S, Deng CH, Sahu SK, Jaiswal P (2023) Exploring pan-genomes: an overview of resources and tools for unraveling structure, function, and evolution of crop genes and genomes. Biomolecules 13. https://doi.org/10.3390/biom13091403

Peláez D, Aguilar PA, Mercado M, López-Hernández F, Guzmán M, Burbano-Erazo E, Denning-James K, Medina CI, Blair MW, De Vega JJ, Cortés AJ (2022) Genotype selection, and seed uniformity and multiplication to ensure common bean (*Phaseolus vulgaris* L.) var. Liborino Agronomy 12:2285. https://doi.org/10.3390/agronomy12102285

Pritchard JK, Stephens M, Donnelly P (2000) Inference of population structure using multilocus genotype data. Genetics 155:945–959. https://doi.org/10.1093/genetics/155.2.945

Reyes-Herrera PH, Muñoz-Baena L, Velásquez-Zapata V, Patiño L, Delgado-Paz OA, Díaz-Diez CA, Navas-Arboleda AA, Cortés AJ (2020) Inheritance of rootstock effects in avocado (*Persea americana* Mill.) cv. Hass Front Plant Sci 11:555071. https://doi.org/10.3389/fpls.2020.555071

Romero Navarro JA, Willcox M, Burgueño J, Romay C, Swarts K, Trachsel S et al (2017) A study of allelic diversity underlying flowering-time adaptation in maize landraces. Nat Genet 49:476–480. https://doi.org/10.1038/ng.3784

Satterlee JW, Alonso D, Gramazio P, Jenike KM, He J, Arrones A et al (2024) Convergent evolution of plant prickles by repeated gene co-option over deep time. Science 385:eado1663. https://doi.org/10.1126/science.ado1663

Sibbesen JA, Eizenga JM, Novak AM, Sirén J, Chang X, Garrison E et al (2023) Haplotype-aware pantranscriptome analyses using spliced pangenome graphs. Nat Methods 20:239–247. https://doi.org/10.1038/s41592-022-01731-9

Spindel JE, Begum H, Akdemir D, Collard B, Redoña E, Jannink JL et al (2016) Genome-wide prediction models that incorporate de novo GWAS are a powerful new tool for tropical rice improvement. Heredity (Edinb) 116:395–408. https://doi.org/10.1038/hdy.2015.113

Tester M, Langridge P (2010) Breeding technologies to increase crop production in a changing world. Science 80-.). 327:818–822. https://doi.org/10.1126/science.1183700

Tibshirani R (1996) Regression shrinkage and selection via the Lasso. J. R. Stat Soc Ser B Stat Methodol 58:267–288. https://doi.org/10.1111/j.2517-6161.1996.tb02080.x

Usadel B (2022) Solanaceae pangenomes are coming of graphical age to bring heritability back. aBIOTECH 3:233–236. https://doi.org/10.1007/s42994-022-00087-0

Varshney RK, Terauchi R, McCouch SR (2014) Harvesting the promising fruits of genomics: applying genome sequencing technologies to crop breeding. PLoS Biol 12:e1001883. https://doi.org/10.1371/journal.pbio.1001883

Wolf JBW, Ellegren H (2017) Making sense of genomic islands of differentiation in light of speciation. Nat Rev Genet 18:87–100. https://doi.org/10.1038/nrg.2016.133

Wu X, Islam FASM, Limpot N, Mackasmiel L, Mierzwa J, Cortés AJ, Blair MW (2020) Genome-wide SNP identification and association mapping for seed mineral in mung bean. Front Genet 11:656. https://doi.org/10.3389/fgene.2020.00656

Wu D, Lao S, Fan L (2021) De-domestication: an extension of crop evolution. Trends Plant Sci 26:560–574. https://doi.org/10.1016/j.tplants.2021.02.003

Wu X, Michael VN, López-Hernández F, Cortés AJ, Morris JB, Wang M, Tallury S, Miller Ii MC, Blair MW (2024) Genetic diversity and genome-wide Association in Cowpeas (*Vigna unguiculata* L. Walp). Agronomy 14(5). https://doi.org/10.3390/agronomy14050961

Xu Y, Li P, Zou C, Lu Y, Xie C, Zhang X et al (2017) Enhancing genetic gain in the era of molecular breeding. J Exp Bot 68:2641–2666. https://doi.org/10.1093/jxb/erx135

Yuan M, Jiang Z, Bi G, Nomura K, Liu M, Wang Y et al (2021) Pattern-recognition receptors are required for NLR-mediated plant immunity. Nature 592:105–109. https://doi.org/10.1038/s41586-021-03316-6

Zhao W, Lai X, Liu D, Zhang Z, Ma P, Wang Q et al (2020) Applications of support vector machine in genomic prediction in pig and maize populations. Front Genet 11:1–7. https://doi.org/10.3389/fgene.2020.598318

Zhou Y, Zhang Z, Bao Z, Li H, Lyu Y, Zan Y et al (2022) Graph pangenome captures missing heritability and empowers tomato breeding. Nature 606:527–534. https://doi.org/10.1038/s41586-022-04808-9

Phenomic Selection and Digital Imaging for Characterizing Plant Genetic Resources: A Case Study from the Colombian Maize Collection

17

Diego Felipe Conejo-Rodríguez, Creuci Maria Caetano, Héctor Estrada Marin, Daniel Orlando Osorio Garcia, Ana María Serna Landaeta, Vanesa Diaz Giraldo, Amanda Ortiz Escobar, Carlos Iván Cardozo Conde, and Milan Oldřich Urban

Abstract

High-throughput phenotyping and digital image processing have generated new digital trait acquisition that can be used in the characterization and evaluation of plant genetic resources conserved in genebanks. Maize is of importance to world agriculture and has been fundamental to the food security of peasant and indigenous communities in Colombia. In Colombia, there are maize races that have been dispersed and conserved by communities leading to an increase in phenotypic variation in ear shape and kernel color, driven by domestication processes and random crosses among the diversity of races. In this chapter, we relate the morphological and colorimetric variation of ears of 24 Colombian maize races

D. F. Conejo-Rodríguez (✉)
Universidad Nacional de Colombia sede Palmira, Facultad de Ciencias Agropecuarias, Palmira, Valle del Cauca, Colombia

Grupo de estudio Fenómica, Ciencia de datos e Inteligencia Artificial, Universidad Nacional de Colombia sede Palmira, Facultad de Ciencias Agropecuarias, Palmira, Valle del Cauca, Colombia

Grupo de Investigación en Recursos Fitogenéticos Neotropicales (GIRFIN), Universidad Nacional de Colombia sede Palmira, Facultad de Ciencias Agropecuarias, Palmira, Valle del Cauca, Colombia

Genetic Resources Program, International Center for Tropical Agriculture (CIAT), Palmira, Valle del Cauca, Colombia
e-mail: dfconejoro@unal.edu.co

C. M. Caetano
Universidade Federal de Rondônia UNIR, Presidente Médici, Rondônia, Brazil

P. M. Priyadarshan, R. Ortiz (eds.), *Plant Breeding 2050*,
https://doi.org/10.1007/978-981-95-0583-8_17

using digital image processing, machine learning, and functional diversity indices that allow characterization and selection of potential accessions for genetic improvement from digital images. We generate a methodological approach to understand and support the use of digital images in processes of characterization of genetic resources, in addition to integrating functional diversity indexes in the exploration of functional traits and selection of potential accessions in genebanks as well as in working collections in breeding programs.

Keywords

Morphological characterization · Genebank · Prebreeding · Image analysis · Digital colorimetric · Geometric morphometric · Random Forest · Functional diversity

17.1 Introduction

Maize is a valuable crop genetic resource for food and agriculture in South America, particularly among traditional communities and indigenous people. Colombia is considered one of the important centers of crop diversification. Due to maize importance, the Ministry of Agriculture of Colombia, in collaboration with the Rockefeller Foundation, made a significant effort in 1950 for the collection and classification of maize varieties. At that time, they managed to collect about 2000 introductions of

H. E. Marin
Universidad Nacional de Colombia sede Palmira, Facultad de Ciencias Agropecuarias, Palmira, Valle del Cauca, Colombia

Grupo de estudio Fenómica, Ciencia de datos e Inteligencia Artificial, Universidad Nacional de Colombia sede Palmira, Facultad de Ciencias Agropecuarias, Palmira, Valle del Cauca, Colombia

D. O. O. Garcia · A. M. S. Landaeta
Universidad Nacional de Colombia sede Palmira, Facultad de Ciencias Agropecuarias, Palmira, Valle del Cauca, Colombia

V. D. Giraldo
Grupo de estudio Fenómica, Ciencia de datos e Inteligencia Artificial, Universidad Nacional de Colombia sede Palmira, Facultad de Ciencias Agropecuarias, Palmira, Valle del Cauca, Colombia

A. O. Escobar · C. I. C. Conde
Universidad Nacional de Colombia sede Palmira, Facultad de Ciencias Agropecuarias, Palmira, Valle del Cauca, Colombia

Grupo de Investigación en Recursos Fitogenéticos Neotropicales (GIRFIN), Universidad Nacional de Colombia sede Palmira, Facultad de Ciencias Agropecuarias, Palmira, Valle del Cauca, Colombia

M. O. Urban
International Center for Biosaline Agriculture (ICBA), Near Zayed University, Dubai, United Arab Emirates

this cereal, which were studied and described by Roberts et al. (1957), resulting in the identification of 23 races of maize. The races were divided into three categories: (I) Primitive (*Pollo, Pira*), (II) Probably introduced (*Amagaceño, Andaquí, Cariaco, Clavo, Güirua, Harinoso dentado, Imbricado, Maíz Dulce, Pira naranja, Sabanero*), and (III) Colombian hybrids (*Cabuya, Cacao, Capio, Chococeño, Común, Costeño, Montaña, Negrito, Puya, Puya Grande*). The races are additionally classified based on their tolerance to distinct altitudes: highlands (more than 1800 meters above sea level) and lowlands (0–1800 meters above sea level).

To continue with a previous effort and bring new data, from 2007 to 2014, the Research Group on Recursos geneticos neotropicales (GIRFIN) of the Universidad Nacional de Colombia—Palmira and instituto colombiano agropecuario (ICA) collected maize races and indigenous germplasms in locations specified by Roberts et al. (1957). Colombia currently recognizes 42 maize races, which are grouped into four categories: (I) Primitive (Pollo, Pira, Imbricado); (II) Possibly introduced (Andaquí, Amagaceño, Berrendo, Cabuya, Canguil, Capia, Cariaco, Chulpi, Clavo, Costeño, Güirua, Huevito, Morochillo, Morochillo Rosado Cordobés, Morocho, Negro Peruano, Ojo de Gallo, Sangre Toro, Yucatán); (III) Colombian hybrids (Azul Vallecaucano, Brisa, Caucano, Chococeño, Común, Harinoso Dentado, Montaña, Negrito, Negro de Tabanok, Pira Naranja, Puya, Puya Grande, Rojo del Valle, Rojo Sureño, Sabanero, Sikuani, Timbrado); and (IV) Colombian hybrids Strictu sensu (Cacao, Capio, Maíz Dulce). The collection is currently conserved at the Universidad Nacional de Colombia sede Palmira (UNAL-PAL) (Caetano et al. 2015).

Due to the quantification of high heritability traits associated with functional diversity indexes, high-throughput phenotyping and the use of images in the evaluation and characterization of genetic resources have become available tools during selection processes in genetic improvement (Conejo-Rodríguez et al. 2024). By using diversity indicators, like the Colombian maize landraces, we will carry out image processing and phenomic selection. To deal with this task, the Colombian maize landraces from the "*Creuci Maria Caetano*" collection of UNAL-PAL were used (Fig. 17.1).

17.1.1 Digital Image Processing and Extraction of Digital Descriptors in the Colombian Races Maize Collection

Methodological approaches that include standardizing image acquisition and extracting features related to digital morphometry and colorimetry are necessary for digital image processing. The methodological approaches are based on six steps: (I) Capturing an image (Fig. 17.2a): This requires controlled lighting to ensure that the object's distance from the camera is uniform. (II) Binarization and color image extraction (Fig. 17.2b and c): This procedure involves removing the object of interest. We have extracted the corn from the picture using the R Pliman library (Olivoto 2022).

The Pliman simplifies digital image processing by incorporating algorithms that enhance feature extraction and detection as well as the ability to adapt various

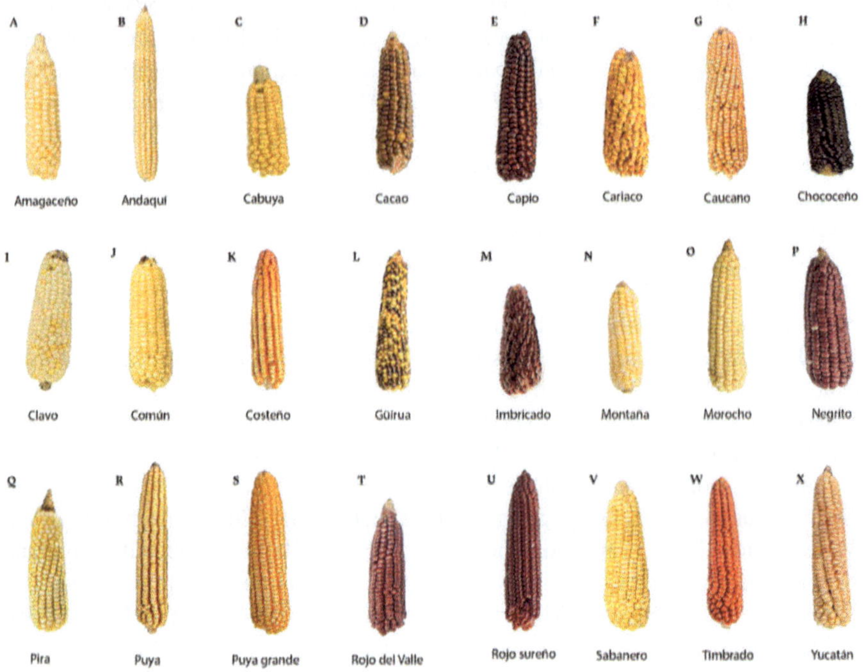

Fig. 17.1 Colombian maize races evaluated. (**a**) *Amagaceño*, (**b**) *Andaquí*, (**c**) *Cabuya*, (**d**) *Cacao*, (**e**) *Capio*, (**f**) *Cariaco*, (**g**) *Caucano*, (**h**) *Chococeño*, (**i**) *Clavo*, (**j**) *Común*, (**k**) *Costeño*, (**l**) *Güirua*, (**m**) *Imbricado*, (**n**) *Montaña*, (**o**) *Morocho*, (**p**) *Negrito*, (**q**) *Pira*, (**r**) *Puya*, (**s**) *Puya grande*, (**t**) *Rojo del Valle*, (**u**) *Rojo sureño*, (**v**) *Sabanero*, (**w**) *Timbrado*, and (**x**) *Yucatán*

mathematical models with RGB color plate to recognize the desired object. (III) Geometric morphometry (Fig. 17.2d): The Momocs library is used to perform contour analysis using the binarized image (Bonhomme et al. 2014). This software makes possible to determine the morphospaces using the Fourier elliptic method. (IV) The fourth method is digital colorimetry (Fig. 17.2e), which involves extracting the color groups with the highest representation at the target cluster number from the original color image. To find the colors in RGB space that have the largest representation in the image, we use Colordistance library via cluster analysis with k means (Weller and Westneat 2019). (V) Digital descriptors (Fig. 17.2f): The aim of this process is to pick the descriptors that best differentiate the evaluated accessions by combining the morphometric and colorimetric digital descriptors that were already extracted. The methodological approaches and the code to be used are accessible in the supplementary data repository at https://github.com/dfconejoro/Phenomics.

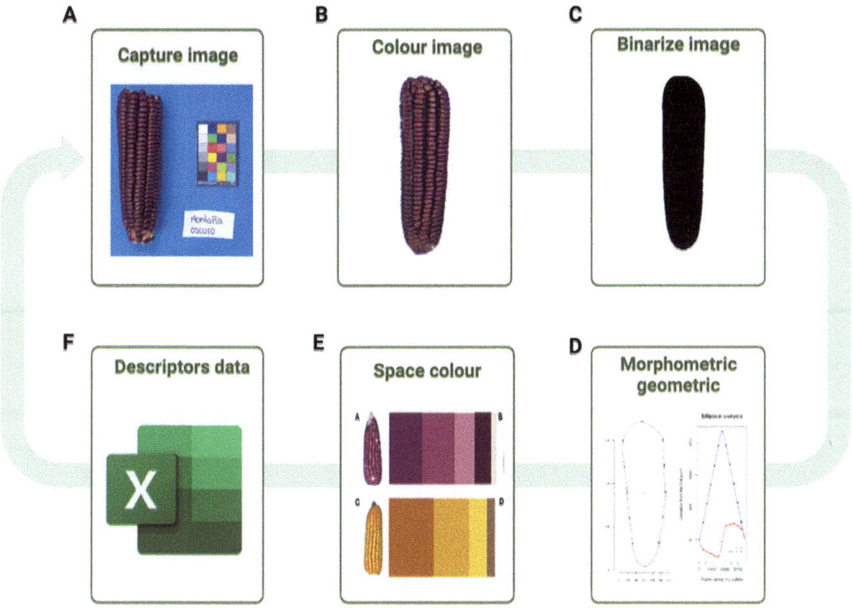

Fig. 17.2 Digital image processing and extraction of digital descriptors. (**a**) Image capture with standard white led light and color card, (**b**) Image with white background for colorimetric analysis, (**c**) Binarized image for morphometric analysis, (**d**) Extraction of morphospaces from the variation of Fourier elliptic, (**e**) Extraction of RGB color spaces for each cluster, and (**f**) Database related to morphometric and colorimetric digital descriptors

17.1.2 Morphometric and Colorimetric Digital Descriptors of the Variation of the Colombian Races Maize Collection

The variations in seed, leaf, and fruit morphology of the most important crops have served as a basis for the creation of descriptors used in its characterization and evaluation (IBPGR 1991). Numerous researches that present morphological variation quantitatively have used geometric morphometrics (Morales-Alonso et al. 2023; Conejo-Rodríguez et al. 2024). In the Colombian maize collection "Creuci Maria Caetano" of UNAL - PAL, we carried out contour analysis of ear shape variation. Contour analysis provides insight into the morphological variations in the maize collection and how they relate to the landraces identified (Revelo Portilla et al. 2015). The range of the X-axis is used to calculate the morphometric variation in the assessed accessions (Fig. 17.3a, b). The similar impact can be seen in the difference between the ear shape's mean and standard deviation. Race has a direct correlation with ear width and length (Fig. 17.3c). The phenotypic diversity of the ear shape is determined based on the 24 morphological spaces corresponding to the principal components (PC1-PC24) (Table 1S: Morphological spaces). These are digital traits that represent the morphological variation of the Colombian races evaluated. The digital colorimetric descriptors show the heterogeneity found in the

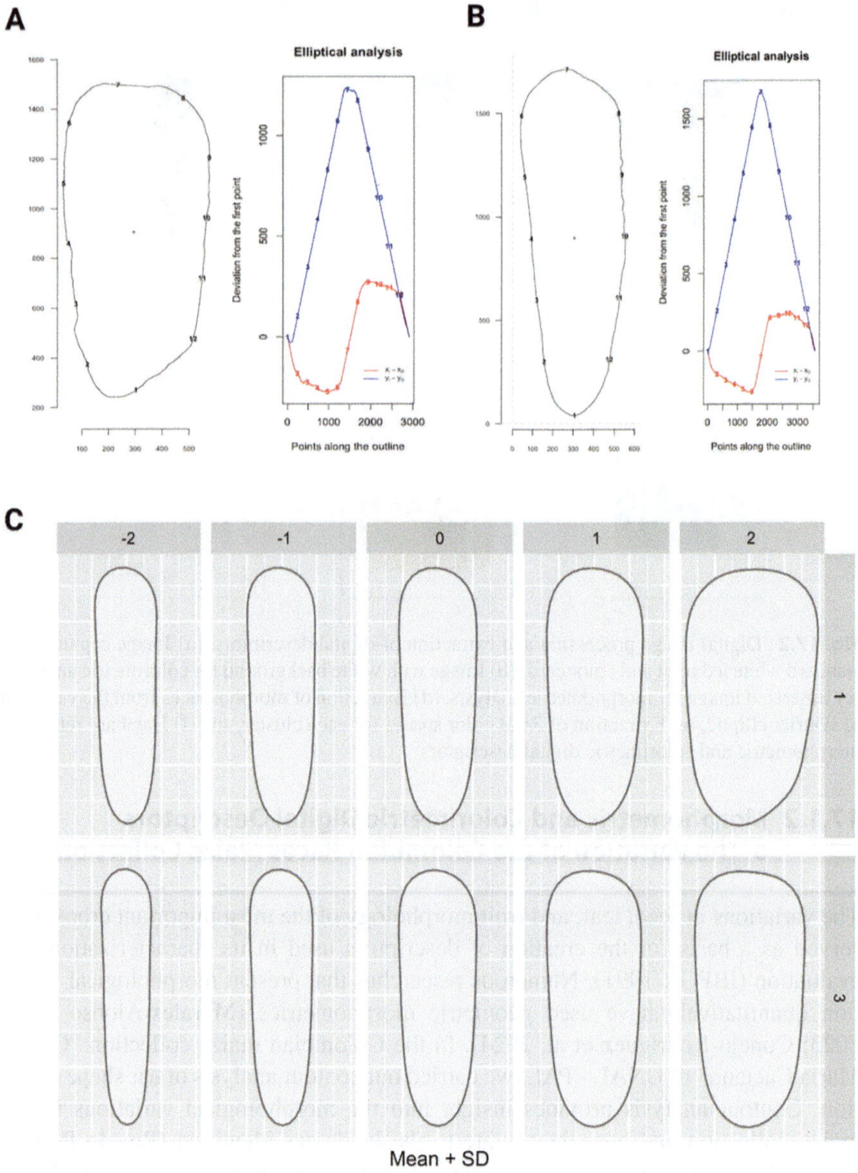

Fig. 17.3 Colombian races maize morphometric digital descriptors: (**a**, **b**) Fourier elliptic variation in the ear form of the corn, (**c**) Standard and average deviations of the variation in the outlines of the corn in the functional spaces PCs 1 through 3

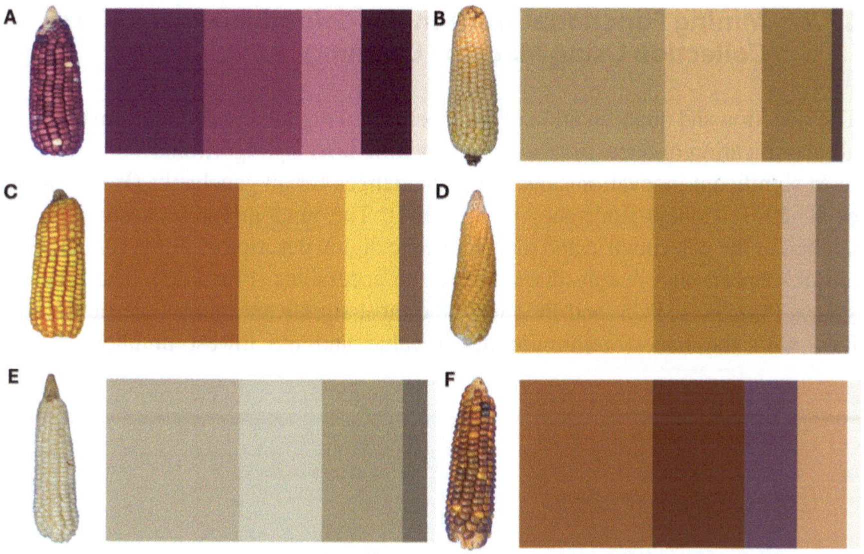

Fig. 17.4 Colorimetric variation of digital colorimetric descriptors of ears from selected Colombian races collection. (**a**) *Negrito* race, (**b**) *Clavo* race, (**c**) *Costeño* race, (**d**) *Chococeño* race, (**e**) *Common* race, (**f**) *Cacao* race. Each race has five main colors that are a grouping, for a total of five colors per ear evaluated

Colombian races that were evaluated (Fig. 17.4). Colorimetric analysis allows quantifying color from RGB spaces; however, Colombian races present similar colorimetric variations (IBPGR 1991; Revelo Portilla et al. 2015).

In this study, we demonstrate how to use digital images from the RGB color space values for quantification of ears' color distribution (Fig. 17.4). Kernel color is directly related to the indigenous populations selection in relation to the place of its introduction. The variations within kernel colors can be found in the majority of maize races (Revelo Portilla et al. 2015; Caetano et al. 2015), and genetically, it is connected to variations in carotenoid/anthocyanin pathways, usually with the presence of dominant alleles for red/purple and recessive alleles for yellow/white color. RGB values are extracted for every processed image for each defined cluster. The five colors which most effectively characterize each ear are represented by the RGB values per cluster (qualitative data), which are transposed into colorimetric digital descriptors (quantitative information; Fig. 17.4). It is worth noting that there exists multiple RGB colors within a single tone not easily recognizable for a human eye. For instance, the yellow of the *costeño* race (Fig. 17.4c) exhibits conventional color combinations ranging from orange to light yellow. These colors can also be employed as a means of choosing specific desired traits, like sweetness, flour production, carotenoids (for yellow seeds, Fig. 17.4b–d) or antioxidant ratio (for purple and red kernels of *negrito* races, Fig. 17.4a), proteins, and fiber (white kernels; Fig. 17.4e) as demonstrated by the colored kernels in selected races in Fig. 17.4.

17.2 Mining Functional Traits in the Colombian Races Maize Collection Using Machine Learning

The selection and quantification of the most useful digital descriptors from the overall classification of accessions—using machine learning algorithms—is one of the most significant innovations and future challenges for all genebanks (Henao-Rojas et al. 2021; Conejo-Rodríguez et al. 2024). The random forests technique was applied to the colorimetric and morphometric digital descriptors of the ears to relate ten descriptors that deeply discriminate the accessions (Fig. 17.5). The morphospaces PC9, PC1, PC3, and PC2 were the most significant morphometric descriptors, with the lowest minimum depth value and the lowest number of trees

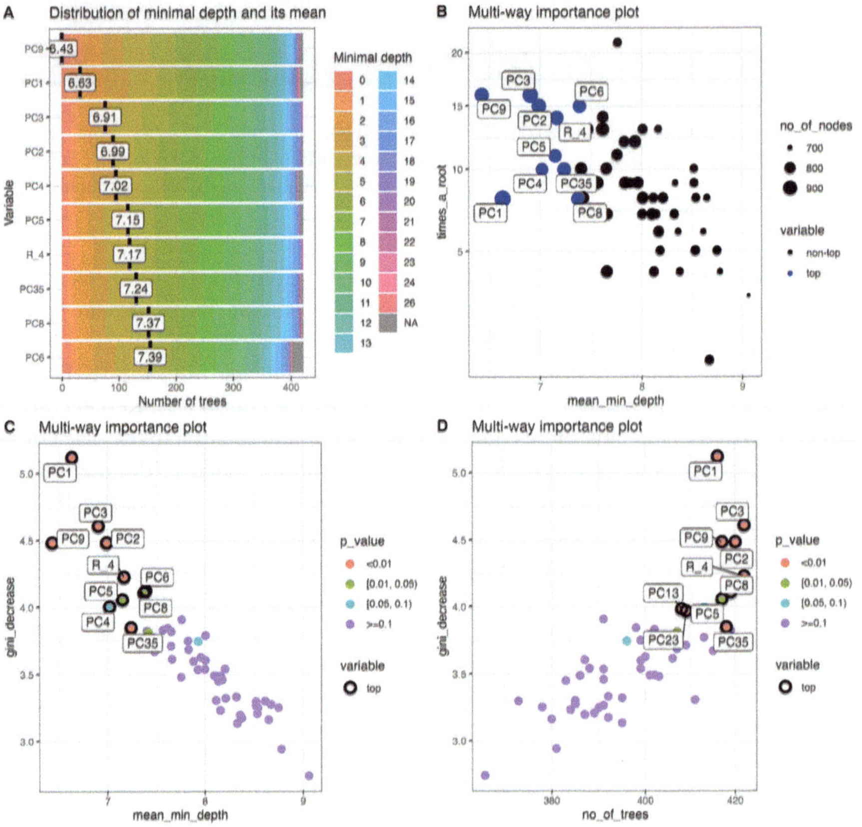

Fig. 17.5 Selection of important descriptors from random forest analyses. (**a**) Relationship of the importance variable number of trees and minimum depth, (**b**) Relationship of time roots and minimum depth. Blue color indicates important variables, (**c**) Gini index, minimum depth, and p-value, and (**d**) Gini index, number of trees, and p-value. The colors indicate the p-value of the first 10 variables of importance in the classification of maize landraces

(Fig. 17.5a). The colorimetric descriptor R_4 (Red_cluster 4) was the only one that significantly contributed to the classification of used maize accessions.

In addition to linking the Gini index that establishes the significance of the descriptor, the morphometric descriptors present a greater number of trees at lower minimum depths, which—importantly—results in a more accurate classification than using the sum of descriptors (Fig. 17.5b, c, d).

17.3 Selection of Potential Colombian Maize Accessions for Genetic Improvement Using Functional Diversity Indexes

In order to evaluate how individual differences in morphology and physiology affect ecosystem functions and their capacity to adapt to various environments, functional diversity is a crucial method in the characterization and selection of genetic resources (Mathur et al. 2023). This diversity in Colombian races is an invaluable resource for agronomic traits that can be used to further genetic development. Not only does the discovery and selection of accessions exhibiting exceptional functional traits support the preservation of this genetic diversity, but it also enhances prospects for the creation of more resilient and productive cultivars. Accessions that occupy extreme spaces of the convex area in the morphometric space can be chosen using functional diversity indices (Fig. 17.6), suggesting that they may provide unique traits to breeding programs. The variation of the functional traits of the accessions across different morphospaces is highlighted by the functional diversity of Colombian maize. As indicated by the morphospaces (PC) and their various dimensions (PC1, PC2, PC3, PC4, and PC5), there are multiple relationships among the various accessions that constitute the vertices of the two-dimensional functional spaces (Fig. 17.7). The accessions of Colombian races with assignments at the vertices of the morphospaces, referred to as vertices accessions, exhibit the highest degree of phenotypic variation among the accessions analyzed. In order to correct desired traits, these accessions can be employed in population improvement procedures. Particularly based on ear shape, races as *Chococeño*, *Costeño*, and *Puya* exhibit traits with the most obvious functional diversity (Fig. 17.7a–d).

It is believed that accessions that occupy functional extremes tend to provide greater genetic variability. Despite the fact that these accessions are not very representative in the genebanks, the analysis of the vertices accessions allows evaluation of its potential contribution for conservation and agronomic value for well-defined pre-breeding strategies. When evaluating the various races of maize, functional specialization (FSpe) analyzes the functional divergence between them, emphasizing the accessions at the vertices of the functional spaces that exhibit distinct morphometric traits. It is possible that these races and accessions, which are far from the functional space centroid, have evolved unique adaptations to their specific environments (Fig. 17.8). The degree to which the various races of maize phenotypically resemble one another is measured by a metric considered functional identity, or FIde. Races that exhibit distinctive traits and/or those that share characteristics are

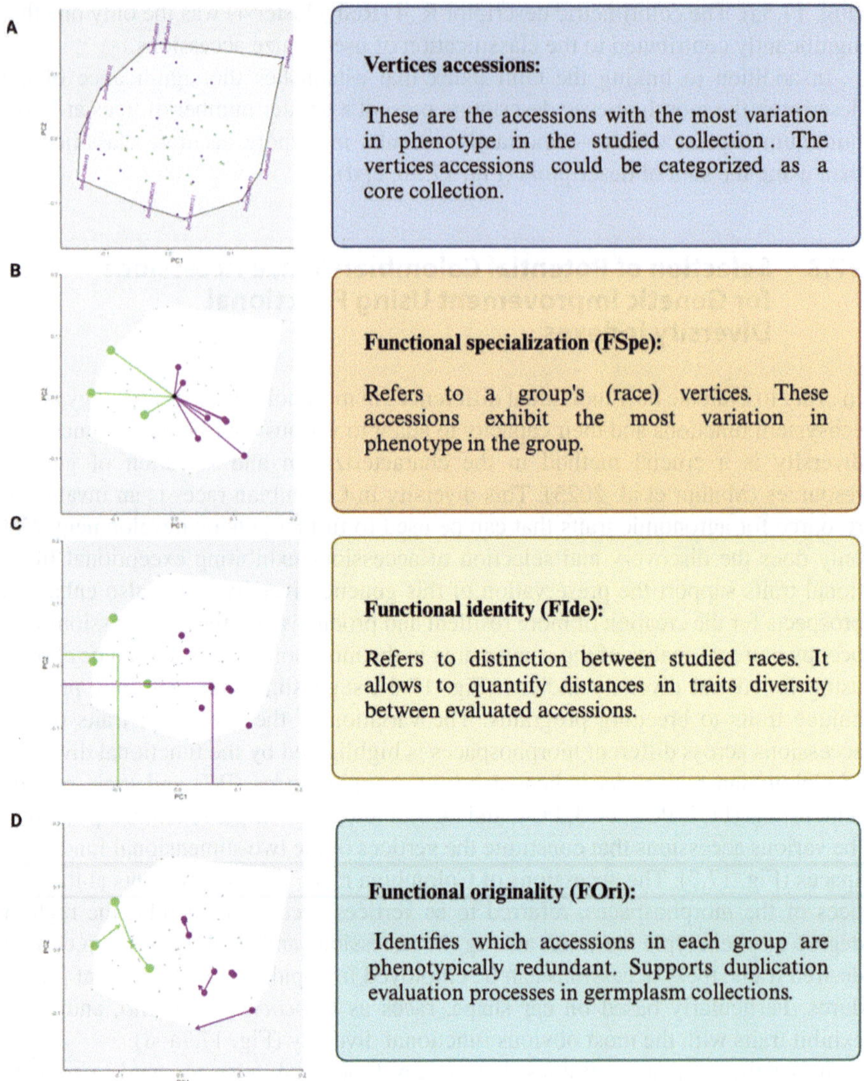

Fig. 17.6 Indices of functional diversity used for the evaluation of plant genetic resources. (**a**) Vertices accessions, (**b**) Functional specialization (FSpe), (**c**) Functional identity (FIde), and (**d**) Functional originality (FOri)

highlighted by this index, which measures the functional aggregate of each particular race. Landraces classified as functionally unique or redundant within the race group are known as functionally original (FOri). The arrows that highlight similarities between the accessions are suggestive of their redundant functional traits, which may facilitate the identification of duplicates in working collections and genebanks. By classifying races for adaptable functional traits, performing image-based

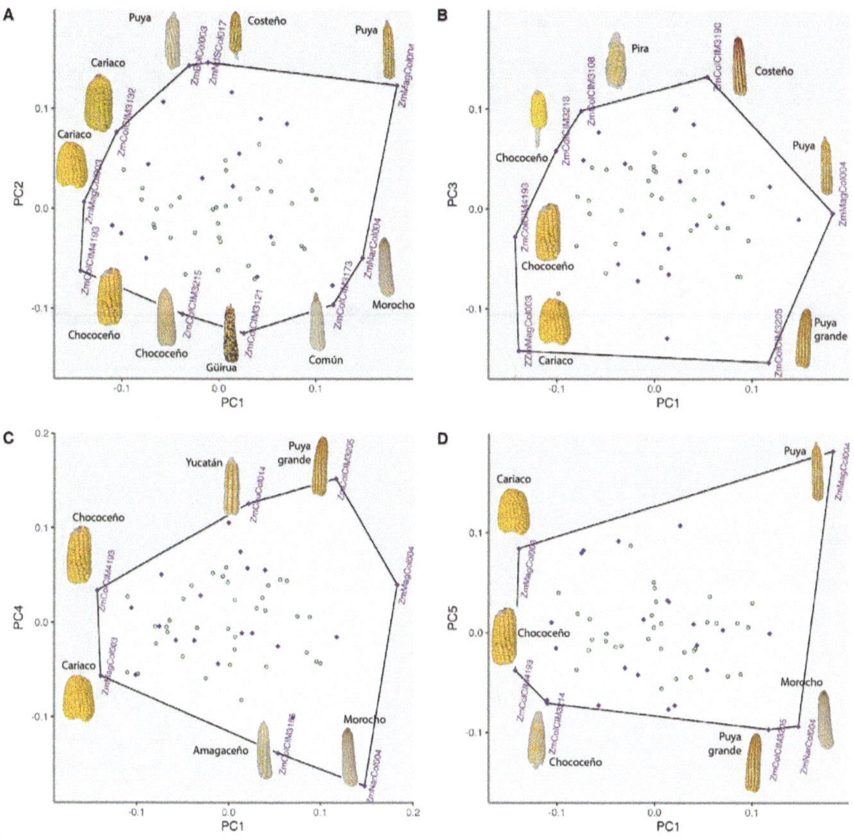

Fig. 17.7 Vertices accessions of functional trait variation in the ear of Colombian maize races. (**a**) Functional spaces PC1—PC2. (**b**) Functional spaces PC1—PC3. (**c**) Functional spaces PC1—PC4, and (**d**) Functional spaces PC1—PC5. Each purple dot represents an accession that is a vertex in some of the two-dimensional relations of the functional spaces

selection processes based on machine learning trait selection and detailed understanding of variation with functional diversity could promote easier but targeted parental selection and thus accelerate population improvement (Conejo-Rodríguez et al. 2024).

17.4 Conclusions and Prospects

We present a machine learning-based methodology for phenomic selection and image processing, utilizing functional trait selection procedures. This approach integrates functional diversity indices to complement biological selection criteria in the breeding process. Our work provides a methodological framework, a comprehensive understanding of the procedures, and a phenome-centric perspective on the

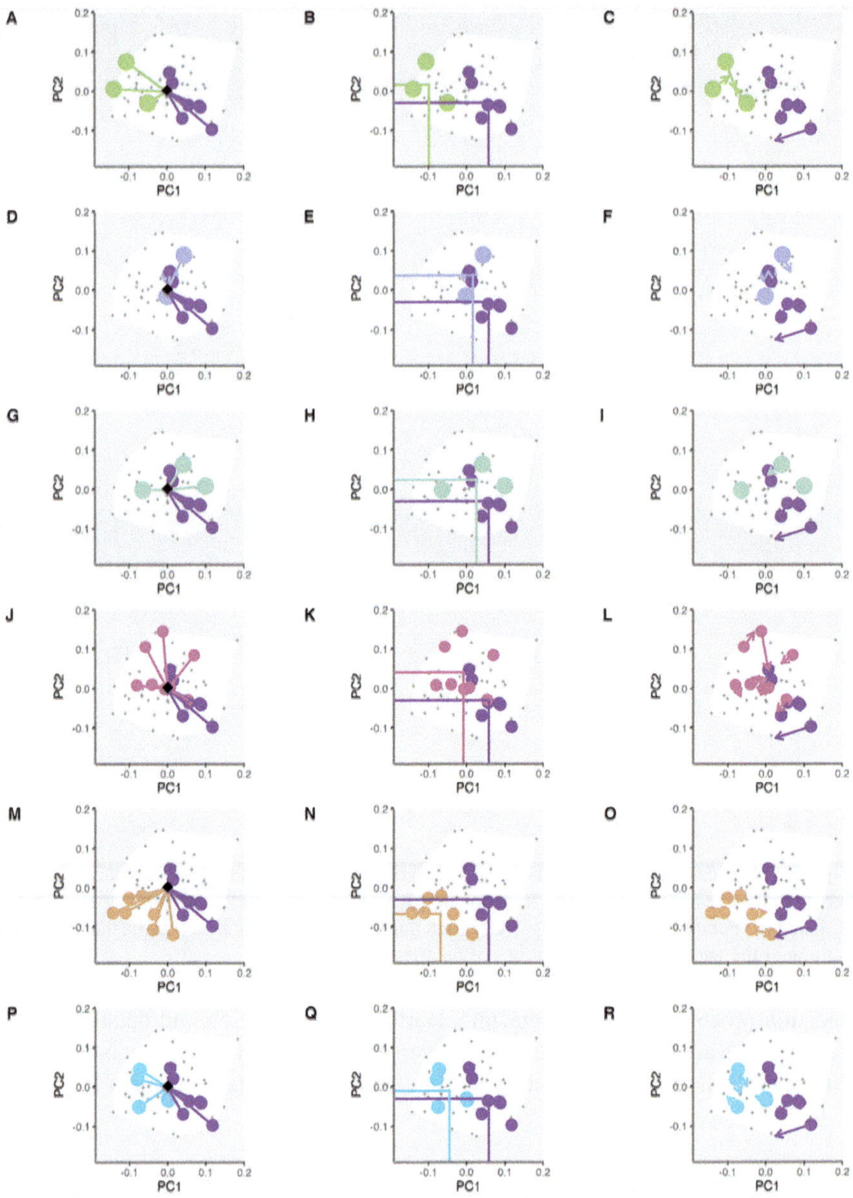

Fig. 17.8 Functional diversity indices related to Colombian maize races. (**a–c**) *Common* race-*Cariaco;* (**d–f**) *Common* race—*Cabuya*; (**g–i**) *Common* race—*Clavo*; (**j–l**) *Common* race—*Costeño*; (**m–o**) *Common* race—*Choconeño*; (**p–r**) *Common* race—*Pira*

crop indices. By applying these analytical techniques, genebank collections can be described via additional quantitative and qualitative trait data, facilitating deeper characterization of germplasm. Additionally, index-based functional diversity techniques aid in the conservation, evaluation, and regeneration of accessions, ensuring the representation of higher phenomic variation within conserved accessions for each protected species.

References

Bonhomme V, Picq S, Gaucherel C, Claude J (2014) Momocs: outline analysis using R. J Stat Softw 56:1–24

Caetano CM, Cuellar RDP, Juajibioy JLM, Ávila LNV, Nunes DGC, de Pazdiora BRCN (2015) Mejoramiento participativo: herramienta para la conservación de cultivos subutilizados y olvidados. Acta Agronómica 64(3sup):307–327

Conejo-Rodríguez DF, Gonzalez-Guzman JJ, Ramirez-Gil JG, Wenzl P, Urban MO (2024) Digital descriptors sharpen classical descriptors, for improving genebank accession management: a case study on Arachis spp. and Phaseolus spp. PLoS One 19(5):e0302158

Henao-Rojas JC, Rosero-Alpala MG, Ortiz-Muñoz C, Velásquez-Arroyo CE, Leon-Rueda WA, Ramírez-Gil JG (2021) Machine learning applications and optimization of clustering methods improve the selection of descriptors in blackberry germplasm banks. Plants 10(2):247

IBPGR, C (1991) Descriptors for maize. International Board for Plant Genetic Resources, Rome and International Maize and Wheat Improvement Center, Mexico City

Mathur S, Mason AJ, Bradburd GS, Gibbs HL (2023) Functional genomic diversity is correlated with neutral genomic diversity in populations of an endangered rattlesnake. Proc Natl Acad Sci 120(43):e2303043120

Morales-Alonso A, Villaverde T, Jiménez-Mejías P (2023) Geometric Morphometrics sheds light on the systematics affinities of two enigmatic dwarf Neotropical sedges (Carex, Cyperaceae). PhytoKeys 232:167

Olivoto T (2022) Lights, camera, pliman! An R package for plant image analysis. Methods Ecol Evol 13(4):789–798

Revelo Portilla EA, Cardozo Conde CI, Caetano CM (2015) Estudio molecular preliminar de accesiones de maíz (Zea maysL.) criollo e indígena Colombiano, utilizando una región de ADN cloroplástico. Acta Agronómica 64(1):72–82

Weller HI, Westneat MW (2019) Quantitative color profiling of digital images with earth mover's distance using the R package colordistance. PeerJ 7:e6398

Part V

New Plant Breeding Techniques

New Plant Breeding Technologies for Climate-Smart Agriculture: Enhancing Abiotic Stress Tolerance

18

Andrés Gatica-Arias and Alejandro Hernández-Soto

Abstract

Global agriculture faces serious difficulties from climate change, which puts food security and agricultural yield at risk. Innovative approaches are required to generate robust crops due to temperature shifts, changing precipitation patterns, and an increase in the frequency of extreme weather events. With cutting-edge plant breeding technologies (NPBTs), including CRISPR/Cas9, farmers may tweak crop genomes with previously unheard-of precision to increase resistance to abiotic stresses like drought, salinity, and high temperatures. By overcoming the time-consuming and genetic diversity-restricted nature of traditional breeding procedures, these technologies expedite the generation of climate-resilient cultivars. The agricultural enhancement process is further accelerated by speed breeding techniques, which allow for many generations in a year. The most recent developments in NPBTs and their uses in climate-smart agriculture to increase resistance to abiotic stresses are summarized in this chapter. Genome editing offers a holistic strategy to ensure agricultural sustainability when combined with climate-smart agricultural techniques, such as improving soil health and water use efficiency. The promise for NPBTs to lessen the effects of climate change and protect global food security lies in their ability to combine contemporary breeding methods with climate-resilient farming practices. A thorough bibliometric study of research trends from 2020 to 2024 highlights important

A. Gatica-Arias (✉)
University of Costa Rica, School of Biology, Plant Biotechnology Laboratory, San José, Costa Rica

Red Latinoamericana en Biotecnología, San José, Costa Rica
e-mail: andres.gatica@ucr.ac.cr; https://redbiotecnologia-latam.org/

A. Hernández-Soto
Technologic Institute of Costa Rica, Biology School, Biotechnology Research Center, Cartago, Costa Rica

P. M. Priyadarshan, R. Ortiz (eds.), *Plant Breeding 2050*,
https://doi.org/10.1007/978-981-95-0583-8_18

areas of concentration, including food security, sustainable development, and crop stress tolerance. The results highlight how important genome editing technologies are to improving agricultural resilience to ongoing environmental problems.

Keywords

Genome editing · CRISPR/Cas9 · Abiotic stress · Food security · Sustainable agriculture

18.1 Introduction

Agriculture depends on the climate and specific weather patterns to produce food and vital goods necessary for human survival (Anderson et al. 2020). Furthermore, agriculture is influenced by climate change, which is anticipated to bring about both beneficial and detrimental effects (Raza et al. 2019). Climate change, characterized by substantial shifts in long-term average values of meteorological elements such as precipitation and temperature, poses a significant threat to agriculture worldwide, impacting crop productivity and food security (Malhi et al. 2021). The consequences of climate change are complex, affecting various aspects of crop growth and development. Shifts in temperature, precipitation patterns, and the frequency of extreme weather events have already shown adverse effects on crop yields and quality (Malhi et al. 2021). For example, elevated atmospheric CO_2 levels, though potentially beneficial for plant growth, can also cause nutrient imbalances, complicating agricultural outcomes (McGrath and Lobell 2013). Higher temperatures can speed up crop maturation, shorten grain filling periods, and induce heat stress, negatively impacting yield and quality (Skendžić et al. 2021). Changes in precipitation patterns can lead to droughts or excessive rainfall, both detrimental to crops by affecting soil moisture and nutrient availability (Engelhardt et al. 2021). Moreover, climate change can increase the prevalence of pests and diseases, further challenging crop management and productivity (Skendžić et al. 2021).

The variability in climate conditions calls for urgent and innovative approaches to ensure the resilience and sustainability of agricultural systems. To address these challenges, breeding strategies have evolved to enhance crop resilience to both abiotic and biotic stresses linked to climate change (Cooper and Messina 2023). While traditional breeding methods are effective, they are time-consuming and often constrained by the genetic diversity available within species (Breseghello and Coelho 2013). Modern breeding techniques, such as marker-assisted and genomic selection, enable more precise and efficient development of crop varieties with desirable traits. These methods leverage advancements in genetic mapping and sequencing technologies to identify and incorporate beneficial genes into breeding programs (Budhlakoti et al. 2022).

Genome editing, especially with CRISPR/Cas9, has revolutionized plant breeding by allowing precise modifications to the plant genome (Yadav et al. 2021). This technology enables targeted manipulation of genes associated with biotic and abiotic stress tolerance (Pandita 2022a, 2022b), yield, and nutritional quality (Chaudhry et al. 2023). Unlike traditional transgenic approaches, genome editing can create plants with specific traits without introducing foreign DNA, addressing some regulatory and public acceptance concerns (Bain et al. 2020; Menz et al. 2020). The rapid development of climate-resilient crops through genome editing holds great promise for mitigating the impacts of climate change on agriculture (Hamdan et al. 2022; Garland and Curry 2022).

Climate-smart agriculture (CSA) includes a set of practices aimed at sustainably increasing agricultural productivity, adapting to and building resilience against climate change, and reducing greenhouse gas emissions (Bai et al. 2019). CSA practices include optimizing water use, enhancing soil health, and integrating agroforestry systems. Farmers can improve their adaptability to changing climatic conditions by adopting CSA principles while contributing to climate change mitigation (De Pinto et al. 2020). Integrating modern breeding techniques and CSA practices can synergistically enhance the sustainability and resilience of agricultural systems (Gentile Alessandra and La Malfa 2019; De Pinto et al. 2020).

Speed breeding is an innovative approach that accelerates the breeding cycle of crops, enabling the development of new varieties in a fraction of the time required by conventional methods (Watson et al. 2017, 2018). By controlling environmental conditions such as light and temperature, speed breeding can shorten the time from seed to seed, allowing multiple generations of crops to be grown within a single year (Samantara et al. 2022). This method is precious for rapidly incorporating beneficial traits into crops, enhancing their ability to withstand climate-related stresses. When combined with genome editing and other advanced breeding techniques, speed breeding can significantly expedite the development of climate-resilient crops (Razzaq et al. 2021).

This review will synthesize recent advancements and insights in genome editing and climate-smart agriculture by analyzing articles published between 2020 and 2024. A comprehensive literature search was conducted in the SCOPUS database (https://www.scopus.com/) using the keywords "genome," "editing," "smart," and "agriculture." Thirty-nine relevant articles were identified and reviewed to provide an overview of current research trends, methodologies, and applications in the field.

18.2 Bibliometric Search Results

The annual production of articles over four years (2020–2024) was analyzed using the keywords "genome," "editing," "smart," and "agriculture." In 2020, two articles were published, increasing to 4 in 2021. This upward trend continued into 2022 with 7 articles, peaking at 16 articles in 2023, and subsequently stabilizing at 10 articles up to the date of this review (Fig. 18.1).

Figure 18.2 shows the frequency of key terms in the literature reviewed on genome editing and climate-smart agriculture. "Agriculture" is the most mentioned term (12 occurrences), highlighting its central role. "Food security" (7) and

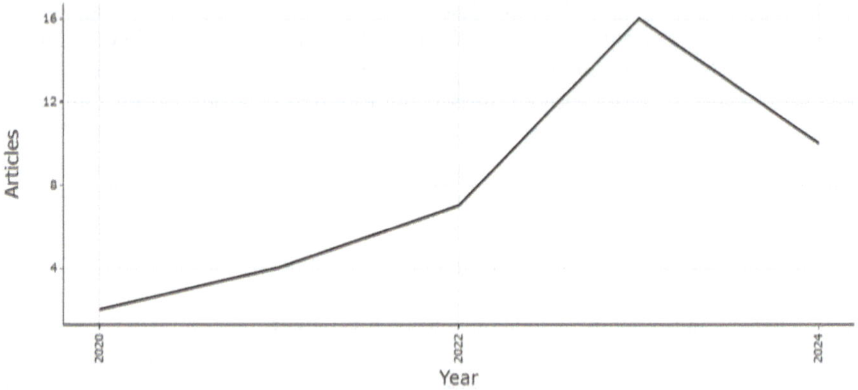

Fig. 18.1 Trends in annual article production from 2020 to 2024 using keywords "genome," "editing," "smart," and "agriculture" in the literature search conducted in the SCOPUS database (https://www.scopus.com/)

Fig. 18.2 Frequency of key terms in the literature on genome editing and climate-smart agriculture based on a search conducted in the SCOPUS database (https://www.scopus.com/)

"sustainable development" (6) are also significant. Terms with 5 mentions, such as "arsenic," "climate change," "crop," "gene editing," "nonhuman," and "plant breeding," reflect environmental and genetic research focuses.

The thematic map presented in Fig. 18.3 illustrates the interconnectedness and thematic clustering within the field of new plant breeding technologies (NPBTs) for climate-smart agriculture. Prominently connected key themes, such as "sustainable development," "nonhuman," and "food security," underscore their central role in this research domain. The clusters are color-coded: green nodes represent topics like "agricultural sustainability" and "cultivation," blue nodes are associated with "crop improvement" and "physiological stress," red nodes signify themes related to

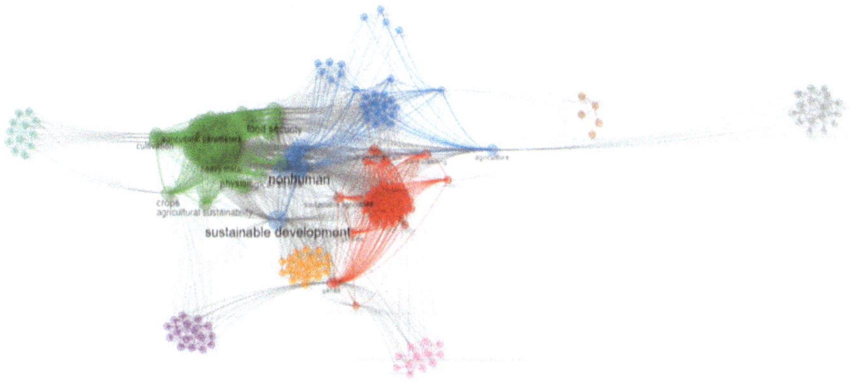

Fig. 18.3 Thematic map of interconnectedness and clustering in New Plant Breeding Technologies (NPBTs) for climate-smart agriculture based on a search conducted in the SCOPUS database (https://www.scopus.com/)

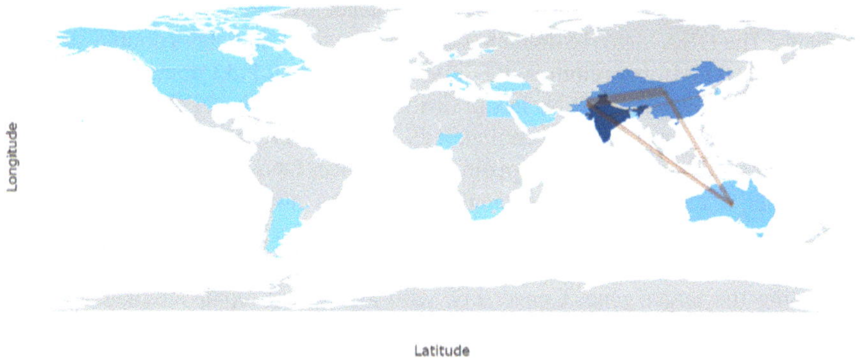

Fig. 18.4 Collaborators world map for international research partnerships based on a search conducted in the SCOPUS database (https://www.scopus.com/)

"plant breeding" and "genetics," and purple nodes indicate emerging areas such as "CRISPR-associated protein." This map visually underscores the multidisciplinary nature of NPBTs, emphasizing their relevance in addressing global challenges related to food security, sustainable agriculture, and environmental resilience.

Figure 18.4 presents a diverse international research network characterized by several key collaborations. Australia partners with Cyprus, Italy, Turkey, and the United Arab Emirates, alongside multiple collaborations with China, India, and Pakistan. China demonstrates a robust network with collaborations spanning Australia, Cyprus, Egypt, Italy, Korea, Lebanon, Turkey, and the United Arab Emirates, as well as collaborations with India. Cyprus collaborates with Australia, China, Italy, Pakistan, and the United Arab Emirates. India collaborates across Australia, China, Denmark, Estonia, Korea, Pakistan, and Saudi Arabia. Italy collaborates with Australia, China, Cyprus, Pakistan, and Turkey. Korea's collaborations include China, Denmark, India, and Pakistan. Pakistan has a broad collaboration network with Australia, Canada, China, Cyprus, Egypt, Italy, Korea, Lebanon, Turkey, and the United Arab Emirates. The USA collaborates notably with Bangladesh and Nigeria with South Africa.

18.3 Basic Genome-Editing Techniques

18.3.1 Zinc Finger Nucleases (ZFNs)

Zinc Finger Nucleases (ZFNs) represent a significant advancement in biotechnology, allowing precise modification of specific DNA sequences through genetic engineering. These synthetic enzymes have two fundamental components: a DNA-binding domain derived from zinc finger proteins (ZFPs) and a DNA-cleaving domain from the FokI restriction enzyme (Gaj et al. 2013).

ZFPs are modular proteins containing zinc finger domains, each capable of recognizing and binding to a distinct triplet of DNA base pairs. Researchers can tailor ZFPs to target virtually any desired DNA sequence by carefully engineering these domains. The DNA-cleaving domain, sourced from FokI, initiates a double-stranded DNA break (DSB), but requires the dimerization of two FokI domains to become active. The innovation of ZFNs lies in fusing these customized ZFPs with the FokI domain. This fusion allows the engineered ZFN to bind to its target DNA sequence and recruit a second ZFN, forming a dimer that induces DNA cleavage (Carroll 2011).

The cell's natural repair mechanisms, such as nonhomologous end joining (NHEJ) or homologous recombination (HR), are activated upon cleavage. NHEJ may introduce small insertions or deletions without a repair template, potentially disrupting the target gene. Conversely, with a supplied repair template, HR can precisely insert a new gene sequence or correct a specific mutation (Carroll 2011).

ZFNs' versatility and precision have revolutionized genetic engineering, facilitating applications across diverse fields, from fundamental research to therapeutic interventions. Researchers have successfully applied ZFNs to edit genomes across various organisms, expanding our ability to understand and manipulate biological systems (Gaj et al. 2013). Nonetheless, ensuring the specificity of DNA-binding domains and minimizing off-target effects remain critical challenges in developing and applying ZFN technology. A detailed review of genome editing by ZFNs in agriculture can be found in (Petolino 2015).

18.3.2 Transcription Activator-like Effector Nucleases—TALENs

Transcription Activator-Like Effector Nucleases (TALENs) are a powerful genome editing tool that allows for precise targeting and modification of DNA sequences (Gaj et al. 2013). They were developed as an alternative to the earlier Zinc Finger Nucleases (ZFNs) technology (Sun and Zhao 2013). TALENs have two key components: a DNA-binding domain and a DNA-cleavage domain. The DNA-binding domain is derived from transcription activator-like effectors (TALEs), proteins secreted by *Xanthomonas* bacteria (Sun and Zhao 2013). TALEs contain a series of 33–34 amino acid repeats, with the 12th and 13th amino acids (known as the repeat variable diresidue or RVD) determining the recognized DNA base. By engineering the sequence of these RVDs, researchers can create TALEs that can bind to virtually any desired DNA sequence.

The DNA-cleavage domain of TALENs is derived from the FokI restriction enzyme, which creates a double-stranded DNA break when two FokI domains dimerize. By fusing the customized TALE DNA-binding domain to the FokI cleavage domain, the resulting TALEN can bind to its target DNA sequence and recruit a second TALEN to form a dimer, triggering DNA cleavage. The DNA breaks created by TALENs can then be repaired by the cell's natural repair mechanisms, such as nonhomologous end joining (NHEJ) or homologous recombination (HR). NHEJ can lead to small insertions or deletions that disrupt the target gene, while HR can be used to precisely insert a new gene or correct a specific mutation if a repair template is provided (Sun and Zhao 2013; Gaj et al. 2013).

TALENs have been widely used for genome editing in various organisms, including plants, animals, and human cells (Sprink et al. 2015). They offer advantages over ZFNs in terms of their ease of design and high targeting specificity. However, challenges remain in ensuring the specificity of TALEN binding and minimizing off-target effects (Mahfouz et al. 2014). Ongoing research continues to improve the efficiency and safety of TALEN-based genome editing technologies. A detailed review of genome editing by TALEN in agriculture can be found in (Khan et al. 2017).

18.3.3 CRISPR/Cas9 System

The CRISPR/Cas9 system is a revolutionary genome editing technology that has transformed genetic research and plant breeding (Doudna and Charpentier 2014; Wang and Doudna 2023). It was initially discovered as a bacterial immune system against invading viruses in about 50% of bacteria and 90% of archaea (Doudna and Charpentier 2014; Hille et al. 2018). However, it has since been adapted for precise gene editing in various organisms (Bao et al. 2019).

The CRISPR/Cas9 system consists of two key components: a CRISPR RNA (crRNA) that recognizes and binds to a specific DNA sequence and a Cas9 enzyme that cuts the DNA at that target site (Doudna and Charpentier 2014). The crRNA is derived from the CRISPR arrays found in bacteria, which serve as a genetic memory to detect and destroy invading viruses. The Cas9 enzyme is an endonuclease that can cleave double-stranded DNA when guided by the crRNA (Tang et al. 2016; Bao et al. 2019).

Scientists have streamlined the CRISPR system by combining the CRISPR RNA (crRNA) and the trans-activating CRISPR RNA (tracrRNA) into a single molecule known as single guide RNA (sgRNA), which simplifies the targeting process. (La

Russa and Qi 2015). This sgRNA is then paired with the Cas9 enzyme to form the CRISPR/Cas9 complex. When this complex binds to the target DNA sequence, the Cas9 enzyme creates a double-stranded break in the DNA (Tang et al. 2016; Mekler et al. 2017; Bao et al. 2019; Wang et al. 2023).

The DNA break can then be repaired by the cell's natural repair mechanisms, such as nonhomologous end joining (NHEJ) or homologous recombination (HR). NHEJ can lead to insertions or deletions (InDels) that tend to disrupt the target gene, resulting in knockout. At the same time, HR can be used to precisely insert a new gene or correct a specific mutation if a repair template is provided (Bao et al. 2019).

The CRISPR/Cas9 system has revolutionized the field of genome editing, offering unparalleled precision and efficiency in gene modification. In agriculture, CRISPR/Cas9 holds promise for accelerating breeding programs and advancing smart agriculture practices by enabling targeted modifications in crops for desired traits like disease resistance, yield enhancement, and nutritional fortification (Rao and Wang 2021; Ali et al. 2022). This transformative technology not only expedites the development of improved crop varieties but also enhances their resilience to environmental stresses, fostering sustainable agricultural practices tailored to meet global food security challenges (González Guzmán et al. 2022). A detailed review of genome editing by CRISPR/Cas9 in agriculture can be found in (Chen et al. 2019; Singh et al. 2023).

18.4 New Developments in Genome-Editing Techniques

18.4.1 Base Editing

Base editing is an advanced genome editing technology that precisely modifies individual DNA bases without generating double-stranded breaks (DSBs)(Li et al. 2020). This technique employs components from the CRISPR system, specifically a catalytically inactive Cas9 protein (dCas9) or a Cas nickase, combined with a deaminase enzyme to achieve targeted genetic alterations (Eid et al. 2018).

The base editing process involves three key components. First, the Guide RNA (gRNA) directs the Cas protein to the DNA sequence targeted for editing. Second, the Cas protein, a modified version of Cas9, binds to the DNA without inducing double-stranded breaks, thereby facilitating the targeted recruitment of the deaminase enzyme. Finally, the deaminase enzyme carries out the chemical modification of the target base. For example, a cytosine base editor (CBE) converts cytosine (C) to uridine (U), which is then repaired to thymine (T) during DNA replication. Similarly, an adenine base editor (ABE) transforms adenine (A) into inosine (I), which pairs with cytosine (C) upon repair, ultimately converting A to guanine (G) in the final DNA sequence. A detailed review of genome editing by base editing in agriculture can be found in (Bharat et al. 2020; Li et al. 2020).

18.4.2 Prime Editing

Prime editing has emerged as a revolutionary tool in genome editing, enabling precise insertion, deletion, and base substitution without inducing double-strand breaks (Molla et al. 2021; Choi et al. 2022). The prime editing process involves several key components and steps. First, the Prime Editing Guide RNA (pegRNA) directs the editing machinery to the target site in the genome. This RNA molecule includes a sequence complementary to the target DNA and an additional segment that encodes the desired edit. Next, the Cas9 Nickase, specifically the H840A variant, introduces a single-strand break (nick) in the DNA, facilitating the editing process. Attached to the Cas9 nickase is the Reverse Transcriptase (RT), which synthesizes a new DNA sequence based on the pegRNA's information. The pegRNA binds to the target DNA, the Cas9 nickase creates a nick, and the RT uses the pegRNA's extended portion to generate a new DNA sequence that includes the desired edit, forming a 3′ flap of DNA. Finally, the cell's natural DNA repair mechanisms incorporate this new sequence into the genome. The cell recognizes the nicked strand and uses the edited strand as a template to repair the unedited strand, ensuring that the genetic change is maintained in both DNA strands (Petrova and Smirnikhina 2023).

Prime editing technology was first developed in rice with the creation of the initial system, known as the plant prime editors (pPE2) toolkit (Xu et al. 2020). This breakthrough has led to the widespread adoption of prime editing in various crops, including rice, peanut, chickpea, and cowpea, serving as a proof of concept for GFP editing in protoplasts (Biswas et al. 2022). A detailed review of genome editing by prime editing in agriculture can be found in (Hassan et al. 2022).

18.5 Genome Editing Tools for Developing Climate-Resilient Crops

18.5.1 Drought Tolerance

Global climate change has accelerated the frequency of adverse and unpredictable weather events, causing substantial yield losses and crop failures across the globe. Drought is identified as a critical limiting factor among these climate-induced challenges. Plants subjected to drought stress experience stunted growth due to photoinhibition and decreased nutrient uptake. This stress also reduces transpiration efficiency, manifesting as lower stomatal conductance, increased leaf abscission, and stomatal closure. Consequently, these physiological responses decrease CO_2 levels, impairing photosynthetic potential. On a biochemical level, drought stress hampers carbon metabolism and interferes with cellular signaling pathways. CRISPR technology provides possibilities for engine drought, salinity, and thermotolerant plants (Chennakesavulu et al. 2022; Shelake et al. 2022; Angon et al. 2023; Rai et al. 2023).

CRISPR-based studies on drought tolerance have identified a variety of genes associated with improved drought resistance across multiple plant species (Zafar et al. 2020; Bhat et al. 2021; Chennakesavulu et al. 2022; Shelake et al. 2022; Angon et al. 2023; Kumar et al. 2023; Rai et al. 2023). In *Arabidopsis*, key genes include *miR169a, OST2, AVP1* (Vacuolar H + -pyrophosphatase), *AREB1, PtoMYB216, UGT79B2, UGT79B3, AtAITR, AtOST2, RGLG2, AtPUB19, ARR1, ARR10, ARR12,* and *FtMYB10*. In cotton, genes such as *GhRDL1* and *GhPIN1–3* have been linked to drought tolerance, while *MeMAPKK* and *MeKUPs* play similar roles in cassava. In chickpeas, genes like *4CL* (4-coumarate ligase) and *RVE7* (Reveille 7) have been noted. In maize, drought resistance is associated with genes such as *ARGOS8, ARGOS*, and *abh2*, and in papaya, *CpDreb2* plays a key role. For rice, several genes, including *OsERA1, OsSRL1, OsPUB67, OsSAPK2, OsADR3, OsDIS1, OsiSAP7, MODD, OMTN2, OMTN3, OMTN4, OMTN6, OsDST, OsNAC14*, and *SRL1, SRL2*, have been identified. In sugarcane, *ScNsLTP* contributes to drought tolerance, while in soybean, *GmMYB118* and *GmHdz4* are important. Tomato has been found to rely on genes such as *SlARF4, SlMPK3, SlLBD40,* and *SlNPR1*, and mitogen-activated kinases like *GID1a*, while in wheat, *TaDREB2, DREDB3*, and *TaERF3* are involved. In *Brassica napus*, the gene *BnaA6 RGA* has been noted. This diversity of drought-tolerance genes across species is significant, as they may have homologs in other orphan crops, providing similar drought-resistance benefits.

A study by Nuñez-Muñoz et al. (2021) focused on enhancing drought tolerance by targeting the trehalase gene. They developed a novel vector platform, incorporating a streptomycin resistance marker to enable genome editing in both monocots and dicots using the hypervirulent *Agrobacterium tumefaciens* AGL1 strain. The guide RNA (gRNA) was designed based on the trehalase sequence from drought-tolerant Selaginella species. In *A. thaliana*, plants with edits in the trehalase substrate-binding domains exhibited increased drought tolerance. Additionally, the researchers characterized a transient transformation system for editing genes in maize leaves. These results demonstrate the power of CRISPR/Cas9 technology to improve drought resistance, with promising applications for agricultural sustainability and food security. Similar results were observed by Hernández-Soto et al. (2022), who induced CRISPR/Cas9 mutations in trehalose-degrading genes (*NTH1* and *NTH2*) in yeast, leading to stress-tolerant traits.

In maize, Yang et al. (2024) identified and characterized *ZmASR1*, a drought-induced ASR gene, revealing its role in drought tolerance. *ZmASR1* transcription was upregulated under drought stress and abscisic acid (ABA) treatment. CRISPR/Cas9 knockout lines showed enhanced drought tolerance, characterized by reduced reactive oxygen species (ROS) accumulation, increased ABA levels, and greater stomatal closure compared to wild-type plants. Transcriptome analysis highlighted the involvement of ABA signaling, antioxidant defense, and photosynthetic pathways in drought response. These results suggest that *ZmASR1* acts as a negative regulator of drought tolerance and presents a promising target for improving drought resistance in maize.

Also, in maize, Niu et al. (2024) investigated the role of β-amylase (BAM) in leaf starch degradation and drought tolerance, focusing on the *ZmBAM8* gene. *ZmBAM8* expression was significantly upregulated in drought-tolerant maize lines and localized in the vascular bundle sheath, the site of transitory starch (TS) turnover.

Overexpression of *ZmBAM8* led to enhanced starch hydrolysis under drought stress, while knockout lines showed reduced starch degradation. *ZmBAM8*-overexpressing plants displayed improved drought tolerance compared to knockout lines. This study highlights the role of starch metabolism in C4 crops and suggests *ZmBAM8* as a target for developing drought-resistant maize varieties.

In rice, Kim et al. (2024) investigated the drought tolerance of a gamma-ray-induced mutant line, ditl4, which exhibited significant improvements in various physiological parameters under drought conditions. Compared to wild-type (WT) plants, ditl4 showed greater fresh weight, survival rates, chlorophyll content, and proline and soluble sugar accumulation, alongside reduced levels of hydrogen peroxide (H2O2) and malondialdehyde (MDA). Enhanced activities of antioxidant enzymes, reduced leaf curling, and decreased water loss further characterized ditl4's drought resilience. Whole-genome resequencing identified a SNP in *OsDIRH2*, a RING-H2 type E3 ligase, leading to a premature stop codon. CRISPR/Cas9-mediated knockout of *OsDIRH2* enhanced drought tolerance by lowering stomatal density. These findings suggest that ditl4 could serve as a valuable genetic resource for breeding drought-resistant rice with improved water-use efficiency. Also, in rice, Rashid et al. (2024) generated a *Osckx2* mutant using CRISPR/Cas9 to investigate the effects of cytokinin accumulation on grain yield and drought tolerance. The Osckx2 mutant exhibited increased cytokinin levels, leading to enhanced panicle branching, grain number, and overall yield. Under drought conditions, the mutant showed improved water conservation, reduced transpiration, and better survival compared to wild-type plants. Additionally, it maintained membrane and chloroplast integrity, boosting photosynthetic efficiency through enhanced antioxidant activity. These findings highlight the potential of *Osckx2* mutants in breeding drought-tolerant, high-yielding rice varieties.

In tomato, Li et al. (2024a, b) explored the role of the WUSCHEL-related homeobox (WOX) transcription factor *SlWOX4* in drought tolerance, utilizing CRISPR/Cas9 to create knockout lines. *SlWOX4*, homologous to *Nicotiana tabacum NtWOX4*, was significantly downregulated under drought and ABA treatments. The knockout lines exhibited improved drought tolerance characterized by reduced water loss rates, enhanced stomatal closure, and lower accumulation of reactive oxygen species (ROS) and malondialdehyde (MDA). Additionally, there were increases in antioxidant enzyme activity, proline, and ABA levels, indicating that *SlWOX4* negatively regulates drought tolerance and may serve as a target for breeding drought-resistant tomato varieties. Also, in tomato, Saikia et al. (2024) employed CRISPR/Cas9 to edit two negative regulatory genes, *SlHyPRP1* and *SlDEA1*, in *S. lycopersicum* to enhance tolerance to drought, salinity, bacterial leaf spot, and bacterial wilt. The CRISPR-edited plants showed improved phenotypic responses, including increased chlorophyll and proline levels under abiotic stress and reduced reactive oxygen species accumulation and cell death under biotic stress. The study highlights the critical role of *SlDEA1*, a poorly understood 8CM family gene, in multi-stress tolerance.

In potato, Karlsson et al. (2024) evaluated CRISPR/Cas9-edited *Stdmr6–1* mutants, demonstrating enhanced resistance to *Phytophthora infestans* (late blight),

Alternaria solani (early blight), and common scab in both field and controlled conditions. Field trials over four years showed improved disease resistance without yield or quality penalties. Additionally, *Stdmr6–1* mutants displayed tolerance to drought and salinity, indicating broader stress resistance. These findings suggest that knocking out susceptibility genes, such as Stdmr6–1, holds promise for sustainable potato cultivation by reducing reliance on fungicides and improving resistance to both biotic and abiotic stresses.

In rice, Ahmad et al. (2024a, b) conducted a comprehensive in silico analysis of the *OsMYB48* transcription factor (TF), comparing it with 21 dicot and 10 monocot MYB TFs. Gene structure, protein domain, motif analysis, and phylogenetic classification identified key conserved sequences involved in cellular functions and stress responses. Promoter analysis revealed cis-elements linked to hormonal signaling and abiotic stress responses. *Agrobacterium*-mediated transformation of *OsMYB48* into tobacco demonstrated enhanced stress tolerance, emphasizing the role of OsMYB TFs in stress resilience and advocating for future functional investigations, including CRISPR/Cas9 genome editing, to develop crops with improved abiotic stress tolerance.

In tomato, the study by Lv et al. (2024) demonstrates the use of CRISPR/Cas9 to edit the trihelix transcription factor *SlGT30*, aiming to enhance both drought resistance and fruit yield. Knockout lines exhibited reduced stomatal density and increased cell size through the endoreduplication pathway, improving drought tolerance by minimizing water loss. Additionally, knockout lines showed larger fruits with increased cell size and number in the pericarp, leading to higher fruit weight. This highlights *SlGT30* as a valuable target for improving agronomic traits in tomato breeding programs.

Table 18.1 summarizes key findings from recent studies investigating various plant species, their target genes, associated phenotypes, and highlights from each study, providing insights into the genetic mechanisms underlying salinity tolerance and adaptation. The search was conducted on October 4 in the PubMed database (https://pubmed.ncbi.nlm.nih.gov/advanced/) for articles from 2024 using the keywords "drought", "plants", and "CRISPR/Cas9".

18.5.2 Salinity Tolerance

Salinity stress constitutes a major environmental constraint that negatively impacts plant growth and agricultural productivity. This stress predominantly arises from the accumulation of soluble salts in the soil, which induces osmotic stress and ion toxicity (Isayenkov and Maathuis 2019; Raza et al. 2023).

Salinity stress impacts plants through a series of sequential mechanisms. Initially, osmotic stress occurs shortly after exposure to salt, as the high concentration of salts in the soil reduces the water potential. This decrease in water potential makes it challenging for roots to absorb water, leading to physiological drought conditions. As a result, stomatal closure occurs, and cell expansion, particularly in the shoot, is restricted. Over time, excessive uptake of sodium (Na^+) and chloride (Cl^-) ions can

Table 18.1 Summary of recent studies on drought tolerance in plants using CRISPR/Cas9 technology

Plant species	Stress type	Target gene	Phenotype	Highlight	Reference
Maize	Drought	*ZmBAM8*	Enhanced starch hydrolysis, improved drought tolerance	Identified as a key regulator of leaf starch degradation and drought response.	Niu et al. (2024)
	Drought	*ZmASR1*	Reduced ROS, increased ABA levels, improved stomatal closure	Acts as a negative regulator of drought tolerance, promising target for drought resistance improvement.	Yang et al. (2024)
Rice	Drought	*OsDIRH2*	Enhanced drought tolerance, reduced stomatal density	Knockout mutant showed improved water conservation and yield potential under drought conditions.	Kim et al. (2024)
	Drought	*Osckx2*	Increased cytokinin levels, enhanced yield	Knockout mutant improved water use efficiency and maintained photosynthetic integrity under drought.	Rashid et al. (2024)
	Abiotic stress	*OsMYB48*	Enhanced stress tolerance in tobacco	Functional role identified in stress resilience; suggests potential for CRISPR/Cas9 editing.	Ahmad et al. (2024a, b)
Tomato	Drought	*SlGT30*	Reduced stomatal density, larger fruit size	Enhanced drought resistance and fruit yield, highlighting potential for tomato breeding programs.	Lv et al. (2024)
	Multi-stress	*SlHyPRP1, SlDEA1*	Improved phenotypic responses under stress	Highlights the role of SlDEA1 in multi-stress tolerance, addressing both abiotic and biotic stress.	Saikia et al. (2024)
Potato	Multi-stress	*Stdmr6–1*	Enhanced resistance to diseases and drought	Demonstrated sustainable cultivation benefits by knocking out susceptibility genes.	Karlsson et al. (2024)

The studies were obtained from the PubMed database using the keywords "drought", "plants", and "CRISPR/Cas9" during a search conducted on October 4, 2024 cause ion toxicity. The accumulation of these ions within plant tissues reaches toxic levels, disrupting metabolic processes, accelerating premature senescence, and potentially resulting in cell death (Angon et al. 2022).

Plants exhibit a range of physiological and biochemical adaptations to cope with salinity stress. They regulate the uptake and compartmentalization of Na^+ and Cl^- ions through various mechanisms, including membrane transporters that sequester these ions into vacuoles, thereby minimizing their toxic effects. Plants synthesize compatible solutes such as proline and glycine betaine to counteract osmotic stress. These solutes help maintain cellular turgor pressure and protect cellular structures (Zhao et al. 2020; Balasubramaniam et al. 2023).

Salinity stress also induces oxidative stress due to the production of reactive oxygen species (ROS). Plants enhance their antioxidant systems to mitigate oxidative damage, producing enzymes like superoxide dismutase and catalase. Additionally, salinity influences the levels of plant hormones, particularly abscisic acid (ABA), which plays a crucial role in regulating stomatal closure and enhancing stress tolerance (Hasanuzzaman and Fujita 2022; Balasubramaniam et al. 2023).

Recent research into salinity tolerance's genetic and molecular basis has yielded valuable insights. Genomic and transcriptomic analyses are used to identify and select traits associated with salt tolerance, aiming to improve crop resilience to salinity stress (Atta et al. 2023). Furthermore, proteomic studies focus on the expression of salt stress-responsive proteins to understand their roles in conferring tolerance, with particular attention to proteins involved in ion transport, osmotic regulation, and stress signaling pathways (Athar et al. 2022).

The study by Zhang et al. (2019) aimed to improve salinity tolerance in rice by using CRISPR/Cas9 to target the *OsRR22* gene. The researchers developed a vector containing Cas9 and OsRR22-gRNA to generate transgenic rice plants. Out of 14 T0 transgenic plants, nine mutants were identified, each with one of six distinct mutation types at the target site. These mutations were stably transmitted to subsequent generations. In the T1 generation, segregation analysis revealed mutant plants without transferred DNA (T-DNA). In the T2 generation, two homozygous mutant lines demonstrated significantly improved salinity tolerance at the seedling stage compared to wild-type plants, with no notable differences in agronomic traits.

Similarly, Wang et al. (2021) employed CRISPR/Cas9 to edit *GmAITR* genes in soybean, aiming to enhance salinity tolerance. Transfection assays in soybean protoplasts showed that *GmAITR*s function as nuclear proteins with transcriptional repression activity. By simultaneously targeting six *GmAITR* genes, the researchers generated Cas9-free gmaitr36 double and gmaitr23456 quintuple mutants. These mutants displayed increased sensitivity to ABA (abscisic acid) and altered ABA responses in key signaling genes. Improved salt tolerance was observed in seed germination and seedling growth assays, and these results were validated in field trials, confirming that CRISPR/Cas9-induced mutations in GmAITR genes effectively enhance salinity tolerance in soybeans.

In *Solanum lycopersicum* L. (tomato), the study by Saikia et al. (2024) investigates the role of CRISPR/Cas9 genome editing in enhancing multi-stress tolerance by focusing on the genes *SlHyPRP1* and *SlDEA1*, identified as potential negative regulators of responses to both abiotic (drought, salinity) and biotic (bacterial leaf spot, bacterial wilt) stresses. Using CRISPR/Cas9, the researchers generated both single and dual-gene knockouts, which exhibited improved phenotypic responses compared to wild-type (WT) plants. The edited plants showed higher chlorophyll and proline levels under abiotic stress and reduced reactive oxygen species accumulation and cell death under biotic stress. Notably, *SlDEA1* was found to play a critical role alongside *SlHyPRP1* in conferring multi-stress tolerance. The successful replication of transient CRISPR/Cas9 editing events in stably generated parent-genome-edited lines (GEd0) and first-generation lines (GEd1) underscores the potential for stable, heritable stress tolerance traits. This study highlights the potential of CRISPR/Cas9 as a tool for crop improvement, contributing to sustainable strategies for enhancing stress tolerance in tomatoes.

In *Solanum tuberosum* (potato), the study by Karlsson et al. (2024) addresses challenges from various diseases and harsh abiotic conditions. While frequent fungicide applications are common, the researchers explored a more sustainable solution through genome editing by knocking out susceptibility genes. They examined CRISPR/Cas9-generated StDMR6–1 mutants, where the *DMR6* gene was disrupted to increase resistance to *Phytophthora infestans* (late blight). Over four years of field trials demonstrated that StDMR6–1 mutants exhibited enhanced resistance to late blight without any yield penalty or reduction in tuber quality. Additionally, the mutants showed resistance to common scab and early blight (*Alternaria solani*) in controlled conditions, as well as improved tolerance to abiotic stresses like drought and salinity. These findings highlight the broad-spectrum resistance conferred by StDMR6–1 mutations, making them a promising tool for future sustainable potato cultivation.

In *Oryza sativa* (rice), the study by Ly et al. (2024) explores the role of the *OsDSG1* gene, associated with the ubiquitination pathway, in regulating biochemical processes related to stress. By utilizing CRISPR/Cas9 to introduce targeted mutations in *OsDSG1*, the researchers confirmed that the mutant rice lines exhibited enhanced salt tolerance during both the germination and seedling stages. The improvements were reflected in increased plant height, root length, total fresh weight, chlorophyll content, and relative water content under saline conditions, alongside reduced levels of proline and malondialdehyde (MDA). Importantly, the CRISPR/Cas9-induced mutations did not negatively impact growth under normal conditions, demonstrating the utility of *OsDSG1* mutations for enhancing salt tolerance in rice.

In another study on *Oryza sativa*, Yan et al. (2024) investigated the function of *OsNF-YC5*, a subunit of the NF-Y transcription factor. Using CRISPR/Cas9 to disrupt *OsNF-YC5*, the researchers observed significantly enhanced salinity tolerance and ABA hypersensitivity in rice. The *OsNF-YC5* mutants displayed reduced levels of hydrogen peroxide (H_2O_2) and malondialdehyde (MDA), higher catalase (CAT) activity, and increased expression of stress marker genes under salt stress. These results suggest that *OsNF-YC5* negatively regulates salinity tolerance, and its disruption enhances stress resistance by modulating gene expression and increasing CAT enzyme activity.

In a different approach with *Oryza sativa*, Rashid et al. (2024) developed *OsCKX2*-deficient mutants using CRISPR/Cas9 to enhance drought tolerance. The mutants exhibited increased cytokinin accumulation, resulting in improved panicle branching, grain yield, and enhanced drought resistance. Under drought conditions, the OsCKX2 mutants showed better water conservation traits, improved survival rates, and reduced transpiration compared to wild-type plants. The study highlights the potential of the *OsCKX2* allele in breeding high-yielding, drought-tolerant rice varieties, offering a valuable tool for developing climate-resilient crops.

In another investigation involving *Oryza sativa*, Xie et al. (2024) conducted a genome-wide identification and functional analysis of K-homologous (KH) family proteins, identifying 31 *KH* genes. Among these, *OsKH12* demonstrated a significant decrease in expression under salt stress. The researchers generated an Oskh12 knockout mutant using CRISPR/Cas9, which showed increased sensitivity to salinity compared to wild-type Nipponbare rice. These findings shed light on the role of KH proteins in rice's response to salt stress, providing a foundation for future research on abiotic stress tolerance.

In the study by Akter et al. (2024) on *Oryza sativa*, the researchers focused on glutamate decarboxylase 4 (OsGAD4), a gene linked to GABA production. Using CRISPR/Cas9, they removed the calmodulin binding domain (CaMBD) from *OsGAD4*, which enhanced its activity and significantly increased gamma-aminobutyric acid (GABA) levels. The genome-edited line (#14–1) produced 11.26 mg GABA/100 g of grain, nearly nine times higher than the wild type. Under abiotic stresses such as salinity, flooding, and drought, GABA accumulation was significantly higher in the #14–1 mutant compared to the wild-type and another mutant with a shorter deletion. This study is the first to demonstrate that endogenous GABA accumulation can enhance stress tolerance in rice, providing new insights into its role in plant stress responses.

Table 18.2 summarizes key findings from recent studies investigating various plant species, their target genes, associated phenotypes, and highlights from each study, providing insights into the genetic mechanisms underlying salinity tolerance and adaptation. The search was conducted on October 4 in the PubMed database (https://pubmed.ncbi.nlm.nih.gov/advanced/) for articles from 2024 using the keywords "salinity", "plants", and "CRISPR/Cas9".

18.5.3 Extreme Temperature Tolerance

High-temperature plant stress often results in significant cellular damage, characterized by protein denaturation, membrane instability, and enzyme inactivation. These detrimental effects impair metabolic processes, reducing photosynthetic efficiency and compromising energy production. Elevated temperatures also exacerbate water loss through increased evapotranspiration, causing drought stress, wilting, and diminished turgor pressure. Furthermore, reproductive functions are adversely affected, with heat stress impairing flower development and pollen viability, ultimately leading to lower fruit sets and reduced yields. High temperatures can also

Table 18.2 Summary of recent studies on salinity tolerance in plants using CRISPR/Cas9 technology

Plant species	Stress type	Target gene	Phenotype	Highlight	Reference
G. max	Salinity	*GmAITR*	Increased salt tolerance in mutants	Confirmed CRISPR/Cas9 editing effectiveness for salinity tolerance	Wang et al. (2021)
O. sativa	Salinity	*OsCKX2*	Improved panicle branching and drought tolerance	Highlighted potential for developing climate-resilient rice varieties	Rashid et al. (2024)
		OsDSG1	Enhanced salt tolerance; increased growth parameters	Confirmed role of OsDSG1 in salt resistance mechanisms	Ly et al. (2024)
		OsGAD4	Enhanced GABA production; improved stress tolerance	First evidence that endogenous GABA accumulation enhances stress tolerance	Akter et al. (2024)
		KH proteins	Increased sensitivity to salinity in Oskh12 mutant	Provided insights into KH proteins' role in abiotic stress responses	Xie et al. (2024)
		OsRR22	Improved salinity tolerance in homozygous mutants	Demonstrated successful CRISPR/Cas9 editing for enhanced salt tolerance	Zhang et al. (2019)
S. lycopersicum L.	Drought, salinity, biotic	*SlHyPRP1, SlDEA1*	Enhanced chlorophyll and proline levels; reduced cell death	Identified SlDEA1 as a critical negative regulator for multi-stress tolerance	Saikia et al. (2024)
S. tuberosum	Disease (late blight, scab)	StDMR6–1	Increased resistance to late blight and common scab	Demonstrated broad-spectrum disease resistance and abiotic stress tolerance	Karlsson et al. (2024)

The studies were obtained from the PubMed database using the keywords "salinity", "plants", and "CRISPR/Cas9" during a search conducted on October 4, 2024 alter growth patterns, resulting in stunted growth or excessive elongation depending on the species (Hasanuzzaman et al. 2013; Awasthi et al. 2015; Nievola et al. 2017).

Conversely, chilling temperature stress disrupts metabolic activities by slowing or halting essential processes, affecting overall plant growth and development. The integrity of cell membranes is compromised due to lipid peroxidation and leakage, which can lead to cell injury or death. Seed germination and seedling growth are particularly vulnerable to chilling stress, often exhibiting reduced rates of both. Additionally, chilling temperatures can weaken plant defenses, making them more susceptible to pathogenic attacks. Photosynthesis is similarly impaired under chilling conditions through mechanisms such as reduced enzyme activity and altered chlorophyll content. Plants utilize various adaptive mechanisms to mitigate these stresses, including producing heat shock proteins and antioxidants and adjusting membrane lipid composition and cryoprotectant synthesis. Advances in breeding and genetic engineering aim to enhance plant tolerance to these extreme temperatures, thereby improving resilience and productivity in changing climates (Ding et al. 2020; Aslam et al. 2022).

Table 18.3 summarizes key findings from recent studies investigating various plant species, their target genes, associated phenotypes, and highlights from each study, providing insights into the genetic mechanisms underlying heat and cold tolerance and adaptation. The search was conducted on October 4 in the PubMed database (https://pubmed.ncbi.nlm.nih.gov/advanced/) for articles from 2024 using the keywords "cold", "heat", "high temperature", "plants", and "CRISPR/Cas9".

18.5.3.1 Heat Tolerance

High temperature (HT) stress poses a major challenge to plant growth, development, and crop productivity by disrupting physiological processes such as photosynthesis, stress signaling, and reproductive timing. Recent studies have focused on identifying the molecular mechanisms underlying HT tolerance in various species, including rice, soybean, tomato, and leaf lettuce. Using advanced techniques like genome assembly, CRISPR/Cas9-mediated gene editing, and transcriptome analysis, researchers have uncovered key regulatory genes and pathways that enable plants to withstand heat stress, providing valuable insights for breeding heat-tolerant crop varieties.

In rice (*O. sativa*), Ahmad et al. (2024a, b) identified two allelic mutants exhibiting pale-green or variegated leaves under high temperature (HT) stress. The mutants, controlled by the *PGL10* gene encoding NADPH oxidoreductase B, showed impaired chlorophyll production. Functional and transcriptome analyses revealed that *PGL10* regulates chloroplast function, photosynthesis, and ROS detoxification under HT.

In soybean (*Glycine max*), Hou et al. (2024) investigated the roles of *GmBSK1* and *GmBES1.5* in heat tolerance. CRISPR/Cas9 knockout mutants showed increased sensitivity to HT due to impaired ROS scavenging. The study identified a regulatory interaction between *GmBSK1*, *GmGSK1*, and *GmBES1.5*, which could be targeted for breeding heat-tolerant soybean cultivars.

Table 18.3 Summary of recent studies on heat and cold tolerance in plants using CRISPR/Cas9 technology

Plant species	Stress type	Target gene	Phenotype	Highlight	Reference
G. max	Heat	*GmBSK1, GmBES1.5*	Increased heat sensitivity in gmbsk1 mutants	GmBSK1-GmGSK1-GmBES1.5 module improves heat tolerance	Hou et al. (2024)
L. sativa	Heat	*LsFUL, LsSMU2, LsCOL5*	Early bolting in HT-sensitive cultivars	*LsFUL* interacts with *LsSMU2* and *LsCOL5* to control bolting under HT	Zhong et al. (2024)
O. sativa	Heat	*PGL10*	Pale-green or variegated leaves under high temperature (HT)	*PGL10* regulates chloroplast function, photosynthesis, and ROS detoxification	Ahmad et al. (2024a, b)
S. americanum	Heat	*SaHSF17*	Reduced thermotolerance in knockout mutants	*SaHSF17* regulates heat shock protein genes, crucial for heat tolerance	Yu et al. (2024)
O. sativa	Cold	*OsTIL1*	Improved cold tolerance	Protects cell membrane integrity and enhances glycerolipid biosynthesis under cold stress.	Ji et al. (2024)
		OsWRKY70	Enhanced susceptibility to fungal disease	Negatively contributes to blast resistance but positively regulates cold tolerance.	Li et al. (2024a, b)
		OsCAO1	Reduced chilling tolerance	Knockout mutants show weakened chilling tolerance and decreased yield; important for photosynthesis.	Xiong et al. (2024)
T. aestivum	Cold	*TaTPS11-6D*	Enhanced cold resistance	Increases sugar content and improves cold tolerance in winter wheat.	Lu et al. (2024)

(continued)

Table 18.3 (continued)

Plant species	Stress type	Target gene	Phenotype	Highlight	Reference
S. lycopersicum	Cold	*SlMAPK3*	Enhanced cold resistance	Regulates raffinose synthesis and expression of SlWRKY46; enhances cold tolerance in tomatoes.	Shu et al. (2024)
		SlHB8	Improved pollen cold tolerance	Regulates tapetal programmed cell death; essential for pollen development under cold stress.	Guan et al. (2024)
V. amurensis	Cold	*VaBAM3*	Increased cold tolerance	Enhances soluble sugar accumulation and scavenging of reactive oxygen species in transgenic plants.	Liang et al. (2024)

The studies were obtained from the PubMed database using the keywords "cold," "plants," and "CRISPR/Cas9" during a search conducted on October 4, 2024

In tomato (*Solanum americanum*), Yu et al. (2024) produced a chromosome-level genome assembly and identified *SaHSF17*, a heat-inducible transcription factor critical for thermotolerance. Knockout of *SaHSF17* reduced heat tolerance and downregulated heat shock protein genes. Transcriptome analysis also highlighted delphinidins as major anthocyanins in the fruit, regulated by key biosynthetic genes.

In leaf lettuce (*Lactuca sativa*), Zhong et al. (2024) explored HT-induced bolting mechanisms in two cultivars. The study revealed that *LsFUL*, a MADS-box gene, plays a key role in regulating bolting under HT. Overexpression of *LsFUL* accelerated bolting, while its knockout delayed the process. *LsSMU2* and *LsCOL5* were found to interact with LsFUL, influencing bolting time based on temperature conditions.

18.5.3.2 Cold Tolerance

In rice (*O.sativa*), Zhou et al. (2024) investigated the critical agronomic trait of heading date, essential for seasonal and regional adaptation. By employing a CRISPR-Cas9-based multiplex promoter-targeting method (HMP), the researchers created novel alleles in key regulatory genes (*Hd1*, *Ghd7*, *DTH8*). Field trials identified germplasms suitable for various regions, including the elite cultivar Ningjing8, demonstrating HMP's potential for improving rice adaptability. *O. sativa* also featured in Ji et al. (2024), who examined the lipocalin gene *OsTIL1*'s role under cold stress. The study revealed that *OsTIL1* enhances cold tolerance by maintaining cell membrane integrity against oxidative damage and facilitating glycerolipid remodeling. Overexpression of *OsTIL1* led to increased levels of unsaturated fatty acids, supporting its potential as a candidate gene for enhancing cold tolerance.

In the research conducted by Li et al. (2024a, b), *O. sativa* was used to study the transcription factor OsWRKY70. The findings indicated that knockout mutants exhibited heightened resistance to fungal pathogens but reduced cold tolerance. OsWRKY70 was identified as a negative regulator of blast resistance and a positive influencer of cold tolerance, offering valuable insights for breeding stress-tolerant rice cultivars. *O. sativa* was further explored by Hu et al. (2024), focusing on the role of *OsRBCS3* in chilling tolerance at the booting stage. The study found a positive correlation between *OsRBCS3* expression and chilling tolerance. Overexpression lines displayed enhanced chilling tolerance and improved antioxidant and photosynthetic capacities compared to knockout lines, highlighting OsRBCS3's potential for breeding chilling-tolerant rice. Xiong et al. (2024) investigated the chlorophyllide, an oxygenase gene *OsCAO1* in *O. sativa*, focusing on its role in chilling stress responses. Knockout mutants exhibited significantly reduced chilling tolerance and severe chloroplast damage. The findings suggested that enhancing *OsCAO1* expression could improve both yield and chilling tolerance, providing a strategy for developing high-yield, cold-resistant rice varieties.

In *Vitis amurensis,* Liang et al. (2024) studied the *VaBAM3* gene and its impact on enhancing cold tolerance. Overexpressing *VaBAM3* in various plant species resulted in less wilting and improved enzyme activity under cold stress. The findings indicated that *VaBAM3* enhances cold tolerance by promoting soluble sugar accumulation and scavenging reactive oxygen species, laying a foundation for developing cold-resistant grape cultivars.

In tomato (*Solanum lycopersicum*), Shu et al. (2024) investigated the role of *SlMAPK3* in raffinose synthesis under cold stress. Overexpression of *SlMAPK3* resulted in increased levels of galactinol and raffinose, while knockouts exhibited reduced accumulation. This study highlighted the regulatory pathway involving *SlMAPK3* and *SlWRKY46*, which together enhance cold resistance through raffinose synthesis. In research conducted by Guan et al. (2024), *S. lycopersicum* was used to study the gene *SlHB8*, which regulates tapetum degradation in anthers during cold conditions. *SlHB8* knockouts showed improved pollen cold tolerance. The study demonstrated that *SlHB8* activates critical genes in response to cold, shedding light on the molecular mechanisms underlying anther adaptation to temperature fluctuations.

18.6 Conclusions

With the ability to precisely develop climate-resilient crops, new plant breeding technologies (NPBTs), including CRISPR/Cas9, TALENs, and base editing, are revolutionizing modern agriculture. These technologies can address the increasing problems brought about by climate change, such as drought, salt, and high-temperature strains that jeopardize global food security. More resilient and sustainable agricultural systems can be achieved by combining NPBTs with climate-smart techniques like improving soil health and conserving water. Genome editing and speed breeding work together to create new crop types more quickly while reducing the time needed for conventional breeding. NPBTs will remain essential in improving crop resilience, productivity, and sustainability as research advances, providing a viable approach to mitigating the complex effects of climate change on agriculture. The long-term sustainability of agricultural systems worldwide is contingent upon the persistent utilization and innovation of these biotechnological instruments, which will determine the future of global food security.

Acknowledgments A. Gatica-Arias expresses gratitude to the Humboldt 2023 Program at the University of Costa Rica for its support in promoting biotechnology for the conservation and sustainable use of biodiversity.

Author Contributions All authors contributed to the study conception and design, material preparation, data collection, and analysis equally. All authors read and approved the final manuscript.

Conflict of Interest AGA declares no financial interests. AHS is employed by CropLifeLA, funded by companies developing GM crops.

Funding This study was financed by the "Espacio de Estudios Avanzados de la Universidad de Costa Rica" (UCREA; Project No. 801-B7–294). G.H. was supported by funding from the Deutsche Forschungsgemeinschaft under Germany's Excellence Strategy—EXC-2048/1 (project ID 390686111 and grants 426557363 and 458717903).

References

Ahmad S, Tabassum J, Sheng Z et al (2024a) Loss-of-function of PGL10 impairs photosynthesis and tolerance to high-temperature stress in rice. Physiol Plant 176(3):e14369. https://doi.org/10.1111/ppl.14369

Ahmad Y, Haider S, Iqbal J et al (2024b) In-silico analysis and transformation of OsMYB48 transcription factor driven by CaMV35S promoter in model plant—Nicotiana tabacum L. conferring abiotic stress tolerance. GM Crops Food 15(1):130–149. https://doi.org/10.1080/21645698.2024.2334476

Akter N, Kulsum U, Moniruzzaman M, Yasuda N, Akama K (2024) Truncation of the calmodulin binding domain in rice glutamate decarboxylase 4 (OsGAD4) leads to accumulation of γ-aminobutyric acid and confers abiotic stress tolerance in rice seedlings. Mol Breed 3:21. https://doi.org/10.1007/s11032-024-01460-1

Ali Q, Yu C, Hussain A et al (2022) Genome engineering technology for durable disease resistance: recent progress and future outlooks for sustainable agriculture. Front Plant Sci 13

Anderson R, Bayer PE, Edwards D (2020) Climate change and the need for agricultural adaptation. Curr Opin Plant Biol 56:197–202. https://doi.org/10.1016/j.pbi.2019.12.006

Angon PB, Tahjib-Ul-Arif M, Samin SI et al (2022) How do plants respond to combined drought and salinity stress?—a systematic review. Plants 11. https://doi.org/10.3390/plants11212884

Angon PB, Mondal S, Akter S et al (2023) Roles of CRISPR to mitigate drought and salinity stresses on plants. Plant Stress 8:100169. https://doi.org/10.1016/j.stress.2023.100169

Aslam M, Fakher B, Ashraf MA et al (2022) Plant low-temperature stress: signaling and response. Agronomy 12. https://doi.org/10.3390/agronomy12030702

Athar HUR, Zulfiqar F, Moosa A et al (2022) Salt stress proteins in plants: an overview. Front. Plant Sci 13

Atta K, Mondal S, Gorai S et al (2023) Impacts of salinity stress on crop plants: improving salt tolerance through genetic and molecular dissection. Front. Plant Sci 14

Awasthi R, Bhandari K, Nayyar H (2015) Temperature stress and redox homeostasis in crops. Front Environ Sci 3. https://doi.org/10.3389/fenvs.2015.00011

Bai X, Huang Y, Ren W et al (2019) Responses of soil carbon sequestration to climate-smart agriculture practices: a meta-analysis. Glob Chang Biol 25:2591–2606. https://doi.org/10.1111/gcb.14658

Bain C, Lindberg S, Selfa T (2020) Emerging sociotechnical imaginaries for gene edited crops for foods in the United States: implications for governance. Agric Human Values 37:265–279. https://doi.org/10.1007/s10460-019-09980-9

Balasubramaniam T, Shen G, Esmaeili N, Zhang H (2023) Plants' response mechanisms to salinity stress. Plants 12. https://doi.org/10.3390/plants12122253

Bao A, Burritt DJ, Chen H, Zhou X, Cao D, Tran L-SP (2019) The CRISPR/Cas9 system and its applications in crop genome editing. Crit Rev Biotechnol 39:321–336. https://doi.org/10.1080/07388551.2018.1554621

Bharat SS, Li S, Li J et al (2020) Base editing in plants: current status and challenges. Crop J 8:384–395. https://doi.org/10.1016/j.cj.2019.10.002

Bhat MA, Mir RA, Kumar V et al (2021) Mechanistic insights of CRISPR/Cas-mediated genome editing towards enhancing abiotic stress tolerance in plants. Physiol Plant 172:1255–1268. https://doi.org/10.1111/ppl.13359

Biswas S, Bridgeland A, Irum S et al (2022) Optimization of prime editing in rice, peanut, chickpea, and cowpea protoplasts by restoration of GFP activity. Int J Mol Sci 23:9809. https://doi.org/10.3390/ijms23179809

Breseghello F, Coelho ASG (2013) Traditional and modern plant breeding methods with examples in Rice (Oryza sativa L.). J Agric Food Chem 61:8277–8286. https://doi.org/10.1021/jf305531j

Budhlakoti N, Kushwaha AK, Rai A et al (2022) Genomic selection: a tool for accelerating the efficiency of molecular breeding for development of climate-resilient crops. Front Genet 13. https://doi.org/10.3389/fgene.2022.832153

Carroll D (2011) Genome engineering with zinc-finger nucleases. Genetics 188:773–782. https://doi.org/10.1534/genetics.111.131433

Chaudhry A, Hassan AU, Khan SH et al (2023) The changing landscape of agriculture: role of precision breeding in developing smart crops. Funct Integr Genomics 23:167

Chen K, Wang Y, Zhang R et al (2019) CRISPR/Cas genome editing and precision plant breeding in agriculture. Annu Rev Plant Biol 70:667–697. https://doi.org/10.1146/annurev-arplant-050718-100049

Chennakesavulu K, Singh H, Trivedi PK et al (2022) State-of-the-art in CRISPR technology and engineering drought, salinity, and Thermo-tolerant crop plants. Plant Cell Rep 41:815–831. https://doi.org/10.1007/s00299-021-02681-w

Choi J, Chen W, Suiter CC et al (2022) Precise genomic deletions using paired prime editing. Nat Biotechnol 40:218–226. https://doi.org/10.1038/s41587-021-01025-z

Cooper M, Messina CD (2023) Breeding crops for drought-affected environments and improved climate resilience. Plant Cell 35:162–186

De Pinto A, Cenacchi N, Kwon H-Y et al (2020) Climate-smart agriculture, and global food-crop production. PLoS One 15:e0231764. https://doi.org/10.1371/journal.pone.0231764

Ding Y, Shi Y, Yang S (2020) Molecular regulation of plant responses to environmental temperatures. Mol Plant 13:544–564. https://doi.org/10.1016/j.molp.2020.02.004

Doudna JA, Charpentier E (2014) The new frontier of genome engineering with CRISPR-Cas9. Science 1979:346. https://doi.org/10.1126/science.1258096

Eid A, Alshareef S, Mahfouz MM (2018) CRISPR base editors: genome editing without double-stranded breaks. Biochem J 475:1955–1964. https://doi.org/10.1042/BCJ20170793

Engelhardt IC, Niklaus PA, Bizouard F et al (2021) Precipitation patterns and N availability alter plant-soil microbial C and N dynamics. Plant Soil 466:151–163. https://doi.org/10.1007/s11104-021-05015-7

Gaj T, Gersbach CA, Barbas CF III (2013) ZFN, TALEN, and CRISPR/Cas-based methods for genome engineering. Trends Biotechnol 31:397–405. https://doi.org/10.1016/j.tibtech.2013.04.004

Garland S, Curry HA (2022) Turning promise into practice: crop biotechnology for increasing genetic diversity and climate resilience. PLoS Biol 20. https://doi.org/10.1371/journal.pbio.3001716

Gentile Alessandra, La Malfa S (2019) New breeding techniques for sustainable agriculture. In: Muhammad F, Pisante M (eds) Innovations in sustainable agriculture. Springer International Publishing, Cham, pp 411–437

González Guzmán M, Cellini F, Fotopoulos V et al (2022) New approaches to improve crop tolerance to biotic and abiotic stresses. Physiol Plant 174:e13547. https://doi.org/10.1111/ppl.13547

Guan H, Yu C, Zeng Z et al (2024) SlHB8 is a novel factor in enhancing cold resistance in tomato anthers by modulating tapetal cell death. Int J Mol Sci 25(17):9336. https://doi.org/10.3390/ijms25179336

Hamdan MF, Karlson CKS, Teoh EY et al (2022) Genome editing for sustainable crop improvement and mitigation of biotic and abiotic stresses. Plants 11. https://doi.org/10.3390/plants11192625

Hasanuzzaman M, Fujita M (2022) Plant responses and tolerance to salt stress: physiological and molecular interventions. Int J Mol Sci 23

Hasanuzzaman M, Nahar K, Alam MM et al (2013) Physiological, biochemical, and molecular mechanisms of heat stress tolerance in plants. Int J Mol Sci 14:9643–9684. https://doi.org/10.3390/ijms14059643

Hassan MM, Yuan G, Liu Y et al (2022) Precision genome editing in plants using gene targeting and prime editing: existing and emerging strategies. Biotechnol J 17:2100673. https://doi.org/10.1002/biot.202100673

Hernández-Soto A, Delgado-Navarro JP, Benavides-Acevedo M, et al (2022) NTH2 1271_1272delTA Gene Disruption Results in Salt Tolerance in Saccharomyces cerevisiae. Fermentation 2022, Vol 8, Page 166 8:166. doi:https://doi.org/10.3390/FERMENTATION8040166

Hille F, Richter H, Wong SP et al (2018) The biology of CRISPR-Cas: backward and forward. Cell 172:1239–1259. https://doi.org/10.1016/j.cell.2017.11.032

Hou ZH, Gao Y, Zheng JC et al (2024) GmBSK1-GmGSK1-GmBES1.5 regulatory module controls heat tolerance in soybean. J Adv Res 3. https://doi.org/10.1016/j.jare.2024.09.004

Hu Y, Tian C, Song S, Li R (2024) Insights on the enhancement of chilling tolerance in rice through over-expression and knock-out studies of OsRBCS3. Plant Signal Behav 19(1):2318514. https://doi.org/10.1080/15592324.2024.2318514

Isayenkov SV, Maathuis FJM (2019) Plant salinity stress: many unanswered questions remain. Front Plant Sci 10. https://doi.org/10.3389/fpls.2019.00080

Ji L, Zhang Z, Liu S et al (2024) The OsTIL1 lipocalin protects cell membranes from reactive oxygen species damage and maintains the 18:3-containing glycerolipid biosynthesis under cold stress in rice. Plant J117(1):72–91. https://doi.org/10.1111/tpj.16470

Karlsson M, Kieu NP, Lenman M et al (2024) CRISPR/Cas9 genome editing of potato StDMR6-1 results in plants less affected by different stress conditions. Hortic Res 11(7):uhae130. https://doi.org/10.1093/hr/uhae130

Khan Z, Khan SH, Mubarik MS et al (2017) Use of TALEs and TALEN technology for genetic improvement of plants. Plant Mol Biol Report 35:1–19. https://doi.org/10.1007/s11105-016-0997-8

Kim JH, Cho AY, Lim SD, Jang CS (2024) Mutation of a RING E3 ligase, OsDIRH2, enhances drought tolerance in rice with low stomata density. Physiol Plant 76(5):e14565. https://doi.org/10.1111/ppl.14565

Kumar M, Prusty MR, Pandey MK et al (2023) Application of CRISPR/Cas9-mediated gene editing for abiotic stress management in crop plants. Front Plant Sci 14:1–22. https://doi.org/10.3389/fpls.2023.1157678

La Russa MF, Qi LS (2015) The new state of the art: Cas9 for gene activation and repression. Mol Cell Biol 35:3800–3809. https://doi.org/10.1128/mcb.00512-15

Li C, Zhang R, Meng X et al (2020) Targeted, random mutagenesis of plant genes with dual cytosine and adenine base editors. Nat Biotechnol 38:875–882. https://doi.org/10.1038/s41587-019-0393-7

Li H, Ma W, Wang X et al (2024a) A WUSCHEL-related homeobox transcription factor, SlWOX4, negatively regulates drought tolerance in tomato. Plant Cell Rep 43(11):253. https://doi.org/10.1007/s00299-024-03333-5

Li J, Chen Y, Zhang R et al (2024b) OsWRKY70 plays opposite roles in blast resistance and cold stress tolerance in rice. Rice (NY) 7(1):61. https://doi.org/10.1186/s12284-024-00741-9

Liang G, Wang H, Gou H et al (2024) Overexpression of VaBAM3 from Vitis amurensis enhances seedling cold tolerance by promoting soluble sugar accumulation and reactive oxygen scavenging. Plant Cell Rep 43(6):151. https://doi.org/10.1007/s00299-024-03236-5

Lu X, Zhang F, Zhang C et al (2024) TaTPS11 enhances wheat cold resistance by regulating source-sink factor. Plant Physiol Biochem 211:108695. https://doi.org/10.1016/j.plaphy.2024.108695

Lv H, Wang X, Dong X et al (2024) CRISPR/Cas9 edited SlGT30 improved both drought resistance and fruit yield through endoreduplication. Plant Cell Environ 48. https://doi.org/10.1111/pce.14927

Ly LK, Ho TM, Bui TP et al (2024) CRISPR/Cas9 targeted mutations of OsDSG1 gene enhanced salt tolerance in rice. Funct Integr Genomics 24(2):70. https://doi.org/10.1007/s10142-024-01347-6

Mahfouz MM, Piatek A, Stewart CN Jr (2014) Genome engineering via TALENs and CRISPR/Cas9 systems: challenges and perspectives. Plant Biotechnol J 12:1006–1014. https://doi.org/10.1111/pbi.12256

Malhi GS, Kaur M, Kaushik P (2021) Impact of climate change on agriculture and its mitigation strategies: a review. Sustainability (Switzerland) 13:1–21

Mcgrath JM, Lobell DB (2013) Reduction of transpiration and altered nutrient allocation contribute to the nutrient decline of crops grown in elevated CO2 concentrations. Plant Cell Environ 36:697–705. https://doi.org/10.1111/pce.12007

Mekler V, Minakhin L, Severinov K (2017) Mechanism of duplex DNA destabilization by RNA-guided Cas9 nuclease during target interrogation. Proc Natl Acad Sci USA 114:5443–5448. https://doi.org/10.1073/pnas.1619926114

Menz J, Modrzejewski D, Hartung F et al (2020) Genome edited crops touch the market: a view on the global development and regulatory environment. Front Plant Sci 11. https://doi.org/10.3389/fpls.2020.586027

Molla KA, Sretenovic S, Bansal KC, Qi Y (2021) Precise plant genome editing using base editors and prime editors. Nat Plants 7:1166–1187. https://doi.org/10.1038/s41477-021-00991-1

Nievola CC, Carvalho CP, Carvalho V, Rodrigues E (2017) Rapid responses of plants to temperature changes. Temperature 4:371–405

Niu L, Wu X, Liu H, Hu X, Wang W (2024) Leaf starch degradation by β-amylase ZmBAM8 influences drought tolerance in maize. Carbohydr Polym 345:122555. https://doi.org/10.1016/j.carbpol.2024.122555

Nuñez-Muñoz L, Vargas-Hernández B, Hinojosa-Moya J, Ruiz-Medrano R, Xoconostle-Cázares B (2021) Plant drought tolerance provided through genome editing of the trehalase gene. Plant Signal Behav 16:1877005. https://doi.org/10.1080/15592324.2021.1877005

Pandita D (2022a) CRISPR/Cas-mediated genome editing technologies in plants for stress resilience. In: Antioxidant defense in plants: molecular basis of regulation. Springer Nature, pp 285–303

Pandita D (2022b) CRISPR/Cas-mediated genome engineering for abiotic stress resilience in plants. Metabolic Engineering in Plants. Springer Nature, pp 263–283

Petolino JF (2015) Genome editing in plants via designed zinc finger nucleases. In Vitro Cell Dev Biol—Plant 51:1–8. https://doi.org/10.1007/s11627-015-9663-3

Petrova IO, Smirnikhina SA (2023) The development, optimization and future of prime editing. Int J Mol Sci 24. https://doi.org/10.3390/ijms242317045

Rai GK, Khanday DM, Kumar P et al (2023) Enhancing crop resilience to drought stress through CRISPR-Cas9 genome editing. Plants 12:1–14. https://doi.org/10.3390/plants12122306

Rao MJ, Wang L (2021) CRISPR/Cas9 technology for improving agronomic traits and future prospective in agriculture. Planta 254:68. https://doi.org/10.1007/s00425-021-03716-y

Rashid A, Achary VMM, Abdin MZ et al (2024) Cytokinin oxidase2-deficient mutants improve panicle and grain architecture through cytokinin accumulation and enhance drought tolerance in indica rice. Plant Cell Rep 43(8):207. https://doi.org/10.1007/s00299-024-03289-6

Raza A, Razzaq A, Mehmood SS et al (2019) Impact of climate change on crops adaptation and strategies to tackle its outcome: a review. Plants 8

Raza A, Tabassum J, Fakhar AZ, Sharif R, Chen H, Zhang C, Ju L, Fotopoulos V, Siddique KHM, Singh RK, Zhuang W, Varshney RK (2023) Smart reprogramming of plants against salinity stress using modern biotechnological tools. Crit Rev Biotechnol 43:1035–1062. https://doi.org/10.1080/07388551.2022.2093695

Razzaq A, Kaur P, Akhter N et al (2021) Next-generation breeding strategies for climate-ready crops. Front. Plant Sci 12

Saikia BSR, Debbarma J, Maharana J, Sastry GN, Chikkaputtaiah C (2024) CRISPR/Cas9-based genome editing and functional analysis of SlHyPRP1 and SlDEA1 genes of Solanum lycopersicum L. in imparting genetic tolerance to multiple stress factors. Front. Plant Sci 15:1304381. https://doi.org/10.3389/fpls.2024.1304381

Samantara K, Bohra A, Mohapatra SR et al (2022) Breeding more crops in less time: a perspective on speed breeding. Biology (Basel) 11:275. https://doi.org/10.3390/biology11020275

Shelake RM, Kadam US, Kumar R et al (2022) Engineering drought and salinity tolerance traits in crops through CRISPR-mediated genome editing: targets, tools, challenges, and perspectives. Plant Commun 3:100417. https://doi.org/10.1016/j.xplc.2022.100417

Shu P, Li Y, Sheng J, Shen L (2024) Tomato SlMAPK3 modulates cold resistance by regulating the synthesis of raffinose and the expression of SlWRKY46. J Agric Food Chem 72(10):5185–5196. https://doi.org/10.1021/acs.jafc.3c09066

Singh P, Thakur S, Kumar S et al (2023) Genetic engineering for enhancing abiotic stress tolerance in pulses. In: Legumes: physiology and molecular biology of abiotic stress tolerance. Springer Nature, pp 345–367

Skendžić S, Zovko M, Živković IP et al (2021) The impact of climate change on agricultural insect pests. Insects 12

Sprink T, Metje J, Hartung F (2015) Plant genome editing by novel tools: TALEN and other sequence-specific nucleases. Curr Opin Biotechnol 32:47–53. https://doi.org/10.1016/j.copbio.2014.11.010

Sun N, Zhao H (2013) Transcription activator-like effector nucleases (TALENs): a highly efficient and versatile tool for genome editing. Biotechnol Bioeng 110:1811–1821. https://doi.org/10.1002/bit.24890

Tang X, Zheng X, Qi Y et al (2016) A single transcript CRISPR-Cas9 system for efficient genome editing in plants. Mol Plant 9:1088–1091. https://doi.org/10.1016/j.molp.2016.05.001

Wang T, Xun H, Wang W et al (2021) Mutation of GmAITR genes by CRISPR/Cas9 genome editing results in enhanced salinity stress tolerance in soybean. Front Plant Sci 12. https://doi.org/10.3389/fpls.2021.779598

Wang J, Arantes PR, Ahsan M et al (2023) Twisting and swiveling domain motions in Cas9 to recognize target DNA duplexes, make double-strand breaks, and release cleaved duplexes. Front Mol Biosci 9

Watson A, Ghosh S, Williams MJ et al (2017) Speed breeding: a powerful tool to accelerate crop research and breeding. bioRxiv 16:161182. https://doi.org/10.1101/161182

Watson A, Ghosh S, Williams MJ et al (2018) Speed breeding is a powerful tool to accelerate crop research and breeding. Nat Plants 4:23–29. https://doi.org/10.1038/s41477-017-0083-8

Xie Q, Zhang Y, Wu M et al (2024) Identification and functional analysis of KH family genes associated with salt stress in rice. Int J Mol Sci 25(11):5950. https://doi.org/10.3390/ijms25115950

Xiong J, Wen G, Song J et al (2024) Knockout of the chlorophyll and oxygenase gene OsCAO1 reduces chilling tolerance in Rice seedlings. Genes (Basel) 15(6):721. Published 2024 Jun 2. https://doi.org/10.3390/genes15060721

Xu R, Li J, Liu X et al (2020) Development of plant prime-editing systems for precise genome editing. Plant Commun 1. https://doi.org/10.1016/j.xplc.2020.100043

Yadav R, Thankappan R, Kumar A (2021) Novel approaches for genome editing to develop climate smart crops. In: Microbiomes and the global climate change. Springer Nature, pp 267–291

Yan X, Han M, Li S et al (2024) A member of NF-Y family, OsNF-YC5 negatively regulates salt tolerance in rice. Gene 892:147869. https://doi.org/10.1016/j.gene.2023.147869

Yang Y, Li A, Liu Y et al (2024) ZmASR1 negatively regulates drought stress tolerance in maize. Plant Physiol Biochem 211:108684. https://doi.org/10.1016/j.plaphy.2024.108684

Yu S, Wang Y, Li T et al (2024) Chromosome-scale assembly and gene editing of Solanum americanum genome reveals the basis for thermotolerance and fruit anthocyanin composition. Theor Appl Genet 137(1):15. https://doi.org/10.1007/s00122-023-04523-7

Zafar K, Sedeek KE, Rao GS, Khan MZ, Amin I, Kamel R (2020) Genome editing technologies for rice improvement: progress, prospects, and safety concerns. Front Genome Ed 2. https://doi.org/10.3389/fgeed.2020.00005

Zhang A, Liu Y, Wang F et al (2019) Enhanced rice salinity tolerance via CRISPR/Cas9-targeted mutagenesis of the OsRR22 gene. Mol Breed 39:47. https://doi.org/10.1007/s11032-019-0954-y

Zhao C, Zhang H, Song C et al (2020) Mechanisms of plant responses and adaptation to soil salinity. The Innovation 1. https://doi.org/10.1016/j.xinn.2020.100017

Zhong M, Wang H, Hao X et al (2024) LsFUL-LsSMU2 module positively controls bolting time in leaf lettuce (Lactuca sativa L.) under high temperature. Plant Sci 347:112195. https://doi.org/10.1016/j.plantsci.2024.112195

Zhou S, Cai L, Wu H et al (2024) Fine-tuning rice heading date through multiplex editing of the regulatory regions of key genes by CRISPR-Cas9. Plant Biotechnol J 22(3):751–758. https://doi.org/10.1111/pbi.14221

Gene Editing to Ensure Food Supply Under Global Warming

19

Akshaya Unnikrishnan and Sneha Yogindran

Abstract

Climate change is already at its peak, and the world population will continue to increase; in this situation, we face significant challenges in the food system. Anthropogenic actions are the main cause of climatic change. One of the major risks to world health in the twenty-first century, according to the World Health Organization, is climate change. The environment is gradually being affected by climate change. People are now facing increasing heat every year, flooding, more diseases, increased water and food scarcity, droughts, sea level rise, etc., which are the impacts of climate change.

All these impacts will directly affect the agriculture sector; when the agriculture sector is affected, food production decreases, thereby increasing the price of the available food products. Introduction of climate-resilient crops is hence essential. The field of gene editing is helpful in this situation for improving the agriculture sector. It includes adding new genes, changing how genes are regulated, or modifying or removing existing genes. Zinc finger nucleases (ZFNs), transcription activator-like effector nucleases (TALENs), and CRISPR-Cas9 are a few of the techniques that can be used to do this. The invention of CRISPR/Cas systems has greatly increased the accuracy and productivity of producing modifications. Since the discovery of CRISPR/Cas-9, it has been used to genetically alter food to increase its nutritional content, shelf life, drought tolerance, and disease resistance; it is an existing addition to the field. With the help of the CRISPR technique, drastic changes can be made in the agriculture sector. It is also reported that many changes can be made in the agriculture sector. To mitigate the effects of climate change, gene editing has already been successfully

A. Unnikrishnan · S. Yogindran (✉)
Plant Biotechnology Laboratory, Department of Biotechnology, Cochin University of Science and Technology, Kochi, Kerala, India
e-mail: snehayogindran@cusat.ac.in

P. M. Priyadarshan, R. Ortiz (eds.), *Plant Breeding 2050*,
https://doi.org/10.1007/978-981-95-0583-8_19

used on plants. Successful gene editing in various crops like rice, wheat, cassava, potato, tomato, etc., gives the idea that gene editing can help crop improvement for sustainable agriculture. Both the biotic and abiotic factors affect the plants. CRISPR can help the plants in all conditions for their improvement. Scientists are attempting to genetically modify these crops to make them resistant to these obstacles and challenges.

The global food crisis is primarily being resolved in three ways by CRISPR. It has the power to replenish food sources, support plant survival in stressful conditions, and enhance plant health by promoting growth. Hence, this chapter discusses how CRISPR is going to solve the food crisis in the scenario of global warming.

Keywords

Gene editing · Global warming · CRISPR-Cas system · Food supply

19.1 Introduction

The high quality of life, growth, nutrition, and development of all living things directly or indirectly depends on their natural environment. The effects of climate change have already started to emerge and will undoubtedly worsen. The earth's climate system is being negatively impacted by global warming due to the fast emissions of greenhouse gases (GHGs). The average temperature of the Earth's climate system is continuously rising, referred to as global warming. The main reason for climate change is the emission of greenhouse gases, mainly CO2. The increasing population day by day, and increasing anthropogenic activities, increasing transportation rate, industries, etc., are responsible for releasing greenhouse gases. Temperatures are rising, hotter seas, acrid waters, and weather events like floods and droughts are becoming more frequent and more intense as a result of the rapid rise in greenhouse gas emissions over the past century (Kabir et al. 2023; Priya et al. 2023; Lieber et al. 2020). Climate change will cause more negative changes in the world and have a huge negative impact on all living creatures which also affects the agriculture sector. There is a strong link between agricultural production and climate change.

Climate change poses a severe threat to agricultural systems and food security. Agricultural production is under threat due to climate change in food-insecure regions, especially in Asian countries. Various climate-driven extremes, i.e., drought, heat waves, erratic and intense rainfall patterns, storms, floods, and emerging insect pests, have adversely affected the livelihood of the farmers (Nunez et al. 2019). Effects of environmental change are turning out to be more remarkable in the present, as opposed to being far away later. In the twenty-first century, climate change has been recognized as possibly the best well-being challenge and it mainly affects the agriculture sector by decreasing global productivity, crop growth, and quality of grains. The increasing population requires more agricultural production which is now actually decreasing due to climate change. The growth rate, improvement, and yield are affected by the increase of the suboptimal range of temperature for a long duration at any significant stage (Burgess 2024; Campbell-Lendrum et al. 2023). Temperatures that remain consistently above those that are best suited for plant growth may cause heat stress (HS), which restricts the development of flowers and fruits and significantly reduces yields. High temperatures may kill plants at a certain point. The findings of the temperature threshold of important crop plants and how they influence plant yield are essential under the condition of global warming (Janni et al. 2024).

The total population of the world is anticipated to come to an expected 9.7 billion by 2050 and food production worldwide ought to rise by at least 70%. This necessitates production, sustainability, and resilience of the food systems for the rapid increase in global food demand. However, the practice of simply increasing food production without increasing food system efficiency is outdated. Soil biodiversity and ecosystems are put at risk by traditional intensive agricultural systems that heavily rely on resource inputs, monocultures, and chemical use. Researchers can completely alter crop characteristics, enhancing their resistance to pests, diseases, and environmental stresses by introducing specific genes or utilizing advanced gene-editing tools (Galanakis 2024). Zinc finger nucleases (ZFNs), transcription activator-like effector nucleases (TALENs), and CRISPR-based technologies are the most popular and commonly used nuclease-based gene editing methods (Wei and Li 2023). ZFNs were a huge breakthrough in gene editing technology, but they have several drawbacks, including expensive cost, intricate design, high off-target rate, and challenges in doing multi-target editing (Chandrasegaran and Carroll 2016). Although ZFNs and TALENs can increase the effectiveness of gene knockout, they also depend on proteins that have been engineered for specialized DNA recognition, which means that new proteins must be created for various target sites. Their use in gene editing is restricted primarily by their time-consuming operation and complex technical requirements (Wei and Li 2023). The field of Clustered Regularly Interspaced Short Palindromic Repeat (CRISPR)/CRISPR-associated protein (Cas) has seen a significant surge in research, paving the way for the current era of more precise and targeted genome editing (Bhardwaj and Nain 2021).

In 2012, Jennifer Doudna and Emmanuelle Charpentier developed CRISPR/Cas9, a precise genome-editing tool for modifying DNA in animals, plants, and microbes, earning them the 2020 Nobel Prize in Chemistry (Jinek et al. 2012; Mali et al. 2013; Westermann et al. 2020). Due to its simplicity and multiplexed gene editing capabilities, this system has revolutionized crop improvement, enabling rapid enhancements in less than a year instead of traditional methods that take six to seven years (Galanakis 2024). The established methods for improving plants are more expensive than the CRISPR technology, and more importantly, the CRISPR-Cas system is faster, simpler, and more accurate. Crop quality can be enhanced by CRISPR technology by increasing nutrition, resistance to pests and diseases, and drought tolerance. Through simple modifications at specific genetic loci, CRISPR has been used to edit the genomes of several crops, including grains, vegetables, and fruits. Improved nutrient profiles, including higher levels of carbohydrate, protein, vitamin, and oleic acid in cereal crops like rice, maize, and wheat, are the outcome of biofortified crops with CRISPR-Cas9 (Ndudzo et al. 2024).

Since the CRISPR genome editing system was invented, various Cas nucleases related to it, including Cas9, Cas12, Cas13, and Cas14, have been found and are utilized in various contexts (Ceasar and Kavas 2024). The CRISPR/Cas system was initially discovered in two important food crops: rice and wheat (Akram et al. 2022). Till now, the CRISPR system is used for the gene editing in different plants: tomato, potato, cassava, soybean, cowpea, watermelon, etc. (Veillet et al. 2019; Wang et al. 2022; Zhang et al. 2022; Che et al. 2021; Tian et al. 2018). The use of CRISPR/Cas9 technology in a variety of crops has enormous promise for transforming agriculture and ensuring food and nutrition security for the world's rapidly growing population in the scenario of climate change due to global warming that is of much importance (Kafle 2023). Hence, this chapter focuses on the efficient use of CRISPR-Cas mediated gene editing to address food security under climate change.

19.2 Challenges in Twenty-First-Century Food Systems

Compared to 2005, worldwide food needs are supposed to increase by 50–100% by 2050 due to an expansion in per capita utilization (Fanzo et al. 2022). Global warming causes a range of negative impacts on plants especially due to rapid changes in temperatures, alterations of rainfall patterns, foods or drought conditions, and outbreaks of pests and diseases. These problems directly affect the agriculture sector, bringing down the quantity and quality of food products. World food consumption will increase due to increasing population and loss of food products due to climatic changes, notably creating huge problems in the agriculture sector (Munaweera et al. 2022).

Food safety has faced numerous issues as a consequence of climate change, such as changing precipitation patterns, extreme weather events of higher frequency, and extreme temperatures. These modifications have an effect on the food supply chain at many points, from production to consumption, increasing already-existing risks to food safety and generating new ones (Michael Alurame Eruaga 2024). The consequences of food safety issues include millions of people in disaster-prone areas losing their food security and suffering fatalities (Varzakas and Smaoui 2024). The effects of climate change, such as increased carbon dioxide, have a great impact on the quality of micronutrients, especially zinc, iron, and vitamin A in the agriculture yield (Loladze 2014; Mcgrath and Lobell 2012; Semba et al. 2022). The micronutrient deficiencies affected an estimated 2 billion people worldwide, mainly including young children and women (Bailey et al. 2015). Malnutrition and hunger continue to be unacceptable worldwide issues, affecting every nation to some degree (Fanzo et al. 2022). Plant diseases due to climate change conditions reduce output and cause plant mortality, which has an impact on food supply and, additionally, on food security and it also leads to malnutrition (Rizzo et al. 2021; Fanzo et al. 2024).

Plant biotechnology has advanced significantly in recent decades with the help of genome editing technologies, making it possible to produce genome-edited crops for food production and biomass (Li et al. 2024). It is very significant for the present and future generations to attain food security in climatic extreme and crop productivity-reducing situations. The improvement of biotechnological approaches such as genetic engineering, genome and gene editing using CRISPR-Cas system, RNA-mediated gene silencing armored with next-generation sequencing, and genome mapping opened the way for faster and more precise modifications of plants (Munaweera et al. 2022).

19.3 Why Gene Editing Is a Climate Change Solution?

When the Sustainable Development Goals (SDGs) were introduced in 2015, the elimination of poverty, enhanced food security, and improved human health were ranked as the top three objectives. The target year for attaining each of the 17 SDGs is 2030. Since the middle of the twentieth century, advances in plant breeding have greatly enhanced the efficiency of food production. New developments in the current millennium have shown increased potential to improve crop yields, the nutritional content of food crops, and the effects on the environment (Smyth 2022).

In order to attain sustainable development in the agriculture field, the conventional practices have not given the expected results in the climate change conditions. Therefore, the agriculture sector needs modern biotechnical practices like gene editing to improve the agriculture sector as a climate change solution. Gene editing using CRISPR has shown promising results in many plants for their improvement in various stress conditions (Tariq et al. 2023). Humankind throughout the world uses wheat and wheat products as a food source, and globally, 1 in 170 to 1 in 100 people are affected by celiac disease, an autoimmune condition brought on by the glutenin and prolamin found in wheat (Gujral 2012). It is not feasible to produce wheat cultivars using conventional breeding methods for the production of lower prolamin levels. However, the editing of 35 genes simultaneously in the wheat gluten family might be achieved through the application of gene-editing technologies, resulting in transgene-free cultivars that could drastically lower immunological reactivity (Sánchez-León et al. 2017).

Potato is an important crop for maintaining the nutritional and food security of the world. It is vulnerable to different types of stress and thereby the production of yield is affected. However, conventional breeding techniques are insufficient to achieve significant trait improvement in tetraploid potato cultivars in a reasonably short amount of time due to several characteristics of the potato, including large levels of heterozygosity, inbreeding depression, tetrasomic inheritance, and self-incompatibility of diploid potato. The use of CRISPR/Cas genome editing has given new avenues for the production of innovative potato varieties with significant commercial potential (Chincinska et al. 2022).

Fruits after harvesting are ripened fast and decayed which cause a huge loss which can be avoided using gene editing techniques. In tomatoes, CRISPR/Cas9 editing of the ripening-related lncRNA1459 (long noncoding RNA) yielded homozygous indel mutants with delayed ripening (Ueta et al. 2017). There are abiotic and biotic factors that cause huge loss of tomatoes and biotic stresses are the major constraint to the production of tomatoes (Wai et al. 2020). Viruses are the major biotic stress for plant growth; the CRISPR technology can help in the development of tomato plants against viruses. Tashkandi et al. (2018) showed the potential of CRISPR-Cas technology for the development of Tomato Yellow Leaf Curl Virus (TYLCV) resistance in *N. benthamiana* and tomato. Grain yield is an attribute of quality and can be increased with the help of CRISPR technology. In Zhonghua 11 variety of rice, this technology has helped to eliminate genes such as GS3, DEP1, IPA1, and Gn1a, which operate as negative regulators of grain size and number, panicle architecture, and plant architecture, respectively (Li et al. 2016). The above studies show the potential of CRISPR gene editing technology, for bringing improvements in the agriculture sector.

The primary obstacle to global agriculture policy and the economy is the requirement to provide food for the increasing world population. Innovative solutions are needed to solve the food scarcity issues. In this situation, genetic engineering has received a lot of attention and it provides several tools to boost crop resilience, raise yield, and improve nutrition. However, public opinion, decision-making, and regulations determine how applicable they are to industrial uses.

Genome editing strategies and techniques like CRISPR-Cas9 are of much importance for effective agricultural improvement, providing a remedy for the crop failure issue (Nguyen and Do 2024). Appropriate policy strategies that integrate new technology with consumer and social expectations are necessary when utilizing genome editing as a technological solution. To transform technological innovations into policy strategies that could guarantee ethical innovation in agricultural biotechnology, social license is an essential component (Anyshchenko 2022).

19.4 The Evolution of Gene Editing Technologies

CRISPR-Cas system has been getting a lot of attention since it is an effortless and uncomplicated process (Chaudhuri et al. 2022). Before these CRISPR-Cas systems, other site-directed nucleases (SDN) facilitated gene editing techniques for plant genome engineering such as zinc finger nucleases (ZFNs) and transcription activator-like effector nucleases (TALENs) (Bhuyan et al. 2023). A significant milestone in genetics and molecular biology was in 1983 with the discovery of the first zinc finger protein (Miller et al. 1985). The use of ZFNs in genome editing has always been difficult and has not been extensively adopted by the research community because occasionally ZFNs go off-target as a result of nonspecific DNA binding and also produce mutant plants by creating DSBs (double-strand breaks) and inducing NHEJ (nonhomologous end joining) repair (Bhuyan et al. 2023).

When compared with ZFNs, TALENs are another protein-based DNA-targeting tool, which have improved sequence precision. However, TALENs have been reported to be more mutagenic than ZFNs (Becker and Boch 2021, Bhuyan et al. 2023).

The CRISPR repeat arrays are special loci in bacterial or archaeal genomes in which the foreign genomes were inserted known as the CRISPR spacers, which is a naturally evolved programming as an adaptive immune system protects bacteria from repeated viral attacks (Koonin et al. 2023). CRISPR was first discovered from the DNA sequences of *Escherichia coli* bacteria in 1987 (Ishino et al. 1987). In different strains of the same species of pathogenic bacteria, CRISPR loci had a high degree of polymorphism identified by genotyping *Mycobacterium tuberculosis* initially in 1993 and later *Streptococcus pyogenes* in 1999 (Groenen et al. 1993, Hoe et al. 1999). Francisco Mojica was one of the first scientists who discovered similar structures in the archaeal genome of *Haloferax mediterranei* which include fragments of foreign DNA with similar biological functions (Mojica et al. 1993; Mojica et al. 2005). In 2007, studies by two French food scientists, Rodolphe Barrangou and Philippe Horvath, in *Streptococcus thermophilus* bacteria using the yogurt culture provided experimental evidence on the mechanism of action of the CRISPR system (Barrangou et al. 2007).

When foreign genetic elements invade the prokaryotes, the foreign DNA can be cut into short fragments by Cas proteins which have endonuclease activity, encoded by CRISPR-associated (cas) genes and then the DNA fragments will be integrated into the CRISPR array as new spacers. CRISPR RNA (crRNA) can rapidly identify the same invader and guides the Cas protein to cleave the target sequences of foreign DNA by pairing with the foreign DNA, hence protecting the host (Makarova et al. 2011; Xu and Li 2020).

19.5 Classification of CRISPR-Cas System

Based on the structural differences between the Cas genes, CRISPR-Cas systems are divided into Class I and Class II, both having a total of 6 types (Type I to VI) with several subtypes (Fig. 19.1). The Class I systems include Type I, III, and IV, which have multi-Cas protein effector complexes, and on the other hand, Class 2 systems have single effector proteins which contain Type II, V, and VI (Koonin et al. 2017; González et al. 2020; Bhuyan et al. 2023). In 2022, Kato et al. (2022) reported a total of 34 subtypes. Due to its versatility and advanced gene editing capabilities, CRISPR-Cas9 which comes under the type II system is more used in various studies. The most studied types of CRISPR-Cas systems are type I, II, and III based on their target sequence recognition and cleavage. *Streptococcus pyogenes* simultaneously possess all these three systems and Cas protein from this organism is an important tool for gene editing because it can precisely target DNA sequences of specific interests (Makarova et al. 2011).

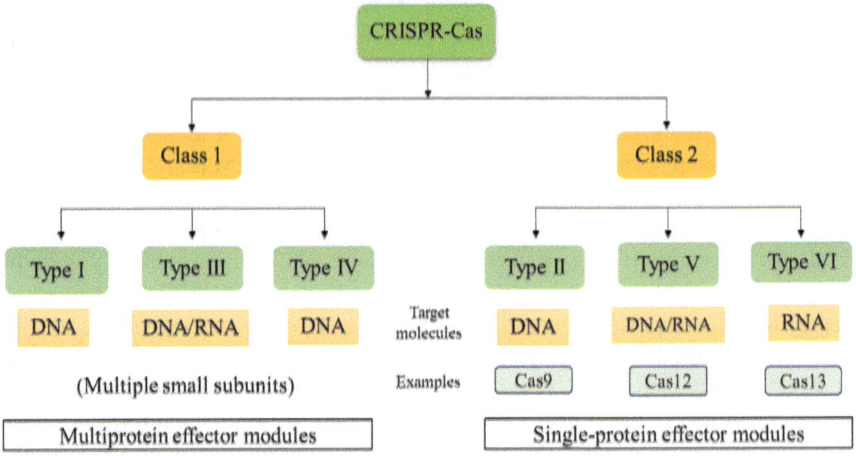

Fig. 19.1 Classification of CRISPR-Cas system

19.6 Mechanism of Action of CRISPR-Cas

CRISPR-Cas system consists of two main components: genomic CRISPR repeat array and the cas gene operon located in close vicinity to the array. One CRISPR locus contains a noncoding leader sequence with promoters and regulatory protein binding sites, followed by a series of direct repeat and spacer units. Unique spacers determine the specificity of the system and the identical direct repeats play a regulatory role. CRISPR-Cas specificity arises from spacers complementary to invading genetic elements, enabling the system to recognize and recall previous infections upon reinvasion.

For the functioning of the system, Cas proteins are essential during each of three distinct phases: adaptation, crRNA maturation, and interference (Zakrzewska and Burmistrz 2023). Figure 19.2 depicts the CRISPR-Cas mechanism exhibited by bacteria during virus infection.

Scientists have found that the CRISPR-Cas system is an adaptive immunity that is acquired by the bacteria or archaea against a viral attack or many other foreign substances entering into the cell. CRISPR-Cas system also has different applications in genome editing in RNA-level studies (Hillary and Ceasar 2022).

19.7 CRISPR-Cas Systems

The range of Cas proteins that can target DNA and RNA has increased with the discovery of additional CRISPR/Cas systems, making them an effective tool for editing plant genomes. The editing capabilities of these tools have been expanded with the help of the use of different Cas orthologues with different PAM specificities. The field of precision genome editing was transformed by the growing CRISPR/

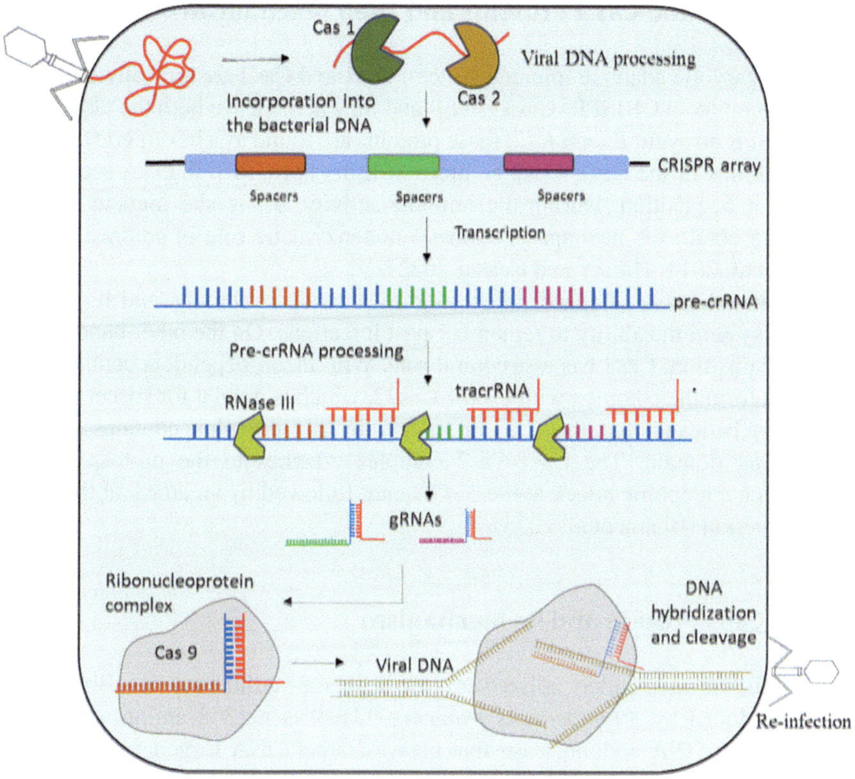

Fig. 19.2 CRISPR-Cas mechanism exhibited by bacteria during virus infection. When the virus infects the bacteria, the bacteria acquire a small piece of viral DNA to generate CRISPR arrays by integrating the viral foreign DNA into the CRISPR locus. The Cas1–Cas2 complex inserts a sequence of the invasive DNA (protospacer) into the CRISPR array. The CRISPR arrays consist of palindromic repeats of bacterial genomes and spacers. The CRISPR array is then transcribed into a long precursor CRISPR RNA (pre-crRNA), which is then cleaved into mature crRNAs. When combined with Cas proteins, these mature crRNAs form effector complexes and tracrRNA, which have sections complementary to the palindromic sequence. Ribonuclease III or RNase III will cleave the strand between the effector complexes. The cell is prepared to fight against the foreign invader when these effector complexes form. When sequences of foreign genetic material match a CRISPR spacer, the corresponding crRNA attaches itself to the invasive strand and uses Cas nuclease to cleave it. Thus, it can degrade the new viral molecules

Cas toolkits, which also made diversified genome modification possible for crop engineering (Atia et al. 2024). There are different CRISPR-Cas systems having a wide range of applications in the agriculture field. Genome editing can be done precisely with the help of these Cas systems which give further insights into new possibilities to improve the agriculture field. The upcoming section will provide a brief understanding of different CRISPR-Cas systems.

19.7.1 Cas 1 and Cas 2 Proteins and Their Mechanism of Action

In the prokaryotic adaptive immune system, Cas1 and Cas2 are typically conserved proteins among all CRISPR–Cas systems and they help cleave both the target DNA and CRISPR array. In *E. coli K12*, these proteins are found as a hexameric complex. The mutations in the Cas1–Cas2 complex will disrupt CRISPR DNA recognition and spacer acquisition. During the immune activity, active site mutants of Cas2 which can obtain the new spacers show a nonenzymatic role in contrast to Cas1 (Nuñez et al. 2014; Hillary and Ceasar 2022).

Cas1 has the role of spacer integration as a catalytic activity, and it gives the CRISPR system the ability to remember past infections. On the other hand, for the spacer acquisition, Cas2 has a structural role. With metal-dependent activity, Cas1 and Cas2 form the complex with Cas14-Cas22, which is crucial for spacer recognition. It is believed that the Cas1 monomer in this complex contains a PAM-recognizing domain. The Cas1-Cas2 complex attaches to the protospacer and initiates a nucleophilic attack at the 3'-OH end, followed by an attack at the leader repeat junction (Bhatia et al. 2023).

19.7.2 Cas9 Protein and Its Mechanism

Cas9 is still the most widely utilized enzyme in genome editing studies (Bhatia et al. 2023) produced by *Streptococcus pyogenes*. Cas-9 is a 1368 amino acids' long multi-domain DNA endonuclease that cleaves target DNA to generate a double-stranded break (Mengstie and Wondimu 2021).

The recognition (REC) lobe and the nuclease (NUC) lobe are the two regions of Cas 9. The binding of guide RNA takes place in the REC lobe which further consists REC1 and REC2 domains. In contrast, the RuvC, HNH, and Protospacer Adjacent Motif (PAM) interacting domains all comprise the NUC lobe. PAM specificity is granted by the PAM interacting domain, which also starts the binding process to target DNA and single-stranded DNA molecule is cut using the RuvC and HNH domains (Mengstie and Wondimu 2021).

Cas-9 nuclease creates double-stranded breaks (DSBs) at a point three base pairs upstream of PAM. The PAM sequence recognized by the Cas-9 protein is 5′-NGG-3′ (N can be any nucleotide base). Finding a target site with the proper PAM initiates local DNA melting by Cas-9, which is followed by the formation of an RNA-DNA hybrid and the Cas-9 protein is then ready to cleave DNA. To produce predominantly blunt-ended DSBs, HNH domain cleaves the complementary strand and the target DNA's noncomplementary strand is cut by the RuvC domain (Mengstie and Wondimu 2021).

19.7.3 Cas 12 Protein and Its Mechanism

Cas12 protein belongs to the type V CRISPR system. With its great degree of versatility, the Cas12 protein can be used for several dynamic applications, including epigenome editing. It has been identified that the Cas12 protein serves as an RNA-guided DNA endonuclease that can be used for editing genomes in place of the Cas9 protein (Agarwal 2023). The CRISPR-Cas12 complex exhibits a bilobed characteristic consisting of nuclease (NUC) lobe and recognition (REC) lobe. The WED domain is the domain in the REC lobe, which displays established PAM-interacting domains for a few effectors. REC domain, which is further described as the REC1 and REC2 domains, is situated in the REC lobe. On the other hand, the NUC lobe is made up of a nuclease domain that contains a ZnF domain that directly regulates cleavage and a RuvC domain comprising conserved amino acid residues (D/E/D) (Badon et al. 2024).

Single effectors Cas12a and crRNA mediate the operation of the CRISPR-Cas12a system. The complex attaches to its recognition site downstream of the PAM sequence by combining Cas12a and crRNA. A 23–25 nucleotide reference sequence of crRNA helps DNA binding (Dronina et al. 2022). Without any requirement for a tracrRNA, Cas12a converts pre-crRNA into mature crRNA on its own, making it a special effector protein with endonuclease and endoribonuclease functions. Cas12a cuts the pre-crRNA 4 nt upstream of the hairpin structures formed by the CRISPR repetitions after it has been transcribed during the expression stage. This results in the formation of intermediate crRNA molecules, which are then further processed in vivo to become mature crRNAs (Paul and Montoya 2020). To effectively cut at ssDNA and dsDNA, the Cas12 protein just needs the crRNAs. Once Cas12 comes across a PAM sequence near to a possible target site, it initiates the R-loop, which helps the target DNA strand and the crRNA to hybridize base-pair. Following the formation of the R-loop, the PAM sequence is utilized by the Cas12 protein's active RuvC domain to cleave the nontarget strand (Hillary and Ceasar 2022). Plant genome editing has been successfully carried out in several species, including rice, Arabidopsis, maize, soybeans, and others, using the temperature-sensitive CRISPR/LbCas12a system (Li et al. 2020; Yang et al. 2023).

19.7.4 Cas 13 Protein and Its Mechanism

The crRNA of 64–66 nucleotides and the Cas13 protein effector make up the CRISPR-Cas13 system. The nuclease (NUC) domain and crRNA-recognition (REC) lobe comprise the Cas13a protein. The N-terminal domain (NTD) and the helix-1 or helical-1 domain are there in the REC lobe. The NTD is a non-conserved region of the Cas13a, which is made up of two subdomains: a smaller one with three α helices, a β-hairpin, and a β-sheet, and a bigger subdomain with an ordered fragment made up of seven α helices and a disordered fragment. The Cas13a-crRNA complex is formed when Cas13a and crRNA bind together and the Cas13a binds the appropriate RNA with the help of crRNA, causing prokaryotes to become defensive against RNA viruses (Huang et al. 2022a).

Cas13 activity is not necessary for crRNA maturation; target RNA identification can be achieved even with unprocessed pre-crRNA. The ribonucleoprotein complex (RNP) exhibits a conformational change following interaction between crRNA and a target RNA. A catalytic site is generated mainly by the close contact of two HEPN domains. The binding of crRNA is anticipated to cause the cleavage of the target RNA along with other ssRNAs that surround the RNP complex, which might involve host RNA, due to the distance between the catalytic site and the crRNA-RNA duplex. As an outcome, the CRISPR-Cas13 system is capable of cutting both nontarget and target RNA at the same time. This ability has been widely used for RNA knockout techniques, disease treatment, viral infection prevention, loss-of-function mutant screening, molecular biological agent detection, and CRISPR-based antimicrobial agents (Huang et al. 2022a).

In the climate change situation for the agriculture sector, the major threat is viral infection. Viral infections are a major constraint to food production, and to overcome these problems, a proper solution is needed. The primary focus of recent research on plant CRISPR/Cas13 systems has been on viral RNA interference. The important feature of CRISPR/Cas13 can also be used to target noncoding RNA in plants without changing the genome and improve plant resilience. Plant transcriptome research can benefit significantly from the adoption of CRISPR/Cas13 systems for RNA targeting and editing in plants. The primary goal of CRISPR/Cas13 use in plants is to provide resistance against plant viruses. CRISPR/Cas13 systems may develop into a strong platform for RNA targeting in plants, opening up new possibilities for relevant agricultural applications (Kavuri et al. 2022).

19.7.5 Cas 14 Protein and Its Mechanism

The CRISPR-Cas Type V-F proteins are commonly known as Cas14 (Ullah et al. 2023). The small operational CRISPR system identified to date is the Cas14 system (Huang et al. 2022b). It contains only ~400–700 amino acids. Researchers identified the absence of a PAM binding requirement for Cas14 in Cas effectors. Activity assays suggest that Cas14 only cleaves single-stranded DNA (ssDNA) and not double-stranded DNA or single-stranded RNA. Cas14 is particularly vulnerable to mismatches in the middle of its target region, which is around 6 bp long (Savage 2019). Similarities can be seen between RuvC-containing bacterial proteins C2c10 and C2c9, which are usually found close to Cas arrays instead of Cas genes, and Cas14 in the phylogenetic analysis. These small effectors don't require a PAM sequence to create RNA-guided single-strand (ssDNA) breaks in the target DNA. Later research, nevertheless, has demonstrated that these effectors can use the T-rich PAM sequence to cause double-stranded breaks (DSB) (Ullah et al. 2023). The crystal structure of Cas14 was discovered, revealing insight into its identification and cleavage mechanism (Takeda et al. 2021). The two subunits of the homodimer Cas14 protein are identical, both containing a helicase–nuclease domain and a nuclease domain. The guide RNA–target DNA complex is situated within the channel that separates the nuclease (NUC) lobe and the REC lobe of the Cas12f dimer, and it takes on a lobed form (Ullah et al. 2023).

Cas14 is the only known DNA-targeted Cas that needs no PAM. Cas14 can identify ssDNA complementary to the recognition arm of a small guided RNA (sgRNA) and use this information to produce an R-loop structure with indiscriminate ssDNA cleaving activity. In contrast to Cas12a, Cas14 has high specificity and sensitivity in detecting ssDNA, making it a valuable tool for identifying pathogens and single-nucleotide polymorphism (SNP) genotyping (Zhou et al. 2023).

19.8 Applications of Gene Editing in Agriculture

Genome editing is vital as of now and it is needed for the future in the agriculture field for the sustainable agriculture in the climate change scenario. Genome editing in crops has the potential to accelerate breeding program development greatly. It has various advantages that influence the evolution of new crop features, enabling increase in production and pest resistance, climate change adaptation, and industrial uses (Ricroch 2019). It is essential to look into particular traits including broad, durable resistance to various diseases, as well as enhanced yield and tolerance of biotic and abiotic stress, with tolerance to heat, salt, and drought being particularly significant. Other target trait qualities include photosynthetic efficiency, plant architecture, so-called output traits, like lipid composition, and agricultural inputs (nitrogen use efficiency and utilization of other nutrients, water use efficiency, herbicide tolerance). The notable thing is that these can be achieved through gene editing (Khaipho-Burch et al. 2023).

CRISPR technology has been widely used to confer stress tolerance to a wide range of crops, such as rice, wheat, tomato, cotton, soybean, cassava, and potato. Table 19.1 provides information about various crops improved with the help of CRISPR technology using different Cas systems.

19.9 The Impact of Gene Editing on Climate Change and Food Security

19.9.1 Advanced Agricultural Resilience

Agriculture is climate-dependent; thus, rising temperatures and carbon dioxide levels can significantly impact optimum nutrient levels, water availability, soil moisture, and a variety of other essential performance factors (Gomez-Zavaglia et al. 2020). Abiotic stressors, including drought and severe temperatures, can cause growth inhibition in crops, which lowers crop vigor, stunts development and growth, and decreases the production of crops (Kim and Lee 2023). High temperature is the primary environmental stressor that reduces plant development and agricultural productivity. Tomatoes are one of the most popular vegetables worldwide in demand

Table 19.1 Examples of CRISPR-Cas technology used for crop improvement

S.No	Crop	Gene-edited	CRISPR-Cas	Trait obtained	References
1	*Solanum tuberosum*	DMR6	Cas9	Resistance to various stress conditions	Karlsson et al. (2024)
2	*Oryza sativa*	GA20ox-2, OsCKX2, and OsSWEET11	Cas9	Early maturing, short stems, high yield, and bacterial leaf blight resistance	Roviqowati (2024)
3	*Nicotiana Benthamiana*	AC1 and AV1	Cas9	Disease resistance	Talakayala et al. (2024)
4	*Manihot esculenta*	MeSWEET10a	Cas9	Reduce susceptibility to disease	Elliott et al. (2024)
5	*Solanum tuberosum*	StPM1	Cas9	Resistance to phytophthora	Bi et al. (2023)
6	*Musa paradisiaca Linn*	ENODL3	Cas9	Disease resistance	Ntui et al. (2023)
7	*Oryza sativa*	xopV	Cas12a from *Francisella novicida*	Yield (disease resistance)	Yan et al. (2023)
8	*Phaseolus vulgaris*	XDH, NSH1, NSH2, XMPP, and GSDA	Cas9	Ureide biosynthesis	Voß et al. (2022)
9	*Glycine max*	GmPIN	Cas9	Crop improvement	Zhang et al. (2022)
10	*Glycine max*	GmNN1/FT2a	Cas 9	Crop improvement	Li et al. (2022)
11	*Glycine max*	GmIPK1	Cas9	Seed improvement with reduced PA content	Song et al. (2022)
12	*Manihot esculenta*	CYP79D1	Cas9	Help to reduce cyanide content in cassava	Juma et al. (2022)
13	*Manihot esculenta*	MeSWEET10a	Cas9	Cassava resistant to bacterial blight	Wang et al. (2022)
14	*Hordeum vulgare*	EIF4E	Cas9	Resistant against barley yellow mosaic virus. Normal growth and grain production.	Hoffie et al. (2021)

(continued)

Table 19.1 (continued)

S.No	Crop	Gene-edited	CRISPR-Cas	Trait obtained	References
15	*Brassica napus*	BnSHP1 and BnSHP2	Cas9	Yield (pod shatter-resistant	Zaman et al. (2021)
16	*Brassica napus*	BnaBP	Cas9	Yield (plant architecture)	Fan et al. (2021)
17	*Glycine max*	Glyma10g42470 and FAD2-A	Cpf1 from Lachnospiraceae bacterium ND 2006 (LbCpf1)	Fatty acid composition and oil content (oil content)	Duan et al. (2021)
18	*Vigna unguiculata*	Vu-SPO11	spCas9	Loss of seed and pod Development	Che et al. (2021)
19	*Glycine max*	FAD2-A and Glyma10g42470	LbCpf1	Improve editing efficiency	Duan et al. (2021)
20	*Glycine max*	GmIPK1 and GmIPK2	Cas9	Successful gene editing	Carrijo et al. (2021)
21	*Glycine max*	GmAITR	Cas9	Improve salinity tolerance	Wang et al. (2021)
22	*Musa paradisiaca Linn.*	MaACO1	Cas9	Promotes the shelf life of banana fruit	Hu et al. (2021)
23	*Saccharum officinarum*	ALS	Cas9	Herbicide tolerance	Oz et al. (2021)
24	*Brassica napus*	BnaEOD3	Cas9	Yield (seed size and seed Number)	Khan et al. (2020)
25	*Brassica napus*	BnaFAD2	Cas9	Fatty acid composition and oil content (oleic acid Content)	Huang et al. (2020)
26	*Gossypium hirsutum*	Deoxyglucose-5-phosphate Synthase (GhCLA)	Cas12b (C2c1) From Alicyclobacillus Acidoterrestris (AacCas12b)	Cotton genome editing Efficiency by Cas 12	Wang et al. (2020)
27	*Solanum lycopersicum*	SlBRI1, SlRIN	XNG-Cas9	Plant architecture and Fruit ripening	Niu et al. (2020)
28	*Arabidopsis thaliana*	PDS3 (phytoene desaturase)	Cas V	Carotenoid biosynthesis	Pausch et al. (2020)
29	*Arachis hypogaea L*	AhNFR1 and AhNFR5	Cas9	Improvement in root nodule formation	Shu et al. (2020)

(continued)

Table 19.1 (continued)

S.No	Crop	Gene-edited	CRISPR-Cas	Trait obtained	References
30	*Oryza sativa*	OsPQT3	Cas9	Crop improvement with stress resistance	Alfatih et al. (2020)
31	*Musa paradisiaca Linn.*	LCYε	Cas9	Nutritional improvement	Kaur et al. (2020)
32	*Brassica napus*	BnALS1	Cas9	Herbicide-resistance	Wu et al. (2020)
33	*Brassica napus*	BnaMAX1	Cas9	Plant yield and architecture	Zheng et al. (2020)
34	Brassica napus	BnLPAT2 and BnLPAT5	Cas9	Fatty acid composition and oil content (oil content)	Zhang et al. (2019a)
35	*Glycine max*	GmF3H1, GmF3H2 and GmFNSII-1	Cas9	Isoflavone content	Zhang et al. (2019b)
36	*Oryza sativa*	OsSWEET11 (PthXo1) and OsSWEET14 (PthXo3/ AvrXa7)	Cas 9	Increased resistance tobacterial blight	Xu et al. (2019b)
37	*Citrus sinensis*	Two CsWRKY22 alleles	Cas 9	Disease resistance	Wang et al. (2019b)
38	*Oryza sativa*	OsALS, NRT1.1B, OsCDC48, and OsWaxy	Three Cas9 variants SpCas9variantsSpCas9-HF2, HypaCas9, andeSpCas9 (1.1)	Crop improvement	Xu et al. (2019a)
39	*Solanum tuberosum and Solanum lycopersicum*	SlALS1	nCas9 cytidine Base editing system	Crop improvement (chlorsulfuron-resistant Plants)	Veillet et al. (2019)
40	*Arachis hypogaea L.*	FAD2	Cas9	Crop improvement	Yuan et al. (2019)
41	*Solanum lycopersicum L.*	CCD8	Cas9	Resistance against parasite	Bari et al. (2019)

(continued)

Table 19.1 (continued)

S.No	Crop	Gene-edited	CRISPR-Cas	Trait obtained	References
42	*Oryza sativa*	OsAAP3	Cas 9	Yield (blocking OsAAP3 Increases grain yield by Increasing tiller number In rice)	Lu et al. (2018)
43	*Oryza sativa*	OsGS3, OsGW2, and OsGn1a	Cas 9	Yield (grain weight, grainnumber, and grain size)	Zhou et al. (2018)
44	*Oryza sativa*	eIF4G	Cas9	Resistance to virus	Macovei et al. (2018)
45	*Solanum lycopersicum*	Coat protein (CP) gene	Cas9	Resistance against tomato yellow leaf curl virus	Tashkandi et al. (2018)
46	*Citrullus lanatus*	ALS	Cas9	Herbicide-resistance	Tian et al. (2018)
47	*Solanum lycopersicum*	SlGAD2 and SlGAD3	Cas9	Increased GABA content in tomato fruits	Nonaka et al. (2017)
48	*Citrus sinensis Osbeck*	CsLOB1	Cas9	Canker-resistant citrus cultivars	Peng et al. (2017)
49	*Zea mays*	ARGOS8	Cas9	Improve maize grain yield	Shi et al. (2016)
50	*Cucumis sativus L.*	eIF4E	Cas9	Virus resistance	Chandrasekaran et al. (2016)
51	*Oryza sativa*	ALS	Cas9	Herbicide-resistance	Sun et al. (2016)

(Taheri et al. 2022). Heat stress suppresses photosynthesis, resulting in difficulties with growth and decreased production (Fahad et al. 2017). As a result, heat-tolerant agricultural plants are in high demand. Mitogen-activated protein kinases (MAPKs) are highly conserved serine and threonine protein kinases that play a role in plant responses to various environmental stressors. CRISPR/Cas9-mediated SlMAPK3 mutants improved heat tolerance in tomato plants (Yu et al. 2019). Lateral Organ Boundaries Domain (LBD)-containing genes are plant-specific genes that play crucial roles in lateral organ development, such as LBD40, which belongs to subfamily II of the LBD gene family in tomato. Knockout of SlLBD40 by CRISPR/Cas9 improved drought tolerance in tomatoes by increasing their water-holding abilities under drought conditions (Liu et al. 2020).

Compared to many other crops, soybeans are particularly vulnerable to water scarcity, and drought stress can significantly reduce soybean yield. Using the CRISPR/Cas9 technology to alter the homeodomain-leucine zipper gene (GmHdz4) in soybeans may increase drought stress and offer a possible target for molecular breeding in soybeans (Zhong et al. 2022). Similarly, soybean salt tolerance is crucial because it can significantly decrease soybean productivity (Phang et al. 2008). The researchers identified abscisic acid (ABA)-induced transcription repressors (AITRs) as a novel family of transcription factors involved in regulating ABA signaling and discovered that knocking down the whole family of AITR genes in Arabidopsis improved salinity and drought tolerance without sacrificing fitness; also, it is reported that CRISPR/Cas9 alteration of GmAITR genes is an effective method for improving soybean salinity tolerance (Wang et al. 2021). Stomatal control of water loss is one adaptation strategy that plants use to cope with water scarcity. Epidermal patterning factors (VvEPFL9–1) knockout in Grapevine using CRISPR/Cas9 suggests potential benefits in lowering stomatal density under future drier environmental conditions (Clemens et al. 2022).

19.9.2 Strengthen Crop Performance

The quality of agricultural products is important, and improving the features of crops is essential to get better yields. Rice is a major staple crop worldwide, and CRISPR/Cas9 genome editing provides a rapid, sustainable, and cost-effective technique for increasing rice grain quality. Rice's OsAGAP gene encodes a protein related to the ADP-ribosylation factor GTPase-activating protein (ArfGAP). It affects the polar movement of auxin in rice, influencing plant morphology and rice grain development. OsPDCD5 encodes the programmed cell death 5 protein, which interacts with OsAGAP and reduces rice plant structure and yield of grains. The targeted alteration of OsPDCD5 with the CRISPR/Cas9 system increased plant structure and yield (Dong et al. 2021). Also, CRISPR/Cas9-guided OsSPL16 mutations can increase rice grain yield (Usman et al. 2020). Knocking out the hexokinase gene (OsHXK1) using the CRISPR/Cas9 gene-editing technology could develop high-photosynthetic efficiency and high-yield rice cultivars. They also discovered that OsHXK1 significantly regulates rice yield and photosynthesis (Zheng et al. 2021).

In the flowering plant, alfalfa CRISPR/Cas9-mediated mutagenesis FLOWERING LOCUS Ta1 (MsFTa1) causes delayed flowering time with better forage biomass yield and quality (Wolabu et al. 2023). Maize ARGOS8 is a negative ethylene response regulator, and CRISPR-Cas9-generated ARGOS8 variations increase maize grain yield in drought-stressed fields (Shi et al. 2016). Strigolactones are a class of small compounds that serve essential functions in plant development and root biotic interactions and mutation of Carotenoid Cleavage Dioxygenase (CCD) gene in Sorghum by CRISPR/Cas9 modifies strigolactone biosynthesis and plant biotic interactions that will help to manipulate strigolactone synthesis in sorghum to recruit beneficial microorganisms and reduce parasitic weeds (Hao et al. 2023). In sweet potatoes, CRISPR/Cas9 technology resulted in improved starch quality and the development of polyploid root crops (Wang et al. 2019a).

Low nitrogen availability is one of the main obstacles to plant growth and development, and substantial N fertilizer doses are required to obtain significant agricultural yields (Lebedev et al. 2021). Apigenin and other flavones promoted biological nitrogen-fixing. CRISPR-based gene editing in rice targets apigenin breakdown, increasing plant apigenin content and root exudation. When grown under nitrogen-limited soil conditions, improved rice plants produced more grain with the help of the CRISPR system (Yan et al. 2022). Also, the CRISPR interference (CRISPRi) technology fills a critical gap in modifying microbial nitrogen fixation for specific uses (Russell et al. 2024).

Late bolting or flowering is directly linked to quality and productivity. Early flowering degrades leaf quality and reduces the harvest period. Thus, delayed flowering is a significant horticultural characteristic for boosting quality. For lettuce, delayed flowering allows for more extensive vegetative development, resulting in higher leaf numbers and probably larger leaves. CRISPR/Cas9-mediated insertion of a single base mutation in SOC1, a transcription factor that controls flowering time, generates late-bolting phenotypes in lettuce (Choi et al. 2022).

Another way the CRISPR system helps the agriculture sector is by mitigating food wastage and increasing the ripening time of crops. Whenever fruits reach the most edible stage, they gradually deteriorate, resulting in a drop in fruit quality (Li et al. 2021). So, it is essential to regulate fruit ripening. Fruit ripening in tomatoes is caused by the selective expression of ripening-related genes controlled by transcription factors (TFs). The NAC (NAM, ATAF1/2, and CUC2) TF family is one of the most prominent groups of plant-specific transcription factors. SlNAM1, a NAC TF, favorably influences the onset of tomato fruit ripening by controlling ethylene production. The commencement of fruit ripening in slnam1-deficient mutants produced using CRISPR/Cas9 was delayed (Gao et al. 2021), which will help to get more harvesting time to mitigate the food wastage.

19.9.3 Lessen Environmental Emissions

Carbon dioxide is the most significant greenhouse gas and is seen as vital to global warming and climate change (Lee et al. 2023). So, the most effective strategies to address the climate change challenge are using sustainable methods to store and reduce CO2 and our carbon footprint (Matinvafa et al. 2023). Plants have been capable of capturing CO2 for billions of years, making it the most reliable carbon capture technology on the planet. Finding a technique to store the carbon absorbed by plants permanently is critical. Plant roots are densely packed with carbon and discharge carbon-rich molecules into the soil around them. Developing future plants with more profound, more extensive root systems, with an improved chemical composition that enhances carbon content and slows the rate of biological degradation, offers a way to store atmospheric carbon captured by plants below ground for years (Busch and Miller 2022). Gene editing allows for precise changes to a plant's genome, increasing the development of new crop varieties, such as those that survive changing environmental conditions. It can also capture and store excess atmospheric carbon dioxide, which can be significantly improved with the help of CRISPR/Cas systems (Admas and Feyissa 2024). CRISPR technology can be used to reduce carbon-based pollutants, which are the primary cause of environmental pollution, and it has the potential to speed up bio-based approaches to environmental rehabilitation, such as bioremediation. CRISPR technology provides sustainable ways and approaches that lead to a sustainable ecosystem (Matinvafa et al. 2023).

19.9.4 Harness a Plant's Natural Potential

The long crop duration limits the production of improved crop types in plant breeding. Given the various phases of crossing, selection, and testing needed to develop new plant types, creating a new cultivar can take up to two decades. Recent developments in genome editing technologies, including programmable nucleases and the CRISPR-Cas system, resulted in a new era of plant breeding. Genome editing technologies have various benefits over traditional agricultural methods due to their simplicity, efficiency, high specificity, and ability to multiplex (Ahmar et al. 2020).

Large-scale crop improvement through genome editing, such as CRISPR-Cas technology, is about to become a reality. Genome editing can accelerate crop improvement strategies and increase their effectiveness, allowing for the development of crops with higher yields, improved nutrient content, pest resistance, and abiotic stress tolerance, which could promote food security in both developed and developing countries (Scheben et al. 2017). CRISPR/Cas systems are simple, versatile, and resilient, making genome editing a potent tool for precise crop improvement via gene knock-in, knockout, point mutations, replacement, fine-tuning of gene regulation, and other alterations at any gene locus in crops (Chen et al. 2019). Gene editing is more precise than conventional crop breeding and genetic engineering procedures. Thus, this technology is a highly potent tool for ensuring the world's food supply (Abdallah et al. 2015).

19.10 Conclusion

Since the discovery of CRISPR technology, several CRISPR-based tools have been designed that have revolutionized crop engineering with their expanded targeting ranges, improved editing specificity and efficiency, and other special features. CRISPR-Cas technology has features such as versatility, ease of design, stability, and the ability to multiplex gene editing, so it will continue to be the most commonly used system.

CRISPR-Cas technology holds progressive potential in addressing the multifaceted challenges posed by climate change on food security. By enabling precise genetic modifications, this far-reaching tool can enhance crop resilience, improve crop yield, and mitigate greenhouse gas emissions through sustainable agricultural practices. Additionally, CRISPR's applications extend to ecosystems restoration and carbon sequestration, further augmenting its application for environmental sustainability.

Gene editing has shown promise in research, but the practical application for mitigating climate change has not yet been followed. Adoption of these technological innovations needs to overcome regulatory barriers, social barriers, and prohibitive policies. However, realizing its full potential requires addressing ethical considerations, regulatory frameworks, and equitable access to this technology. As we steer the complexities of a climate-heating world, CRISPR-Cas emerges as a powerful tool in nurturing innovation and resilience for a sustainable future.

References

Abdallah NA, Prakash CS, McHughen AG (2015) Genome editing for crop improvement: challenges and opportunities. GM Crop Food 6(4):183–205. https://doi.org/10.1080/21645698.2015.1129937

Admas A, Feyissa T (2024) The role of modern biotechnology in the fight against the current and the future climate change. J Adv Biol Biotechnol 27(5):595–610. https://doi.org/10.9734/jabb/2024/v27i5822

Agarwal C (2023) A review: CRISPR/CAS12-mediated genome editing in fungal cells: advancements, mechanisms, and future directions in plant-fungal pathology [preprint]. https://doi.org/10.14293/pr2199.000129.v2

Ahmar S et al (2020) Conventional and molecular techniques from simple breeding to speed breeding in crop plants: recent advances and future outlook. Int J Mol Sci 21(7):2590. https://doi.org/10.3390/ijms21072590

Akram F et al (2022) An insight into modern targeted genome-editing technologies with a special focus on CRISPR/Cas9 and its applications. Mol Biotechnol 65(2):227–242. https://doi.org/10.1007/s12033-022-00501-4

Alfatih A et al (2020) Loss of rice paraquat tolerance 3 confers enhanced resistance to abiotic stresses and increases grain yield in field. Plant Cell Environ 43(11):2743–2754. https://doi.org/10.1111/pce.13856

Anyshchenko A (2022) Aligning policy design with science to achieve food security: the contribution of genome editing to sustainable agriculture. Front Sustain Food Syst 6. https://doi.org/10.3389/fsufs.2022.897643

Atia M et al (2024) Crop bioengineering via gene editing: reshaping the future of agriculture. Plant Cell Rep 43(4). https://doi.org/10.1007/s00299-024-03183-1

Badon IW et al (2024) Recent application of CRISPR-Cas12 and omega system for genome editing. Mol Ther 32(1):32–43. https://doi.org/10.1016/j.ymthe.2023.11.013

Bailey RL, West KP Jr, Black RE (2015) The epidemiology of global micronutrient deficiencies. Ann Nutr Metab 66(Suppl. 2):22–33. https://doi.org/10.1159/000371618

Bari VK et al (2019) CRISPR/Cas9-mediated mutagenesis of carotenoid cleavage dioxygenase 8 in tomato provides resistance against the parasitic weed Phelipanche Aegyptiaca. Sci Rep 9(1). https://doi.org/10.1038/s41598-019-47893-z

Barrangou R et al (2007) CRISPR provides acquired resistance against viruses in prokaryotes. Science 315(5819):1709–1712. https://doi.org/10.1126/science.1138140

Becker S, Boch J (2021) TALE and TALEN genome editing technologies. Gene and Genome Editing, [online] 2(2):100007. https://doi.org/10.1016/j.ggedit.2021.100007

Bhardwaj A, Nain V (2021) Talens—an indispensable tool in the era of CRISPR: a mini review. J Genetic Eng Biotechnol 19(1):125. https://doi.org/10.1186/s43141-021-00225-z

Bhatia S, Pooja, Yadav SK (2023) CRISPR-Cas for genome editing: classification, mechanism, designing and applications. Int J Biol Macromol 238:124054. https://doi.org/10.1016/j.ijbiomac.2023.124054

Bhuyan SJ et al (2023) Progress in gene editing tools, implications and success in plants: a review. Frontiers in Genome Editing 5. https://doi.org/10.3389/fgeed.2023.1272678

Bi W et al (2023) Crispr/cas9-guided editing of a novel susceptibility gene in potato improves phytophthora resistance without growth penalty. Plant Biotechnol J 22(1):4–6. https://doi.org/10.1111/pbi.14175

Burgess M (2024) Five considerations for twenty-first century climate policy. FIU Law Review 18(2):283–312. https://doi.org/10.25148/lawrev.18.2.6

Busch W, Miller C (2022) The greenest revolution—harnessing the power of plants to help combat climate change. Biochemist 44(3):13–18. https://doi.org/10.1042/bio_2022_113

Campbell-Lendrum D et al (2023) Climate change and health: three grand challenges. Nat Med 29(7):1631–1638. https://doi.org/10.1038/s41591-023-02438-w

Carrijo J et al (2021) Two efficient CRISPR/Cas9 systems for gene editing in soybean. Transgenic Res 30(3):239–249. https://doi.org/10.1007/s11248-021-00246-x

Ceasar SA, Kavas M (2024) 'Plant genome editing to achieve food and nutrient security', BMC. Methods 1(1). https://doi.org/10.1186/s44330-024-00003-6

Chandrasegaran S, Carroll D (2016) Origins of programmable nucleases for genome engineering. J Mol Biol 428(5):963–989. https://doi.org/10.1016/j.jmb.2015.10.014

Chandrasekaran J et al (2016) Development of broad virus resistance in non-transgenic cucumber using CRISPR/cas9 technology. Mol Plant Pathol 17(7):1140–1153. https://doi.org/10.1111/mpp.12375

Chaudhuri A, Halder K, Datta A (2022) Classification of CRISPR/CAS system and its application in tomato breeding. Theor Appl Genet 135(2):367–387. https://doi.org/10.1007/s00122-021-03984-y

Che P et al (2021) Developing a rapid and highly efficient cowpea regeneration, transformation and genome editing system using embryonic Axis explants. Plant J 106(3):817–830. https://doi.org/10.1111/tpj.15202

Chen K et al (2019) CRISPR/Cas genome editing and precision plant breeding in agriculture. Annu Rev Plant Biol 70(1):667–697. https://doi.org/10.1146/annurev-arplant-050718-100049

Chincinska IA, Miklaszewska M, Sołtys-Kalina D (2022) Recent advances and challenges in potato improvement using CRISPR/Cas genome editing. Planta 257(1). https://doi.org/10.1007/s00425-022-04054-3

Choi SH et al (2022) Development of late-bolting plants by CRISPR/Cas9-mediated genome editing from mesophyll protoplasts of lettuce. Plant Cell Rep 41(7):1627–1630. https://doi.org/10.1007/s00299-022-02875-w

Clemens M et al (2022) VVEPFL9-1 knock-out via CRISPR/Cas9 reduces stomatal density in grapevine. Front Plant Sci 13. https://doi.org/10.3389/fpls.2022.878001

Dong S et al (2021) ospdcd5 negatively regulates plant architecture and grain yield in Rice. Proc Natl Acad Sci 118(29). https://doi.org/10.1073/pnas.2018799118

Dronina J, Samukaite-Bubniene U, Ramanavicius A (2022) Towards application of CRISPR-CAS12A in the design of modern viral DNA detection tools (review). J Nanobiotechnol 20(1). https://doi.org/10.1186/s12951-022-01246-7

Duan K et al (2021) Large chromosomal segment deletions by CRISPR/LBCPF1-mediated multiplex gene editing in soybean. J Integr Plant Biol 63(9):1620–1631. https://doi.org/10.1111/jipb.13158

Elliott K et al (2024) CRISPR/Cas9-generated mutations in a sugar transporter gene reduce cassava susceptibility to bacterial blight. Plant Physiol 195(4):2566–2578. https://doi.org/10.1093/plphys/kiae243

Fahad S et al (2017) Crop production under drought and heat stress: plant responses and management options. Front Plant Sci 8. https://doi.org/10.3389/fpls.2017.01147

Fan S et al (2021) CRISPR/Cas9-targeted mutagenesis of the bnaa03.bp gene confers semi-dwarf and compact architecture to rapeseed (brassica napus L.). Plant Biotechnol J 19(12):2383–2385. https://doi.org/10.1111/pbi.13703

Fanzo J et al (2022) Sustainable food systems and nutrition in the 21st century: a report from the 22nd annual Harvard nutrition obesity symposium. Am J Clin Nutr 115(1):18–33. https://doi.org/10.1093/ajcn/nqab315

Fanzo J, Schneider KR, Wood S (2024) 'Food insecurity and malnutrition', climate change and public. Health:153–179. https://doi.org/10.1093/oso/9780197683293.003.0008

Galanakis CM (2024) The future of food. Foods 13(4):506. https://doi.org/10.3390/foods13040506

Gao Y et al (2021) A tomato NAC transcription factor, SLNAM1, positively regulates ethylene biosynthesis and the onset of tomato fruit ripening. Plant J 108(5):1317–1331. https://doi.org/10.1111/tpj.15512

Gomez-Zavaglia A, Mejuto JC, Simal-Gandara J (2020) Mitigation of emerging implications of climate change on food production systems. Food Res Int 134:109256. https://doi.org/10.1016/j.foodres.2020.109256

González MN et al (2020) Reduced enzymatic browning in potato tubers by specific editing of a polyphenol oxidase gene via ribonucleoprotein complexes delivery of the CRISPR/cas9 system. Front Plant Sci 10. https://doi.org/10.3389/fpls.2019.01649

Groenen PM et al (1993) Nature of DNA polymorphism in the direct repeat cluster of mycobacterium tuberculosis; application for strain differentiation by a novel typing method. Mol Microbiol 10(5):1057–1065. https://doi.org/10.1111/j.1365-2958.1993.tb00976.x

Gujral N (2012) Celiac disease: prevalence, diagnosis, pathogenesis and treatment. World J Gastroenterol 18(42):6036. https://doi.org/10.3748/wjg.v18.i42.6036

Hao J et al (2023) CRISPR/Cas9-mediated mutagenesis of carotenoid cleavage dioxygenase (ccd) genes in sorghum alters strigolactone biosynthesis and plant biotic interactions. Phytobiom J [Preprint]. https://doi.org/10.1094/pbiomes-08-22-0053-r

Hillary VE, Ceasar SA (2022) A review on the mechanism and applications of CRISPR/Cas9/CAS12/CAS13/CAS14 proteins utilized for genome engineering. Mol Biotechnol 65(3):311–325. https://doi.org/10.1007/s12033-022-00567-0

Hoe N, Nakashima K, Grigsby D, Pan X, Dou SJ, Naidich S, Garcia M, Kahn E, Bergmire-Sweat D, Musser JM (1999) Rapid molecular genetic subtyping of serotype M1 Group A streptococcus strains. Emerg Infect Dis 5(2):254–263. https://doi.org/10.3201/eid0502.990210

Hoffie RE et al (2021) Targeted knockout of eukaryotic translation initiation factor 4E confers bymovirus resistance in winter barley. Front Genome Editing 3. https://doi.org/10.3389/fgeed.2021.784233

Hu C et al (2021) CRISPR/Cas9-mediated genome editing of maaco1 (aminocyclopropane-1-carboxylate oxidase 1) promotes the shelf life of banana fruit. Plant Biotechnol J 19(4):654–656. https://doi.org/10.1111/pbi.13534

Huang H et al (2020) Modifications of fatty acid profile through targeted mutation at BNAFAD2 gene with CRISPR/Cas9-mediated gene editing in brassica napus. Theor Appl Genet 133(8):2401–2411. https://doi.org/10.1007/s00122-020-03607-y

Huang Z et al (2022a) CRISPR-cas13: a new technology for the rapid detection of pathogenic microorganisms. Front Microbiol 13. https://doi.org/10.3389/fmicb.2022.1011399

Huang Z et al (2022b) Fluorescence signal-readout of CRISPR/Cas biosensors for nucleic acid detection. Biosensors 12(10):779. https://doi.org/10.3390/bios12100779

Ishino Y et al (1987) Nucleotide sequence of the IAP gene, responsible for alkaline phosphatase isozyme conversion in escherichia coli, and identification of the gene product. J Bacteriol 169(12):5429–5433. https://doi.org/10.1128/jb.169.12.5429-5433.1987

Janni M et al (2024) Plant responses to climate change, how global warming may impact on food security: a critical review. Front Plant Sci 14. https://doi.org/10.3389/fpls.2023.1297569

Jinek M et al (2012) A programmable dual-RNA–guided DNA endonuclease in adaptive bacterial immunity. Science 337(6096):816–821. https://doi.org/10.1126/science.1225829

Juma BS et al (2022) Targeted mutagenesis of the CYP79D1 gene via CRISPR/Cas9-mediated genome editing results in lower levels of cyanide in cassava. Front Plant Sci 13. https://doi.org/10.3389/fpls.2022.1009860

Kabir M et al (2023) Climate change due to increasing concentration of carbon dioxide and its impacts on environment in 21st century; a mini review. J King Saud Univ—Sci 35(5):102693. https://doi.org/10.1016/j.jksus.2023.102693

Kafle S (2023) CRISPR/Cas9: a new paradigm for crop improvement revolutionizing agriculture. J Agric Food Res 11:100484. https://doi.org/10.1016/j.jafr.2022.100484

Karlsson M et al (2024) CRISPR/Cas9 genome editing of potato stDMR6-1 results in plants less affected by different stress conditions. Hortic Res 11(7). https://doi.org/10.1093/hr/uhae130

Kato K et al (2022) Structure and engineering of the type III-E CRISPR-Cas7-11 effector complex. Cell 185(13). https://doi.org/10.1016/j.cell.2022.05.003

Kaur N et al (2020) CRISPR/Cas9 directed editing of lycopene epsilon-cyclase modulates metabolic flux for β-carotene biosynthesis in banana fruit. Metab Eng 59:76–86. https://doi.org/10.1016/j.ymben.2020.01.008

Kavuri NR et al (2022) Applications of CRISPR/cas13-based RNA editing in plants. Cells 11(17):2665. https://doi.org/10.3390/cells11172665

Khaipho-Burch M et al (2023) Genetic modification can improve crop yields — but stop overselling it. Nature 621(7979):470–473. https://doi.org/10.1038/d41586-023-02895-w

Khan MH et al (2020) Targeted mutagenesis of eod3 gene in brassica napus L. regulates seed production. J Cell Physiol 236(3):1996–2007. https://doi.org/10.1002/jcp.29986

Kim K-H, Lee B-M (2023) Effects of climate change and drought tolerance on maize growth. Plants 12(20):3548. https://doi.org/10.3390/plants12203548

Koonin EV et al (2017) Diversity, classification and evolution of CRISPR-Cas systems. Curr Opin Microbiol 37:67–78. https://doi.org/10.1016/j.mib.2017.05.008

Koonin EV et al (2023) Discovery of diverse CRISPR-cas systems and expansion of the genome engineering toolbox. Biochemistry 62(24):3465–3487. https://doi.org/10.1021/acs.biochem.3c00159

Lebedev VG, Popova AA, Shestibratov KA (2021) Genetic engineering and genome editing for improving nitrogen use efficiency in plants. Cells 10(12):3303. https://doi.org/10.3390/cells10123303

Lee T-M et al (2023) Clustered regularly interspaced short palindromic repeats (CRISPR) technology and genetic engineering strategies for microalgae towards carbon neutrality: a critical review. Bioresour Technol 368:128350. https://doi.org/10.1016/j.biortech.2022.128350

Li M et al (2016) Reassessment of the four yield-related genes GN1A, DEP1, GS3, and IPA1 in rice using a CRISPR/cas9 system. Front Plant Sci 7. https://doi.org/10.3389/fpls.2016.00377

Li B et al (2020) The application of temperature sensitivity CRISPR/LBCPF1 (LBCAS12A) mediated genome editing in allotetraploid cotton (G.& hirsutum) and creation of nontransgenic, gossypol-free cotton. Plant Biotechnol J 19(2):221–223. https://doi.org/10.1111/pbi.13470

Li S, Chen K, Grierson D (2021) Molecular and hormonal mechanisms regulating fleshy fruit ripening. Cells 10(5):1136. https://doi.org/10.3390/cells10051136

Li X et al (2022) Shoot-to-root translocated GMNN1/ft2a triggers nodulation and regulates soybean nitrogen nutrition. PLoS Biol 20(8). https://doi.org/10.1371/journal.pbio.3001739

Li J et al (2024) Application of multiple sgrnas boosts efficiency of CRISPR/Cas9-mediated gene targeting in Arabidopsis. BMC Biol 22(1). https://doi.org/10.1186/s12915-024-01810-7

Lieber M et al (2020) A systematic review and meta-analysis assessing the impact of droughts, flooding, and climate variability on malnutrition. Glob Public Health 17(1):68–82. https://doi.org/10.1080/17441692.2020.1860247

Liu L et al (2020) CRISPR/Cas9 targeted mutagenesis of SLLBD40, a lateral organ boundaries domain transcription factor, enhances drought tolerance in tomato. Plant Sci 301:110683. https://doi.org/10.1016/j.plantsci.2020.110683

Loladze I (2014) 'Hidden shift of the ionome of plants exposed to elevated CO2 depletes minerals at the base of human nutrition', eLife, 3. https://doi.org/10.7554/elife.02245

Lu K et al (2018) Blocking amino acid transporter OSaap3 improves grain yield by promoting outgrowth buds and increasing tiller number in Rice. Plant Biotechnol J 16(10):1710–1722. https://doi.org/10.1111/pbi.12907

Macovei A et al (2018) Novel alleles of rice eif4g generated by CRISPR/Cas9-targeted mutagenesis confer resistance to rice tungro spherical virus. Plant Biotechnol J 16(11):1918–1927. https://doi.org/10.1111/pbi.12927

Makarova KS et al (2011) Evolution and classification of the CRISPR–Cas systems. Nat Rev Microbiol 9(6):467–477. https://doi.org/10.1038/nrmicro2577

Mali P et al (2013) RNA-guided human genome engineering via cas9. Science 339(6121):823–826. https://doi.org/10.1126/science.1232033

Matinvafa MA et al (2023) 'CRISPR-CAS technology secures sustainability through its applications: a review in green biotechnology', 3. Biotech 13(11). https://doi.org/10.1007/s13205-023-03786-7

Mcgrath JM, Lobell DB (2012) Reduction of transpiration and altered nutrient allocation contribute to nutrient decline of crops grown in elevated co2 concentrations. Plant Cell Environ 36(3):697–705. https://doi.org/10.1111/pce.12007

Mengstie MA, Wondimu BZ (2021) Mechanism and applications of CRISPR/cas-9-mediated genome editing. Biologics: Target Therapy 15:353–361. https://doi.org/10.2147/btt.s326422

Michael Alurame Eruaga (2024) Policy strategies for managing food safety risks associated with climate change and agriculture. Int J Scholar Res Rev 4(1):021–032. https://doi.org/10.56781/ijsrr.2024.4.1.0026

Miller J, McLachlan AD, Klug A (1985) Repetitive zinc-binding domains in the protein transcription factor IIIA from xenopus oocytes. EMBO J 4(6):1609–1614. https://doi.org/10.1002/j.1460-2075.1985.tb03825.x

Mojica FJ et al (1993) Transcription at different salinities of haloferax mediterranei sequences adjacent to partially modified pstI sites. Mol Microbiol 9(3):613–621. https://doi.org/10.1111/j.1365-2958.1993.tb01721.x

Mojica FJM et al (2005) Intervening sequences of regularly spaced prokaryotic repeats derive from foreign genetic elements. J Mol Evol 60(2):174–182. https://doi.org/10.1007/s00239-004-0046-3

Munaweera TI et al (2022) Modern plant biotechnology as a strategy in addressing climate change and attaining food security. Agric Food Security 11(1). https://doi.org/10.1186/s40066-022-00369-2

Ndudzo A et al (2024) CRISPR-Cas9 genome editing in crop breeding for climate change resilience: implications for smallholder farmers in Africa. J Agric Food Res 16:101132. https://doi.org/10.1016/j.jafr.2024.101132

Nguyen CX, Do PT (2024) 'CRISPR/Cas9-based technology for functional genomics and crop improvement in soybean', CRISPR and plant functional. Genomics:348–356. https://doi.org/10.1201/9781003387060-21

Niu Q et al (2020) Expanding the scope of CRISPR/Cas9-mediated genome editing in plants using an xcas9 and cas9-ng hybrid. J Integr Plant Biol 62(4):398–402. https://doi.org/10.1111/jipb.12886

Nonaka S et al (2017) Efficient increase of ɣ-aminobutyric acid (GABA) content in tomato fruits by targeted mutagenesis. Sci Rep 7(1). https://doi.org/10.1038/s41598-017-06400-y

Ntui VO et al (2023) Targeted knockout of early nodulin-like 3 (musaenodl3) gene in banana reveals its function in resistance to Xanthomonas wilt disease. Plant Biotechnol J 22(5):1101–1112. https://doi.org/10.1111/pbi.14248

Nunez S, Arets E, Alkemade R, Verwer C, Leemans R (2019) Assessing the impacts of climate change on biodiversity: is below 2 °C enough? Clim Change 154(3–4):351–365. https://doi.org/10.1007/s10584-019-02420-x

Nuñez JK et al (2014) CAS1–CAS2 complex formation mediates spacer acquisition during CRISPR–CAS adaptive immunity. Nat Struct Mol Biol 21(6):528–534. https://doi.org/10.1038/nsmb.2820

Oz MT et al (2021) CRISPR/Cas9-mediated multi-allelic gene targeting in sugarcane confers herbicide tolerance. Front Genome Editing 3. https://doi.org/10.3389/fgeed.2021.673566
Paul B, Montoya G (2020) CRISPR-Cas12a: functional overview and applications. Biom J 43(1):8–17. https://doi.org/10.1016/j.bj.2019.10.005
Pausch P et al (2020) CRISPR-CasΦ from huge phages is a hypercompact genome editor. Science 369(6501):333–337. https://doi.org/10.1126/science.abb1400
Peng A et al (2017) Engineering canker-resistant plants through crispr/cas9-targeted editing of the susceptibility gene cslob1 promoter in citrus. Plant Biotechnol J 15(12):1509–1519. https://doi.org/10.1111/pbi.12733
Phang T, Shao G, Lam H (2008) Salt tolerance in soybean. J Integr Plant Biol 50(10):1196–1212. https://doi.org/10.1111/j.1744-7909.2008.00760.x
Priya AK, Muruganandam M, Rajamanickam S, Sivarethinamohan S, Gaddam MKR, Velusamy P, et al (2023) Impact of climate change and anthropogenic activities on aquatic ecosystem – A review. Environ Res 238:117233. https://doi.org/10.1016/j.envres.2023.117233
Ricroch A (2019) Global developments of genome editing in agriculture. Transgenic Res 28(S2):45–52. https://doi.org/10.1007/s11248-019-00133-6
Rizzo DM et al (2021) Plant health and its effects on food safety and security in a one health framework: four case studies. One Health Outlook 3(1). https://doi.org/10.1186/s42522-021-00038-7
Roviqowati F (2024) Introduction of CRISPR/cas9 with the target genes to improve agronomic traits and leaf blight resistance in Rice. SABRAO 56(1):29–44. https://doi.org/10.54910/sabrao2024.56.1.3
Russell SJ, Garcia AK, Kaçar B (2024) A CRISPR interference system for engineering biological nitrogen fixation. mSystems 9(3). https://doi.org/10.1128/msystems.00155-24
Sánchez-León S et al (2017) Low-gluten, nontransgenic wheat engineered with CRISPR/Cas9. Plant Biotechnol J 16(4):902–910. https://doi.org/10.1111/pbi.12837
Savage DF (2019) CAS14: big advances from small CRISPR proteins. Biochemistry 58(8):1024–1025. https://doi.org/10.1021/acs.biochem.9b00035
Scheben A et al (2017) Towards crispr/cas crops—bringing together genomics and genome editing. New Phytol 216(3):682–698. https://doi.org/10.1111/nph.14702
Semba RD et al (2022) The potential impact of climate change on the micronutrient-rich food supply. Adv Nutr 13(1):80–100. https://doi.org/10.1093/advances/nmab104
Shi J et al (2016) argos8 variants generated by crispr-cas9 improve maize grain yield under field drought stress conditions. Plant Biotechnol J 15(2):207–216. https://doi.org/10.1111/pbi.12603
Shu H et al (2020) The application of CRISPR/Cas9 in hairy roots to explore the functions of AHNFR1 and AHNFR5 genes during peanut nodulation. BMC Plant Biol 20(1). https://doi.org/10.1186/s12870-020-02614-x
Smyth SJ (2022) Contributions of genome editing technologies towards improved nutrition, environmental sustainability and poverty reduction. Front Genome Editing 4. https://doi.org/10.3389/fgeed.2022.863193
Song J et al (2022) Mutation of GMIPK1 gene using CRISPR/Cas9 reduced phytic acid content in soybean seeds. Int J Mol Sci 23(18):10583. https://doi.org/10.3390/ijms231810583
Sun Y et al (2016) Engineering herbicide-resistant rice plants through CRISPR/Cas9-mediated homologous recombination of acetolactate synthase. Mol Plant 9(4):628–631. https://doi.org/10.1016/j.molp.2016.01.001
Taheri S et al (2022) Drought tolerance improvement in solanum lycopersicum: an insight into "OMICS" approaches and genome editing. Biotech 12(3). https://doi.org/10.1007/s13205-022-03132-3
Takeda SN et al (2021) Structure of the miniature type V-F CRISPR-cas effector enzyme. Mol Cell 81(3). https://doi.org/10.1016/j.molcel.2020.11.035
Talakayala A, Ankanagari S, Mallikarjuna G (2024) Augmenting the CRISPR/cas9 system to edit Mungbean yellow mosaic viral genome for enhanced disease resistance in nicotiana benthamiana. https://doi.org/10.2139/ssrn.4683744

Tariq A et al (2023) Putting CRISPR-CAS system in action: a golden window for efficient and precise genome editing for crop improvement. GM Crops Food 14(1):1–27. https://doi.org/10.1080/21645698.2023.2219111

Tashkandi M et al (2018) Engineering resistance against tomato yellow leaf curl virus via the CRISPR/cas9 system in tomato. Plant Signal Behav 13(10). https://doi.org/10.1080/15592324.2018.1525996

Tian S et al (2018) Engineering herbicide-resistant watermelon variety through CRISPR/Cas9-mediated base-editing. Plant Cell Rep 37(9):1353–1356. https://doi.org/10.1007/s00299-018-2299-0

Ueta R et al (2017) Rapid breeding of parthenocarpic tomato plants using CRISPR/Cas9. Sci Rep 7(1). https://doi.org/10.1038/s41598-017-00501-4

Ullah N et al (2023) Comparative analysis and phylogenetic insights of CAS14-homology proteins in bacteria and archaea. Genes 14(10):1911. https://doi.org/10.3390/genes14101911

Usman B et al (2020) Programmed editing of rice (Oryza sativa L.) OSSPL16 gene using CRISPR/cas9 improves grain yield by modulating the expression of pyruvate enzymes and cell cycle proteins. Int J Mol Sci 22(1):249. https://doi.org/10.3390/ijms22010249

Varzakas T, Smaoui S (2024) Global food security and sustainability issues: the road to 2030 from nutrition and sustainable healthy diets to food systems change. Foods 13(2):306. https://doi.org/10.3390/foods13020306

Veillet F et al (2019) Transgene-free genome editing in tomato and potato plants using agrobacterium-mediated delivery of a CRISPR/cas9 cytidine base editor. Int J Mol Sci 20(2):402. https://doi.org/10.3390/ijms20020402

Voß L et al (2022) 'Enzymes and cellular interplay required for flux of fixed nitrogen to Ureides in bean nodules', nature. Communications 13(1). https://doi.org/10.1038/s41467-022-33005-5

Wai AH et al (2020) Molecular genetic approaches for enhancing stress tolerance and fruit quality of tomato. Plant Biotechnol Report 14(5):515–537. https://doi.org/10.1007/s11816-020-00638-1

Wang H et al (2019a) CRISPR/Cas9-based mutagenesis of starch biosynthetic genes in sweet potato (Ipomoea batatas) for the improvement of starch quality. Int J Mol Sci 20(19):4702. https://doi.org/10.3390/ijms20194702

Wang L et al (2019b) CRISPR/Cas9-mediated editing of CSWRKY22 reduces susceptibility to xanthomonas citri subsp. citri in Wanjincheng orange (citrus sinensis (L.) Osbeck). Plant Biotechnol Reports 13(5):501–510. https://doi.org/10.1007/s11816-019-00556-x

Wang Q et al (2020) The application of a heat-inducible CRISPR/CAS12B (C2C1) genome editing system in tetraploid cotton (G.& hirsutum) plants. Plant Biotechnol J 18(12):2436–2443. https://doi.org/10.1111/pbi.13417

Wang T et al (2021) Mutation of GMAITR genes by CRISPR/cas9 genome editing results in enhanced salinity stress tolerance in soybean. Front Plant Sci 12. https://doi.org/10.3389/fpls.2021.779598

Wang Y et al. (2022) Engineering bacterial blight-resistant plants through CRISPR/Cas9-targeted editing of themesweet10apromoter in cassava [preprint]. https://doi.org/10.1101/2022.03.02.482644

Wei J, Li Y (2023) CRISPR-based gene editing technology and its application in microbial engineering. Eng Microbiol 3(4):100101. https://doi.org/10.1016/j.engmic.2023.100101

Westermann L et al (2020) Nobel prize 2020 in chemistry honors CRISPR: a tool for rewriting the code of life. Pflugers Arch—Eur J Physiol 473(1):1–2. https://doi.org/10.1007/s00424-020-02497-9

Wolabu, Tezera W, et al. Multiplex CRISPR/Cas9-mediated mutagenesis of Alfalfa FLOWERING LOCUS Ta1 (MsFTa1) leads to delayed flowering time with improved forage biomass yield and quality. Plant Biotechnol J 21(7), 25 Mar. 2023, pp. 1383–1392, https://doi.org/10.1111/pbi.14042. Accessed 10 Dec 2023

Wu J et al (2020) Engineering herbicide-resistant oilseed rape by CRISPR/Cas9-mediated cytosine base-editing. Plant Biotechnol J 18(9):1857–1859. https://doi.org/10.1111/pbi.13368

Xu Y, Li Z (2020) CRISPR-Cas systems: overview, innovations and applications in human disease research and gene therapy. Comput Struct Biotechnol J 18:2401–2415. https://doi.org/10.1016/j.csbj.2020.08.031

Xu W et al (2019a) Multiplex nucleotide editing by high-fidelity cas9 variants with improved efficiency in Rice. BMC Plant Biol 19(1). https://doi.org/10.1186/s12870-019-2131-1

Xu Z et al (2019b) Engineering broad-spectrum bacterial blight resistance by simultaneously disrupting variable tale-binding elements of multiple susceptibility genes in Rice. Mol Plant 12(11):1434–1446. https://doi.org/10.1016/j.molp.2019.08.006

Yan D et al (2022) Genetic modification of flavone biosynthesis in rice enhances biofilm formation of soil diazotrophic bacteria and biological nitrogen fixation. Plant Biotechnol J 20(11):2135–2148. https://doi.org/10.1111/pbi.13894

Yan F et al (2023) CRISPR/FNCAS12A-mediated efficient multiplex and iterative genome editing in bacterial plant pathogens without donor DNA templates. PLoS Pathog 19(1). https://doi.org/10.1371/journal.ppat.1010961

Yang Y et al (2023) Research progress on nucleic acid detection and genome editing of CRISPR/Cas12 system. Mol Biol Rep 50(4):3723–3738. https://doi.org/10.1007/s11033-023-08240-8

Yu W et al (2019) Knockout of SLMAPK3 enhances tolerance to heat stress involving Ros homeostasis in tomato plants. BMC Plant Biol 19(1). https://doi.org/10.1186/s12870-019-1939-z

Yuan M et al (2019) Mutagenesis of FAD2 genes in peanut with CRISPR/Cas9 based gene editing. BMC Biotechnol 19(1). https://doi.org/10.1186/s12896-019-0516-8

Zakrzewska M, Burmistrz M (2023) Mechanisms regulating the CRISPR-cas systems. Front Microbiol 14. https://doi.org/10.3389/fmicb.2023.1060337

Zaman QU et al (2021) Characterization of shatterproof homoeologs and CRISPR-Cas9-mediated genome editing enhances pod-shattering resistance in brassica napus L. CRISPR J 4(3):360–370. https://doi.org/10.1089/crispr.2020.0129

Zhang K et al (2019a) Effective editing for lysophosphatidic acid acyltransferase 2/5 in allotetraploid rapeseed (brassica napus L.) using CRISPR-Cas9 system. Biotechnol Biofuels 12(1). https://doi.org/10.1186/s13068-019-1567-8

Zhang P et al (2019b) Multiplex CRISPR/Cas9-mediated metabolic engineering increases soya bean isoflavone content and resistance to soya bean mosaic virus. Plant Biotechnol J 18(6):1384–1395. https://doi.org/10.1111/pbi.13302

Zhang Z et al (2022) Gmpin1-mediated auxin asymmetry regulates leaf petiole angle and plant architecture in soybean. J Integr Plant Biol 64(7):1325–1338. https://doi.org/10.1111/jipb.13269

Zheng M et al (2020) Knockout of two BNAmax1 homologs by crispr/cas9-targeted mutagenesis improves plant architecture and increases yield in rapeseed (brassica napus L.). Plant Biotechnol J 18(3):644–654. https://doi.org/10.1111/pbi.13228

Zheng S et al (2021) Improving the rice photosynthetic efficiency and yield by editing OSHXK1 via CRISPR/Cas9 system. Int J Mol Sci 22(17):9554. https://doi.org/10.3390/ijms22179554

Zhong X et al (2022) CRISPR/Cas9 mediated gene-editing of GMHDZ4 transcription factor enhances drought tolerance in soybean (glycine max [L.] merr.). Front Plant Sci 13. https://doi.org/10.3389/fpls.2022.988505

Zhou J et al (2018) Multiplex QTL editing of grain-related genes improves yield in elite rice varieties. Plant Cell Rep 38(4):475–485. https://doi.org/10.1007/s00299-018-2340-3

Zhou B et al (2023) CRISPR/CAS14 provides a promising platform in facile and versatile aptasensing with improved sensitivity. Talanta 254:124120. https://doi.org/10.1016/j.talanta.2022.124120

Beyond Gene Editing in the Plant Breeding Toolbox

20

Jagannath Swain, Kutubuddin A. Molla, K. Jagadis Gupta, and K. C. Bansal

Abstract

In recent years, gene editing technologies, such as CRISPR-Cas9, have revolutionized plant breeding by enabling precise modifications in plant genomes, enhancing traits like yield, nutritional quality, disease resistance, and stress tolerance. However, the future of crop improvement extends beyond the conventional CRISPR system, incorporating a diverse array of innovative strategies such as next-generation genome editing tools based on transposon-associated RNA-guided nucleases (TnpB), CRISPR-mediated regulation of gene expression via CRISPR-activator (CRISPRa) and CRISPR-interference (CRISPRi), and CRISPR-mediated epigenome editing. By fully integrating these methods, it is possible to enhance traits associated with mitigating greenhouse gas emissions, improving nutrient-use efficiency, enabling biological nitrogen fixation, and increasing carbon dioxide capturing capacity in crops. These advancements contribute to the development of climate-resilient crops, addressing challenges posed by climate change to meet growing food demands and ensuring global food security.

J. Swain · K. J. Gupta
BRIC-National Institute of Plant Genome Research, New Delhi, India

K. A. Molla
ICAR-Central Rice Research Institute, Cuttack, Odisha, India

K. C. Bansal (✉)
Department of Biotechnology, Guru Jambheshwar University of Science and Technology, Hisar, India

Centre for Crop and Food Innovation, Murdoch University, Perth, Australia

P. M. Priyadarshan, R. Ortiz (eds.), *Plant Breeding 2050*,
https://doi.org/10.1007/978-981-95-0583-8_20

Keywords

Genome editing · CRISPR/Cas · Climate resilience · Crop improvement · Food security

20.1 Introduction

According to recent estimates, the global population is projected to reach 9.6 billion by 2050, with food demand expected to increase by 100–110%, posing a significant challenge to global food security (Samant et al. 2024). Climate-driven changes, such as rising global temperatures, extreme weather events, flash flooding, and regional conflicts, further threaten global crop production and supply chains (Rajkhowa and Sarma 2021). Additionally, the shrinking availability of arable land and rapid urbanization contribute to declining agricultural productivity (Xiong et al. 2022). Therefore, there is an urgent need to develop robust and efficient breeding methods to mitigate food security threats through the creation of next-generation smart crops—genetically engineered to be more resilient to stress while achieving higher yield potential (Wang et al. 2024).

Plant breeding has long been an effective method for crop improvement, offering the potential to meet increasing food demands. However, traditional breeding methods are labor-intensive and time-consuming. Since the early 1900s, both farmers and researchers have relied on conventional breeding to develop improved crop varieties (Tester and Langridge 2010). The process of developing elite and smart crops through conventional breeding is slow, costly, and resource-intensive (Swarup et al. 2021). Traditional breeding primarily relies on identifying natural genetic variations within large populations, requiring the screening of thousands of individuals to identify desirable traits such as higher yield, stress tolerance, and disease resistance. Once a desirable trait is identified, extensive crossing is required to introgress it into an elite genetic background. This process often takes multiple generations—sometimes decades—to achieve the desired trait combination in a stable and agronomically viable variety (Xiong et al. 2022; Xu et al. 2017). Additionally, conventional breeding is limited by genetic diversity within a species' gene pool and the unpredictability of trait inheritance due to complex genetic interactions (Adhikari et al. 2024). These challenges underscore the need for advanced breeding technologies, such as marker-assisted selection (MAS), genomic-assisted breeding, and genetic engineering, to accelerate the development of improved varieties with enhanced precision and efficiency.

Over the past three decades, molecular breeding techniques, such as molecular marker-assisted selection (MAS) and transgenic approaches, have been widely applied for developing improved crops. However, both approaches have notable limitations (Xu et al. 2012b). For instance, MAS is time-consuming and requires extensive genotyping and phenotyping of large populations, as well as the development of high-density genetic maps. Additionally, the effectiveness of MAS depends on the availability of molecular markers tightly linked to target traits, which may not always be readily accessible. On the other hand, the transgenic approach, which involves the introduction of foreign genes to confer desirable traits, necessitates comprehensive biosafety regulatory evaluations before the environmental release and commercialization of GM crops (Cobb et al. 2019; Hundleby et al. 2018).

Recently, genome editing tools have emerged as breakthrough technologies, revolutionizing plant breeding by enabling precise and predictable genetic modifications directly in elite crop backgrounds (Adeyinka et al. 2023; Ndudzo et al. 2024; Mankotia et al. 2022). Genome editing techniques such as transcription activator-like effector nucleases (TALENs) and zinc-finger nucleases (ZFNs), as well as chromosome engineering, have been widely explored to address global food security concerns (Elsharawy and Refat 2023). However, these tools have limitations, including low efficiency, labor-intensive processes, and complex design need (Kumar et al. 2024). These challenges underscore the need for further refinement of genome editing technologies to enhance their precision, efficiency, and applicability in modern plant breeding.

To overcome the limitations of both conventional and molecular breeding approaches, CRISPR-Cas-based genome editing has emerged as a powerful technology to accelerate crop improvement. The CRISPR method has advanced rapidly, with significant progress made in major crops such as rice, wheat, and tomato, demonstrating immense agricultural potential. Although CRISPR-Cas technology was introduced only about a decade ago, its impact on both basic and applied plant research, including molecular breeding, has been unprecedented (Zhu et al. 2020). The application of the CRISPR-Cas system in plant breeding has facilitated the rapid development of improved crop varieties, allowing researchers and agricultural companies to accelerate trait improvement through targeted genetic modifications (Ndudzo et al. 2024).

CRISPR-Cas is a game-changing technology poised to revolutionize molecular plant breeding, as it enables simultaneous modification of multiple traits through multiplex genome editing (Adeyinka et al. 2023; Ndudzo et al. 2024). Compared with conventional breeding, which may take 10–20 years to generate new allelic combinations or eliminate undesirable traits, the CRISPR-Cas system allows for the precise deletion of negative genetic elements or the introduction of gain-of-function mutations, leading to the rapid development of improved crop varieties (Li et al. 2018). Moreover, CRISPR-based genome editing offers the advantage of producing transgene-free plants, thereby alleviating biosafety concerns associated with transgenic crops in various countries (Zhou et al. 2023). Thus, CRISPR-Cas technology represents the most effective genome editing tool available today for plant breeding, with the potential to develop climate-resilient, disease-resistant, and nutritionally enhanced high yielding crops.

In this chapter, we discuss the potential applications of CRISPR-Cas genome editing in plant breeding for sustainable agriculture, highlighting its role in developing climate-resilient crops to ensure global food security.

20.2 Beyond Conventional Genome Editing

20.2.1 CRISPR/Cas System as an Advanced Tool for Plant Breeding

The CRISPR/Cas system was originally identified as an RNA-mediated adaptive immune system in bacteria and archaea, where it provides defense against foreign genetic elements such as phages and viruses by cleaving their DNA (Jiang and Doudna 2017). The CRISPR/Cas system was first reported in archaea by Mojica et al. (1993), and Jinek et al. (2012) later demonstrated its application for genome editing.

CRISPR/Cas systems are categorized into two main classes, each further subdivided into three subtypes. The key distinction between the two classes lies in the composition of the interference complex. Class 1 CRISPR-Cas systems (types I, III, and IV) are characterized by multi-Cas protein complexes forming the interference complex, whereas Class 2 systems (types II, V, and VI) use a single Cas protein along with CRISPR RNAs (Chen et al. 2019; Jiang and Doudna 2017; Zhu et al. 2020).

The Class 2 CRISPR/Cas system is the most efficient and is rapidly becoming the preferred tool for plant biologists and molecular breeders. Cas9, Cas12 and their homologs with additional functionalities have been identified from diverse bacteria and employed in genome editing applications (Belhaj et al. 2015; Chen et al. 2019; Patra et al. 2024; Molla and Yang 2020).

The type II CRISPR/Cas9 system, originally derived from *Streptococcus pyogenes*, was the first to be used for genetic manipulation in vitro and in vivo in both animals and plants (Bortesi et al. 2016). These systems require two components for genome editing: (1) a DNA endonuclease, CRISPR-associated protein 9 (Cas9), and (2) a single guide RNA (sgRNA), which is an artificial fusion of a noncoding CRISPR RNA (crRNA) and a trans-activating crRNA (tracrRNA) (Chen et al. 2019; Jiang and Doudna 2017). The Cas9/sgRNA complex directs sequence-specific cleavage of the target DNA adjacent to the protospacer-adjacent motif (PAM) sequence, 5′-NGG-3′. Cas9 generates double-stranded breaks in the target DNA, which are repaired by the native repair system via nonhomologous end joining (NHEJ), often leading to insertion or deletion mutations (Belhaj et al. 2015; Langner et al. 2018; Molla and Yang 2020).

Cas9 proteins possess a bilobed structure comprising a recognition (REC) lobe and a nuclease (NUC) lobe. The NUC lobe contains two signature nuclease domains: the HNH nuclease domain, which cleaves the DNA strand complementary to the sgRNA, and the RuvC-like nuclease domain, which cleaves the noncomplementary DNA strand, resulting in double-strand breaks (Nishimasu et al. 2014; Molla and Yang 2020). However, SpCas9 exhibits PAM specificity for NGG sequences, limiting its efficiency in AT-rich genomic regions. Consequently, several engineered Cas9 variants have been developed to recognize different PAM sequences, including EQR-Cas9 (NGAG PAM), SaKKH-Cas9 (NNNRRT PAM), xCas9 (NG, GAA, GTA PAM), VQR-Cas9 (NGA PAM), VRER-Cas9 (NGCG PAM), and SpCas9-NG

(NG PAM) (Belhaj et al. 2015; Xu et al. 2017; Chen et al. 2019; Molla and Yang 2020).

Beyond the type II CRISPR/Cas9 system, another Class 2, type V nuclease, Cas12a (formerly known as Cpf1, CRISPR from *Prevotella* and *Francisella* 1), has shown immense potential in genome editing (Bandyopadhyay et al. 2020). It offers advantages over SpCas9, such as requiring only the RuvC nuclease domain and not needing a tracrRNA sequence for genome editing. Cas12a recognizes an upstream T-rich (5′-TTTV-3′) PAM sequence. To enhance its versatility, several Cas12a variants have been identified from different bacteria and engineered in laboratories, including *Lachnospiraceae bacterium* (LbCas12a), *Acidaminococcus* sp. (AsCas12a), and *Francisella novicida* (FnCas12a), which exhibit relaxed PAM recognition (Gao et al. 2017). All Cas12a variants create sticky-end DNA breaks, making them favorable for homology-directed repair (HDR) (Zetsche et al. 2015).

Cas12b, another type V endonuclease, shares similarities with Cas9, as it requires both crRNA and tracrRNA for cleavage and processing (Strecker et al. 2019). Cas12b demonstrates higher specificity and has been successfully applied to induce targeted mutations in rice and human genomes (Ming et al. 2020; Wu et al. 2020). Furthermore, the recently discovered CRISPR/Cas12j2 and CRISPR/Cms1 systems represent new additions to the type V CRISPR repertoire (Liu et al. 2022).

CRISPR-Cas systems have also been harnessed for RNA editing. CRISPR/Cas13a (formerly C2c2), a type VI nuclease, has been shown to specifically cleave RNA. Cas13 and Cas12a have also been utilized in nucleic acid diagnostics (Gootenberg et al. 2018; Freije et al. 2019; Karmakar et al. 2022). Recent efforts are underway to employ Cas13 for virus interference in plants (Mahas et al. 2019).

20.2.2 Next-Generation Genome Editing Tools with TnpB

Cas9 and Cas12a are large proteins (~1200–1400 amino acids), which make their delivery into target cells challenging, especially via viral vectors. Transposon-associated TnpB has been identified as an RNA-guided nuclease and is considered an evolutionary precursor of Cas12a nucleases. Notably, TnpBs from diverse prokaryotic species are significantly smaller (~400 amino acids) than Cas9 and Cas12a. Recently, TnpB has been repurposed to develop efficient genome-editing tools for both monocot and dicot plants (Karmakar et al. 2024).

Similar to the protospacer adjacent motif (PAM) of Cas proteins, TnpB recognizes a transposon-associated motif (TAM). The IsDra2 TnpB specifically recognizes the 5′TTGAT3′ TAM, located upstream of the 20-nt protospacer (guide). TnpB cleaves DNA at the TAM-distal ends of the protospacer, generating 5′ overhangs (Karvelis et al. 2021). Since TnpB-mediated cleavage produces staggered DNA ends similar to Cas12a, it may be particularly suitable for homology-directed repair (HDR)-based genome modifications. Additionally, a catalytically deactivated version of TnpB (dTnpB) has been engineered to develop an adenine base-editing tool capable of A to G conversions in the plant genome (Karmakar et al. 2024).

Due to their small size, ease of fusion with other effector proteins, and compatibility with viral delivery systems, TnpB and other compact nucleases represent the next generation of genome manipulation tools.

20.2.3 CRISPR-Mediated Regulation of Gene Expression

The catalytic activity of Cas nucleases can be abolished by introducing one or more mutations in the coding sequence. For example, SpCas9 can be rendered catalytically inactive by introducing the D10A and H840A mutations. This modified version, known as dCas9, retains its DNA-binding ability but lacks nuclease activity. Diverse effector proteins can be fused with dCas9 to direct them to specific genomic loci by designing appropriate guide RNAs.

CRISPR-activation (CRISPRa) and CRISPR-interference (CRISPRi) are two approaches developed using dCas9 to upregulate or repress gene expression, respectively. In CRISPRa, dCas9 is fused with an activator domain, such as VP64, to create a programmable gene activation system (Konermann et al. 2015; Pan et al. 2022). The chimeric dCas9-VP64 complex is guided to the promoter region of the target gene using single or multiple guide RNAs. Similarly, a deactivated version of TnpB (dTnpB) has been engineered and fused with a synthetic activator domain (6XTAL-VP128) to activate gene expression in rice (Karmakar et al. 2024).

Conversely, in CRISPRi, dCas9 is fused with a transcriptional repressor domain to suppress gene expression. For instance, dCas9 tethered to the KRAB repressor domain has been employed to inhibit target gene expression (Li et al. 2020). Although CRISPRa and CRISPRi enable precise and reversible regulation of gene expression without altering the underlying DNA sequence, the activator/repressor cassette must be continuously expressed in the plant genome for sustained activity. Consequently, plants engineered using these systems remain transgenic (Mushtaq and Molla 2021).

20.2.4 CRISPR-Mediated Epigenome Editing

Epigenome editing techniques modify the epigenomic status of a target genetic locus by utilizing dCas9 or other programmable DNA-binding domains (TAL/ZF) tethered to epi-effectors. For example, DNA methylation and histone acetylation at a target chromatin site can be altered using epigenome editors to influence gene expression or chromatin accessibility. Epi-effectors such as TET1 (for DNA demethylation), HDAC (for histone deacetylation), DNMT3A (for DNA methylation), PRDM9 or DOT1L (for histone methylation), and P300HAT (for histone acetylation) have been fused with dCas9 to introduce or erase chromatin marks in target organisms (Nakamura et al. 2021). Interestingly, heritable and targeted DNA demethylation was achieved in Arabidopsis, resulting in an observable phenotype (Ghoshal et al. 2020). Additionally, epigenome editing has been demonstrated to enhance disease resistance in cassava (Veley et al. 2023; Karmakar et al. 2022).

20.3 Application of Advanced Genome Editing Tools for Climate Resilience

20.3.1 Potential Use of CRISPR/Cas for Nitrogen Use Efficiency

In the twenty-first century, nitrogen fertilizer plays an indispensable role in ensuring global food security. However, excessive use of nitrogen fertilizers in agriculture poses a major threat to ecosystems and contributes to greenhouse gas emissions, highlighting the urgent need to improve nitrogen use efficiency (NUE) in crops. Achieving sustainable agriculture requires minimizing fertilizer use while maintaining productivity (Roy et al. 2006; Han et al. 2015; Samant et al. 2025).

The production of one ton of nitrogen fertilizer via the Haber-Bosch process requires approximately 873 m^3 of natural gas and a substantial amount of heat, contributing to global warming and ozone depletion. Additionally, the application of nitrogen fertilizers leads to the production of nitrogen oxides NO_x (NO, NO_2, and N_2O) through microbial denitrification in soil, exacerbating climate change and ozone pollution. Indiscriminate fertilizer application further amplifies NO_x emissions, causing environmental issues such as eutrophication and negative impacts on human health. These NO_x pollutants can also reduce nitrate transport by S-nitrosylation-mediated inhibition of high-affinity nitrate transporters (Samant et al. 2025). Thus, improving NUE while mitigating NOx emissions is critical for developing sustainable and cost-effective agricultural practices.

Over the past few decades, conventional breeding approaches have been employed to enhance NUE in various crops, including rice and wheat (Gao 2018). NUE is a highly complex and tightly regulated process involving nitrogen sensing, uptake, assimilation via the glutamine synthetase (GS)-glutamate synthase (GOGAT) cycle, conversion to essential and nonessential amino acids, utilization in metabolic pathways, and mobilization to various tissues (Masclaux-Daubresse et al. 2010).

In recent years, CRISPR/Cas-mediated genome editing has emerged as a powerful tool for improving NUE by precisely targeting and modifying specific genes associated with nitrogen uptake, assimilation, and utilization (Chen et al. 2019; Kumar et al. 2024; Hu et al. 2023). Effective gene targeting for NUE enhancement requires a comprehensive understanding of nitrogen signaling, transporter proteins, enzymes, and regulatory elements such as transcription factors involved in nitrogen metabolism (Han et al. 2015; Hu et al. 2023; Kumar et al. 2024; Xu et al. 2012a). Additionally, multi-omics approaches such as transcriptomics, proteomics, and metabolomics provide valuable insights into the expression and activity of nitrogen-responsive genes and enzymes, facilitating rational genome editing strategies for NUE improvement.

Nitrogen is predominantly taken up by plants as nitrate, whereas ammonium and peptides are absorbed to a lesser extent (Xu et al. 2012b). The uptake of nitrogen is primarily facilitated by specialized nitrate transporters (NRTs), which belong to four major families: Nitrate Transporter 1 (NRT1), Peptide Transporter (PTR) as part of the Nitrate and Peptide Transporter Family (NPF), the Chloride Channel

(CLC) family, and the Slowly Activating Anion Channel (SLAC) family (Xu et al. 2012a; Li et al. 2023; Samant et al. 2025).

Based on their nitrate uptake kinetics, NRTs are classified into high-affinity nitrate transporters (HATs) and low-affinity nitrate transporters (LATs). Under low nitrogen conditions (<1 mM), HATs are upregulated, whereas under high nitrogen availability (>1 mM), LAT expression is enhanced. In rice, five HATs belonging to the NRT2 family and two NAR2s (OsNAR2.1/2.2) have been identified. Among them, OsNRT2.1, OsNAR2.1, OsNRT2.2, and OsNRT2.3 are significantly upregulated in roots under low-nitrate conditions (Fan et al. 2016). Overexpression of OsNRT2.1, OsNRT2.2, and OsNRT2.3a enhances nitrate uptake under nitrogen-limited conditions (Chen et al. 2020).

Rice encodes approximately 93 LATs belonging to the NPF (NRT1/PTR) family. Among them, OsNPF6.5 (OsNRT1.1b) is a dual-affinity nitrate transporter that primarily facilitates nitrate uptake and transport from root to shoot. A study by Fan et al. (2016) demonstrated that overexpression of OsNRT1.1b significantly improves rice growth under nitrogen-limited conditions by enhancing nitrogen accumulation.

To date, eight peptide transporters have been identified in rice. Among them, OsNPF7.3 plays a crucial role in nitrogen use efficiency (NUE) (Li et al. 2017). Overexpression of OsNPF7.3 enhances growth under nitrogen deficiency by increasing ammonium uptake. Additionally, OsNPF2.2 and OsNPF2.4 contribute to long-distance nitrate transport from root to shoot (Xia et al. 2015).

Nitrogen is also taken up in the form of ammonium through ammonium transport proteins (OsAMTs). Rice has ten putative OsAMT genes, classified into four subfamilies: OsAMT1, OsAMT2, OsAMT3, and OsAMT4. Among them, the OsAMT1 gene family is well studied, and its expression is regulated by nitrogen availability (Bao et al. 2015). Targeting regulatory elements of these genes using CRISPR-based approaches could be a potential strategy for improving NUE.

After nitrate uptake, it is converted to nitrite by nitrate reductase (NR), which is subsequently reduced to ammonium in plastids by nitrite reductase (NiR). The resulting ammonium is transported to various tissues by ammonium transporters (AMTs), which belong to the ammonium transporter/methylammonium permease/Rhesus (AMT/MEP/Rh) family (Han et al. 2015; Miller et al. 2007). Nitric oxide (NO), a free radical generated through nitrate and nitrite reduction by nitrate reductase, modulates nitrate transporter expression (Samant et al. 2025).

Ammonia is assimilated into glutamine and glutamate through the glutamine synthetase/glutamate synthase (GS/GOGAT) pathway, ensuring nitrogen mobilization to various sink tissues (Fortunato et al. 2023). In rice, three cytosolic glutamine synthetases—OsGS1.1, OsGS1.2, and OsGS1.3—and one chloroplastic glutamine synthetase—OsGS2—have been identified. OsGS1.1 and OsGS1.2 are primarily expressed in the elongation zone and are differentially regulated by nitrogen availability (Funayama et al. 2013). Knockout of *OsGS1.1* or *OsGS1.2* results in a severe reduction in active tiller number and axillary bud formation, whereas *OsGS1.3* is specifically expressed in spikelets and is essential for grain filling and germination.

Targeting negative regulators of the GS-GOGAT cycle can improve nitrogen uptake, assimilation, and mobilization, thereby enhancing NUE (Urriola and Rathore 2015).

Several key transcription factors regulate NUE, including NLP7, NAC, bZIP, DEP1, HY5, LBD, and MADS25. CRISPR/Cas9 technology has been employed to enhance NUE by targeting negative regulators such as NAC7, OsACTPK1, NIGT1, and bZIP60 (Yağız et al. 2022). Recent studies have used CRISPR/Cas9 to improve NUE in rice and barley. For instance, Sun et al. (2024) successfully improved NUE in rice by targeting *OsELF3*, a gene that inhibits Ghd7, a key regulator of agronomic traits such as grain number and plant height. Knockout of *OsELF3* leads to the expression of flowering-related genes.

OsHHO3 is a GARP/G2-like transcription factor that negatively regulates ammonium transporters, such as OsAMT1.1, OsAMT1.2, and OsAMT1.3, under low-nitrogen conditions. Liu et al. (2023) generated an *oshho3* knockout mutant using the CRISPR/Cas9 system, which significantly enhanced nitrogen use efficiency (NUE), growth, and yield under low-nitrogen conditions.

OsTCP19 is a nitrogen-responsive gene that negatively regulates NUE via brassinosteroid signaling. CRISPR/Cas-mediated knockout of *ostcp19* enhanced NUE under low nitrogen (Liu et al. 2021). *OsSBM1* encodes an oligopeptide transporter domain-containing protein involved in nitrogen utilization by interacting with OsMPK6. Xu et al. (2021) reported that CRISPR/Cas-mediated knockout of *OsSBM1* enhances NUE in rice under low nitrogen.

ARE1 is a repressor of the GS-GOGAT cycle in rice. The *ARE1* knockout mutant in rice, wheat, and barley showed improved NUE and yield under low-nitrogen conditions (Sathee et al. 2022). Homologous genes associated with NUE in rice and Arabidopsis can be targeted in other crops, such as wheat and maize, using CRISPR/Cas to enhance NUE.

Furthermore, CRISPR/Cas9 can be employed to genetically engineer the active sites of various enzymes involved in nitrogen assimilation and nitrate transporters to enhance nitrate uptake from the soil. Phytoglobins can also be modulated to improve NUE (Kim et al. 2017). Several successful attempts have been made to enhance NUE in various crops, predominantly in rice, using genome editing via CRISPR/Cas (Table 20.1).

20.3.2 CRISPR/Cas as a Tool to Improve Biological Nitrogen Fixation

The increased application of chemical fertilizers has led to significant human interference in the nitrogen cycle, raising environmental concerns such as increased emissions of nitrogen oxides (NO_x), soil acidification, and water eutrophication (Cunha-Zeri and Ometto 2021). Biological nitrogen fixation (BNF) is a promising alternative that provides fixed nitrogen, which is less prone to leaching and volatilization, as it is utilized in situ and serves as an important and sustainable input to agriculture (Wang et al. 2016).

Table 20.1 List of the genes targeted to improve NUE by CRISPR/Cas system

Gene name	Crop	Type of mutation	Outcome	References
Os MYB61	Rice	Deletion	Enhanced nitrogen utilization and biomass production in rice under low N	Gao et al. (2018)
OsELF3	Rice	8BP deletion	Showed more grain number, panicle, grain yield, and increases N uptake under low N	Sun et al. (2024)
OsNPF3.1	Rice	20–23 BP deletion	Showed more tiller number, grain number, and grain weight	Hang et al. (2024)
OsTCP19	Rice	29 BP deletion	Showed more tiller number under low N	Liu et al. (2021)
OsMIR396e and *OsMIR396f*	Rice		Improved grain size and elongate panicles under low N	Zhang et al. (2020)
OsDNR1	Rice	Deletion	Improved NUE and grain yield under low N	Zhang et al. (2021b)
OsAAP5	Rice	2BP deletion	Showed increase in tiller number and grain yield and quality under low N	Wang et al. (2019)
OsAAP3	Rice	Deletion	Improved grain yield, and increasing tiller number was observed in rice under low N	Lu et al. (2018)
OsSBM1	Rice	Substitution	Showed enhanced rice yield under low N	Xu et al. (2021)
OsHHO3	Rice	Deletion	Showed more tiller number, grain yield, increased photosynthesis rate under low N	Liu et al. (2023)
OsARE1	Rice	1BP deletion	Promote overall growth and grain yield	Wang et al. (2018)
OsNhd1	Rice		Increases N uptake and NUE, regulate photoperiod and flowering	Zhang et al. (2021a)
HvARE1	Barley	Deletion	Increased yield and overall growth were observed under low N	Karunarathne et al. (2022)

BNF is the process by which atmospheric nitrogen is converted into ammonia by symbiotic, associative, and free-living bacteria (Dixon and Kahn 2004; Mus et al. 2016). Nitrogenase, a complex metalloenzyme, catalyzes biological nitrogen fixation (Kim and Rees 2002). Legume plants belonging to the orders Fabales, Fagales, Cucurbitales, and Rosales have the potential to establish mutualistic associations with soil nitrogen-fixing bacteria known as rhizobia, which form specialized structures called nodules (Downie 2014).

During symbiosis, flavonoid compounds and rhizobial transcriptional activators (NodD) facilitate successful rhizobial-legume associations (Moscatiello et al. 2010). Nod factors bind to a receptor complex consisting of lysin motif (LysM) receptor-like kinases (LYK/LYR) located in the epidermal cells of legume roots.

The binding initiates signaling cascades required for nodule development. A key transcription factor, NODULE INCEPTION (NIN), plays a central role in nodulation by regulating rhizobial infection, nodule organogenesis, transition to nitrogen fixation, and nodule number regulation (Ivanov and Harrison 2024; Zhou et al. 2024). As NIN is essential for these processes, it represents a critical target for optimizing symbiotic nitrogen fixation efficiency and enhancing nodulation. Regulatory elements controlling NIN expression and activity can be strategically manipulated to maximize the symbiotic potential of legumes, thereby improving nitrogen fixation and overall plant productivity (Liu et al. 2019). Consequently, NIN could serve as a valuable target for improving BNF for sustainable agricultural practices.

Recent advancements in genome editing have led to the development of transformation pipelines for legumes, including chickpea, soybean, cowpea, pea, Lotus japonicus, and Medicago truncatula, using particle bombardment and Agrobacterium-mediated transformation (Belhaj et al. 2015; Langner et al. 2018). While the CRISPR/Cas system holds potential for enhancing the efficiency of BNF in legumes, there are prospects of establishing nodule-like structures in nonlegume plants such as rice, wheat, maize, and sugarcane, which require significant nitrogen fertilizer inputs for optimal productivity (Downie 2014).

20.3.3 Improvement of CO_2 Capturing Capacity Through CRISPR/Cas-Mediated Genome Editing

In an era where every country is striving for development, massive industrialization, anthropogenic activities, and extensive combustion of fossil fuels have led to enormous CO_2 emissions. This rise in CO_2 levels is a major contributor to the greenhouse effect, driving global warming and climate change, thereby affecting crop productivity (Barati et al. 2021). Therefore, it is crucial to develop sustainable methods for mitigating CO_2 emissions and enhancing atmospheric CO_2 capture by crops.

To date, various technologies have been developed to improve CO_2 capture in plants, such as activated carbon, and zeolites (Zahed et al. 2021). However, these technologies are highly expensive and lack scalability for commercial applications.

Alternatively, photosynthetic green algae offer a promising biological approach for CO_2 sequestration due to their unique adaptive abilities, including high growth rates, ease of cultivation, and relatively low production costs (Onyeaka and Ekwebelem 2023). Additionally, they possess efficient CO_2 uptake mechanisms, adaptability to fluctuating CO_2 and oxygen levels, and unique physiological traits. Several strategies can be employed to enhance CO_2 fixation in microalgae: 1. Enhancing the efficiency of the CO_2 fixation pathway, 2. Improving the efficiency of energy-harvesting complexes, and 3. Introducing additional CO_2 fixation pathways (Ighalo et al. 2022; Xu et al. 2023). Two pivotal pathways are involved in CO_2 fixation: the Calvin-Benson-Bassham (CBB) cycle, which fixes atmospheric inorganic CO_2 into organic compounds, and carbon concentration mechanisms (CCMs), which supply inorganic carbon to the Calvin cycle (Burnap et al. 2015; Giordano et al. 2005).

Ribulose-1,5-bisphosphate carboxylase/oxygenase (Rubisco) is the most critical photosynthetic enzyme in CO_2 fixation, but has a low intrinsic affinity for CO_2 as compared to oxygen. Therefore, enhancing Rubisco catalytic activity has been proposed as a key strategy to improve CO_2 fixation via photosynthesis (Andrews and Lorimer 1987; Onyeaka and Ekwebelem 2023). CRISPR-Cas can be utilized to modify the catalytic site of the Rubisco enzyme, improving its CO_2 fixation efficiency. This targeted modification could optimize enzyme performance, leading to enhanced CO_2 fixation and greater biomass production in both plants and microalgae. Additionally, Rubisco activity can be further enhanced by modulating the expression of Rubisco activase genes (Amaral et al. 2024; Parry et al. 2013). Several inhibitors suppress Rubisco activase activity; these suppressors can be targeted by the CRISPR-Cas system to enhance Rubisco activity by alleviating inhibition.

Other key enzymes in the CBB cycle include fructose-1,6-bisphosphate aldolase (aldolase) and sedoheptulose-1,7-bisphosphatase (SBPase), where SBPase catalyzes the regeneration of ribulose-1,5-bisphosphate (RuBP), the substrate for Rubisco (Ludwig et al. 2024; Tamoi et al. 2006). A recent study reported that overexpression of the SBPase gene enhanced CO_2 fixation by 1.2-fold in microalgae such as *Dunaliella bardawil* and *Euglena* (Barati et al. 2021). Through precise editing techniques like base editing and prime editing (Molla et al., 2020, 2021), both SBPase and aldolase can be targeted to improve CO_2 fixation efficiency in crops.

CCMs enhance the efficiency of photosynthesis by concentrating CO_2 around Rubisco, thereby minimizing photorespiration and optimizing carbon fixation efficiency. The three principal components of CCMs are: Ci transporters, Carboxysomes, and Carbonic anhydrases (CAs), which catalyze the conversion of HCO_3^- to CO_2 for Rubisco function (Burlacot et al. 2022). Studies have shown that CCM-associated genes are differentially modulated in response to varying CO_2 concentrations and light intensities. Key components of CCMs include multiple CAs (CAH1, CAH2, CAH3, CAH4, CAH5, and CAH6), Ci transporters (HLA3, LCI1, LCIA, CCP1, CCP2, CemA, RHP1, and RHP2), CO_2 uptake proteins (LCIB, LCIC, LCID, and LCIE), a putative methyltransferase (CIA6), and two nuclear transcriptional regulators (CIA5/CCM1 and LCR1). Among these, HLA3 and LCI1 play key roles in CO_2 uptake, particularly under low CO_2 concentrations (Burlacot et al. 2022; Burnap et al. 2015; Parry et al. 2013; Spalding 2008). By precisely modifying genes, with base and prime editing tools, encoding carbonic anhydrases, bicarbonate transporters, and regulators of carboxysome/pyrenoid formation using multiplex genome editing, researchers can optimize CCM function, thereby increasing CO_2 fixation efficiency and mitigating the adverse effects of photorespiration.

20.4 Future Prospects

Over the past decades, global agricultural productivity has been affected significantly due to climate change-induced extreme weather events, including various abiotic stresses such as drought, high temperatures, flooding, salinity, and cold. Concurrently, food demand has surged due to rapid population growth. To meet this increasing demand and achieve sustainable crop production under adverse environmental conditions, the adoption of smart agricultural practices is imperative. Climate-smart and stress-resilient crops can be game changers in overcoming environmental challenges.

Plant breeders are increasingly utilizing advanced biotechnological tools and new breeding technologies, such as CRISPR-based genome editing, to enhance crop productivity. The emergence of CRISPR/Cas technology has revolutionized plant molecular breeding by enabling precise editing of key regulatory genes associated with important agronomic traits while saving time and costs. By identifying and modifying genes that regulate yield attributes, confer multiple stress resistance, and improve nutrient use efficiency, breeders can introduce desirable traits such as increased yield, improved product quality, and enhanced tolerance to biotic and abiotic stresses across various crops.

Climate-smart crops can help meet future food and energy demands. Multiplex genome editing enables simultaneous targeting of multiple metabolic pathway genes, serving as a stepping stone toward developing climate-resilient crops. Recent advancements in genome-editing tools for crop genetic improvement could play a pivotal role in developing high-yielding, climate-resilient crops, ultimately fulfilling the increasing global food demand.

References

Adeyinka OS, Tabassum B, Koloko BL, Ogungbe IV (2023) Enhancing the quality of staple food crops through CRISPR/Cas-mediated site-directed mutagenesis. Planta 257(4):78. https://doi.org/10.1007/s00425-023-04110-6

Adhikari S, Kumar A, Joshi A, Bharati A, Pramod Prasad OP, Gangwar CL, Kumar S, Bhardwaj SC (2024) Speed breeding: rapid generation tool for breeding climate smart crops. In: Climate-resilient agriculture. Apple Academic Press

Amaral J, Lobo AKM, Carmo-Silva E (2024) Regulation of rubisco activity in crops. New Phytol 241(1):35–51. https://doi.org/10.1111/nph.19369

Andrews TJOHN, Lorimer GEORGEH (1987) Chapter 3: Rubisco: structure, mechanisms, and prospects for improvement. In: Hatch MD, Boardman NK (eds) Photosynthesis. Academic, pp 131–218

Bandyopadhyay A, Kancharla N, Javalkote VS, Dasgupta S, Brutnell TP (2020) CRISPR-Cas12a (Cpf1): a versatile tool in the plant genome editing tool box for agricultural advancement. Front Plant Sci 11. https://doi.org/10.3389/fpls.2020.584151

Bao A, Liang Z, Zhao Z, Cai H (2015) Overexpressing of OsAMT1-3, a high affinity ammonium transporter gene, modifies Rice growth and carbon-nitrogen metabolic status. Int J Mol Sci 16(5):9037–9063. https://doi.org/10.3390/ijms16059037

Barati B, Zeng K, Baeyens J, Wang S, Addy M, Gan S-Y, Abomohra AE-F (2021) Recent Progress in genetically modified microalgae for enhanced carbon dioxide sequestration. Biomass Bioenergy 145:105927. https://doi.org/10.1016/j.biombioe.2020.105927

Belhaj K, Chaparro-Garcia A, Kamoun S, Patron NJ, Nekrasov V (2015) Editing plant genomes with CRISPR/Cas9. Curr Opin Biotechnol 32:76–84. https://doi.org/10.1016/j.copbio.2014.11.007

Bortesi L, Zhu C, Zischewski J, Perez L, Bassié L, Nadi R, Forni G, Lade SB, Soto E, Jin X, Medina V, Villorbina G, Muñoz P, Farré G, Fischer R, Twyman RM, Capell T, Christou P, Schillberg S (2016) Patterns of CRISPR/Cas9 activity in plants, animals and microbes. Plant Biotechnol J 14(12):2203–2216. https://doi.org/10.1111/pbi.12634

Burlacot A, Dao O, Auroy P, Cuiné S, Li-Beisson Y, Peltier G (2022) Alternative photosynthesis pathways drive the algal CO_2-concentrating mechanism. Nature 605(7909):366–371. https://doi.org/10.1038/s41586-022-04662-9

Burnap RL, Hagemann M, Kaplan A (2015) Regulation of CO_2 concentrating mechanism in cyanobacteria. Life 5(1):348–371. https://doi.org/10.3390/life5010348

Chen K, Wang Y, Zhang R, Zhang H, Gao C (2019) CRISPR/Cas genome editing and precision plant breeding in agriculture. Ann Rev Plant Biol 70:667–697. https://doi.org/10.1146/annurev-arplant-050718-100049

Chen J, Liu X, Liu S, Fan X, Zhao L, Song M, Fan X, Guohua X (2020) Co-overexpression of OsNAR2.1 and OsNRT2.3a increased agronomic nitrogen use efficiency in transgenic Rice plants. Front Plant Sci 11. https://doi.org/10.3389/fpls.2020.01245

Cobb JN, Biswas PS, Damien Platten J (2019) Back to the future: revisiting MAS as a tool for modern plant breeding. Theor Appl Genet 132(3):647–667. https://doi.org/10.1007/s00122-018-3266-4

Cunha-Zeri G, Ometto J (2021) Nitrogen emissions in Latin America: A conceptual framework of drivers, impacts, and policy responses. Environm Dev 38:100605. https://doi.org/10.1016/j.envdev.2020.100605

Dixon R, Kahn D (2004) Genetic regulation of biological nitrogen fixation. Nat Rev Microbiol 2(8):621–631. https://doi.org/10.1038/nrmicro954

Downie JA (2014) Legume Nodulation. Curr Biol 24(5):R184–R190. https://doi.org/10.1016/j.cub.2014.01.028

Elsharawy H, Refat M (2023) CRISPR/Cas9 genome editing in wheat: enhancing quality and productivity for global food security—a review. Funct Integr Genomics 23(3):265. https://doi.org/10.1007/s10142-023-01190-1

Fan X, Feng H, Tan Y, Yanling X, Miao Q, Guohua X (2016) A putative 6-transmembrane nitrate transporter OsNRT1.1b plays a key role in Rice under low nitrogen. J Integr Plant Biol 58(6):590–599. https://doi.org/10.1111/jipb.12382

Fortunato S, Nigro D, Lasorella C, Marcotuli I, Gadaleta A, Concetta M, de Pinto. (2023) The role of glutamine Synthetase (GS) and glutamate synthase (GOGAT) in the improvement of nitrogen use efficiency in cereals. Biomolecules 13(12):1771. https://doi.org/10.3390/biom13121771

Freije CA, Myhrvold C, Boehm CK, Lin AE, Welch NL, Carter A, Metsky HC, Luo CY, Abudayyeh OO, Gootenberg JS, Yozwiak NL, Zhang F, Sabeti PC (2019) Programmable inhibition and detection of RNA viruses using Cas13. Mol Cell 76(5):826–837.e11. https://doi.org/10.1016/j.molcel.2019.09.013

Funayama K, Kojima S, Tabuchi-Kobayashi M, Sawa Y, Nakayama Y, Hayakawa T, Yamaya T (2013) Cytosolic glutamine Synthetase1;2 is responsible for the primary assimilation of ammonium in Rice roots. Plant Cell Physiol 54(6):934–943. https://doi.org/10.1093/pcp/pct046

Gao C (2018) The future of CRISPR Technologies in Agriculture. Nat Rev Mol Cell Biol 19(5):275–276. https://doi.org/10.1038/nrm.2018.2

Gao L, Cox DBT, Yan WX, Manteiga JC, Schneider MW, Yamano T, Nishimasu H, Nureki O, Crosetto N, Zhang F (2017) Engineered Cpf1 variants with altered PAM specificities increase genome targeting range. Nat Biotechnol 35(8):789–792. https://doi.org/10.1038/nbt.3900

Gao J, Wang F, Hang H, Jiang S, Muhammad A, Shao Y, Sun C, Tian Z, Jiang D, Dai T (2018) Improved leaf nitrogen reutilisation and rubisco activation under short-term nitrogen-deficient conditions promotes photosynthesis in winter wheat (Triticum Aestivum L.) at the Seedling stage. Funct Plant Biol 45(8):840–853. https://doi.org/10.1071/FP17232

Ghoshal B, Vong B, Picard CL, Feng S, Tam JM, Jacobsen SE (2020) A viral guide RNA delivery system for CRISPR-based transcriptional activation and heritable targeted DNA demethylation in Arabidopsis Thaliana. PLoS Genet 16(12):e1008983. https://doi.org/10.1371/journal.pgen.1008983

Giordano M, Beardall J, Raven JA (2005) CO_2 concentrating mechanisms in algae: mechanisms, environmental modulation, and evolution. Ann Rev Plant Biol 56:99–131. https://doi.org/10.1146/annurev.arplant.56.032604.144052

Gootenberg JS, Abudayyeh OO, Kellner MJ, Joung J, Collins JJ, Zhang F (2018) Multiplexed and portable nucleic acid detection platform with Cas13, Cas12a, and Csm6. Science 360(6387):439–444. https://doi.org/10.1126/science.aaq0179

Han M, Okamoto M, Beatty PH, Rothstein SJ, Good AG (2015) The genetics of nitrogen use efficiency in crop plants. Ann Rev Genet 49:269–289. https://doi.org/10.1146/annurev-genet-112414-055037

Hang J, Wu B, Qiu D, Yang G, Fang Z, & Zhang M (2024) OsNPF3.1, a nitrate, abscisic acid and gibberellin transporter gene, is essential for rice tillering and nitrogen utilization efficiency. Journal of Integrative Agriculture, 23(4):1087–1104. https://doi.org/10.1016/j.jia.2023.04.024

Hu B, Wang W, Chen J, Liu Y, Chu C (2023) Genetic improvement toward nitrogen-use efficiency in rice: lessons and perspectives. Mol Plant 16(1):64–74. https://doi.org/10.1016/j.molp.2022.11.007

Hundleby PAC, Sack M, Twyman RM (2018) Biosafety, risk assessment, and regulation of molecular farming. In: Molecular pharming. Wiley, pp 327–351

Ighalo JO, Dulta K, Kurniawan SB, Omoarukhe FO, Ewuzie U, Eshiemogie SO, Ojo AU, Abdullah SRS (2022) Progress in microalgae application for CO_2 sequestration. Cleaner Chem Eng 3:100044. https://doi.org/10.1016/j.clce.2022.100044

Ivanov S, Harrison MJ (2024) Receptor-associated kinases control the lipid provisioning program in plant–fungal symbiosis. Science 383(6681):443–448. https://doi.org/10.1126/science.ade1124

Jiang F, Doudna JA (2017) CRISPR–Cas9 structures and mechanisms. Ann Rev Biophys 46:505–529. https://doi.org/10.1146/annurev-biophys-062215-010822

Jinek M, Chylinski K, Fonfara I, Hauer M, Doudna JA, Charpentier E (2012) A programmable dual-RNA–guided DNA endonuclease in adaptive bacterial immunity. Science 337(6096):816–821. https://doi.org/10.1126/science.1225829

Karmakar, S., Das, P., Panda, D., Xie, K., Baig, M. J., & Molla, K. A. (2022). A detailed landscape of CRISPR-Cas-mediated plant disease and pest management. Plant Science, 323, 111376. https://doi.org/10.1016/j.plantsci.2022.111376

Karmakar S, Panda D, Panda S, Dash M, Saha R, Priya Das SP, Avinash JS, Yinong Yang AK, Nayak MJ, Baig, and Kutubuddin A. Molla. (2024) A miniature alternative to Cas9 and Cas12: transposon-associated TnpB mediates targeted genome editing in plants. Plant Biotechnol J 22(10):2950–2953. https://doi.org/10.1111/pbi.14416

Karunarathne SD, Han Y, Zhang X.-Q, & Li C (2022) CRISPR/Cas9 gene editing and natural variation analysis demonstrate the potential for HvARE1 in improvement of nitrogen use efficiency in barley. Journal of Integrative Plant Biology, 64(3):756–770. https://doi.org/10.1111/jipb.13214

Karvelis T, Druteika G, Bigelyte G, Budre K, Zedaveinyte R, Silanskas A, Kazlauskas D, Venclovas Č, Siksnys V (2021) Transposon-associated TnpB is a programmable RNA-guided DNA endonuclease. Nature 599(7886):692–696. https://doi.org/10.1038/s41586-021-04058-1

Kim J, Rees DC (2002) Nitrogenase and biological nitrogen fixation. ACS Publications. Retrieved February 3, 2025 https://pubs.acs.org/doi/pdf/10.1021/bi00168a001.

Kim H, Kim S-T, Ryu J, Kang B-C, Kim J-S, Kim S-G (2017) CRISPR/Cpf1-mediated DNA-free plant genome editing. Nat Commun 8(1):14406. https://doi.org/10.1038/ncomms14406

Konermann S, Brigham MD, Trevino AE, Joung J, Abudayyeh OO, Barcena C, Hsu PD, Habib N, Gootenberg JS, Nishimasu H, Nureki O, Zhang F (2015) Genome-scale transcriptional activation by an engineered CRISPR-Cas9 complex. Nature 517(7536):583–588. https://doi.org/10.1038/nature14136

Kumar A, Subbaiah M, Roy J, Phogat S, Kaushik M, Saini MR, Madhavan J, Sevanthi AM, Mandal PK (2024) Strategies to utilize genome editing for increasing nitrogen use efficiency in crops. Nucleus 67(1):205–225. https://doi.org/10.1007/s13237-024-00475-5

Langner T, Kamoun S, Belhaj K (2018) CRISPR crops: plant genome editing toward disease resistance. Ann Rev Phytopathol 56:479–512. https://doi.org/10.1146/annurev-phyto-080417-050158

Li H, Bin H, Chu C (2017) Nitrogen use efficiency in crops: lessons from Arabidopsis and rice. J Exp Bot 68(10):2477–2488

Li H, Rasheed A, Hickey LT, He Z (2018) Fast-forwarding genetic gain. Trends Plant Sci 23(3):184–186. https://doi.org/10.1016/j.tplants.2018.01.007

Li K, Liu Y, Cao H, Zhang Y, Zhimin G, Liu X, Andy Y, Kaphle P, Dickerson KE, Ni M, Jian X (2020) Interrogation of enhancer function by enhancer-targeting CRISPR epigenetic editing. Nat Commun 11(1):485. https://doi.org/10.1038/s41467-020-14362-5

Li Z, Xu Na W, Jacquot A, Chaput V, Adamo M, Neuhäuser B, Straub T, Lejay L, Schulze WX (2023) Phosphoregulation in the N-terminus of NRT2.1 affects nitrate uptake by controlling the interaction of NRT2.1 with NAR2.1 and kinase HPCAL1 in Arabidopsis. J Exp Bot:erad490. https://doi.org/10.1093/jxb/erad490

Liu C-W, Breakspear A, Guan D, Cerri MR, Jackson K, Jiang S, Robson F, Radhakrishnan GV, Roy S, Bone C, Stacey N, Rogers C, Trick M, Niebel A, Oldroyd GED, de Carvalho-Niebel F, Murray JD (2019) NIN acts as a network hub controlling a growth module required for Rhizobial infection. Plant Physiol 179(4):1704–1722. https://doi.org/10.1104/pp.18.01572

Liu Y, Wang H, Jiang Z, Wang W, Ruineng X, Wang Q, Zhang Z, Li A, Liang Y, Shujun O, Liu X, Cao S, Tong H, Wang Y, Zhou F, Liao H, Bin H, Chu C (2021) Genomic basis of geographical adaptation to soil nitrogen in Rice. Nature 590(7847):600–605. https://doi.org/10.1038/s41586-020-03091-w

Liu S, Sretenovic S, Fan T, Cheng Y, Li G, Aileen Qi X, Tang YX, Guo W, Zhong Z, He Y, Liang Y, Han Q, Zheng X, Xiaofeng G, Qi Y, Zhang Y (2022) Hypercompact CRISPR–Cas12j2 (CasΦ) enables genome editing, gene activation, and epigenome editing in plants. Plant Commun 3(6). https://doi.org/10.1016/j.xplc.2022.100453

Liu K, Sakuraba Y, Ohtsuki N, Yang M, Ueda Y, Yanagisawa S (2023) CRISPR/Cas9-mediated elimination of OsHHO3, a transcriptional repressor of three AMMONIUM TRANSPORTER1 genes, improves nitrogen use efficiency in Rice. Plant Biotechnol J 21(11):2169–2172. https://doi.org/10.1111/pbi.14167

Lu K, Wu B, Wang J, Zhu W, Nie H, Qian J, Huang W, & Fang Z (2018) Blocking amino acid transporter Os3 improves grain yield by promoting outgrowth buds and increasing tiller number in rice. Plant Biotechnology Journal, 16(10):1710–1722. https://doi.org/10.1111/pbi.12907

Ludwig M, Hartwell J, Raines CA, Simkin AJ (2024) The calvin-benson-bassham cycle in C4 and Crassulacean acid metabolism species. Semin Cell Dev Biol 155:10–22. https://doi.org/10.1016/j.semcdb.2023.07.013

Mahas A, Aman R, Mahfouz M (2019) CRISPR-Cas13d mediates robust RNA virus interference in plants. Genome Biol 20(1):263. https://doi.org/10.1186/s13059-019-1881-2

Mankotia S, Swain J, Satbhai SB (2022) Chapter 17: Biotechnological approaches for generating iron-rich crops. In: Kumar V, Srivastava AK, Suprasanna P (eds) Plant nutrition and food security in the era of climate change. Academic, pp 437–451

Masclaux-Daubresse C, Daniel-Vedele F, Dechorgnat J, Chardon F, Gaufichon L, Suzuki A (2010) Nitrogen uptake, assimilation and remobilization in plants: challenges for sustainable and productive agriculture. Ann Bot 105(7):1141–1157. https://doi.org/10.1093/aob/mcq028

Miller AJ, Fan X, Orsel M, Smith SJ, Wells DM (2007) Nitrate transport and Signalling. J Exp Bot 58(9):2297–2306. https://doi.org/10.1093/jxb/erm066

Ming M, Ren Q, Pan C, He Y, Zhang Y, Liu S, Zhong Z, Wang J, Malzahn AA, Jun W, Zheng X, Zhang Y, Qi Y (2020) CRISPR–Cas12b enables efficient plant genome engineering. Nat Plants 6(3):202–208. https://doi.org/10.1038/s41477-020-0614-6

Mojica FJM, Juez G, Rodriguez-Valera F (1993) Transcription at different salinities of Haloferax Mediterranei sequences adjacent to partially modified PstI sites. Mol Microbiol 9(3):613–621. https://doi.org/10.1111/j.1365-2958.1993.tb01721.x

Molla KA, Yang Y (2020) Predicting CRISPR/Cas9-induced mutations for precise genome editing. Trends Biotechnol 38(2):136–141. https://doi.org/10.1016/j.tibtech.2019.08.002

Molla KA, Shih J, Yang Y (2020) Single-nucleotide editing for zebra3 and wsl5 phenotypes in rice using CRISPR/Cas9-mediated adenine base editors. Abiotech 1(2):106–118. https://doi.org/10.1007/s42994-020-00018-x

Molla KA, Sretenovic S, Bansal KC et al (2021) Precise plant genome editing using base editors and prime editors. Nat Plants 7:1166–1187. https://doi.org/10.1038/s41477-021-00991-1

Moscatiello R, Squartini A, Mariani P, Navazio L (2010) Flavonoid-induced calcium signalling in rhizobium leguminosarum Bv. Viciae. New Phytol 188(3):814–823. https://doi.org/10.1111/j.1469-8137.2010.03411.x

Mus F, Crook MB, Garcia K, Costas AG, Geddes BA, Kouri ED, Paramasivan P, Ryu M-H, Oldroyd GED, Poole PS, Udvardi MK, Voigt CA, Ané J-M, Peters JW (2016) Symbiotic nitrogen fixation and the challenges to its extension to nonlegumes. Appl Environ Microbiol 82(13):3698–3710. https://doi.org/10.1128/AEM.01055-16

Mushtaq M, Molla KA (2021) CRISPR technologies for plant biotechnology innovation. In: Ricroch A, Chopra S, Kuntz M (eds) Plant biotechnology: experience and future prospects. Springer International Publishing, Cham, pp 51–67

Nakamura M, Gao Y, Dominguez AA, Lei SQ (2021) CRISPR technologies for precise epigenome editing. Nat Cell Biol 23(1):11–22. https://doi.org/10.1038/s41556-020-00620-7

Ndudzo A, Makuvise AS, Moyo S, Bobo ED (2024) CRISPR-Cas9 genome editing in crop breeding for climate change resilience: implications for smallholder farmers in Africa. J Agric Food Res 16:101132. https://doi.org/10.1016/j.jafr.2024.101132

Nishimasu H, Ann Ran F, Hsu PD, Konermann S, Shehata SI, Dohmae N, Ishitani R, Zhang F, Nureki O (2014) Crystal structure of Cas9 in complex with guide RNA and target DNA. Cell 156(5):935–949. https://doi.org/10.1016/j.cell.2014.02.001

Onyeaka H, Ekwebelem OC (2023) A review of recent advances in engineering bacteria for enhanced CO2 capture and utilization. Int J Environ Sci Technol 20(4):4635–4648. https://doi.org/10.1007/s13762-022-04303-8

Pan W, Cheng Z, Han Z, Yang H, Zhang W, Zhang H (2022) Efficient genetic transformation and CRISPR/Cas9-mediated genome editing of watermelon assisted by genes encoding developmental regulators. J Zhejiang Univ Sci B 23(4):339–344. https://doi.org/10.1631/jzus.B2200119

Parry MAJ, John Andralojc P, Scales JC, Salvucci ME, Elizabete Carmo-Silva A, Alonso H, Whitney SM (2013) Rubisco activity and regulation as targets for crop improvement. J Exp Bot 64(3):717–730. https://doi.org/10.1093/jxb/ers336

Patra S, Chatterjee D, Basak S, Sen S, Mandal A (2024) CRISPR/Cas9 opens new horizon of crop improvement under stress condition. Biochim Biophys Acta Gen Subj 1868(10):130685. https://doi.org/10.1016/j.bbagen.2024.130685

Rajkhowa S, Sarma J (2021) Chapter 14: Climate change and flood risk, global climate change. In: Singh S, Singh P, Rangabhashiyam S, Srivastava KK (eds) Global climate change. Elsevier, pp 321–339

Roy RN, Finck A, Blair GJ, Tandon HLS (2006) "Plant nutrition for food security." A guide for integrated nutrient management. FAO Fertil Plant Nutr Bull 16(368)

Samant SB, Yadav N, Swain J, Joseph J, Kumari A, Praveen A, Sahoo RK, Manjunatha G, Seth CS, Singla-Pareek SL, Foyer CH, Pareek A, Gupta KJ (2024) Nitric oxide, energy, and redox-dependent responses to hypoxia. J Exp Bot 75(15):4573–4588. https://doi.org/10.1093/jxb/erae139

Samant SB, Swain J, Yadav N, Yadav R, Singh P, Rai P, Sheri V, Sreeman S, Subramanyam R, Pareek A, Gupta KJ (2025) Overexpression of Phytoglobin1 in rice leads to enhanced nitrogen use efficiency via modulation of nitric oxide. Plant Cell Environ. https://doi.org/10.1111/pce.15289

Sathee L, Jagadhesan B, Pandesha PH, Barman D, Sandeep Adavi B, Shivani Nagar GK, Krishna ST, Jha SK, Chinnusamy V (2022) Genome editing targets for improving nutrient use efficiency and nutrient stress adaptation. Front Genet 13. https://doi.org/10.3389/fgene.2022.900897

Spalding MH (2008) Microalgal carbon-dioxide-concentrating mechanisms: chlamydomonas inorganic carbon transporters. J Exp Bot 59(7):1463–1473. https://doi.org/10.1093/jxb/erm128

Strecker J, Jones S, Koopal B, Schmid-Burgk J, Zetsche B, Gao L, Makarova KS, Koonin EV, Zhang F (2019) Engineering of CRISPR-Cas12b for human genome editing. Nat Commun 10(1):212. https://doi.org/10.1038/s41467-018-08224-4

Sun Q, Zhiwen Y, Wang X, Chen H, Jiahao L, Zhao C, Jiang L, Li F, Quan X, Ma D (2024) EARLY FLOWERING3-1 represses grain number, plant height, and heading Date7 to promote ABC1

REPRESSOR1 and regulate nitrogen uptake in Rice. Plant Physiol 196(3):1857–1868. https://doi.org/10.1093/plphys/kiae416
Swarup S, Cargill EJ, Crosby K, Flagel L, Kniskern J, Glenn KC (2021) Genetic diversity is indispensable for plant breeding to improve crops. Crop Sci 61(2):839–852. https://doi.org/10.1002/csc2.20377
Tamoi M, Nagaoka M, Miyagawa Y, Shigeoka S (2006) Contribution of Fructose-1,6-Bisphosphatase and Sedoheptulose-1,7-Bisphosphatase to the photosynthetic rate and carbon flow in the Calvin cycle in transgenic plants. Plant Cell Physiol 47(3):380–390. https://doi.org/10.1093/pcp/pcj004
Tester M, Langridge P (2010) Breeding technologies to increase crop production in a changing world. Science 327(5967):818–822. https://doi.org/10.1126/science.1183700
Urriola J, Rathore KS (2015) Overexpression of a glutamine Synthetase gene affects growth and development in sorghum. Transgenic Res 24(3):397–407. https://doi.org/10.1007/s11248-014-9852-6
Veley KM, Elliott K, Jensen G, Zhong Z, Feng S, Yoder M, Gilbert KB, Berry JC, Lin Z-JD, Ghoshal B, Gallego-Bartolomé J, Norton J, Motomura-Wages S, Carrington JC, Jacobsen SE, Bart RS (2023) Improving cassava bacterial blight resistance by editing the epigenome. Nat Commun 14(1):85. https://doi.org/10.1038/s41467-022-35675-7
Wang L, Wang L, Tan Q, Fan Q, Zhu H, Hong Z, Zhang Z, Duanmu D (2016) Efficient inactivation of symbiotic nitrogen fixation related genes in lotus Japonicus using CRISPR-Cas9. Front Plant Sci 7. https://doi.org/10.3389/fpls.2016.01333
Wang Q, Nian J, Xie X, Yu H, Zhang J, Bai J, Dong G, Hu J, Bai B, Chen L, Xie Q, Feng J, Yang X, Peng J, Chen F, Qian Q, Li J, & Zuo J (2018) Genetic variations in ARE1 mediate grain yield by modulating nitrogen utilization in rice. Nature Communications, 9(1):735. https://doi.org/10.1038/s41467-017-02781-w
Wang J, Wu B, Lu K, Wei Q, Qian J, Chen Y, & Fang Z (2019) The Amino Acid Permease 5 (OsAAP5) Regulates Tiller Number and Grain Yield in Rice. Plant Physiology, 180(2):1031–1045. https://doi.org/10.1104/pp.19.00034
Wang P, Li Z, Li H, Zhang D, Wang W, Xiaodong X, Xie Q, Duan Z, Xia X, Guo G, Shaheen A, Zhou Y, Wang D, Guo S, Zhubing H, Galbraith DW, Song C-P (2024) Smart crops. New Crops 1:100007. https://doi.org/10.1016/j.ncrops.2023.12.002
Wu F, Qiao X, Zhao Y, Zhang Z, Gao Y, Shi L, Haokun D, Wang L, Ya-Jie Zhang Y, Zhang LL, Wang Q, Kong D (2020) Targeted mutagenesis in Arabidopsis Thaliana using CRISPR-Cas12b/C2c1. J Integr Plant Biol 62(11):1653–1658. https://doi.org/10.1111/jipb.12944
Xia X, Fan X, Wei J, Feng H, Hongye Q, Xie D, Miller AJ, Guohua X (2015) Rice nitrate transporter OsNPF2.4 functions in low-affinity acquisition and long-distance transport. J Exp Bot 66(1):317–331. https://doi.org/10.1093/jxb/eru425
Xiong W, Reynolds M, Yunbi X (2022) Climate change challenges plant breeding. Curr Opin Plant Biol 70:102308. https://doi.org/10.1016/j.pbi.2022.102308
Xu G, Fan X, Miller AJ (2012a) Plant nitrogen assimilation and use efficiency. Ann Rev Plant Biol 63:153–182. https://doi.org/10.1146/annurev-arplant-042811-105532
Xu Y, Li Z-K, Thomson MJ (2012b) Molecular breeding in plants: moving into the mainstream. Mol Breed 29(4):831–832. https://doi.org/10.1007/s11032-012-9717-8
Xu Y, Li P, Zou C, Yanli L, Xie C, Zhang X, Prasanna BM, Olsen MS (2017) Enhancing genetic gain in the era of molecular breeding. J Exp Bot 68(11):2641–2666. https://doi.org/10.1093/jxb/erx135
Xu J, Shang L, Wang J, Chen M, Xue F, He H, Wang Z, Zeng D, Zhu L, Jiang H, Zhang C, Chen G, Gao Z, Zou W, Ren D, Dong G, Shen L, Zhang Q, Li Q, Guo L, Qian Q, Zhang G (2021) The SEEDLING BIOMASS 1 allele from Indica Rice enhances yield performance under low-nitrogen environments. Plant Biotechnol J 19(9):1681–1683. https://doi.org/10.1111/pbi.13642
Xu P, Li J, Qian J, Wang B, Liu J, Rui X, Chen P, Zhou W (2023) Recent advances in CO2 fixation by microalgae and its potential contribution to carbon neutrality. Chemosphere 319:137987. https://doi.org/10.1016/j.chemosphere.2023.137987

Yağız AK, Yavuz C, Naeem M, Dangol SD, Aksoy E (2022) Genome editing for nutrient use efficiency in crops. In: Prakash CS, Fiaz S, Fahad S (eds) Principles and practices of OMICS and genome editing for crop improvement. Springer International Publishing, Cham, pp 347–383

Zahed MA, Movahed E, Khodayari A, Zanganeh S, Badamaki M (2021) Biotechnology for carbon capture and fixation: critical review and future directions. J Environ Manag 293:112830. https://doi.org/10.1016/j.jenvman.2021.112830

Zetsche B, Gootenberg JS, Abudayyeh OO, Slaymaker IM, Makarova KS, Essletzbichler P, Volz SE, Joung J, van der Oost J, Regev A, Koonin EV, Zhang F (2015) Cpf1 is a single RNA-guided endonuclease of a class 2 CRISPR-Cas system. Cell 163(3):759–771. https://doi.org/10.1016/j.cell.2015.09.038

Zhang J, Zhou Z, Bai J, Tao X, Wang L, Zhang H, & Zhu J-K (2020) Disruption of MIR396e and MIR396f improves rice yield under nitrogen-deficient conditions. National Science Review, 7(1):102–112. https://doi.org/10.1093/nsr/nwz142

Zhang S, Zhang Y, Li K, Yan M, Zhang J, Yu M, Tang S, Wang L, Qu H, Luo L, Xuan W, & Xu G (2021a) Nitrogen Mediates Flowering Time and Nitrogen Use Efficiency via Floral Regulators in Rice. Current Biology, 31(4):671–683.e5. https://doi.org/10.1016/j.cub.2020.10.095

Zhang S, Zhu L, Shen C, Ji Z, Zhang H, Zhang T, Li Y, Yu J, Yang N, He Y, Tian Y, Wu K, Wu J, Harberd NP, Zhao Y, Fu X, Wang S, & Li S (2021b) Natural allelic variation in a modulator of auxin homeostasis improves grain yield and nitrogen use efficiency in rice. The Plant Cell, 33(3):566–580.https://doi.org/10.1093/plcell/koaa037

Zhou X, Zhao Y, Ni P, Ni Z, Sun Q, Zong Y (2023) CRISPR-mediated acceleration of wheat improvement: advances and perspectives. J Genet Genomics 50(11):815–834. https://doi.org/10.1016/j.jgg.2023.09.007

Zhou M, Li Y, Yao X-L, Zhang J, Liu S, Cao H-R, Bai S, Chen C-Q, Zhang D-X, Ao X, Lei J-N, Qian-Zhuo Mao Y, Zhou D-QD, Guan Y-F, Chen Z-C (2024) Inorganic nitrogen inhibits symbiotic nitrogen fixation through blocking NRAMP2-mediated iron delivery in soybean nodules. Nat Commun 15(1):8946. https://doi.org/10.1038/s41467-024-53325-y

Zhu H, Li C, Gao C (2020) Applications of CRISPR–Cas in agriculture and plant biotechnology. Nat Rev Mol Cell Biol 21(11):661–677. https://doi.org/10.1038/s41580-020-00288-9

Nanotechnology in Crop Breeding

21

Revanth Kishore Pulamolu, Pradeep Kumar Shukla, Atul Kumar Singh, and Pragati Shukla nee Misra

Abstract

This chapter explores the role of nanotechnology in revolutionizing crop breeding and advancing sustainable agriculture. Nanotechnology, operating at the nanoscale, offers innovative solutions for enhancing crop productivity, improving nutrient delivery, and enhancing crop protection. The various applications include Nano-fertilizers, nano-pesticides, and nano-herbicides which have demonstrated significant potential in improving nutrient uptake, controlling pests and diseases, and managing weeds while minimizing environmental impact and reducing several other pressing challenges in modern agriculture. Furthermore, nanotechnology is revolutionizing genetic modification by enabling precise and efficient gene delivery systems that accelerate breeding processes. Using nanoparticles (NPs), desirable traits like disease resistance and drought tolerance can be introduced into crops with greater precision and effectiveness. On the other hand, nanosensors play a crucial role in monitoring plant health, detecting pathogens, and optimizing resource use, leading to more efficient and sustainable farming practices. Despite challenges such as environmental safety, regula-

R. K. Pulamolu
University of Illinois, Urbana-Champaign, Urbana, IL, USA

P. K. Shukla
Department of Biological Sciences, Sam Higginbottom University of Agriculture, Technology and Sciences, Prayagraj, Uttar Pradesh, India

A. K. Singh
Department of Plant Physiology, KL University, Guntur, Andhra Pradesh, India

P. Shukla nee Misra (✉)
Centre for Tissue Culture Technology, Jacob Institute of Biotechnology and Bioengineering, Sam Higginbottom University of Agriculture, Technology and Sciences, Prayagraj, Uttar Pradesh, India
e-mail: pragati.misra@shiats.edu.in

P. M. Priyadarshan, R. Ortiz (eds.), *Plant Breeding 2050*,
https://doi.org/10.1007/978-981-95-0583-8_21

tory issues, and economic viability, continued research and innovation in nanotechnology are set to propel progress in crop breeding, growing a more sustainable and resilient future for agriculture.

Keywords

Nanotechnology · Agriculture · Breeding · Nanoparticles · Nanofertilizers

21.1 Introduction

The concept of nanotechnology was first introduced by the distinguished Nobel Laureate Richard Feynman in his renowned lecture at the California Institute of Technology on December 29, 1959. Today, nanotechnology is widely applied across various fields, including biology, chemistry, physics, engineering, medicine, textiles, painting manufacturing, cosmetics, and agriculture (Singh et al. 2011). At its core, nanotechnology refers to technologies that operate on the nanoscale, utilizing individual atoms and molecules to create functional structures with real-world applications. Nanotechnology involves manipulation and control of matter at the atomic and molecular scale, typically with one or more dimensions in the nanometer range, usually between 1 and 100 nm (Chellaram et al. 2014). Due to their increased surface area and improved mass transfer rates, nanoparticles exhibit enhanced chemical and biological activity, greater enzymatic reactivity, higher penetrability, catalytic properties, and quantum characteristics when compared to larger particles of the same material (Avella et al. 2007). The uses of NPs are endless; this chapter is going to discuss the role of NPs in agriculture, particularly in crop improvement and breeding. One of the most promising areas where nanotechnology can contribute to sustainable crop management is in agriculture. Nanomaterials in agriculture effectively reduce nutrient losses and enhance crop yields by enabling controlled and targeted chemical release, which promotes eco-friendly pest management systems (Usman et al. 2020; Benzon et al. 2015). These advancements improve agricultural productivity and sustainability, supporting efficient resource use. Moreover, nanomaterials in agriculture have demonstrated significant benefits, such as promoting efficient plant growth, increasing crop output, and improving nutritional quality. Agri-nanotechnology offers effective solutions for enhancing crop performance across diverse environmental conditions, leveraging advancements in science and technology to address key agricultural challenges and drive innovation. Subsequently, the rise of nanotechnology has led to the development of nano formulations for sustainable agriculture, which are proving to be more effective due to their reduced contamination levels (Fraceto et al. 2016). In the future, agriculture may increasingly rely on nano-enabled fertilizers, as nanotechnologies gain widespread acceptance for enhancing crop yields while maintaining a healthy agro-ecosystem, even in the face of environmental challenges (Gogos et al. 2012; Rajput et al. 2021; Singh et al. 2021; Lowry et al. 2019; Bandala and Berli 2019). As global food demand continues to rise, the role of nanotechnology in ensuring agricultural productivity and sustainability cannot be overstated (Lowry et al. 2019; Gogos et al. 2012). The various applications, benefits, and challenges of nanotechnology in crop breeding and agriculture in general are discussed in this chapter.

21.2 Applications of Nanotechnology in Crop Breeding

21.2.1 Nano-fertilizers

Nano-fertilizers are emerging as a game-changing innovation in agriculture, offering precise and efficient nutrient delivery systems. Unlike traditional fertilizers, which often lead to nutrient loss and environmental contamination, nano-fertilizers ensure targeted nutrient uptake and enhanced plant growth. This precision delivery is made possible through the unique morphological properties of nanocarriers, such as their high surface area-to-volume ratio and enhanced permeability (Hussein and Abou-Baker 2018) (Fig. 21.1).

Studies have demonstrated the efficiency of various nano-fertilizers across multiple crops. For instance, nano-Zinc (nano-Zn) has significantly improved biomass

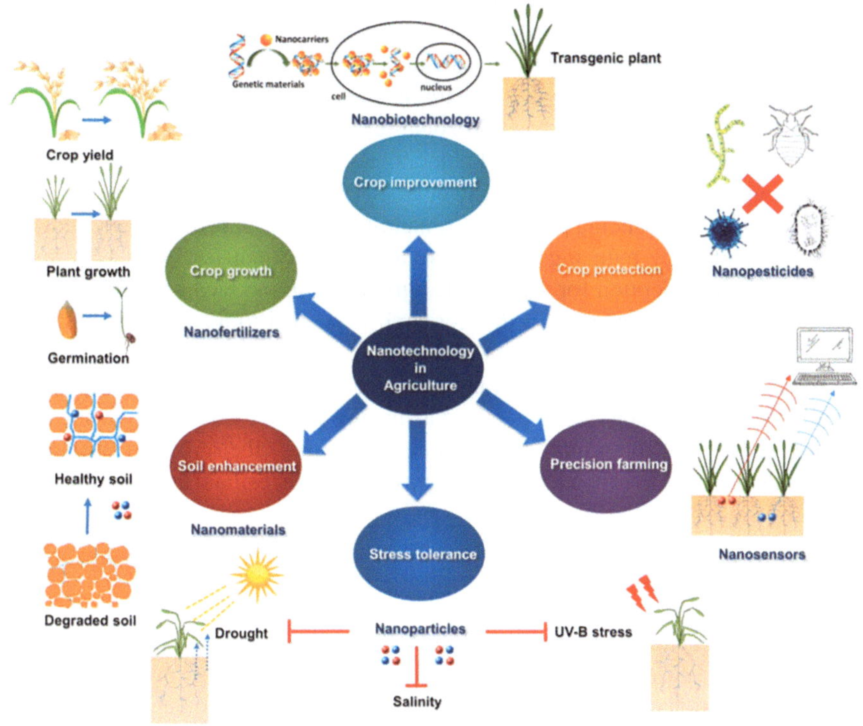

Fig. 21.1 Applications of nanotechnology in plant growth and crop protection. (Source: Shang et al. 2019)

production in rice and sunflower while enhancing drought resilience in maize by boosting nitrogen use efficiency (NUE) (Upadhyaya et al. 2015; Torabian et al. 2016; Farnia et al. 2015). Similarly, nano-chitosan (nCHT) has been shown to improve seed germination, seedling vigor, and biomass in crops like chickpea and tomato, emphasizing its potential for sustainable crop improvement (Saharan et al. 2015, 2016).

The mechanism of nutrient delivery via nano-fertilizers is both intricate and efficient. Nanoparticles penetrate plant tissues through stomata and root systems, facilitating better nutrient absorption and distribution. Once absorbed, they interact with cellular components, such as chloroplasts, to enhance photosynthetic activity. For example, mesoporous silica nanoparticles (MSNs) have optimized light-harvesting processes in chlorophyll–protein complexes, improving photosynthesis without inducing stress in plants (Rajput et al. 2021; Bhat et al. 2021).

Another significant advantage of nano-fertilizers is their ability to prime seeds for enhanced germination and growth. Nano-priming, as this process is known, enables nanoparticles to penetrate hard seed coatings, allowing better water uptake and accelerating the germination process. This technique has shown promising results in improving the yield potential of high-value crops (Acharya et al. 2020; Anand et al. 2020).

Despite these benefits, the application of nano-fertilizers is not without challenges. Understanding the translocation and accumulation of nanoparticles within plants is crucial for optimizing their use. Studies indicate that the uptake mechanism is influenced by factors such as nanoparticle size, composition, and cell wall pore diameter. For instance, smaller nanoparticles exhibit higher rates of internalization and translocation through xylem and phloem pathways, as seen in the case of negatively charged CeO_2 nanoparticles (Liu et al. 2019; Zhu et al. 2012).

In addition to enhancing productivity, nano-fertilizers contribute to environmental sustainability by reducing chemical runoff and nutrient leaching. This makes them an attractive option for sustainable agriculture in the face of global environmental challenges. However, further research is needed to assess their long-term impact on soil health and ecosystem balance.

By combining scientific innovation with environmental stewardship, nano-fertilizers represent a pivotal step toward achieving global food security while minimizing ecological footprints. As research progresses, their role in sustainable agriculture is likely to expand, paving the way for a more resilient agricultural future.

21.2.2 Nano-pesticides

Nano-pesticides encompass pesticide formulations that incorporate nanoscale active ingredients or utilize nanocarriers to deliver active substances more effectively. These formulations can be categorized into various types, including nanocapsules, nanospheres, nanoemulsions, and nanogels, each designed to improve the stability, solubility, and controlled release of pesticides which depicted their superiority over conventional pesticides.

For instance, copper oxide nanoparticles (CuO NPs) have proven effective against fungal pathogens like *Alternaria alternata* and *Botrytis cinerea*, significantly reducing disease incidence with lower application rates compared to traditional fungicides (Itodo et al. 2017). Similarly, silver nanoparticles (AgNPs) have demonstrated broad-spectrum antimicrobial activity, effectively controlling bacterial and fungal pathogens in crops such as tomato and wheat (Kumar et al. 2020). Insect pests, which often develop resistance to conventional pesticides, have also been successfully managed using nano-pesticides. Zinc oxide nanoparticles (ZnO NPs) have shown significant larvicidal activity against *Helicoverpa armigera*, a major pest in cotton production, with minimal environmental residue (Ghafariyan et al. 2013).

Moreover, nano-emulsions of neem oil encapsulated within nanoparticles have provided prolonged pest control against aphids and whiteflies by ensuring slow and sustained release of active ingredients (Koul et al. 2019). Another example is the use of silica nanoparticles to control the red flour beetle (*Tribolium castaneum*), where their abrasive nature damages the exoskeleton, leading to dehydration and death (Barik et al. 2020).

Polymeric nanoparticles have been used to deliver insecticides like pyrethroids, improving stability and reducing the risk of resistance development in pests such as *Plutella xylostella* (Sharma et al. 2019).

Nano-pesticides have also proven effective in combating vector-borne diseases that affect crops. For instance, titanium dioxide nanoparticles (TiO_2 NPs) have shown larvicidal effects against mosquito vectors, reducing populations that can transmit plant diseases like rice yellow mottle virus (Sundararajan et al. 2021). Furthermore, carbon-based nanomaterials, such as graphene oxide, have demonstrated potential in targeting nematodes like *Meloidogyne incognita*, a pest responsible for significant crop yield losses (Albanese et al. 2020).

As per a recent study conducted by da Silva Pontes et al. (2023), it is found that nano-pesticides act through two interconnected processes such as Toxicokinetics and Toxicodynamics. In Toxicokinetics, the process essentially governs the uptake, distribution, biotransformation, and excretion of nano-pesticides wherein there is an enhancement in adhesion and absorption rates of nano-pesticides as compared to traditional pesticides. However, in Toxicodynamics the biochemical and physiological responses of the organism towards nano-pesticides have been observed. For example, when nanoatrazine is used in lettuce, it induced oxidative stress by generating reactive oxygen species (ROS) which disrupted antioxidant enzyme cycles. However, the understanding of modes of action is still limited due to the diversity in nano-pesticide formulations.

Overall, by reducing pesticide usage, improving pest targeting, and minimizing environmental contamination, nano-pesticides represent a sustainable and innovative approach to modern agriculture. However, future research is needed to address potential risks such as nanoparticle toxicity, environmental persistence, and regulatory challenges to ensure their safe and widespread adoption.

21.2.3 Nano-herbicides

Nano-herbicides are an innovative advancement in weed management, offering precision-targeting capabilities that minimize environmental contamination and enhance efficacy compared to traditional herbicides. Recent studies have shown that nano-formulations improve the efficiency of active ingredients, allowing for lower application rates while achieving the desired weed control. Another significant development involves the use of mesoporous silica nanoparticles to encapsulate herbicides like glyphosate, ensuring slow and controlled release, thus preventing leaching and groundwater contamination (Khan and Yadav 2021).

Copper oxide nanoparticles (CuO NPs) have also demonstrated effectiveness as nano-herbicides. These nanoparticles selectively target weeds by entering through the root system and disrupting critical metabolic pathways, leading to stunted growth and eventual death of the weed. Studies have highlighted their success in managing invasive species such as *Echinochloa crus-galli* in rice fields, with minimal impact on nontarget crops (Itodo et al. 2017).

Additionally, polymeric nanocarriers have been used to deliver herbicides like atrazine and 2,4-Dichlorophenoxyacetic acid (2,4-D), enhancing their stability and bioavailability. These formulations have shown significant promise in controlling broadleaf weeds and grasses in maize and wheat crops (Barik et al. 2020; Chauhan et al. 2022). Another example is the application of zinc oxide nanoparticles (ZnO NPs) in combination with natural extracts, which has successfully inhibited weed growth in legume crops by interfering with photosynthesis and nutrient uptake (Rastogi and Singh 2021).

Recent advancements also include biodegradable nanocarriers designed to minimize residue accumulation in the environment. For instance, chitosan-based nanoparticles loaded with preemergent herbicides have provided prolonged weed control while breaking down into nontoxic by-products (Kumar and Jain 2023). Similarly, graphene oxide-based nano-herbicides have emerged as a promising solution for weed management by enhancing adhesion to leaf surfaces, thus increasing absorption and efficacy (Sundararajan et al. 2023).

21.2.4 Nano-sensors

Nano-sensors have emerged as a vital tool in modern agriculture, enabling precise monitoring of soil conditions, plant health, and environmental contaminants. By utilizing advanced nanotechnology, these sensors provide real-time data that can significantly improve decision-making processes in crop management. For example, nanosensors embedded with cadmium telluride (CdTe) nanoparticles have been developed to detect heavy metals like cadmium in soil, ensuring safer agricultural practices (Wang et al. 2016). Similarly, gold nanoparticles have been utilized in surface plasmon resonance (SPR) sensors to detect lead contamination in crops and soil (Ikram et al. 2019).

The versatility of nanosensors extends beyond detecting contaminants. They are also instrumental in identifying plant diseases and pathogens, offering early detection methods that can prevent widespread crop damage. For instance, gold nanorods immobilized on fiber optic cores have been employed to detect *Cymbidium mosaic virus* (CymMV) and *Odontoglossum ringspot virus* (ORSV) with high sensitivity (Lin et al. 2014). Moreover, iron oxide nanoparticles have been used to develop sensors for identifying viruses such as *Tomato Ring Spot Virus* (ToRSV) and *Arabis Mosaic Virus* (ArMV), showcasing their adaptability in various agricultural contexts (Zhang et al. 2013).

In addition to pathogen detection, nanosensors play a crucial role in monitoring pesticide residues. For example, copper oxide/multi-walled carbon nanotube (CuO/MWCNT) sensors have been designed for glyphosate detection, offering a high level of accuracy and sensitivity (Chang et al. 2016). Similarly, electrochemical sensors incorporating iron oxide nanoparticles have been employed to detect malathion residues in agricultural products, ensuring food safety and regulatory compliance (Prabhakar et al. 2016; Xie et al. 2018).

The integration of nanosensors into precision agriculture systems enhances sustainability by reducing resource wastage and optimizing inputs like water, fertilizers, and pesticides. Furthermore, advances in nanosensor technology have enabled the development of portable devices, allowing farmers to conduct on-site analyses without the need for extensive laboratory infrastructure. This democratization of advanced diagnostics paves the way for more inclusive and efficient agricultural practices.

Looking ahead, nanosensors are poised to play an even greater role in sustainable agriculture, particularly as innovations continue to reduce costs and improve accessibility. However, addressing challenges such as sensor durability, environmental impact, and data integration will be essential for their widespread adoption.

21.3 Genomic Interventions

Genomic interventions using nanoparticles for gene delivery in crops have emerged as a promising approach to introduce desirable traits such as drought resistance and pest tolerance. This technology offers significant potential to enhance crop productivity and resilience in the face of climate change and increasing global food demand (Jat et al. 2020; Demirer et al. 2019). Nanoparticle-mediated gene delivery overcomes many limitations of conventional plant transformation methods by enabling efficient transport of genetic cargo across the rigid plant cell wall barrier (Wu et al. 2023; Cunningham et al. 2018).Various types of nanoparticles have been explored for plant genetic engineering, including mesoporous silica nanoparticles (MSNs), carbon nanotubes (CNTs), metal nanoparticles, and polymeric nanoparticles (Wu et al. 2023; Torney et al. 2007). These nanocarriers can be functionalized to bind and protect DNA, RNA, or protein cargo and facilitate its uptake into plant cells and subcellular compartments (Cunningham et al. 2018; Kwak et al. 2019).

For example, MSNs with their large surface area and tunable pore size have been used to deliver plasmid DNA encoding reporter genes into tobacco and maize cells. Similarly, single-walled CNTs complexed with plasmid DNA achieved efficient chloroplast-targeted gene delivery in mature plant leaves. One of the key advantages of nanoparticle-mediated delivery is the ability to introduce genetic cargo into intact plant tissues without tissue culture or regeneration steps (Cunningham et al. 2018; Zhao et al. 2017). This allows rapid screening of gene function and trait improvement in mature plants.

Carbon nanotubes have successfully delivered CRISPR-Cas9 components into mature wheat leaves and roots, protecting the Cas9 protein and guide RNA from degradation while facilitating their co-delivery into plant cells. Similarly, magnetic nanoparticles carrying herbicide resistance genes enabled stable transformation in cotton plants via pollen magnetofection, with transgenes inherited across generations. Additionally, mesoporous silica nanoparticles (MSNs) have been used to deliver a drought tolerance gene to tomato plants, significantly enhancing water use efficiency.

For pest resistance, RNA interference (RNAi) approaches using nanocarriers have been explored. Clay nanosheets loaded with double-stranded RNA targeting essential insect genes provided sustained crop protection against viral infections when sprayed on plant leaves. The nanocarriers protected the RNA from degradation and facilitated its uptake by both plant cells and insect pests feeding on the plants. Despite significant advancements, challenges persist in optimizing nanoparticle design for efficient cargo delivery and expression across various plant species and tissues (Wu et al. 2023; Cunningham et al. 2018). The mechanisms governing nanoparticle uptake, intracellular trafficking, and cargo release in plants remain incompletely understood, necessitating further research.

Nanotechnology offers transformative benefits in crop improvement by revolutionizing genetic modification precision, accelerating breeding processes, and fostering the development of stress-resistant varieties.

In the realm of genetic modification, nanotechnology enables precise and efficient delivery of genetic material into plant cells, overcoming limitations of traditional methods. Nanoparticle-mediated gene delivery has been shown to significantly enhance transformation efficiency for both transient and stable genetic modifications across various plant species (Ahmar et al. 2021; Ijaz et al. 2023). For instance, carbon nanotubes and other nanoparticles penetrate plant cell walls to deliver DNA, RNA, or proteins directly to specific subcellular targets, such as chloroplasts (Raza et al. 2023). These nanoparticles not only protect genetic cargo from degradation but also allow for controlled release, improving stability and gene expression (Soutter 2019; Raza et al. 2023).

Nanotechnology also accelerates crop breeding by streamlining phenotyping and genotyping processes. Nanomaterials facilitate rapid screening and selection of desired traits, enabling high-throughput analysis that increases breeding efficiency (Ahmar et al. 2021). The integration of nanoparticle-mediated delivery of CRISPR/Cas9 components further enhances the precision and speed of genome editing in plants (Ijaz et al. 2023). Combined with speed breeding techniques, these advancements allow researchers to drastically shorten breeding cycles, potentially achieving multiple generations per year (Ahmar et al. 2021).

Moreover, nanotechnology is pivotal in developing stress-resistant crop varieties to combat abiotic and biotic challenges. Nanoparticles such as zinc oxide, silver, and carbon nanotubes have demonstrated the ability to enhance plant growth, photosynthesis, and stress tolerance when applied directly or as carriers for other compounds (Wu et al. 2023; Khan et al. 2022). Nanomaterials also enable precise delivery of stress-responsive genes or protective compounds to targeted plant tissues, amplifying their effectiveness (Liu et al. 2021; Khan et al. 2022). Additionally, nanosensors play a crucial role in early stress detection, providing timely insights for effective intervention and management (Liu et al. 2021).

By leveraging these advanced approaches, researchers are accelerating the development of crop varieties with enhanced resilience to drought, salinity, extreme temperatures, and other environmental stresses. These nanotechnology-driven innovations have the potential to revolutionize sustainable agriculture, contributing significantly to global food security in the face of climate change and increasing population demands (Raza et al. 2023; Khan et al. 2022).

21.4 Current Research and Developments

Current research and developments on the role of nanotechnology in crop breeding are advancing rapidly, offering innovative solutions to enhance crop improvement and address global agricultural challenges. Nanotechnology is revolutionizing genetic modification techniques, accelerating breeding processes, and facilitating the development of stress-resistant crop varieties while improving nutrient delivery and crop protection strategies.

In the area of precision genetic modification, nanoparticle-mediated systems have significantly enhanced transformation efficiency for both transient and stable genetic modifications across various plant species (Soutter 2012). Carbon nanotubes and other nanoparticles penetrate plant cell walls, delivering DNA, RNA, or proteins directly to specific subcellular targets, such as chloroplasts, to improve the precision of genetic modifications (Fig. 21.2) (Soutter 2012). Furthermore, integrating nanotechnology with CRISPR/Cas9 systems has enabled more accurate genome editing, exemplified by advancements in enhancing disease resistance in wheat (Naik et al. 2022).

Nanotechnology has also accelerated crop breeding by streamlining phenotyping and genotyping processes. Nanomaterials enable high-throughput screening and selection of desired traits, improving breeding efficiency (Soutter 2012). The nanoparticle-mediated delivery of CRISPR/Cas9 components further increases the speed and precision of genome editing in plants (Naik et al. 2022). Researchers are also employing artificial intelligence and machine learning to develop "digital twins," simulating interactions between nanocarrier molecules and plants to design more efficient targeted delivery systems (Bernstein 2024).

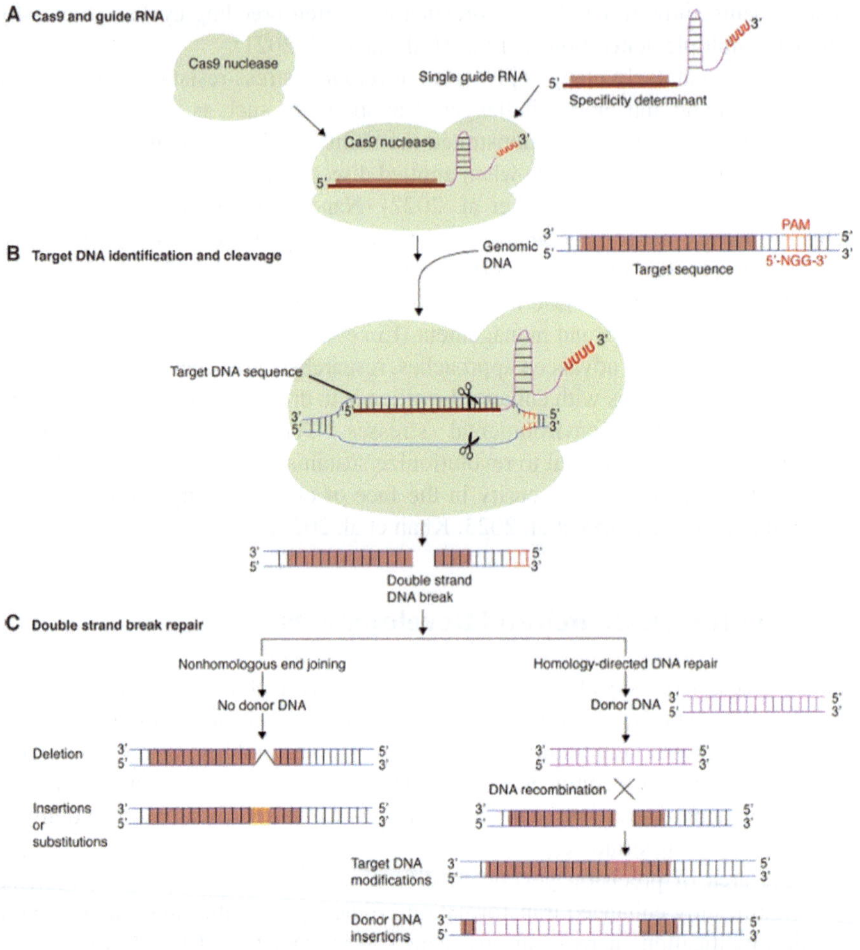

Fig. 21.2 Targeted genome editing using CRISPR-Cas9. (**a**) The CRISPR Cas9 system consists of a Cas9 protein and one or several guide RNA. Guide RNAs determine target DNA specificity by sequence complementarity. (**b**) Guide RNA and Cas9 protein form a binary complex that specifically cleaves target DNA creating a double-strand DNA break. (**c**) Cellular DNA repair mechanisms, nonhomologous end joining (NHEJ) and homology-directed repair (HDR), repairs the double strand DNA break. In the process, short insertions, deletions, nucleotide substitutions, or gene insertion may occur. (El-Mounadi et al. 2020)

In developing stress-resistant crops, nanoparticles such as zinc oxide, silver, and carbon nanotubes improve plant growth, photosynthesis, and stress tolerance when applied directly or as carriers for other compounds (Soutter 2012). Innovative nanoparticles have been designed to release stress-soothing hormones like spermidine into plant tissues upon detecting chemical stress signals, aiding plants in coping with extreme weather, and protecting yields in a changing climate (Bryce 2024). Nano-enabled precision delivery systems are transforming agriculture by improving

Table 21.1 Effect of nanoparticles on plant species under abiotic stress

Plant species	Nanoparticles	Size	Concentration	Stress type	Response	Citations
Zea mays L.	Cu	30–40 nm	3.33, 4.44, 5.55 mg L^{-1}	Drought	Higher biomass and grain yield	Van Nguyen et al. (2021)
Oryza sativa L.	ZnO	30 nm	50 mg L^{-1}	Chilling	Regulated antioxidative system and chilling response transcription factors	Song et al. (2021)
Solanum melongena L.	ZnO	–	50 and 100 ppm	Drought stress	Improved growth characteristics and increased fruit yield	Semida et al. (2021)
Triticum aestivum L.	TiO_2	10–25 nm	60 ppm	Salinity stress	Enhancement of seed vigor, leaf water status, and antioxidant enzyme activities	Shah et al. (2021)
Trigonella foenum-graecum	ZnO	10–30 nm	1000, 3000 ppm	Salinity	Upregulation of protein and proline levels, enhancement of antioxidant activities, reduced H_2O_2 and MDA levels	Noohpisheh et al. (2021)
Dracocephalum moldavica	TiO_2	20–30 nm	50, 100, 200 mg L^{-1}	Salinity stress	Improved agronomic traits and increased antioxidant enzyme activity and essential oil content	Gohari et al. (2020)
Glycine max	Ag NP	15 nm	5 ppm	Flooding stress	Enhanced root length/weight and hypocotyl length/weight	Hashimoto et al. (2020)
Abelmoschus esculentus L.	ZnO	16–35 nm	10 mg L^{-1}	Salt stress	Enhanced photosynthetic pigments, antioxidant activity, reduced proline, and soluble sugar accumulation	Alabdallah and Alzahrani (2020)
Musa acuminata	Si	–	200, 400, 600 mg L^{-1}	Salinity stress	Mitigated oxidative stress	Mahmoud et al. (2020)
Zea mays L.	ZnO	37.7 ± 15.5 nm	100 mg L^{-1}	Drought	Enhanced melatonin synthesis and metabolism	Sun et al. (2020)

Table 21.2 Effect of nanoparticles on plant species under biotic stress

Plant species	Nanoparticles	Size	Concentration	Stress type	Response	Citations
Bougainvillea	CuO	5–20 nm	80, 100 ppm	*A. niger*	Antifungal	Maisa'a and Awwad (2021)
Malus species	CuO	80 nm	0.05–1 mg mL^{-1}	*Alternaria mali, Diplodia seriata, Botryosphaeria dothidea*	Antifungal	Ahmad et al. (2020)
Triticum aestivum L.	Ag NP	19.8–44.9 nm	5–20 μg mL^{-1}	*F. Graminearum*	Antifungal	Ibrahim et al. (2020)
Lycopersicon esculentum	Au-chitosan	80 nm	25–75 μg mL^{-1}	*Fusarium oxysporum*	Antifungal	Lips et al. (2020)
Saccharum officinarum	ZnO	72–183 nm	3–20 ppm	*Holotrichia* sp.	Insecticidal	Shukla et al. (2020)
Oryza sativa	Ag NP	100–250 nm	–	*R. solani, F. moniliforme*	Antifungal	Manikandaselvi et al. (2020)
Gossypium sp.	Ag NP	63–85 nm	1 mM	*Earias insulana*	Insecticidal	Al Shater et al. (2020)
Pongamia pinnata	Ag NP	10–25 nm	0.0062–1.6 mg mL^{-1}	*P. ultimum*	Antifungal	Khatua et al. (2020)
Oryza sativa	Chitosan–Fe_2O_3	50–70 nm	0.25–1%	*Rhizopus oryzae*	Antifungal	Saqib et al. (2019)

crop resilience to biotic and abiotic stresses (Koppes 2024). The various types and effects of nanoparticles in combating several abiotic and biotics stresses are mentioned in Tables 21.1 and 21.2 of this chapter.

Nanotechnology is also advancing nutrient delivery and crop protection. Nano or smart fertilizers encapsulate nutrients in nanomaterials for controlled and targeted release, enhancing efficiency and minimizing environmental impact compared to traditional fertilizers (Soutter 2012). Similarly, nanocarriers are being developed to deliver pesticides, herbicides, and fungicides to specific biological targets in plants, reducing off-target effects and environmental contamination (Bernstein 2024). Nanoscale carriers, such as clay nanotubes, further optimize the delivery of fertilizers and pesticides, achieving higher efficacy while mitigating environmental risks (Soutter 2012).

By leveraging these advancements, nanotechnology is playing an essential role in developing sustainable solutions to meet global food security challenges. These transformative innovations enable more precise genetic modifications, efficient breeding processes, and stress-resistant varieties while supporting environmentally conscious nutrient delivery and crop protection practices.

21.5 Challenges and Limitations

Nanotechnology offers transformative potential in crop breeding, enhancing agricultural productivity and sustainability. However, several challenges must be addressed to ensure its successful implementation. Among these, potential environmental and health risks stand out as critical concerns. The long-term effects of nanoparticles on plants, ecosystems, and human health remain insufficiently understood, raising questions about their safety (Soumya et al. 2023; Chaud et al. 2021). For instance, bioaccumulation of nanoparticles in soil, water, and food chains poses threats to environmental integrity and human well-being (Chaud et al. 2021). Specific examples include the toxic effects of carbon nanotubes, which can induce cell death in plants and disrupt soil microbial populations by generating reactive oxygen species (Chaud et al. 2021). Additionally, nanomaterials' ability to cross biological barriers, such as the blood-brain barrier, further exacerbates exposure risks, potentially causing irreversible damage to vital organs (Chaud et al. 2021). To address these risks, comprehensive research is needed to balance the potential benefits and unintended consequences of nanotechnology in agriculture, ensuring sustainable practices (Soumya et al. 2023).

Regulatory hurdles and public acceptance also pose significant challenges to the adoption of nanotechnology in agriculture. The regulatory landscape is inconsistent across regions, with some nations adopting lenient approaches for genome-edited crops, while others, like the European Union and New Zealand, apply stringent regulations akin to those for genetically modified organisms (GMOs) (Conrow 2021). Furthermore, many countries lack specific frameworks tailored to nano-enabled agricultural products (Arya and Manghwar 2023). This regulatory ambiguity is compounded by low public acceptance of biotech crops, particularly in regions such as Europe (Lassoued et al. 2018). Bridging this gap requires evidence-based regulations that consider ethical and socioeconomic factors, as well as targeted efforts to increase public awareness of biotechnological advancements (Conrow 2021). Promoting understanding and fostering rational discourse can pave the way for amendments to current regulatory laws, aligning them with modern agricultural needs (Conrow 2021).

Economic and scalability issues present additional barriers to the widespread implementation of nanotechnology in agriculture. The high cost of nanofertilizers compared to conventional fertilizers limits their adoption among farmers (Su et al. 2022). Moreover, the scalability of nanotechnology-based solutions remains a challenge, as their feasibility in real-world farming practices and compatibility with existing agricultural infrastructure require further analysis (Soumya et al. 2023). Addressing these issues involves optimizing the dosage and efficiency of nanomaterials to maximize returns while minimizing costs (Su et al. 2022). Additionally, developing cost-effective production methods and improved formulations can enhance the economic viability of nanotechnology for farmers (Soumya et al. 2023).

In conclusion, while nanotechnology holds immense promise for revolutionizing crop breeding and agriculture, tackling challenges related to environmental and health risks, regulatory hurdles, and economic feasibility is essential. By addressing these limitations through research, regulation, and public engagement, nanotechnology can contribute to sustainable agricultural development and global food security.

21.6 Future Prospects of Nanotechnology in Crop Breeding

Nanotechnology holds immense potential for revolutionizing crop breeding and addressing global food security challenges. By integrating with advanced breeding technologies, nanotechnology is paving the way for transformative changes in agricultural practices. For instance, the combination of CRISPR/Cas9 genome editing with nanoparticle-mediated delivery systems has demonstrated remarkable efficiency, overcoming limitations of traditional transformation methods and ensuring precise gene delivery and expression in plants (Ahmar et al. 2021; Chaud et al. 2021). Additionally, researchers are exploring the synergy between nanotechnology and speed breeding techniques, which could significantly accelerate crop improvement cycles and facilitate the rapid development of new varieties with desirable traits (Ahmar et al. 2021). Furthermore, the integration of nanotechnology-enabled sensors and smart delivery systems with precision agriculture techniques is optimizing resource use and enhancing crop management (University of Birmingham, June 2021).

Nanotechnology offers promising solutions to global food security challenges through several innovative avenues. Nano-enabled fertilizers and growth promoters enhance nutrient uptake and plant growth, potentially boosting yields per unit area (Sharma and Kumar 2024). Nanoparticles designed to improve crop resilience against biotic and abiotic stresses are crucial in combating the adverse impacts of climate change (Liu et al. 2021). Moreover, nano-biofortification techniques are being developed to enrich the nutritional content of crops, addressing malnutrition among vulnerable populations (Khan et al. 2022).

Emerging trends in nano-assisted crop breeding showcase exciting possibilities. Advanced nanobiosensors are enabling early detection of plant stresses and pathogens, facilitating timely interventions and reducing crop losses (Wu et al. 2023). Targeted gene delivery systems using nanocarriers are revolutionizing genetic engineering by ensuring precise and efficient delivery of genetic material to specific plant tissues (Liu et al. 2021). Researchers are also developing smart nanomaterials that respond to environmental cues, releasing nutrients or protective compounds as needed, thereby enhancing crop performance (Naik et al. 2022).

However, the adoption of nanotechnology in crop breeding raises important ethical and societal considerations. Concerns about the long-term effects of nanomaterials on human health and the environment necessitate ongoing research and transparent risk assessments (Seleiman et al. 2022). Developing appropriate regulatory frameworks for nano-enabled agricultural products is another critical challenge that must be addressed to ensure safety and public trust (Sharma and Kumar 2024). Additionally, equitable access to these innovations is essential to prevent widening the gap between resource-rich and resource-poor farming communities (Seleiman et al. 2022).

Compared to conventional breeding methods, nanotechnology-assisted breeding offers significant advantages. Nano-enabled techniques provide unparalleled precision in genetic modifications and targeted delivery of agrochemicals (Chaud et al. 2021). Nanoparticle-mediated gene delivery can improve transformation efficiency across various plant species (Wu et al. 2023). The integration of nanotechnology with other breeding technologies has the potential to drastically reduce breeding cycle durations, allowing for multiple generations of crops within a year (Ahmar et al. 2021). Moreover, nano-enabled agricultural practices hold promise for reducing the environmental footprint of crop production compared to traditional methods (Naik et al. 2022; Sharma and Kumar 2024).

21.7 Conclusion

Nanotechnology is poised to revolutionize crop breeding, offering transformative solutions to enhance agricultural productivity and addressing global food security challenges. By integrating with advanced breeding technologies such as CRISPR/Cas9 and speed breeding, nanotechnology is accelerating the development of improved crop varieties with enhanced traits. Nano-enabled fertilizers, pesticides, and herbicides enhance nutrient delivery, improving crop protection and minimizing environmental impact. Furthermore, nanotechnology is enabling the development of stress-resistant crops, optimizing resource use, and improving the nutritional quality of food. While challenges related to environmental safety, regulatory hurdles, and economic feasibility remain, ongoing research and innovation in nanotechnology will continue to drive advancements in crop breeding. By addressing these challenges and fostering responsible development and implementation, nanotechnology can play a pivotal role in ensuring food security and sustainable agricultural practices for a growing global population.

References

Acharya P, Jayaprakasha GK, Crosby KM, Jifon JL, Patil BS (2020) Nanoparticle-mediated seed priming improves germination, growth, yield, and quality of watermelons (*Citrullus lanatus*) at multi-locations in Texas. Sci Rep 10:5037

Ahmad H, Venugopal K, Bhat AH, Kavitha K, Ramanan A, Rajagopal K, Srinivasan R, Manikandan E (2020) Enhanced biosynthesis of copper oxide nanoparticles (CuO-NPs) for their antifungal activity toxicity against major phytopathogens of apple orchards. Pharm Res 37(1):1–12

Ahmar S, Mahmood T, Fiaz S, Mora-Poblete F, Shafique MS, Chattha MS, Jung KH (2021) Advantage of nanotechnology-based genome editing system and its application in crop improvement. Front Plant Sci 12:663849

Al Shater H, Fouad H, Alwasel S, Al-Otaibi S, Al-Ghamdi A, El-Sheikh T (2020) Synthesis, phytochemical screening and toxicity measuring against *Earias insulana* (Boisd.) (*Lepidoptera: Noctuidae*) of silver nanoparticles from *Origanum marjorana* extract in the field. Egypt Acad J Biol Sci F Toxicol Pest Control 12(1):175–184

Alabdallah NM, Alzahrani HS (2020) The potential mitigation effect of ZnO nanoparticles on [*Abelmoschus esculentus* L. Moench] metabolism under salt stress conditions. Saudi. J Biol Sci 27(12):3132–3137

Albanese A, Tang PS, Chan WC (2020) The effect of nanoparticle size, shape, and surface chemistry on biological systems. Annu Rev Biomed Eng 14:1–16

Anand KV, Anugraga A, Kannan M, Singaravelu G, Govindaraju K (2020) Bio-engineered magnesium oxide nanoparticles as nano-priming agent for enhancing seed germination and seedling vigour of green gram (*Vigna radiata* L.). Mater Lett 8:127792

Arya S, Manghwar H (2023) Regulation and safety measures for nanotechnology-based agri-products. Front Genome Ed 5:1200987

Avella M, Bruno G, Errico ME, Gentile G, Piciocchi N, Sorrentino A, Volpe MG (2007) Innovative packaging for minimally processed fruits. Packag Technol Sci 20:325–335

Bandala ER, Berli M (2019) Engineered nanomaterials (ENMs) and their role at the nexus of food, energy, and water. Mater Sci Energy Technol 2:29–40

Barik TK, Sahu B, Swain V (2020) Nano-silica—from medicine to pest control. Parasitol Res 103(2):253–258

Benzon HR, Rubenecia MR, Ultra VU Jr, Lee SC (2015) Nano-fertilizer affects the growth, development, and chemical properties of rice. Int J Agron Agric Res 7:105–117

Bernstein J (2024) Better farming through nanotechnology. UCR News

Bhat JA, Rajora N, Raturi G, Sharma S, Dhiman P, Sanand S, Shivaraj SM, Sonah H, Deshmukh R (2021) Silicon nanoparticles (SiNPs) in sustainable agriculture: major emphasis on the practicality, efficacy and concerns. Nanoscale Adv 3:4019

Bryce E (2024) Can medical advancements make agriculture more sustainable? Anthropocene Magazine

Chang YC, Lin YS, Xiao GT, Chiu TC, Hu CC (2016) A highly selective and sensitive nanosensor for the detection of glyphosate. Talanta 161:94–98

Chaud M, Souto EB, Zielinska A, Severino P, Batain F, Oliveira-Junior J, Alves T (2021) Nanopesticides in agriculture: benefits and challenge in agricultural productivity, toxicological risks to human health and environment. Toxics 9(6):131

Chauhan D, Gupta S, Sharma AK (2022) Polymeric nanocarriers for controlled delivery of 2,4-D in weed management. J Weed Sci Technol 50(3):198–207

Chellaram C, Murugaboopathi G, John AA, Sivakumar R, Ganesan S, Krithika S, Priya G (2014) Significance of nanotechnology in food industry. APCBEE Procedia 8:109–113

Conrow J (2021) Ease regulatory burden to support 'new revolution' in crop breeding, researchers urge. Alliance for Science

Cunningham FJ, Goh NS, Demirer GS, Matos JL, Landry MP (2018) Nanoparticle-mediated delivery towards advancing plant genetic engineering. Trends Biotechnol 36(9):882–897

da Silva Pontes M, de Lima Forini MM, Caires ARL, Santiago EF, Tripathi DK, Kah M, Grillo R (2023) Nanopesticides: modes of action and impacts. In: Nano-enabled sustainable and precision agriculture. Elsevier, pp 79–103

Demirer GS, Zhang H, Matos JL, Goh NS, Cunningham FJ, Sung Y, Chang R, Aditham AJ, Chio L, Cho MJ (2019) High aspect ratio nanomaterials enable delivery of functional genetic material without DNA integration in mature plants. Nat Nanotechnol 14(5):456–464

El-Mounadi K, Morales-Floriano ML, Garcia-Ruiz H (2020) Principles, applications, and biosafety of plant genome editing using CRISPR-Cas9. Front Plant Sci 11:56

Farnia A, Omidi MM, Farnia A (2015) Effect of nano-zinc chelate and nano-biofertilizer on yield and yield components of maize (*Zea mays* L.) under water stress condition. Indian J Nat Sci 5:4614–4646

Fraceto LF, Grillo R, de Medeiros GA, Scognamiglio V, Rea G, Bartolucci C (2016) Nanotechnology in agriculture: which innovation potential does it have? Front Environ Sci 4:20

Ghafariyan MH, Malakouti MJ, Dadpour MR, Stroeve P, Mahmoudi M (2013) Effects of magnetite nanoparticles on soybean chlorophyll. Environ Sci Technol 47(19):10645–10652

Gogos A, Knauer K, Bucheli TD (2012) Nanomaterials in plant protection and fertilization: current state, foreseen applications, and research priorities. J Agric Food Chem 60:9781–9792

Gohari G, Mohammadi A, Akbari A, Panahirad S, Dadpour MR, Alirezalu A, Fotopoulos V (2020) Titanium dioxide nanoparticles (TiO_2 NPs) promote growth and ameliorate salinity stress effects on essential oil profile and biochemical attributes of *Dracocephalum moldavica*. Sci Rep 10:912

Hashimoto T, Miura K, Takahashi F, Yamaguchi-Shinozaki K, Shinozaki K, Tanaka K, Nakamura S, Yamamoto Y, Matsui M, Shimada H et al (2020) Comparative analysis of the effect of inorganic and organic chemicals with silver nanoparticles on soybean under flooding stress. Int J Mol Sci 21(4):1300

Hussein MM, Abou-Baker NH (2018) The contribution of nano-zinc to alleviate salinity stress on cotton plants. R Soc Open Sci 5:171809

Ibrahim E, Fouad H, Zhang M, Zhang Y, Qiu W, Yan C, Li B, Mo J, Chen J (2020) Green-synthesization of silver nanoparticles using endophytic bacteria isolated from garlic and its antifungal activity against wheat Fusarium head blight pathogen *Fusarium graminearum*. Nanomaterials 10(2):219

Ijaz M, Khan F, Ahmed T, Noman M, Zulfiqar F, Rizwan M, Chen J, Siddique KH, Li B (2023) Nanobiotechnology to advance stress resilience in plants: current opportunities and challenges. Mater Today Bio 22:100759

Ikram F, Qayoom A, Aslam Z, Shah MR (2019) Epicatechin-coated silver nanoparticles as highly selective nanosensors for the detection of Pb^{2+} in environmental samples. J Mol Liq 277:649–655

Itodo HU, Nnamonu LA, Wuana RA (2017) Green synthesis of copper chitosan nanoparticles for controlled release of pendimethalin. Asian J Chem Sci 2:1–10

Jat SK, Bhattacharya J, Sharma MK (2020) Nanomaterial based gene delivery: a promising method for plant genome engineering. J Mater Chem B 8(19):4165–4175

Khan MA, Yadav SK (2021) Mesoporous silica nanoparticles for controlled herbicide delivery. Agriculture 13(10):2031

Khan MN, Khan S, Al-Ghamdi AA, Al-Huqail AA (2022) Role of nanoparticles in enhancing crop tolerance to abiotic stress: a comprehensive review. Front Plant Sci 13:946717

Khatua A, Priyadarshini E, Rajamani P, Patel AK, Jha B, Suar M, Paul S, Singh SK (2020) Phytosynthesis, characterization and fungicidal potential of emerging gold nanoparticles using *Pongamia pinnata* leaf extract: a novel approach in nanoparticle synthesis. J Clust Sci 31(1):125–131

Koppes S (2024) Jul 23. Purdue University News, Researchers examine novel methods for improving agriculture

Koul O, Walia S, Dhaliwal GS (2019) Essential oils as green pesticides: potential and constraints. Biopestic Int 15(1):1–12

Kumar S, Jain RK (2023) Biodegradable chitosan nanoparticles for pre-emergent herbicide delivery in crop protection. Front Agron 5:1497041

Kumar S, Sachdeva S, Chaudhary S, Chaudhary GR (2020) Assessing the potential application of bio-compatibly tuned nanosensor of Yb2O3 for selective detection of imazapyr in real samples. Colloids Surf A Physicochem Eng Asp 593:124612

Kwak SY, Lew TT, Sweeney CJ, Koman VB, Wong MH, Bohmert-Tatarev K, Snell KD, Seo JS, Chua NH, Strano MS (2019) Chloroplast-selective gene delivery and expression in planta using chitosan-complexed single-walled carbon nanotube carriers. Nat Nanotechnol 14(5):447–455

Lassoued R, Smyth SJ, Phillips PW, Hesseln H (2018) Regulatory uncertainty around new breeding techniques. Front Plant Sci 9:1291

Lin CW, Tsai CH, Lai YH, Chang HC (2014) Highly sensitive and specific detection of plant viruses using nanorod-based optical fiber plasmon resonance immunosensors. Biosens Bioelectron 61:55–61

Lips A, Albishi T, Alharbi N, Alshahrani M, Alzahrani F, Alqahtani A, Alghamdi S (2020) Evaluation of the antifungal activity of gold-chitosan and carbon nanoparticles on *Fusarium oxysporum*. Agronomy 10(8):1143

Liu M, Feng S, Ma Y, Xie C, He X, Ding Y, Zhang J, Luo W, Zheng L, Chen D et al (2019) Influence of surface charge on the phytotoxicity, transformation, and translocation of CeO_2 nanoparticles in cucumber plants. ACS Appl Mater Interfaces 11(30):16905–16913

Liu C, Zhou H, Zhou J (2021) The applications of nanotechnology in crop production. Molecules 26(23):7070

Lowry GV, Avellan A, Gilbertson LM (2019) Opportunities and challenges for nanotechnology in the agri-tech revolution. Nat Nanotechnol 14(6):517–525

Mahmoud LM, Abd El-Gawad HG, El-Borhamy AM, El-Badri AM, El-Mogy MM, Abdel-Salam EF, Al-Sayed E (2020) Silicon nanoparticles mitigate oxidative stress of in vitro-derived banana (*Musa acuminata* 'Grand Nain') under simulated water deficit or salinity stress. S Afr J Bot 132:155–163

Maisa'a W, Awwad AM (2021) A novel route for the synthesis of copper oxide nanoparticles using *Bougainvillea* plant flowers extract and antifungal activity evaluation. Chem Int 7(1):71–78

Manikandaselvi S, Sathya V, Vadivel V, Sampath N, Brindha P (2020) Evaluation of bio-control potential of AgNPs synthesized from *Trichoderma viride*. Adv Nat Sci Nanosci Nanotechnol 11(3):035004

Naik BJ, Shimoga G, Kim SC, Manjulatha M, Reddy CS, Palem RR, Kumar M, Kim SY, Lee SH (2022) CRISPR/Cas9 and nanotechnology pertinence in agricultural crop refinement. Front Plant Sci 13:843575

Noohpisheh Z, Aynehband A, Ghorbani A, Zarei M, Mojaddam M, Akbari GA, Mahmoudi M (2021) Effect of the foliar application of zinc oxide nanoparticles on some biochemical and physiological parameters of *Trigonella foenum-graecum* under salinity stress. Plant Biosyst 155(2):267–280

Prabhakar N, Thakur H, Bharti A, Kaur N (2016) Chitosan-iron oxide nanocomposite based electrochemical aptasensor for determination of malathion. Anal Chim Acta 939:108–116

Rajput VD, Singh A, Minkina T, Rawat S, Mandzhieva S, Sushkova S, Shuvaeva V, Nazarenko O, Rajput P, Komariah et al (2021) Nano-enabled products: challenges and opportunities for sustainable agriculture. Plants 10:2727

Rastogi A, Singh PK (2021) Application of zinc oxide nanoparticles for enhancing weed management in legume crops. Environ Sci Pollut Res 28(14):17627–17639

Raza A, Charagh S, Salehi H, Abbas S, Saeed F, Poinern GE, Siddique KH, Varshney RK (2023) Nano-enabled stress-smart agriculture: can nanotechnology deliver drought and salinity-smart crops? J Sustain Agric Environ 2(3):189–214

Saharan V, Sharma G, Yadav M, Choudhary MK, Sharma SS, Pal A, Raliya R, Biswas P (2015) Synthesis and in vitro antifungal efficacy of Cu–chitosan nanoparticles against pathogenic fungi of tomato. Int J Biol Macromol 75:346–353

Saharan V, Kumaraswamy RV, Choudhary RC, Kumari S, Pal A, Raliya R, Biswas P (2016) Cu-chitosan nanoparticle mediated sustainable approach to enhance seedling growth in maize by mobilizing reserved food. J Agric Food Chem 64:6148–6155

Saqib S, Naqvi ZA, Rizwan M, Shahzad T, Bashir S, Nadeem M (2019) Organometallic assembling of chitosan-iron oxide nanoparticles with their antifungal evaluation against *Rhizopus oryzae*. Appl Organomet Chem 33(6):e5190

Seleiman MF, Al-Suhaibani N, Ali N, Akmal M, Alotaibi M, Refay Y, Dindaroglu T, Abdul-Wajid HH, Battaglia ML (2022) Harnessing the known and unknown impact of nanotechnology on enhancing food security and reducing postharvest losses: constraints and future prospects. Agronomy 12(7):1657

Semida WM, Abdelkhalik A, Rady MM, Marey RA, Abd El-Mageed TA, Howladar SM, Abd El-Mageed SA, Ali EF (2021) Foliar application of zinc oxide nanoparticles promotes drought stress tolerance in eggplant (*Solanum melongena* L.). Plants 10(2):421

Shah T, Latif S, Khan MN, Muhammad N, Ali A, Rizwan M, Shahid MJ, Alkahtani J, Alfarraj S (2021) Seed priming with titanium dioxide nanoparticles enhances seed vigor, leaf water status, and antioxidant enzyme activities in maize (*Zea mays* L.) under salinity stress. J King Saud Univ Sci 33:101207

Shang Y, Hasan MK, Ahammed GJ, Li M, Yin H, Zhou J (2019) Applications of nanotechnology in plant growth and crop protection: a review. Molecules 24(14):2558

Sharma A, Kumar V (2024) Ethical and regulatory perspective on nanotechnology in agriculture. Just Agriculture

Sharma A, Kumar V, Shahzad B (2019) Worldwide pesticide usage and its impacts on ecosystem. Front Environ Sci 7:43

Shukla G, Gaurav SS, Singh A (2020) Synthesis of mycogenic zinc oxide nanoparticles and preliminary determination of its efficacy as a larvicide against white grubs (*Holotrichia* sp.). Int. Nano Lett 10(2):131–139

Singh RP, Shukla VK, Yadav RS, Sharma PK, Singh PK, Pandey AC (2011) Biological approach of zinc oxide nanoparticles formation and its characterization. Adv Mater Lett 2:313–317

Singh H, Sharma A, Bhardwaj SK, Arya SK, Bhardwaj N, Khatri M (2021) Recent advances in the applications of nanoagrochemicals for sustainable agricultural development. Environ Sci Process Impacts 23:213–239

Song Y, Jiang M, Zhang H, Li R (2021) Zinc oxide nanoparticles alleviate chilling stress in rice (*Oryza sativa* L.) by regulating antioxidative system and chilling response transcription factors. Molecules 26(8):2196

Soumya D, Delvadiya IR, Dinesh MS, Ginoya AV (2023) Nanotechnology-enabled advancements in plant breeding. Biol Forum Int J 15(5):585–592

Soutter W (2012) Nanotechnology in agriculture. AZoNano

Soutter W (2019) Nanotechnology in agriculture: an overview. AZoNano

Su Y, Zhou X, Meng H, Xia T, Liu H, Rolshausen P, Roper C, McLean JE, Zhang Y, Keller AA et al (2022) Cost–benefit analysis of nanofertilizers and nanopesticides emphasizes the need to improve the efficiency of nanoformulations for widescale adoption. Nat Food 3(12):1050–1059

Sun L, Song F, Guo J, Zhu X, Liu S, Liu F, Li X (2020) Nano-ZnO-induced drought tolerance is associated with melatonin synthesis and metabolism in maize. Int J Mol Sci 21(3):782

Sundararajan B, Thomas J, Palanisamy M (2021) Ecofriendly management of vector-borne diseases using titanium dioxide nanoparticles. Environ Sci Nano 8(3):689–705

Sundararajan B, Thomas J, Palanisamy M (2023) Graphene oxide-based nano-herbicides for efficient weed control. Environ Sci Nano 10(3):502–518

Torabian S, Zahedi M, Khoshgoftarmanesh A (2016) Effect of foliar spray of zinc oxide on some antioxidant enzymes activity of sunflower under salt stress. J Agric Sci Technol 18:1013–1025

Torney F, Trewyn BG, Lin VS-Y, Wang K (2007) Mesoporous silica nanoparticles deliver DNA and chemicals intoplants. Nat Nanotechnol 2:295–300

University of Birmingham (2021) Nanotech and AI could hold key to unlocking global food security challenge. Phys.org

Upadhyaya H, Shome S, Tewari S, Bhattacharya MK, Panda SK (2015) Effect of Zn nanoparticles on growth responses of rice. In: Singh B, Kaushik A, Mehta SK, Tripathi SK (eds) Nanotechnology: novel perspectives and prospects. New Delhi, McGraw Hill Education, pp 508–512

Usman M, Farooq M, Wakeel A, Nawaz A, Alam Cheema SA, Rehman HU, Ashraf I, Sanaullah M (2020) Nanotechnology in agriculture: current status, challenges and future opportunities. Sci Total Environ 721:137778

Van Nguyen D, Nguyen HM, Le NT, Nguyen KH, Nguyen HT, Le HM, Nguyen AT, Dinh NTT, Hoang SA, Ha C (2021) Copper nanoparticle application enhances plant growth and grain yield in maize under drought stress conditions. J Plant Growth Regul 40:1–12

Wang J, Jiang C, Wang X, Wang L, Chen A, Hu J, Luo Z (2016) Fabrication of an "ion-imprinting" dual-emission quantum dot nanohybrid for selective fluorescence turn-on and ratiometric detection of cadmium ions. Analyst 141(21):5886–5892

Wu K, Xu C, Li T, Ma H, Gong J, Li X, Sun X, Hu X (2023) Application of nanotechnology in plant genetic engineering. Int J Mol Sci 24(19):14836

Xie Y, Yu Y, Lu L, Ma X, Gong L, Huang X, Liu G, Yu Y (2018) CuO nanoparticles decorated 3D graphene nanocomposite as non-enzymatic electrochemical sensing platform for malathion detection. J Electroanal Chem 812:82–89

Zhang B, Zheng X, Li H, Lin J (2013) Optical fiber sensors for plant disease detection using functionalized iron oxide nanoparticles. Anal Chem 85(13):6332–6337

Zhao X, Meng Z, Wang Y, Chen W, Sun C, Cui B, Cui J, Yu M, Zeng Z, Guo S et al (2017) Pollen magnetofection for genetic modification with magnetic nanoparticles as gene carriers. Nat Plants 3(12):956–964

Zhu ZJ, Wang H, Yan B, Zheng H, Jiang Y, Miranda OR, Rotello VM, Xing B, Vachet RW (2012) Effect of surface charge on the uptake and distribution of gold nanoparticles in four plant species. Environ Sci Technol 46:12391–12398

Systems Biology Approach for Plant Breeding

22

Shivani Tiwari, Lumbini Devi Darapu, and Prateek Gupta

Abstract

Systems biology is a rapidly evolving field that integrates various scientific disciplines to understand the complex interactions within biological systems. It employs computational modeling, quantitative analysis, and experimental techniques to explore how biological components, from molecules to entire organisms, interact and function. This holistic approach is crucial for deciphering the intricate networks that govern biological processes, making it a key tool in modern plant breeding.

Systems biology emerged in response to the limitations of traditional reductionist approaches in biology, which often focus on individual components without considering their interactions. By adopting a systems-level perspective, researchers can better understand emergent properties and characteristics that arise from the collective behavior of components within a system. This shift has been facilitated by advancements in screening techniques and computational approaches, enabling the integration of vast amounts of biological data from proteomics, genomics and metabolomics. The primary focus of this book chapter is to explain how systems biology serves as a vital tool in modern plant breeding by providing a holistic framework for understanding complex biological processes and guiding targeted genetic improvements. This approach not only enhances crop performance, but also contributes to sustainable agricultural practices in response to global challenges such as climate change and food security.

S. Tiwari · L. D. Darapu · P. Gupta (✉)
Department of Biological Sciences, SRM University AP, Amaravati, Andhra Pradesh, India

P. M. Priyadarshan, R. Ortiz (eds.), *Plant Breeding 2050*,
https://doi.org/10.1007/978-981-95-0583-8_22

Keywords

Systems biology · Genomics · Transcriptomics · Proteomics · Metabolomics · Multi-omics

22.1 Introduction

Systems biology uses a holistic approach to recognize and understand how biological systems function. It combines insights from biology, computer science, engineering, and mathematics to predict and develop models for biological processes and their responses to environmental changes. Many agronomically important traits are polygenic and influenced by numerous environmental factors (Houle et al. 2010). Systems biology helps elucidate the molecular and genetic basis of these traits by analyzing the interactions within biological networks. By integrating data from various omics technologies, it provides a comprehensive view of plant biology (Jamil et al. 2020). This integrative approach aids in identifying key regulatory pathways and metabolic processes that can be targeted for crop improvement (Yagi and Shiina 2014). It enables researchers to create predictive models that simulate plant growth under different conditions (Hammer et al. 2005). These models can inform breeding strategies to develop varieties resilient to climate change and other environmental stresses (Messina et al. 2009). The insights gained from systems biology can streamline breeding efforts by identifying genetic markers associated with desirable traits (Zhu et al. 2008). This accelerates the selection process and improves the efficiency of developing new plant varieties (Alvarado Beltrán et al. 2020).

22.2 Traditional Breeding and Systems Biology

The origins of plant breeding can be traced back to early 10,000 years ago beginning with the domestication of wild plants. This process involved selecting and cultivating plants with desirable traits, while screening for specific phenotypic characteristics. Over time, people began domesticating plants to select important traits, like larger seeds or better taste, from their wild ancestors, leading to the development of staple crops (Purugganan and Fuller 2009). By the early twentieth century, with the rediscovery of Mendel's inheritance laws, breeding became more scientific, focusing on hybridization and selection based on phenotypes. The mid-twentieth century saw the rise of green revolution technologies, such as the development of high-yielding varieties through cross-breeding, mutation breeding, and marker-assisted selection (MAS) (Pingali 2012).

Traditional breeding methods, despite their wide utility, exhibit several limitations. These approaches are time-intensive, often requiring more than 10 years for the release of a new cultivar (Bharti and Chimata 2019). It relies predominantly on phenotypic traits to identify superior cultivars; however, phenotypic expressions are influenced by genotype-environment interactions, leading to potential errors in selection (Lema 2018). Environmental factors can also impact phenotypic traits, resulting in the selection of poor breeding materials (Oladosu et al. 2019). Moreover, the overreliance on phenotypic selection can lead to a narrowing of the gene pool, which may hinder the efficiency of crop improvement programs. The variability and inconsistency of results from conventional breeding are largely due to subjective analysis. To overcome these challenges, modern breeding techniques have been increasingly integrated into plant breeding practices.

Systems biology significantly enhances plant breeding by providing a comprehensive framework that integrates genetic, metabolic, and environmental interactions. This approach allows for more precise breeding outcomes compared to traditional methods, which often focus on individual traits in isolation. Systems biology enhances plant breeding by providing a holistic approach to understanding how genes, proteins, and metabolites interact within the plant system. This is essential for addressing complex traits that are influenced by multiple genes and environmental factors (Mochida and Shinozaki 2011).

Advancements in genomics, transcriptomics, proteomics, and metabolomics have led to the production of large amounts of data in plant biology. By integrating these multi-omic datasets, systems biology provides valuable insights into the genetic basis of complex traits. This integration helps identify key genes and pathways that can be targeted for improvement through breeding (Weckwerth 2011). It also predicts models that help breeders anticipate how specific genetic changes will impact plant performance under different conditions, allowing for more effective selection strategies (Messina et al. 2018). It improves Marker-Assisted Selection (MAS) by identifying markers linked to complex traits through genomic mapping and functional analysis. This allows for more accurate selection in breeding programs by using molecular markers associated with specific traits (Xu and Crouch 2008).

Systems biology accelerates trait discovery and validation by utilizing high-throughput phenotyping (Cabrera-Bosquet et al. 2012) and advanced genomic techniques such as CRISPR/Cas9 for gene editing (Jaganathan et al. 2018). These technologies enable targeted changes in plants, allowing specific traits to be introduced or enhanced without the unintended consequences often associated with traditional breeding methods. Systems biology transforms plant breeding by offering a more integrated approach that accounts for the complexities of genetic, metabolic, and environmental interactions (Weckwerth 2011). By leveraging advanced technologies and holistic frameworks, systems biology is poised to significantly enhance the efficiency and effectiveness of modern plant breeding programs (Hirayama and Shinozaki 2010).

22.3 The Multi-omics Approach

22.3.1 Genomics

Genomics is the study of genes and genomes, emphasizing their structure, function, mapping, epigenomics, metagenomics, and genome editing (Muthamilarasan et al. 2019). Genomic sequencing allows researchers to explore the genetic diversity within and between plant species. By comparing the genomes of different varieties or species, scientists can identify quantitative trait loci- (QTL) specific regions of the genome that correlate with phenotypic traits (Huang and Han 2014). This information is essential for understanding how genetic variation contributes to observable characteristics in plants, such as growth patterns or resistance to diseases.

Genome-Wide Association Studies (GWAS) leverage genomic sequencing data to associate specific genetic markers with traits of interest. GWAS can also uncover associations between genetic variants and phenotypes within a population of any organism by examining SNPs in the sequence data (Challa and Neelapu 2018). For example, studies have successfully identified genes linked to root morphology (Zhao et al. 2011) and fruit quality by analyzing the genomes of various cultivars and correlating them with phenotypic data (Huang and Han 2014). GWAS plays a crucial role in genomics approaches to improve crop tolerance to abiotic stress. This method provides a powerful tool for pinpointing the genetic basis of complex traits that are influenced by multiple genes.

Genomic sequencing supports precision breeding methods like genomic selection (GS) and marker-assisted selection (MAS). MAS involves selecting individuals based on specific molecular markers linked to desirable traits, while GS employs genome-wide markers to predict the performance of breeding candidates without the need for extensive phenotyping (Varshney et al. 2005) This approach helps identify key regulatory genes involved in stress responses or developmental processes, thereby enhancing our understanding of how plants adapt to their environment.

22.3.2 Transcriptomics

Transcriptomics is a branch of molecular biology that focuses on studying the transcriptome, which includes all RNA molecules transcribed from the genome at a specific time under conditions (Wang et al. 2009). This field provides valuable insights into gene expression patterns during various developmental stages and in response to environmental stresses, making it an essential tool in plant biology and breeding. This approach allows researchers to observe gene expression differences in vitro to understand the primary function of specific genes. Initially, gene expression analysis relied on traditional techniques like cDNA-AFLP, differential display-PCR (DD-PCR), and SSH, though these provided limited resolution (Nataraja et al. 2017). Later, robust methods like microarrays, digital gene expression profiling, next-generation sequencing (NGS), and RNA-seq made RNA profiling much more precise (Kawahara et al. 2012; De Cremer et al. 2013; Duque et al. 2013). Advances in high-throughput sequencing, especially RNA sequencing (RNA-seq), have transformed transcriptomics by enabling accurate measurement of RNA levels, offering a snapshot of gene expression (Mortazavi et al. 2008). RNA-seq also helps identify differentially expressed genes (DEGs), comparing gene expression under various conditions (e.g., stressed vs. non-stressed plants) to reveal genes that are significantly upregulated or downregulated (Conesa et al. 2016). Transcriptomic studies have facilitated the identification of key regulatory networks involved in plant stress responses (Wang et al. 2020). For example, signaling pathways related to abscisic acid (ABA), ethylene, and jasmonic acid have been shown to mediate drought and salinity tolerance (Hong et al. 2019). By mapping out these networks, researchers can better understand the molecular mechanisms underlying plant resilience and identify targets for genetic improvement.

RNA sequencing (RNA-seq) has become a pivotal tool in plant research, particularly for identifying regulatory networks that govern gene expression. Through the analysis of transcriptomic data, researchers can uncover the intricate interactions between genes and their regulatory elements, offering insights into how plants respond to various environmental conditions and developmental cues. RNA-seq allows for the comprehensive assembly of the transcriptome, capturing both known and novel transcripts. This data is the foundation for further analysis of gene interactions and regulatory mechanisms (Tu et al. 2022). After establishing gene expression profiles, researchers can build gene co-expression networks by grouping genes with similar expression patterns across various conditions, suggesting possible functional relationships. Using clustering analysis, they can identify modules or clusters within the network (Cassan et al. 2021).

By analyzing gene co-expression modules, researchers can identify transcription factors (TFs) that are likely to regulate these genes. TFs bind to specific DNA sequences in the promoters of target genes, influencing their expression. RNA-seq data can be combined with other omics techniques like Chromatin Immunoprecipitation Sequencing (ChIP-seq), which maps protein-DNA interactions to identify direct targets of TFs. Integrating ChIP-seq with RNA-seq provides a more precise picture of regulatory interactions. Additionally, advances in Single-Cell RNA Sequencing (scRNA-seq) allow researchers to examine gene regulatory networks at the single-cell level, revealing how individual cells respond to stimuli and contribute to overall plant functions (Tripathi and Wilkins 2021). As methodologies continue to advance, RNA-seq will play an increasingly vital role in enhancing our understanding of plant biology and improving breeding strategies for crop resilience and productivity.

22.3.3 Proteomics

The term "proteome" originates from "proteins expressed by a genome," and much like genomics, "proteomics" refers to the study and analysis of the complete set of proteins present in a cell at any given time (Wilkins et al. 1995). As a relatively recent tool, proteomics is becoming indispensable for modern plant breeders, offering cellular-level insights that reveal molecular mechanisms vital to plant growth, development, and interactions with both physical and biological environments (Das et al. 2015). Proteomics serves plant biologists and breeders by identifying novel genes and DNA markers valuable for breeding. Marker-Assisted Selection (MAS), a widely used technique that employs DNA markers in breeding programs to select desirable genes or quantitative trait loci (QTLs), has significantly improved breeding lines (Collard and Mackill 2008). For example, Damerval et al. (1994) integrated proteomics with MAS to identify protein quantity loci (PQLs), which explained variations in protein expression, aiding breeders in correlating responsive genes with stress tolerance traits. These correlations can be further confirmed using advanced mapping populations, such as near-isogenic lines (NILs), recombinant inbred lines (RILs), and double haploid lines (Salekdeh and Komatsu 2007).

However, MAS is restricted to species with cross-compatible parents to effectively transfer superior genes/alleles using this molecular breeding approach. To enhance crop productivity, it is essential to identify stress tolerance and disease resistance genes and proteins continuously. Proteomics provides snapshots of all proteins in a cell under specific conditions, helping researchers observe how protein levels change over time (Hashiguchi et al. 2010). Advances in mass spectrometry (MS)-based proteomics, often termed "New Genomics," have made MS a powerful tool for studying protein posttranslational modifications and interactions, providing unique insights into how cells make decisions, thereby strengthening the role of proteomics in understanding whole biological systems (Cox and Mann 2007). Integrating proteomics with crop science will significantly enrich genome annotation and accelerate the development of crop models, enhancing our understanding of gene functions that shape phenotypic traits. However, to apply proteomics in biotechnology programs, genetic modifications must be reflected at the protein level. The progress achieved through various omics approaches, alongside the expansion of the gene pool via modern biotechnology, offers a promising path for improving crop productivity to meet food demands by 2050 (Eldakak et al. 2013).

22.3.4 Metabolomics

Metabolomics is the comprehensive study of metabolites within a biological system, providing insights into metabolic processes and their regulation. In plants, metabolites serve as biomarkers that reflect physiological conditions, stress responses, and overall health, making them essential for understanding plant growth, stress resistance, and yield (Hall 2006). Plant metabolomics is a valuable tool in both basic and applied science. For instance, metabolomics is already being used to monitor crop quality traits (Hall et al. 2005) and to identify biochemical markers for detecting product contamination and adulteration (Reid et al. 2004). A significant advantage of metabolomics is its ability to identify silent plant phenotypes—traits that may not be visually apparent but show distinct differences at the metabolic level (Weckwerth et al. 2004). Metabolomics can also help link functions to orphan genes—genes that have been sequenced but whose functions are still unknown

(Goodacre et al. 2003). Through the analysis of metabolites, these genes can be associated with specific functions. When metabolomics is combined with large-scale robotic enzyme analysis (Gibon et al. 2004), it becomes even more powerful, offering a detailed view of the biochemical pathways and processes in plants and providing deeper insights into plant metabolism.

Incorporating systems biology into plant breeding offers a transformative approach to understanding complex traits and optimizing breeding strategies. By integrating data from multiple biological levels, genes, proteins, metabolites, and environmental factors systems biology helps breeders identify crucial interactions that govern plant traits like yield, stress resistance, and nutrient efficiency. This holistic perspective is essential in addressing the growing challenges of food security and climate change. However, realizing the full potential of systems biology in plant breeding depends on advanced tools and technologies that can capture, analyze, and interpret vast datasets. The next section will explore the key tools and technologies—such as high-throughput omics, computational modeling, and bioinformatics platforms that drive innovation in this field, enabling breeders to accelerate crop improvement through precise and data-driven strategies.

22.4 Systems Biology Integration

Systems biology integration is defined as the process of combining and integrating data from various biological studies to create an extensive understanding of complex biological systems. It prioritizes integration strategies over reduction approaches (Senapati et al. 2020). It enables the researcher to interpret complex biological systems and their interaction, which helps uncover the process underlying growth development, stress response, and adaptation to any changes in the environment. The following sections overview the various steps in a systems biology approach and its applications in plant breeding.

22.5 Multi-omics Data Integration

The primary challenge in integrating omics data lies in the processing, scaling, and analyzing multidimensional information. Integration and evaluation of the datasets created from various interfaces encompass data acquisition, preprocessing, suitable normalization, and integration into a unified matrix. Most statistical methods involve reducing dimensionality and analyzing co-expressed clusters among the various variables measured in the same samples. It is hypothesized that biological entities showing similar expression patterns within the samples may have potential functional relationships (González et al. 2012). Several important sequence libraries, such as the European Molecular Biology Laboratory (EMBL) (Kanz et al. 2005), Phytozome (Goodstein et al. 2012), GenBank (Benson et al. 2012), and Plant GDB (Duvick et al. 2007), play a crucial role in collecting and organizing genomic data. However, these libraries do not include variant or phenotype information from other sources, making it more challenging for breeders and plant biologists to build a bridge between phenotype and genotype which generally needs data on genomics, epigenomics, phenotype, and environment (Hu et al. 2018).

Various platforms are available for unifying multidimensional datasets, for example, mixOmics, OnPLS, Integromics, PANOMICS, and COVAIN (Pazhamala et al. 2021). mixOmics provides a suite of methods for multivariate analysis to explore relationships between different omics datasets (Rohart et al. 2017). OnPLS modeling enables the decomposition of multiblock data to uncover shared variation across multiple datasets (Reinke et al. 2018). Integromics focuses on integrative data analysis to reveal connections across genomics, proteomics, and metabolomics (Lê Cao et al. 2009). COVAIN is a versatile tool for statistical analysis and visualization of complex omics data, often used for metabolomics and other high-dimensional datasets (Sun and Weckwerth 2012). Additionally, PANOMICS is a platform developed for integrating multi-omics data, merging insights from genomics, epigenomics, transcriptomics, proteomics, posttranslational modifications, metabolomics, and phenomics (Weckwerth et al. 2020). These platforms aid researchers in identifying correlations and patterns across diverse omics layers, advancing systems biology and plant breeding studies. PLINK is a popular bioinformatics tool utilized for Genome-Wide Association Studies (GWAS), which facilitates standard regression analysis to correlate genotypes with phenotypes (Purcell et al. 2007).

A systematic approach for integrating multi-omics data has emerged, combining diverse methodologies, software tools, web applications, and databases specifically designed for plant systems biology (Jamil et al. 2020; Pinu et al. 2019). These techniques are particularly valuable in crop plants, as they help investigate plant metabolism and reveal molecular mechanisms linked to agriculturally significant traits (Pazhamala et al. 2021). For example, an integrated analysis of transcriptomics, proteomics, and metabolomics has been used to study how rice seedling leaves respond to ozone exposure (Kyoungwon et al. 2008). Additionally, recent studies using multi-omics integration have shown a relationship between sucrose metabolism genes and carbohydrate content in poplar roots infected by Fusarium wilt, highlighting complex interactions at the metabolic level (Xu et al. 2024). Such integrated multi-omics data approaches also help link phenotype with genomic variations, enhancing predictive models for genomics-based applications (Azodi et al. 2020; Weckwerth et al. 2020). Recent research integrates metabolomics with epigenomic QTL, proteomic QTL, and metabolomic QTL mapping to dissect genetic variation across mRNA, protein, and metabolic levels using comprehensive multi-omics datasets (Raza 2022). A brief list of plant research studies using an integrated omics approach is given in Table 22.1.

Table 22.1 Examples of crop research studies using an integrated omics approach

S. No.	Plant species	Study	Omics methodologies	References
1.	*Oryza sativa* (Rice)	Unraveling the light-specific metabolic and regulatory signatures of rice	Genomics, transcriptomics, proteomics	Lakshmanan et al. (2015)
2.	*Oryza sativa* (Rice)	Rice metabolic activities of salt-stressed flag leaf were systematically characterized	Transcriptomics and metabolomics	Wanichthanarak et al. (2020)
3.	Soybean	Primary metabolism regulation in response to Rhizoctonia foliar blight disease	Transcriptomics and metabolomics	Copley et al. (2017)
4.	*Solanum lycopersicum* (Tomato)	Pollen development	Transcriptomics and proteomics	Keller and Simm (2018)
5.	Poplar Tree	Synergistic link between sucrose metabolic isogenes and carbohydrates in poplar roots infected by Fusarium Wilt	Transcriptomics and metabolomics	Xu et al. (2024)
6.	*Quercus suber* (Oak)	Genome-scale metabolic model of the *Q. suber*	Transcriptomics and metabolomics	Cunha et al. (2023)
7.	Strawberry	Effects of blue and red lights on the metabolism and gene expression of strawberry using metabolomics combined with transcriptomics	Transcriptomics and metabolomics	Chen et al. (2020)
8.	*Gynostemma pentaphyllum*	Investigation of the response mechanism of *G. pentaphyllum* under Cd stress to develop varieties with low Cd accumulation and high tolerance	Transcriptomics and metabolomics	Zhou et al. (2023)
9.	Pepper	Fruit development	Transcriptomics and proteomics	Liu et al. (2019)
10.	*Arabidopsis thaliana*	Identification of reactions and pathways that are highly altered in a mutant relative to the wild type.	Metabolics and transcriptomics	Pries et al. (2021)
11.	Common bean	Characterizing variability	Transcriptomics, proteomics, and metabolomics	Mensack et al. (2010)

22.6 Computational Network Analysis

The study of biological networks focuses on understanding the complex relationships within biological systems through various types of networks, including gene regulatory networks (GRNs), protein-protein interaction networks (PPIs), transcriptional regulatory networks (TRNs), metabolic networks, signaling transduction networks, and gene co-expression networks (GCNs). These networks are central to understanding how plants respond to and adapt to stress, with many of these processes being regulated by specific network structures. Biological networks can be represented mathematically as nodes (representing biological components like genes, proteins, RNAs, etc.) connected by edges that signify interactions or relationships, such as co-expression. Edges can be weighted to indicate the strength of these interactions, and networks can be directed or undirected based on their structural properties (Langfelder and Horvath 2008).

There are several approaches for constructing biological networks. The first method involves downloading existing network data from curated online databases such as TRANSFAC, STRING, KEGG, and others (Szklarczyk et al. 2021; Peri et al. 2003), which provide experimentally validated interactions for model organisms. However, data for many organisms are still limited. The second approach uses computer algorithms to simulate biomolecular networks. These algorithms often rely on the duplication-divergence (DD) model, where biomolecule duplication is simulated through node reiteration and divergence is modeled through edge deletion or dimerization (Wang et al. 2015; Vázquez et al. 2003). The third approach employs complex network theory to analyze network topology and dynamics (Mei et al. 2015; Wu and Jiang 2021). The fourth, and most advanced, method involves data-driven techniques that infer biological networks from experimental data using mathematical and statistical models, including correlation analysis, regression, Bayesian inference, and the Gaussian graphical model (Langfelder and Horvath 2008). These approaches help uncover the intricate molecular mechanisms governing plant traits and behaviors, facilitating advances in systems biology.

Network construction forms the basis of network biology and can be approached through either experimental detection or data-driven/model-driven inference. In crop plants, integrated omics approaches have been used to study complex network interactions. For instance, RiceNet combined with proteomics data helped predict proteins involved in abiotic stress resistance, such as XA21-mediated immunity in rice (Lee InSuk et al. 2011). In tobacco (*Nicotiana tabacum*), combining co-expression gene modules with metabolite data uncovered gene-metabolite relationships and identified key genes involved in the carotenoid metabolism pathway (Liu et al. 2019). Similarly, Mounet et al. (2009) used integrated transcriptomic and metabolomic data to uncover genes involved in tomato fruit development and metabolism.

Gene co-expression networks have shown significant promise in exploring specialized metabolite pathways at high throughput. By applying network biology, researchers can uncover the genetic mechanisms that enable plants to adapt to changing environmental conditions (Wisecaver et al. 2017). To fully exploit the potential of network biology, it is critical to employ appropriate temporal designs for omics data collection, followed by detailed analysis through network-based frameworks. This approach has proven essential for modeling plant responses to stress and other environmental challenges.

22.7 Predictive Modeling and Simulations

Computational modeling plays a vital role in simulating biological processes and understanding how changes within a system can impact the whole, such as in protein-protein interactions, disease mechanisms, or responses to mutations and drugs. These models help test biological hypotheses, which are further developed to predict system responses over time, supported by dynamic simulations and machine learning techniques. Biological systems are modelled based on hypotheses formed by researchers, and these models are tested over time, with results compared to experimental data to refine the hypotheses and better predict the system's behavior under dynamic conditions (Macklin 2019; Muthuramalingam et al. 2019). The rapid progress in molecular biology and biotechnology has made large datasets available for genetic studies, especially in plant breeding. However, these data have not been fully utilized for crop improvement due to a lack of suitable tools. Simulation approaches can effectively use this genetic information to predict performance across different breeding strategies. For instance, QuLine, a software tool for simulating breeding processes, allows breeders to modify methods and improve breeding efficiency (Jian-kang and Pfeiffer 2007). Computer simulation in plant breeding serves four key purposes: (1) comparing breeding methods to identify the most effective strategy, (2) gene mapping and evaluating mapping techniques, (3) plant-breeding platforms that replicate the entire breeding process, and (4) crop modeling that integrates genetic traits with environmental data to link genotype and phenotype (Li et al. 2012) (Fig. 22.1).

Genome-scale metabolic models (GEMs) are essential in systems biology, enabling the simulation of metabolic states to understand cellular functions. GEMs integrate all metabolic reactions within a mathematical framework, aiding in biological discoveries and metabolic engineering applications (Reed et al. 2006; Park et al. 2007). Context-specific models, tailored to specific conditions or diseases, offer more precise insights than generic GEMs (Opdam et al. 2017). Over 6000 GEMs have been created for various organisms, including multiple models for rice, which have been used to study chlorophyll biosynthesis and rice responses to environmental changes (Chatterjee and Kundu 2015; Lakshmanan et al. 2016). To support systems modelling, tools like Cell Illustrator, COBRA Toolbox, and Acorn are widely used to simulate biological pathways (Nagasaki et al. 2010; Schellenberger et al. 2011). Machine learning has also become a key tool in analyzing metabolite association networks, helping to describe biological responses to environmental stress (Jahagirdar and Saccenti 2020; Jahagirdar et al. 2019). In plant research, machine learning models are used to interpret phenotypic traits such as stress responses and crop quality. For instance, random forest regression was used to

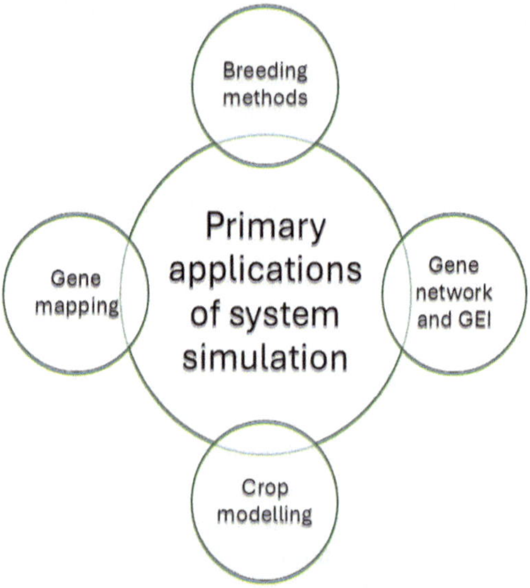

Fig. 22.1 Application of computer simulations in plant breeding (1) breeding methods (2) gene mapping (3) gene network and GEI (4) Crop modeling

analyze multi-omics data and identify key genes and metabolites affecting potato tuber quality (Acharjee et al. 2016). Predictive modeling, particularly through ensemble neural networks, has gained popularity for phenotype prediction, significantly contributing to crop development (Kim and Tagkopoulos 2018; Tong et al. 2020).

22.8 Systems Biology in Crop Improvement and Breeding

Systems biology offers great potential for sustainable agriculture by helping us to understand the complex traits that connect genotype and phenotype. It can be applied to model multigenic traits affecting agricultural productivity, such as plant structure, nitrogen and water use efficiency, and resistance to abiotic and biotic stresses. Advances in plant genetics and molecular biology have significantly boosted breeding programs. High-throughput experimental techniques, machine learning, and simulation now make it easier to explore biological complexities. The availability of whole-genome sequencing and rapidly advancing technologies, tools, and software (Table 22.2) have enabled the study of the molecular and biochemical responses of plants to stress and the related signaling pathways. The interactions between biomolecules, including nucleic acids, transcription factors, enzymes, proteins, and metabolites, form the basis of molecular pathways. These interactions, collectively known as the interactome, help connect genotype to phenotype and can also assist in disease detection. Therefore, analyzing these pathways and interactions is crucial for understanding diseases and their mechanisms (Cantu et al. 2008; Seo et al. 2011).

Table 22.2 List of computational tools and databases for systems biology research

S. No.	Name of computational tools	Function	References
1.	OmicsAnalyzer	Network mapping	Xia et al. (2010)
2.	OmicKriging	Data integration & visualization	Wheeler et al. (2014)
3.	OmicsPLS	OmicsPLS	el Bouhaddani (2018)
4.	JSBML 1.0	Provides a smorgasbord of options to encode systems biology models	Rodriguez et al. (2015)
5.	PlantSEED	Metabolic reconstruction	Seaver et al. (2018)
6.	Plant Regulomics	Functional characterization of data, Retrieve upstream regulators, Data mining	Ran et al. (2020)
7.	PathVisio 3	Pathway editor and data visualization & Pathway analysis	Kutmon et al. (2015)
8.	Systems biology markup language (SBML)	A medium for representation and exchange of biochemical network models	Hucka et al. (2003)
9.	VirtualPlant	A software platform to support systems biology research	Katari et al. (2010)
10.	AtPID (*Arabidopsis thaliana* protein interactome database)	An integrative platform for plant systems biology	Cui et al. (2007)
11.	TEPIC 2	A framework for fast, accurate, and versatile prediction, and analysis of TF binding from epigenetics data	Schmidt et al. (2019)
12.	MapMan4	Enrichment analysis & visualization of data expression	Schwacke et al. (2019)
13.	Crops in silico	An integrative platform for plant systems biology research	Marshall-Colon et al. (2017)
14.	Mergeomics	Multidimensional data integration to identify pathogenic perturbations to biological systems	Shu et al. (2016)
15.	VANTED (Visualization and Analysis of Network	Metabolic mappings, correlation networks analysis	Hartmann and Jozefowicz (2018)
16.	KBCommons (Knowledge Base Commons)	Universal framework for data management and retrieve	Zeng et al. (2019)
17.	KaPPA-View	Pathway mapping & data visualization	Sakurai et al. (2011)
18.	Babelomics	A complete suite of web tools for functional analysis of genome-scale experiments	Al-Shahrour et al. (2006)

(continued)

Table 22.2 (continued)

S. No.	Name of computational tools	Function	References
19.	VESPER 1.5	Spatial prediction software for precision agriculture	Whelan et al. (2002)
20.	COVAIN (Covariance inverse)	Transcriptomics, proteomics and metabolomics	Sun and Weckwerth (2012)

Systems biology approaches have already been used to identify important genes for agriculture, providing new insights into plant growth at different stages, such as roots, stems, leaves, fruits, and seeds, ultimately helping to increase productivity (Kumar et al. 2015). Integrating multi-omics and systems biology using top-down (phenotype to genotype) and bottom-up (genotype to phenotype) approaches will be useful in improving crop traits under environmental stress (Yang et al. 2021). Additionally, integrating multi-omics data with high-throughput phenotyping in comprehensive models will improve breeding accuracy for complex, agronomically important traits in the future.

22.9 Future of Systems Biology in Crop Breeding

Systems biology makes it possible to interpret crops at a systemic level, presenting new molecular insights that can increase agricultural productivity in the future (Kumar et al. 2015). The practice of various omics techniques, such as genomics, transcriptomics, proteomics, metabolomics, epigenomics, and single-cell omics, is picking upstream in plant sciences as sequencing costs decrease and expertise in the field increases. The generation of sequencing and resequencing data for various crops has facilitated the identification of novel alleles from diverse sources, regardless of the availability of complete genome sequences. Additionally, rapid advancements in omics technologies present circumstances to generate new and informative datasets across different plant species. Proficiently unifying and combining these genomic and functional omics data with genetic and phenotypic information will lead to the uncovering of genes and pathways connected with important agronomic traits.

The development in systems biology continues to grow, with accessible repository database resources being updated regularly with new information. Researchers are actively developing programs with enhanced algorithms focused on understanding the root, stem, leaf, fruit, and seed systems of crops, or crop systems altogether, applying both bottom-up and top-down approaches. This focused development in systems biology will aid in interpreting the temporal and spatial molecular changes that occur during different stages and conditions of plant development. The resulting molecular structure will make it possible to engineer crops to meet future food and nutritional security demands.

Acknowledgments Dr. Prateek Gupta would like to acknowledge the SRM University-AP for providing the seed grant (Sanction No. SRMAP/URG/SEED/2023-24/010 for financial assistance).

References

Acharjee A, Kloosterman B, Visser RG, Maliepaard C (2016) Integration of multi-omics data for prediction of phenotypic traits using random forest. BMC Bioinf 17:363–373

Al-Shahrour F, Minguez P, Tárraga J, Montaner D, Alloza E, Vaquerizas JM, Conde L, Blaschke C, Vera J, Dopazo J (2006) BABELOMICS: a systems biology perspective in the functional annotation of genome-scale experiments. Nucleic Acids Res 34(suppl_2):W472–W476

Alvarado Beltrán G, Rodríguez FM, Pacheco Gil RA, Burgueño J, Crossa J, Vargas-Hernández M, Perez-Rodriguez P, Lopez-Cruz M (2020) META-R: A software to analyze data from multi-environment plant breeding trials. Crop Sci 60(1):313–317

Azodi CB, Pardo J, Van Buren R, de Los Campos G, Shiu SH (2020) Transcriptome-based prediction of complex traits in maize. Plant Cell 32(1):139–151

Benson DA, Cavanaugh M, Clark K, Karsch-Mizrachi I, Lipman DJ, Ostell J, Sayers EW (2012) GenBank. Nucleic Acids Res 41(D1):D36–D42

Bharti G, Chimata MK (2019) Review on new plant breeding techniques. Int J Sci Res 8(4):723–730

Bouhaddani, S.E., Uh, H.W., Jongbloed, G., Hayward, C., Klarić, L., Kiełbasa, S.M. and Houwing-Duistermaat, J., 2018. Integrating omics datasets with the OmicsPLS package. BMC bioinformatics, 19(1), p.371

Cabrera-Bosquet L, Crossa J, von Zitzewitz J, Serret MD, Luis Araus J (2012) High-throughput phenotyping and genomic selection: The frontiers of crop breeding converge. J Integr Plant Biol 54(5):312–320

Cantu D, Vicente AR, Greve LC, Dewey FM, Bennett AB, Labavitch JM, Powell ALT (2008) The intersection between cell wall disassembly, ripening, and fruit susceptibility to Botrytis cinerea. Proc Natl Acad Sci 105(3):859–864

Cassan O, Lèbre S, Martin A (2021) Inferring and analyzing gene regulatory networks from multi-factorial expression data: a complete and interactive suite. BMC Genomics 22(1):387

Challa S, Neelapu NR (2018) Genome-wide association studies (GWAS) for abiotic stress tolerance in plants. In: Biochemical, physiological and molecular avenues for combating abiotic stress tolerance in plants. Academic, pp 135–150

Chatterjee A, Kundu S (2015) Revisiting the chlorophyll biosynthesis pathway using genome scale metabolic model of Oryza sativa japonica. Sci Rep 5(1):14975

Chen X, Cai W, Xia J, Yu H, Wang Q, Pang F, Zhao M (2020) Metabolomic and transcriptomic analyses reveal that blue light promotes chlorogenic acid synthesis in strawberry. J Agric Food Chem 68(44):12485–12492

Cho K, Shibato J, Agrawal GK, Jung YH, Kubo A, Jwa NS, Tamogami S, Satoh K, Kikuchi S, Higashi T, Kimura S (2008) Integrated transcriptomics, proteomics, and metabolomics analyses to survey ozone responses in the leaves of rice seedling. J Proteome Res 7(7):2980–2998

Collard BC, Mackill DJ (2008) Marker-assisted selection: an approach for precision plant breeding in the twenty-first century. Philo Trans R Soc B: Biol Sci 363(1491):557–572

Conesa A, Madrigal P, Tarazona S, Gomez-Cabrero D, Cervera A, McPherson A, Szcześniak MW, Gaffney DJ, Elo LL, Zhang X, Mortazavi A (2016) A survey of best practices for RNA-seq data analysis. Genome Biol 17:1–19

Copley TR, Aliferis KA, Kliebenstein DJ, Jabaji SH (2017) An integrated RNAseq-1 H NMR metabolomics approach to understand soybean primary metabolism regulation in response to Rhizoctonia foliar blight disease. BMC Plant Biol 17:1–18

Cox J, Mann M (2007) Is proteomics the new genomics? Cell 130:395–398

Cui J, Li P, Li G, Xu F, Zhao C, Li Y, Yang Z, Wang G, Yu Q, Li Y, Shi T (2007) AtPID: Arabidopsis thaliana protein interactome database—an integrative platform for plant systems biology. Nucleic Acids Res 36(suppl_1):D999–D1008

Cunha E, Silva M, Chaves I, Demirci H, Lagoa DR, Lima D, Rocha M, Rocha I, Dias O (2023) The first multi-tissue genome-scale metabolic model of a woody plant highlights suberin biosynthesis pathways in Quercus suber. PLoS Comput Biol 19(9):e1011499

Damerval C, Maurice A, Josse JM, De Vienne D (1994) Quantitative trait loci underlying gene product variation: a novel perspective for analyzing regulation of genome expression. Genetics 137:289–301

Das A, Paudel B, Rohila JS (2015) Potentials of proteomics in crop breeding. In: Advances in plant breeding strategies: breeding, biotechnology and molecular tools. Springer, pp 513–537

De Cremer K, Mathys J, Vos C, Froenicke L, Michelmore RW, Cammue BPA, De Coninck B (2013) RNA-seq-based transcriptome analysis of Lactuca sativa infected by the fungal necrotroph Botrytis cinerea. Plant Cell Environ 36(11):1992–2007

Duque AS, de Almeida AM, da Silva AB, da Silva JM, Farinha AP, Santos D, Fevereiro P, de Sousa Araújo S (2013) Abiotic stress responses in plants: unraveling the complexity of genes and networks to survive. In: Abiotic stress-plant responses and applications in agriculture. Academic, pp 49–101

Duvick J, Fu A, Muppirala U, Sabharwal M, Wilkerson MD, Lawrence CJ, Lushbough C, Brendel V (2007) PlantGDB: a resource for comparative plant genomics. Nucleic Acids Res 36(suppl_1):D959–D965

Eldakak M, Milad SI, Nawar AI, Rohila JS (2013) Proteomics: a biotechnology tool for crop improvement. Front Plant Sci 4:35

Gibon Y, Blaesing OE, Hannemann J, Carillo P, Hohne M, Hendriks JH, Palacios N, Cross J, Selbig J, Stitt M (2004) A robot-based platform to measure multiple enzyme activities in Arabidopsis using a set of cycling assays: comparison of changes of enzyme activities and transcript levels during diurnal cycles and in prolonged darkness. Plant Cell 16(12):3304–3325

González I, Cao KAL, Davis MJ, Déjean S (2012) Visualising associations between paired 'omics' data sets. BioData Mining 5:1–23

Goodacre R, York EV, Heald JK, Scott IM (2003) Chemometric discrimination of unfractionated plant extracts analyzed by electrospray mass spectrometry. Phytochemistry 62(6):859–863

Goodstein DM, Shu S, Howson R, Neupane R, Hayes RD, Fazo J, Mitros T, Dirks W, Hellsten U, Putnam N, Rokhsar DS (2012) Phytozome: a comparative platform for green plant genomics. Nucleic Acids Res 40(D1):D1178–D1186

Hall RD (2006) Plant metabolomics: from holistic hope, to hype, to hot topic. New Phytol 169(3):453–468

Hall RD, de Vos CR, Verhoeven HA, Bino RJ (2005) Metabolomics for the assessment of functional diversity and quality traits in plants. In: Metabolome analyses: strategies for systems biology. Springer US, Boston, pp 31–44

Hammer GL, Chapman S, van Oosterom E, Podlich DW (2005) Trait physiology and crop modelling as a framework to link phenotypic complexity to underlying genetic systems. Aust J Agric Res 56(9):947–960

Hartmann A, Jozefowicz AM (2018) VANTED: a tool for integrative visualization and analysis of-omics data. In: Plant membrane proteomics: methods and protocols, pp 261–278

Hashiguchi A, Ahsan N, Komatsu S (2010) Proteomics application of crops in the context of climatic changes. Food Res Int 43(7):1803–1813

Hirayama T, Shinozaki K (2010) Research on plant abiotic stress responses in the post-genome era: past, present and future. Plant J 61(6):1041–1052

Hong ZH, Zhou Y, Hong ZH, He S, Ning Z, Liu Q (2019) Transcriptome profiling reveals insights into the molecular mechanism of drought tolerance in sweetpotato. J Integr Agric 18(1):9–23

Houle D, Govindaraju DR, Omholt S (2010) Phenomics: the next challenge. Nat Rev Genet 11(12):855–866

Hu H, Scheben A, Edwards D (2018) Advances in integrating genomics and bioinformatics in the plant breeding pipeline. Agriculture 8(6):75

Huang X, Han B (2014) Natural variations and genome-wide association studies in crop plants. Annu Rev Plant Biol 65(1):531–551

Hucka M, Finney A, Sauro HM, Bolouri H, Doyle JC, Kitano H, Arkin AP, Bornstein BJ, Bray D, Cornish-Bowden A, Cuellar AA (2003) The systems biology markup language (SBML): a medium for representation and exchange of biochemical network models. Bioinformatics 19(4):524–531

Jaganathan D, Ramasamy K, Sellamuthu G, Jayabalan S, Venkataraman G (2018) CRISPR for crop improvement: an update review. Front Plant Sci 9:985

Jahagirdar S, Saccenti E (2020) On the use of correlation and MI as a measure of metabolite–metabolite association for network differential connectivity analysis. Metabolites 10(4):171

Jahagirdar S, Suarez-Diez M, Saccenti E (2019) Simulation and reconstruction of metabolite–metabolite association networks using a metabolic dynamic model and correlation based algorithms. J Proteome Res 18(3):1099–1113

Jamil IN, Remali J, Azizan KA, Nor Muhammad NA, Arita M, Goh HH, Aizat WM (2020) Systematic multi-omics integration (MOI) approach in plant systems biology. Front Plant Sci 11:944

Jian-kang WANG, Pfeiffer WH (2007) Simulation modeling in plant breeding: principles and applications. Agric Sci China 6(8):101–105

Kanz C, Aldebert P, Althorpe N, Baker W, Baldwin A, Bates K, Browne P, van den Broek A, Castro M, Cochrane G, Duggan K (2005) The EMBL nucleotide sequence database. Nucleic Acids Res 33(suppl_1):D29–D33

Katari MS, Nowicki SD, Aceituno FF, Nero D, Kelfer J, Thompson LP, Cabello JM, Davidson RS, Goldberg AP, Shasha DE, Coruzzi GM (2010) VirtualPlant: a software platform to support systems biology research. Plant Physiol 152(2):500–515

Kawahara Y, Oono Y, Kanamori H, Matsumoto T, Itoh T, Minami E (2012) Simultaneous RNA-Seq analysis of a mixed transcriptome of rice and blast fungus interaction. PLoS One 7:e49423

Keller M, Simm S (2018) The coupling of transcriptome and proteome adaptation during development and heat stress response of tomato pollen. BMC Genomics 19:1–20

Kim M, Tagkopoulos I (2018) Data integration and predictive modeling methods for multi-omics datasets. Mol Omics 14(1):8–25

Kumar A, Pathak RK, Gupta SM, Gaur VS, Pandey D (2015) Systems biology for smart crops and agricultural innovation: filling the gaps between genotype and phenotype for complex traits linked with robust agricultural productivity and sustainability. Omics 19(10):581–601

Kutmon M, van Iersel MP, Bohler A, Kelder T, Nunes N, Pico AR, Evelo CT (2015) PathVisio 3: an extendable pathway analysis toolbox. PLoS Comput Biol 11(2):e1004085

Lakshmanan M, Lim SH, Mohanty B, Kim JK, Ha SH, Lee DY (2015) Unraveling the light-specific metabolic and regulatory signatures of rice through combined in silico modeling and multiomics analysis. Plant Physiol 169(4):3002–3020

Lakshmanan M, Cheung CM, Mohanty B, Lee DY (2016) Modeling rice metabolism: from elucidating environmental effects on cellular phenotype to guiding crop improvement. Front Plant Sci 7:1795

Langfelder P, Horvath S (2008) WGCNA: an R package for weighted correlation network analysis. BMC Bioinf 9:1–13

Lê Cao KA, González I, Déjean S (2009) integrOmics: an R package to unravel relationships between two omics datasets. Bioinformatics 25(21):2855–2856

Lee InSuk, L.I., Seo, Y.S., Coltrane, D., Hwang SoHyun, H.S., Oh TaeYun, O.T., Marcotte, E.M. and Ronald, P.C. 2011. Genetic dissection of the biotic stress response using a genome-scale gene network for rice

Lema M (2018) Marker assisted selection in comparison to conventional plant breeding. Agric Res Technol 14:555914

Li X, Zhu C, Wang J, Yu J (2012) Computer simulation in plant breeding. Adv Agron 116:219–264

Liu Z, Lv J, Zhang Z, Li H, Yang B, Chen W, Dai X, Li X, Yang S, Liu L, Ou L (2019) Integrative transcriptome and proteome analysis identifies major metabolic pathways involved in pepper fruit development. J Proteome Res 18(3):982–994

Macklin P (2019) Key challenges facing data-driven multicellular systems biology. Gigascience 8(10):giz127

Marshall-Colon A, Long SP, Allen DK, Allen G, Beard DA, Benes B, Von Caemmerer S, Christensen AJ, Cox DJ, Hart JC, Hirst PM (2017) Crops in silico: generating virtual crops using an integrative and multi-scale modeling platform. Front Plant Sci 8:786

Mei S, Zarrabi N, Lees M, Sloot PM (2015) Complex agent networks: An emerging approach for modeling complex systems. Appl Soft Comput 37:311–321

Mensack MM, Fitzgerald VK, Ryan EP, Lewis MR, Thompson HJ, Brick MA (2010) Evaluation of diversity among common beans (Phaseolus vulgaris L.) from two centers of domestication using 'omics' technologies. BMC Genomics 11:1–11

Messina C, Hammer G, Dong Z, Podlich D, Cooper M (2009) Modelling crop improvement in a G× E× M framework via gene-trait-phenotype relationships. In: Crop physiology: applications for genetic improvement and agronomy. Elsevier, pp 235–265

Messina CD, Technow F, Tang T, Totir R, Gho C, Cooper M (2018) Leveraging biological insight and environmental variation to improve phenotypic prediction: Integrating crop growth models (CGM) with whole genome prediction (WGP). Eur J Agron 100:151–162

Mochida K, Shinozaki K (2011) Advances in omics and bioinformatics tools for systems analyses of plant functions. Plant Cell Physiol 52(12):2017–2038

Mortazavi A, Williams BA, McCue K, Schaeffer L, Wold B (2008) Mapping and quantifying mammalian transcriptomes by RNA-Seq. Nat Methods 5(7):621–628

Mounet F, Moing A, Garcia V, Petit J, Maucourt M, Deborde C, Bernillon S, Le Gall G, Colquhoun I, Defernez M, Giraudel JL (2009) Gene and metabolite regulatory network analysis of early developing fruit tissues highlights new candidate genes for the control of tomato fruit composition and development. Plant Physiol 149(3):1505–1528

Muthamilarasan M, Singh NK, Prasad M (2019) Multi-omics approaches for strategic improvement of stress tolerance in underutilized crop species: a climate change perspective. Adv Genet 103:1–38

Muthuramalingam P, Jeyasri R, Krishnan SR, Pandian STK, Sathishkumar R, Ramesh M (2019) Integrating the bioinformatics and omics tools for systems analysis of abiotic stress tolerance in Oryza sativa (L.). In: Advances in plant transgenics: methods and applications, pp 59–77

Nagasaki M, Saito A, Jeong E, Li C, Kojima K, Ikeda E, Miyano S (2010) Cell Illustrator 4.0: a computational platform for systems biology. In Silico Biol 10(1–2):5–26

Nataraja KN, Madhura BG, Parvathi MS (2017) Omics: modern tools for precise understanding of drought adaptation in plants. In: Plant OMICS and crop breeding. Apple Academic Press, pp 263–294

Oladosu Y, Rafii MY, Samuel C, Fatai A, Magaji U, Kareem I, Kamarudin ZS, Muhammad II, Kolapo K (2019) Drought resistance in rice from conventional to molecular breeding: a review. Int J Mol Sci 20(14):3519

Opdam S, Richelle A, Kellman B, Li S, Zielinski DC, Lewis NE (2017) A systematic evaluation of methods for tailoring genome-scale metabolic models. Cell Syst 4(3):318–329

Park JH, Lee KH, Kim TY, Lee SY (2007) Metabolic engineering of Escherichia coli for the production of L-valine based on transcriptome analysis and in silico gene knockout simulation. Proc Natl Acad Sci 104(19):7797–7802

Pazhamala LT, Kudapa H, Weckwerth W, Millar AH, Varshney RK (2021) Systems biology for crop improvement. Plant Genome 14(2):e20098

Peri S, Navarro JD, Amanchy R, Kristiansen TZ, Jonnalagadda CK, Surendranath V, Niranjan V, Muthusamy B, Gandhi TKB, Gronborg M, Ibarrola N (2003) Development of human protein reference database as an initial platform for approaching systems biology in humans. Genome Res 13(10):2363–2371

Pingali PL (2012) Green revolution: impacts, limits, and the path ahead. Proc Natl Acad Sci 109(31):12302–12308

Pinu FR, Beale DJ, Paten AM, Kouremenos K, Swarup S, Schirra HJ, Wishart D (2019) Systems biology and multi-omics integration: viewpoints from the metabolomics research community. Metabolites 9(4):76

Pries C, Razaghi-Moghadam Z, Kopka J, Nikoloski Z (2021) Integration of relative metabolomics and transcriptomics time-course data in a metabolic model pinpoints effects of ribosome biogenesis defects on Arabidopsis thaliana metabolism. Sci Rep 11(1):4787

Purcell S, Neale B, Todd-Brown K, Thomas L, Ferreira MA, Bender D, Maller J, Sklar P, De Bakker PI, Daly MJ, Sham PC (2007) PLINK: a tool set for whole-genome association and population-based linkage analyses. Am J Hum Genet 81(3):559–575

Purugganan MD, Fuller DQ (2009) The nature of selection during plant domestication. Nature 457(7231):843–848

Ran X, Zhao F, Wang Y, Liu J, Zhuang Y, Ye L, Qi M, Cheng J, Zhang Y (2020) Plant Regulomics: a data-driven interface for retrieving upstream regulators from plant multi-omics data. Plant J 101(1):237–248

Raza A (2022) Metabolomics: a systems biology approach for enhancing heat stress tolerance in plants. Plant Cell Rep 41(3):741–763

Reed JL, Patel TR, Chen KH, Joyce AR, Applebee MK, Herring CD, Bui OT, Knight EM, Fong SS, Palsson BO (2006) Systems approach to refining genome annotation. Proc Natl Acad Sci 103(46):17480–17484

Reid LM, O'Donnell CP, Downey G (2004) Potential of SPME-GC and chemometrics to detect adulteration of soft fruit purées. J Agric Food Chem 52(3):421–427

Reinke SN, Galindo-Prieto B, Skotare T, Broadhurst DI, Singhania A, Horowitz D, Djukanović R, Hinks TS, Geladi P, Trygg J, Wheelock CE (2018) OnPLS-based multi-block data integration: a multivariate approach to interrogating biological interactions in asthma. Anal Chem 90(22):13400–13408

Rodriguez N, Thomas A, Watanabe L, Vazirabad IY, Kofia V, Gómez HF, Mittag F, Matthes J, Rudolph J, Wrzodek F, Netz E (2015) JSBML 1.0: providing a smorgasbord of options to encode systems biology models. Bioinformatics 31(20):3383–3386

Rohart F, Gautier B, Singh A, Lê Cao KA (2017) mixOmics: An R package for 'omics feature selection and multiple data integration. PLoS Comput Biol 13(11):e1005752

Sakurai N, Ara T, Ogata Y, Sano R, Ohno T, Sugiyama K, Hiruta A, Yamazaki K, Yano K, Aoki K, Aharoni A (2011) KaPPA-View4: a metabolic pathway database for representation and analysis of correlation networks of gene co-expression and metabolite co-accumulation and omics data. Nucleic Acids Res 39(suppl_1):D677–D684

Salekdeh GH, Komatsu S (2007) Crop proteomics: aim at sustainable agriculture of tomorrow. Proteomics 7:2976–2996

Schellenberger J, Que R, Fleming RM, Thiele I, Orth JD, Feist AM, Zielinski DC, Bordbar A, Lewis NE, Rahmanian S, Kang J (2011) Quantitative prediction of cellular metabolism with constraint-based models: the COBRA Toolbox v2. 0. Nat Protoc 6(9):1290–1307

Schmidt F, Kern F, Ebert P, Baumgarten N, Schulz MH (2019) TEPIC 2—an extended framework for transcription factor binding prediction and integrative epigenomic analysis. Bioinformatics 35(9):1608–1609

Schwacke R, Ponce-Soto GY, Krause K, Bolger AM, Arsova B, Hallab A, Gruden K, Stitt M, Bolger ME, Usadel B (2019) MapMan4: a refined protein classification and annotation framework applicable to multi-omics data analysis. Mol Plant 12(6):879–892

Seaver SM, Lerma-Ortiz C, Conrad N, Mikaili A, Sreedasyam A, Hanson AD, Henry CS (2018) PlantSEED enables automated annotation and reconstruction of plant primary metabolism with improved compartmentalization and comparative consistency. Plant J 95(6):1102–1113

Senapati M, Dash SS, Lenka B, Panda D (2020) An overview on systems biology driven plant breeding. Int J Curr Microbiol App Sci 9(7):1956–1964

Seo YS, Chern M, Bartley LE, Han M, Jung KH, Lee I, Walia H, Richter T, Xu X, Cao P, Bai W (2011) Towards establishment of a rice stress response interactome. PLoS Genet 7(4):e1002020

Shu L, Zhao Y, Kurt Z, Byars SG, Tukiainen T, Kettunen J, Orozco LD, Pellegrini M, Lusis AJ, Ripatti S, Zhang B (2016) Mergeomics: multidimensional data integration to identify pathogenic perturbations to biological systems. BMC Genomics 17:1–16

Sun X, Weckwerth W (2012) COVAIN: a toolbox for uni-and multivariate statistics, time-series and correlation network analysis and inverse estimation of the differential Jacobian from metabolomics covariance data. Metabolomics 8:81–93

Szklarczyk D, Gable AL, Nastou KC, Lyon D, Kirsch R, Pyysalo S, Doncheva NT, Legeay M, Fang T, Bork P, Jensen LJ (2021) The STRING database in 2021: customizable protein–protein networks, and functional characterization of user-uploaded gene/measurement sets. Nucleic Acids Res 49(D1):D605–D612

Tong H, Madison I, Long TA, Williams CM (2020) Computational solutions for modeling and controlling plant response to abiotic stresses: A review with focus on iron deficiency. Curr Opin Plant Biol 57:8–15

Tripathi RK, Wilkins O (2021) Single cell gene regulatory networks in plants: Opportunities for enhancing climate change stress resilience. Plant Cell Environ 44(7):2006–2017

Tu M, Zeng J, Zhang J, Fan G, Song G (2022) Unleashing the power within short-read RNA-seq for plant research: Beyond differential expression analysis and toward regulomics. Front Plant Sci 13:1038109

Varshney RK, Graner A, Sorrells ME (2005) Genomics-assisted breeding for crop improvement. Trends Plant Sci 10(12):621–630

Vázquez A, Flammini A, Maritan A, Vespignani A (2003) Modeling of protein interaction networks. Complexus 1(1):38–44

Wang Z, Gerstein M, Snyder M (2009) RNA-Seq: a revolutionary tool for transcriptomics. Nat Rev Genet 10(1):57–63

Wang P, Chen Y, Lü J, Wang Q, Yu X (2015) Graphical features of functional genes in human protein interaction network. IEEE Trans Biomed Circuits Syst 10(3):707–720

Wang X, Li N, Li W, Gao X, Cha M, Qin L, Liu L (2020) Advances in transcriptomics in the response to stress in plants. Glob Med Genet 7(02):030–034

Wanichthanarak K, Boonchai C, Kojonna T, Chadchawan S, Sangwongchai W, Thitisaksakul M (2020) Deciphering rice metabolic flux reprogramming under salinity stress via in silico metabolic modeling. Comput Struct Biotechnol J 18:3555–3566

Weckwerth W (2011) Green systems biology—from single genomes, proteomes and metabolomes to ecosystems research and biotechnology. J Proteome 75(1):284–305

Weckwerth W, Loureiro ME, Wenzel K, Fiehn O (2004) Differential metabolic networks unravel the effects of silent plant phenotypes. Proc Natl Acad Sci 101(20):7809–7814

Weckwerth W, Ghatak A, Bellaire A, Chaturvedi P, Varshney RK (2020) PANOMICS meets germplasm. Plant Biotechnol J 18(7):1507–1525

Wheeler HE, Aquino-Michaels K, Gamazon ER, Trubetskoy VV, Dolan ME, Huang RS, Cox NJ, Im HK (2014) Poly-omic prediction of complex traits: OmicKriging. Genet Epidemiol 38(5):402–415

Whelan BM, McBratney AB, Minasny B (2002) Vesper 1.5–spatial prediction software for precision agriculture. In: Precision Agriculture, Proceedings of the 6th international conferences on precision agriculture, vol 179. ASA/CSSA/SSSA, Madison

Wilkins MR, Sanchez JC, Gooley AA, Appel RD, Humphery-Smith I, Hochstrasser DF et al (1995) Progress with proteome projects: why all proteins expressed by a genome should be identified and how to do it. Biotechnol Genet Eng Rev 13:19–50

Wisecaver JH, Borowsky AT, Tzin V, Jander G, Kliebenstein DJ, Rokas A (2017) A global coexpression network approach for connecting genes to specialized metabolic pathways in plants. Plant Cell 29(5):944–959

Wu R, Jiang L (2021) Recovering dynamic networks in big static datasets. Phys Rep 912:1–57

Xia T, Hemert JV, Dickerson JA (2010) OmicsAnalyzer: a Cytoscape plug-in suite for modeling omics data. Bioinformatics 26(23):2995–2996

Xu Y, Crouch JH (2008) Marker-assisted selection in plant breeding: from publications to practice. Crop Sci 48(2):391–407

Xu X, Wei H, Yao K, Wu H, Huang T, Han M, Su T, Cao F (2024) Integrative omics studies revealed synergistic link between sucrose metabolic isogenes and carbohydrates in poplar roots infected by Fusarium wilt. Plant Mol Biol 114(2):29

Yagi Y, Shiina T (2014) Recent advances in the study of chloroplast gene expression and its evolution. Front Plant Sci 5:61

Yang Y, Saand MA, Huang L, Abdelaal WB, Zhang J, Wu Y, Li J, Sirohi MH, Wang F (2021) Applications of multi-omics technologies for crop improvement. Front Plant Sci 12:563953

Zeng S, Lyu Z, Narisetti SRK, Xu D, Joshi T (2019) Knowledge Base Commons (KBCommons) v1. 1: a universal framework for multi-omics data integration and biological discoveries. BMC Genomics 20:1–16

Zhao K, Tung CW, Eizenga GC, Wright MH, Ali ML, Price AH, Norton GJ, Islam MR, Reynolds A, Mezey J, McClung AM (2011) Genome-wide association mapping reveals a rich genetic architecture of complex traits in Oryza sativa. Nat Commun 2(1):467

Zhou Y, Yao L, Huang X, Li Y, Wang C, Huang Q, Yu L, Pan C (2023) Transcriptomics and metabolomics association analysis revealed the responses of Gynostemma pentaphyllum to cadmium. Front Plant Sci 14:1265971

Zhu C, Gore M, Buckler ES, Yu J (2008) Status and prospects of association mapping in plants. Plant Genome 1(1):5–20

Computational Biology in Plants: Technological Innovations in the Post-Genomic Era

23

Justine K. Kitony, Lillian K. Padgitt-Cobb, Nolan T. Hartwick, and Todd P. Michael

Abstract

Computational biology is reshaping plant science, opening new possibilities for analyzing and interpreting vast biological datasets with precision. The advent of single cell and third-generation sequencing (TGS) has propelled plant science into the digital era where genomes are routinely assembled, annotated, and characterized. Digitized genomes reveal genetic and epigenetic variation, highlighting the limitations of single reference genomes and enable pangenome construction to capture diversity for breeding and editing. In parallel, advances in computational power have driven the development of high-throughput phenotyping platforms, capable of measuring complex plant traits with unprecedented accuracy. From drones to environmental sensors, these technologies generate data that bridge the gap between genomes and phenotypes, enabling the dissection of genotype-environment interactions. Today, plant science intersects with artificial intelligence (AI) and synthetic biology. AI-driven models are revealing new insights into growth, stress responses, and metabolisms, while CRISPR-

Supplementary Information The online version contains supplementary material available at https://doi.org/10.1007/978-981-95-0583-8_23.

J. K. Kitony · L. K. Padgitt-Cobb · N. T. Hartwick
Plant Molecular and Cellular Biology Laboratory, The Salk Institute for Biological Studies, La Jolla, CA, USA

T. P. Michael (✉)
Plant Molecular and Cellular Biology Laboratory, The Salk Institute for Biological Studies, La Jolla, CA, USA

University of California San Diego, La Jolla, CA, USA

San Diego Botanical Garden, Encinitas, CA, USA
e-mail: tmichael@salk.edu

P. M. Priyadarshan, R. Ortiz (eds.), *Plant Breeding 2050*,
https://doi.org/10.1007/978-981-95-0583-8_23

based gene editing and plant artificial chromosomes (PACs) are pushing the boundaries of crop design and domestication. Collectively, these innovations point toward crops that are climate-resilient, resource-efficient, and capable of addressing global challenges in food security, and climate change. the next frontier lies in weaving together multi-omics data, predictive analytics, and biodesign, positioning plant sciences at the forefront of sustainable agriculture and biotechnological innovation.

Keywords

Weighted Gene Co-expression Network Analysis (WGCNA) · Telomere-to-Telomere (T2T) · Single-cell RNA sequencing (scRNA-seq) · Multiplexed Error-Robust Fluorescence In Situ Hybridization (MERFISH) · Plant Artificial Chromosomes (PACs) · Artificial Intelligence (AI) · Multi-omics integration · Genome-Wide Association Studies (GWAS) · Genomic Selection (GS) · Clustered Regularly Interspaced Short Palindromic Repeats (CRISPR) · Genome editing · Climate-resilient crops

23.1 Introduction

Computational biology has emerged as a cornerstone of modern plant science, transforming vast genome sequences into actionable insights that decode the complexities of biological systems. It harnesses advanced data analytics, machine learning, and predictive modeling to uncover hidden patterns within raw genomic data and elucidates the intricate networks of genes, proteins, and metabolic pathways that drive plant growth and adaptation. Furthermore, computational biology is accelerating breakthroughs in crop resilience, climate adaptation, and sustainable agriculture, offering innovative solutions to address global food security ("Computation and Biology: A Partnership" 2021).

The rise of computational biology in plant science can be traced back to the early days of the genomics era, when the shift from traditional sequencing methods to high-throughput technologies marked a pivotal turning point. Early genome sequencing relied heavily on chemical-based approaches, but as the scale of sequencing expanded, it became clear that computational methods were essential to manage, assemble, and interpret the vast amounts of raw data. The development of tools for sequence alignment, error correction, and de novo genome assembly laid the foundation for what would become digital biology—where plant genomes are not only sequenced but also modeled to reveal their structural and functional properties.

As sequencing advanced from short-reads to long-reads, computational biology evolved to manage the massive data volumes and the complexity of highly repetitive plant genomes. This technological shift enabled the assembly of complete, telomere-to-telomere (T2T) genomes, capturing the full extent of genomic structural variation and resolving regions previously inaccessible with short reads. The rise of pangenome assemblies further expanded this capability by constructing graph-based genomic references that reflect the genetic diversity within species, offering a more comprehensive view of presence/absence variations (PAVs), copy number variations (CNVs), tandem duplications (TDs), and structural variations (SVs). These advancements have driven the sequencing of non-model plants and wild relatives, uncovering novel alleles and adaptive traits critical for climate resilience and crop improvement. By integrating T2T to pangenomes, researchers can better explore underutilized species, unlocking their potential for agricultural innovation and biodiversity conservation.

A significant area of advancement in plant science is the rise of correlation biology, which maps gene networks and co-expression modules underlying complex traits. Correlation biology dissects the intricate relationships between genes, metabolic pathways, and environmental signals, offering a deeper understanding of polygenic traits that drive yield, stress resilience, and disease resistance. Tools such as weighted gene co-expression network analysis (WGCNA) and genome-wide association studies (GWAS) play a critical role in this process, enabling researchers to link genomic variation to phenotypic outcomes across diverse plant populations. The integration of genetic and phenotypic data from large plant populations has facilitated genomic selection (GS), allowing breeders to identify, track, and accelerate the deployment of favorable alleles linked to critical agronomic traits such as drought tolerance, nutrient efficiency, and pathogen resistance. This convergence enhances the accuracy of predictive models, enabling breeders to make informed decisions early in the breeding cycle and fast-track the development of high-performing, resilient crop varieties. By leveraging multi-environment trials and big data analytics, GS captures the complexity of genotype-environment interactions, providing a powerful framework for breeding crops that can thrive under variable climatic conditions and contribute to sustainable agriculture. This systems-level approach accelerates the identification of candidate genes for CRISPR-based genome editing and facilitates the design of next-generation crops capable of thriving in variable climates and resource-limited environments. As a result, correlation biology serves as a cornerstone for developing climate-resilient, high-yield plant varieties, addressing the pressing challenges of global food security and sustainable agriculture.

Looking to the future, computational biology is poised to integrate artificial intelligence (AI), generative AI (GenAI), and synthetic biology to redesign biological processes for sustainable agriculture. AI-powered models are already being used to analyze multi-omics datasets, predict gene function, and simulate plant responses to environmental stress. Emerging technologies such as plant artificial chromosomes (PACs), genome-scale digital twins, and large language models (LLMs) hold the promise of redefining crop design, enabling the creation of entirely new plant varieties optimized for climate resilience, nutrient use efficiency, and sustainable production.

As computational biology continues to evolve, its influence will extend beyond plant science, shaping the future of agriculture, biodiversity conservation, and global food systems (Skobelev et al. 2021).

23.2 Digital Biology Emerges in the Post-Genomic Era

Since the development of early sequence alignment algorithms in the 1970s (Needleman and Wunsch 1970), computational biology has become an integral part of plant biology research and applications (Fogg and Kovats 2015). This evolution has been driven by advancements in computing resources leading to the creation of digital germplasm information, portals like GRIN (Supplementary Table 23.1), and genomic databases such as NCBI GenBank, Phytozome, Ensembl Plants, UniProt, InterPro, REVIGO, Gramene, Pfam, Panache, TEdb, and TAIR (Supplementary Table 23.2) (Kanehisa and Bork 2003; Li et al. 2013). These repositories, combined with recent cloud-based computational platforms designed around FAIR (Findable, Accessible, Interoperable, and Reusable) principles, facilitate data storage, retrieval, and large-scale analysis. They enable sequence similarity searches, structural predictions, evolutionary studies, and functional annotation for both model and non-model species. This infrastructure serves as the backbone of modern genomics, empowering researchers to unravel the complexities of gene networks, metabolic pathways, and single-cell processes (VanBuren et al. 2020; Fusaro et al. 2011).

The advent of advanced sequencing technologies (Supplementary Table 23.3), such as Oxford Nanopore Technologies (ONT) and Pacific Biosciences (PacBio HiFi), has revolutionized the resolution of complex genomic regions, including structural variations like duplications, translocations, inversions, and repetitive elements that are key to understanding genomic plasticity and evolutionary dynamics. These long-read platforms, paired with complementary technologies like Hi-C for 3D chromosomal interactions and bisulfite sequencing (WGBS) for epigenomic profiling, enable chromosome-scale genome assemblies and insights into subgenome complexities in polyploid plants. High-resolution transcriptomics, including RNA sequencing (RNA-seq), single-cell RNA-seq (scRNA-seq), and 10× Genomics Chromium, alongside chromatin accessibility tools such as ATAC-seq, ChIP-seq, CUT&RUN, and Tn5-based library prep (Nextera), offer unparalleled detail on gene expression, regulation, and chromatin states. Together, these innovations transform raw sequencing data into comprehensive genomic blueprints, fueling breakthroughs in crop improvement, conservation genomics, and synthetic biology and ushering in a new era of digital biology to develop future-ready plants.

23.2.1 Complete Plant Genomes

Recent advancements in genome sequencing, assembly, and annotation have accelerated the generation of high-quality, complete plant genomes. This progress is driven by the refinement of sequencing technologies and the development of faster, more accurate computational tools that enable the assembly of complex plant genomes at chromosome-level resolution. Genome assembly tools have evolved to handle the size, heterozygosity, and repetitive content typical of plant genomes. Modern tools like Minimap2, CANU, Flye, VERKKO, and HIFIasm handle the size, heterozygosity, and repetitive content of plant genomes, while polishing tools such as Racon and Pilon improve assembly accuracy. Tools like BUSCO, Jellyfish, and GenomeScope2 assess genome completeness, size, ploidy, and repeat content, and HiC-Pro enables spatial organization of contigs into chromosome-length scaffolds (Durand et al. 2016). These methods (Table 23.1), combined with PacBio HiFi, ONT long reads, and Hi-C scaffolding, now make it possible to assemble complete, gap-free sequences of a chromosome (Naish et al. 2021), the so-called T2T genomes, even for polyploid and highly repetitive species like wheat and conifers.

23.2.2 Pangenomics

Pangenomics has transformed plant genomics by shifting from the limitations of single-reference genomes to capturing the full genomic diversity of a species. This paradigm shift began with the first plant pangenome assembled for soybean in 2014 (Li et al. 2014), paving the way for a deeper understanding of structural variations (SVs), gene PAVs, and rare alleles that underpin critical agronomic traits.

Table 23.1 Genome assembly, completeness assessment, and pangenomics

Tool	Description	Reference
Minimap2	Lightweight tools for mapping and assembling long-read sequencing data.	Li (2018)
Canu	A tool for de novo assembly of long reads, optimized for PacBio and ONT data.	Koren et al. (2017)
Flye	Assembler for long-read sequencing data, often used with ONT reads.	Kolmogorov et al. (2019)
Verkko	A hybrid assembler leveraging long-read, hi-C, and optical mapping data.	Rautiainen et al. (2023)
Hifiasm	Assembler specifically designed for highly accurate HiFi reads from PacBio sequencing.	Cheng et al. (2024)
Racon	Polishing tool that improves genome assemblies using long reads.	Vaser et al. (2017)
Pilon	Assembly improvement tool that uses Illumina data to correct assemblies.	Walker et al. (2014)
Busco	Completeness assessment tool that evaluates genome assemblies based on conserved genes.	Seppey et al. (2019)
Jellyfish	K-mer counting tool, used for estimating genome size and heterozygosity.	Marçais and Kingsford (2011)
Genomescope2	Tool to estimate genome size, ploidy, and repeat content from k-mer data.	Ranallo-Benavidez et al. (2020)
Hic-pro	Hi-C data processing pipeline, used for genome scaffolding and structural assembly validation.	Servant et al. (2015)
Pggb	Tool for progressive genome graph construction and visualization.	Garrison et al. (2024)
Pankmer	Tool for pangenome analysis using k-mers	Aylward et al. (2023)
Panx	Pangenome analysis platform for visualizing gene content variability across multiple genomes.	Ding et al. (2018)

Table 23.2 Variant calling, structural variation, evolutionary analysis, and functional genomics

Tool	Description	Reference
Helixer	Tool for ab initio prediction of gene structure using Deep Neural Networks and a Hidden Markov Model.	Holst et al. (2023)
Maker	A portable and easily configurable genome annotation pipeline.	Cantarel et al. (2008)
Augustus	Tool for finding protein-coding genes and their exon-intron structure in genomic sequences.	Stanke et al. (2008)
Edta	Repeat annotation and identification tool for LTR retrotransposons and other repeats.	Ou et al. (2024)
Eggnog mapper	Functional annotation of large sets of sequences based on fast orthology assignments	Cantalapiedra et al. (2021)
Orthofinder	Comparative genomics tool for orthology inference and phylogenetic analysis.	Emms and Kelly (2019)
Orthodb	Database for hierarchical orthology information, often used in orthology analysis in pangenomes.	Kriventseva et al. (2019)
Cafe	Tool for gene family evolution analysis, used to infer gene gain/loss rates.	Mendes et al. (2021)
Goatools	Python library for Gene Ontology (GO) enrichment analysis and annotation.	Klopfenstein et al. (2018)
Coge	Platform for synteny and comparative genomics analysis with interactive visualizations.	Nelson et al. (2018)
Mcscan	Synteny analysis tool that identifies collinear gene pairs and generates syntenic plots.	Y. Wang et al. (2024)
Freebayes	Variant calling software for calling SNPs and short indels, optimized for diploid genomes.	Garrison and Marth (2012)
Syri	Tool for detecting structural variants in genomes using paired-end and long-read data.	Goel et al. (2019)
Gblup	Genomic Best Linear Unbiased Prediction, used for breeding value prediction based on genomic data.	Meuwissen et al. (2001)
Bglr	Variety of tools for high-dimensional regression and variable selection like BayesB and RKHS	Pérez and de los Campos (2014)

Table 23.3 Transcriptomics tools

Tool	Description	Reference
Salmon	Tool for transcript quantification from RNA-Seq data, known for high-speed and accuracy.	Patro et al. (2017)
Seurat	Single-cell RNA-Seq analysis tool enables dimensionality reduction, clustering, and cell-type annotation.	Hao et al. (2024)
Merfish	Multiplexed Error-Robust Fluorescence In Situ Hybridization, enabling spatially resolved transcriptomics at single-cell resolution.	Fang et al. (2023), Moffitt et al. (2016)
Phytomap	Spatial transcriptomics tool exemplified by Arabidopsis thaliana root tips, allowing validation of clusters and genes across an entire genome.	Nobori et al. (2023)

Pangenomes consolidates genomic data from multiple accessions and uncovers novel alleles often missed in single-reference approaches, which expands the genomic toolbox for crop improvement, stress resilience, and adaptive evolution (Lei et al. 2021). Recent breakthroughs in graph-based pangenome construction have further elevated the field, providing high-resolution, gapless representations of plant genomes. Tools like PGGB and PanX (Table 23.1) generate graph-based structures that integrate T2T assemblies, enabling the visualization of entire genomic landscapes and capturing complex SVs across populations (Garrison et al. 2024). PanKmer, an ultra-fast k-mer-based pangenome analysis tool, facilitates the detection of population-level genetic signatures by efficiently processing thousands of genomes, making it particularly useful for GWAS and trait discovery in large datasets (Aylward et al. 2023). The application of these tools extends beyond major crops, catalyzing research into non-model plants and underutilized species. For instance, the exploration of duckweed (Acosta et al. 2021) and the sequencing of baobab genomes (Kitony et al. 2024) highlight how pangenomics can unlock the genetic potential of wild and orphan species for addressing climate resilience, carbon sequestration, and food security. As pangenomics continues to evolve, it is reshaping our understanding of plant biodiversity, evolution, and the genomic architecture that underlies complex traits. This transformative approach not only enhances breeding programs, but also serves as a vital tool for biodiversity conservation and the sustainable management of genetic resources in the face of global environmental challenges.

23.2.3 Next-Generation Genome Annotation

Genome annotation has significantly advanced with the integration of structural and functional analyses, enhancing gene prediction and characterization. Tools like AI-enabled Helixer, MAKER, and Augustus (Table 23.2) leverage RNA-seq and proteomic data to automate gene model prediction (Holst et al. 2023; Holt and Yandell 2011; Stanke et al. 2006; Mahood et al. 2020), refine gene boundaries, detect alternative splicing, and annotate noncoding RNAs (ncRNAs) critical for regulatory functions in plant development and stress responses. Repeat annotation with EDTA further supports genome assembly by identifying LTR retrotransposons and other repetitive elements, which are integral to structural and functional genome analysis. Functional annotation is enriched by tools like OrthoFinder, which infers orthology and paralogy (Emms and Kelly 2019), and CAFE, which analyzes gene family evolution to uncover gene gain and loss rates. Comparative genomics platforms such as CoGe and MCScan facilitate synteny and structural conservation analysis, while GOATOOLS perform Gene Ontology (GO) enrichment analysis and annotation. For variant calling, commonly used tools are FreeBayes, which identifies SNPs and INDELs, and Syri for SV. Together, these tools enable a comprehensive annotation process, illuminating conserved and novel gene functions, evolutionary divergence, and adaptive traits, even in non-model species.

23.2.4 Epigenetics

Epigenomic tools such as ATAC-seq and bisulfite sequencing (Supplementary Table 23.3) offer tools to interrogate the chromatin landscape and DNA methylation patterns that regulate gene expression. By mapping open chromatin regions, histone modifications, and methylation marks, these techniques reveal the epigenetic mechanisms underlying plant responses to environmental stimuli. This epigenomic profiling plays a pivotal role in breeding for resilience, as it identifies regulatory marks associated with stress tolerance, development, and yield stability. Notably, epigenetic modifications can prime gene expression to respond more robustly to recurrent stress, a phenomenon known as stress memory or transgenerational epigenetic inheritance (Lämke and Bäurle 2017; Yong et al. 2016). By leveraging these insights, researchers can develop crops that "remember" past environmental stressors and exhibit enhanced resistance to drought, salinity, and pathogen attacks. Moreover, the integration of epigenomic data with genomic selection and gene-editing technologies (e.g., CRISPR-dCas9-epigenetic editors) opens new avenues for modulating gene expression without altering DNA sequences (Papikian et al. 2019). This approach holds promise for generating stress-adaptive crops that maintain high yields while minimizing genetic disruption, contributing to the development of sustainable agricultural systems. As epigenomics continues to expand, it offers transformative potential for climate-resilient breeding programs, enabling the fine-tuning of plant responses at the regulatory level. This innovation with newer basecalling and methylation detection tools like ONT's Dorado paves the way for the design of crops that are not only genetically optimized but also epigenetically conditioned to thrive in fluctuating and extreme environments.

23.2.5 Single Cell and Spatial Transcriptomics

Plant transcriptomics has expanded from traditional bulk RNA-seq quantified by tools like Salmon to single-cell and spatial transcriptomics (Table 23.3). These deeper-level transcriptomics are redefining our understanding of plant gene expression by providing unprecedented insights into cellular heterogeneity and the spatial organization of tissues. scRNA-seq enables the profiling of gene expression at the resolution of individual cells, uncovering rare cell types, developmental trajectories, and dynamic responses to environmental cues. In plants, scRNA-seq has been pivotal in mapping root, shoot, and leaf cell lineages, shedding light on how gene regulatory networks orchestrate growth, differentiation, and stress adaptation (Ryu et al. 2019). A critical challenge in non-model plants lies in the annotation of scRNA-seq clusters, where reference genomes and well-curated gene databases are often limited (Abramson et al. 2021). Computational tools, such as Seurat, enable cluster annotation by integrating gene ontology (GO) enrichment, marker gene prediction, and cross-species comparisons. In these systems, researchers increasingly rely on de novo transcript assembly and comparative genomics to identify cluster-specific markers, bridging the annotation gap in underexplored species.

Meanwhile, spatial transcriptomics bridges gene expression profiling with spatial context, preserving the architectural integrity of plant tissues and revealing how gene activity is distributed across specific tissue regions (Zahedi et al. 2024). This technology is critical for dissecting complex developmental processes and understanding how plants respond to biotic and abiotic stress at the cellular level (Shahan et al. 2022; Giolai et al. 2019).

Recent advances in multiplexed error-robust fluorescence in situ hybridization (MERFISH), which integrates 3D imaging, confocal microscopy, and deep learning algorithms, expand the capabilities of spatial transcriptomics. This approach enables the detection and quantification of thousands of transcripts in tissue sections up to 200 μm thick, allowing for comprehensive analysis of gene expression within plant organs (Fang et al. 2023). The application of PHYTOMAP to *Arabidopsis thaliana* root tips exemplifies the power of spatial transcriptomics in plant biology (Nobori et al. 2023), while it also enables the validation of clusters and genes across an entire non-model plant, *Wolffia australiana* (Denyer et al. 2024). Together, single-cell and spatial transcriptomics provide a multidimensional view of plant gene expression, unlocking new frontiers in developmental biology, systems genetics, and precision breeding. These technologies are poised to drive breakthroughs in crop improvement, stress resilience, and plant-environment interactions, ultimately enhancing agricultural sustainability and productivity.

23.2.6 Proteomics

Computational biology has changed proteome analysis in plant biology, enabling the identification, quantification, and functional characterization of proteins on a genome-wide scale. Tools such as MaxQuant and Proteome Discoverer (Table 23.4) facilitate label-free and isotopic labeling-based proteomics, providing insights into protein abundance, posttranslational modifications (PTMs), and protein-protein interactions (Tyanova et al. 2016). InterPro and Pfam databases annotate protein domains, while OrthoFinder and EggNOG mapper assist in identifying orthologous and paralogous protein families, shedding light on evolutionary conservation across species (Table 23.2 and Supplementary Table 23.2) (Mitchell et al. 2019; Emms and Kelly 2019; Cantalapiedra et al. 2021). STRING and Cytoscape map protein interaction networks, revealing the regulatory pathways that govern plant development and stress responses (Szklarczyk et al. 2021; Shannon et al. 2003). Additionally, AlphaFold2 and RoseTTAFold predict protein structures with near-experimental accuracy, aiding in the functional analysis of novel or uncharacterized proteins (Jumper et al. 2021; Baek et al. 2021). Computational tools allow researchers to explore how proteins dynamically respond to environmental stimuli, pathogen attack, and nutrient availability by integrating mass spectrometry (MS) data with multi-omics approaches. This holistic view of the plant proteome is essential for engineering stress-resilient crops, understanding complex signaling networks, and advancing precision agriculture.

Table 23.4 Proteomics tools

Tool	Description	Reference
Maxquant	Tool for label-free and isotopic labeling-based proteomics, enabling insights into protein abundance, PTMs, and interactions.	Tyanova et al. (2016), Sinitcyn et al. (2018)
Proteome discoverer	Software for comprehensive proteomics data analysis, supporting PTM discovery and quantification.	Tyanova et al. (2016)
String	Database and tool for mapping protein interaction networks, elucidating regulatory pathways involved in development and stress responses.	Szklarczyk et al. (2021)
Cytoscape	Visualization and analysis tool for protein interaction and biological networks.	Shannon et al. (2003), Su et al. (2014)
Alphafold2	Deep learning-based tool for predicting protein structures with near-experimental accuracy.	Jumper et al. (2021)
Rosettafold	Tool for protein structure prediction using a three-track deep learning architecture.	Baek et al. (2021)

Table 23.5 Metabolomics tools

Tool	Description	Reference
Xcms	Software for processing mass spectrometry data, including peak detection, retention time alignment, and metabolite annotation.	Smith et al. (2006)
Mzmine	Open-source software for mass spectrometry data processing and visualization, supporting peak alignment and compound identification.	Pluskal et al. (2010),Heuckeroth et al. (2024)
Metaboanalyst	Platform for statistical analysis, visualization, and functional interpretation of metabolomics data.	Pang et al. (2021)
Kegg	Database for metabolic pathway reconstructions, linking metabolites to biosynthetic and regulatory networks.	Kanehisa et al. (2025)
Plantcyc	Database for plant-specific metabolic pathways, connecting genes, enzymes, and their metabolic products.	Schläpfer et al. (2017), Hawkins et al. (2025)
Gnps	Platform for clustering structurally related metabolites and discovering novel phytochemicals.	Wang et al. (2016)

23.2.7 Metabolomics

Computational biology has become indispensable in metabolomics, enabling the comprehensive analysis of metabolites and their roles in plant biology. Tools such as XCMS, MZmine, and MetaboAnalyst process and interpret mass spectrometry (MS) and nuclear magnetic resonance (NMR) data, facilitating peak detection, alignment, and compound identification (Smith et al. 2006; Pluskal et al. 2010; Pang et al. 2021). These platforms support untargeted and targeted metabolomics, allowing researchers to profile thousands of metabolites and uncover biomarkers linked to stress responses, developmental stages, and secondary metabolism. KEGG and PlantCyc provide metabolic pathway reconstructions, mapping metabolites to biosynthetic networks and highlighting connections between genes, enzymes, and their metabolic products (Kanehisa et al. 2023; Schläpfer et al. 2017). Molecular networking tools like GNPS (Global Natural Products Social Molecular Networking) cluster structurally related metabolites, enhancing the discovery of novel phytochemicals and specialized metabolites in plants (Wang et al. 2016). Integration with transcriptomics and proteomics further enables correlation analyses between gene expression, enzyme activity, and metabolite levels, offering insights into how plants adapt to environmental stressors, nutrient availability, and pathogen interactions. These computational frameworks not only advance fundamental research but also drive innovations in crop improvement, bioengineering, and natural product discovery (Table 23.5).

23.2.8 CRISPR (Clustered Regularly Interspaced Short Palindromic Repeats)

Building on high-quality genome assemblies, CRISPR technology serves as a transformative bridge between genome sequencing and functional annotation, enabling precise, targeted manipulation of genes. This gene-editing platform accelerates the validation of gene functions, facilitates the dissection of complex traits, and empowers the development of climate-resilient, high-yield crops. For example, CRISPR-Cas9 has been applied to edit genes associated with traits like disease resistance, drought tolerance, and yield (Pickar-Oliver and Gersbach 2019). Single base and prime editing "search-and-replace" modifications improve specificity, as shown in crops like maize and wheat (Vats et al. 2024). CRISPR-modified sorghum exhibits resistance to the parasitic witchweed, promising new solutions for global food security (Zhou et al. 2023; Ledford 2024). As CRISPR technology evolves, its convergence with machine learning, multi-omics data, and synthetic biology will unlock new frontiers in crop improvement. This holistic approach will enable the design of next-generation crops tailored to withstand climate change, enhance nutrition, and contribute to sustainable agriculture, ultimately redefining global food production systems.

The advent of digital biology has fundamentally transformed plant science, reshaping not only how experiments are designed but also expanding the scope and complexity of questions that can be addressed. By integrating high-throughput sequencing, single-cell and spatial transcriptomics, multi-omics approaches, and AI-driven analytics, researchers can now explore plant systems at an unprecedented resolution, from the dynamics of individual cells to the evolution of entire genomes. This shift has enabled deeper insights into genotype-environment interactions, regulatory networks, and adaptive traits, paving the way for precision breeding, synthetic biology, and the domestication of non-model species. As computational tools grow more sophisticated and accessible, plant scientists are empowered to tackle grand challenges in food security, climate resilience, and biodiversity conservation, marking the post-genomic era as an exciting time of discovery and innovation that is redefining the future of agriculture and ecological sustainability.

23.3 The Coming of Age of Correlation Biology in the Post-Genomic Era

Since the discovery of Mendelian genetics and the development of statistical methodologies for trait analysis in the early twentieth century, biological correlations have become a cornerstone of genomics research (Sukhija et al. 2023). Advancements in sequencing and high-throughput phenotyping have transformed correlation biology, enabling the integration of multi-omic datasets with large-scale phenotypic assessments to link genetic variants to traits with unprecedented precision. Techniques such as GWAS, GS, and co-expression analysis are essential for dissecting the genetic architecture of complex traits, modeling gene-environment interactions, and accelerating the development of resilient, high-yielding, and resource-efficient crops. This integrated framework is redefining data-driven breeding strategies to address global challenges in food security and sustainable agriculture.

23.3.1 High-Throughput Phenotyping

In the post-genomic era, phenotyping serves as the foundation of correlation biology, precisely connecting genetic data to observable traits. Advances in technology and computational power have enabled high-throughput phenotyping systems to capture critical traits like drought tolerance, photosynthetic efficiency, and root system architecture (RSA). Indoor platforms, such as SLEAP-Roots, use AI-driven tools to rapidly phenotype root systems, offering new insights into root morphology and growth dynamics (Berrigan et al. 2024). Field phenotyping has expanded through tools like drones, LiDAR, and hyperspectral imaging, which noninvasively monitor crop performance under real-world conditions over time (Gano et al. 2024; Moghimi et al. 2019; Dhami et al. 2020). Gantry systems equipped with multi-sensor arrays further track dynamic trait expression in response to environmental changes, bridging the gap between lab experiments and field applications (Bascon et al. 2022; Sadeghi-Tehran et al. 2017). By integrating machine learning and multi-sensor data, modern phenotyping systems uncover complex genotype-phenotype-environment interactions, allowing for the accurate prediction of traits like yield and disease resistance (Lee et al. 2018). These advancements streamline breeding pipelines, reducing selection cycles and accelerating the development of climate-resilient, high-yield crops.

23.3.2 Genome-Wide Association Studies (GWAS)

GWAS is a powerful research approach for linking genomic variants to specific traits by identifying statistical associations across diverse populations. In the context of plants, GWAS surveys the genomes of large collections of individuals to uncover genetic variants associated with critical agronomic traits such as drought tolerance, disease resistance, or yield. By pinpointing these variants, researchers can identify genes or regions in the genome that contribute directly to these traits, offering precise targets for crop improvement via gene editing or conventional marker-assisted breeding. In maize (*Zea mays*), for example, GWAS coupled with transcriptome analyses identified *ZmNAC111*, a key gene regulating drought tolerance. This discovery highlighted the potential of GWAS to uncover genetic factors controlling water use efficiency, a vital trait for resilience in water-limited environments (Mao et al. 2015). Similarly, GWAS integrated with WGCNA has mapped loci associated with seed dormancy, a complex trait critical for ensuring stable germination under variable conditions (Ma et al. 2023). Such multi-omics approaches accelerate the discovery of genes underlying complex agronomic traits, guiding breeders in the development of high-performing crop varieties.

23.3.3 Genomic Selection (GS)

GS is a breeding methodology that leverages genomic data to predict the performance of untested individuals for specific traits (Fig. 23.1). Unlike traditional marker-assisted selection (MAS), GS uses genome-wide markers to estimate the genetic value of a genotype, making it particularly effective for improving complex traits controlled by many genes (Table 23.2). This approach has revolutionized plant breeding by enabling early selection of superior genotypes, thereby reducing breeding cycles and accelerating crop improvement. In rice (*Oryza sativa*), GS models that incorporate genotypic main effects (G) and environmental covariates (W) have demonstrated robust predictive accuracy across diverse environments. For example, traits like grain yield and drought resilience have been predicted with high confidence, enabling breeders to focus resources on the most promising candidates (Nguyen et al. 2023). However, the inclusion of genotype-by-environment (G × E) or genotype-by-covariate (G × W) interactions yielded mixed results, reflecting the context-specific nature of these models. This contrasts with successes in barley (*Hordeum vulgare*) and wheat (*Triticum aestivum*), where environmental factors significantly enhanced predictive accuracy (Jarquín et al. 2014; Malosetti et al. 2016). These example findings illustrate the adaptability of GS and its ability to tailor breeding strategies to the unique genetic and environmental dynamics of each crop.

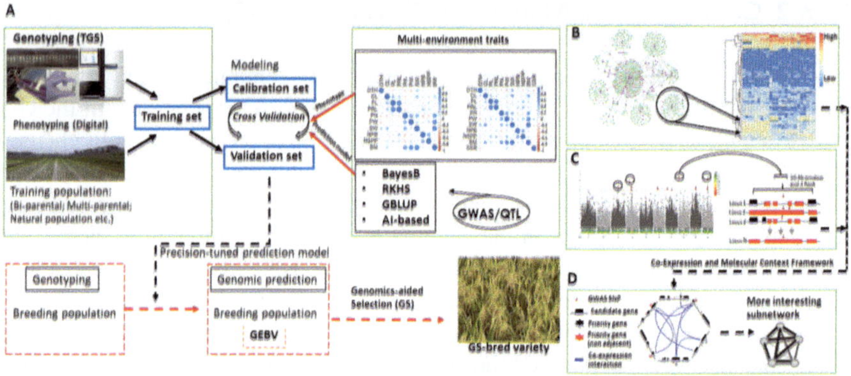

Fig. 23.1 Integration of genomic selection (GS), co-expression networks, and GWAS in the post-genomic era for trait discovery and gene prioritization. (**a**) Genomic Selection (GS): Leveraging third-generation sequencing (TGS) and high-throughput phenotypic data collected across diverse environments; statistical and AI-driven models are trained to predict genomic breeding values (GEBVs). This accelerates the development of climate-smart plant varieties by enabling cost-effective and precise genomics-assisted selection. (**b**) Co-Expression Networks: Gene co-expression networks uncover functional relationships by analyzing gene expression profiles across diverse conditions and accessions. These networks reveal gene subnetworks with highly correlated expression patterns, providing unbiased insights into putative biological co-function and regulatory interactions. (**c**) Genome-Wide Association Studies (GWAS): GWAS identifies SNPs associated with complex traits, pinpointing genomic regions of interest. Candidate genes are mapped based on proximity to significant SNPs, using user-defined parameters such as flanking regions (e.g., 50 kb and two neighboring genes). (**d**) Integration of Co-Expression and GWAS Evidence: By combining GWAS results with co-expression networks, genes with strong functional connectivity are prioritized. This approach identifies key candidate genes, including those distant from GWAS SNPs (nonadjacent), that might be missed using proximity-based mapping alone. Network interactions are statistically validated, highlighting gene subnetworks with significant co-expression and trait relevance. Together, these frameworks enhance precision in trait discovery and enable the identification of next-generation candidate genes, driving innovation in climate-smart breeding programs. (Figure adapted from Schaefer et al. (2018))

23.3.4 Co-Expression Analysis

Co-expression analysis investigates gene expression patterns to uncover networks of genes that work together to regulate traits. By identifying groups of genes with correlated expression profiles, researchers can reveal the molecular pathways underlying stress responses, growth, and development in plants. This systems biology approach is invaluable for dissecting complex traits and identifying key regulatory genes that can serve as targets for breeding or genetic engineering. In wheat (*Triticum aestivum*), WGCNA identified drought-responsive networks centered around key regulatory genes such as *PP2C* (*PROTEIN PHOSPHATASE 2C*), *ERF* (*ETHYLENE RESPONSE FACTOR*), and *HSP90* (*HEAT SHOCK PROTEIN 90*). These genes regulate essential processes like stress signaling and protein stability under water-deficit conditions, contributing to enhanced drought tolerance (Lv et al. 2020). Similarly, co-expression analysis in peanut (*Arachis hypogaea*) highlighted antioxidant defense pathways involving *SOD* (*SUPEROXIDE DISMUTASE*), *CAT* (*CATALASE*), and *APX* (*ASCORBATE PEROXIDASE*), which protect cells from oxidative stress under salinity conditions (Wang et al. 2024). Beyond stress tolerance, the integration of co-expression analysis with scRNA-seq in soybean (*Glycine max*) revealed complex gene networks controlling specific cell types in the root system for carbon storage (Sun et al. 2024). By dissecting these intricate networks, co-expression analysis provides actionable insights for breeding resilient and high-yielding crops.

23.3.5 Whole Genome Multiplication (WGM)

WGM, also known as polyploidy, is a pivotal evolutionary mechanism driving genetic diversity and adaptability in plants. Sequence analyses in crops such as wheat (*Triticum aestivum*) have revealed a reticulate evolutionary history, showcasing how ancient polyploidy events have facilitated the expansion of gene families, enabling resilience to biotic and abiotic stresses and adaptability across diverse environments (Marcussen et al. 2014; Chen et al. 2024). In *Arabidopsis thaliana*, genome-wide duplications have been linked to ecological success, with genetic redundancy and subfunctionalization fostering the evolution of novel traits and environmental flexibility (Raes et al. 2003). Advances in computational biology and comparative genomics now enable precise analysis of WGM events, shedding light on how they contribute to trait innovation, regulatory network evolution, and species diversification. By incorporating WGM into correlation biology frameworks such as GWAS and co-expression analysis, researchers can better understand the genetic basis of resilience and productivity in polyploid crops, paving the way for targeted breeding strategies that harness the power of duplication events.

Collectively, modern correlational biology underscores the transformative role of computational biology in decoding evolutionary pathways and deepening our understanding of plant resilience, gene function, and adaptive strategies. The techniques like GWAS have moved beyond merely identifying genetic variants; they now offer precise insights into the complex interplay between genotype, phenotype, and environment. By integrating high-throughput data with advanced computational tools, these methods enable researchers to uncover novel genetic drivers, predict trait performance with remarkable accuracy, and design resilient, high-yielding crops tailored to the challenges of climate change and global food security.

23.4 The Future of Computational Biology in a Post-Genomic World

Agriculture is entering an era of unprecedented transformation, driven by breakthroughs in AI, synthetic biology, and CRISPR-Cas technologies. By 2050, agriculture is projected to evolve into a precision-driven, data-centric ecosystem, where crops are genetically tailored to thrive in specific climates, and resources are optimized in real time to maximize yield and minimize environmental impact. This convergence of AI, quantum computing, and synthetic biology will redefine how we cultivate food, ensuring that global food systems sustainably support a population nearing 10 billion people while preserving biodiversity and ecosystem health (Mukerjee et al. 2024; United Nations n.d.). The future of agriculture will be characterized by the widespread deployment of *Integrated Multi-Omics*, *Gene Network Modeling*, *PACs*, *Genome-Scale Digital Twins*, *AI-Driven Crop Management,* and *quantum computing*. These innovations will empower breeders to develop climate-resilient crops, optimize resource use with unprecedented precision, and simulate plant growth at the cellular level to predict and mitigate environmental stressors. This shift not only aligns agriculture with global sustainability goals, but also lays the foundation for a regenerative, resilient food system capable of addressing the complex challenges posed by climate change and population growth.

23.4.1 Integrated Multi-Omics

Next-generation computational tools will drive the seamless integration of multi-omics data to generate new hypotheses (Fig. 23.2a). This holistic integration will revolutionize precision agriculture, enabling the development of climate-resilient crops tailored for specific environments, optimizing growth under conditions of drought, heat stress, and nutrient imbalances (Cembrowska-Lech et al. 2023). The synchronization of omics-driven insights with phenomic logs of continuous data collected by advanced sensor networks at various growth stages will facilitate dynamic, real-time plant health monitoring. This capacity to track crop performance and physiological responses at the cellular level will shrink the yield gap and enhance resource efficiency, ensuring that crops receive precisely what they need at the right moment. Artificial intelligence and machine learning will play a pivotal role in analyzing complex omics datasets, identifying hidden correlations between genetic markers and phenotypic traits. These algorithms will predict how specific gene edits or breeding strategies influence plant resilience, accelerating the development of next-generation cultivars optimized for fluctuating climates. Additionally, metabolic modeling driven by integrated omics will unlock novel pathways for nutrient uptake, carbon sequestration, and pest resistance. By analyzing how genome-wide expression patterns interact with external stressors, researchers can design crops that self-regulate nutrient flow, optimize water use, and thrive in degraded soils. In practical terms, sensor-enabled phenomics will allow farmers to monitor plant health, detect early signs of disease, nutrient deficiency, or environmental stress, and deploy corrective actions with minimal input. This approach enhances sustainability by reducing fertilizer runoff, minimizing pesticide use, and conserving water resources. The convergence of omics technologies, AI, and high-throughput phenotyping marks the dawn of predictive agriculture, where decisions are driven by data-driven insights and molecular precision. This transformative framework will boost global food security while promoting ecological resilience and advancing sustainable farming practices for future generations.

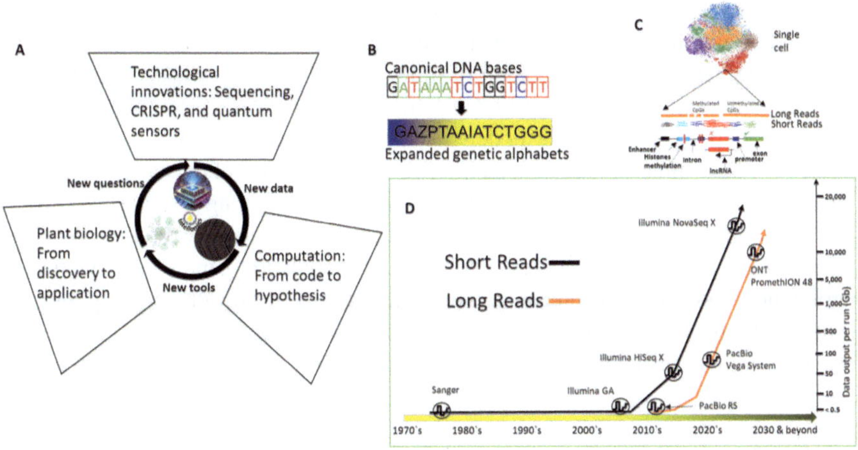

Fig. 23.2 Overview of technological innovations, computational advances, and biological applications driving plant science toward digital biology in the post-genomic era, enabling exploration of DNA elements down to single-cell epigenetic marks. (**a**) A continuous and iterative cycle illustrating the interplay among advancements in sequencing, CRISPR, and quantum sensors (technological innovations); the transformation of these datasets into hypotheses and computational tools (computation); and their application to biological discovery and solutions in plant science (plant biology). At the heart of this cycle is Intellomics, integrating AI (artificial intelligence), ML (machine learning), LLMs (large language models), and quantum computing to dynamically enhance each stage, fueling progress in post-genomic research and applications. (**b**) A conceptual depiction of the transition from canonical DNA bases to expanded genetic alphabets, emphasizing how quantum computing could revolutionize future sequencing paradigms and redefine our understanding of subtle genetic variations through expanded genetic codes. (**c**) The integration of long-read and short-read sequencing technologies for generating gapless T2T (telomere-to-telomere) genome assemblies. Initially, short-read sequencing transformed genomics, producing numerous low-quality draft genomes, but quickly became the primary platform for studying DNA elements such as enhancers (black), histones (cyan), DNA/histone methylation (dark red), introns (black lines), promoters (blue), silenced gene exons (red), antisense long noncoding RNA (lncRNA, red), and expressed gene exons (green). Advances now enable single-cell applications, providing high-resolution spatial data essential for in silico modeling, synthetic circuit design, and development of synthetic plants. (**d**) A timeline of sequencing technology evolution, showcasing the exponential transition from low-throughput Sanger sequencing to high-throughput short-read and long-read sequencing platforms, highlighting milestones that shaped modern genomics

23.4.2 Gene Network Modeling

Enhanced versions of WGCNA are evolving to incorporate machine learning (ML) and AI techniques, significantly advancing the identification of key driver genes and functional modules (Langfelder and Horvath 2008). By integrating deep learning, random forest, and support vector machines (SVMs), these enhanced models can more accurately predict gene-gene interactions and prioritize biologically relevant pathways. A major breakthrough is the ability to connect quantitative trait loci (QTL) to dynamic gene networks, enabling the discovery of regulatory hubs and actionable gene modules that drive complex traits (Liu 2024; Sun et al. 2024). This approach not only accelerates crop improvement through the identification of yield-related, stress-resilient, and disease-resistant genes, but also enhances functional genomics across diverse species. Furthermore, the integration of multi-omics data (e.g., transcriptomics, proteomics, and metabolomics) into WGCNA frameworks provides a comprehensive systems biology perspective. These models leverage unsupervised learning to detect subtle patterns in high-dimensional data, facilitating the exploration of epistatic interactions, gene pleiotropy, and noncoding regulatory elements. By adopting scalable cloud-based platforms and high-throughput sequencing technologies, next-generation WGCNA pipelines are positioned to address challenges in population genomics, personalized agriculture, and precision breeding. This convergence of network biology and ML promises transformative insights into genetic architecture and trait heritability, paving the way for more resilient and productive crop varieties.

23.4.3 Synthetic Biology

Synthetic biology, driven by computational biology, is unlocking unprecedented opportunities to engineer plants with novel traits that range from enhanced pest resistance to biofuel production by designing and integrating synthetic pathways. This field now enables the transfer of entire biosynthetic pathways across species, exemplified by efforts to enhance nitrogen fixation in nonleguminous crops like rice and maize. Such innovations aim to reduce reliance on synthetic fertilizers, lowering agricultural inputs while promoting sustainable farming practices (Roell and Zurbriggen 2020). Beyond enhancing productivity, synthetic biology has demonstrated its potential to address global health and energy challenges by engineering plants for the high-value production of metabolites such as artemisinin, a key antimalarial compound, and terpenes for renewable biofuels (Paddon and Keasling 2014; Xu et al. 2020; Hao et al. 2020; Mewalal et al. 2017). These efforts expand the role of plants from traditional food sources to biomanufacturing platforms capable of producing pharmaceuticals and sustainable materials. Advances in synthetic CO_2 fixation pathways and the engineering of enhanced photosynthetic efficiency are also being pursued to boost crop yields and mitigate the impacts of climate change (Luo et al. 2023). Scientists are exploring ways to increase biomass production, improve water-use efficiency, and future-proof agriculture against environmental stressors by redesigning the fundamental processes of carbon assimilation. A rapidly growing frontier is the development of AI-driven virtual cells (AIVC), which model and simulate cellular processes to guide the design of optimized plant systems for resilience, productivity, and environmental sustainability (Bunne et al. 2024). By integrating machine learning, protein engineering, and gene circuit design, synthetic biology is pushing the boundaries of what is possible in plant science, laying the groundwork for next-generation crops tailored for a rapidly changing world.

23.4.4 Plant Artificial Chromosomes (PACs)

PACs represent a transformative tool for introducing multigenic traits that enhance water-use efficiency carbon sequestration nutrient uptake and pest resistance in a single targeted step (Dymond et al. 2011; Coradini et al. 2023). By enabling the transfer of entire gene clusters or networks PACs bypass the limitations of single-gene edits accelerating the development of climate-resilient crops optimized for extreme environments. PACs are particularly promising for the domestication of wild and underutilized plant species offering a platform to rapidly introduce traits that transform these species into viable nutrient-dense food sources. This technology holds immense potential for expanding agricultural biodiversity introducing orphan crops that thrive in marginal soils and require fewer resources addressing both food security and ecological sustainability. Recent advances in chromosome engineering provide a blueprint for PAC development. In *Saccharomyces cerevisiae* CRISPR-Cas9-mediated chromosome fusion through the deletion of centromeres and telomeres has created novel chromosomal configurations serving as a model for synthetic chromosome construction in plants (Shao et al. 2018). Moreover CRISPR-induced double-strand breaks (DSBs) have been shown to trigger large-scale chromosomal rearrangements including crossovers and chromothripsis-like events a process involving extensive genomic shuffling highlighting new pathways for engineering stable PACs (Samach et al. 2023). The ability to construct customized synthetic chromosomes introduces opportunities to fine-tune gene expression stacking regulatory elements that control plant growth stress tolerance and secondary metabolite production. PACs could facilitate the simultaneous introduction of genes encoding biotic and abiotic stress resistance nitrogen fixation pathways and enhanced photosynthetic capacity creating super-crops capable of withstanding global climate pressures. As agriculture faces the dual challenges of feeding a growing population and mitigating environmental impacts PACs offer a forward-looking solution that bridges synthetic biology genome engineering and crop improvement. By reimagining how genetic material is introduced and organized PACs pave the way for sustainable high-yield agricultural systems that drive regenerative farming practices and carbon-negative agriculture (Fig. 23.1b)

23.4.5 Genome-Scale Digital Twins

Genome-scale digital twins, virtual models that replicate crops at the cellular and molecular levels, are set to revolutionize plant breeding and agricultural innovation. These digital twins will simulate growth dynamics, developmental processes, and stress responses across diverse environmental conditions, enabling in silico testing of gene-editing strategies and accelerating the breeding cycle (Peladarinos et al. 2023). By integrating high-throughput phenotypic data and multi-omics layers (genomics, transcriptomics, and metabolomics), digital twins will drive the breeding of crops with complex, polygenic traits such as drought resilience, nutrient use efficiency, and pathogen resistance (Robles et al. 2023). These models will not only predict gene-environment interactions but also map the cascading effects of genetic edits across entire biological networks, offering breeders a systems-level understanding of crop performance. Advanced simulation platforms like the Agricultural Production Systems sIMulator (APSIM) (Holzworth et al. 2014) and the Decision Support System for Agrotechnology Transfer (DSSAT) (Hoogenboom et al. 2019) are already laying the groundwork for this innovation. The integration of these tools with large language models such as CodonBERT and BioGPT will refine the accuracy of trait prediction and multi-trait selection, enhancing the precision of gene-editing targets (Ren et al. 2024). This fusion of AI and crop modeling will enable the rapid identification of candidate genes linked to critical agronomic traits, optimizing breeding strategies with unprecedented speed and precision. Moreover, genome-scale digital twins will facilitate scenario testing by simulating the impact of climate change, soil variability, and pest pressures on crop yields. Researchers and breeders will be able to optimize genetic modifications for specific agroecosystems, tailoring crops to perform under region-specific challenges. This predictive capacity will enhance climate resilience, ensuring stable yields in the face of increasing environmental volatility. The convergence of digital twins, AI-driven language models, and predictive agronomic tools heralds a new era of sustainable agriculture, where crop improvement is driven by data-driven insights and precision breeding. This transformative approach promises to reduce breeding timelines, conserve resources, and unlock genetic potential at a scale never before possible, paving the way for next-generation crops that are resilient, productive, and adaptive to the future of farming.

23.4.6 AI-Driven Crop Management

AI and ML are set to transform agriculture by seamlessly integrating climate, soil health, and crop growth data into dynamic, predictive systems. These AI-driven platforms will anticipate environmental challenges in real-time, optimizing key agricultural processes such as irrigation, nutrient delivery, and pest management. By harnessing vast datasets from satellite imagery, IoT sensors, and on-field analytics, AI will enable precision farming at an unprecedented scale, reducing resource use while maximizing yields. Digital climate twins, which are virtual models that replicate entire farm ecosystems, will allow farmers to simulate crop responses to drought, heat stress, and shifting weather patterns, facilitating proactive decision-making. Similarly, satellite-enabled decision-support systems will monitor field conditions continuously, providing insights into soil moisture, disease spread, and pest activity. This convergence of remote sensing, machine learning, and real-time feedback loops will empower farmers to enhance productivity, improve soil health, and minimize ecological footprints. Machine learning models have already demonstrated remarkable capabilities, such as predicting disease outbreaks weeks in advance by analyzing climatic trends and pathogen behavior (Mana et al. 2024). In pest management, AI tools can classify insect species, recommend biocontrol solutions, and track population dynamics, significantly reducing the reliance on chemical pesticides. Additionally, automated irrigation systems driven by AI adjust water delivery based on soil sensor data and weather forecasts, ensuring optimal hydration while conserving water resources. As these technologies mature, they pave the way for fully automated, self-regulating farms where AI continuously fine-tunes agricultural operations. By integrating robotic systems for planting and harvesting with AI-driven crop monitoring, farms of the future will operate with minimal human intervention, fostering a sustainable, regenerative agricultural paradigm. This shift promises not only to boost food security and resilience to climate change, but also to drive carbon-smart farming practices, where AI optimizes land use to sequester carbon, enrich biodiversity, and restore degraded ecosystems. Ultimately, the fusion of AI, climate modeling, and sustainable agriculture holds the key to feeding a growing global population while safeguarding the planet's natural resources.

23.4.7 Quantum Computing

Quantum computing is poised to advance plant science by overcoming computational limitations inherent to classical systems, addressing complex biological challenges such as protein folding, metabolic pathway optimization, and single-cell gene regulatory networks (Cheng et al. 2020; Ma et al. 2024). Unlike traditional bits, qubits leverage superposition, allowing them to exist in a probabilistic state between 0 and 1, such as 20% 0 and 80% 1, or 50% 0 and 50% 1, allowing quantum algorithms to explore vast solution spaces in parallel. Additionally, entanglement enhances processing power by enabling deeply interconnected computations, vastly increasing the speed and efficiency of problem-solving.

In plant science, these capabilities will facilitate precise molecular simulations of dynamic cellular processes, offering unprecedented insights into metabolic flux, epigenetic modifications, and gene-environment interactions. For example, quantum models could simulate the intricate folding of plant proteins under various environmental stresses, accelerating the discovery of novel traits linked to drought tolerance, nutrient efficiency, and pathogen resistance. Figure 23.2b illustrates the hypothetical transition from canonical DNA bases to expanded genetic alphabets, highlighting how quantum computing could redefine future sequencing paradigms.

Beyond molecular biology, quantum computing holds promise for ecosystem-level modeling and breeding optimization. By simulating entire plant populations and their genomic variability, researchers can predict polygenic trait inheritance patterns and identify ideal genomic configurations for enhanced yield and resilience. Quantum-driven optimization of photosynthetic pathways and carbon fixation mechanisms could lead to plants with significantly improved biomass production and reduced water usage. By 2050, the integration of quantum computing into plant genomics and precision agriculture is expected to drive the development of next-generation crops capable of thriving under extreme climate conditions, supporting global food security while minimizing environmental footprints (Maraveas et al. 2024). This technological leap will not only accelerate crop breeding cycles but also open new frontiers in synthetic biology, biopharmaceuticals, and carbon sequestration, reinforcing agriculture's role in addressing pressing global sustainability challenges.

23.5 Concluding Remarks

The integration of digital biology, correlation biology, AI, and quantum computing is reshaping plant science, enabling precision-driven approaches to address critical challenges such as food security, biodiversity loss, and climate change. With the rapid expansion of computational biology tools and their applications (Fig. 23.2), ensuring equitable access and addressing ethical concerns will require collaborative efforts, standardized frameworks, and active stakeholder engagement. Striking a balance between technological advancement, social equity, and environmental stewardship will be pivotal in building a sustainable and inclusive agricultural landscape for 2030 and beyond.

References

Abramson BW, Novotny M, Hartwick NT, Colt K, Aevermann BD, Scheuermann RH, Michael TP (2021) The genome and preliminary single-nuclei transcriptome of Lemna Minuta reveals mechanisms of invasiveness. Plant Physiol 188. https://doi.org/10.1093/plphys/kiab564

Acosta K, Appenroth KJ, Borisjuk L, Edelman M, Heinig U, Jansen MAK, Oyama T et al (2021) Return of the Lemnaceae: duckweed as a model plant system in the genomics and postgenomics era. Plant Cell 33(10):3207–3234. https://doi.org/10.1093/plcell/koab189

Aylward AJ, Petrus S, Mamerto A, Hartwick NT, Michael TP (2023) PanKmer: K-Mer-based and reference-free pangenome analysis. Bioinformatics 39(10). https://doi.org/10.1093/bioinformatics/btad621

Baek M, DiMaio F, Anishchenko I, Dauparas J, Ovchinnikov S, Lee GR, Wang J et al (2021) Accurate prediction of protein structures and interactions using a three-track neural network. Science (New York, N.Y.) 373(6557):871–876. https://doi.org/10.1126/science.abj8754

Bascon MV, Nakata T, Shibata S, Takata I, Kobayashi N, Kato Y, Inoue S, Doi K, Murase J, Nishiuchi S (2022) Estimating yield-related traits using UAV-derived multispectral images to improve rice grain yield prediction. Agriculture 12(8):1141. https://doi.org/10.3390/agriculture12081141

Berrigan EM, Wang L, Carrillo H, Echegoyen K, Kappes M, Torres J, Ai-Perreira A et al (2024) Fast and efficient root phenotyping via pose estimation. Plant phenomics 6(April):0175. https://doi.org/10.34133/plantphenomics.0175

Bunne C, Roohani Y, Rosen Y, Gupta A, Zhang X, Roed M, Alexandrov T et al (2024) How to build the virtual cell with artificial intelligence: priorities and opportunities. Cell 187(25):7045–7063. https://doi.org/10.1016/j.cell.2024.11.015

Cantalapiedra CP, Hernández-Plaza A, Letunic I, Bork P, Huerta-Cepas J (2021) EggNOG-Mapper v2: functional annotation, orthology assignments, and domain prediction at the metagenomic scale. bioRxiv. https://doi.org/10.1101/2021.06.03.446934

Cantarel BL, Korf I, Robb SM C, Parra G, Ross E, Moore B, Holt C, Sánchez Alvarado A, & Yandell M (2008) MAKER: an easy-to-use annotation pipeline designed for emerging model organism genomes. Genome Research, 18(1):188–196. https://doi.org/10.1101/gr.6743907

Cembrowska-Lech D, Krzemińska A, Miller T, Nowakowska A, Adamski C, Radaczyńska M, Mikiciuk G, Mikiciuk M (2023) An integrated multi-omics and artificial intelligence framework for advance plant phenotyping in horticulture. Biology 12(10):1298. https://doi.org/10.3390/biology12101298

Chen Y, Wang W, Yang Z, Peng H, Ni Z, Sun Q, Guo W (2024) Innovative computational tools provide new insights into the polyploid wheat genome. aBIOTECH 5(1):52–70. https://doi.org/10.1007/s42994-023-00131-7

Cheng H-P, Deumens E, Freericks JK, Li C, Sanders BA (2020) Application of quantum computing to biochemical systems: a look to the future. Front Chem 8(November):587143. https://doi.org/10.1007/s42994-023-00131-7

Cheng, H., Asri, M., Lucas, J. et al (2024) Scalable telomere-to-telomere assembly for diploid and polyploid genomes with double graph. Nat Methods 21, 967–970. https://doi.org/10.1038/s41592-024-02269-8

Computation and Biology: A Partnership (2021) Nat Methods 18(7):695. https://doi.org/10.1038/s41592-021-01215-2

Coradini ALV, Ville CN, Krieger ZA, Roemer J, Hull C, Yang S, Lusk DT, Ehrenreich IM (2023) Building synthetic chromosomes from natural DNA. Nat Commun 14(1):8337. https://doi.org/10.1101/2023.05.09.540074

Denyer T, Pin-Jou W, Colt K, Abramson BW, Pang Z, Solansky P, Mamerto A et al (2024) Streamlined spatial and environmental expression signatures characterize the minimalist duckweed Wolffia Australiana. Genome Res 34(7):1106–1120. https://doi.org/10.1101/gr.279091.124

Dhami H, Yu K, Xu T, Zhu Q, Dhakal K, Friel J, Li S, Tokekar P (2020) Crop height and plot estimation for phenotyping from unmanned aerial vehicles using 3D LiDAR. In: 2020 IEEE/RSJ International Conference on Intelligent Robots And Systems (IROS). IEEE. https://doi.org/10.1109/iros45743.2020.9341343

Ding W, Baumdicker F, Neher RA (2018) panX: pan-genome analysis and exploration. Nucleic Acids Res 46(1):e5. https://doi.org/10.1093/nar/gkx977

Durand NC, Robinson JT, Shamim MS, Machol I, Mesirov JP, Lander ES, Aiden EL (2016) Juicebox provides a visualization system for Hi-C contact maps with unlimited zoom. Cell Syst 3(1):99–101. https://doi.org/10.1016/j.cels.2015.07.012

Dymond JS, Richardson SM, Coombes CE, Babatz T, Muller H, Annaluru N, Blake WJ et al (2011) Synthetic chromosome arms function in yeast and generate phenotypic diversity by design. Nature 477(7365):471–476. https://doi.org/10.1038/nature10403

Emms DM, Kelly S (2019) OrthoFinder: phylogenetic orthology inference for comparative genomics. Genome Biol 20(1):238. https://doi.org/10.1186/s13059-019-1832-y

Fang R, Halpern AR, Rahman MM, Huang Z, Lei Z, Hell SJ, Dulac C, Zhuang X (2023) Three-dimensional single-cell transcriptome imaging of thick tissues. bioRxiv. https://doi.org/10.1101/2023.07.21.550124

Fogg CN, Kovats DE (2015) Computational biology: moving into the future one click at a time. PLoS Comput Biol 11(6):e1004323. https://doi.org/10.1371/journal.pcbi.1004323

Fusaro VA, Patil P, Gafni E, Wall DP, Tonellato PJ (2011) Biomedical cloud computing with Amazon web services. PLoS Comput Biol 7(8):e1002147. https://doi.org/10.1371/journal.pcbi.1002147

Gano B, Bhadra S, Vilbig JM, Ahmed N, Sagan V, Shakoor N (2024) Drone-based imaging sensors, techniques, and applications in plant phenotyping for crop breeding: a comprehensive review. Plant Phenome J 7(1). https://doi.org/10.1002/ppj2.20100

Garrison E, & Marth G (2012) Haplotype-based variant detection from short-read sequencing. arXiv:1207.3907. https://doi.org/10.48550/ARXIV.1207.390

Garrison E, Guarracino A, Heumos S, Villani F, Bao Z, Tattini L, Hagmann J et al (2024) Building pangenome graphs. Nat Methods 21:2008–2012. https://doi.org/10.1038/s41592-024-02430-3

Giolai M, Verweij W, Lister A, Heavens D, Macaulay I, Clark MD (2019) Spatially resolved transcriptomics reveals plant host responses to pathogens. Plant Methods 15(1):114. https://doi.org/10.1186/s13007-019-0498-5

Goel M, Sun H, Jiao WB, & Schneeberger K (2019) SyRI: finding genomic rearrangements and local sequence differences from whole-genome assemblies. Genome Biology, 20(1):277. https://doi.org/10.1186/s13059-019-1911-0

Hao Y, Ma Q, Liu X, Fan X, Men J, Heyun W, Jiang S, Tian D, Xiong B, Xie X (2020) High-yield production of L-valine in engineered Escherichia Coli by a novel two-stage fermentation. Metab Eng 62(November):198–206. https://doi.org/10.1016/j.ymben.2020.09.007

Hao, Y., Stuart, T., Kowalski, M. H., Choudhary, S., Hoffman, P., Hartman, A., Srivastava, A., Molla, G., Madad, S., Fernandez-Granda, C., & Satija, R (2024) Dictionary learning forintegrative, multimodal and scalable single-cell analysis. Nature Biotechnology, 42(2), 293–304. https://doi.org/10.1038/s41587-023-01767-y

Heuckeroth, S., Damiani, T., Smirnov, A., Mokshyna, O., Brungs, C., Korf, A., Smith, J. D., Stincone, P., Dreolin, N., Nothias, L.-F., Hyötyläinen, T., Orešič, M., Karst, U., Dorrestein, P. C., Petras, D., Du, X., van der Hooft, J. J. J., Schmid, R., & Pluskal, T (2024) Reproducible mass spectrometry data processing and compound annotation in MZmine 3. Nature Protocols, 19(9), 2597–2641. https://doi.org/10.1038/s41596-024-00996-y

Holst F, Bolger A, Günther C, Maß J, Triesch S, Kindel F, Kiel N et al (2023) Helixer–*de Novo*prediction of primary eukaryotic gene models combining deep learning and a hidden Markov Model bioRxiv 2023.02.06.527280. https://doi.org/10.1101/2023.02.06.527280

Holt C, Yandell M (2011) MAKER2: an annotation pipeline and genome-database management tool for second-generation genome projects. BMC Bioinform 12(1):491. https://doi.org/10.1186/1471-2105-12-491

Holzworth DP, Huth NI, deVoil PG, Zurcher EJ, Herrmann NI, McLean G, Chenu K et al (2014) APSIM—evolution towards a new generation of agricultural systems simulation. Environ Model Softw 62(December):327–350. https://doi.org/10.1016/j.envsoft.2014.07.009

Hoogenboom G, Porter CH, Boote KJ, Shelia V, Wilkens PW, Singh U et al (2019) The DSSAT crop modeling ecosystem. In: Advances in crop modelling for a sustainable agriculture. Burleigh Dodds Science Publishing, pp 173–216. https://doi.org/10.19103/AS.2019.0061.10

Jarquín D, Crossa J, Lacaze X, Du Cheyron P, Daucourt J, Lorgeou J, Piraux F et al (2014) A reaction norm model for genomic selection using high-dimensional genomic and environmental data. Theor Appl Genet 127(3):595–607. https://doi.org/10.1007/s00122-013-2243-1

Jumper J, Evans R, Pritzel A, Green T, Figurnov M, Ronneberger O, Tunyasuvunakool K et al (2021) Highly accurate protein structure prediction with AlphaFold. Nature 596(7873):583–589

Kanehisa M, Bork P (2003) Bioinformatics in the post-sequence era. Nat Genet 33(3):305–310

Kanehisa M, Furumichi M, Sato Y, Kawashima M, Ishiguro-Watanabe M (2023) KEGG for taxonomy-based analysis of pathways and genomes. Nucleic Acids Res 51(D1):D587–D592

Kanehisa, M., Furumichi, M., Sato, Y., Matsuura, Y., & Ishiguro-Watanabe, M (2025) KEGG: biological systems database as a model of the real world. Nucleic Acids Research, 53(D1), D672–D677. https://doi.org/10.1093/nar/gkae909

Kitony JK, Colt K, Abramson BW, Hartwick NT, Petrus S, Konozy EHE, Karimi N, Yant L, Michael TP (2024) Chromosome-level baobab genome illuminates its evolutionary trajectory and environmental adaptation. Nat Commun 15(1):8833

Klopfenstein DV, Zhang L, Pedersen BS, Ramírez F, Warwick Vesztrocy A, Naldi A, Mungall CJ, Yunes JM, Botvinnik O, Weigel M, Dampier W, Dessimoz C, Flick P, Tang H (2018) GOATOOLS: A Python library for Gene Ontology analyses. Sci Rep 8(1):10872. https://doi.org/10.1038/s41598-018-28948-z

Kolmogorov M, Yuan J, Lin Y, & Pevzner PA (2019) Assembly of long, error-prone reads using repeat graphs. Nature Biotechnology, 37(5):540–546. https://doi.org/10.1038/s41587-019-0072-8

Kriventseva EV, Kuznetsov D, Tegenfeldt F, Manni M, Dias R, Simão FA, & Zdobnov EM (2019) OrthoDB v10: sampling the diversity of animal, plant, fungal, protist, bacterial and viral genomes for evolutionary and functional annotations of orthologs Nucleic Acids Research, 47(D1):D807–D811. https://doi.org/10.1093/nar/gky1053

Koren S, Walenz BP, Berlin K, Miller JR, Bergman NH, & Phillippy AM (2017) Canu: scalable and accurate long-read assembly via adaptive k-mer weighting and repeat separation. Genome Research, 27(5):722–736. https://doi.org/10.1101/gr.215087.116

Lämke J, Bäurle I (2017) Epigenetic and chromatin-based mechanisms in environmental stress adaptation and stress memory in plants. Genome Biol 18(1):124

Langfelder P, Horvath S (2008) WGCNA: an R package for weighted correlation network analysis. BMC Bioinform 9(1):559

Ledford H (2024) CRISPR-edited crops break new ground in Africa. Nature 626(7998):245–246 https://doi.org/10.1038/d41586-024-00176-8

Lee U, Chang S, Putra GA, Kim H, Kim DH (2018) An automated, high-throughput plant phenotyping system using machine learning-based plant segmentation and image analysis. PLoS One 13(4):e0196615

Lei L, Goltsman E, Goodstein D, Guohong Albert W, Rokhsar DS, Vogel JP (2021) Plant pangenomics comes of age. Annu Rev Plant Biol 72(1):411–435

Li M-W, Qi X, Ni M, Lam H-M (2013) Silicon era of carbon-based life: application of genomics and bioinformatics in crop stress research. Int J Mol Sci 14(6):11444–11483

Li Y-H, Zhou G, Ma J, Jiang W, Jin L-G, Zhang Z, Guo Y et al (2014) De novo assembly of soybean wild relatives for pan-genome analysis of diversity and agronomic traits. Nat Biotechnol 32(10):1045–1052

Li H (2018) Minimap2: pairwise alignment for nucleotide sequences. Bioinformatics (Oxford, England), 34(18):3094–3100. https://doi.org/10.1093/bioinformatics/bty191

Liu, Y (2024) CWGCNA: an R package to perform causal inference from the WGCNA framework. NAR Genomics and Bioinformatics, 6(2):lqae042. https://doi.org/10.1093/nargab/lqae042

Luo S, Diehl C, He H, Bae Y, Klose M, Claus P, Cortina NS et al (2023) Construction and modular implementation of the THETA cycle for synthetic CO2 fixation. Nat Catal 6(12):1228–1240

Lv L, Zhang W, Sun L, Zhao A, Zhang Y, Wang L, Liu Y, Li Z, Li H, Chen X (2020) Gene co-expression network analysis to identify critical modules and candidate genes of drought-resistance in wheat. PLoS One 15(8):e0236186

Ma X, Feng L, Tao A, Zenda T, He Y, Zhang D, Duan H, Tao Y (2023) Identification and validation of seed dormancy loci and candidate genes and construction of regulatory networks by WGCNA in maize introgression lines. Theor Appl Genet 136(12):259

Ma P, Chen Y, Lu H, Zhong W (2024) Bisection Grover's search algorithm and its application in analyzing CITE-Seq data. J Am Stat Assoc 120(549):1–12

Mahood EH, Kruse LH, Moghe GD (2020) Machine learning: a powerful tool for gene function prediction in plants. Appl Plant Sci 8(7):e11376

Malosetti M, Bustos-Korts D, Boer MP, van Eeuwijk FA (2016) Predicting responses in multiple environments: issues in relation to genotype × environment interactions. Crop Sci 56(5):2210–2222

Mana AA, Allouhi A, Hamrani A, Rehman S, el Jamaoui I, Jayachandran K (2024) Sustainable AI-based production agriculture: exploring AI applications and implications in agricultural practices. Smart Agric Technol 7:100416

Mao H, Wang H, Liu S, Li Z, Yang X, Yan J, Li J, Tran L-SP, Qin F (2015) A transposable element in a NAC gene is associated with drought tolerance in maize seedlings. Nat Commun 6(1):8326

Marçais G, & Kingsford C (2011) A fast, lock-free approach for efficient parallel counting of occurrences of k-mers. Bioinformatics (Oxford, England), 27(6):764–770. https://doi.org/10.1093/bioinformatics/btr011

Maraveas C, Konar D, Michopoulos DK, Arvanitis KG, Peppas KP (2024) Harnessing quantum computing for smart agriculture: empowering sustainable crop management and yield optimization. Comput Electron Agric 218:108680

Marcussen T, Sandve SR, Heier L, Spannagl M, Pfeifer M, International Wheat Genome Sequencing Consortium, Jakobsen KS et al (2014) Ancient Hybridizations among the Ancestral Genomes of Bread Wheat. Science (New York, N.Y.) 345(6194):1250092

Mendes FK, Vanderpool D, Fulton B, & Hahn MW (2021) CAFE 5 models variation in evolutionary rates among gene families. Bioinformatics (Oxford, England), 36(22–23):5516–5518. https://doi.org/10.1093/bioinformatics/btaa1022

Meuwissen TH, Hayes BJ, Goddard ME (2001) Prediction of total genetic value using genome-wide dense marker maps. Genetics 157(4):1819–1829. https://doi.org/10.1093/genetics/157.4.1819

Mewalal R, Rai DK, Kainer D, Chen F, Külheim C, Peter GF, Tuskan GA (2017) Plant-derived terpenes: a feedstock for specialty biofuels. Trends Biotechnol 35(3):227–240

Mitchell AL, Attwood TK, Babbitt PC, Blum M, Bork P, Bridge A, Brown SD et al (2019) InterPro in 2019: improving coverage, classification and access to protein sequence annotations. Nucleic Acids Res 47(D1):D351–D360

Moffitt JR, Hao J, Wang G, Chen KH, Babcock HP, & Zhuang X (2016) High-throughput single-cell gene-expression profiling with multiplexed error-robust fluorescence in situ hybridization. Proceedings of the National Academy of Sciences of the United States of America, 113(39):11046–11051. https://doi.org/10.1073/pnas.1612826113

Moghimi A, Yang C, Anderson JA (2019) Aerial hyperspectral imagery and deep neural networks for high-throughput yield phenotyping in wheat. arXiv [eess.IV]. https://doi.org/10.48550/ARXIV.1906.09666

Mukerjee A, Mukherjee G, Mallik BB (2024) Quantum active learning for vegetable plant phenology annotation. In: Information systems engineering and management. Springer Nature Switzerland, Cham, pp 13–23

Naish M, Alonge M, Wlodzimierz P, Tock AJ, Abramson BW, Schmücker A, Mandáková T et al (2021) The genetic and epigenetic landscape of the arabidopsis centromeres. Science 374(6569):eabi7489

Needleman SB, Wunsch CD (1970) A general method applicable to the search for similarities in the amino acid sequence of two proteins. J Mol Biol 48(3):443–453

Nelson AD L, Haug-Baltzell AK, Davey S, Gregory BD, Lyons E (2018) EPIC-CoGe: managing and analyzing genomic data. Bioinformatics (Oxford, England), 34(15):2651–2653. https://doi.org/10.1093/bioinformatics/bty106

Nguyen VH, Morantte RIZ, Lopena V, Verdeprado H, Murori R, Ndayiragije A, Katiyar SK et al (2023) Multi-environment genomic selection in rice elite breeding lines. Rice (New York, N.Y.) 16(1):7

Nobori T, Oliva M, Lister R, Ecker JR (2023) Multiplexed single-cell 3D spatial gene expression analysis in plant tissue using PHYTOMap. Nat Plants 9(7):1026–1033

Ou S, Scheben A, Collins T, Qiu Y, Seetharam AS, Menard CC, Manchanda N, Gent JI, Schatz MC, Anderson SN, Hufford MB, & Hirsch CN (2024) Differences in activity and stability drive transposable element variation in tropical and temperate maize. Genome Research, 34(8):1140–1153. https://doi.org/10.1101/gr.278131.123

Paddon CJ, Keasling JD (2014) Semi-synthetic artemisinin: a model for the use of synthetic biology in pharmaceutical development. Nat Rev Microbiol 12(5):355–367

Pang Z, Chong J, Zhou G, de Lima Morais DA, Le Chang MB, Gauthier C, Jacques P-É, Li S, Xia J (2021) MetaboAnalyst 5.0: narrowing the gap between raw spectra and functional insights. Nucleic Acids Res 49(W1):W388–W396

Papikian A, Liu W, Gallego-Bartolomé J, Jacobsen SE (2019) Site-specific manipulation of arabidopsis loci using CRISPR-Cas9 SunTag systems. Nat Commun 10(1):729

Patro R, Duggal G, Love MI, Irizarry RA, & Kingsford C (2017) Salmon provides fast and bias-aware quantification of transcript expression. Nature Methods, 14(4):417–419. https://doi.org/10.1038/nmeth.4197

Peladarinos N, Piromalis D, Cheimaras V, Tserepas E, Munteanu RA, Papageorgas P (2023) Enhancing smart agriculture by implementing digital twins: a comprehensive review. Sensors (Basel, Switzerland) 23(16). https://doi.org/10.3390/s23167128

Pérez P, de los Campos G (2014) Genome-wide regression and prediction with the BGLR statistical package. Genetics 198(2):483–495. https://doi.org/10.1534/genetics.114.164442

Pickar-Oliver A, Gersbach CA (2019) The next generation of CRISPR-Cas technologies and applications. Nat Rev Mol Cell Biol 20(8):490–507

Pluskal T, Castillo S, Villar-Briones A, Oresic M (2010) MZmine 2: modular framework for processing, visualizing, and analyzing mass spectrometry-based molecular profile data. BMC Bioinform 11(1):395

Raes J, Vandepoele K, Simillion C, Saeys Y, Van de Peer Y (2003) Investigating ancient duplication events in the arabidopsis genome. J Struct Funct Genom 3(1/4):117–129

Rautiainen M, Nurk S, Walenz BP, Logsdon GA, Porubsky D, Rhie A, Eichler EE, Phillippy AM, Koren S (2023) Telomere-to-telomere assembly of diploid chromosomes with Verkko. Nat Biotechnol 41(10):1474–1482. https://doi.org/10.1038/s41587-023-01662-6

Ren Z, Jiang L, Di Y, Zhang D, Gong J, Gong J, Jiang Q et al (2024) CodonBERT: a BERT-based architecture tailored for codon optimization using the cross-attention mechanism. Bioinformatics (Oxford, England) 40(7). https://doi.org/10.1093/bioinformatics/btae330

Ranallo-Benavidez TR, Jaron KS, & Schatz MC (2020) GenomeScope 2.0 and Smudgeplot for reference-free profiling of polyploid genomes. Nature Communications, 11(1):1432. https://doi.org/10.1038/s41467-020-14998-3

Robles J, Martín C, Díaz M (2023) OpenTwins: an open-source framework for the design, development and integration of effective 3D-IoT-AI-powered digital twins. arXiv [cs.SE]. arXiv. http://arxiv.org/abs/2301.05560

Roell M-S, Zurbriggen MD (2020) The impact of synthetic biology for future agriculture and nutrition. Curr Opin Biotechnol 61(February):102–109

Ryu KH, Huang L, Kang HM, Schiefelbein J (2019) Single-cell RNA sequencing resolves molecular relationships among individual plant cells. Plant Physiol 179(4):1444–1456

Sadeghi-Tehran P, Sabermanesh K, Virlet N, Hawkesford MJ (2017) Automated method to determine two critical growth stages of wheat: heading and flowering. Front Plant Sci 8(February):252

Samach A, Mafessoni F, Gross O, Melamed-Bessudo C, Filler-Hayut S, Dahan-Meir T, Amsellem Z, Pawlowski WP, Levy AA (2023) CRISPR/Cas9-induced DNA breaks trigger crossover, chromosomal loss, and chromothripsis-like rearrangements. Plant Cell 35(11):3957–3972

Schaefer RJ, Michno J-M, Jeffers J, Hoekenga O, Dilkes B, Baxter I, Myers CL (2018) Integrating coexpression networks with GWAS to prioritize causal genes in maize. Plant Cell 30(12):2922–2942

Schläpfer P, Zhang P, Wang C (2017) CycADS: an annotation database system to integrate and analyze plant metabolism. Plant Physiol 173(1):620–630

Seppey M, Manni M, & Zdobnov EM (2019). BUSCO: Assessing genome assembly and annotation completeness. Methods in Molecular Biology (Clifton, N.J.), 1962:227–245

Servant N, Varoquaux N, Lajoie BR, Viara E, Chen CJ, Vert JP, Heard E, Dekker J, & Barillot E (2015) HiC-Pro: an optimized and flexible pipeline for Hi-C data processing. Genome Biology, 16(1):259. https://doi.org/10.1186/s13059-015-0831-x

Sinitcyn P, Tiwary S, Rudolph J, Gutenbrunner P, Wichmann C, Yılmaz Ş, Hamzeiy H, Salinas F & Cox J (2018). MaxQuant goes Linux. Nature Methods, 15(6):401. https://doi.org/10.1038/s41592-018-0018-y

Shahan R, Hsu C-W, Nolan TM, Cole BJ, Taylor IW, Greenstreet L, Zhang S et al (2022) A single-cell arabidopsis root atlas reveals developmental trajectories in wild-type and cell identity mutants. Dev Cell 57(4):543–60.e9

Shannon P, Markiel A, Ozier O, Baliga NS, Wang JT, Ramage D, Amin N, Schwikowski B, Ideker T (2003) Cytoscape: a software environment for integrated models of biomolecular interaction networks. Genome Res 13(11):2498–2504

Shao Y, Ning L, Qin Z, Xue X (2018) CRISPR-Cas9 facilitated multiple-chromosome fusion in saccharomyces cerevisiae. ACS Synth Biol 7(11):2706–2708

Skobelev P, Mayorov I, Simonova E, Goryanin O, Zhilyaev A, Tabachinskiy A, Yalovenko V (2021) Development of digital twin of plant for adaptive calculation of development stage duration and forecasting crop yield in a cyber-physical system for managing precision farming. In: Cyber-physical systems, Studies in systems, decision and control. Springer International Publishing, Cham, pp 83–96

Smith CA, Want EJ, O'Maille G, Abagyan R, Siuzdak G (2006) XCMS: processing mass spectrometry data for metabolite profiling using nonlinear peak alignment, matching, and identification. Anal Chem 78(3):779–787

Stanke M, Keller O, Gunduz I, Hayes A, Waack S, Morgenstern B (2006) AUGUSTUS: Ab Initio prediction of alternative transcripts. Nucleic Acids Res 34(Web Server issue):W435–W439

Stanke M, Diekhans M, Baertsch R, Haussler D (2008) Using native and syntenically mapped cDNA alignments to improve de novo gene finding. Bioinformatics (Oxford, England), 24(5):637–644. https://doi.org/10.1093/bioinformatics/btn013

Su G, Morris JH, Demchak B , & Bader GD (2014) Biological network exploration with Cytoscape 3. Et Al [Current Protocols in Bioinformatics], 47(1):8.13.1–24. https://doi.org/10.1002/0471250953.bi0813s47

Sukhija N, Kanaka KK, Purohit PB, Ganguly I, Malik AA, Sanjeev Singh SP, Dixit AV, Dash A (2023) Mendelism: connecting the dots across centuries. Cytol Genet 57(5):500–516

Sun Y, Miller C, Rajurkar AB, Lynch RC, Alyward A, Zhang L, Shaner M et al (2024) Genome-Wide Association Study reveals influence of cell-specific gene networks on soybean root system architecture. https://doi.org/10.1101/2024.02.27.581071

Szklarczyk D, Gable AL, Nastou KC, Lyon D, Kirsch R, Pyysalo S, Doncheva NT et al (2021) Correction to 'The STRING Database in 2021: customizable protein-protein networks, and functional characterization of user-uploaded gene/measurement sets. Nucleic Acids Res 49(18):10800

Tyanova S, Temu T, Cox J (2016) The MaxQuant computational platform for mass spectrometry-based shotgun proteomics. Nat Protoc 11(12):2301–2319

United Nations (2024) World Population Projected to Reach 9.8 Billion in 2050, and 11.2 Billion in 2100 | United Nations. Accessed 30 Nov 2024. https://www.un.org/en/desa/world-population-projected-reach-98-billion-2050-and-112-billion-2100

VanBuren R, Wai CM, Wang X, Pardo J, Yocca AE, Wang H, Chaluvadi SR et al (2020) Exceptional subgenome stability and functional divergence in the allotetraploid ethiopian cereal Teff. Nat Commun 11(1):1–11

Vaser R, Sović I, Nagarajan N, & Šikić M (2017) Fast and accurate de novo genome assembly from long uncorrected reads. Genome Research, 27(5):737–746. https://doi.org/10.1101/gr.214270.116

Vats S, Kumar J, Sonah H, Zhang F, Deshmukh R (2024) Prime editing in plants: prospects and challenges. J Exp Bot 75(17):5344–5356

Walker BJ, Abeel T, Shea T, Priest M, Abouelliel A, Sakthikumar S, Cuomo CA, Zeng Q, Wortman J, Young SK, & Earl AM (2014) Pilon: an integrated tool for comprehensive microbial variant detection and genome assembly improvement. PloS One, 9(11):e112963. https://doi.org/10.1371/journal.pone.0112963

Wang M, Carver JJ, Phelan VV, Sanchez LM, Garg N, Peng Y, Nguyen DD et al (2016) Sharing and community curation of mass spectrometry data with global natural products social molecular networking. Nat Biotechnol 34(8):828–837

Wang F, Miao H, Zhang S, Xiaohui H, Chu Y, Yang W, Wang H, Wang J, Shan S, Chen J (2024) Weighted gene co-expression network analysis reveals hub genes regulating response to salt stress in Peanut. BMC Plant Biol 24(1):425

Wang Y, Tang H, Wang X, Sun Y, Joseph PV, & Paterson AH (2024) Detection of colinear blocks and synteny and evolutionary analyses based on utilization of MCScanX. Nature Protocols, 19(7):2206–2229. https://doi.org/10.1038/s41596-024-00968-2

Xu N, Liu Y, Jiang H, Liu J, Ma Y (2020) Combining protein and metabolic engineering to construct efficient microbial cell factories. Curr Opin Biotechnol 66(December):27–35

Yong W-S, Hsu F-M, Chen P-Y (2016) Profiling genome-wide DNA methylation. Epigenetics Chromatin 9(1):26

Zahedi R, Ghamsari R, Argha A, Macphillamy C, Beheshti A, Alizadehsani R, Lovell NH, Lotfollahi M, Alinejad-Rokny H (2024) Deep learning in spatially resolved transcriptfomics: a comprehensive technical view. Brief Bioinform 25(2). https://doi.org/10.1093/bib/bbae082

Zhou J, Luan X, Liu Y, Wang L, Wang J, Yang S, Liu S, Zhang J, Liu H, Yao D (2023) Strategies and methods for improving the efficiency of CRISPR/Cas9 gene editing in plant molecular breeding. Plants 12(7). https://doi.org/10.3390/plants12071478

Correction to: The Power of *De Novo Domestication* of Wild Solanaceae for a Changing Climate

Karla Gasparini, Lucía Rodríguez Gargantini,
Carlos Federico Marfil, and Agustin Zsögön

Correction to:
Chapter 5 in: P. M. Priyadarshan, R. Ortiz (eds.), *Plant Breeding 2050*, https://doi.org/10.1007/978-981-95-0583-8_5

Owing to an unfortunate oversight on the part of production, the original version of this chapter was inadvertently published with an incorrect authorship.

The authorship has now been updated as below:

Karla Gasparini, Lucía Rodríguez Gargantini, Carlos Federico Marfil and Agustin Zsögön

The updated version of this chapter can be found at
https://doi.org/10.1007/978-981-95-0583-8_5

P. M. Priyadarshan, R. Ortiz (eds.), *Plant Breeding 2050*,
https://doi.org/10.1007/978-981-95-0583-8_24

MIX
Papier aus verantwortungsvollen Quellen
Paper from responsible sources
FSC® C105338

If you have any concerns about our products,
you can contact us on
ProductSafety@springernature.com

In case Publisher is established outside the EU,
the EU authorized representative is:
Springer Nature Customer Service Center GmbH
Europaplatz 3, 69115 Heidelberg, Germany

Printed by Libri Plureos GmbH
in Hamburg, Germany